Mathematicians in Bologna 1861–1960

The old "Aula Magna" of the University of Bologna and the reading room of the University Library. Photograph courtesy of the Historical Archive of the University of Bologna, Alma Mater Studiorum.

Salvatore Coen
Editor

Mathematicians in Bologna
1861–1960

 Birkhäuser

Editor
Salvatore Coen
Dipartimento di Matematica
Università di Bologna
Bologna
Italy

ISBN 978-3-0348-0226-0   e-ISBN 978-3-0348-0227-7
DOI 10.1007/978-3-0348-0227-7
Springer Basel Dordrecht Heidelberg London New York

Library of Congress Control Number: 2012935226

Mathematics Subject Classification (2010): 01-06, 01Axx

© Springer Basel AG 2012
This work is subject to copyright. All rights are reserved, whether the whole or part of the material is concerned, specifically the rights of translation, reprinting, re-use of illustrations, recitation, broadcasting, reproduction on microfilms or in other ways, and storage in data banks. For any kind of use, permission of the copyright owner must be obtained.

Cover photograph: Palazzo Poggi, the official site of the University of Bologna. Used with the kind permission of the Archivio storico dell'Università di Bologna.

Printed on acid-free paper

Springer Basel AG is part of Springer Science+Business Media
(www.birkhauser-science.com)

A Short Overview on Mathematicians in Bologna in the First Century after the Establishment of Italy

Salvatore Coen

The present volume considers the lives and achievements of mathematicians who studied and worked in various roles at the Bologna University in the century following Italian unification. Most contributions to this volume are historical in character; the few which more closely focus on mathematical research do so strictly in relation to a discussion of the mathematicians concerned.

Without claiming to be exhaustive, the volume deals with many of the most representative mathematicians who worked in Bologna in the period 1860-1960, namely with Luigi Cremona (b. 1830), Eugenio Beltrami (b. 1836), Salvatore Pincherle (b. 1853), Pietro Burgatti (1868), Federigo Enriques (b. 1871), Ugo Amaldi (b. 1875), Beppo Levi (b. 1875), Giuseppe Vitali (b. 1875), Enrico Bompiani (b. 1889), Beniamino Segre (b. 1903), Tullio Viola (b. 1904), Dario Graffi (b. 1905), Gianfranco Cimmino (b. 1908) Bruno Pini (b. 1918) and Lamberto Cattabriga (b. 1930). Sometimes the same mathematician is presented from different points of view by different authors. The work of a number of other mathematicians also comes in for perusal, most notably the contributions made by Giulio Vivanti (b. 1859), Ettore Bortolotti (b. 1866) and Filippo Sibirani (b. 1880).

The University of Bologna is one of the oldest universities in the world (indeed, it is traditionally said to have been founded in 1088) and many well-known mathematicians have studied and taught there. We may recall Luca Pacioli, Domenico Maria Novara, Scipione del Ferro, Girolamo Cardano, Ludovico Ferrari, Rafael Bombelli, Bonaventura Cavalieri, Pietro Mengoli, Giandomenico Cassini, Domenico Guglielmini and Gabriele Manfredi, to mention just a few. It is, however, hard to come up with a corresponding list for the first half of the nineteenth century, a dark and difficult period for mathematics in Bologna, although studies of astronomy and hydraulics continued developing along traditional lines.

In order to overcome the local weakness of mathematical research at Bologna, in 1860, soon after Bologna became part of the Kingdom of Sardinia (by the plebiscite of March 1860), the Minister of Education appointed Luigi Cremona as Professor of Higher Geometry and Matteo Fiorini (b. 1827) as Professor of Geodesy. The presence of Luigi Cremona and Eugenio Beltrami (starting from 1862) brought about a dramatic improvement in the state of mathematical studies and education at the university. Unfortunately, they both left the university a few years after their nomination so that mathematical studies at Bologna went through a new period of difficulty as far as pure mathematics was concerned. Starting from 1880, however, a new generation of mathematicians revitalized the study of mathematics at the university; they attracted colleagues of high quality and the number of students in pure mathematics grew year after year. About forty years later, the founding of the Unione Matematica Italiana (UMI), which has since continued to maintain its headquarters at the University of Bologna, the launching of the *Bollettino dell' Unione Matematica Italiana* by Salvatore Pincherle and his election as President of the IMU in 1924 were a sign of the new improved standing of mathematics in Bologna, while the holding of the International Congress of Mathematicians in Bologna in 1928 testified to the international recognition that the school had acquired.

Starting from the twenties, however, the effect of the central government policy on the university was already becoming evident. The so-called Gentile reform of education embodied a changed conception of the university's duties and objectives. Universities became more and more authoritarian; Jewish university professors were sacked as a result of the Race Laws of 1938. At Bologna two professors in the College of Engineering (Facoltà di Ingegneria), Giulio Supino and Emanuele Foà lost their posts. At that time, in 1938, there were four professors of mathematics in the School of Science (Facoltà di Scienze MM.FF.NN) at Bologna University: Pietro Burgatti, Luigi Fantappiè, Beppo Levi, Beniamino Segre. Pietro Burgatti died in May; Segre and Levi were dismissed and Fantappiè was working abroad. For a few months the School of Science had no active professor of Mathematics in Bologna. So, once again, mathematics in Bologna had to face difficult times which profoundly affected the direction and nature of the research that was carried out.

This book is about the protagonists of this arduous and eventful story with all its highs and lows.

What follows are some selected sources for general information about mathematical studies at Bologna (of course, many other interesting papers were written on particular mathematicians). Reference [1] is a beautiful booklet by Ettore Bortolotti, distributed to the participants at the 1928 International Congress of Mathematicians held in Bologna; [2] is a revised and 1947 updated version of [1]. For a short introduction to the history of Mathematics in Bologna, see [3].

The proceedings of the international conference held on the occasion of the IX centennial of the University of Bologna are published in [4]: the volume contains more than thirty contributions inspired by research carried out in Bologna. Reference [5] deals with teaching of mathematics at Bologna from 1860 to 1940.

For a more in-depth study of the period 1860- 1960, later on we list volumes of collected works by mathematicians who worked in Bologna during those years.

LIST OF VOLUMES OF SELECTED WORKS OF MATHEMATICIANS WHO HAVE WORKED IN BOLOGNA IN THE FIRST CENTURY AFTER THE ESTABLISHMENT OF THE KINGDOM ITALY.

Arzelà, C. 1992. *Opere,* edited by Unione Matematica Italiana, I, XXXIX + 348; II, VII + 357. Bologna: Cremonese
Beltrami, E. 1902-1920. *Opere matematiche di Eugenio Beltrami,* edited by University of Rome Science Faculty, Tomo primo (1902), VII + 437; Tomo secondo (1904) 465; Tomo terzo (1911), 488; Tomo quarto (1920), 554. Milano: Ulrico Hoepli
Bompiani, E. 1978. *Opere scelte,* edited by Unione Matematica Italiana and published with the support of the Consiglio Nazionale delle Ricerche, I, 404; II, 468; III, 436. Bologna: Unione Matematica Italiana
Burgatti, M. 1951. *Opere scelte* (published under the auspices of the universities of Bologna and Ferrara, the Accademia delle Scienze di Bologna and the Unione Matematica Italiana), VI+354. Bologna: Zanichelli
Cimmino, G. 2002. *Opere Scelte* (published by Accademia delle Scienze Fisiche e Matematiche della Società Nazionale di Scienze Lettere e Arti in Napoli), Giannini, Napoli, XV +668.
Cremona, L. 1914–1917. *Opere matematiche di Luigi Cremona,* published under the auspices of the R. Accademia dei Lincei, Tomo primo (1914), con note dei revisori, VII+492, Tomo secondo (1915), con note dei revisori, 459, Tomo terzo (1917), con note dei revisori, con notizie della opere e della vita dell' autore e con indice analitico per materie, 520. Milano: Ulrico Hoepli
Donati, L. 1925. *Memorie e note scientifiche: elasticità, vettori, elettrologia, correnti alternative, argomenti vari,* XIII +350. Bologna: Zanichelli
Enriques, F. 1956-1966. *Memorie scelte di Geometria,* edited by the Accademia Nazionale dei Lincei, Volume I (1956), 1893-1898, XXII + 541; Volume II, (1959) 1899-1910, 527; Volume III, (1966), 1911-1940, VII+456. Bologna: Zanichelli
Fantappiè, L. 1973. *Opere scelte,* edited and published by Unione Matematica Italiana with the support of the Consiglio Nazionale delle Ricerche, I, 646; II, 420. Bologna: Unione Matematica Italiana
Graffi, D. 1999. *Opere scelte,* edited and published by Consiglio Scientifico del G.N.F.M., XXXI + 458. Roma: Gruppo Nazionale per la Fisica Matematica
Levi, B. 1999. *Opere 1897–1926,* edited by Unione Matematica Italiana and published with the support of the Consiglio Nazionale delle Ricerche, I, il primo decennio, 1897/1906, CXXII + 498; II, il ventennio, 1907/1926, VIII+ 434. Roma: Cremonese
Pincherle, S. 1954. *Opere scelte,* edited by Unione Matematica Italiana, I, VI + 396; II, 494. Roma: Edizioni Cremonese
Segre, B. 1987-2000. *Opere scelte,* edited by Unione Matematica Italiana and published with the support of the Consiglio Nazionale delle Ricerche, I (1987), periodo 1923-1937, LI+420; II (1987) periodo 1937-1970 ,VIII+ 44; III (2000), 460. Bologna: Cremonese
Tonelli, L. 1960-1963. *Opere scelte,* edited by Unione Matematica Italiana and published with the support of the Consiglio Nazionale delle Ricerche, Volume I

(1960), Funzioni di variabile reale, VI + 604; Volume II (1961), Calcolo delle Variazioni 1911–1924, VI+534; Volume III (1962), Calcolo delle Variazioni 1926–1950, 508; Volume IV (1963), Argomenti vari, VI+330. Roma: Cremonese

Vitali, G. 1984. *Opere sull'analisi reale e complessa. Carteggio*, edited by Unione Matematica Italiana and published with the support of the Consiglio Nazionale delle Ricerche, XII + 524. Roma: Cremonese

Acknowledgments

The present volume has been prepared on occasion of the nineteenth Congress of the Unione Matematica Italiana (UMI), held at the University of Bologna from the 12th to the 17th of September, 2011. We thank the University of Bologna and the Carisbo Foundation for their support and the Organizing Committee of the Congress. Appreciation also goes to Cristina Chersoni and Chiara Cocchi archivists of the historical archivals of the University of Bologna, for their skill and patience in my regard. Finally, special recognition goes to Claudia Benedetti for her competence and dedication in the preparation of this work.

References

1. Bortolotti, Ettore. 1928. *L' école mathématique de Bologne. Aperçu historique.* Congrès international des Mathématiciens, Bologna, 75: N. Zanichelli
2. Bortolotti, Ettore. 1947. *La storia della matematica nella università di Bologna*, Bologna, 226: Nicola Zanichelli
3. Coen, Salvatore. 1991. Preface. In *Geometry and Complex Variables: Proceedings of an international Meeting on the Occasion of the IX Centennial of the University of Bologna.* Lecture Notes in Pure and Applied Mathematics, 132, iii–vii. Dekker
4. Coen, Salvatore. (ed.). 1991. *Geometry and Complex Variables: Proceedings of an international Meeting on the Occasion of the IX Centennial of the University of Bologna.* Lecture Notes in Pure and Applied Mathematics, 132, 494. Dekker
5. Francesconi, Stefano. 1991. L'insegnamento della Matematica nell'Università di Bologna dal 1860 al 1940. In *Geometry and Complex Variables: Proceedings of an international Meeting on the Occasion of the IX Centennial of the University of Bologna.* Lecture Notes in Pure and Applied Mathematics, 132, 415–474. Dekker

Dipartimento di Matematica Salvatore Coen
Università di Bologna
Alma Mater Studiorum
salvatore.coen@unibo.it

Contents

A Short Overview on Mathematicians in Bologna in the First Century after the Establishment of Italy v
Salvatore Coen

Beltrami's Models of Non-Euclidean Geometry 1
Nicola Arcozzi

Giuseppe Vitali: Real and Complex Analysis and Differential Geometry 31
Maria Teresa Borgato

Pincherle's Early Contributions to Complex Analysis 57
Umberto Bottazzini

Luigi Cremona's Years in Bologna: From Research to Social Commitment 73
Aldo Brigaglia and Simonetta Di Sieno

Federigo Enriques: The First Years in Bologna 105
Ciro Ciliberto and Paola Gario

Enrico Bompiani: The Years in Bologna 143
Ciro Ciliberto and Emma Sallent Del Colombo

Dario Graffi in a Complex Historical Period 179
Mauro Fabrizio

Pietro Burgatti and His Studies on Mechanics 197
Paolo Freguglia and Sandro Graffi

Federigo Enriques (1871–1946) and the Training of Mathematics Teachers in Italy 209
Livia Giacardi

Beppo Levi and Quantum Mechanics 277
Sandro Graffi

Leonida Tonelli: A Biography 289
Angelo Guerraggio and Pietro Nastasi

Bruno Pini and the Parabolic Harnack Inequality: The Dawning of Parabolic Potential Theory 317
Ermanno Lanconelli

Federigo Enriques as a Philosopher of Science 333
Gabriele Lolli

The *Enciclopedia delle Matematiche elementari* and the Contributions of Bolognese Mathematicians 343
Erika Luciano

The Role of Salvatore Pincherle in the Development of Fractional Calculus 373
Francesco Mainardi and Gianni Pagnini

Tullio Viola and his *Maestri* in Bologna: Giuseppe Vitali, Leonida Tonelli and Beppo Levi 383
Clara Silvia Roero and Michel Guillemot

Developement of the Theory of Lie Groups in Bologna (1884–1900) 415
Enrico Rogora

Difference Equations in Spaces of Regular Functions: a tribute to Salvatore Pincherle 427
Irene Sabadini and Daniele C. Struppa

The Work of Beniamino Segre on Curves and Their Moduli 439
Edoardo Sernesi

Lamberto Cattabriga and the Theory of Linear Constant Coefficients Partial Differential Equations 451
Daniele C. Struppa

New Perspectives on Beltrami's Life and Work – Considerations Based on his Correspondence 465
Rossana Tazzioli

On Cimmino Integrals as Residues of Zeta Functions 519
Sergio Venturini

Index of Names and Locations 539

Beltrami's Models of Non-Euclidean Geometry

Nicola Arcozzi

Abstract In two articles published in 1868 and 1869, Eugenio Beltrami provided three models in Euclidean plane (or space) for non-Euclidean geometry. Our main aim here is giving an extensive account of the two articles' content. We will also try to understand how the way Beltrami, especially in the first article, develops his theory depends on a changing attitude with regards to the definition of surface. In the end, an example from contemporary mathematics shows how the boundary at infinity of the non-Euclidean plane, which Beltrami made intuitively and mathematically accessible in his models, made non-Euclidean geometry a natural tool in the study of functions defined on the real line (or on the circle).

1 Introduction

In two articles published in 1868 and 1869, Eugenio Beltrami, at the time professor at the University of Bologna, produced various models of the hyperbolic version non-Euclidean geometry, the one thought in solitude by Gauss, but developed and written by Lobachevsky and Bolyai. One model is presented in the *Saggio di interpretazione della geometria non-euclidea [5] [Essay on the interpretation of non-Euclidean geometry]*, and other two models are developed in *Teoria fondamentale degli spazii di curvatura costante* [6] [*Fundamental theory of spaces with constant curvature*]. One of the models in the *Teoria*, the so-called Poincaré disc, had been briefly mentioned by Riemann in his *Habilitationschrift [26]*, the text of which was posthumously published in 1868 only, after Beltrami had written his first

---

N. Arcozzi (✉)
Dipartimento di Matematica, Università di Bologna, Piazza di porta San Donato, 5 40127 Bologna, Italy
e-mail: arcozzi@dm.unibo.it

paper [5],[1] and it served as a lead for the second one [6]. Riemann was not so much interested in getting involved in a *querelle* between Euclidean and non-Euclidean geometry, which he had in fact essentially solved in his remark, as he was interested in developing a broad setting for geometry, a "library" of theories of space useful for the contemporary as well as for the future developments of sciences. At the time Beltrami wrote his *Saggio*, however, Riemann's *Habilitationschrift* was not widely available and although its connection to non-Euclidean geometry was rather clear, it was not explicitely stated. It should be mentioned that the second model in the *Teoria* had been previously considered by Liouville, who used it as an example of a surface with constant, negative curvature.

Beltrami's papers were widely read and promptly translated into French by Jules Hoüel. Their impact was manifold. (1) They clearly showed that the postulates of non-Euclidean geometry described the simply connected, complete surfaces of negative curvature (surfaces which, however, only locally could be thought of as surfaces in $\mathbb{R}^3$). (2) Hence, it was not possible proving the Postulate of the Paralles using the remaining ones, as J. Hoüel explicited in [17].[2] (3) It was then possible to consider and use non-Euclidean geometry without having an opinion, much less a faith, concerning the "real geometry" (of space, of Pure Reason). (4) More important, and lasting, the universe of non-Euclidean geometry was not anymore the counter-intuitive world painted by Lobachevsky and Bolyai: any person instructed in Gaussian theory of surfaces could work out all consequences of the non-Euclidean principles directly from Beltrami's models; this legacy is quite evident up to the present day. (5) In all of Beltrami's models, the non-Euclidean plane (or n-space) is confined to a portion of the Euclidean plane (or n-space), whose boundary encodes important geometric features of the non-Euclidean space it encloses.

The presence of an important "boundary at infinity" in non-Euclidean geometry had been realized before. In Beltrami's models, this boundary is (from the Euclidean viewpoint of the "external observer") wholly within reach, easy to visualize, complete with natural coordinate systems (spherical, when the boundary is seen as a the limit of a sphere having fixed finite center and radius tending to infinity; Euclidean-flat when it is seen as the limit of horocycles: distinguished spheres having infinite radius and center at infinity). With this structure in place, it was possible to radically change viewpoint and to see the non-Euclidean space as the "filling" of its boundary, spherical or Euclidean. To wit, some properties of functions defined on the real line

---

[1] In a letter to Genocchi in 1868 ([15] p. 578–579), Beltrami says that he had the manuscript of the *Saggio* ready in 1867, but that, faced with criticism from Cremona, he had postponed its publication. After reading Riemann's *Habilitationschrift*, he felt confident in submitting the article to the Neapolitan *Giornale di matematiche*, emended of a statement about three dimensional non-Euclidean geometry and with some integration "which I can hazard now, because substantially agreeing with some of Riemann's ideas."

[2] In another letter to Genocchi ([15] p. 588), Beltrami writes that this fact clearly follows from his *Saggio*, and that "in the note of Hoüel I do not find further elements to prove it". Beltrami being generally rather unassuming about his own work, it is likely that he had already thought of this consequence of his model, but that he thought it prudent to leave it to state explicitely to the reader.

or on the unit circle become more transparent when we consider their extensions to the non-Euclidean plane of which the line or the circle are the boundary in Beltrami's models. This line of reasoning is not to be found in Beltrami's articles, but it relies on Beltrami's models, and it is the main reason why non-Euclidean (hyperbolic) geometry has entered the toolbox of such different areas as harmonic and complex analysis, potential theory, electrical engineering and so on. The pioneer of these kind of applications is Poincaré [25]. It is still matter of discussion whether or not Poincaré had had any exposure to Beltrami's work, or if he re-invented one of Beltrami's models ([19], p. 277–278). Even if he had not had first-hand knowledge of Beltrami's work, however, I find it unlikely that he had not heard of the debate about non-Euclidean geometry in which Beltrami's work was so central, and of the possibility of having concrete models of it. It is nowadays very common, for mathematicians from all branches, to start with a class of objects (typically, but not only, functions) naturally defined on some geometric space, and to look for a "more natural" geometry which might help in understanding some properties of those same objects. The new geometry, this way, has a "model" built on the old one.

Let me end these introductory remarks by reminding the reader that, unfortunately, in the mathematical pop-culture the name of Beltrami is seldom attached to his models. The model of the *Saggio* is generally called the *Klein model* and the two models of the *Teoria* are often credited to Poincaré. There are reasons for this. Klein made more explicit the connections between the model in the *Saggio* and projective geometry, which Beltrami had just mentioned in his article. Poincaré, as I said above, was the first to use the other two models in order to understand phenomena apparently far from the non-Euclidean topic. More informed sources refer to the projective model as the *Beltrami–Klein (projective) disc model;* the other two should perhaps be called *Riemann–Beltrami–Poincaré (conformal) disc model* and *Liouville–Beltrami (conformal) half-plane model*.

The aim of this note is mostly expository. There are excellent accounts of how non-Euclidean geometry developed: from the scholarly and influential monograph of Roberto Bonola [11], which is especially interesting for the treatment of the early history, to the lectures of Federigo Enriques [16], to the recent, flamboyant book by Jeremy Gray [19], in which the development of modern geometry is treated in all detail. To have a taste of what happened after Beltrami, Klein, and Poincaré, I recommend the beautiful article [21] by Milnor, which is also historically accurate, and the less historically concerned, but equally useful article [14] by Cannon, Floyd, Kenyon, and Parry. An extensive account of the modern view of hyperbolic spaces (from the metric space perspective) is in Bridson and Haefliger's beautiful monograph [13]. I will just summarize the well known story up to Beltrami for ease of the reader. Then, I will describe in some detail the main mathematical content of the *Saggio* and of the *Teoria*. Not only such content is a masterful piece of mathematics, but it was also in the non-pretentious style of Beltrami to present his work as double faced: his reader could think of it as an investigation on the foundations of geometry, but, if skeptical, he could also give it *a purely analytic meaning* ([6] p. 406; Beltrami refers to the geometric terms he uses, but the same distinction applies to the two papers as a whole). We can appreciate the analytic content in itself. With this at hand, we will try to understand what Beltrami claimed to have achieved in geometric (and logical) terms.

To end on a different tune, I will describe how, starting from a reasonable problem about functions defined on the real line (looking at a signal at different scales) one is naturally led to consider non-Euclidean geometry in the upper-half space, the last of Beltrami's models. My aim here is giving a simple example of one of the most important legacies of Beltrami's models, which is point (v) above. For understandable reasons, this aspect is seldom mentioned in historical accounts, while it is central (and popular) in the research literature and in textbooks.

A disclaimer is due. I am neither trained in geometry, nor in history of mathematics. This surely accounts for the bibliography, which is probably not the one an historian of science would have used, and for other naiveties I can not be aware of. I have tried, however, to be as historically correct as I could and to be honest about anachronisms. I have been using Beltrami's models for many years, as a tool or as an inspiring metaphor, while working in harmonic analysis and complex function theory, and this is my only title to discuss them. I thank Salvatore Coen, who entrusted me with writing this note and for investing so much energy to edit the volume.

Note on bibliography. *Beltrami's papers in the bibliography are given with the coordinates of their publication, except for the page numbers, which refer to the edition of his collected works [2].*

2 Non-Euclidean Geometry Before Beltrami

As in many other scientific revolutions, at the roots of the non-Euclidean one we find an orthodox theory and a disturbing asymmetry. The orthodox theory is Euclid's *Elements,* in which the science of measurement and space (ideal space from a Platonic viewpoint, real from an Aristotelian one: it does not matter here) is given a hierarchical structure (axioms, postulates, definitions, theorems), held together by logics. The postulates should encode unquestionable truths about space, from which other truths are deduced. The asymmetry consists in the Fifth Postulate (or Parallel Postulate), concerning properties of parallel lines, whose use Euclid postpones until after Proposition XXVIII; where he proves that two straight lines a and b in the plane do not meet, if the internal angles they make on the same side of a third line c meeting both of them sum to a straight angle. The Fifth Postulate states that the converse is true: if the sum is less than a straight angle, then a and b eventually meet on that same side of c. The main disturbing feature of the Fifth Postulate is that, in order to verify the property it states, one has to consider the straight lines in their infinite extension. It was soon realized that the Fifth Postulate is equivalent to the *uniqueness* of the straight line through a given point P, which is parallel to a given straight line a not containing P. Very early, attempts were made to prove it, based on the remaining Postulates and Axioms.

In the effort, several properties were found which, given the other Postulates, were equivalent to the Fifth. Also, the critical thought unleashed in search for a proof of the Fifth Postulate helped in finding a number of hidden assumption (namely, hidden postulates) in Euclid's work: the line divides the plane into two parts, for instance, or the Archimedean property of lengths.

Trying without success to prove the Postulate of Parallels by contradiction, mathematicians went deeper and deeper into a geometric world in which the Postulate did not hold, finding increasingly counterintuitive properties of figures. This kind of research reached maturity with the work if Girolamo Saccheri[3] [1667–1733]. In his *Euclides ab omni naevo vindicatus* (see [11], Chap. II), Saccheri considered a fixed quadrilateral $ABCD$ with right angles in A and B and equal sides $AC = BD$. He considered the three possibilities for the angles $\widehat{C} = \widehat{D}$: (r) the angles are both right (then the Fifth Postulate holds); (o) the angles are both obtuse; (a) the angles are bothe acute. The *obtuse hypothesis* (o) leads to contradiction[4] It remained the *acute angle hypothesis,* which Saccheri developed at length. I again refer to Bonola's monograph [11] for more on this early history, but let me mention a few among the properties which, in the effort of proving the Fifth Postulate by contradiction, were found to hold in the fictional geometry where the Fifth would be false.

(i) The set of points equidistant from a straight line a (on one side of a) *is not* a straight line (this is somehow implicit in Posidonius [I cent. a.C.], who *defines* two straight lines to be parallel if they are equidistant).

---

[3]Interestingly, the work of Saccheri, which had an indirect role in the development of non-Euclidean geometry and was then forgotten, was re-discovered by Beltrami [10].

[4]The obtuse hypothesis holds on a sphere, using geodesics (great circles) as straight lines; but on a sphere we do not have uniqueness of the geodesic through two points. This was considered to be a major problem by Beltrami, who was looking for a geometry in which all principles of Euclidean geometry hold true, but the uniqueness of parallels, but not for Riemann. In the *Habilitationschrift,* Riemann offered the sphere as a model for a geometry in which no parallel existed. Some of the principles used by Saccheri had to be abandoned: uniqueness of the line through two points, it seemed; but also the infinite extension of lines (Riemann seems to be the first to point out that the right geometric requirement is not that the straight lines have infinite length – what he calls "infinite extent of the line", translated in "unboundedness" in modern netric space theory –, but that one finds no obstructions while following a straight line – a property he calls "unboundedness", translated nowadays in "metrically complete and without boundary"). Clifford used the two-to-one covering of the real projective plane by the sphere to exhibit a geometry with positive constant curvature in which (1) there was just one line through two points; (2) space was homogeneous and isotropic; and (3) there are no distinct, parallel straight lines. Kelin realized the role of this model in the discussion of non-Euclidean geometries. Of course, in the projective-space model, often called *Riemann's non-Euclidean geometry,* one not only has to give up the infinite extension of straight lines, but also the fact that a line divides the plane into two parts; or, which is about the same, orientability of the plane.

(ii) For each planar figure F, there is not a similar figure F' (i.e., with its points having reciprocal distances in proportion to distances between corresponding points on F) of arbitrary size (J. Wallis [1616–1703]).

(iii) The sum of the internal angles for one (hence, of all) triangle ABC is less than a straight angle (G. Saccheri).

(iv) More, the quantity $\pi - (\hat{A} + \hat{B} + \hat{C})$ is proportional to the area of ABC (J.E. Lambert [1728–1777]); hence, the proportionality constant is an *absolute quantity*.

(v) There are distinct parallel lines a and b having a common point and a common orthogonal line at infinity (G. Saccheri).

(vi) There are points P, Q, R which are not collinear, such that no circle passes through all of them (W. Bolyai [1775–1856], father of J. Bolyai).

Lambert went on to observe that (iv) holds, with reversed signs, for geodesic triangles on a sphere of radius r:

$$\text{Area}(ABC) = r^2(\hat{A} + \hat{B} + \hat{C} - \pi);$$

hence, that (iv) could be seen as a phenomenon taking place on a sphere of imaginary radius. This viewpoint played a major role in the development of non-Euclidean geometry.

Finally, independently of each other, N.I. Lobachevsky [1793–1856], J. Bolyai [1802–1860] and C.F. Gauss [1777–1855] totally changed perspective. Starting from the assumption that the Parallel Postulate could not be proved based on the remaining ones, they assumed it did not hold and went on developing the corresponding theory, trusting that no contradiction could possibly arise. They all shared the view that geometry was a description of physical space, and Gauss and Lobachevsky even compared their non-Euclidean geometry (as Gauss called it) and the absolute quantity already pointed out by Lambert, with available astronomical evidence. They deduced that, in real space, the value of r must be very large. Lobachevsky and Bolyai published their foundings, but Gauss did not: because he did not want to be involved in a philosophical–mathematical struggle, but also because he learned of the research of Bolyai and Lobachevsky and was satisfied that others had gone public with the new geometry.

At this stage we have two competing geometries: Euclidean and non-Euclidean, with and without the Fifth Postulate. There also was a substantial body of *absolute geometry*, which was the intersection of the two. Bonola suggests that Kant's doctrine of space, according to which Euclidean Geometry was the main concrete example of *synthetic a priori knowledge*, acted against the acceptance of, or even the debate about non-Euclidean geometry. J. Gray [18] argues that, at the time Bolyai and Lobachevsky published their work, Kantian philosophy was not hegemonic as it had been a generation before. On the mathematical side, Lobachevsky proposed to model non-Euclidean geometry using as analytic model hyperbolic trigonometry, the non-Euclidean version of trigonometry (or, better, the "imaginary" version of spherical trigonometry). He was logically on firm ground, but the mathematical community seem not to have reacted to his proposal.

Another fact has to be taken into account. At the beginning of nineteenth century, Euclidean geometry was considered the least questionable of all sciences (an ancient state of affairs, of course, that had also directed Kant to exemplify by it the notion of "synthetic a priori"), while the conceptual foundations of calculus were still matter of debate and controversy. It was through the work of Cauchy, Bolzano and, especially, Weierstrass, that mathematical analysis, hence the differential geometry of surfaces, reached a level of philosophical reliability comparable (for the standards of the time) to that of geometry [12]. Weierstrass lectured on the foundations of calculus in 1859–1860. The 1860s were an excellent decade for developing analytic models for geometry.

Starting with Gauss work on surfaces, a substantial body of knowledge had been accumulating on surfaces of constant curvature. The reason is clear: on these surfaces only could figures be moved freely, at least locally, without alteration of their metric properties. Gauss published his *Theorema egregium* in 1827 and it was already clear that, if figures could be moved isometrically, curvature had to be constant. Minding observed that the converse was true in the 1830s, and he found various surfaces of constant negative curvature in Euclidean space, the tractroid among them. Liouville found other examples in 1850, one of which – the Liouville-Beltrami half plane- will be discussed below. Riemann, in 1854, revolutioned the concepts of geometry in his *Habilitationschrift*, in which surfaces of constant curvature played a distinguished and exemplary role. Codazzi, in 1857, found that the trigonometric formulae on the tractroid could be obtained by those on the sphere by considering the radius as imaginary. With the exception of Riemann's work, at that time unpublished. See [19] for an extensive and very readable account of the early history of differential geometry. Beltrami knew all of this literature, which he brought to synthesis, and more, in the *Saggio*.

3 The Models of Beltrami

Between 1868 and 1869, in two influential articles, Beltrami provided models of the non-Euclidean geometry of Lobachevsky and Bolyai. One of these models is better known as the *Klein model*, another one as the *Poincaré disc model*, the third as the *Poincaré half plane model*. A fourth one, which Beltrami worked out, like the disc model, directly from Riemann's *Habilitationschrift*, has had a much minor impact. In a sense, the first article, *Saggio di interpretazione della Geometria non-euclidea* [5], was written under the influence of Gauss, and the second, *Teoria fondamentale degli spazi di curvatura costante* [6], under the influence of Riemann. In this section we first give an exposition of the mathematical content of the two papers, then we will try to clarify some points of methodology and philosophy.

3.1 The "Projective" Model

In his *Saggio* ([5], p. 377), Beltrami introduces a family of Riemannian metrics on the disc of radius $a > 0$ centered at the origin by letting

$$ds^2 = R^2 \frac{(a^2 - v^2)du^2 + 2uv\,du\,dv + (a^2 - u^2)dv^2}{(a^2 - u^2 - v^2)^2}. \tag{1}$$

The Gaussian curvature of ds^2 is the constant $-1/R^2$. The advantage of such metric is that its geodesics are straight lines.

Where does the metric ds^2 come from? In 1865 ([3], p. 262–280), Beltrami had considered the problem of finding all surfaces which could be parametrized in such a way that all geodesics were given by straight lines. The starting point was Lagrange's rather obvious observation that, by projecting the sphere S onto a plane Π from the center of S, the geodesics of S (great circles) were mapped $2-1$ onto straight lines in Π, and viceversa. To his surprise, Beltrami found out that the only surfaces which had such distinguished parametrizations where the surfaces of constant curvature. In the positive curvature case, he computed the Riemannian metric

$$ds^2 = R^2 \frac{(a^2 + v^2)du^2 - 2uv\,du\,dv + (a^2 + u^2)dv^2}{(a^2 + u^2 + v^2)^2}, \tag{2}$$

which is in fact the spherical metric on a sphere of radius R, written after the sphere has been projected from its center C to a plane Π having distance a from C. He had probably worked out the analogous expression (1) for the negative curvature case, but he did not write it in the article *per abbreviare il discorso*, "for the sake of brevity" ([3], p. 276). Beltrami also mentions ([5], Nota I, p. 399) that (1) can be obtained from (2) by changing a and R into their imaginary counterparts ia and iR, according to the general euristic principle we have already mentioned, that hyperbolic geometry is spherical geometry on a sphere of imaginary radius. Since the equations of the geodesics are the same in both the real and in the imaginary case, the geodesics with respect to the coordinates u, v are straight lines. Hence, they are chords of the limiting circle.

The angle θ between two "extrisically perpendicular" geodesics $u = $ const and $v = $ const is computed:

$$\cos(\theta) = \frac{uv}{\sqrt{(a^2 - u^2)(a^2 - v^2)}}; \sin(\theta) = \frac{a\sqrt{a^2 - u^2 - v^2}}{\sqrt{(a^2 - u^2)(a^2 - v^2)}}. \tag{3}$$

Form the expression of $sin(\theta)$ we see that, when the two geodesics meet at a point on the circle at infinity, the angle they make vanishes.[5] This is exactly the

---

[5] This is true also for geodesics not in the form $u = const$ and $v = const$. See the formulae at p. 381 in the *Saggio*.

phenomenon (recalled in Sect. 2, (v)) that Saccheri had found "repugnant to the nature of the straight line".

Beltrami's first aim is proving two basic properties he singles out as crucial in the introduction to the *Saggio* ([5], p. 375–376):

(A1) Given two distinct points A and B, there is exactly one line passing through them ([5], p. 381).
(A2) Given points A and B and oriented lines $l \ni A$ and $m \ni B$, there is a rigid movement of the plane which maps A to B and l into m, preserving the orientation ([5], Nota II, p. 400–405: I summarize in twentieth century language what Beltrami writes as "principle of superposition" at p. 376).

The first property does not hold, he says, in the positive curvature case. He is thinking of the sphere model and the statement, as such, is not wholly accurate. In fact, as Clifford had shown, but not linked to the dibate on the Parallel Postulate, the real projective space has the properties (A1) and (A2). It was known that properties (A1) and (A2) locally hold on surfaces of constant negative curvature: Beltrami's goal is realizing a model in which they hold globally.

Here, Beltrami explains, "point" means "point on the surface of constant curvaturem parametrized by (u, v) in the disc $\{(u, v) : u^2 + v^2 < a^2\}$" and "line" is a geodesic for the metric (1). I use the expression "rigid movement" for any transformation which preserves the metric (Beltrami writes of the possibility of superposing without restrictions a figure onto another). To have a complete picture of the model, it is useful to know:

(A3) Lines can be indefinitely prolonged ([5], p. 380 for geodesics passing through the origin; extend to the general case using (A3)) and they have infinite length in both directions.
(A4) The plane is simply connected ([5], p. 378).
(A5) There are more lines passing through a point $A \notin l$, which do not meet l ([5], p. 382).

Beltrami calls *psedudospheres* the surfaces bijectively parametrized by the coordinates (u, v) and endowed with the metric (1) (below, we will use in differently *pseudosphere, hyperbolic plane, non-Euclidean plane*). At p. 381, he correctly says that the theorems of non-Euclidean geometry apply (translating geometric terms into differential-geometric terms as above) to the pseudospheres. We know that the implication goes in the other direction as well: if a question is posed in terms of non-Euclidean geometry and it is answered working on the pseudosphere model, we can consider that answer as belonging to non-Euclidean geometry; the same way questions in Euclidean geometry can be answered in terms of the Cartesian plane.[6] How aware was Beltrami of this?

---

[6]This amounts to say that, for each given value of R there is, but for isometries, exactly one non-Euclidean plane having "radius of curvature" equal to R (where R is connected to the universal geometric constants envisioned by Lobachevsky, Bolyai, and Gauss).

It is my opinion that he was personally convinced, but hesitant to state it explicitely. He surely lacked the conceptual frame which was going to be provided, much later, by Peano and Hilbert [27]. He also did not have the philosophical confidence of Riemann, who simply avoided this kind of questions, to go directly to the general science of space. There is little doubt, I think, that, for Riemann, non-Euclidean, hyperbolic geometry *was* that of the model he had mentioned in the *Habilitationschrift* and that Beltrami was going to study in depth in his *Teoria fondamentale* [6].

Instead of simply saying that his model was, rather than contained, non-Euclidean geometry, Beltrami further proceeds in two directions. On the one hand, he goes on to deduce from his model a number of important theorems in non-Euclidean geometry, as to convince the reader and himself that his model correctly answers all reasonable questions in non-Euclidean geometry. On the other hand, he seems to worry that the surface with the metric (1) might not be considered wholly "real", since it is not clear in which relation it stands with respect to Euclidean three space (the strictest measure of "reality"). Then, he will show that, after cutting pieces of it, the pseudosphere can be isometrically folded onto a "real" constant curvature surface in Euclidean space.[7] If it is true that both directions seem to be suggested by logical and philosophical hesitation, this hesitation induces Beltrami to do some really elegant mathematics, and to unravel some very interesting features of his model.

After one knows that the geodesics for (1) are chords of the unit disc, it is obvious that the geodesics in this metric satisfy the incidence properties of straight lines in non-Euclidean geometry, and Beltrami is very coscious in proving this with all details ([5], p. 382). In the *Nota II* ([5], Nota II, p. 400–405), as we said in (A2), Beltrami shows that the metric ds^2 is invariant under rotations around the origin and, for each fixed (u_0, v_0), there is a metric preserving transformation of the unit disc, which maps $(0, 0)$ to (u_0, v_0): that is, with respect to the metric (1), the disc is–in contemporary terms–homogeneous and isotropic; figures can be moved around with no more restrictions than the ones they are subjected to in Euclidean plane. Being the geodesics straight lines, one is not surprised in learning that the isometries are the projective transformations of the plane fixing the circle at infinity. The isometries of the pseudosphere are compositions of (Euclidean with respect to u, v coordinates) rotations around the origin, reflections in the coordinate axis and maps of the form

$$(u, v) \mapsto \left(\frac{a^2(u - r_0)}{a^2 - r_0 u}, \frac{a\sqrt{a^2 - r_0^2}\, v}{a^2 - r_0 u} \right), \quad 0 < r_0 < a.$$

---

[7] As a matter of fact, later Hilbert showed that no regular surface in Euclidean three space is isometric to the whole pseudosphere. Much later, Nash showed that any surface can be isometrically imbedded in a Euclidean space having dimension large enough.

This is one of the two points in which Beltrami makes explicit the connection between his model and projective geometry, which will be later developed in depth by Felix Klein.

The distance ρ between the origin and (u, v) is computed by straightforward integration as

$$\rho = \frac{R}{2} \log \frac{a + \sqrt{u^2 + v^2}}{a - \sqrt{u^2 + v^2}}, \tag{4}$$

which implies (A3). Also, using another integration of (1) on a Euclidean (hence, by (4), non-Euclidean) circle centered at the origin, one deduces that the semiperimeter of the non-Euclidean circle having radius ρ is

$$\pi R \sinh(\rho/R),$$

a formula already known to Gauss ([5], p. 380 and 384).

Beltrami now turns his attention to the *angle of parallelism*. Consider two geodesics α and β meeting at right angles at a point P, a point Q on β at non-Euclidean distance δ from P and let ζ, ξ be the two geodesics passing through Q which are *parallel* to α (that is, which separate the geodesics through Q which meet, and the ones which do not, α). Let Δ be the angle ζ and ξ make with β (by reflection invariance, it is the same). Assume Q is the origin in the model. Using the facts that (1) is conformal to the Euclidean metric at the origin and that geodesics are straight lines, and (4), one easily computes

$$\tan(\Delta) = 1/\sinh(\delta/R),$$

which had been previously found by Battaglini on surfaces of constant negative curvature. After standard manipulation, this becomes Lobachevsky's pivotal formula

$$\tan(\Delta/2) = e^{-\frac{\delta}{R}}. \tag{5}$$

Among the other theorems Beltrami deduces from his models we find the main formulae of non-Euclidean trigonometry and the formula for the area of a geodesic triangle having angles $\hat{A}$, $\hat{B}$, $\hat{C}$;

$$R^2(\pi - \hat{A} - \hat{B} - \hat{C}).$$

When the angles vanishe, we obtain the single value πR^2 for the area of a geodesic triangle with all vertices at infinity. It is curious that ([5], p. 389) when mentioning that the value "is independent of its [the triangle's] shape", Beltrami fails to observe that all such triangles are in fact isometric and hence, from the pseudospherical point of view, they have in fact the same shape.

At p. 387, between the discussion of trigonometry and that about areas, Beltrami writes: "The preceeding results seem to fully show the correspondence between non-Euclidean (two-dimensional) geometry and geometry of the pseudosphere."

Again, this is evidence that he thought of his model as essentially identical to Lobachevsky's geometry, but he lacked a conceptual frame for defending his belief in front of possible criticism; hence, he preferred to keep a reasonably low profile.

We now come to the second appearance of projective geometry in the article ([5], p. 391–392). In the Nota II, Beltrami finds that the equation of the pseudospherical circle having radius ρ and center (u_0, v_0) is

$$\frac{a^2 - u_0 u - v_0 v}{\sqrt{(a^2 - u^2 - v^2)(a^2 - u_0^2 - v_0^2)}} = \cosh(\rho/R) \geq 1. \tag{6}$$

From Gauss Lemma, such circles are orthogonal to the geodesics issuing from (u_0, v_0). In Beltrami's model, however, it makes sense to find the curves which are orthogonal to the geodesics passing through (u_0, v_0) even when $u_0^2 + v_0^2 = a^2$ (a *point at infinity*) or $u_0^2 + v_0^2 > a^2$ (an *ideal point*). The equation of the orthogonal curve is deduced the same way as (6), and it has a similar form,

$$\frac{a^2 - u_0 u - v_0 v}{\sqrt{a^2 - u^2 - v^2}} = C, \ u^2 + v^2 < a^2. \tag{7}$$

We might be called *generalized (metric) circles* the curves γ_C given by (7). The curves γ_C corresponding to the same (u_0, v_0), but to different values of C, are equidistant as it is easy to show. When the center (u_0, v_0) is ideal, the value $C = 0$ is admissible and γ_0 is a geodesic, in pseudospherical terms, and it is the polar of the point (u_0, v_0) with respect to the limiting circle. For $C \neq 0$, γ_C is a set of points having a fixed distance from the geodesic γ_0 and γ_C *is not* a geodesic: a fact which reminds us of early attempts to prove Euclid's Fifth Postulate.

While the case of the ideal points seems to have been forgotten in contemporary hyperbolic pop-culture, the case of the points at infinity has not. Suppose that $u_1^2 + v_1^2 = a^2$, rewrite (6) with $\rho' - \rho$ instead of ρ,

$$\frac{a^2 - u_0 u - v_0 v}{\sqrt{a^2 - u^2 - v^2}} = \sqrt{a^2 - u_0^2 - v_0^2} \cosh\left(\frac{\rho' - \rho}{R}\right), \tag{8}$$

and let $\rho' \to \infty$ and $(u_0, v_0) \to (u_1, v_1)$ in such a way that

$$\frac{1}{2}\sqrt{a^2 - u_0^2 - v_0^2} e^{\rho'/R} \to a.$$

Then, the geodesic circle $S((u_0, v_0), \rho' - \rho)$ having equation (9) stabilizes to the *horocicle* $H((u_1, v_1), \rho)$ having equation

$$\frac{a^2 - u_1 u - v_1 v}{\sqrt{a^2 - u^2 - v^2}} = a e^{-\frac{\rho}{R}}. \tag{9}$$

The distance between two concentric horocicles $H((u_1, v_1), \rho_0)$ and $H((u_1, v_1), \rho_1)$ is clearly constant and has value $|\rho_1 - \rho_0|$. By a limiting argument, it is also easily deduced that the horocycle $H((u_1, v_1), \rho)$ is normal to all geodesics issuing from its center (u_1, v_1); that is, to all geodesics converging to the point at infinity (u_1, v_1).

Beltrami has, this way a unifying interpretation in terms of generalized circles having center in the Euclidean plane for three important different non-Euclidean objects: the (usual) metric circle, the horocycle, the set of the points having a fixed distance from a geodesic. (The reader will figure out without difficulty what the corresponding objects are in Euclidean and in spherical geometry).

Another important feature of the *Saggio*, as said before, is that Beltrami "folds" his pseudosphere onto surfaces of constant curvature in Euclidean space. His clever construction, however, is easier if translated in the "conformal" Liouville–Beltrami and Riemann–Beltrami–Poincaré models; hence, we will postpone it after we have discussed the second article, *Teoria fondamentale*.

It seems evident that the projective nature of the model was perfectly clear to Beltrami: (1) the metric is invariant under projective transformations; (2) important geometric objects (the circles with ideal center, for instance) had to do with projective concepts (the polar of a point with respect to a circle, to remain in the example). Probably, Beltrami did not pursue this line of investigation because it was not to his taste. Klein's main contribution was that the Riemannian distance produced by the metric had a projective interpretation (as the logarithm of a bi-ratio; see [11, 16] for extensive accounts of this), hence that the model could be fully developed within *synthetic* geometry, without resorting to infinitesimal calculus. The preference for synthetic arguments goes in a direction which is opposite to the one taken by Riemann in the footsteps of Gauss, and by Beltrami in the footsteps of both. It has its own merits and it had a great influence in the discussion on the foundation of mathematics, but, as far as geometry is concerned, the major conceptual developments of the past century see synthetic geometry in an ancillary position.

3.1.1 Anisotropy

Let us spend some words for some remarks on the metric (1) itself. We will let $a = 1$, since the particular value of a has no importance for us. We can rewrite (1) as

$$\mathrm{d}s^2 = R^2 \frac{\mathrm{d}u^2 + \mathrm{d}v^2}{1 - u^2 - v^2} + R^2 \frac{(u\mathrm{d}u + v\mathrm{d}v)^2}{(1 - u^2 - v^2)^2}.$$

In this form, it is easy to see that the infinitesimal discs centered at $(r, 0)$ (this is no loss of generality in view of rotational invariance) and having radius $\mathrm{d}r$ are ellipsis having a semiaxis $\frac{(1-r^2)}{R}\mathrm{d}r$ in the u direction and $\frac{(1-r^2)^{1/2}}{R}\mathrm{d}r$ in the v direction. In fact, the metric at $(r, 0)$ becomes

$$ds^2 = R^2 \frac{du^2 + dv^2}{1-r^2} + R^2 \frac{(r\,du)^2}{(1-r^2)^2} = R^2 \frac{du^2}{(1-r^2)^2} + R^2 \frac{dv^2}{1-r^2}.$$

This important eccentricity says that the metric is very far from being conformal to the Euclidean metric in (u, v) coordinates (a metric ds^2 ic *conformal* to the metric $du^2 + dv^2$ if $ds^2 = f(u, v)(du^2 + dv^2)$ for some strictly positive function $f(u, v)$). It is interesting that, in a foundational article published the year before, *Delle variabili complesse su una superficie curva, 1867* [4] p. 318–373, Beltrami himself had considered and solved the problem of finding conformal coordinates on a (Riemannian) surface. It is possible that, at the time of writing the *Saggio*, Beltrami had worked out conformal coordinates for the metric (1), possibly the ones associated with the so-called Poincaré disc model. In fact, Beltrami ([5], p. 378) says that "the form of this expression [i.e, ds^2 as in (1)], although *less simple than other equivalent forms which might be obtained by introducing different variables*, has the peculiar advantage [...] that any linear equation in the u, v variables represents a geodesic line and that, conversely, each geodesic line is represented by a linear equation" [my translation]. Since at the time he sent the article to the journal he had already read Riemann's *Habilitationschrift*, however, Beltrami could also have found there a disc model he had not thought of before.

3.2 The "Conformal" Models

When he learned of Riemann's *Habilitationschrift*, Beltrami was in the position of developing a remark contained in it, that the length element of an n-dimensional manifold having constant (Riemannian) curvature α ($\alpha \in \mathbb{R}$) could be written in the form

$$ds^2 = \frac{\sum_j dx_j^2}{1 + \frac{\alpha}{4} \sum_j x_j^2}. \tag{10}$$

More important than this, Riemann had given solid philosophical and scientific, as well as geometric, foundations for a theory of manifolds in which Euclidean spaces played no priviledge role (except for the infinitesimal structure of space, about which, by the way, he had interesting questions and observations). To wit: (1) geometry was the study of n-dimensional manifolds endowed with a length structure (an idea which is still actual; see for instance [13]); (2) the "geometricity" of a space was not, then, less so because the space was not a subspace of Euclidean three-space. Backed by this philosophy, which he however did not fully endorse in the *Teoria*, Beltrami could present his "purely analytic" calculations as (almost) sound geometry.

The starting point of the *Teoria* is a different way to write down the (n-dimensional) length element ds^2 in (1),

$$ds = R\frac{\sqrt{dx^2 + dx_1^2 + \cdots + dx_n^2}}{x}, \tag{11}$$

subjected to the constraint

$$x^2 + x_1^2 + \cdots + x_n^2 = a^2 \tag{12}$$

([6], p. 407). A formal calculation, in fact, shows that this is the same as the n-dimensional version of (1), already cited in the *Saggio*:

$$ds^2 = R^2\left(\frac{|d\underline{x}|^2}{a^2 - |\underline{x}|^2} - \frac{\underline{x} \cdot d\underline{x}}{(a^2 - |\underline{x}|^2)^2}\right), \tag{13}$$

with $\underline{x} = (x_1, \ldots, x_n)$ and obvious vector notation. The interest of the new way to write the metric is that the metric ds in (11) can be thought of as a metric living in the "right" half-space $H_{n+1}^+ = \{(x, x_1, \ldots, x_n) : x > 0\}$, restricted to the half-sphere $S_{n+1}^+(a)$ given by the constraint. The space H_{n+1}^+ with the metric (11) is itself a model for the $(n+1)$-dimensional non-Euclidean geometry, as we shall see shortly.

We will use the notation $B_n(a) = \{(x_1, \ldots, x_n) : x_1^2 + \cdots + x_n^2 < a^2\}$. With Beltrami, we take x to be a variable dependent on $(x_1, \ldots, x_n)$ in $B_n(a)$. A variational argument shows that the geodesics for the metric (11) are chords of the ball $B_n(a)$, as in the 2-dimensional case. This also guarantees that the Riemannian manifold here introduced is simply connected and that it has the property that there is exactly one geodesic passing through any two given points. Next, Beltrami shows that his metric is invariant under a (Euclidean) rotation around the center of $B_n(a)$. This would immediately follow from the expression (13) for (11), but Beltrami prefers to point out that each of the following is invariant under such rotations: (1) the metric (11) when thought of as a metric on H_{n+1}^+ (here, the rotation acts on the coordinates $x_1, \ldots, x_n$ and leaves x fixed); (2) both sets H_{n+1}^+ and $S_{n+1}(a)$. He then shows by a lengthy argument that metric isometries act transitively on $B_n(a)$. We skip here the argument and only give its conclusion ([6], p. 416): the isometries for (11)–(12) are those projective transformations of Euclidean n-space which map $B_n(a)$ onto itself.

Next, the metric (11) is written in polar coordinates $\underline{x} = r\Lambda = r(\lambda_1, \ldots, \lambda_n)$, with $0 \le r < a$ and $|\Lambda| = 1$:

$$ds^2 = R^2\frac{a^2 dr^2}{a^2 - r^2} + R^2\frac{|d\Lambda|^2}{(a^2 - r^2)^2}.$$

Changing from r to ρ, the non-Euclidean distance from the origin, $d\rho = R a dr/(a^2 - r^2)$, one obtains

$$ds^2 = d\rho^2 + R^2 \sinh^2(\rho/R)|d\Lambda|^2. \tag{14}$$

Introducing new coordinates by radially stretching the metric,

$$\xi_j = 2R\tanh(\rho/2R)\lambda_j, \quad j = 1,\ldots,n,$$

and doing elementary calculations with hyperbolic functions, one finally obtains that

$$ds^2 = \frac{\sum_j d\xi_j^2}{1 - \frac{\sum_j \xi_j^2}{4R^2}}, \qquad (15)$$

which is Riemann's choice for the coordinates of the n-dimensional space of constant curvature. From here, closely following Riemann's exposition in [26], Beltrami computes that the curvature of the space metrized by (15) (hence, the metric (11) itself) is $-1/R^2$. In the two-dimensional case, (15) was independently rediscovered by Poincaré in [25].

The third model is obtained by a fractional transformation of the coordinates $(x, x_1, \ldots, x_{n-1})$ in (11) (while keeping the constraint (12)):

$$(\eta, \eta_1, \ldots, \eta_{n-1}) = \left(\frac{Rx}{a - x_n}, \frac{Rx_1}{a - x_n}, \ldots, \frac{Rx_{n-1}}{a - x_n}\right).$$

In the new coordinates,

$$ds^2 = R^2 \frac{d\eta^2 + d\eta_1^2 + \cdots + \eta_{n-1}^2}{[\eta^2]}. \qquad (16)$$

In dimension two, as Beltrami notes, the metric had been computed by Liouville in his *Note IV* to Monge's *Application de l'Analyse à la Géométrie* ([22], p. 600), at the end of a discussion on Gauss' *Theorema Egregium*. What is interesting, Beltrami notes, is that the metric (16) is no other than the metric (11) in the ambient space (without the constraint (12)), with one less dimension.

The original metric (11) (hence, the two-dimensional metric (1) of the *Saggio*) might then so be interpreted. Start with the (conformal) metric (11) on the right half-space H_{n+1}^+. The metric (11) restricted to the half n-dimensional sphere $S_{n+1}^+(a)$, makes $S_{n+1}^+(a)$ in a model of n-dimensional pseudo-sphere. Beltrami proves so, but we know that $S_{n+1}^+(a)$ contains the whole geodesic for (11) connecting any two of its points (this is clear, once we know that the geodesics for (11) are arcs of circles or straight lines, which are perpendicular to the boundary of H_{n+1}^+). The projective metric (1), and its n-dimensional generalizations, is the projection of the conformal metric (11) on the disc cut by $S_{n+1}(a)$ on the boundary of H_{n+1}^+.

A special feature of the metric (16) is that the horocycles Γ_k having center at the point at infinity of H_{n+1}^+ have equation $\eta = k$, for a positive constant k. It is immediate, then, that the restriction of the pseudospherical metric to Γ_k is (a rescaled version of) the Euclidean metric; a fact already noted by Bolyai in the three-dimensional case.

3.2.1 Imbedding the Pseudosphere in Euclidean Space

In his *Saggio,* Beltrami found three surfaces in Euclidean space which carried his metric (1), by cleverly folding the pseudosphere.[8] One at least of these surfaces was previously known. What Beltrami was interested in was finding "real" models for non-Euclidean geometry; i.e. surfaces of constant curvature in Euclidean space. This he did in the *Saggio,* written before reading Riemann; at a time when Beltrami kept a rather conservative profile about "reality" of geometric objects.

The construction of the three surfaces is easier to see, I believe, if carried out in the conformal models of the later *Teoria*; I will then translate them in that context. I will also let $R = 1$, leaving to the reader to figure out the elementary modifications for different values of the curvature. The general frame adopted by Beltrami is the following. The metric

$$ds^2 = d\rho^2 + G(\rho)^2 d\theta^2$$

is the length element on a surface of revolution S, θ being the angular coordinate and $d\rho$ being the length element on the generatrix of S, provided $|G'(\rho)| \leq 1$, because dG is the projection of $d\rho$ in the direction orthogonal to the axis of revolution.

We start from the *horocycle construction*, which also is the best known. Consider the Liouville–Beltrami metric

$$ds^2 = \frac{dx^2 + dy^2}{y^2}, \; y > 0. \tag{17}$$

Let $d\rho = dy/y$ (then, ρ is hyperbolic length on any geodesic $x = $ const) and consider $0 \leq x \leq 2\pi$. We set $\rho = \log y$ (we will recover the other primitives using an isometry). The metric ds^2 becomes

$$ds^2 = d\rho^2 + e^{-2\rho} dx^2. \tag{18}$$

The condition $|dG| \leq d\rho$ becomes $\rho \geq 0$; i.e. $y \geq 1$. The surface of revolution S_1 so obtained is the one often finds in articles and web pages concerning Beltrami's models. It is rather confusing the use of calling such surface "pseudosphere", the term Beltrami uses for the (simply connected) model he has for the non-Euclidean plane. Being the generating curve of S_1 a *tractrix*, the term *tractroid*,

---

[8] At this point only I disagree with Gray's account of Beltrami's work. He writes, p. 208 in [19]: "...but all Beltrami did was hint that the pseudosphere [the tractroid, in this paper's terminology] must be cut open before there is any chance of a map between it and the disc [Beltrami's projective model]". In fact, Beltrami's hint at p. 390 follows his formula (12), which can be used to find (after a cut) a surface of revolution in Euclidean space, what we call here S_3. Formula (14) at p. 393 leads to our S_2 and, finally, formula (17) at p. 394 leads to the tractroid, our S_1 below. The tractrix, called by Beltrami *linea dalle tangenti costanti* (constant tangent curve) is explicitly mentioned at p. 395. I find it indicative of Beltrami virtuosism, that he managed to exhibit three different surfaces of revolution by using the projective model, which is not especially amenable to this sort of calculations.

which is sometimes used, is better (but ugly). The part of the pseudosphere which can be mapped onto the surface S_1 in a one-to-one way (but for a curve) is $\{(x, y) : y \geq 1, 0 \leq x \leq 2\pi\}$. The vertical sides $x = 0$, $y \geq 1$ and $x = 2\pi$, $y \geq 1$ are half-geodesics which have to be glued together. The resulting surface in Euclidean space looks like an exponentially thin thorn as $y \to +\infty$ and as a trumpet near $y = 1$. At the aperture of the trumpet, the metric of S_1 can not be carried by Euclidean space anymore.

The Euclidean dilation $(x, y) \mapsto (\lambda x, \lambda y)$, $\lambda > 0$, is clearly an isometry for ds^2 in (17). By composing with these dilations, we find other regions in the pseudosphere which can be mapped onto the surface S_1; regions which might be otherwise be found by choosing the primitives $\rho = \log(y/y_0)$ with $y_0 > 0$, as Beltrami did in the *Saggio*. By isometry, however, the tractroid does not change with y_0.

The contemporary way to think of S_1 would be to take a quotient of the pseudosphere with the parameters x, y with respect to the action of the group of isometries generated by $(x, y) \mapsto (x + 2\pi, y)$. The resulting Riemannian surface, then, can be embedded in Euclidean space for the part where $y \geq 1$ as a tractroid, as we have just seen.

We now give the corresponding construction with two *geodesic circles having ideal center* playing the role that was above played by a horocycle. Again in the upper plane model, consider the geodesic γ_∞ having equation $x = 0$. The lines γ_m with equation $y = mx$, $m \in \mathbb{R}$, are then equidistant from γ_0 (proof: the lines $y = \pm mx$ are the envelope of a family of circles having constant hyperbolic radius and centers on γ_0). Writing $x = r\cos(\theta)$, $y = r\sin(\theta)$ in polar coordinates, the metric becomes

$$ds^2 = \frac{d\theta^2}{\sin^2(\theta)} + \frac{dr^2}{r^2 \sin^2(\theta)}.$$

Let $d\theta/\sin(\theta) = d\rho$, $\rho = \log(\tan(\theta/2))$, and $dr/r = dt$, $r = e^t$. Then,

$$ds^2 = d\rho^2 + z(\rho)^2 dt^2, \tag{19}$$

where $z(\rho) = 1/\sin(\theta)$. It follows that $|dz/d\rho| = |1/\tan(\theta)| \leq 1$ if and only if $|\theta - \pi/2| \leq \pi/4$. Keeping in mind that we want the "angle" coordinate t in $[A, A + 2\pi]$, we have that any Euclidean circular sector

$$\Omega = \{(x, y) : y \geq |x|, r_0 \leq \sqrt{x^2 + y^2} \leq r_0 e^{2\pi}\},$$

with $r_0 > 0$, is isometric to a surface of revolution S_2 in Euclidean space. The two arcs of circles (which are geodesics for the metric) are glued together (identifying points lying on the same line $y = mx$). The segments of the lines $y = \pm x$ lying on the boundary of Ω are apertures corresponding to the aperture of the trumpet in the horocycle construction. While in S_1 the aperture is a piece of horocycle, in this case it is made by two pieces of circle having ideal center. Contrary to the case of the surface S_1, in this case we can construct a one-parameter family of non-isometric surfaces $S_2(k)$ by integrating $dr/r = k dt$ ($k > 0$), having then $z(\rho) = 1/[k\sin(\theta)]$: as k grows larger, the interval of the allowable r's becomes thinner, while the circles with ideal center $y = mx$ can range over a greater range of m's. In this case, reverting to $k = 1$, group of isometries generated by $(x, y) \mapsto (e^{2\pi} x, e^{2\pi} y)$ (Euclidean dilations) is the one giving (an extension of) S_2 as quotient.

Lastly, we consider the conformal disc model instead,

$$ds^2 = \frac{dx^2 + dy^2}{\left(1 - \frac{x^2+y^2}{4}\right)^2} = \frac{dr^2 + r^2 d\theta^2}{\left(1 - \frac{r^2}{4}\right)^2},$$

in usual polar coordinates (r, θ). We set $d\rho = dr/(1 - r^2/4)$ (ρ is then non-Euclidean distance from the origin), so that $\tanh(\rho/2) = r/2$ and $r/(1 - r^2/4) = \sinh(\rho)$. The metric then becomes

$$ds^2 = d\rho^2 + \sinh^2(\rho) d\theta^2. \tag{20}$$

Since $(d/d\rho)\sinh(\rho) = \cosh(\theta) \geq 1$, (20) can not be the length element of a surface of revolution S_3 whose angular coordinate is θ. In fact, Beltrami points out that, if such were the case, then S_3 would have, by symmetry, equal principal curvatures at the point corresponding to $(x, y) = (0, 0)$; it could not then have negative curvature there. The remedy is considering $0 \leq \theta \leq 2\pi\epsilon$ only, with $\epsilon < 1$, and letting $t = \theta/\epsilon$ be the angular coordinate. This way, the metric becomes $ds^2 = d\rho^2 + \epsilon^2 \sinh^2(\rho) dt^2$, which defines the metric for a surface of revolution S_3^ϵ provided that $\cosh(\rho) \leq 1/\epsilon$. The surface S_3^ϵ has a cusp at the point corresponding to the origin, where the total angle is only $2\pi\epsilon$. The aperture at its other hand is an arc of the geodesic circle centered at the origin and having radius $\rho_\epsilon = \cosh^{-1}(1/\epsilon)$. It is intuitive (and can be proved) that S_3^ϵ converges to S_1 as ϵ tends to zero. The isometry group producing S_3^ϵ as a quotient, in the case $\epsilon = 1/n$ ($n \geq 2$ integer), is the one generated by a rotation of an angle $2\pi/n$.

In all the three types of surfaces we have seen the gluing is done along geodesic segments. The apertures due to the partial imbeddability in Euclidean space are always arcs of generalized circles: a horocycle for S_1; two circles with ideal center for S_2 and a metric circle for S_3. In this case as well we see the tripartition of the "projective" generalized circles emerging very elegantly from Beltrami's *Saggio*.

3.2.2 Anachronism (for the Complex Reader)

If we replace the unit disc in the real plane $\mathbb{R}^2$ by the unit ball $\mathbb{B}_{\mathbb{C}^2} = \{(z, w) : |z|^2 + |w|^2 < 1\}$ in the complex 2-space $\mathbb{C}^2$ and we replace real by complex variables, we might consider (1) as the restriction to the real disc $\mathbb{B}_{\mathbb{R}^2} = \{(x, y) : x^2 + u^2 < 1\} \subset \mathbb{B}_{\mathbb{C}^2}$ of the metric

$$d\sigma^2 = R^2 \frac{|dz|^2 + |dw|^2}{1 - |z|^2 - |w|^2} + R^2 \frac{|\bar{z}dz + \bar{w}dw|^2}{(1 - |z|^2 - |w|^2)^2}. \tag{21}$$

People working several complex variables will immediately recognize $d\sigma^2$ as the *Bergman metric* on $\mathbb{B}_{\mathbb{C}^2}$; i.e. of the (essentially) unique bi-holomorphically invariant Riemannian metric $\mathbb{B}_{\mathbb{C}^2}$. Beltrami's first model, then, can be interpreted as a totally geodesic surface in Bergman's complex ball. Easy considerations show that the automorphism group of $\mathbb{B}_{\mathbb{C}^2}$ acts transitively on the tangent space of the real disc $\mathbb{B}_{\mathbb{R}^2} = \{(z, w,) \in \mathbb{B}_{\mathbb{C}^2} : z, w \in \mathbb{R}\}$; hence, that by a theorem of Gauss, as noted by Beltrami– the metric ds^2 has constant curvature. It is easy to see that another totally geodesic surface in $B_{\mathbb{C}^2}$ can be obtained by letting $w = 0$ in. This gives the metric

$$d\tau^2 = R^2 \frac{|dz|^2}{(1 - |z|^2)^2} \tag{22}$$

on the complex disc $\mathbb{B}_{\mathbb{C}^1}$. The metric $d\tau^2$ makes the unit complex disc $\mathbb{B}_{\mathbb{C}^1} = \{z \in \mathbb{C} : |z| < 1\}$ into the Riemann–Beltrami–Poincaré disc model. Note that the projective model $\mathbb{B}_{\mathbb{R}^2}$ has here curvature $K = -1$, while the conformal model $\mathbb{B}_{\mathbb{C}^1}$ has curvature $K = -4$: the two surfaces in $\mathbb{B}_{\mathbb{C}^2}$ *are not isometric*. It is also interesting to note that, if we pass to higher (complex) dimension, we still find this way higher dimensional (real) projective models of non-Euclidean geometry, but we always find two-dimensional only conformal disc models of it.

The bi-holomorphisms of the unit disc are exactly the projective transformations of the complex two-space, which map the unit ball into itself: in the light of Beltrami's discussion of his projective model, this is almost obvious. Among these transformations, the subgroup formed by the ones having real coefficients also map the unit real disc $\mathbb{B}_{\mathbb{R}^2}$ onto itself, and exhaust the projective self maps of $\mathbb{B}_{\mathbb{R}^2}$ which preserve orientation. This are the (sense preserving) isometries of the Beltrami–Klein (projective) model. On the other hand, the projective self maps of $\mathbb{B}_{\mathbb{C}^2}$ fixing the second coordinate w, which we can identify with projective transformations of the projective line $\mathbb{C} \cup \{\infty\}$ (i.e. fractional linear transformations) fixing the unit disc in $\mathbb{C}$, are the isometries of the Riemann–Beltrami–Poincaré (conformal) model. They have the form

$$z \mapsto e^{i\lambda} \frac{a - z}{1 - \bar{a}z}$$

for fixed a in $\mathbb{C}$, $|a| < 1$, and λ in $\mathbb{R}$. It interesting that the two-dimensional disc model and the (eventually) higher dimensional projective model studied by Beltrami find a unification in the context of complex projective geometry.

The metric (21) has not constant curvature in the sense of Riemann (if it had, we could not have found in it geodesic surfaces having different curvature), it is not, then, by itself a model for non-Euclidean geometry. It is interesting to see in one of its features how much it departs from having constant curvature. We saw before that the projective metric (1) exhibits an extrinsic (hence, illusory), but important anisotropy. The metric (21), which is defined similarly, exhibits a similar anisotropy, which is not, this time, illusory at all. A way to see this is the following. Consider the non-Euclidean plane in the Riemann–Beltrami–Poincaré three-dimensional model

$$ds^2 = \frac{dx^2 + dy^2 + dz^2}{(1 - x^2 - y^2 - z^2)^2}$$

considered before and the family of circles $\Gamma_r = \{(x, y, z) : x^2 + y^2 + z^2 = r^2\}$, $0 < r < 1$. The metric ds^2 restricted to each Γ_r is a spherical metric and all the metrics so obtained are just rescalings of each other. (Keeping track of the rescaling factors, we can this way give spherical coordinates to the boundary at infinity of the non-Euclidean space). The situation is much different with the metric (21). Its restrictions to spheres centered at the origin (which are spheres for the Bergman metric itself!) are not rescaled versions of each other: there is not a map Λ from one of the spheres (S_1) to another one (S_2) such that the $d_2(\Lambda(A), \Lambda(B))/d_1(A, B)$ is constant (here, d_j is the distance on S_j which comes from the restriction of the Bergman metric). Moreover, if we normalize the metrics on the spheres of (Bergman) radius ρ in order that they stabilize to a metric on the boundary at infinity as $\rho \to \infty$, such metric turns out *not to be a Riemannian metric*. In fact, it becomes what is called a *sub-Riemannian metric,* in which, for instance, the uniqueness of the geodesic through two points fails on any (arbitrary small) small open set.[9]

Beltrami's projective model coupled with complex variable opens the way for a world which is outside (but in the limit of) Riemannian geometry.

3.3 What was Beltrami's Interpretation of His Own Work?

I will try to argue here that Beltrami had a model of non-Euclidean geometry, that this model was geometrically rather "real" and that he thought so. What is uncertain is whether, at the time he wrote the *Saggio,* he claimed (or believed) that this model was a faithful representation of non-Euclidean geometry; that is, if he thought that a property which could be proved in his model was necessarily true in the non-Euclidean (synthetic) theory of Bolyai and Lobachevsky. In his [27], M.J. Scanlan argues otherwise. In particular, he maintains that Beltrami's pseudosphere *is* the imbedded tractroid S_1 we have met before; rather, two of such surfaces glued together, which would not be a model for non-Euclidean plane (try to see what such union becomes on the Liouville–Beltrami model).

---

[9]The main reason for this is that, infinitesimal balls centered at the point $(r, 0)$ in $\mathbb{B}_{\mathbb{C}^2}$ ($0 < r < 1$), have two (real) large directions and two (real) small directions. Let $z = x + iy$: end consider $(z, 0)$ in the small ball the x (normal) direction is small exactly as in the Beltrami–Klein model. Due to the holomorphic coupling of the variables in the metric (21), the (tangential) direction y is small as well. We have then two large and one small tangential directions, causing an interesting degeneracy of the metric in the limit.

Beltrami begins, we have seen, by giving the metric (1), where (u, v) are coordinates satisfying $u^2 + v^2 < a^2$. The coordinates u, v are meant to bijectively parametrize the points of a surface S. A this point, after all properties of the metric (1) are proved, we surely have an *analytic model:* a class of objects verifying the assumptions of non-Euclidean (hyperbolic) geometry.

What is this surface and where does it live? Beltrami does not say, but it is clear that the question has for him the greatest relevance. If the model has to *geometric,* then the surface S has to be "real". The strictest standard for "reality" is being a surface in Euclidean three-space.

That Beltrami considers S a surface, not merely an analytic fiction, becomes evident at p. 379, in a passage which might be confusing for the contemporary reader, who is accustomed to the abstract definition of manifolds (i.e.: set theory coupled with analytic considerations). Beltrami considers the *geometric* disc $x^2 + y^2 < a^2$ in a Euclidean plane endowed with Cartesian coordinates x, y. Setting $x = u$ and $y = v$ we have a map from the pseudosphere S (parametrized by u and v) to a region of the Euclidean plane. We have then two geometric objects (the pseudosphere, a Euclidean disc), endowed with coordinate systems, and we use the equality of the coordinates' values to established a map between these two geometric objects. The confusion for us is that we are not used anymore to consider Euclidean discs geometrically, and equating the coordinates makes for us little sense.

The surface S exists as an unquestionably "real object" if it lives in Euclidean three space (otherwise, someone could object that it is more like an "analytic object", at least before Riemann's paper). It lives there, however, not as we are used to think of it nowadays, as an imbedded submanifold; but, rather, as a physical surface, which remains the same if moved without alteration of the mutual metric relations within the surface; without taking into account self intersections. This idea of surface in space is shortly explained in the introductory remarks to [4] (p. 318): *Each couple of values for u, v determine a point of the surface, which remains essentially distinct from that corresponding to a different couple of values. The fact that, in same place in space, two points having different curvilinear coordinates might coincide, does not matter unless we consider a given configuration of the surface.* That is to say: the contemporary imbedded surface is for Beltrami a *configuration* of his own surface, and it is possible that parts of the surface overlap. What Beltrami calls (real) *surface,* then, is the (equivalence) class of all "configurations" which are isometric; i.e. for which a parametrization u, v such that the length element on the surface can be expressed by

$$ds^2 = E du^2 + 2F du dv + G dv^2$$

exists (the couple u, v ranging on a fixed set and E, F, G being given functions of u and v). This definition of Riemannian surface is perfectly consistent with Gauss theory and its motivations, and with physical intuition. Of course, in our highly formalized mathematics this definition is hard to work with. Beltrami, however, felt perfectly at ease with it.

Being such the definition, Beltrami does not find any problem when he has to fold his pseudosphere. Rather, he might have problems when the imbedding can not be locally done, due to the obstruction that $|G'(\rho)| > 1$ on his surfaces of revolution. The tractroid in Euclidean space, for instance, can only model the part of the pseudosphere which is internal to the horocycle bounding it (i.e., between the horocycle and its point at infinity).

However, having freed the surface from its particular "configuration" in space, it is possible to identify it with the range of the parameters u, v and with the length element ds^2: what we would call a patch of (abstract) surface with a Riemannian metric. Namely, further extending the family of the possible "configurations" to those in which the imbedding is just locally possible, we come to a definition of Riemannian surface which is almost equivalent to the contemporary one. Beltrami appears to be open to this possibility in the first few sentences of [4]: *It is not useless to remind that, when considering a surface as defined solely by its linear element [i.e., ds^2], one has to abstract from any concept or image which implies a concrete determination of its shape which is related to external objects; e.g. [its position with respect to] a system of orthogonal axis [in three space].* A surface of this kind, which is intermediate between "real" and "purely analytic" is the pseudosphere in his *Saggio*. In the *Teoria*, relying –at least as two-dimensional manifolds are concerned– on Riemann's authority, Beltrami does not even mention the problem of finding a Euclidean space where he can imbed his pseudospheres.

Let now get to the second problem: the relation between pseudospherical and non-Euclidean geometry. Beltrami shows that, properly interpreted, terms and postulates of non-Euclidean geometry hold on the pseudosphere. Hence, all theorems of non-Euclidean geometry hold on the pseudosphere.[10]

As a consequence, as Hoüel made explicit in [17], the Postulate of Parallels can not be proved from the remaining ones (on this point, I disagree with Scanlan). In fact, any proposition which can be proved by the remaining postulates would also hold on the pseudosphere (no matter if it is real-geometric or purely-analytic), but on the pseudosphere we have not uniqueness of parallels, so we have a contradiction. Scanlan says that Beltrami and Hoüel did not have enough mathematical logics behind to fully justify this. In my opinion, on this point Scanlan is close to the anachronisms he is castigating.

Let me use a metaphor. Suppose a child has realized that Odd and Even satisfy, with respect to sum, the same relation that Minus and Plus signs satisfy with respect to product. Lacking the idea of group isomorphism, she might be hesitant in describing this discovery. Challenged to prove that an expression with one hundred sums of Odds and Evens ends correspondingly to the corresponding expression of one hundred products of Minuses and Pluses, she will probably compute some shorter expressions and then she will use (as we sometimes do when teaching) the non-rigorous expression "and so on". Now, it is true that the higher

---

[10] A similar remark is in *Teoria*, p. 427.

level of formalism provided by isomorphisms would give her a tool to extend this observation and to be more fully aware of its meaning; but, in my opinion, the clever observation of the child remains solid even in its naïf form.

Another point of interest is the following: was Beltrami convinced that the pseudosphere had to the non-Euclidean plane the same relation that Cartesian coordinates have to the Euclidean one? That is, as I was saying above: if a question in non-Euclidean geometry is answered doing analytic calculations on the pseudosphere, was Beltrami certain that the answer was the correct one in non-Euclidean geometry as well? In the *Saggio,* there seems to be some hesitation on this point. In fact, Beltrami proves a number of known results in non-Euclidean geometry, to conclude (p. 387): *The preceeding results seem to us to fully show the correspondence between non-Euclidean planimetry and pseudospherical geometry.* But he is not yet satisfied, and goes on: *To verify the same thing from a differen viewpoint, we also want to directly establish, by our analysis, the theorem about the sum of the three angles of a triangle.* These passages, hesitant as they are, seem to show that Beltrami was personally convinced, but that, like the child of the metaphor, he did not have the language to write it down convincingly. Beltrami certainly knew that the pseudosphere and the non-Euclidean plane are isometric (he introduced the coordinates u, v in non-Euclidean plane in place of Cartesian coordinates, and using known metric properties of non-Euclidean triangles the isometry follows rather easily) Then, all geometric quantities (angles, areas, lengths) which can be expressed in terms of distances correspond. For carrying out this program in all details, however, the mathematical technology available at the time was probably not sufficient (defining the area of a generic figure through distance alone, for instance, requires concepts like Hausdorff measure, which were developed only much later). Notwithstanding, Beltrami seemed rather confident that he could answer one by one all possible problems in non-Euclidean geometry by his model, even if he did not have a general *logical* recipe to do so, but rather a *method.*

4 From the Boundary to the Interior: An Example from Signal Processing

In this section, I would like to give the reader a hint of what it means that, in Beltrami's models, the non-Euclidean space can be usefully thought of as the "filling" of the Euclidean space. I make no attempt to trace the history of this idea, which is probably rather old. In the context of complex analysis, an idea with this flavor it was first formalized by Georg Pick [24] (I have learned the details of this story from a nice article by Robert Osserman [23]) by noting that the classical Schwarz Lemma could be interpreted as saying that holomorphic functions from the unit disc D in the complex plane to itself, decrease hyperbolic distance. Let me remind the reader that Schwarz Lemma states that if $f : D \to D$ is holomorphic and $f(0) = 0$, then $|f(0)| \leq |z|$ in D; equality occurring for some $z \neq 0$ if and only if for some λ with $|\lambda| = 1$ one has that $f(z) = \lambda z$ holds for all z in D. Changing variables by means of a fractional transformation of the unit disc, Pick showed that, if $f : D \to D$ is holomorphic, then $\frac{2|df(z)|}{(1-|f(z)|^2)} \leq \frac{2|dz|}{(1-|z|^2)} = ds$, but ds is the length element in the Riemann–Beltrami–Poincaré model (corresponding to curvature $K = -1$), proving the remark. Lars Valerian Ahlfors [1] extended this observation to holomorphic maps from D to Riemann surfaces S having curvature $K \leq -1$. Bounded holomorphic functions can be reconstructed from their boundary values by means of an integral reproducing formula, which is obviously invariant under rotations of D: it was not a priori obvious that the underlying geometry was much richer (the first step in this direction had been taken by Poincaré in [25]).

It would be interesting, I think, knowing more about the history of how non-Euclidean (hyperbolic) geometry came to be recognized as the geometry underlying classes of objects arising in a Euclidean or spherical setting. Here, instead, I offer a small sample of this relation, showing how non-Euclidean geometry naturally arises from the problem of "looking at a signal at a given scale". What I am going to say is not at all original, perhaps except for the order in which the argument is presented.

Suppose we want to analyze *signals* which can be modeled as functions in $L^\infty(\mathbb{R}) \cup L^1(\mathbb{R})$ (Lebesgue spaces with respect to the Lebesgue measure in $\mathbb{R}$; where $\mathbb{R}$ could be thought of as time). Suppose that we have a device $\mathcal{T} = \{T_t \ t > 0\}$ which looks at the signal *at the scale t* for each $t > 0$; i.e., dropping in an orderly fashion the information at smaller scales, while keeping information at larger scales. Here, is a short list of reasonable (not all independent) *desiderata* for the device.

(L) **Linearity.** $T_t(af + bg) = aT_t f + bT_t g$.
(T) **Translation invariance.** If

$$\tau_a f(x) = f(x-a)$$

is (forward) translation in time, then

$$\tau_a \circ T_t = T_t \circ \tau_a. \tag{23}$$

The obvious meaning of the requirement is that the device acts uniformly in time.
(D) **Dilations.** Let

$$\delta_\lambda f(x) = \lambda^{-1} f(\lambda^{-1} x)$$

be a *dilation* in time by a factor $\lambda > 0$ (normalized to keep invariant the $L^1(\mathbb{R})$-norm of the signal). Then,

$$T_t \circ \delta_\lambda = \delta_\lambda \circ T_{t/\lambda}. \tag{24}$$

The meaning is: dilating a signal by λ an then looking it at the scale $t > 0$ is the same as looking at it at the scale t/λ, then dilating it by λ: in both cases we loose the same amount of detail.

(S) **Stability.** We want the device $\mathcal{T}$ to give an output which can be put in the same device. For instance,

$$\|T_t f\|_{L^p} \leq C(t)\|f\|_{L^p},$$

for all $p \in [1, \infty]$.

(C) **Continuity.** As t goes to 0, $T_t f$ should be "close" to f. For instance,

$$\lim_{t \to 0^+} T_t f = f \text{ in } L^p(\mathbb{R}), \ 1 \leq p < \infty, \text{ and in } C_0(\mathbb{R}),$$

where $C_0(\mathbb{R})$ is the space of the continuous functions vanishing at infinity.

(SG) **Semigroup.** Looking at the signal at the scale s, then at the scale t, we loose all detail at scale t, then s. It is reasonable that, overall, we have lost details at scale $t + s$:

$$T_t \circ T_s = T_{s+t}$$

(P) **Positivity.** $f \geq 0 \implies T_t f \geq 0$.

None of the requirements is arbitray, but we can indeed think of useful devices which do not satisfy some of them.

It is easy to see that (L), (T), (D), (S), (C), (SG), (P) are satisfied by the operators $T_t = P_t$, $f \mapsto P_t f$, where

$$P_t f(x) = \int_{-\infty}^{+\infty} \frac{1}{\pi} \frac{t}{y^2 + t^2} f(x - y) \mathrm{d}y, \qquad (25)$$

which is the *Poisson integral* of f. Using elementary arguments involving Fourier transforms, it can also be shown that such operators are essentially the only ones satisfying all the requirements. More precisely, either $T_t f = f$ for all $t > 0$ (but this way we would not loose any detail!), or there is some $k > 0$ such that $T_t = P_{kt}$.

We now look at all of the scaled versions of a signal together:

$$u(x, t) = P_t f(x),$$

where $u : H_2^+ = \mathbb{R} \times [0, +\infty) \to \mathbb{R}$. It is easy to see that u is *harmonic* on H_2^+, $\partial_{xx} u + \partial_{tt} u = 0$; but will not use this fact. We ask, now, how much can the (scaled) signal change with respect to x or t. We consider a positive signal $f \geq 0$. The answer, we will see, can be interpreted in terms of non-Euclidean geometry.

Theorem 1. *Harnack's inequality. Let $u(x, t) = P_t f(x)$, with $f \geq 0$, $f \in L^1$. If (x, t), $(x', t') \in \mathbb{R}_+^2$ and*

$$1/2 \leq t'/t \leq 2, \ |x - x'| \leq \max(t, t\prime),$$

then
$$u(x,t) \leq C u(x',t'),$$
where *C* is a universal constant.

The geometric meaning of Harnack's inequality is this: *u* is essentially constant (in multiplicative terms) on squares having the form

$$Q(x_0, t_0) = \{(x,t) : |t - t_0| \leq t_0/2, \ |x - x_0| \leq t_0/2\}.$$

It is immediate from (17), the expression of the non-Euclidean distance in the Liouville–Beltrami model,

$$ds^2 = \frac{dx^2 + dt^2}{t^2}$$

(with $R = 1$) that $Q(t_0, x_0)$ is comparable with a disc having fixed hyperbolic radius; i.e. that there are positive constants $c_1 < c_2$ such that $D_{\text{hyp}}((x_0, t_0), c_1) \subseteq Q(x_0, t_0) \subseteq D_{\text{hyp}}((x_0, t_0), c_2)$. As a consequence, if (x_1, t_1) and (X_2, t_2) lie a hyperbolic distance *d* apart, then $u(x_2, t_2) \leq C^d u(x_1, t_1)$. The usual proof of Harnack's inequality makes use of the harmonicity of *u*, which might be thought of as an infinitesimal version of the mean value property on circles, which implies a Poisson integral formula like (25), with discs contained in H_2^+ instead of H_2^+ itself. We have preferred a different approach to show how the inequality directly follows for the geometric requirements we have made.

Proof.

$$u(0,t) = \frac{1}{\pi} \int_{\mathbb{R}} f(y) \frac{t}{t^2 + y^2} dy$$

$$\approx \frac{1}{t} \left[\int_{|y/t| \leq 1} f(y) dy + \sum_{n=1}^{\infty} \frac{1}{1 + 2^{2n}} \int_{2^{n-1} \leq |y/t| \leq 2^n} f(y) dy \right]$$

$$\approx u(0, 2t).$$

For the next calculation, consider that, if $|x/t| \leq 1$ e $2^{n-1} \leq |(x-y)/t| \leq 2^n$, then

$$(2^n - 1)t \leq |x - y| - |x| \leq |y| \leq |x - y| + |x| \leq (2^n + 1)t.$$

$$u(x,t) \approx \frac{1}{t} \left[\int_{|(x-y)/t| \leq 1} f(y) dy + \sum_{n=1}^{\infty} \frac{1}{1 + 2^{2n}} \int_{2^{n-1} \leq |(x-y)/t| \leq 2^n} f(y) dy \right]$$

$$\lesssim \frac{1}{t} \left[\int_{|y/t| \leq 1} f(y) dy + \sum_{n=1}^{\infty} \frac{1}{1 + 2^{2n}} \int_{2^{n-1} \leq |y/t| \leq 2^n} f(y) dy \right]$$

$$\approx u(0, t).$$

In order to better understand Harnack's inequality and its proof, divide H_2^+ in dyadic squares:

$$Q_{n,j} = \left\{(x,t) : \frac{1}{2^{n+1}} \leq t < \frac{1}{2^n}, \frac{j-1}{2^n} \leq x \leq \frac{1}{2^{n+1}}\right\}.$$

The inequality implies that u changes by at most a multiplicative constant when we move from a square $Q = Q_{n,j}$ to a square $Q' = Q_{n',j'}$ such that $Q \cap Q' \neq \emptyset$. Let $\mathcal{Q}$ be the set of the dyadic squares and define a *graph structure* G on $\mathcal{Q}$ by saying that there is an edge of the graph between the squares Q_1 and Q_2 if $Q \cap Q' \neq \emptyset$. We can define a distance d_G on the graph G by saying that $d_G(Q_1, Q_2)$ is the minimum number of edges of the graph G one has to cross while going from Q_1 to Q_2. Let d be non-Euclidean distance (in the Liouville–Beltrami model). It is clear that, if (x_j, t_j) lis in Q_j, $j = 1, 2$, then

$$d((x_1,t_1), (Q_1, Q_2)) + 1 \approx d_G(Q_1, Q_2) + 1 : \tag{26}$$

the graph and the hyperbolic distance are, *in the large,* comparable. This is the main reason why Cannon *et al* ([14]) call the dyadic decomposition of the upper half space *the Fifth model* of non-Euclidean (hyperbolic) geometry, after the three models by Beltrami and the "Fourth", commonly used model living in Minkovsky space. In the language of the metric d_G, Harnack's inequality reads as

$$|\log u(x_1, t_1) - \log u(x_2, t_2)| \leq C' d_G(Q_1, Q_2) + K'. \tag{27}$$

for some constants C' and K'. More precisely, it could be proved that there is a universal constant $K > 0$ such that

$$|\log u(x_1, t_1) - \log u(x_2, t_2)| \leq K d((x_1, t_1), (x_2, t_2)). \tag{28}$$

and that, for given (x_1, t_1), (x_2, t_2), the constant on the right of (28) is best possible for Harnack's inequality. The proof is in the books and exploits the harmonicity of u and a version of the Poisson integral on circles. Also, Harnack's inequality hold as well for positive harmonic functions u, which do not arise as Poisson integrals of positive, integrable functions on the real line. Both (28) and (27) can be read as Lipschitz conditions for $|\log u|$ in terms of the discretized, or of the continuous hyperbolic distance.

The example above reflects the modern view that "geometry" has not to be related to "real space". Such view was endorsed by Riemann, but it found a coherent and rather complete logical and mathematical framework only much later, with the work of Peano and Hilbert on the axiomatization of mathematical theories. Although Beltrami seems to have felt safer on the old ground of reality, his models are nonetheless an essential step in the direction of the contemporary, freer way to think and use geometry.

References

1. Ahlfors, Lars Valerian. 1938. An extension of Schwarz's lemma. *Trans. Amer. Math. Soc.* 43: 3 359–364.
2. Beltrami, Eugenio. 1904. Opere matematiche di Eugenio Beltrami, pubblicate per cura della Facoltà di Scienze dalla Regia Università di Roma. vol. I (1902) vii+437 p.; vol. II, 468 p., Hoepli: Milano.
3. Beltrami, Eugenio. 1865. Risoluzione del problema: "Riportare i punti di una superficie sopra un piano in modo che le linee geodetiche vengano rappresentate da linee rette". *Ann. Mat. Pura App.* I no. 7: 185–204.
4. Beltrami, Eugenio. 1867. Delle variabili complesse sopra una superificie qualunque, *Ann. Mat. Pura App.* Ser. II vol. I 329–366.
5. Beltrami, Eugenio. 1868. Saggio di interpretazione della geometria non-euclidea, Giornale di Matematiche. VI 284–312; English transl., Essai d'interpretation de la géométrie noneuclidéenne, Annales scientifiques de l'É.N.S. Ire série, 6, 1869, 251–288.
6. Beltrami, Eugenio. 1868–1869. Teoria fondamentale degli spazii di curvatura costante, Ann. Mat. Pura App., Serie II, vol. II, (Unknown Month 1868), 232–255; English transl., Théorie fondamentale des espace de courbure constante, Annales scientifiques de l'É.N.S. Ire série, 6, 1869, 347-375-288.
7. Beltrami, Eugenio. 1871–1873. Osservazione sulla nota del prof. L. Schlaefli alla memoria del sig. Beltrami "Sugli spazi di curvatura costante". *Ann. Mat. Pura Appl.* V, 194–198.
8. Beltrami, Eugenio. 1872. Teorema di geometria pseudosferica. *Giorn. Mat.* X, 53.
9. Beltrami, Eugenio. 1872. Sulla superficie di rotazione che serve di tipo alle superficie pseudosferiche. *Giorn. Mat.* X, 147–159.
10. Beltrami, Eugenio. 1889. Un precursore italiano di Legendre e di Lobatschewsky. Rendiconti della R. Accad. dei Lincei, V. I semestre, 441–448.
11. Bonola, Roberto. 1906. La geometria non-euclidea. Zanichelli: Bologna.
12. Boyer, Carl B. 1959. The History of the Calculus and Its Conceptual Development. Dover Publications, Inc.: New York.
13. Bridson, Martin R. 1999. André, Häfliger, Metric Spaces of Non-Positive Curvature, Grundlehren der mathematischen Wissenschaften, Vol. 319, p. Springer: Berlin.
14. Cannon, James W. Floyd, William J., Kenyon, Richard, Walter R. 1997. Hyperbolic Geometry. Math. Sci. Res. Inst. Publ., Cambridge Univ. Press: Cambridge.
15. Cicenia, Salvatore. 1989. Le lettere di E. Beltrami ad A. Genocchi sulle geometrie non euclidee. Nuncius, 567–593.
16. Enriques, Federigo. 1918. Conferenze sulla geometria non-euclidea. Zanichelli: Bologna.
17. Hoüel, Jules. 1870. Note sur l'impossibilité de démotrer par une construction plane le principe des parallèles, dit Postulatum d'Euclide, Nouvelles annales de mathématiques 2e série. tome 9: 93–96.
18. Gray, Jeremy. 1979. Non-Euclidean geometry: a re-interpretation. *Historia Mathematica* 6: 236–258.
19. Gray, Jeremy. 2007. Worlds out of nothing: a course in the history of geometry in the 19th Century. Springer Undergraduate Mathematics Series. Springer: Berlin.
20. Merrick, Teri. 2006. What Frege Meant When He Said: Kant is Right about Geometry. *Philophia Math.* (III), 1–32.
21. Milnor, John. 1982. Hyperbolic Geometry: the first 150 years. *Bull. Am. Math. Soc.* 6: 19–24.
22. Monge. 1850. Application de l'Analyse à la Géométrie, 5. éd., rev., cor. et annotée par m. Liouville. Bachelier: Paris.
23. Osserman, Robert. 1999. From Schwarz to Pick to Ahlfors and beyond. *Notices Amer. Math. Soc.* 46: no. 8, 868–873.
24. Pick, Georg. 1915. Über eine Eigenschaft der konformen Abbildung kreisförmiger Bereiche. *Math. Ann.* 77: 1 1–6.
25. Poincaré, Henri. 1882. Théorie des groupes Fuchsiens. *Acta Math.* 1: 1 1–76.

26. Riemann, Bernhard. 1873. Ueber die Hypothesen, welche der Geometrie zu Grunde liegen, Abhandlungen der Kniglichen Gesellschaft der Wissenschaften zu Göttingen, 13, 1867, On the Hypotheses which lie at the Bases of Geometry, Nature, no. VIII, 183: 14–17.
27. Scanlan, Michael J. 1988. Beltrami's Model and the Independance of the Parallel Postulate. Hist. Philos. Logic, 9: no. 1, 13–34.
28. Winger, R.M. 1925. Gauss and non-Euclidean geometry. *Bull. Amer. Math. Soc.* 31: 7 356–358.

Giuseppe Vitali: Real and Complex Analysis and Differential Geometry

Maria Teresa Borgato

Abstract Giuseppe Vitali's mathematical output has been analysed from various points of view: his contributions to real analysis, celebrated for their importance in the development of the discipline, were accompanied by a more correct evaluation of his works of complex analysis and differential geometry, which required greater historical investigation since the language and themes of those research works underwent various successive developments, whereas the works on real analysis maintained a collocation within the classical exposition of that theory.

This article explores the figure of Vitali and his mathematical research through the aforementioned contributions and, in particular, the edition of memoirs and correspondence promoted by the *Unione Matematica Italiana*, which initiated and encouraged the analysis of his scientific biography.

Vitali's most significant output took place in the first 8 years of the twentieth century when Lebesgue's measure and integration were revolutionising the principles of the theory of functions of real variables. This period saw the emergence of some of his most important general and profound results in that field: the theorem on discontinuity points of Riemann integrable functions (1903), the theorem of the quasi-continuity of measurable functions (1905), the first example of a non-measurable set for Lebesgue measure (1905), the characterisation of absolutely continuous functions as antiderivatives of Lebesgue integrable functions (1905), the covering theorem (1908). In the complex field, Vitali managed to establish fundamental topological properties for the functional spaces of holomorphic functions, among which the theorem of compacity of a family of holomorphic functions (1903–1904).

Vitali's academic career was interrupted by his employment in the secondary school and by his political and trade union commitments in the National Federation of Teachers of Mathematics (*Federazione Nazionale Insegnanti di Matematica*,

---

M.T. Borgato (✉)
Dipartimento di Matematica, Università di Ferrara, via Machiavelli, 35, 44121 Ferrara, Italy
e-mail: bor@unife.it

FNISM), which brought about a reduction, and eventually a pause, in his publications.

Vitali took up his research work again with renewed vigour during the national competition for university chairs and then during his academic activity firstly at the University of Modena, then Padua and finally Bologna. In this second period, besides significant improvements to his research of the first years, his mathematical output focussed on the field of differential geometry, a discipline which in Italy was long renowned for its studies, and particularly on some leading sectors like connection spaces, absolute calculus and parallelism, projective differential geometry, and geometry of the Hilbertian space.

Vitali's connection with Bologna was at first related to his origins and formation. Born in Ravenna, Vitali spent the first 2 years of university at Bologna where he was taught by Federigo Enriques and Cesare Arzelà. Enriques commissioned him with his first publication, an article for the volume *Questioni riguardanti la geometria elementare* on the postulate of continuity. Vitali then received a scholarship for the *Scuola Normale Superiore* and later completed his university studies at Pisa under the guidance of Ulisse Dini and Luigi Bianchi.

From the end of 1902 until the end of 1904, when he was teaching in a secondary school in Voghera, Vitali returned to mix with the Bologna circles as he, at times, resided there. During his last years, after spending some time at the Universities of Modena and Padua, Vitali returned to teach at Bologna University, a role he carried out with energy and generosity but unfortunately not long enough to be able to found a school.

1 Formation and Career

Giuseppe Vitali was born in Ravenna on 24th August 1875, first of five children. His father worked in the Railway Company and his mother was a housewife. Before going to the university, he attended the Dante Alighieri Secondary School in Ravenna where he distinguished himself in mathematics, and his teacher, Giuseppe Nonni, strongly advised his father to allow him to continue his studies in this discipline.[1]

---

[1] A useful source for Vitali's biography is the Vitali Archive, donated by his daughter, Luisa, to the *Unione Matematica Italiana* in 1984. It included original documents of interest for his biography, and numerous letters sent to Vitali from various mathematicians, among which 100 letters judged to be the most significant, were chosen and inserted into the edition of his works on real and complex analysis: [1], which also includes a complete list of Vitali's publications. The correspondence, in particular, was edited by Maria Teresa Borgato and Luigi Pepe. Unedited letters and documents have contributed to the new biography of Giuseppe Vitali by Luigi Pepe, inserted into the volume: [1]: pp. 1–24. Specific studies on Vitali's contributions to real and complex analysis and differential geometry, in: [2, 4, 5]. On Vitali's treatises and production related to mathematics teaching see also: [6]. On Vitali's life and work see also Vitali's obituary [7]. For a general overview on the development of measure theory and integration, see: [8, 9].

So it was that Giuseppe attended the first 2 years of the degree course in Bologna University where he was taught by Federigo Enriques, who had just arrived from Rome to teach descriptive and projective geometry, and by Cesare Arzelà, who had the chair in infinitesimal analysis. Both Enriques and Arzelà recognised his ability and did their best to help him gain a scholarship. In the academic year 1897–98 he was accepted in the Scuola Normale of Pisa where, in July 1899, he graduated summa cum laude under the guidance of Luigi Bianchi. He wrote his thesis on analytic functions on Riemann surfaces. The teachers who had most influence over him were undoubtedly Ulisse Dini, who taught infinitesimal calculus, and Luigi Bianchi analytical geometry, other teachers were Eugenio Bertini for higher geometry, Gian Antonio Maggi for rational mechanics, and Cesare Finzi for algebra.

After graduating, Vitali became Dini's assistant for 2 years, during which time he took a teaching diploma with a thesis, under the guidance of Bianchi, on: "Le equazioni di Appell del 2° ordine e le loro equazioni integrali", (Appell's 2nd order equations and their integral equations) which was later published in a first work in 1902, and later in a more in-depth work in 1903.

In 1902, Vitali left university since he had accepted a position as a mathematics teacher in a secondary school (better paid than the post of assistant at university), first in Sassari and then in Voghera (until 1904) and finally in the "C. Colombo" Secondary School of Genoa. At that time it was not unusual for a person to begin a university career with a period of secondary school teaching (other examples are Luigi Cremona and Cesare Arzelà), but it was a great pity that it took the university up till 1922 to concede full research activity to such talent like that of Vitali.

Vitali, however, kept in contact with his mentors, and Arzelà in particular remained a constant point of reference for him, through exchange of letters and also meetings above all during the holiday periods (Vitali returned to Bologna during his teaching years in Voghera, and Arzelà, in his turn, spent long periods in Santo Stefano Magra in Liguria). He always maintained a friendship with Enriques, who had followed his transfer from the University of Bologna to the Scuola Normale Superiore of Pisa, but his lifelong friend was Guido Fubini, who had been a fellow student in Pisa.

In these 20 years mild attempts were made to procure him a position, including a post as temporary lecturer (there were offers by Dini of a free course in 1907, an offer from Mario Pieri of a temporary position in Parma, a post teaching infinitesimal analysis in Genoa for the academic year 1917–1918) but when it came to making the decision a preference was shown for younger scholars mostly within the university in consolidated areas of research.

Vitali's biography [10] provides an excellent description of the academic situation and the widespread opinion which the university establishment held of Vitali's research and the new real analysis in general; the fruits of the review of the foundations of analysis initiated by Dini and his school were reaped above all in the field of integral equations thanks to Vito Volterra. It was along these lines that Leonida Tonelli found great success and his results gave ample space for application. In contrast, Vitali's problems appeared very limited to the discipline, above all those on the "groups of points", and only his works on integrability and the convergence of series, which fell within traditional interest, aroused general attention. On the other hand, also Lebesgue's research was initially received with some diffidence, so much so that his famous thesis was published in Italy not in France. Only when one considers that Italy and France were the countries in which this type of research was given most space, in an international setting of relative indifference, do we realise the initial difficulties in earning justified recognition. Lebesgue, in fact, reminds us that the new discipline had not yet received either definition or application: "Jusqu'à ces dernières temps, la plus part des travaux sur les fontions réelles, ceux concernant les séries trigonométriques exceptés, se réduisaient à des remarques, parfois très élégantes, mais sans lien, ne formant nul corps de doctrine et n'ayant servi pratiquement à rien... On en était encore à la phase d'observation: on explorait l'amas désordonné des fonctions pour y découvrir des catégories intéressantes, mais, comme on ne savait pas expérimenter sur les fonctions, c'est-à-dire calculer avec elles, s'en servir, on manquait totalement de critères pour juger qu'une catégorie était intéressante." In spite of this, a change of mind was swift in occurring abroad: Lebesgue's integral became part of courses on calculus as early as 1910 and Vitali was frequently cited in works by Montel and De La Vallée Poussin (1912, 1916), whereas in Italy Vitali continued to be excluded from academic research...

Let me quote a well-known passage from Giovanni Sansone' repertoire on famous mathematicians graduated from the Scuola Normale Superiore in the years 1860–1929. Sansone related that Vito Volterra, perhaps in 1922, had occasion to meet Lebesgue in Paris. Lebesgue asked Volterra for news about Vitali and when he heard that Vitali was teaching mathematics in secondary school in Genoa, surprised, he said: "I am pleased that Italy has the possibility to maintain mathematicians like Vitali in secondary schools" [11].

Finally in 1922, Vitali was one of the successful applicants of a national call for a university chair and on that occasion he had begun to publish once more: six works between the end of 1921 and 1922. Vitali came second, preceded by Gustavo Sannia (the Examining Board consisted of Guido Fubini, Tullio Levi-Civita, Salvatore Pincherle, Leonida Tonelli, and Gabriele Torelli. Only Fubini and Levi-Civita favoured Vitali for first place). Since Sannia did not take up the position, Vitali was called to the University of Modena where he remained for 3 years, then in the December of 1925, he transferred to the University of Padua, where he took up the chair in mathematical analysis which had previously been occupied by Gregorio Ricci Curbastro. His teaching, research and organisational activities over these 5 years in Padua were intense, in spite of the partial paralysis which afflicted him down the right side of his body in 1925. He explored new areas of research with various young students, founded the *Seminario matematico* of the University of Padua and was elected as Director: the aim of this initiative was to promote interest in mathematics in Padua by organising activities such as general conferences, elementary preparatory lessons for the teaching of mathematics, and seminars regarding ongoing research. The journal connected to this, the *Rendiconti*, founded in 1930, still continues publishing as the journal of the University of Padua. In 1928, Vitali was asked to return to Pisa to take up the chair previously occupied by Luigi Bianchi, but he refused. In order to remain close to his mother, he accepted the offer of transfer to Bologna in 1930, for the chair of infinitesimal calculus, which Tonelli was leaving in order to go to Pisa. Even in this brief time spent in Bologna he devoted himself with enthusiasm to his academic commitments until his sudden death under the portico of the city as he was going home from the university with Ettore Bortolotti after his lessons and speaking on recent mathematical papers.

2 First Research Works in Complex Analysis

Both in Bologna and Pisa, not surprisingly, Vitali's mathematical research was initially guided by his teachers, who influenced him in different ways. To Enriques, Vitali owes the focus on the foundations and didactics of mathematics. Enriques commissioned Vitali to write a chapter on the postulate of continuity for the volume: *Questioni riguardanti la geometria elementare*, of formative and cultural interest to teachers of mathematics, which was also his first published work [12].

Moreover, Vitali may have considered research themes in algebraic geometry: Enriques was developing his theory of algebraic curves by means of projective models which was to allow him a complete classification, and there are hints, in Enriques' correspondence with Vitali, of a few ideas or, rather, research plans, on algebraic varieties with elliptic section curves of higher dimensions, which were never developed.

The first two memoirs that Vitali published in 1900, in the same volume of the journal of the *Circolo Matematico* of Palermo, take their origin from his degree thesis and so the line of research carried out by Bianchi. The first one contains an extension of Mittag-Leffler's theorem to holomorphic functions with given singularities on Riemann surfaces [13], the second one, a property of holomorfic functions of one variable for which the limit of the n-derivatives at a regular point, exists as n goes to infinity [14].

Other important memoirs of 1902 and 1903 with the same title are linked to the teaching diploma thesis, thus still under the influence of Bianchi and on complex analysis [15, 16]: *On linear homogeneous differential equations with algebraic coefficients* where Abelian integrals and Abelian functions are applied to a class of fuchsian differential equations introduced by Appel [17] and called by Vitali "Appell equations" (with singular points in the Fuchs class). After a classical part derived from the Appell-Goursat treatise, Vitali concentrates on second order equations, and even if his results do not give a conclusion to the topic, they are still nowadays considered to be of interest and value, as the questions treated at that time, and then abandoned, by Appell and Vitali have had a successive revival.[2]

---

[2]Problems closely linked to those of Vitali were treated by André Weil: [18], and more recently in [137] much space is also given to differential equations of the 2nd order in the book: [138]. A description of this research work by Vitali, using the sheaf-theoretical notations, and its collocation in the research of his day, in [5]: 376–377.

Vitali's later research took a different direction, under the influence of Cesare Arzelà, who, in 1899 and 1900, had published two extensive memoirs [19], in which he gave a systematic exposition of his results in the field of the theory of functions: on the sum of a series of continuous functions, on integrability of series, on developability in Fourier series, giving them the prominence they deserved [20]. Arzelà continued his research in the memoir "on the series of analytic functions" [21] in 1903 and, in the same year, over a short space of time, following Arzelà's research and using his techniques, Vitali published, under the same title, three memoirs that demonstrated four original theorems which in themselves were sufficient to guarantee him international renown and a place in the history of mathematical analysis [22–24].

The question, on which also Arzelà, Osgood and Montel worked, was to find the minimal sufficient conditions for a series of analytic functions in some range of the complex plane to converge to a holomorphic function. In the first memoir, using Arzelà's techniques, Vitali extended a theorem by Osgood and established that a sequence of holomorphic functions, uniformly bounded on a connected open set T and convergent at each point of a subset of T, with a cluster point in T, will converge uniformly, on compact subsets of T, to a holomorphic function.

The most famous result that Vitali obtained was the theorem of compacity, later taken up and generalised by Montel, in which, from any uniformly bounded sequence of holomorphic functions (on a connected open set), one can extract a convergent subsequence to a holomorphic function. Vitali, moreover, discovered another more important theorem (called "tautness theorem" by A. Vaz Ferreira),[3] even if his demonstration is incomplete, according to which a pointwise convergent sequence of holomorphic functions which omit two values is normally convergent.[4] In the last of the three memoirs, Vitali arrives at the conditions necessary and sufficient for the convergence of a series of holomorphic functions.

3 Research on Real Analysis

Still today the name of Giuseppe Vitali is mainly linked to his first research works on real analysis, whose basic topological concepts were not immediately successful due to the innovation in approach and language. The fact that his work fell within a line of research that had not yet attained complete international recognition may also account for the delay in his academic career.

---

[3]For a detailed analysis of Vitali's results on this topic, Vitali's priority with respect to Montel, Arzelà and Osgood, and the attribution of the discovery of these theorems in following works (Carathéodory-Landau) and treatises see: [5]: 386–391.

[4]"Se $u_1, u_2, \ldots$ è una successione di funzioni analitiche finite e monodrome convergenti in ogni punto di un campo semplicemente connesso C, e se inoltre le funzioni suddette non assumono mai i valori 0, 1 la successione converge verso una funzione analitica finita e monodroma in C."

His work was linked to the research into the theory of measure and integration which, in France, was being developed by Borel and Lebesgue. As for the Italian tradition, Ulisse Dini's work constituted the reference point, in particular the volume entitled *Fondamenti per la teorica delle funzioni di variabili reali* [25]. In it the basic concepts of the theory of functions of a real variable show organisation and precision, and provide a coherent and comprehensible structure to the entire theory which contains the preceding or contemporary research carried out by Weierstrass, Schwarz, Heine, Cantor, Du Boys-Reymond, Hankel, and Riemann as well as his own. Dini's *Fondamenti* were translated into German by J. Lüroth and A. Schepp, and then even 14 years after the Italian edition, as the editors stated, it was still considered to be "the only book of modern theory of a real variable" [26].

The volume includes the following topics: Dedekind's theory on real numbers, derived set and set of first and second category, least upper bound and greatest lower bound, the theory of the limits of sequences, continuous and various types of discontinuous functions (Weierstrass), uniform continuity (Cantor, Schwarz), pointwise discontinuous and totally discontinuous functions (Hankel), derivatives and derivability, intermediate value theorem (correctly demonstrated), types of discontinuity of the derivate function, series and series of functions, uniform convergence and its applications to continuity, differentiability, and integrability, Hankel's principle of condensation of singularities, continuous everywhere but nowhere derivable functions, functions of bounded variation, oscillation of a function, etc. Then in the section on finite integrals Dini introduced Riemann's integral and interesting criteria of integrability, correcting Hankel's condition, and based on the concept of negligible set. Continuing this line of research, one of his pupils, Giulio Ascoli published, in 1875, a work which gave a demonstration of a necessary and sufficient condition equivalent to Riemann integrability, based on the concept of oscillation intended as a point function. Vito Volterra, in his turn, in a work of 1881, presented an example of a nowhere dense not negligible set, to complete Dini's discussion, as well as new criteria of integrability through use of jump function.

The research carried out by Dini and his school formed the basis of successive development in France by Borel, Baire, Lebesgue and Fréchet. It is to be remembered that Baire was in Italy in 1898 when he worked with Volterra, and his thesis ("Sur les fonctions de variables réelles") was published in the *Annali di matematica pura ed applicata* of which Dini was the editor. Lebesgue's thesis, too, ("Integral, longueur, aire") was published in the *Annali di matematica* in 1902.

At the beginning of the twentieth century, after the premature death of Ascoli and Volterra who was particularly engaged in the field of integral equations, among Dini's pupils there remained Cesare Arzelà who could carry forth research in the field of real and complex analysis, and the *Fondamenti* by Dini as well as the long memoir *Sulle serie di funzioni* by Arzelà constituted the reference texts of the young Vitali.

Many of the most important results for which Vitali is still remembered, derived from the period between 1903 and 1908, after he had left his position as assistant in Pisa for the better paid job of secondary school teacher in Sassari and then in Voghera. Until his transfer to the Liceo Colombo in Genoa in 1904, Vitali continued to live in Bologna during the summer and Christmas holidays, as can be seen from his correspondence at that time.

It was during those years that the legacy left by Dini and Arzelà tied up with the new research that Borel and Lebesgue were carrying out on the theory of measure; in 1898 Borel had provided a vast class of measurable sets [27] and, in 1901, Lebesgue had come up with the definition of the integral which carries his name. In 1902 [28], in his famous doctoral thesis, Lebesgue had developed the properties of the measure and observed that the points of discontinuity of a Riemann integrable function form a set of measure zero. Vitali arrived on his own at the same concept of measure in 1903, and, in 1904,[5] demonstrated that the necessary and sufficient condition for a function to be Riemann-integrable was that its set of points of discontinuity is of measure zero, in a sense that coincided with Lebegue's, thus releasing the integrability from the behaviour of the function in these points. The same year Lebesgue also demonstrated the sufficiency of the condition.

In order to clarify the interconnection of these results we can examine Vitali's works in particular. In the first one, starting from one of Osgood's theorems, redemonstrated in an elementary form, Vitali provided his criterion for *Riemann integrabiliy*, that is that the *minimal extension* ("estensione minima") of the points of discontinuity must be equal to zero,[6] which he reformulated and clarified in the second note, in which he also observed that countable sets had minimal extension zero and that minimal extension coincides, for closed sets and only on these, with the *Inhalt* of Cantor. In the third note, Vitali, inspired by Borel [27], defined the measurable sets and studied the properties that make the *minimal extension* a measure for measurable linear sets of points: finite and countable additivity, that the family of measurable sets is closed under the operations of countable union, and intersection, comparing it with the measures of Jordan[7] and Borel (he demonstrated that both Jordan's and Borel's families of measurable sets were extended). While this memoir was being published, (it was presented on the 5th November 1803), Pincherle brought to the attention of Vitali Lebesgue's work of 1901, in which "mention is made to the concepts treated herein... for the construction of integrals of derived functions which are not integrable" ([31] p. 126).

It may be said that Vitali's *minimal extension* coincides with Lebesgue outer measure, but while he also introduced the inner measure by means of closed sets, and defined measurable sets as those for which inner and outer measure coincide, Vitali did not introduce an inner measure and defined measurable sets by means of a scalar function ("allacciamento" - *linking*) which links a set E of an interval (a, b)

---

[5] Three of Vitali's memoirs refer to Riemann's integrability of a function in relation to the set of his points of discontinuity: [29–31].

[6] The *minimal extension* of a linear set of points in Vitali's theory is the least upper bound of the series of measures of (finite or countable) families of pairwise disjoint open intervals, which covers the set.

[7] One should say: "Peano-Jordan", see: [32]. See also Peano's letter of 29th February 1904 to Vitali, in which Peano compares the *minimal extension* with his outer measure.

to its complement E^*: $Z(E, E^*) = (\text{mis}_e E + \text{mis}_e E^*) - (b - a)$. Those sets for which the linking is zero are defined measurable.[8]

In 1905, Vitali published as many as seven articles, as well as two more up to the end of 1908, most of them originated from problems connected to the theory of measure and the integral of Lebesgue: the existence or otherwise of non measurable sets, the characterisation of measurable functions, the characterisation of the integrals of summable functions (Lebesgue integrable), the extension of the fundamental theorem of calculus to Lebesgue integrals, integration of series term by term, integrability on unlimited intervals, the extension to functions of two or more variables.

A brief note of 1905 [33] is famous as it contains the first example of non Lebesgue measurable set, usually cited in every critical presentation of measure theory. The counter-example is constructed using the axiom of choice in the form of the well-ordering theorem, which Vitali immediately accepted, whereas the mathematical milieu was divided over the matter. In Italy, for example, Tonelli wanted to work out a version of the calculus of variations independent from it. The fact that Vitali's counter-example was not inserted into a journal but was published separately is also significant. At the end of the note, Vitali concluded that, in any case, the possibility of measuring the sets of points of a straight line and that of well ordering the continuum could not coexist.[9] Vitali's statement is not, however, justified for the inverse part, or rather that, by denying the axiom of choice every set would be measurable. Several later research works refer to this problem, among which those of F. Hausdorff and W. Sierpiński.[10]

Naturally, the problem of extending the fundamental theorem of calculus involved functions of bounded variation. Lebesgue had demonstrated that the indefinite integral of a summable function has as a derivative this function except for a zero measure set ([35] pp. 123–125). Vitali introduced the concept of absolute continuity, and therefore demonstrated that the absolute continuity is the necessary and sufficient condition so that a function is the indefinite integral of a summable function [36]. In the same memoir, Vitali also provided the first example of a continuous function of bounded variation, but not absolutely continuous.[11] This work was contested by Legesgue who claimed authorship of various theorems in two letters of 16th and 18th February 1807 to Vitali[12] and pointed out his works were not adequately quoted in Vitali's memoirs. On the other hand, he admitted

---

[8] A description of some of Vitali's works on real analysis of the first period in [3].

[9] "la possibilità del problema della misura dei gruppi di punti di una retta e quella di bene ordinare il continuo non possono coesistere".

[10] In more recent times, to this question is related the famous result [34].

[11] It corresponds to the function called *Cantor function* or *Devil's staircase*. Vitali presented this counterexample, with some more details, also in: [37].

[12] [1], pp. 457–462. Lebesgue also claimed authorship of the demonstration that the set of points in which a continuous function has a finite derived number is measurable, that Beppo Levi had criticised and that Vitali had redemonstrated. See [38, 39].

he had not read many of Vitali's works before Vitali sent them to him because "ici à Poitiers nous n'avons aucun periodique italien ainsi je ne connaissais aucun de vos notes (sauf "Sui gruppi di punti..." que Borel m'avait communiqué)". In reality, Lebesgue had only indicated[13] the result which was completely demonstrated by Vitali in this work of 1905, and was then redemonstrated by Lebesgue in 1907 [40] and once again by Vitali in 1908 with a different method, which could also be extended to functions of more than one variable (multiple integrals) [41]. Here Vitali gave, for the first time, a definition of bounded variation for functions of two variables, obtained from the variations of a function in four vertices of a rectangle.[14] Vitali adopted the following formula:

$$F(X, Y) = \int \int_\rho f(x, y) dx\, dy - F(0, 0) + F(X, 0) + F(0, Y)$$

to define an indefinite integral, where f is a summable function, and ρ the rectangle with vertices $(0, 0), (X, Y), (X, 0), (0, Y)$. The incremental ratio of any function $f(x, y)$:

$$\frac{f(x + h, y + k) + f(x, y) - f(x + h, y) - f(x, y + k)}{hk}$$

is used to define derived numbers, and its numerator to define functions of bounded variation and absolutely continuous functions.

Vitali's priority and contribution was then recognized by Lebesgue in many points of his *Notice* [45], in particular, with reference to the fundamental theorem of calculus ("proposition... la plus feconde" according to Lebesgue) Lebesgue says: "Toute intégrale indéfinie est continue et à variation bornée, mais la réciproque n'est pas vraie. Pour qu'une fonction soit une intégrale indéfinie, il faut que, de plus, la somme des valeurs absolues de ses accroissements dans des intervalles extérieurs les uns aux autres et de mesure ε, tende vers zéro avec ε. On dit alors, avec M. Vitali, qui a été le premier à publier une démonstration de cet énoncé que j'avais formulé, que la fonction est *absolument continue*... C'est aussi M. Vitali qui a publié, le premier, des résultats sur la dérivation des intégrales indéfinies des plusieurs variables".

Linked to the fundamental theorem of the calculus is also another memoir [46] where it was shown that every summable function is "of null integral" (term coined by Dini), that is f is such that $\int_a^x f(t)\, dt = 0$ for every x in (a, b), if and only if $f = 0$ almost everywhere. Lebesgue had demonstrated the theorem for bounded functions, whereas Vitali's demonstration (of the non trivial part, the necessary one) extended to the unbounded functions by using a covering theorem of Lindelöf [47]. In another memoir Vitali dealt with the extension of integrability to unbounded

---

[13][35] see footnote p. 129.

[14]Vitali's priority was also noted by E.W. Hobson [42]. Another definition was given by Lebesgue in 1910 in: [43] p. 364. See: [44].

intervals [48] in which he provided a characterisation of integrable functions in terms of their behaviour at infinity.

In his thesis of 1899, René-Louis Baire had provided the well-known classification of functions [49, 50], but Borel and Lebesgue had managed to construct functions that had escaped this classification. The problem of classifying Baire's functions arose.[15] In 1905 Vitali and Lebesgue demonstrated that all and only the functions of Baire are Borel measurable [52–54]. The same year, however, Vitali also proved that every Borel measurable function can be decomposed in the sum of a Baire function of first or second class, and a function equal to zero almost everywhere [55]. In the same memoir the so-called Luzin's theorem was also demonstrated [56], by which if a function f is finite and measurable on an interval (a, b) of length l, for every ε there exists a closed set in which f is continuous and whose measure is greater than $l - \varepsilon$. The theorem had been indicated by Borel and Lebesgue,[16]; it inspired various analysts in their research into new definitions of integral, in particular we may remember Leonida Tonelli, who posed the functions he called "quasi-continuous" at the basis of his definition of integral [59].

An important problem in relation to the theory of functions of real variables concerns term by term integration of series, which Vitali dealt with in two memoirs [60, 61]. In the first one, he generalised some results of Lebesgue, Borel and Arzelà managing to characterise term by term integrable series under suitable conditions. In the second more important memoir of 1907, Vitali considered the integration extended not only over an interval, but also to any measureable set and, to this purpose, introduced the concept of equi-absolute continuity of a sequence of functions and complete integrability of series.[17] He, therefore, provides the characterisation of term by term integrable series on the basis of the equi-absolute continuity of the integrals of its partial sums. Complete integrability of series implies term by term integrability of series in the ordinary sense but not vice versa unless the partial sums are positive; Vitali, also derived theorems demonstrated by Beppo Levi in 1906 [62] on positive term series, and, therefore, also the necessary and sufficient

---

[15] See, for example, Note II of Lebesgue on the functions of class 1: "Démonstration d'un théorème de M. Baire", in: [51]:149–155 and Note III of Borel "Sur l'existence des fonctions de classe quelconque" [51]: 156–158 "On peut se demander si la classification de M. Baire n'est pas purement idéale, c'est-à-dire *s'il existe* effectivement des fonctions dans les diverses classes définies par M. Baire. Il est claire, en effet, que si l'on prouvait, par exemple, que toutes les fonctions sont de classe 0, 1, 2 ou 3, la plus grande partie de la classification de M. Baire serait sans intérêt. Nous allons voir qu'il n'en rien:... Le raisonnement précédent ne permet pas d'exclure l'hypothèse où un théorème tel que le suivant serait exact: *toute fonction effectivement définie est nécessairement de classe* 0, 1, 2 *ou* 3. Nous allons, au contraire, montrer qu'il est possible de définir effectivement une fonction dont la classe dépasse un nombre donné d'avance".

[16] [57, 58], [35] footnote p. 125.

[17] The equi-absolute continuity corresponds to the uniformity, in relaion to the family of functions, of the condition of absolute continuity of functions. Complete integrability of series extends, to measurabe sets, the ordinary concept of integrability of series (or rather when on every measurable subset Γ of G (measurable) the series of integrals and the integral of the series exist and can be exchanged: $\sum_{n=1}^{\infty} \int_{\Gamma} u_n(x)\,dx = \int_{\Gamma} \sum_{n=1}^{\infty} u_n(x)\,dx)$.

condition for integrability of series over an interval by generalising the result of 1905. Vitali's characterisation turned out to be one of the basic results of the measure theory, extended by Hahn (1922), Nikodym (1931), Saks (1933), Dieudonné (1951) and Grothendieck (1953), and today a great deal of reformulations are inserted into general measure theory.

The results that Vitali produced in the first period are crowned by the so-called Vitali covering theorem, that Arnaud Denjoy evaluated 50 years later as one of the most important theorems in measure theory of Euclidean spaces.[18]

The covering theorem was demonstrated by Vitali as an intermediate result in a memoir written at the end of 1907[19] and was enunciated firstly for the points of the real straight line, and later extended to the plane and then, by analogy but without enunciating explicitly, to higher dimensions:

"Se Σ è un gruppo di segmenti, il cui nucleo abbia misura finita m_1, esiste un gruppo finito o numerabile di segmenti di Σ a due a due distinti, le cui lunghezze hanno una somma non minore di m_1."

This theorem is preceded by another one, known as the Vitali covering lemma:

"Se Σ è un gruppo di segmenti [quadrati] il cui corpo abbia misura finita μ e se ε è un numero maggiore di zero, esiste un numero finito di segmenti [quadrati] di Σ a due a due distinti, le cui lunghezze hanno una somma maggiore di $\mu/3-\varepsilon$ [$\mu/9-\varepsilon$]."

As early as the note of 1904 "Sulla integrabilità delle funzioni", Vitali had provided a generalised version of the covering theorem known as the Heine – Pincherle – Borel theorem,[20] but the reach of this new covering theorem is much vaster; the aim is to cover, up to a measure zero set, a given set E by a disjoint sub-collection extracted from a *Vitali covering* for E: a *Vitali covering* ζ for E is a collection of sets such that, for every $x \in E$ and $\delta > 0$, there is a set U in the collection ζ such that $x \in U$ and the diameter of U is non-zero and less than δ. In the original version of Vitali the collection ζ was composed of intervals, squares, cubes, and so on.

The lemma and theorem of Vitali have been extended to other measures besides that of Lebesgue, and to more general spaces. Mention is to be made of the formulation which Costantin Carathéodory [66] gave a few years later, as well as Stefan Banach's extension, in a fundamental memoir in 1924 [67].

In the second chapter of the memoir, Vitali himself provided interesting applications of the covering theorem, on derived numbers of functions of bounded variation and on the integrals of summable functions, extending many of the results of his previous memoirs to functions of two variables, among which we may recall[21]:

The set of points in which a derived number of a continuous function is finite is measurable.

---

[18] A. Denjoy devoted several notes to Vitali's covering theorem and its generalizations: [63–65].

[19] [41]. Presented at the Academy meeting of 22nd December 1907.

[20] [30] p. 71: "Se si ha un'infinità numerabile di intervalli, tali che ogni punto di un gruppo lineare chiuso P sia dentro ad uno di essi, esiste un numero limitato di intervalli scelti tra gli intervalli dati e aventi la stessa proprietà."

[21] I have sometimes replaced the condition "with the exception of a measure zero set" with the modern terminology "almost everywhere".

The set of points in which a derived number of a continuous function of bounded variation is not finite is of measure zero.

A derived number of a function of bounded variation is summable and if the function is absolutely continuous the integral of the derived number coincides with the function with the exception of a set of measure zero.

A function of null integral is equal to zero with the exception of a set of measure zero.

Two functions with the same integral are equal almost everywhere.

Two derived numbers of the same absolutely continuous function are equal almost everywhere.

A summable function and a derived number of its integral are equal almost everywhere.

4 The Years of Secondary School Teaching

In the first years of his secondary school teaching Vitali managed to continue his research work which may be considered his best production thirteen: papers, seven of which in 1905 alone, the year he was transferred to Genoa and began a family. In Genoa, he taught at the "C. Colombo" Secondary School up to 1922. In 1906, he was promoted to a tenured position. Very soon after becoming a teacher he became involved in politics and the Trade Unions.

This was a period in which the Italian school system underwent many attempts at reform and experimentation which involved teachers in heated debates within the *Società Mathesis*, founded in 1895, and the *Federazione Nazionale Insegnanti Scuola Media* (FNISM) (National federation of Middle School Teachers), founded in 1902 by Giuseppe Kirner and Gaetano Salvemini.

In 1906, the Royal Commission for secondary schools issued a detailed questionnaire to the teachers in view of a reform of state education, which aimed to investigate: 1. the results and defects of the present system, 2. the type of system desired. 3. the value to be attributed to the final exams. Numerous documents and proposals were produced, above all, concerning the teaching of mathematics, and a Bill of Law was presented in 1908 which provided for a single middle school and the institution of a "liceo moderno".[22] The debate on the teaching of mathematics was still a European issue: the CIEM (Commission Internationale de l'Einseignement Mathématique) was also set up in 1908 with Felix Klein as its President.

Vitali was strongly committed to the FNISM and held the position of president of the association of Genoa and its province from 1908 to 1922. He supported positions

---

[22]On mathematics teaching in Italy, from Political Unification to Gentile Reform (1932) see [68]. The "liceo moderno", was a new type of secondary school, which did not substitute the classical one, with more foreign languages and a scientific vocation.

opposite to those of the idealists and Catholics led by Giovanni Gentile. Vitali was also elected town councillor of Genoa from 1910 to 1914.

From 1908 on Vitali's mathematical output was fairly scarce and up to 1921 he published only four short works: one on the mean value property of harmonic functions, already demonstrated in less general hypotheses by Eugenio Elia Levi, Leonida Tonelli and Vito Volterra [69], two on a generalization of Rolle's theorem to additive set functions, linked to contemporary works of Guido Fubini [70, 71], and a didactic note [72].

5 Between the Two Wars: Differential Geometry

The mathematical production of Vitali, which had considerably slowed down owing to his commitments in the secondary school, especially in concomitance with the first world war, was now taken up again with renewed energy on the occasion of examinations for chairs and successive academic activity.

Right from the very first works two trends in his scientific research may be distinguished since the works on real analysis are accompanied with those on absolute calculus and differential geometry.

The works on real analysis, about ten in all, did not achieve the important results which typified those of his youth, but were, nonetheless, carried out with the great insight, elegance and formal agility which constituted the hallmark of his works.

In the first memoir of 1921 [73], Vitali demonstrated how the condition of "closure" of an orthonormal system of square summable functions (in modern terms it corresponds to "completeness" namely, that an orthogonal function to the system is equal to zero almost everywhere) may be verified on the continuous functions alone, a result that turned out to be "precious for applications".[23] A brief note was linked to the research of Tonelli [59] in which the representation by means of absolutely continuous functions of a continuous rectifiable curve is made without resorting to the preventive rectification of the arc [74]. Some interest arose from the demonstration of the equivalence between the new definition of integral provided by Beppo Levi and that of Lebesgue [75]: Beppo Levi had already observed the coincidence for Lebesgue measurable functions [76] and Vitali concluded the matter by demonstrating that every limited function which is Beppo Levi integrable is Lebesgue measurable. Following Vitali's observations, Beppo Levi once more turned his attention to the matter.

The most interesting works in this group of Vitali's memoirs, however, concern the analysis of functions of bounded variation (now BV functions), and are those most closely linked to the research he carried out in his younger days. In the first one [37] Vitali showed that a continuous BV function may be divided into the sum of two functions: the absolutely continuous part, which is represented as

---

[23][11] p. 41.

an integral, and the singular part (the "scarto") whose derivative vanishes almost everywhere, similarly to that which occurs for the functions of bounded variation that are the sum of a continuous function, and one that absorbs the discontinuities (the "saltus function" or "jump function"). A similar decomposition had already been obtained by De La Vallée Poussin in 1909,[24] and used by Fréchet in his Stieltjes integral [78], but the representation of the "scarto" given by Vitali as the sum of a countable infinity of elementary differences was considered "nouvelle et intéressante",[25] allowing classification of functions in accordance with the "scarto" (equal to zero, greater than zero, infinite) into: absolutely continuous, bounded variation, and infinite variation. The relation between his result and De La Vallée Poussin's theorem, is clarified by Vitali himself in a short note [79].

The second memoir [80] displays an even more incisive result, demonstrating a characteristic property of BV functions, thereby introducing an equivalent definition of total variation which may be extended to functions of more variables, different from the one proposed years before [41], with a view to establishing results for surfaces similar to those already obtained by Tonelli for the rectification of curves [81, 82]. It was demonstrated that if f is a continuous function of (a, b) on (c, d) and Γ_r is the set of points of (c, d) that f assumes at least r times, the condition necessary and sufficient in order that f is of bounded variation, is that the series of measures of Γ_r is convergent and in any case this series coincides with the total variation of f:

$$V = \sum_{r=1}^{\infty} \mu(\Gamma_r)$$

Moreover, if G_∞ is the subset of values in (c, d) that f assumes infinite times, and G_r represents the set of values assumed exactly r times, the necessary and sufficient condition so that f is of bounded variation is that G_∞ has zero measure and the series

$$\sum_{r=1}^{\infty} r\,\mu(G_r)$$

is convergent: in this case the sum of this series equals the total variation of f.

In his own studies, Stefan Banach had obtained similar results with different procedures [84]. This memoir is connected to other research works of the Polish School as Waclaw Sierpiński, Stefan Mazurkiewicz and the Russian, Nikolai Luzin, were, at the same time, studying the measurability and cardinality of the set of values that a continuous function assumes a number of times equal to an assigned cardinal.[26] It is to be remembered that, in 1924, from one of Banach's articles he had summarised for the *Bollettino UMI* [86], Vitali drew inspiration for an

---

[24][77] I: 277, Vitali himself had recognised the priority of De La Vallée Poussin in a letter addressed to Fréchet (Vitali Archive, Bologna, 1-V-23).

[25]See two letters from Fréchet to Vitali of 30th March and 4th May 1923, in [1]: 483–485.

[26][85]. See the letter of 3rd October 1942 from Sierpiński to Vitali in [1]: 490–491.

interesting note of measure theory [87]. In the same year, Banach had also published a simplified demonstration of Vitali's covering theorem [67].

Relationships with the Polish researchers were frequent and imbued with esteem; Vitali corresponded not only with Sierpiński, but also with Otton Nikodym, with whom he discussed questions of projective geometry (collineations of the complex projective plane, subsets of the complex projective plane which intersect each line in a certain number of points)[27] and whom he had occasion to meet at the International Congress of Bologna in 1829 (Nikodym held a lecture on the principles in local reasoning of classical analysis). Later, Vitali became a member (presented by Nikodym and his wife) of the *Société Polonaise de Mathématique* and published some of his memoirs in the *Annales de la Société Polonaise de Math.* as well as in *Fundamenta Mathematicae* [80, 88, 89]. At the *Congrès des Mathématiciens des Pays Slaves*, held in Warsaw in 1829, Vitali gave a speech on the definitions of measurable sets and summable functions which he had used to introduce Lebesgue integral in his treatise *Geometria nello spazio hilbertiano* [90] to include, from the beginning, non limited sets and functions [91].

Notwithsatanding this, Vitali's output on real analysis in the interim period between the two wars remains an interesting complement to the research works of his first years (a new demonstration of the Lebesgue-Vitali theorem is to be remembered [92]) whereas his work on differential geometry, a discipline with a strong tradition in Italy, provides much wider and more significant results as it became his main field of investigation.

Even if studies on differential geometry were well established in Italy by the middle of the nineteenth century, there is no doubt that this line of research was boosted by the works of Eugenio Beltrami in the years 1864–1884, after he had entered into contact with Betti and Riemann in Pisa. From then on differential geometry in Italy went through a period of a frenetic development which lasted until the beginning of the Second World War. Several basic techniques of the discipline were honed by Luigi Bianchi, Gregorio Ricci-Curbastro and Tullio Levi-Civita between 1880 and 1920, and later the applications achieved in the field of theoretical physics and, above all, in theory of relativity were so successful that a wide range of researchers were drawn to these studies, so much so that no mathematician of a certain level in Italy had not been involved in the study of differential geometry at some time in the years from 1920 to 1940. Thus alongside the specialists in the discipline, like Enrico Bompiani and Enea Bortolotti, questions of differential geometry were also being tackled by analysts, algebraic geometers, and mathematical physicists such as Pia Nalli, Guido Fubini, Giuseppe Vitali, Francesso Severi, Beniamino Segre, Enrico Fermi and many more. As pointed out in a previous paper, works of a certain importance alone, involving differential geometry produced in Italy from 1880 to 1940 numbered several hundreds.[28] These may be divided into various groups:

---

[27] See the letter of 11th May 1928 from Nikodym to Vitali, in [1]: 494–496.
[28] See [4] p. 49.

- Connection spaces and relative absolute calculus and parallelism, differential geometry of immersed varieties)
- Projective differential geometry
- Theory of surfaces, questions of applicability, transformations and deformations
- Geometric variational problems

Giuseppe Vitali worked mainly in the first two fields, publishing about thirty works from 1922 to 1932. Some of his most important contributions include the discovery of a covariant derivative associated with n covariant systems of first order [93, 94], later named Weitzenböck-Vitali [95–98] also used by Einstein in one of his works in 1928 [99]. The Weitzenböck–Vitali parallel transport was later characterised by Enea Bortolotti as the most general integrable Euclidean transport [100, 101]. Mention must also be made of the generalised absolute calculus, introduced by Vitali as a generalisation of the Ricci calculus [90, 102, 103] as well as the geometric applications made by Vitali himself [88, 89, 104–108] and his followers like Angelo Tonolo and Vitali's less known pupils (Aliprandi, Baldoni, Sacilotto, and Liceni), using the techniques of immersion of varieties in Hilbertian space (of an infinite number of dimensions), in the absence of modern concepts of tangent bundle, normal bundle and so on. Vitali's absolute calculus was used in the thirties by Enea Bortolotti [109–112], in order to arrange Bompiani's work in the domain of what was then called "geometrie riemanniane di specie superiore", and to study the immersed subvarieties in a Riemannian space (Gauss, Codazzi and Ricci equations).

Vitali also provided a direct contribution to projective differential geometry, whose main author was his old university friend Guido Fubini, using the generalised absolute calculus [113, 114]. Finally, Vitali devoted a few works to issues of variational geometry, still within the subvarieties of a Hilbertian space [115–117] Other geometric works were published by Vitali on ruled surfaces and geodesics before the introduction of his geometry in the Hilbertian space [118, 119].

The essential parts of Vitali's research works concerning differential geometry and absolute differential calculus was later re-elaborated and re-exposed in an organic way in a monography: *Geometria nello spazio hilbertiano*.[29] It is divided into five parts: Lebesgue Integral, Developments in series of orthogonal functions and first notions on the Hilbertian space, Complements of algebra, Absolute differential calculus, Differential geometry. In the fourth part, the most original, an extremely general definition of absolute systems is given. In the last part the applications were given, above all, to metric differential geometry, we may particularly remember the notion of principal systems of normals and the study of minimal varieties.

---

[29][90] "Geometry in Hilbertian space is not only an expansion of the field of research, it is not only the passage from finite to countable of the number of dimensions of the ambient space. That would be too little. It is a method of geometric representation that, substituting the usual Cartesian, allows more simple formulas, more concise demonstrations and a clearer and vaster view of the problems" (from the Preface).

The *Geometria nello spazio hilbertiano* was conceived as a basic text for courses on higher analysis: in fact Vitali always had a great interest in the teaching of mathematics. For the courses on analysis held in the first 2 years of university he devoted much time to producing a text which was published in lithograph form in 1930.[30] Another treatise, a monography on functions of one real variable, destined to enter a collection on mathematical topics edited by the National Research Committee (CNR), remained unfinished and was published posthumously in 1935, completed, for the second volume, by Giovanni Sansone [121].

In his last years, Vitali held some important conferences and reports addressed not only to specialists, in which he expressed his strong desire to communicate the ideas that formed the basis of his method and how he envisaged mathematical research [122,123]. Still within this context are to be found some notes published in the *Bollettino della'Unione Matematica Italiana* or in the *Periodico di Matematiche* [72, 74, 92, 124–128] as well as the article on "Limits, series, continued fractions, infinite products" inserted into the *Enciclopedia delle matematiche elementari e complementi* published from 1929 on and edited by L. Berzolari, G. Vivanti and D. Gigli [129].

Finally, we may remember that after having devoted his research to pure mathematics, in the end Vitali also turned his attention to problems of analytical mechanics, structure of matter, and astrophysics, trying to use refined techniques of measure theory and differential geometry [130–134].

I should underline the fact that it was Vitali's fate to be considered something of an outsider. Even his research on differential geometry during his life did not attain great recognition within university circles. No one doubted his worth, but these research works were not considered of great importance to the discipline. Beniamino Segre's study on the development of geometry in Italy starting from 1860 [135], gives evidence of this when, presenting differential geometry, he only cited Vitali as the author of the treatise *Geometria nello spazio hilbertiano*. Further confirmation to this regard comes from the report on the Competition for the Royal Award for Mathematics (which expired on 31st December 1931). The commission (composed of S. Pincherle, G. Castelnuovo, E. Pascal, F. Severi, and G. Fubini) unanimously decided to share the prize between A. Commessatti and L. Fantappié. As for Vitali, whose untimely death occurred during the competition, the Commission said that, were Vitali's entire research works to have been judged, he would have been the winner, but since their judgement was limited to the last 10 years, they had to admit that the introduction Vitali produced of absolute systems and their derivates,

---

[30][120] Concerning Vitali's treatises and his production linked to the teaching of mathematics, see: [6].

although of great value, did not justify its great formal complication in relation to the importance of the results obtained.[31]

It was only after some time, in a long report on differential geometry, that Enea Bortolotti, following the applications and developments that he, himself, had produced, gave formal recognition to Vitali's work [136].

6 Conclusions

In hindsight, we may say that Vitali's absence from university constituted a significant loss to mathematics in Italy; according to contemporary evidence Vitali emerges as a generous person and enthusiastic teacher, with a real genius for identifying key problems and results in mathematical research which he skilfully expressed without excessive formalism in elegant, comprehensible language showing rigour and sobriety and clarity in the exposition of the ideas which formed the basis of his output. The depth of affection and sorrow expressed when news of his sudden death arrived is surprising.[32] Tullio Viola, one of his pupils in Bologna, who followed different lines of research, made reference to his grief during a seminar in Paris, as the words "Vitali est mort!" spread through the audience, and he gave tribute to his mentor [3]: "The Maestro was struck down,... under the portico of the city in which I had spent the best days of my student life..." In a moving obituary, which was also an accurate description of Vitali's research, Angelo Tonolo, who had been one of his colleagues in Padua, defined his personality [7]: gifted with discerning intuition, he not only had great skills in algorithm, but was also able to predict the truth of a proposition even without having logical proof, as well as identifying the general outline of the problem. Original and independent in his approach to a topic, he felt the need to elaborate even well-known results on his own. Devoted to his role as advisor for his students' research, his pupils repaid him with admiration and affection. Tonolo also pointed out, along with his mental attributes,

---

[31] [4] p. 54: "Certainly, if the entire output of a competitor were to be judged, there is no doubt that the prize should go to Vitali, whose work has given the highest honour to mathematics in Italy. The Commission, however, regrets that judgement must be limited exclusively to the works presented in this competition and is not able to take into consideration the research carried out before the last decade. Having these limitations posed on its jurisdiction, the Commission has to admit that the introduction Vitali produced of absolute systems and their derivatives, although of great value, did not justify its great formal complication in relation to the importance of the results obtained".

[32] Let me quote the letter sent by N. Luzin to the Seminario Matematico of Padua University on 18th March 1932 ([2] p. 201): "Messieurs et chers Collègues, C'est avec la plus vive douleur que j'ai appris la mort inattendue de notre cher et inoubliable confrère, le professeur Giuseppe Vitali. Permettez moi de vous exprimer ma condoléance profonde sur cette perte irréparable d'un grand savant dont la vie consacré tout entière sans mélange et sans partage aux recherches scientifiques et aux travaux de l'enseignement et dont les belles découvertes dans la Théorie des Fonctions font l'honneur et la gloire de notre Science".

his great sensitivity, kindness and gentleness, qualities which had prevented him from being assertive in obtaining the role that he deserved.

References

1. Vitali, G. 1984. *Opere sull'analisi reale e complessa, carteggio.* ed. L. Pepe. Bologna: Cremonese.
2. Pepe, L. Giuseppe Vitali e l'analisi reale. 1984. *Rendiconti del Seminario matematico e fisico di Milano* 54: 187–201.
3. Viola, T. 1984. Ricordo di Giuseppe Vitali a 50 anni dalla sua scomparsa. In *Atti del Convegno La Storia delle Matematiche in Italia*, 535–544 (Cagliari 1982). Bologna, Monograf.
4. Borgato, M.T. and Vaz Ferreira, A. 1987 Giuseppe Vitali: ricerca matematica e attività accademica dopo il 1918. In *La matematica italiana tra le due guerre mondiali, Atti del Convegno*, ed. A. Guerraggio, 43–58. Bologna: Pitagora.
5. Vaz Ferreira, A. 1991. Giuseppe Vitali and the Mathematical Research at Bologna. In *Geometry and complex variables proceedings of an international meeting on the occasion of the IX centennial of the University of Bologna.* ed. S. Coen. Lecture Notes in Pure and Appl. Math. 132, 375–395. New York: Dekker.
6. Pepe, L. 1983. Giuseppe Vitali e la didattica della matematica. *Archimede* 35(4): 163–176.
7. Tonolo, A. 1932. Commemorazione di Giuseppe Vitali. *Rendiconti del Seminario Matematico della Università di Padova* 3: 67–81.
8. Pier, J.-P. 1996. *Histoire de l'intégration.* Paris: Masson.
9. Dunham, W. 2005. *The Calculus gallery. Masterpieces from Newton to Lebesgue.* Princeton: Princeton University Press.
10. Pepe, L. Una biografia di Giuseppe Vitali. In [1]: 1–24.
11. Sansone, G. 1977. *Algebristi, analisti, geometri differenzialisti, meccanici e fisici matematici ex-normalisti del periodo 1860–1920*, 40–42. Pisa: Scuola Normale Superiore di Pisa.
12. Vitali, G. 1900. Sulle applicazioni del Postulato della continuità nella geometria elementare. In *Questioni riguardanti la geometria elementare.* Bologna: Zanichelli.
13. Vitali, G. 1900. Sulle funzioni analitiche sopra le superficie di Riemann. *Rendiconti del Circolo Matematico di Palermo* 14: 202–208.
14. Vitali, G. 1900. Sui limiti per $n = \infty$ delle derivate $n.^{me}$ delle funzioni analitiche. *Rendiconti del Circolo Matematico di Palermo* 14: 209–216.
15. Vitali, G. 1902. Sopra le equazioni differenziali lineari omogenee a coefficienti algebrici. *Rendiconti del Circolo Matematico di Palermo* 16: 57–69.
16. Vitali, G. 1903. Sopra le equazioni differenziali lineari omogenee a coefficienti algebrici. *Annali della Scuola Normale Superire di Pisa* 9: 1–57.
17. Appell, P. 1890. Sur les intégrales des fonctions à multiplicateurs et leur application au développement des fonctions abéliennes en séries trigonométriques. *Acta Mathematica* 13: 3–174.
18. Weil, A. 1938. Généralisation des fonctions abéliennes. *Journal de Mathématiques Pures et Appliquées* (9) 17: 47–87.
19. Arzelà, C. 1899. Sulle serie di funzioni. *Memorie dell'Accademia delle scienze di Bologna* (5), parte I, 8: 131–186; parte II, 9 (1900), 701–744.
20. Letta, G., Papini, P.L., and Pepe, L. 1992. Cesare Arzelà e l'Analisi reale in Italia. In Cesare Arzelà, *Opere complete*, I, xiii–xxxvii. Bologna: Cremonese.
21. Arzelà, C. 1903. Sulle serie di funzioni analitiche. *Rendiconti della Accademia delle scienze di Bologna* n. s., 7: 33–42.
22. Vitali, G. 1903. Sopra le serie di funzioni analitiche. *Rendiconti dell'Istituto Lombardo* (2) 36: 772–774.

23. Vitali, G. 1904. Sopra le serie di funzioni analitiche. *Annali di Matematica Pura ed Applicata* (3) 10: 65–82.
24. Vitali, G. 1903–1904. Sopra le serie di funzioni analitiche. *Atti dell'Accademia delle Scienze di Torino* 39: 22–32.
25. Dini, U. 1878. *Fondamenti per la teorica delle funzioni di variabili reali*. Pisa: Nistri.
26. Dini, U. 1892. *Grundlagen für eine Theorie der Funktionen einer veränderlichen reellen Grösse*, Deutsch bearbeitet von Jacob Lüroth und Adolf Schepp. Leipzig: Teubner.
27. Borel, É. 1898. *Leçons sur la théorie des fonctions*. Paris: Gauthier-Villars.
28. Lebesgue, H. 1902. Intégrale, longueur, aire. *Annali di Matematica Pura ed Applicata* 7/1: 231–359.
29. Vitali, G. 1903. Sulla condizione di integrabilità delle funzioni. *Bollettino dell'Accademia Gioenia di Catania* 79: 27–30.
30. Vitali, G. 1904. Sulla integrabilità delle funzioni. *Rendiconti dell'Istituto Lombardo* (2) 37: 69–73.
31. Vitali, G. 1904. Sui gruppi di punti. *Rendiconti del Circolo Matematico di Palermo* 18: 116–126.
32. Borgato, M.T. 1993. Giuseppe Peano: tra analisi e geometria. In *Peano e i fondamenti della matematica. Atti del Convegno* 139–169. Modena: Accademia Nazionale di Scienze Lettere ed Arti.
33. Vitali, G. 1905. *Sul problema della misura dei gruppi di punti di una retta*. Bologna: Gamberini e Parmeggiani.
34. Solovay, R.M. 1970. A model of set-theory in which every set of reals is Lebesgue measurable. *Annals of Mathematics* (2) 92: 1–56.
35. Lebesgue, H. 1904. *Leçons sur l'intégration et la recherche des fonctions primitives*. Paris: Gautier-Villars.
36. Vitali, G. 1904–1905. Sulle funzioni integrali. *Atti dell' Accademia delle Scienze di Torino* 40: 1021–1034.
37. Vitali, G. 1922. Analisi delle funzioni a variazione limitata. *Rendiconti del Circolo Matematico di Palermo* 46: 388–408.
38. Lebesgue, H. 1906. Sur les fonctions dérivées. *Atti della Accademia Nazionale dei Lincei Classe Scienze Fis. Mat. Natur.* 15: 3–8.
39. Lebesgue, H. 1907. Encore une observation sur les fonctions dérivées. *Atti della Accademia Nazionale dei Lincei Classe Scienze Fis. Mat. Natur.* 16: 92–100.
40. Lebesgue, H. 1907. Sur la recherche des fonctions primitives par l'intégration. *Atti della Accademia Nazionale dei Lincei Classe Scienze Fis. Mat. Natur.* 16: 283–290.
41. Vitali, G. 1907–1908. Sui gruppi di punti e sulle funzioni di variabili reali. *Atti dell'Accademia delle Scienze di Torino* 43: 229–246.
42. Hobson, E.W. 1907. *The Theory of Functions of a Real Variable and the Theory of Fourier's Series*, Cambridge: Cambridge University Press.
43. Lebesgue, H. 1910. Sur l'intégration des fonctions discontinues. *Annales Scientifiques de l'Ecole Normale Supérieure* 26: 361–450.
44. Clarkson, J.A. and Adams, C.R. 1933. On definitions of bounded variation for functions of two variables. *Transactions of the American Mathematical Society* 35/4: 824–854.
45. Lebesgue, H. 1922. *Notice sur les travaux scientifiques de M. Henri Lebesgue*. Toulouse: Privat.
46. Vitali, G. Sulle funzioni a integrale nullo. 1905. *Rendiconti del Circolo matematico di Palermo* 20: 136–141.
47. Lindelöf, E. 1903. Sur quelques points de la théorie des ensembles. *Comptes Rendus de l'Académie des Sciences*, 137: 697–700
48. Vitali, G. 1905. *Sugli ordini di infinito delle funzioni reali*. Bologna: Gamberini e Parmeggiani.
49. Baire, R. 1899. Sur les fonctions de variables réelles. *Annali di Matematica pura ed applicata* (3) 3: 1–123.
50. Baire, R. 1905. *Leçons sur les fonctions discontinues*. Paris: Gauthier-Villars.

51. Borel, É. 1905. *Leçons sur les fonctions de variables réelles et les développements en séries des polynômes* Paris: Gauthier-Villars.
52. Vitali, G. 1905. Un contributo all'analisi delle funzioni. *Atti dell'Accademia Nazionale dei Lincei Classe Scienze Fis. Mat. Natur.* (5) 14: 189–198.
53. Lebesgue, H. 1904. Sur les fonctions représentables analytiquement. *Comptes Rendus de l'Académie des Sciences* 139: 29.
54. Lebesgue, H. 1905. Sur les fonctions représentables analytiquement. *Journal de Mathématiques Pures et Appliquées* (6) 1: 139–216
55. Vitali, G. 1905. Una proprietà delle funzioni misurabili. *Rendiconti dell'Istituto Lombardo* (2) 38: 599–603.
56. Luzin, N. 1912. Sur les propriétés des fonctions mesurables. *Comptes Rendus de l'Académie des Sciences* 154: 1688–1690.
57. Borel, E. 1903. Un théorème sur les ensembles mesurables. *Comptes Rendus de l'Académie des Sciences* 137: 966–967.
58. Lebesgue, H. 1903. Sur une proprieté des fonctions. *Comptes Rendus de l'Académie des Sciences* 137: 1228–1230.
59. Tonelli, L. 1921–1923. *Fondamenti di calcolo delle variazioni.* Bologna: Zanichelli.
60. Vitali, G. 1905. Sopra l'integrazione di serie di funzioni di una variabile reale. *Bollettino dell'Accademia Gioenia di Catania* 86: 3–9.
61. Vitali, G. 1907. Sull'integrazione per serie. *Rendiconti del Circolo Matematico di Palermo* 23: 137–155.
62. Levi, B. 1906. Sopra l'integrazione delle serie. *Rendiconti dell'Istituto Lombardo* (2) 39: 775–780.
63. Denjoy, A. 1950. Le veritable théorème de Vitali. *Comptes Rendus de l'Académie des Sciences* 231: 560–562.
64. Denjoy, A. 1950. Le théorème de Vitali. *Comptes Rendus de l'Académie des Sciences* 231: 600–601.
65. Denjoy, A. 1950. Les applications du Théorème général de Vitali. *Comptes Rendus de l'Académie des Sciences* 231: 737–740.
66. Carathéodory, C. 1918. *Vorlesunghen über reelle Funktionen*, 299. Leipzig und Berlin: Teubner.
67. Banach, S. 1924. Sur le théorème de M. Vitali. *Fundamenta Mathematicae* 5: 130–136.
68. Giacardi, L. (ed.). 2006. *Da Casati a Gentile: momenti di storia dell'insegnamento secondario della matematica in Italia.* Centro Studi Enriques, Lugano, Lumières Internationales.
69. Vitali, G. 1912. Sopra una proprietà caratteristica delle funzioni armoniche. *Atti dell' Accademia Nazionale dei Lincei Classe Scienze Fis. Mat. Natur.* (5) 21: 315–320.
70. Vitali, G. 1915–1916. I teoremi della media e di Rolle. *Atti dell'Accademia delle Scienze di Torino* 51: 143–147.
71. Vitali, G. 1916. Sui teoremi di Rolle e della media per le funzioni additive. *Atti dell'Accademia Nazionale dei Lincei Classe Scienze Fis. Mat. Natur.* (5) 25: 684–688.
72. Vitali, G. 1915. Sostituzioni sopra una infinità numerabile di elementi. *Bollettino Mathesis* 7: 29–31.
73. Vitali, G. 1921. Sulla condizione di chiusura di un sistema di funzioni ortogonali. *Atti dell'Accademia Nazionale dei Lincei Classe Scienze Fis. Mat. Natur.* (5) 30: 498–501.
74. Vitali, G. 1922. Sulle rettificazione delle curve. *Bollettino dell'Unione Matematica Italiana* 1/2-3, 47–49.
75. Vitali, G. 1925. Sulla definizione di integrale delle funzioni di una variabile. *Annali di Matematica Pura ed Applicata* (4) 2: 111–121.
76. Levi, B. 1923–1924. Sulla definizione dell'integrale. *Annali di Matematica Pura ed Applicata* (4) 1: 58–82.
77. De La Vallée Poussin, C.-J. 1914. *Cours d'analyse infinitésimale*, 3rd ed. Louvain: Dieudonné - Paris: Gauthier-Villars.

78. Fréchet, M. 1913. Sur les fonctionnelles linéaires et l'intégrale de Stieltjes. *Comptes rendus du Congrès des Sociétés savantes*, 45–54. Paris.
79. Vitali, G. 1923. Sulle funzioni a variazione limitata. *Rendiconti del Circolo Matematico di Palermo* 47: 334–335.
80. Vitali, G. 1926. Sulle funzioni continue. *Fundamenta Mathematicae* 8: 175–188.
81. Tonelli, L. 1907–1908. Sulla rettificazione delle curve. *Atti dell'Accadem delle Scienze di Torino* 43: 783–800.
82. Tonelli, L. 1907–1908. Sulla lunghezza di una curva. *Atti dell'Accademia delle Scienze di Torino* 43: 783–800.
83. Tonelli, L. 1911–1912. Sulla lunghezza di una curva. *Atti dell'Accademia delle Scienze di Torino* 47: 1067–1085.
84. Banach, S. 1925. Sur les lignes rectifiables et ls surfaces dont l'aire est finie. *Fundamenta Mathematicae* 7: 226–236.
85. Mazurkiewicz. S. and Sierpiński, W. 1924. Sur un problème concernant les fonctions continues. *Fundamenta Mathematicae* 6: 161–169.
86. Banach, S. 1923. Sur le problème de la mesure. *Fundamenta Mathematicae* 4: 7–23.
87. Vitali, G. 1924. Sulla misura dei gruppi di punti di una retta. *Bollettino dell'Unione Matematica Italiana* 2: 8–23.
88. Vitali, G. 1928. Sopra alcuni invarianti associati ad una varietà e sopra i sistemi principali di normali delle superficie. *Annales de la Société Polonaise de Mathématique* 7: 43–67.
89. Vitali, G. 1928. Sistemi principali di normali ad una varietà giacenti nel suo σ_2. *Annales de la Société Polonaise de Mathématique* 7: 242–251.
90. Vitali, G. 1929. *Geometria nello spazio hilbertiano*. Bologna: Zanichelli.
91. Vitali, G. 1929. Sulle definizioni di aggregati misurabili e di funzioni sommabili. *Comptes Rendus I Congrès des Mathématiciens des Pays Slaves*, Warszawa, 282–286.
92. Vitali, G. 1927. Sulla condizione della integrabilità riemanniana lungo un dato intervallo delle funzioni limitate di una variabile reale. *Bollettino dell'Unione Matematica Italiana* anno 6: 253–257.
93. Vitali, G. 1224. Una derivazione covariante formata coll'ausilio di n sistemi covarianti del 1° ordine. *Atti della Società Ligustica* 2/4: 248–253.
94. Vitali, G. 1925. Intorno ad una derivazione nel Calcolo assoluto. *Atti della Società Ligustica* 4: 287–291.
95. Weitzenböck, R. 1923. *Invariantentheorie*. Groningen: P. Noordhoff.
96. Schouten, J.A. 1924. *Der Ricci-Kalkül*. Berlin: Springer.
97. Struik, D.J. 1934. *Theory of linear connections*. Berlin: Springer.
98. Schouten, J.A., and Struik, D.J. 1935–1938. *Einfürung in die neueren Methoden der Differentialgeometrie*, 2 vols. Groningen-Batavia: P. Noordhoff.
99. Einstein, A. 1928. Riemann-Geometrie mit Aufrechterhaltung des Begriffes des Fernparallelismus. *Sonderabdruck aus den Sitzungsberichten der Preußischen Akademie der Wissenschaften Physikalisch-mathematischen Klasse* 17: 217–221.
100. Bortolotti, E. 1928. Scostamento geodetico e sue generalizzazioni. *Giornale di Matematiche di Battaglini* 66: 153–191.
101. Bortolotti, E. 1929–1930. Leggi di trasporto sui campi di vettori applicati ai punti di una curva di una V_m in una V_n riemanniane. *Memorie della Reale Accademia delle Scienze dell'Istituto di Bologna* (8) 7: 11–20.
102. Vitali, G. 1930. Nuovi contributi alla nozione di derivazione covariante. *Rendiconti del Seminario Matematico della Università di Padova* 1: 46–72.
103. Vitali, G. 1932. Sulle derivazioni covarianti. *Rendiconti del Seminario Matematico della Università di Padova* 3: 1–2.
104. Vitali, G. 1929. Le identità di Bianchi per i simboli di Riemann nel calcolo assoluto generalizzato. *Atti della Accademia Nazionale dei Lincei Classe di Scienze Mat. Fis. Natur.* (6) 9: 190–192.
105. Vitali, G. 1929. Sui centri di curvatura delle geodetiche di una varietà. *Atti della Accademia Nazionale dei Lincei Classe di Scienze Mat. Fis. Natur.* (6) 9: 391–394.

106. Vitali, G. 1929–1930. Sopra alcune involuzioni delle tangenti ad una superficie. *Atti dell'Istituto Veneto di Scienze Lettere ed Arti* 89: 107–112.
107. Vitali, G. 1930. Evoluta(?) di una qualsiasi varietà dello spazio hilbertiano. *Annali di Matematica Pura ed Applicata* (4) 8: 161–172.
108. Vitali, G. 1931. Sulle relazioni lineari tra gli elementi di un ricciano. *Bollettino dell'Unione Matematica Italiana* 10: 265–269.
109. Bortolotti, E. 1930–1931. Calcolo assoluto generalizzato di Pascal-Vitali e intorni dei vari ordini di un punto su una varietà riemanniana. *Atti dell'Istituto Veneto di Scienze Lettere ed Arti* 90: 461–478.
110. Bortolotti, E. 1931. Vedute geometriche sul calcolo assoluto del Vitali e applicazioni. *Rendiconti del Seminario della Facoltà di Scienze della R. Università di Cagliari* 1: 10–12.
111. Bortolotti, E. 1931. Nuova esposizione, su basi geometriche, del calcolo assoluto del Vitali, e applicazioni alle geometrie riemanniane di specie superiore. *Rendiconti del Seminario Matematico della Università di Padova* 2: 164–212.
112. Bortolotti, E. 1939. Contributi alla teoria delle connessioni. II. Connessioni di specie superiore. *Memorie dell'Istituto Lombardo Cl. Sc. Mat. Nat.* 24: 1–39.
113. Vitali, G. 1928–1929. Forme differenziali a carattere proiettivo associate a certe varietà. *Atti del R. Istituto Veneto* 88: 361–368.
114. Vitali, G. 1929–1930. Saggio di ricerche geometrico-differenziali. *Atti del R. Istituto Veneto* 89: 378–381.
115. Vitali, G. 1929. Sopra i problemi di massimo o di minimo riguardanti le varietà nello spazio hilbertiano. *Rendiconti dell'Istituto Lombardo* (2) 62: 127–137.
116. Vitali, G. 1930. Determinazione della superficie di area minima nello spazio hilbertiano. *Rendiconti del Seminario Matematico dell'Università di Padova* 1: 157–163.
117. Vitali, G. 1931. Alcuni elementi di meccanica negli spazi curvi. *Annali di Matematica Pura ed Applicata* (4) 9: 75–89.
118. Vitali, G. 1922. Sulle superfici rigate e sulle congruenze. *Atti della Società Ligustica* 1: 36–40.
119. Vitali, G. 1925. Superficie che ammettono una famiglia di geodetiche, di cui sono note le proiezioni ortogonali sopra un determinato piano. *Atti del R. Istituto Veneto* 84: 641–644.
120. Vitali, G. 1930. *Analisi matematica*. Bologna: La Grafolito Editrice.
121. Vitali, G. 1933. *Moderna teoria delle funzioni di variabile reale*, Part I. ed. G. Sansone. Bologna: Zanichelli.
122. Vitali, G. 1928. Rapporti inattesi su alcuni rami della matematica. *Atti del Congresso Internazionale dei Matematici*. Bologna 3–10 Sept, 299–302.
123. Vitali, G. 1930. Un trentennio di pensiero matematico. *Atti della XIX Riunione SIPS* I: 315–327.
124. Vitali, G. 1928. Sul teorema fondamentale dell'algebra. *Periodico di Matematiche* (4) 8: 102–105.
125. Vitali, G. 1929. Un teorema sulle congruenze. *Periodico di Matematiche* (4) 9: 193–194.
126. Vitali, G. 1928. Sulle sostituzioni lineari ortogonali. *Bollettino dell'Unione Matematica Italiana* 7: 1–7.
127. Vitali, G. 1929. Sulle equazioni secolari. *Atti della XVIII Riunione SIPS* II: 3–7.
128. Vitali, G. 1933. Del ragionare. *Bollettino dell'Unione Matematica Italiana* 12: 89–94.
129. Vitali, G. 1932. Limiti, serie, frazioni continue, prodotti infiniti. In *Enciclopedia delle matematiche elementari*, I, pars II, 391–439. Milan: Hoepli.
130. Vitali, G. 1926. Sul principio di Hamilton. *Atti della Accademia Nazionale dei Lincei Classe Scienze Fis. Mat. Natur.* (6) 9: 44–48.
131. Vitali, G. 1931. Un risultato di Hausdorff e la compressibilità della materia. *Atti della Accademia Nazionale dei Lincei Rendiconti Classe Scienze Fis. Mat. Natur.* (6) 12: 903–905.
132. Vitali, G. 1931. Alcuni elementi di meccanica negli spazi curvi. *Annali di matematica pura ed applicata* (4) 9: 75–89.
133. Vitali, G. 1931. Una nuova interpretazione del fenomeno gravitazionale. *Bollettino dell'Unione Matematica Italiana* 10: 113–115.

134. Vitali, G. 1931. Una nuova interpretazione del fenomeno della gravitazione universale. *Memorie della Società Astronomica Italiana* (2) 5: 405–421.
135. Segre, B. 1933. La geometria in Italia, dal Cremona ai giorni nostri. *Annali di Matematica Pura ed Applicata* (4) 11: 1–16.
136. Bortolotti, E. 1939. Geometria differenziale. In *Un secolo di progresso scientifico italiano*, I. Roma: SIPS.
137. Gunning, R.C. 1967. *Lectures on Vector Bundles over Riemann Surfaces*, Mathematical Notes 6, Princeton. Princeton University Press.
138. Deligne, P. 1970. *Equations différentielles à points singuliers réguliers*, Lecture Notes in Mathematics 163, Springer.

Pincherle's Early Contributions to Complex Analysis

Umberto Bottazzini

Abstract A young graduate from the Scuola Normale in Pisa, Salvatore Pincherle got his first position as a teacher in Pavia where he got in touch with Casorati. At the latter's suggestion he attended Weierstrass's lectures in Berlin, which marked a lasting influence on his approach to complex function theory. Beginning with Pincherle's own report on his stay in Berlin, the paper offers a survey of his early contributions to complex analysis on the basis of his manuscript notebooks and printed papers as well. They set up the background of his pioneering work in functional analysis.

1 Introduction

One of the outstanding Italian analysts in his days, Salvatore Pincherle (1853–1936) mainly contributed to complex function theory. After graduating in 1874 from the Scuola Normale Superiore in Pisa as a student of Betti's with a thesis on capillarity, Pincherle obtained a professorship at the higher secondary school (*liceo*) in Pavia where he got in touch with Felice Casorati, a professor at the University there. This marked a turning point in Pincherle's scientific career. Indeed, Casorati exerted a decisive influence on his young colleague by turning his research interest to complex function theory.

After the publication of the first volume of his *Teorica* [7] – a mixture of techniques and ideas of Cauchy, Riemann, and Weierstrass – Casorati had planned to let it be followed by a second one, entirely devoted to Riemann's approach to complex function theory. However, being unable to overcome the problems raised by Weierstrass's criticism of Dirichlet's principle – "Riemann's true method for creating

---

U. Bottazzini (✉)
Dipartimento di Matematica 'Federigo Enriques', Università degli Studi di Milano, Via Saldini 50, 20133, Milano, Italy
e-mail: umberto.bottazzini@unimi.it

functions", as Prym called it in a letter to Casorati [16, p. 61] – the latter was forced to delay the publication of the planned volume (which in fact never appeared).[1]

In the hope of getting first-hand information about methods and results of the German mathematicians, in 1877 Casorati supported Pincherle's successful application for a grant to spend the academic year 1877/1878 in Berlin, where he attended Kronecker's and Weierstrass's lectures. Back in Italy, at Casorati's suggestion he was offered the teaching a course at the University of Pavia on the principles of Weierstrass's function theory. Pincherle gave a systematic presentation of it up to the application of the theory of analytic functions to elliptic functions. This was the first time that such a course was taught in Italy. Following it, in 1880 Pincherle summarized the basic elements of Weierstrass's theory of analytic functions in a paper, the *Saggio* [17], that offered the first presentation in Italy of Weierstrass's function theory, and became in some respect the Weierstrassian counterpart of Casorati's *Teorica*.

The *Saggio* made a name of Pincherle as a faithful follower of Weierstrass's approach to complex function theory. In spring 1880, Pincherle was appointed to a chair at the University of Palermo, but in the autumn of the very same year he got a chair at the University of Bologna where he taught until his retirement in 1928. In addition to his scientific achievements – some of which will be discussed below – Pincherle played a major role on the institutional level both in the Italian and the international mathematical community. In 1922, he was the founding President of the Unione Matematica Italiana, and from 1924 to 1928 he served as the President of the International Mathematical Union. In this capacity he organized and chaired the ICM in 1928 in Bologna, the first after World War I that was open to all mathematicians irrespective of nationality, thus terminating the discrimination against the mathematicians of the Central Powers.[2]

In his plenary lecture [11] at the ICM in Bologna Jacques Hadamard hailed Pincherle as a founder, jointly with Volterra, of the new "functional calculus." The first relevant memoir [24] (in Italian) by Pincherle was followed by a second one [26] where he presented his results to the broader readership of Mittag-Leffler's *Acta mathematica*. Those pioneering papers represented the natural outcome of the research work on complex analysis that Pincherle tackled in the first decade of his scientific activity essentially following Weierstrass's approach.

2 Attending Weierstrass's Lectures

Introduced by Casorati's letters to Kronecker and Weierstrass, in November 1877 Pincherle began to attend their "private" courses. As he reported in a letter to Casorati [4, pp. 205–206], Kronecker was lecturing on the application of analysis to

---

[1] For a detailed account of this matter and, more generally, of the history of complex function theory, see [6].

[2] For Pincherle's role in the IMU, see [15], 36–50.

number theory. Pincherle was not very excited by that course, and he limited himself to provide Casorati with a scanty summary of it. Nonetheless, he was pleased to add that Kronecker, occasionally needing Cauchy's integral theorem, offered a proof of it that was essentially the same given by Casorati in his *Teorica*.

In Pincherle's opinion, Weierstrass's course had "a higher character" than Kronecker's. It was devoted to Abelian functions, or "more properly, on hyperelliptic functions". Being familiar with Clebsch and Gordan's treatment of the subject, Pincherle confessed that in the beginning Weierstrass's course – essentially inspired to Weierstrass's 1856 paper [40] – turned out to be very difficult to him with respect to both the terminology and the methods.

At Casorati's request, in a letter on April 21, 1878, Pincherle offered him a short synopsis of Weierstrass's course that is worth quoting in detail:

His [Weierstrass's] course can be divided into three parts. The first of these discusses differential equations of the form

$$u_\lambda = \sum_{\alpha=1}^{\rho} \frac{P(x_\alpha)}{y_\alpha} \frac{\mathrm{d}x_\alpha}{x - a_\alpha} \quad (\lambda = 1, 2, 3, \ldots, \rho),$$

where the u_λ are independent variables and the (x_α, y_α) are pairs of points on the fundamental curve

$$y^2 = (x - a_1)(x - a_2)\ldots(x - a_{2\rho-1}).$$

He shows that the x_α satisfying the above system of differential equations are the roots of an algebraic equation of degree ρ, whose coefficients are shown to be single-valued functions of the independent variables $u_1, u_2, \ldots, u_\rho$. These functions are denoted by $\wp(u_1, u_2, \ldots, u_\rho)$ and he calls them by the name of abelian functions. He then studies the addition theorem for these functions and the relations that obtain between them and their partial derivatives. (Quoted in [4, pp. 206–207]. Transl. in [6])

Pincherle commented that this part did not differ very much from the first part of Weierstrass's 1856 paper. In fact, the major change with respect to that paper was the introduction of the $\wp$-functions of several variables instead of the Als. In the remaining parts of the course, Weierstrass presented some of his more recent material, including the fundamental concept of prime function he had just published in his great paper [41].

Thus, Pincherle went on:

In the second part he treats hyperelliptic integrals, the decomposition of a general integral into integrals in normal form, their infinities and moduli of periodicity; then he shows how to write the normal integrals as functions of the $u_1, u_2, \ldots, u_\rho$. He does this by means of the sums of logarithms of certain single-valued functions which he calls E and which play a major role in the theory. These functions (which have the property of being zero or infinite at only a single point and discontinuous at infinity) he calls prime functions because of the following property: every rational function of x and y, such as $y = R(x)$, can be regarded as function of the independent variables u, whence, as such, it can always be written as a product of the functions E. I do not know whether there exists already some literature about these functions.

In the third part he finally defines Θ-functions of several variables by means of a differential equation, establishes the monodromy and the periodicity of such functions, and their analytic expression; finally, he shows that both the functions $\wp$ and the functions E are expressible as quotients of Θ-functions, and thus finds an analytic expressions for these $\wp$ and E where before he had only proved their existence. This course in Abelian functions is the third in a cycle of courses that Weierstrass gives over three semesters. (Quoted in [4, p. 207]. Transl. in [6])

In order to become acquainted with Weierstrass's theory of analytic functions, and his theory of elliptic functions as well, Pincherle resorted to the lecture notes of the relevant lectures taken by some students of Weierstrass's. Those manuscripts provided Pincherle with the material that he was to include in his *Saggio*, a paper that offered a fresh, first-hand summary of Weierstrass's lectures on analytic functions.[3]

Given Weierstrass's notorious reluctance to publication, and especially to the publication of lecture notes of his courses, the historical importance of Pincherle's *Saggio* can hardly be overestimated. In particular, as Pincherle himself stated, his paper was enough to enable one to read Weierstrass's recent (1876) memoir [41]. This, and Weierstrass's 1880 memoir [42] as well, were the inspiring sources of Pincherle's early papers devoted to complex function theory [2, pp. 29–30].

3 Expansions in Series

Looking back at the beginning of his scientific career Pincherle [33, p. 341] remembered that around 1880 the study of the expansion of an arbitrary function (in the general sense of Dirichlet) of a real variable already constituted a substantial chapter of analysis. As for the expansion of analytic functions in series, on the contrary, by then only special cases had been considered such as the expansions in Legendre polynomials or Bessel functions, as given by the works of Neumann, Heine, Frobenius and others. This suggested to Pincherle to tackle the general problem of the expansion of analytic functions in series of functions belonging to a given system.

In spring 1882, Pincherle began approaching the problem with some preliminary studies on the expansion of analytic functions in spherical functions.[4] The first, related results he published were some theorems that he stated (without any proof) in a one-page note [18]. There he stated that, given a system of analytic functions $p_n(z)$, regular within a disk $|z| < R$, then the powers z^m could be expanded in series of the form

$$z^m = \sum_{n=0}^{\infty} \alpha_{m,n} p_n(z),$$

---

[3] See e.g. [1, pp. 287–289]. For a more detailed account, see [6].

[4] Drafts of them are contained in a workbook by him dated January 1882–August 1882. This is the first of 55 manuscript volumes by Pincherle, in which all along his life he used to record drafts of his own research work as well as notes and comments on the papers and books he occurred to study. The volumes are kept in the library of the Mathematical Department of the Bologna University. See [5].

where $\alpha_{m,n}$ are suitable constants. In addition, if positive numbers K and σ can be assigned such that

$$\sum_{n=0}^{\infty} |\alpha_{m,n} p_n(z)| < \sigma^m K,$$

then any function $F(z)$ regular within a disk $|z| < R'$ ($R' > \sigma$) can be expanded in series as

$$F(z) = \sum_{n=0}^{\infty} c_n p_n(z),$$

where c_n are suitable constants.

Under the same hypotheses, Pincherle went on, there exists a system of "conjugate" functions $q_n(z)$ such that in a suitable domain one has

$$\frac{1}{1-z'z} = \sum_{n=0}^{\infty} q_n(z') p_n(z).$$

In particular, if $p_n(z)$ are polynomials of the degree n, the "conjugate" functions $q_n(z)$ are expressed by power series beginning with z^n, and vice versa [18, p. 225].

Starting from these results Pincherle was led to investigate the convergence conditions of series of the form

$$\sum_{n=0}^{\infty} \varphi_n(z) f_n(z),$$

where the analytic functions $\varphi_n(z)$ are such that the series $\sum \varphi_n(z)$ converges unconditionally and uniformly in a given domain J, and $f_n(z)$ are analytic functions in a connected domain of the complex plane. In addition, he asked the question about the conditions under which given analytic functions can be expanded in series of the above form.

Pincherle collected his results in a paper [19] that represents his first important contribution to the theory of analytic functions. There he began by considering an infinite sequence of meromorphic functions $f_n(z)$ defined in a connected domain A. Let $i_n, i'_n, \ldots, i_n^{(r)}$ ($n = 1, 2, 3, \ldots$) be the poles of the functions $f_n(z)$, and denote by j their (finite or infinite in number) accumulation points. By considering arbitrarily small disks centered at them, or arbitrarily thin strips according to whether the points j were isolated or were building lines, and by taking all of them off of the domain A, Pincherle was able to obtain a (generally not simply connected) domain J. By acting analogously with respect to the remaining (isolated and finite in number) points i_n, Pincherle eventually obtained a domain K where the functions $f_n(z)$ "have a rational, entire character" – as he stated [19, p. 65] by resorting to a Weierstrassian terminology.

At this point Pincherle asked the question: what could be said in general about the limit $\lim_{n \to \infty} f_n(z)$ ($z \in K$) ? In the lack of known "general considerations", in order to overcome the difficulty Pincherle required the functions $f_n(z)$ to be bounded in an appropriate sense. His requirement turned out to be a consequence of a general theorem that he stated in the following terms: If to every point x_0 of a connected domain C, including its boundary, corresponds one and only one value of a function (in the more general sense of the term) $X(x)$, and if a neighborhood U of x_0 can be assigned such that $\limsup |X(x)| = L(x_0)$ with $x \in U$ and $L(x_0)$ a finite number, then there exists a finite number N such that $|X(x)| < N$ in C. In the particular case of plane domains this theorem corresponds to the theorem usually named after Heine and Borel, as Pincherle himself later recognized when claiming priority in the matter.[5]

As a consequence of it, Pincherle stated and proved a theorem that generalized the theorem recently established by Weierstrass, according to which a series of rational functions converging uniformly inside a disconnected domain may represent different monogenic functions on disjoint regions of the domain (see [42, p. 221]).

Let the functions of the infinite system $f_n(z)$ $(n = 1, 2, 3, \ldots)$ be bounded within J, except a finite, fixed number n of them at the most, and let $\varphi_n(z)$ $(n = 1, 2, 3, \ldots)$ be a sequence of analytic functions in J such that the series $\sum \varphi_n(z)$ converges unconditionally and uniformly in J, then "the series

$$S(z) = \sum_{n=0}^{\infty} \varphi_n(z) f_n(z)$$

represents within J a single-valued branch of an analytic, monogenic function if J is connected; if J is disconnected the series represents as many branches of (even different) analytic functions as are the disjoint pieces of which J is built up; these functions have no essential singularities at the points of J, but at the most a finite number of poles at points where the functions $f_n(z)$ have poles" [19, pp. 68–69]

Then he considered the consequences one could draw from this theorem under some particular hypotheses both on the domain A and the system of functions $\varphi_n(z)$. Pincherle showed that all the more important particular cases he studied, including the expansions in Bessel functions and in generalized Lambert functions, could be reduced to the case where $\varphi_n(z) = c_n z^n$ and $f_n(z)$ are regular functions at $z = 0$.

He chose to include this material in the subjects he planned to deal with in his 1882/1883 course of higher analysis, and asked Poincaré for advice in a long letter on June 10, 1882, where he sketched his program. Excerpts from that letter are worth quoting in some detail for they offer an interesting survey of open problems in complex function theory.[6]

Pincherle collected them in four classes. Given an element of an analytic function, the first class (A) of problems asked to establish the properties of

---

[5]"Dans ce Mémoire [19] se trouve, peut être pour la première fois, le concept de 'système de fonctions limitées' dans 'leur ensemble', depuis si commun, et aussi un théorème qui correspond, pour les aires planes, à la célebre proposition de Heine-Borel" [33, p. 46].
[6]For the correspondence between Pincherle and Poincaré see [10, pp. 210–217].

the function defined by the element. Following Weierstrass's approach, Pincherle observed, it is known that from the element one obtains the value(s) of the function in its "whole domain of validity" by means of analytic continuation, but this method is "not well suited to establish (1) the limits of the domain of validity; (2) whether the function is single- or multi-valued; (3) whether it satisfies either an algebraic equation or an algebraic-differential equation or whether it belongs to any known class of functions. "Thus", Pincherle concluded, "it seems that one of the main problems of function theory should be the following: 'to recognise the three above characteristics of a function from the laws of the coefficients of the element'. I do not know whether this problem has been solved apart from the cases of the recurrent, and the hypergeometric series."

The second class (B) of Pincherle's problems was related to the general question: "What functions can be expressed by means of determined, arithmetical operations?." Of course, if the arithmetical form is given by a finite number of rational operations the function is a rational one, and there nothing more to say. But, Pincherle asked, in the case of infinite series, infinite products or continued fractions, what about the domain of convergence of the arithmetical form? These questions, he wrote, have been partly answered in the case of series, but what about the other cases? As for series, Pincherle went on, uniform convergence had provided a sufficient condition, not a necessary one. As shown by Weierstrass, the construction of functions with prescribed zeros and infinities belongs to this class of problems, as does the fact that the very same arithmetical form can represent different functions in disjoint domains. Also included in this class of problems is the study of series of the form $\sum a_n P_n(x)$ where the functions $P_n(x)$ belong to a given system (Fourier series, spherical functions, etc.).

Pincherle's third class (C) of problems dealt with the "research into functions satisfying a given property in their domain of validity," and included the solution of differential as well as functional equations, and above all the solution of Cauchy's problem by means of analytic functions. Finally, the last class (D) of problems involved the study of the behaviour of a function on the boundary of its "domain of validity", including Weierstrass's and Riemann's results on essential singularities, and resp., branch points, and Weierstrass's (and Poincaré's) results on gap series.

In his prompt reply[7] on June 15, Poincaré agreed with Pincherle's approach for, he wrote, "this way is best suited to expound the general theory of functions if one wants the true sense of the problems that one is going to deal with to be understood." Poincaré agreed with Pincherle that the problem of recognizing the essential properties of a function represented by a power series from its coefficients "is far from being solved" and "there is still a lot to do". As for recurrence and hypergeometric series, "I think – Poincaré added – that under the latter name you include not only Gauss's series but all the series that represent the integrals of linear, differential equations with rational coefficients; in fact, there is a linear recurrence relation among p consecutive coefficients of such a series (completely analogous

---

[7]This letter is kept in Pincherle's above mentioned manuscript volume. See also [10, pp. 215–217].

to Gauss's series) involving the rank n of the first of these p coefficients. So there is a condition that allows one to recognize from the law of the coefficients whether the series satisfies a linear equation; and, consequently, *whether it represents an algebraic function."* [Poincaré's emphasis]. There are other cases in which the law of coefficients "immediately shows what is the domain of validity of the function", Poincaré added. As examples he gave the series

$$\sum \frac{1}{2^n} x^{3^n} \text{ or } \sum \varphi_p(n) x^n,$$

where $\varphi_p(n)$ represents the sum of the p-powers of the divisors of n, whose "domain of validity" is the unit disk.

As for Pincherle's second class of problems, Poincaré admitted that the case of continuous fractions had not yet been studied as deeply as it deserved.

Finally, Poincaré added a new class of problems dealing with the conformal mapping and Dirichlet principle that "unfortunately have been treated for a long time without a sufficient rigour", and referred to [37] for "a rigorous solution of them."[8]

Some time later, in August 1882, Pincherle began the study of a problem belonging to the class (B) of problems he had listed in his letter to Poincaré. In the manuscript volumes where he recorded his research work one can find drafts of his preliminary results. His starting point was provided by Neumann's theorem stating that an analytic function within an ellipse with foci ± 1 can be expanded there in a unconditionally and uniformly convergent series of spherical functions. In his later recollections Pincherle recorded that he conjectured that Neumann's result depends on a general property which connects the boundary curves of the domain of convergence of the series expanded in functions $f_n(x)$ of a given system with the nature of the generating function of the $f_n(x)$.[9] His work aimed at clarifying this property, thus re-obtaining Neumann's theorem as a particular case.

Pincherle collected his results in Part I of the memoir [20] which appeared in print in April 1883. After summarizing some basic properties of the spherical functions, and of the series of spherical functions as well, he began by considering an analytic function $T(u, v)$ that in a suitable neighborhood of any point (u_0, v_0) can be expanded in series

$$T(u, v) = \sum_m \sum_n c_{m,n} (v - v_0)^m (u - u_0)^n$$

except at points u, v satisfying an entire (rational or transcendental) equation $f(u, v) = 0$. For $v = 0$ the function $T(u, v)$ will be regular at all values of u except

---

[8] Almost 20 years later Poincaré himself was to take up this subject in [35].

[9] "Ce théorème doit dépendre d'une proprieté générale qui met en rapport les courbes de convergence des séries ordonnées suivant les fonctions d'une suite donnée $f_n(x)$ avec la nature de la fonction génératrice des $f_n(x)$" [33, p. 342].

at (separate) points that are the roots of $f(u, 0) = 0$. Thus one has

$$T(u, v) = \sum_m \sum_n a_{m,n}(u - u_0)^n v^m$$

that is unconditionally convergent for suitable small values of $|u - u_0|$ and $|v|$. Its coefficient $\sum_n a_{m,n}(u - u_0)^n$ is convergent for any m within a suitable disk centered at u_0, and represents an element of the analytic function $p_m(u)$ that can be obtained from it through analytical continuation. Thus, Pincherle [20, pp. 17–18] stated that at every point of the plane, except at points that are roots of $f(u, v) = 0$, one has $T(u, v) = \sum_m v^m p_m(u)$, the function $T(u, v)$ being the generating function of the $p_m(u)$. Then he tackled the problem of determining the convergence domain of the series

$$\sum c_m p_m(u),$$

where c_m are constants, and succeeded in showing that the determination of the boundary curves for its convergence depends on the moduli of the roots of the equation $f(u, v) = 0$. He exemplified this with the case $f(u, v) = u^2 + v^2 - 1 = 0$ where the corresponding boundary curves are Cassini's curves with foci ± 1. Neumann's theorem turns out to be the particular case where $f(u, v) = 1 - 2uv + v^2$.

By resuming and generalizing the results he had only stated in [18], next Pincherle [20, p. 25] considered systems of functions $p_m(u)$ to which a system of function $P_m(v)$ can be associated so that for x, y varying in bounded domains the series $\sum p_m(x) P_m(y)$ converges absolutely and uniformly, and takes the value $\frac{1}{y-x}$. Under these hypotheses, Pincherle stated that an analytic function $f(x)$, regular in a connected domain E bounded by a close curve γ, can be expanded in series of functions of a given system $p_m(x)$ by resorting to Cauchy's integral formula so that

$$f(x) = \frac{1}{2\pi i} \int_\gamma \frac{f(y) dy}{y - x} = \sum C_m p_m(x),$$

where

$$C_m = \frac{1}{2\pi i} \int_\gamma f(y) P_m(y) dy.$$

Alternatively, if there exist a second system of functions $q_n(x)$, and a constant C such that the integral

$$\int_\gamma p_m(x) q_n(x)$$

equals C if $m = n$, 0 otherwise, then the expansion of $f(x)$ can be obtained by resorting to the method of the indeterminate coefficients.

In the remaining part of the memoir Pincherle showed that the relations among integrals occurring in the second method could be obtained by appropriate limit processes. In doing this he succeeded in determining the expansions of an analytic function in series of functions of a given system he was looking for, and in proving their uniqueness as well.

Pincherle sent an offprint of this paper to Poincaré who apparently appreciated it and in turn send Pincherle offprints of his recent papers on fuchsian functions. "I am happy that my memoir has been of some interest to you", Pincherle wrote him on November 3, 1883 [2, pp. 37–38]. The development of his "first essay" in this "promising" domain, Pincherle continued, dealt with the following questions: (1) What can be said in general about the expansion of zero (*Nullentwickelung*) $\sum c_n p_n(x)$? (2) Given a system $p_n(x)$, can the conditions be determined under which a function $f(x)$ – holomorphic, if the $p_n(x)$ are holomorphic in a given domain, or having singularities depending on the ones of the $p_n(x)$ – can be expanded in series $\sum c_n p_n(x)$?

In his letter to Poincaré Pincherle limited himself to state that he was able to find "a satisfying answer" to the first question. As for the "difficult" second one, he hoped to answer it through a generalization of Cauchy's formula

$$f(x) = \frac{1}{2\pi i} \int_\gamma \frac{f(y) dy}{y - x}$$

namely, by substituting $\frac{1}{y-x}$ with a suitable function of x and y. However, Part II of his (1883–1884) memoir, dated December 1883, dealt essentially with the first question. There Pincherle presented examples of expansions of zero, i.e. convergent series of functions whose sum is zero in finite domains, and then generalized his results by showing that such expansions *a priori* existed for the systems of associated functions he had introduced in part I of his memoir. (In particular, he showed that m linearly independent expansions of zero correspond to m singular points (poles) of the associated functions.)

As for the problems related to question (2), we will see in the next section that they became the subject of Pincherle's extensive research since January 1884.

It is worth emphasizing here that Pincherle's 1883–1884 paper shows a remarkable eclecticism from the methodological point of view. It is well known that Weierstrass made a point in avoiding the "transcendental" methods in complex function theory, including Cauchy's integral theorem and formula in particular. In spite of this, although Pincherle was strongly drawn to the Weierstrassian approach, as a research mathematician apparently he could not see any reason to avoid resorting to Cauchy's methods and results. Thus, his paper was a mixture of Cauchy's techniques merged into a Weierstrassian context. In addition, after reading Klein's 1882 booklet on Riemann's theory of algebraic functions [13], Pincherle could see the force of Riemann's ideas, and recorded in his notebook: "I believe that in order to acquire strong knowledge in analysis an in-depth study of the subject of Riemann surfaces is needed."[10]

---

[10] It is interesting to remark that according to the lecture notes of his courses Pincherle apparently followed the same "eclectic" approach in his teaching. In 1892/1893 he lectured on the theory

4 Functional Operations

As his manuscript notes show, in January 1884 Pincherle occurred to consider the integral

$$\sum a_n p_n(x) = \frac{1}{2\pi i} \int_\gamma E(x,y)\varphi(y)\mathrm{d}y.$$

Related to this he asked himself the following question: "given the functions $f(x)$ and $E(x,y)$, is it possible to determine a $\varphi(y)$ such that $f(x) = \int_\gamma E(x,y)\varphi(y)\mathrm{d}y$?". Commenting on this he observed that answering the question would mean establishing a kind of correspondence between the functions f and φ. In other words, he added, $f(x) = \int_\gamma E(x,y)\varphi(y)\mathrm{d}y$ could be regarded as a functional equation. In his view, this kind of problems belonged to a "special part of analysis" that he called "analysis of the functional forms".

Later on he resumed the subject to state that "the studies on functions can be of two kinds, namely studies of value and studies of form". The former ones are more elementary in character, for what matters is the study of the variations occurring in the values of a function when the (independent) variable varies. Functions enjoying similar properties can be collected in classes, whose study can be called a "study of form". In other words, "in a study of form one considers the function as a whole, as an element of the question, and the values of the independent variable essentially play no role". Actually, Pincherle went on, every problem of value on functions led to a problem of form. Thus, for instance, given an element of an analytic function the study of the properties of the represented function is a problem of value that led to the problem: which functions can be represented by series of a given nature? There Pincherle referred to his first letter to Poincaré (see Sect. 3 above) to remark that the problems listed in the classes (B) and (C) were problems of form, whereas the problems listed under (D) belonged to a class of problems of value.

Accordingly, Pincherle proceeded to characterize the "linear group" and, respectively, the "algebraic group" of functions, and published the related results in the papers [21] and [22] before resuming his favorite subject, namely the research on systems of analytic functions. In particular, in his manuscript volume dated September 1884 – October 1885 he referred to his (1883–1884) memoir to state that, if $T(x,y)$ is an analytic function such that

$$T(x,y) = \sum_n y^n p_n(x),$$

$p_n(x)$ being a given system of functions, $|y| < \alpha$ and $x \in E_\alpha$, E_α being a suitable domain, and if $\varphi(\frac{1}{y})$ is analytic in the complementary domain $|y| > \alpha$, then the

---

of analytic functions, and the lecture notes were published as [27]. There he offered a wide-ranging account of the theory according to Weierstrass, Cauchy, and Riemann. Essentially the same approach was followed in both his [29] and [32].

integral along a circle (c) of radius greater than α

$$\int_{(c)} T(x,y)\varphi(\frac{1}{y})\frac{dy}{y}$$

"represents a series of functions $p_n(x)$ suited to represent an analytic function". In addition, under the same hypotheses "any series $\sum_n c_n p_n(x)$ can be transformed into an integral $\int T(x,y)\varphi(\frac{1}{y})\frac{dy}{y}$."

Given this, Pincherle added, "the problem of the inversion of a definite integral, namely the problem of finding a function $\varphi(\frac{1}{y})$ such that

$$f(x) = \int T(x,y)\varphi(\frac{1}{y})\frac{dy}{y}$$

coincides with the problem of checking whether the function $f(x)$ can be expanded in series $\sum_n c_n p_n(x)$".

More precisely, Pincherle proved that

given a function $T(x,y)$ that is regular in a neighborhood of $(0,\infty)$, namely for $|y| > \alpha$ and $|x| < \beta$, and given a function $\varphi(y)$ regular for $|y| > \alpha' > \alpha$, then the integral $\int_{(c)} T(x,y)\varphi(y)dy$ where (c) is a circle of radius r, $\alpha < r < \alpha'$, represents an element of a regular function $f(x)$ for $|x| < \beta$. Consequently, definite integration is a generating method of elements of analytic functions. Thus, the operation

$$\frac{1}{2\pi i}\int T(x,y)\varphi(y)dy$$

for the sake of brevity will be denoted with $T\varphi$. [...] The element $f(x)$ generated by $T\varphi$ defines completely an analytic function, and consequently with the symbol $T\varphi$ one can represent not only the element $f(x)$, but also the whole analytic function defined by this element.

Pincherle exemplified this with the case of functions $A(\frac{x}{y})$ and $\varphi(y)$ that can be expanded in Laurent series within suitable, partly overlapping annuli centered at the origin. Then, Pincherle commented, "one can consider the expression

$$I(\varphi) = \frac{1}{2\pi i}\int_c A(\frac{x}{y})\varphi(y)\frac{dy}{y}$$

as an algorithm applied to the object, namely a new Laurent series", and proved this under various hypotheses on the functions $A(\frac{x}{y})$ and $\varphi(y)$. Pincherle published part of this material in the short note [23] that appeared in *Acta mathematica*. Following this, Pincherle collected his results in his first memoir [24] on "functional operations".

As we have seen, the concept of "functional operation" occurred at various stages of his preliminary work. In the opening lines of [24] Pincherle defined it in general terms as follows: "I call *functional operation* any operation that produces an analytic function when it is performed on an analytic function" [24, p. 92]. Among the "most remarkable algorithms" for functional operations Pincherle mentioned the integration

$$\int_{(c)} f(x,y) dy,$$

where (c) is a suitable path of the y plane, and produced various examples of it, including Cauchy integral formula and his own results that connected integration with the expansion of a function in series of functions of a given system. All these examples were particular cases of the general problem of the inversion of an integral, more precisely the problem of solving an integral equations of the form

$$\psi(x) = \frac{1}{2\pi i} \int_{(c)} A(x,y) \varphi(y) dy,$$

where $A(x, y)$ and $f(x)$ are given functions, and (c) is a suitable path. As Pincherle observed, this equation establishes a functional operation between the functions φ and ψ that he denoted as $\psi = \mathbf{a}\varphi$.

In the first part of his paper he established the properties of the functional in abstract terms then applied them to some special cases where $A(x, y)$ is of the form $g(x/y)$, g being a transcendental entire function, including the Abel transform where $g(x/y) = e^{x/y}$. In the second part he considered the case in which $A(x, y)$ is a rational function $A(x, y) = \frac{g(x,y)}{f(x,y)}$, and showed that $\int_\varrho A(x, y) \varphi(y) dy$, where ϱ is a circle, represents different analytic functions in the different domains of the plane x bounded by branches of the curve C_ϱ that corresponds to the circle ϱ under the transformation $f(x, y) = 0$.

The Laplace transform, corresponding to the particular case where $A(x, y) = e^{xy}$, was studied by Pincherle from his new functional point of view in [25], where he showed that the Laplace transform allowed him to build new classes of entire, transcendental functions and, as a special application, to re-obtain Neumann's and Heine's theory of cylindrical functions.

In the same year, 1887, Pincherle presented his functional approach in the paper [26] published in *Acta mathematica*. Given as usual

$$f(x) = \int_l A(x,y) \varphi(y) dy,$$

($A(x, y)$ being a given function and l a fixed path in the y plane) and denoted it by $f = A\varphi$, Pincherle asked the question whether the problem of the inversion of the integral could be solved by a similar integral; more precisely, he asked the question whether there exists a function $\mathbf{A}(y, t)$ and a path λ such that one has

$$\varphi(x) = \int_\lambda \mathbf{A}(y,t) f(t) dt$$

as a solution of $f = A\varphi$ (at least for certain classes of functions f and φ). Thus, the inversion problem reduced to the search for the function $\mathbf{A}(y, t)$ that he called the reciprocal function of $A(x, y)$.

Pincherle considered first the special kind of functional operations satisfying the conditions

$$f = A\varphi \quad f' = A\varphi'$$

and established their properties in general, formal terms. Then, in Part II of his memoir he produced a detailed study of the case in which the integration path is a circle, and the singularities of $A(x, y)$ satisfy an algebraic equation $f(x, y) = 0$.

5 A Glance Ahead

The papers on "functional operations" announced Pincherle's entry in the new, promising field of the functional calculus. In the early 1890s he begun to elaborate a wide research programm on the general theory of functional, linear (distributive, in his terminology) operations that he expounded in various papers culminating in the memoir *Sur le calcul fonctionnel distributif* [28] that appeared in 1897 in the *Mathematische Annalen*. As he stated in the introduction to it, under the term "functional calculus" he included "the chapters of analysis where the variable element is not a number any more but the function as such" [28, p. 1].

Pincherle's primary aim was the study of distributive operations that can be applied to analytic functions (and in particular that transform power series into power series). It is interesting to remark that he chose to embed this into a geometrical setting. Noteworthy, he considered the set of all power series as a functional space of (denumerably) infinitely many dimensions, where any series can be considered as a point whose coordinates are the coefficients of the series. Having introduced the concept of an $(n-1)$-dimensional, linear vector space generated by the linear combinations of n linearly independent functions $\alpha_1, \alpha_2, \ldots, \alpha_n$, he was able to establish the main properties of distributive operations, including the concept of continuity and of functional derivation as well.

Some years later he gave an expository presentation of his results in the book [30], where he introduced linear spaces in axiomatic terms. This pioneering work not only made Pincherle best suited to contribute the German *Encyklopädie* with a paper on functional operations (see [31]), but also explains why in 1928 in his lecture at the ICM in Bologna Hadamard aptly hailed him as one of the founders of functional analysis.

References

1. Bottazzini, U. 1986. *The higher calculus. A history of real and complex analysis from Euler to Weierstrass*. New York: Springer.
2. Bottazzini, U. 1991. *Pincherle e la teoria delle funzioni analitiche*. In 8: [25–40].
3. Bottazzini, U. 1992. *The influence of Weierstrass's analytical methods in Italy*. In 9: [67–90].

4. Bottazzini, U. 1994. *Va' pensiero. Immagini della matematica nell'Italia dell'Ottocento.* Bologna: Il Mulino.
5. Bottazzini, U. and S. Francesconi. 1989. Manuscript volumes and lecture notes of Salvatore Pincherle. *Historia Mathematica* 16: 379–380.
6. Bottazzini, U. and J.J. Gray. *Hidden Harmony – Geometric Fantasies; The rise of complex function theory.* AMS and LMS (forthcoming).
7. Casorati, F. 1868. *Teorica delle funzioni di variabili complesse.* Pavia: Fusi.
8. Coen, S. (ed.). 1991. *Geometry and complex variable.* New York: Dekker.
9. Demidov, S.S. et al. (eds.). 1992. *Amphora. Festschrift for H.Wussing.* Basel: Birkhäuser.
10. Dugac, P. (ed.). 1989. Lettres de Salvatore Pincherle. *Cahiers du Séminaire d'Histoire des Mathématiques* 10: 210–217.
11. Hadamard, J. 1929. Le développement et le rôle scientifique du Calcul fonctionnel. *Atti del Congresso Internazionale dei Matematici, Bologna 1928.* Bologna: Zanichelli. Vol. 1: 143–161.
12. Hadamard, J. 1968. *Œuvres de Jacques Hadamard*, 4 vols. Paris: Editions du CNRS.
13. Klein, F. 1882a. *Ueber Riemanns Theorie der algebraischen Funktionen und ihrer Integrale.* Teubner, Leipzig. In *Ges. Math. Abh.* 3: 499–573. Engl. trl. as *On Riemann's theory of algebraic functions and their integrals.* Cambridge: Macmillan and Bowes; Rep. New York: Dover 1963.
14. Klein, C.F. 1921–1923. *Gesammelte mathematische Abhandlungen.* ed. Fricke, R. et al., 3 vols. Berlin: Springer.
15. Lehto, O. 1998. *Mathematics without borders. A history of the International Mathematical Union.* New York: Springer.
16. Neuenschwander, E. 1978. Der Nachlass von Casorati (1835–1890) in Pavia. *Archive for History of Exact Sciences* 19: 1–89.
17. Pincherle, S. 1880. Saggio di una introduzione alla teoria delle funzioni analitiche secondo i principi del prof. Weierstrass. *G. di Mat.* 18: 178–154; 317–357 [Not in *Opere scelte*].
18. Pincherle, S. 1882a. Alcuni teoremi sopra alcuni sviluppi in serie per funzioni analitiche. *Rend. Ist. Lombardo* (2) 15: 224–225 [Not in *Opere scelte*].
19. Pincherle, S. 1882b. Sopra alcuni sviluppi in serie per funzioni analitiche. *Mem. Bologna* (4) 3: 149–180. In *Opere scelte* 1: 64–91.
20. Pincherle, S. 1883–1884. Sui sistemi di funzioni analitiche e le serie formate coi medesimi. Memoria I. *Ann. di Mat.* (2) 12: 11–41 and Memoria II. *Ann. di Mat.* (2) 12: 107–133 [Not in *Opere scelte*].
21. Pincherle, S. 1884. Sui gruppi lineari di funzioni di una variabile. *Mem. Bologna* (4) 6: 101–118 [Not in *Opere scelte*].
22. Pincherle, S. 1885a. Alcune osservazioni generali sui gruppi di funzioni. *Mem. Bologna* (4) 6: 205–214 [Not in *Opere scelte*].
23. Pincherle, S. 1885b. Note sur une intégrale définie. *Acta mathematica* 7: 381–386 [Not in *Opere scelte*].
24. Pincherle, S. 1886. Studi sopra alcune operazioni funzionali. *Mem. Bologna* (4) 7: 393–442. In *Opere scelte* 1: 92–141.
25. Pincherle, S. 1887a. Della trasformazione di Laplace e di alcune sue applicazioni. *Mem. Bologna* (4) 8: 125–143. In *Opere scelte* 1: 173–192.
26. Pincherle, S. 1887b. Sur certains opérations fonctionelles représentées par des intégrales définies. *Acta mathematica* 10: 153–182. In *Opere scelte* 1: 142–172.
27. Pincherle, S. 1893. *Lezioni sulla teoria delle funzioni.* ed. E. Maccaferri, Bologna (lith.).
28. Pincherle, S. 1897. Mémoire sur le calcul fonctionnel distributif. *Math. Ann.* 49: 325–382. In *Opere scelte* 2: 1–70.
29. Pincherle, S. 1900. *Lezioni sulla teoria delle funzioni analitiche.* ed. A. Bottari, Bologna (lith.).
30. Pincherle, S. 1901. *Le operazioni distributive e le loro applicazioni all'analisi.* Bologna: Zanichelli.
31. Pincherle, S. 1906. Funktionaloperationen und–Gleichungen. *EMW* II A 11: 761–817. French trl. as Équations et opérations fonctionelles. *ESM* II–26 (1912), 1–81 [Not in *Opere scelte*].

32. Pincherle, S. 1922. *Gli elementi della teoria delle funzioni analitiche*. Bologna: Zanichelli.
33. Pincherle, S. 1925. Notice sur les travaux. *Acta mathematica* 46: 341–362. In *Opere scelte* 1: 45–63.
34. Pincherle, S. 1954. *Opere scelte*, 2 vols. Roma: Edizioni Cremonese.
35. Poincaré, H. 1890. Sur les équations aux dérivées partielles de la physique mathématique. *Amer. J. Math.* 12: 211–294. In *Œuvres* 9: 28–113.
36. Poincaré, H. 1912–1955. *Œuvres*, 11 vols. Paris: Gauthier–Villars.
37. Schwarz, H.A. 1870. Ueber die integration der partiellen Differentialgleichung $\frac{\partial u}{\partial x^2} + \frac{\partial^2 u}{\partial y^2} = 0$ unter vorgeschriebenen Grenz– und Unstetigkeitsbedingungen. *Monatsberichte der K. Preussischen zu Berlin*, 767–795. In *Ges. Math. Abh.* 2: 144–171.
38. Schwarz H.A. 1890. *Gesammelte mathematische Abhandlungen*, 2 vols. Berlin: Springer. Rep. in one volume Chelsea, New York, 1972.
39. Weierstrass K.T.W. 1840. Über die Entwicklung der Modular–Functionen. Ms. In *Math. Werke* 1: 1–49.
40. Weierstrass K.T.W. 1856. Theorie der Abel'schen Functionen. *Journal für die reine und angewandte Mathematik* 52: 285–379. In *Math. Werke* 1: 297–355 [Pages 339–379 of the original 1856 paper are not reproduced because their content essentially coincides with (Weierstrass 1840)].
41. Weierstrass, K.T.W. 1876. Zur Theorie der eindeutigen analytischen Functionen. *Berlin Abh.*, 11–60 (Separate pagination). Rep. in (Weierstrass 1886, 1–52). In *Math. Werke* 2: 77–124. French trl. as Mémoire sur les fonctions analytiques uniformes. *Annales ENS* (2) 8 (1879) 111–150.
42. Weierstrass, K.T.W. 1880. Zur Funktionenlehre. *Monatsberichte Berlin*, 719–743. Nachtrag. *Monatsberichte Berlin* (1881) 228–230. Rep. in (Weierstrass 1886, 67–101, 102–104). In *Math. Werke* 2: 201–233. French trl. as: Remarques sur quelques points de la théorie des fonctions analytiques. *Bull. sci. math.* (2) 5 (1881) 157–183.
43. Weierstrass, K.T.W. 1886. *Abhandlungen aus der Funktionenlehre*. Berlin: Springer.
44. Weierstrass K.T.W. 1894–1927. *Mathematische Werke von Karl Weierstrass*, 7 vols. Berlin: Mayer and Müller. Rep. Hildesheim: Olms.

Luigi Cremona's Years in Bologna: From Research to Social Commitment

Aldo Brigaglia and Simonetta Di Sieno

Abstract Luigi Cremona (1830–1903), unanimously considered to be the man who laid the foundations of the prestigious Italian school of Algebraic Geometry, was active at the University of Bologna from October 1860, when assigned by the Minister Terenzio Mamiani (1799–1885) to cover the Chair of Higher Geometry, until September 1867 when Francesco Brioschi (1824–1897) called him to the Politecnico di Milano.

The "Bolognese years" were Cremona's richest and most significant in terms of scientific production, and, at the same time, were the years when he puts the basis for its most important interventions in the social and political life of the "newborn" kingdom of Italy.

In this article we present these different aspects of Cremona's life, with particular emphasis on the relationship of the geometer of Pavia with the academic life in Bologna, with students and colleagues.

Most Esteemed Professor,
In yesterday's issue the Gazzetta ufficiale *made amends for the omission in a previous article, adding your nomination to those* [announced] *on the 3rd and 10th of this month. I therefore congratulate you as well as the University of Bologna, which has acquired a fine professor, and if my offices have been of help to you, I am pleased, but you owe nothing to me, since, when interpellated by the Ministry, I could not have responded in any other way.*

A. Brigaglia (✉)
Dipartimento di Matematica, Università degli Studi di Palermo, Via Archirafi 34, 90123 Palermo, Italy
e-mail: brig@math.unipa.it

S. Di Sieno
Dipartimento di Matematica, Università degli Studi di Milano, Via Saldini 50, 20133 Milano, Italy
e-mail: simonetta.disieno@unimi.it

In your name I have thanked Prof. Gherardi, who will give you the letter he promised for Bologna, and I will be quite happy to see you ...

A[ngelo] Genocchi[1]

This letter of 22 June 1860 both brings a long period of preparation to its conclusion, and marks the beginning of one of the most interesting, intense, rich and fruitful periods in Luigi Cremona's life. Both the conclusion and the new beginning are surprising. The professor who arrived to the chair of higher geometry in Bologna had very much desired this to happen, and had overcome substantial obstacles – both personal and institutional – to see that it did. His own words, written for a biography of his brother Tranquillo, a celebrated painter of the Milanese Scapigliatura movement, best describe the personal difficulties he faced:

> *The Cremonas of Novara were quite well-off, but overspending had already much impoverished them by the time of Giuseppe Cremona, the father of Gaudenzio* [i.e., the grandfather of Tranquillo and Luigi Cremona]. *Gaudenzio was not capable of improving the conditions of the patrimony that remained; indeed, he was soon forced to seek employment in order to survive. However, his children Giuseppe, Giovanni, and Giovannina received a good education.*
>
> *Giuseppe earned a doctorate in law with honours in Pavia in 1817 - and he soon began a judicial career in which his rise was rapid and distinguished.*
>
> *He was very generous towards his step-brothers* [Gaudenzio, following the death of his first wife, remarried in 1829 and had four more children by his second wife], *who would have had a difficult life and would not have been able to study without him.*[2]

It can be seen that Luigi Cremona's early years were not easy. He lost his father at 12, and his mother at 19, and he faced the hardships of being responsible for his three younger brothers, but at the same time he was blessed with an extended family which did its best to make it possible for him to emerge as he deserved to. In particular, he was able to count on the strong support of his step-brother

---

[1] *Carissimo Professore, la Gazzetta ufficiale nel suo N. di ieri ha emendato l'omissione fatta in un precedente articolo aggiungendo la vostra nomina a [quelle] del 3 e 10 corrente. Me ne congratulo adunque con voi e con l'Università di Bologna che acquista un buon professore, e se i miei uffici vi hanno giovato ne sono lieto, ma voi non mi dovete nulla, poiché interpellato dal Ministro non poteva rispondere in modo diverso. Ho ringraziato a nome vostro il Prof. Gherardi che vi darà la lettera promessa per Bologna e sarà contento assai di vedervi* ... (Angelo Genocchi to Luigi Cremona, dated Torino, 22 June 1860).
This letter is kept in the "Legato Itala Cremona Cozzolino" at the Mazzini Institute, Genoa, with reference number 12863; a digital reproduction is available at the website www.luigi-cremona.it with code 054-12863.

[2] *La famiglia Cremona di Novara era stata molto agiata, ma per lo spendere troppo largo era già molto declinata al tempo di Giuseppe Cremona, padre di Gaudenzio. Gaudenzio non seppe migliorare le condizioni del patrimonio rimasto, anzi fu ben presto costretto a cercare un impiego per vivere. I suoi figli Giuseppe, Giovanni, Giovannina ebbero però una buona educazione. Giuseppe si addottorò in leggi con molto onore a Pavia nel 1817 - entrò ben presto nella carriera giudiziaria e la percorse rapidamente e con distinzione... Fu molto benefico verso i fratelli del secondo letto, i quali senza di lui avrebbero dovuto stentare la vita e non avrebbero potuto fare un corso di studi.* In www.luigi-cremona.it with marks 052-12129, 052-12130, 052-12131.

Giuseppe,[3] who wrote to him on 12 September 1849 – when Luigi, at that time not even 18 years old, had just returned to Pavia after having taken part in the battle in defense of Venice – saying:

> *I have not changed my mind about your future; I thus think that you should continue your studies in Pavia as best you can, dedicating yourself especially to the study of mathematics and to all that regards the construction of railway lines, because I have no doubt that you will find in this, once you are qualified as an engineer, ample means of support.*
> *Then beginning this coming November, you will receive from month to month the sum of fifty Milanese lire that I will assign to you for your needs, and which is the maximum amount I can spare in present circumstances. . . .*
> *But going back to you, I must tell you that I desire you to enrich not only your mind with knowledge but also that you take care to refine your behaviour and your manners, correcting above all an overwhelming impetuosity of character that I have remarked in you, because living and being nice in good society means having an appearance that is smooth, not rough, and is pleasing to be with.*[4]

To all this attention Cremona had first replied in angry defense of his decision not to occupy himself with railway lines,[5] and then taking pains to build a career that would make it possible for him to participate fully in the construction of the new Italy that was then being defined following the devastating battles in the first War of Independence. Thus, though he tried right away to obtain – unsuccessfully – a chair in a high school in either Venice, Milan or Pavia, and had to be content with going to Cremona, he worked to build credibility and a name in the academic world and in the institutions where, moreover, there were many others who, like him, dreamed of

---

[3] Before Luigi was 8 years old, his father had already asked his oldest son Giuseppe for advice about him: *... I desire to ask your council what route to follow, whether to start him in a useful career or not, since in truth he shows uncommon talent (... bramo di consultarti quale partito abbia a prendere, se incamminarlo in una carriera utile giacché a dir vero egli manifesta non comuni talenti.* See 053-12395 in www.luigi-cremona.it) Giuseppe would become the true guardian of his step-brother.

[4] *Non ho cangiato divisamento circa il tuo avvenire; penso quindi che abbi a continuare i tuoi studj in Pavia alla meglio che potrai dedicandoti specialmente allo studio delle matematiche ed a tutto ciò che riguarda alla costruzione delle strade ferrate mentre non dubito che troverai in queste, appena ottenuta la qualificazione d'ingegnere, sufficienti mezzi di sostentamento. . . Incominciando poi col successivo novembre in avanti riceverai di mese in mese anticipatamente la somma di milanesi lire cinquanta che ti assegno per i tuoi bisogni e che è il massimo sacrificio ch'io possa fare nelle presenti circostanze. . . Ma tornando a te debbo manifestarti il desiderio che ti arrichisca non soltanto la mente di cognizioni ma che tu ponga altresì ogni cura nell'ingentilire le tue maniere e la tua educazione correggendo soprattutto una soverchia impetuosità di carattere che ho in te rimarcato, mentre per vivere ed essere gradito nella buona società vuolsi avere una scorza liscia e non ruvida che offende al contatto.* See 052-12153 in www.luigi-cremona.it.

[5] We can deduce this from Giuseppe's letter written from Venice on 16 May 1858: *Because my advice which might have been useful in more than one way has not been followed, it does not follow that it has to be the subject of a permanent censure and permanent rancor. (Perché i miei consigli quando potevano essere utili sotto più di un aspetto, non furono ascoltati, non ne viene che ciò debba formare il soggetto di una permanente censura e di un perpetuo rancore.* See 052-12159 in www.luigi-cremona.it.)

a future where a united Italy would once again become a place of civility and liberty for all its citizens, and where science was assigned a central role in the undertaking.

Given this situation, it is not surprising that on 21 February 1860 (4 months before he would take over the chair in Bologna), Angelo Genocchi wrote to him to explain the government of Piedmont's ambitious plan for higher mathematics, and inquiring as to not only his opinion but his willingness to take over one of the new chairs.

> *The Minister for Public Instruction* [Terenzio Mamiani] *has named several commissions to prepare the programmes for university examinations. Especially in that for the examinations for the natural, physical and mathematical sciences, to which I belong, there is doubt that it is possible to compile the programmes for the examinations in higher analysis, higher geometry, and higher mechanics, since these are vast sciences whose limits are not and cannot be defined, since they are in a state of continual and limitless development. On the other hand, the law requires that the themes of the examinations be published by the government and be the same for all universities and all students, whether they come from state schools or private schools. . . . I wish to know your opinion about all this, even more so because there is hope that you will be entrusted with the teaching one of these very subjects, that of higher geometry at the University of Torino.*[6]

To the problems of making higher education uniform throughout the Kingdom of Italy – which, to make matters even more complicated, seemed to expand from one day to the next – was added that of defining the level of higher education: did it mean teaching the latest developments in the sciences, which had not yet been distilled into treatises, or the contrary, treating the topics which had already been known for some time and were ripe for teaching?

The model unhesitatingly proposed by Cremona in his reply was that of the Germans:

> *I think that higher education should not be subject to the same rules that apply to elementary university instruction. I think that it should not serve so much to complete knowledge acquired by young people who have just come out of the faculties, as much as to set forth in their integrity higher theories never before explained to young people . . . who wish to venture to the very limits of science, in order to be able to become mathematicians themselves. While the elementary course has to have a universal chair, because it has to create all kinds of engineers, the higher course must have a highly scientific stamp, since it must create men who are to carry science forward. We must not be afraid that up to now only a very small number of young people among us wish to dedicate themselves to higher mathematics. Let us look at what is taking place in Germany, and we can be sure that*

---

[6] *Il Ministro della pubblica istruzione ha nominato parecchie commissioni per preparare i programmi degli esami universitarj. In quella degli esami sopra le scienze naturali, fisiche e matematiche, alla quale io appartengo, si è dubitato se sia possibile il compilare i programmi per gli esami di analisi superiore, di geometria superiore, di meccanica superiore, trattandosi di scienze vastissime i cui confini non sono né possono essere determinati e che sono in istato di continuo e indefinito progresso. D'altra parte la legge richiede che i temi degli esami siano pubblicati dal governo e siano comuni a tutte le università e a tutti gli studenti, vengano questi da scuole ufficiali o da scuole private. . . . Bramerei ora di conoscere intorno a ciò il parere della S. V., tanto più perché è da sperarsi che Le sia affidato uno dei medesimi insegnamenti, quello della geometria superiore nella Università di Torino.* See 054-12860 in www.luigi-cremona.it.

the same thing and even better will take place here, when thanks to the changed political conditions the country has emerged from the inertia in which foreign despotism has plunged it, and education will have taken reacquired the energy to which the happy minds of the Italians are so susceptible.
There cannot be programmes, because the subject is vast, indefinite, variable, nor is it possible to confine it to within a course of one, two, or three years. ... How can a programme be made? Do you want pure geometry, or a mention as well of the thousand methods of analytical geometry? Do you want a general theory of the transformation of figures? Do you want the special theory of the conics, the cubics, second-order or third-order surfaces, etc., etc.? Any single one of these theories is enough for a year, or at least half of one. Which should be given first, and which later? None, in my opinion. The professor should be left free ...
There remains the question of the examinations. The exams for higher education are not at all necessary. But if they are desired just the same, then each student should take them in the university where they attended the course. It is impossible to subject higher and elementary education to the same rules. If the [Casati] *laws of 13 November have absurd requirements and are impossible to satisfy, I believe it would be a good deed to make this evident to the new minister* [for public instruction] *who can remedy the situation.*[7]

This shows admirable fighting spirit for a 29-year-old man who is looking for a place in the sun! His design for higher education is outlined with remarkable clarity.

Then the events of this extraordinary period began to happen one after the other in rapid succession: on 12 March Tuscany, under the leadership of Bettino Ricasoli, voted in favour of annexation to Piedmont; on 18 March it was the turn of Emilia, led by the dictator Luigi Carlo Farini; on 11 May Garibaldi landed in Marsala; on

---

[7]*Io penso che gli insegnamenti superiori non debbano essere soggetti alle stesse norme che vincolano l'istruzione elementare universitaria. Penso che essi non debbano tanto servire a completare le cognizioni acquistate da giovani appena usciti dalle facoltà, quanto ad esporre nella loro integrità teorie superiori non mai prima iniziate a giovani ... che desiderano portarsi ai limiti della scienza, per divenire essi medesimi matematici. Se il corso elementare deve avere un carattere universale, perché dee creare ingegneri di tutte le specie, invece il corso superiore dee avere un'impronta altamente scientifica, dovendo creare uomini che poi facciano progredire la scienza. Non dobbiamo spaventarci pel piccolissimo numero di giovani che fra noi sinora si dedicano alle alte matematiche. Guardiamo a ciò che avviene in Germania, e saremo certi che lo stesso e anche meglio avverrà fra noi, quando per le mutate condizioni politiche il paese sarà uscito dalla inerzia in cui l'aveva immerso il dispotismo straniero, e gli studi avranno preso quello slancio di cui sono suscettibilissimi i felici ingegni italiani.*
Non ci ponno essere programmi, perché la materia è vasta, indefinita, variabile, né è possibile restringerla in un corso di uno, né due, né tre anni. Come fare un programma? Volete geometria pura, o anche un cenno sui mille metodi di geometria analitica? Volete la teoria generale della trasformazioni delle figure? Volete le teorie speciali delle coniche, delle cubiche, delle superficie del 2° ordine, di quelle del 3° ordine, etc. etc.? Ciascuna di queste teorie basta ad un anno, o almeno a mezzo anno. Quale si vorrà preferire, quale posporre? Nessuna, a mio credere. Il professore deve essere lasciato libero. ...
Resta la questione degli esami. Gli esami per gli insegnamenti superiori non sono niente necessari. Ma se si vuole che ci siano, si facciano da ciascuno in quella Università ove ha seguito il corso. È impossibile assoggettare alle stesse norme gli studi superiori e gli elementari. Se la legge [Casati] *del 13 novembre ha esigenze assurde e impossibili a soddisfarvi, credo sarebbe opera buona mostrarle al nuovo ministro che potrebbe mettervi rimedio.* Cf. letter no. 1 in [8].

27 May he entered Palermo, and on 7 September Naples; finally, on 11 September the Battle of Castelfidardo marked the liberation of the Marches and Umbria.

In parallel, preparations leading up to the chair mentioned by Genocchi in February were also going forward. Cremona had well understood that the question was one that would decide his future, and so he began to contact all his acquaintances: he wrote to his friend the philosopher Luigi Ferri (then special secretary to Terenzio Mamiani, Minister for Public Instruction) asking him if he might hope to obtain the chair in higher geometry in Torino,[8] and on 7 May he received a reply that regarded the chair of Bologna instead:

> Not only Torino but Bologna as well is lacking a professor of Geometry. The Minister has already taken note and has noted your name with particular favour ... you could send a request that he assign you a chair in one of the state universities. However, I believe you would have a great possibility of success if you were to set your hopes on Bologna. The faculty of law and that of philosophy are beginning to be well constructed, but that of mathematics still needs to be completed. ... Bologna is a city where studies have always flourished, but where the Minister intends to restore them to their ancient splendour.[9]

By 16 May the machine was in motion. There were still problems of overall strategy to be dealt with[10]: for the moment, the minister had no intention of establishing chairs in higher education! But the urgings of Ferri as well as of Brioschi, Genocchi, Tardy and Silvestro Gherardi (professor of mechanics, hydraulics and physics in Bologna) quickly changed the - minister's opinion: the creation of chairs in higher mathematics thus became the axis around which policies regarding research in Italy revolved. Cremona's prospects changed.

Cremona became an official member of the Italian scientific élite. The chair in Bologna was of central importance, not only because of that city's glorious academic tradition – Bologna, along with Torino, Pavia and Pisa (which would soon be joined by Naples and Palermo), was destined to assume the central role in the university system then being created in Italy – but also because it had been the main university of the Papal State. Cremona was probably named to the chair not only thanks to his academic merits, but also thanks to his well-founded fame as a liberal patriot; perhaps it was not by chance that, along with Cremona, Mamiani also nominated the then 25-year-old Giosuè Carducci, not yet famous, who would later also become Cremona's brother in the Masons.

What is more, besides these two very respectable names, between 1860 and 1862 a host of young, and sometimes very young, people were appointed to chairs in

---

[8]Cf. letter no. 4 in [8].

[9]*Il professore di Geometria superiore manca non solo a Torino ma anche a Bologna. Il Ministro ha già preso nota e nota favorevolissima del vostro nome ... voi potreste domandargli una cattedra in una delle Università dello stato. Credo tuttavia che avreste molte probabilità di riuscita se i vostri desideri si fermassero sopra Bologna. La facoltà legale e la filosofica cominciano ad esservi assai bene costituite, ma la matematica ha bisogno d'esservi completata .... Bologna è una città dove gli studi furono sempre floridi, e in cui il Ministero intende rialzarli al grado loro antico.* Cf. 053-12497 in www.luigi-cremona.it.

[10]Cf. letter no. 5 in [8].

Bologna, all of whom were destined to make their mark and profoundly influence their sphere of action. These are the men who have sometimes been called the "Mamiani boys", after the Minister for Public Instruction, and to whose intellectual daring the Italy of those years owed much: Francesco Magni (1828–1887), then 32 years old, professor of clinical ophtamology, later rector of the University of Bologna; Emilio Teza (1831–1912), then 29, a great linguist who was particularly expert in Sanskrit; Giambattista Gandino (1827–1905), then 33, professor of Latin and author of some amongst the most important manuals for teaching it; Pietro Ellero (1833–1933), then 27, philosopher of law, fiery champion of the abolition of the death penalty; Eugenio Beltrami (1835–1900), then 25, who would become a great mathematician; Francesco Fiorentino (1834–1884), then 26, philosopher and author of the books studied by Giovanni Gentile.

1 The Beginnings

Startingin October 1860 Cremona began working in his new offices. His wife, Elisa Ferrari, and little Nina, moved with him. Elisa was expecting a son, Vittorio, who was born in the early months of 1861:

> *I have found myself here since 3 [October] ... Here the studies do not open until 5 November, but who knows when lessons will begin, because they have to be preceded by enrolment and the admission examinations.*[11]

Of course, the most revealing part of his opening lecture was reserved for scientific questions. The situation – indecorous – of teaching in the universities had made it necessary for the government to intervene by seeing that scientific research was given new impetus. This had been done by establishing new chairs for higher education, but as far as mathematics was concerned, it was necessary to take things a step further and regain the time that had been lost. In order to achieve this end, it was necessary to allow pure mathematics to develop fully, and to choose – carefully, but also with broad vision – what higher mathematics was to be taught. Cremona made an attempt at this by proposing a brief synthesis of what was meant by higher analysis and higher geodesy (that is, differential geometry), and thus outlining the contents of possible courses:

> *... the principal results of the theory of determinants, a marvellous instrument of algebraic calculus, which works wonders never dreamed of; the theory of binary forms, which contributed much to the solution of equations; the theory of ternary and quarternary forms, an extremely powerful auxiliary for the geometry of curves and surfaces; transcendent arithmetic, which earned lasting fame for Gauss, Dirichlet, Hermite, Kummer, Eisenstein, Genocchi ... ; the theory of elliptic and hyperelliptic functions, where the genius of the Norwegian Abel and the Prussian Jacobi shone and where today there are the admirable*

---

[11]*Io mi trovo qui sin dal 3* [ottobre] ... *Qui gli studi non si riaprono che con il 5 novembre, ma le lezioni chi sa quando incominceranno, poiché devono precedere le iscrizioni e gli esami d'ammissione.* Cf. letter no. 9 in [8].

> works of Weierstrass, Hermite, Brioschi, Betti and Casorati, a stupendous theory that was at one time linked to the most advanced parts of integral calculus, with the solution of equations, the theory of series and the very arduous and attractive theory of numbers. Well, in the future each of these magnificent branches of science could be examined in rotation by the professor of higher analysis.[12]

And further:

> The theory of curvilinear coordinates, begun by Bordoni and Gauss and then largely promoted by Lamé; research on surfaces presumed to be flexible and inextensible which turn out to be applicable to a given [surface]; the problem of drawing under certain conditions on one surface an image of a given figure drawn on another surface, that is, the problem of the construction of geographic maps; spherical trigonometry; the theory of geodesics: all that from now on will be explained in the school of higher geodesy along with the theory of least squares and other very dense subjects.[13]

After this, of course, he explains the programme of his course. Here, we won't go into the details[14] but will only say that he enunciated an extremely general programme for the study of projective geometry entirely based on the concept of transformation. For some time Cremona had been searching for the most general transformations, which provided the precise geometric conception that led to the birth of the "Cremona transformations" which were to come.

2 The Academic Environment

From the very beginning Cremona noticed that the academic atmosphere in Bologna, the enthusiasm generated by the Risorgimento notwithstanding, was very different from that of Pavia, which was much livelier, and especially from that of

---

[12] ... i principali risultati della teorica de' determinanti, meraviglioso stromento di calcolo algebrico, che opera prodigi non mai sospettati; della teorica delle forme binarie che tanto promosse la risoluzione delle equazioni; della teorica delle forme ternarie e quaternarie, potentissimo ausilio per la geometria delle curve e delle superfici; dell'aritmetica trascendente, per cui s'acquistarono fama non peritura Gauss, Dirichlet, Hermite, Kummer, Eisenstein, Genocchi ... ; della teorica delle funzioni ellittiche ed iperellittiche nella quale brillò il genio del norvego Abel e del prussiano Jacobi, ed or ora apparvero mirabili lavori di Weierstrass, di Hermite, di Brioschi, di Betti e di Casorati, teorica stupenda che si collega a un tempo colle parti più elevate del calcolo integrale, colla risoluzione delle equazioni, colla dottrina delle serie e con quella, sì ardua e sì attraente, de' numeri. Ebbene, ciascuno di questi magnifici rami di scienza potrà in avvenire essere svolto con alternata successione dal professore di analisi superiore. Cf. [10].

[13] La teorica delle coordinate curvilinee, iniziata da Bordoni e da Gauss e poi grandemente promosse da Lamé; la ricerca delle superficie che supposte flessibili e inestensibili riescano applicabili sopra una data; il problema di disegnare con certe condizioni sopra una superficie l'imagine di una figura data su di un'altra superficie, il problema insomma della costruzione delle carte geografiche; la trigonometria sferoidica; la teorica delle linee geodetiche: tutto ciò sarà quind'innanzi esposto nella scuola di alta geodesia insieme colla dottrina de' minimi quadrati e con altri gravissimi argomenti [10].

[14] Cremona's *Prolusione* has been widely studied. We refer to [4, 22].

Milan. Bologna and its university were still in a kind of torpor. The university still seemed to be wrapped in the somnolent cloud of the Papal State, dominated by hoary professors who regarded their younger colleagues' eagerness for reform with mistrust, interposing obstacles directly or indirectly to block their foolish ambitions. Hence, with no surprise we find that on Feb 8th 1861 an impatient Cremona writes to Placido Tardy:

> My work here is almost useless, not because of the students who are eager to learn and are very thankful for my attention, but because of those who should have prepared them and because of the organization of this university.[15]

And again, on June 16th:

> An outrageous forgiveness rules here. The new professors are not examiners: imagine how one can examine in superior geometry. I do not know how we will proceed without great changes; but does the government know this? I have never seen any inspector here.[16]

The regent (the equivalent of today's rector) of the University of Bologna was Antonio Montanari. Once a minister in the government of Pope Pius IX at the time when Pellegrino Rossi was president, with the advent of the Roman Republic he had followed the Pope in his brief exile in Gaeta. After 1849, he had distanced himself from Pius IX's shift to anti-liberalism, while managing to maintain a moderate stance and continue teaching at the University of Bologna. He had then served as Minister for the Interior and Minister for Education during the dictatorship of Farini. Now, regent of the university and senator since 1860, he was very close to Marco Minghetti, who during Cremona's years in Bologna was, from 1860 to 1863, first Minister of the Interior and then Minister of Finance, and finally Prime Minister (1863–1864). How did this moderate and cautious regent receive his young colleague, at that time still very strongly tied to Mazzini and Garibaldi? How did he, still involved in many ways with the designs of Pius IX, receive the appeal of a single, supreme hero, Garibaldi, the senator?

The two opposing political visions of those years could not help from being reflected in academic life in Bologna during the 1860s. On one side there was Montanari, who some years later, as mayor of Meldola, would have difficulties inaugurating a monument to Felice Orsini because he could not take part to that celebration without compromising the principles to which he had dedicated a lifetime. On the other side was Cremona, who knew of Orsini's extremely close ties with Nicolao, his wife Elisa's beloved brother and his own companion in arms! We do not have any documents recording Cremona's political thoughts, nor any notes of

---

[15] *Qui l'opera mia è quasi inutile, non già per colpa de' giovani che sono animati da gran desiderio di imparare e fanno ogni sforzo per mostrarsi grati alle mie cure; ma per colpa di chi avrebbe dovuto istruirli, o per colpa degli attuali ordini di quest'università.* Cf. letter no. 6 in [9].

[16] *Qui regna la più scandalosa indulgenza. I professori nuovi non sono esaminatori; immaginatevi dunque come si esaminerà in geometria superiore. Io non so più come si andrà avanti, senza un gran provvedimento; e d'altronde sa il governo queste cose? Io non ho mai veduto nessuno che fosse mandato per ispezionare.* Cf. letter no. 9 in [9].

his conversations with Elisa, the vestal of the family's allegiance to Mazzini. We can well imagine them, however, just as we can imagine how differently the professor and the regent reacted to Garibaldi's exhortation "Rome, or death!" (as well as their different relationships with the students).

But there were also other, perhaps more important, reasons for conflicts between Cremona and Montanari. Montanari belonged to the old academic guard, which was used to resolving all questions through infinite compromises, both within the teaching staff and with the students. The innumerable, almost feudal privileges of the staff and student associations, the very long vacation periods, all seemed to be part of the mental habitus of the old academic guard.

In contrast, Cremona proposed to turn the university into a support structure for the education of a ruling class that would be capable of modernising, within a short time, the entire nation, bringing it into the heart of European development, and thus could not help but regard the gothic regulations of the university with impatience. Above all he couldn't stand the viscose atmosphere that kept the old Bologna professors tied to each other, and which seemed to him to enwrap even Silvestro Gherardi, who had worked so hard to help him obtain the chair in Bologna: *for heaven's sake, don't say anything about this to Gherardi, who was still close to his old colleagues!*[17]

With obvious and not few exceptions, the "old colleagues" were an unbearable burden for Cremona, and he transferred onto his timid and inept regent all of the disappointment he felt over the slowness and compromise of the process of unity during those years of revolutionary fervor, whether Mazzinian or Garibaldian. It was a conflict that was only one aspect of a broader spiritual travail that many other young patriots like himself were caught up in a mixture of joy over goals achieved and delusion over hopes disappointed.

The first conflict in which Cremona found himself directly opposed to his regent regarded a colleague, the geodesist Matteo Fiorani, a friend of Cremona and Genocchi, who was hotly contested by the students because of his – at least according to Cremona – excessive severity over exams. Cremona took Fiorani's side. He is not prone to compromises, which is one of his traits of character throughout his time in Bologna. Beltrami describes him to Chelini in the following sentence on April 24th 1865:

> *Our Cremona is always struggling with those who, unlike him, fail to see reasonable discipline in the university ... I approve the actions suggested by his zeal for education, but I hope they will not make him regret nor will they be source of concerns and physical consequences.*[18]

---

[17] *Per carità non parlatene a Gherardi che ha qualche affezione a' suoi vecchi colleghi!* Cf. letter no. 15 in [8].

[18] *Il nostro Cremona è sempre alle prese con chi non vorrebbe, come lui, vedere una ragionevole disciplina nelle scuole universitarie. ... Io approvo gli atti che gli vengono suggeriti dal suo zelo per l'istruzione, ma desidero che non gli diano cagione di dispiacersi, o sorgenti di inquietudini e di fisiche conseguenze.* Cf. letter no. 3 in [20].

Nevermind that he had to square off against the direct and obsessive (and not at all favourable) presence of the Ministry, and the Minister himself. Making matters worse, in those years the number of university students was greatly decreased, there were few professors, and the minister could directly intervene even in minor questions [8].

A second conflict between Cremona and Montanari regarded the theme of the exams and the programmes [8, 24]. Then, in 1864, Cremona went up against the entire faculty of mathematics. This time the question was of more general importance, dealing with the creation, in Bologna, of an engineering training school, a kind of special post-graduate degree for mathematics students. However, in 1863 the Politecnico had been founded in Milan; an icon of technological progress in the new Italy and very much desired by Brioschi, it had been envisioned as the culmination of a whole group of initiatives which, beginning with the technical schools, passed through the faculties of mathematics and then the training schools. What was therefore prefigured was a very hierarchical structure which presupposed that there would be a limited number of training schools.

This is where Cremona moved against everyone.

But Cremona's academic life was not made up of just conflicts. As we learned earlier from Genocchi's message of 22 June, Cremona arrived in Bologna armed with letters of introduction, with which he was able to build a first network of acquaintances. These were written by Silvestro Gherardi, who had at one time been a professor at the University of Bologna.[19] To hear him tell it, some of the acquaintances that he made were priceless, as for example, that with Domenico Piani,[20] perpetual secretary of Bologna's Academy of Sciences, by then almost 80:

> For me the memories of the six years in Bologna are filled with the name of Domenico Piani. He welcomed me with open arms upon my first arriving in that city; and he paved the way to such a cordial reception on the part of his colleagues and the most important men to know, that where I thought I was going in as a stranger, I met with only kindness and benevolence. Nor did his support ever diminish; rather, with the passing of time it became increasingly affectionate and profound, and ended only with his death ... I would like to say, if only I had the gift of words, how fruitful and beneficial PIANI's friendship was for me. The inexhaustible goodness of that man naturally led him to put his whole mind and vast erudition at the service of his friends: I never turned to him in vain.[21]

---

[19] Silvestro Gherardi (1802–1879) had been a professor in Bologna from 1827 to 1848, when he left the city to flank the battalions composed of university students. He was then Minister for Public Instruction of the short-lived Roman Republic of 1849. After its defeat, he took refuge in Genoa, finally moving to Torino where he held the chair in physics until 1861. Then he went back to Bologna and became the headmaster of a Technical Institute.

[20] Domenico Piani (1782–1870) was first professor of "sublime", or higher, calculus, at the University of Ferrara, and then director of the Bologna observatory.

[21] ... *Per me i ricordi de' sei anni vissuti a Bologna sono tutti pieni del nome di Domenico Piani. Egli mi ricevette a braccia aperte al mio primo giungere in cotesta città; e mi preparò, presso i colleghi e gli uomini più chiari per sapere, sì cortesi accoglienze che, dove credevo di entrare uomo*

Cremona enjoyed cordial relations with others in the Faculty as well, if not as scientific equals, at least as good friends. One was Matteo Fiorini, professor of geodesy; another was Lorenzo Respighi, (*good man that Respighi, who teaches optics and astronomy*, Cremona wrote to Placido Tardy on 8 February 1861,[22]) who later, in 1864, lost his chair for having refused to swear allegiance to the new Italian state; yet another was Giovanni Capellini, geologist and palaeontologist, and director of the museum of geology.

But Cremona's closest relationship was with Domenico Chelini, professor of rational mechanics, according to Cremona *the only real mathematician* at the University of Bologna.[23] As Cremona tells it, Chelini, a Piarist priest,

> became a professor of mechanics and hydraulics in October 1851 at the University of Bologna; on 24 May 1860 he was dismissed from that position because he had not taken part in the religious function during the celebration of the constitution; on 5 November of that same year an exceptional measure, in the form of a ministerial decree appointing him professore straordinario [an untenured lecturer], reinstated him to the chair of rational mechanics for an unlimited period of time, with no obligation to swear allegiance, and with the same salary that he had earlier enjoyed as a full professor. However, by October 1863 there began to be less willingness to respect Chelini's extraordinary position; he was sent a decree that named him a lecturer for the coming academic year, as is the case for all untenured lecturers. This caused him no small bitterness, because Chelini sincerely loved the nation of Italy and was absolutely as far as could be from associating himself with any act hostile to the Italian government, as many of his intimate friends can amply testify. And a year later the Ministry requested that he swear political allegiance, and following his declaration that he could not swear because of his status as an ecclesiastic, a decree of 18 December 1864 deprived him of his position. When that happened the professors and students of the University of Bologna showed how much esteem and affection they nourished for Chelini, and how pained they were to deprived of all hope to keep him at that University.[24]

---

nuovo, trovai indulgenza e benevolenza. Né il suo patrocinio mi venne mai meno; anzi coll'andar del tempo si fece sempre più affettuoso e intimo, e non cessò che colla vita. ... *Vorrei dire, se avessi l'arte della parola, quanto l'amicizia di PIANI mi sia stata profittevole e benefica. L'inesauribile bontà di quell'uomo lo traeva naturalmente a porre tutto il suo ingegno e la vasta sua erudizione a servizio degli amici: io non ebbi mai ricorso a lui invano*. Cf. [17].

[22] *buono è Respighi che insegna ottica e astronomia*, cf. letter no. 6 in [9].

[23] *il solo vero matematico*, cf. letter no. 6 in [9].

[24] ... *nell'ottobre 1851 andò professore di meccanica e idraulica all'Università di Bologna; il 24 maggio 1860 fu tolto dall'ufficio perché s'era astenuto dall'intervenire alla funzione religiosa della festa dello Statuto; ed il 5 novembre dello stesso anno fu restituito alla cattedra di meccanica razionale con un provvedimento eccezionale sotto forma di decreto ministeriale che lo nominava professore straordinario, senza limite di tempo, senz'obbligo di giuramento e collo stesso stipendio di cui godeva prima come ordinario. Però nell'ottobre 1863 si cominciò a non voler più rispettare la posizione eccezionale del Chelini; gli fu mandato un decreto che lo nominava professore straordinario per l'anno scolastico imminente, come è pratica per gli straordinari. La qual cosa gli recò non poca amarezza, perché il Chelini amava sinceramente la patria italiana ed era assolutamente alieno dall'associarsi a qualunque atto ostile al governo italiano, dei quali suoi sentimenti gli amici intimi possono fare ampia testimonianza. E un anno dopo il Ministero chiese ch'egli prestasse il giuramento politico; e dietro la sua dichiarazione di non lo poter dare*

Cremona was undoubtedly among those who showed their esteem and affection for Chelini, and strove to avoid his expulsion. As early as 8 February 1862, when there were only rumours about the possibility of re-hiring as "new" professors those who had been dismissed in 1860 – a possibility that became reality in 1863 – he wrote to Genocchi:

> *They want him to swear* [allegiance], *while none of the other professors of the defunct system had sworn. They want Chelini to swear because he was dismissed in 1860, and now they want to consider him like a new man. But that very dismissal of 1860 was as extremely unfair as its pretext was ridiculous: a Tedeum. The political question aside, Chelini could not swear without acknowledging the fairness of his dismissal, and thus without humbling himself. Have a little tolerance, you gentlemen who call yourself liberals!*[25]

Cremona not only defended his friend, he defended his right/duty to not swear an oath, and he continued ever after to admire Chelini's consistency of behaviour. On 21 October 1864 he asked Placido Tardy, then a Senator in the Kingdom of Italy, to do everything possible to help Chelini,[26] and a month later, in order to avoid irremediable decisions, he even accepted Brioschi's proposal to give up the course in higher geometry in order to take over teaching one in mechanics that had been taught by Chelini.[27] But nothing was done, and Cremona never ceased to lament Chelini's successor to the course in mechanics, the engineer Luigi Venturi, who, having participated in a competition in Pavia, *came in last but one*:

> *When you have the chance to write to Natoli, I would love for you to let him know what an unhappy gift he has given to the University of Bologna by naming Venturi lecturer in rational mechanics! If the Ministers won't deign to consult with those who are competent, then they should at least hear their judgment about their acts.*[28]

---

per la sua condizione di ecclesiastico, venne destituito con decreto del 18 dicembre 1864. In quell'occasione i professori e gli studenti dell'Università di Bologna in diversi modi mostrarono quanta stima ed affetto nutrissero pel Chelini e con quanto dolore si vedessero privati d'ogni speranza di conservarlo a quell'Ateneo, Cremona quoted by Eugenio Beltrami, in [2].

[25]*Si vuole ch'egli giuri, mentre tutti gli altri professori del cessato sistema non hanno prestato giuramento. Si vuole che Chelini giuri perché egli fu destituito nel 1860, ed ora si vuol considerarlo come un uomo nuovo. Ma appunto quella destituzione fu ingiustissima, come ridicolo ne fu il pretesto: un Tedeum. A parte la quistione politica, il Chelini non può prestarsi al giuramento senza riconoscere la giustizia della sua destituzione, e quindi senza avvilirsi. Un po' di tolleranza, signori che vi dite liberali!.* Cf. letter no. 18 in [8].

[26]Cf. letter no. 17 in [9].

[27]*Incaricato di ciò dal ministro, Brioschi scrisse allora a me pregandomi che, come sacrificio utile allo stesso Chelini, accettassi di insegnare per questo anno la meccanica in luogo della geometria superiore* (Charged to do so by the minister, Brioschi then wrote to me asking me, as a sacrifice to help Chelini, to agree to teach mechanics this year instead of higher geometry), cf. letter no. 22 in [9].

[28]*Quando voi avrete occasione di scrivere al Natoli, amerei che gli faceste osservare quale infelice dono egli abbia fatto all'Università di Bologna destinandovi a professore straord. di meccanica razionale il Venturi! Se i Ministri non si degnano di consultare le persone competenti, sappiano almeno il giudizio che queste portano sui loro atti.* Cf. letter no. 39 in [9].

And shortly afterwards:

> Here we have Venturi teaching rational mechanics. You can't imagine what a enormous ass: the minister made a huge mistake. Even the students realise that he does not know anything: but they are all the happier, because then they have that much less to study, since he proceeds with incredible slowness, he repeats himself, he tries it again, he corrects himself... Losing Chelini to get Venturi ... what a horror! [29]

Even after Chelini's move to Rome, the friendship between the two was carried on with the same closeness and respect. For example, it was to Chelini that on 12 June 1865 Cremona could write of familiar matters:

> I have only good news to tell you about my family. We are all in the country, in a house in the hills not very far from the city, which I have rented from Prof. Emiliani. Here, in addition to my own little brigade, I am in the company of the Emiliani family and professors Magni and Monti, because the latter has likewise rented a house close to mine. Every day, of course, I go to Bologna for the class at the university, and I return to the country late.[30]

We are thus introduced to a world which, though far from the world of mathematics, still left a strong impression on Cremona's life in Bologna. Of particular interest is his mention of Francesco Magni,[31] the friend with whose intentions and experiences his own were very closely tuned for a long time. On the one hand, Magni was a family friend, godfather to little Itala born on 31 March 1865, and later, guardian of Luisa, the youngest child of Elisa and Luigi Cremona, who remained with a wet nurse when the family moved to Milan [4]. On the other hand, he was one of the founders in February 1866, along with Carducci among others, of the Masonic lodge in Bologna, of which Cremona would become Venerable Master. He shared in the hopes and disappointments brought about by the political situation,[32] and was a companion in the attempts to improve the national educational system, especially that of higher education.[33]

Eugenio Beltrami[34] also arrived in Bologna during the academic year 1862–1863, and Cremona's circle was enlarged. Cremona seemed more confident in the

---

[29] *Qui abbiamo il Venturi a insegnare meccanica razionale. Non vi potete immaginare un asino così madornale: il ministro l'ha fatta ben grossa. Gli stessi studenti s'accorgono che egli non ne sa nulla: ma i più si rallegrano che in tal modo avranno poco da studiare, giacché egli procede con una lentezza incredibile, si ripete, rifà, si corregge. ... Perdere il Chelini per acquistare il Venturi ... che orrore!*. Cf. letter no. 41 in [9].

[30] *Della mia famiglia non ho che buone nove da darvi. Siamo tutti in campagna, in un casino in collina, poco discosto dalla città, che ho preso in affitto dal prof. Emiliani. Ivi, oltre alla mia brigatella, ho la compagnia della famiglia Emiliani e dei professori Magni e Monti, perché quest'ultimo ha preso del pari in affitto un casino prossimo al mio. Tutti i giorni, naturalmente, vengo in Bologna, per la lezione all'università, e del tardi torno in campagna*. Cf. letter no. 8 in [21].

[31] For a biography of Magni, see [27].

[32] See [5] and [6].

[33] One example is the profound involvement of the Magni and Cremona in favour of the enactment of the Matteucci Regulations in Bologna; cf. letters no. 1, 2, 3 and 4 in [5].

[34] See, in this volume, the paper by Rossana Tazzioli.

possibility of turning Bologna into a first-rate centre for research: effectively, at the time, the contemporary presence of Cremona, Chelini and Beltrami made Bologna one of the most prestigious universities in Italy for mathematics. Unfortunately for Bologna, but fortunately for the future of Italian mathematics as a whole, this triumvirate ended as soon as it had begun. Beltrami was called the following year to Pisa, at Betti's suggestion, to take over the chair of mathematical physics in the place of Ottaviano Fabrizio Mossotti, who had died in March 1863. Cremona encouraged Beltrami to set aside his doubts and accept the chair of high prestige and responsibility offered by Betti. He himself wrote as much to Betti:

> *On the morning of 16 August Beltrami received your letter of the 11th ... He wrote to me immediately to ask my advice regarding your offer, and I, while replying just as immediately to him exhorting him to accept, am also writing to you. ... Beltrami is worth much, much more than he appears from the few papers he has published so far. He is a amiable young man who unites the most penetrating intellect to a solid commitment and assiduousness in his work. He is also very cultured, gentile and has a truly golden temperament. Thus the minister would have many councillors similar to you.*
>
> *I am very sorry indeed to lose Beltrami as a colleague. ... But I love him too well to postpone his good for the sake of my own. On the other hand, I am sure that, set on this other path, he will excel. If Beltrami goes to Pisa, I beg you to convince the Minister to find a suitable replacement for him in Bologna. Our poor faculty is in the saddest mire: the prof. of higher calculus is a beast, and a great demoraliser of our students.*[35]

Cremona's prophecy was on the mark: in Pisa Beltrami met – in addition to Betti and Novi – Bernhard Riemann. Riemann's intense exchanges with Betti, and especially with the young Beltrami, had decisive consequences for the ulterior development of Italian mathematics and the emergence of an new generation of researchers.

---

[35] *Il Beltrami ricevette nel mattino del 16 agosto il tuo foglio del'11 ... Egli ne scrisse immediatamente a me per chiedere il mio consiglio circa la tua proposta, ed io, mentre rispondo immediatamente a lui esortandolo ad accettare, scrivo anche a te ... Il Beltrami vale assai più di quanto appaja dai pochi scritti finora pubblicati. È un giovane amabile che unisce l'ingegno più penetrante alla salda costanza ed assiduità nel lavoro. Inoltre è assai colto, gentile e di un carattere veramente aureo. Così il ministro avesse molti consiglieri simili a te!*
A me duole assai di perdere il Beltrami come collega ... Ma lo amo troppo per posporre il suo bene al mio. D'altronde son certo che messo in quest'altra via, si farà molto onore.
Se il Beltrami va a Pisa, ti prego di impegnare il Ministro a surrogarlo convenientemente in Bologna. La nostra povera facoltà è in tristissime acque: il prof. di calcolo sublime è una bestia e di più un demoralizzatore di giovani. Cf. letter no. 14 in [24].
For his part, Cremona set his sights high for a replacement for Beltrami: *In Munich there is an excellent young man, Johan Nikol Bischoff, known for distinguished works in advanced analytical geometry published in Crelle's journal ... How wonderful it would be to attract him to us, giving him a post at our university! Do you want to speak to the Minister about him?* (*A Monaco* [di Baviera] *c'è un bravissimo giovane, Johan Nikol Bischoff, noto per egregi lavori d'alta geometria analitica inseriti nel giornale di Crelle. ... Che bella cosa se si potesse attirarlo fra noi, dandogli un posto in una nostra università! Vuoi tu parlarne al ministro?*, ibidem). But he only succeeded in obtaining the transfer of Pietro Boschi from Cagliari, who taught complementary algebra and analytical geometry and who, although he showed himself to be a good teacher, could hardly have made up for the absence of a mathematician of the calibre of Beltrami. See also [36].

3 Teaching and Institutional Responsabilities

We have already seen how the conditions of mathematics teaching in Bologna were one of the main causes of vexation for Cremona. He was bluntly critical:

> *Bologna is not at all improved. Here the introduction to the calculus is taught by a toothless and impudent old man (Ramenghi) who is the quintessence of ignorance. And calculus is taught by a Saporetti who, besides being extremely ignorant, is also the most impudent cheap-jack of tutoring. Who can rid us of the likes of these?*[36]

And again:

> *I have found the strangest things here. For mathematics students there are separate chairs for zoology, botany, mineralogy, civil institutions and political economy, while there is none at all for descriptive geometry!*[37]

His course as well is fraught with various instances of incomprehension: *The scholastic authorities here, even before I arrived, decided that my subject was to be mandatory for third-year students, thinking that higher geometry should mean Monge's* Application de l'analyse [à la géométrie].[38] In consequence, he could certainly not fly rapidly towards the goals described in his opening lecture, but had instead to begin *with the most elementary things about anharmonic ratios.*[39] Still, according to many,[40] he was a fine teacher: clear, well ordered, enthralling, well versed in *the art of presenting difficult subjects in a simple and plain form.* In the course in descriptive geometry:

> *... he had a class of about 12 and lectured on the Theory of the Sun's Dial in connection with his Descriptive Geometry. He is evidently a good lecturer: everything was explained with perfect clearness. One peculiarity of the lecture arrangements was that, instead of a black board on the side of the room, the top of the table before the professor was made of slate and on it he wrote and made figures in chalk. The figures were of course inverted to the audience.*[41]

---

[36]*Bologna non sta punto meglio. Qui l'introduzione al calcolo è insegnata da un vecchio sdentato e impudente (Ramenghi) che è la quintessenza degli ignoranti. E il calcolo è insegnato da un Saporetti che oltre all'essere ignorantissimo fa inoltre il più impudente mercato di ripetizioni. Chi ci libera da costoro!* Cf. letter no. 15 in [8].

[37]*Ho trovato qui cose stranissime. Per gli studenti di matematica vi sono cattedre separate di zoologia, botanica, mineralogia, istituzioni civili ed economia politica, mentre manca quella di geometria descrittiva!* Cf. letter no. 9 in [8].

[38]*L'autorita' scolastica di qui, ancor prima ch'io arrivassi, ha deciso che la mia materia fosse obbligatoria per gli studenti del 3° anno, pensando che per geometria superiore si debba intendere l'*Application à l'Analyse di Monge, cf. letter no. 4 in [9].

[39]*colle cose più elementari de' rapporti anarmonici.* Cf. letter no. 11 in [8].

[40]See for example [3, 40].

[41]*... aveva una classe di circa 12 allievi e faceva lezioni sulla teoria dei quadranti solari in connessione con il corso di geometria descrittiva. Egli è manifestamente un buon insegnante; tutto era spiegato con chiarezza perfetta. Una peculiarità delle condizioni della lezione era che, invece*

In this situation, assembling suitable supporting materials for the lessons and searching for good collaborators naturally became issues of central importance. As regards the first of these, for example, Cremona saw to the publication of the texts of the lessons, but as far as the search for valid collaborators was concerned, the times were obviously much longer than he either could or cared to wait: *the livid Jesuit and the thuggish Austrian*[42] were not the only obstacles to the rebirth of Italian science! It was necessary to find capable young people, and Cremona looked for them in the fertile territory of Pavia:

> Following your last letter I wrote to Gherardi to convince him to propose Dr. Boschi, a student of Brioschi, as the professor of the Introduction to Infinitesimal Calculus in this university. At the same time I wrote to Cav. Ferri, and perhaps Brioschi will have spoken to the Minister about it. The nomination seems probable, I have it from a good source.[43]

But the longed-for nomination would not become a reality for several years. Obviously the minister was pressed by questions of much greater urgency, and Cremona could not even count on his friends taking an interest.

This in spite the fact that in the meantime his involvement in the construction of a national educational system was growing,[44] and thus his dealings with the national authorities had become more intense: on 16 November 1859 he was made a member of the commission charged with establishing the regulations for secondary schools; on 4 November 1862 he became a member of the commission for the examination of school textbooks; on 4 January 1863 he entered that charged with inspecting the secondary schools of Puglia and Calabria; on 23 April 1864 he entered an analogous commission for the schools in Sicily. He thus found himself involved in inspections that kept him far from Bologna for long periods of time, but which also made it possible for him to "see" the situation in a country that was struggling to create an educational system that was adequate to meet the requirements of the new industrial society.

---

di una lavagna in un angolo dell'aula, il piano della cattedra di fronte al professore era di ardesia e su di essa egli scriveva e disegnava le figure con il gesso. Le figure erano ovviamente girate verso il pubblico. Cf. [23].

[42] *il livido gesuita e lo sgherro austriaco* [10].

[43] *In seguito alla vostra ultima lettera scrissi a Gherardi per impegnarlo a proporre il dr. Boschi, allievo di Brioschi, come professore di Introduzione al Calcolo Infinitesimale in questa università. Ne scrissi contemporaneamente al cav. Ferri e forse Brioschi ne avrà parlato al Ministro. La nomina sembra probabile, lo so da buona fonte.* Cf. letter no. 12 in [8].

[44] This is shown, for example, by the fact that in 1865 Cremona had begun the translation for secondary schools of Baltzer's *Elemente der Mathematik*, with the aim of bringing pre-university teaching closer to some of the most recent results of research. The text showed itself to be much more suitable for the training of teachers, and the undertaking, which took a great deal of time away from research, was finally completed only in 1867, although it owes its genesis to Cremona's Bologna period.

4 Cremona's Studies and Research

The torpor of the Bologna environment was such that Cremona immediately and "most ardently" wanted to escape it. He was always tormented of this kind of anxiety: later he wanted just as ardently to leave Milan, and later still, Rome; it reflects the conflict between desires and reality, not only on the political front, but also that of research, which was often set aside for more pressing organisational or institutional requirements.

As early as February 1861, he felt suffocated and asked (as he would continue to ask persistently for the next 6 years) to leave:

> The degradation (with respect to mathematical studies) into which this university has fallen, for many reasons, is such that I see my work as almost futile, in spite of the excellent quality of the students. Moreover, the miserable conditions and gothic regulations of the library do not make it possible for me to study as I used to do in Pavia and Milan. If you add to that the extraordinary abundance of thieves and assassins that make it extremely dangerous to go out at night, you will have the main reasons why I want most ardently to get out of here.[45]

It is perhaps symptomatic of this restlessness that he constantly moved house. The family moved three times in 6 years. They first lived in Via di Mezzo S. Martino, just a short distance from the university; at the beginning of 1862 they moved to Piazza S. Pietro, Strada Galliera, 474/475; they moved once more in the spring of 1866, before Cremona knew that he would shortly move to Milan, to Via S. Stefano 14.

Yet, from many points of view the rhythm of Bologna was perfectly in sync with Cremona's need to dedicate himself to his studies. At least for the first 6 years, his teaching duties were not very taxing (three lessons a week to a small number of students,[46]) and left him ample time to concentrate on his research:

> As for me, I thrive and study. The only good thing about this place is that I can study, meaning with my books. I am not lacking for time.[47]

Of course, after the first years his responsibilities increased, but except for a short time, they remained much less time-consuming than those he would have in any later period of his life.

In any case, Cremona was all too aware that these years were decisive for establishing himself in the scientific world in the way he longed to. To achieve this, he threw himself headlong into his studies. Veronese described it thus:

---

[45] *L'abbiezione in cui è caduta, per tante cause, questa università (rispetto agli studi matematici) è tale ch'io vedo l'opera mia quasi inutile, malgrado l'ottimo buon valore de' giovani. Di più le misere condizioni e i gotici regolamenti della biblioteca non mi consentono quegli studi, cui m'ero abituato a Pavia ed a Milano. Se a ciò aggiungete la straordinaria abbondanza di ladri e d'assassini che rende pericolosissimo l'uscire di sera, avrete le principali cagioni, per cui desidero ardentissimamente andarmene di qui,* Cf. letter no. 3 in [24]).

[46] Cf. letter no. 10 in [8].

[47] *In quanto a me, vegeto e studio. L'unico bene che ho in questo paese è di poter studiare, ben inteso coi miei libri. Il tempo non mi manca.* Cf. letter no. 12 in [8].

> *For many years in Bologna he rose at midnight, after a short nap, to attend assiduously to research in science until daybreak, after which he restored himself with another short nap; since the work often gave him very strong migraines, and to correct the damage due to having neglected exercise in his youth, as an adult he wanted to swim, mountain climb, play billiards and cycle, and he achieved his end. He always wanted something, strongly wanted.*[48]

We can picture him pedalling the new-fangled bicycles of the 1860s (the bicycle as we know it would only come into use in the late 1870s), see him in the billiards parlour, and swimming at the seashore near Genoa, at Elisa's relatives, where he spent long months in summers of 1863 and 1864, or hiking in the mountains with his friends.

Of course, while he might not have lacked time, he did lack access to mathematics journals: the university library boasted 130,000 volumes, but when it came to mathematics, it had little or nothing dating before 1800. Of journals it had only Gergonne, Liouville, the *Annali di Roma*, and the *Philosophical Magazine*; there were only the latest volumes of Crelle. There was not a trace of the *Transactions* of London, nor the publications of the academies of Berlin, Leipzig, and Göttingen, only the *Comptes Rendus* of Paris and the things from Brussels, and these only from a certain point forward.[49]

In this difficult situation, one part of his sleepless nights was used to weave a net of international relations that would soon make him well-known throughout Europe and break the isolation in which the conditions of the Bologna university had plunged him.

The year 1861 began with a good omen: on 9 January he began corresponding with Arthur Cayley, who Brioschi had often described as *the best mathematician in Europe*. It began with very short letter, a simple exchange of publications, but one that promised interesting developments, especially regarding third-order curves in space: *Je m'enteresse beaucoup à cette autre théorie dont vous vous êtes occupé, les Courbes du troisième ordre en espace; et j'éspère toujours pouvoir l'étudier avec plus d'attention* (*I am very interested in other theory with which you are occupied, the curves of third order in space; and I hope to be able to study it with increasing attention*).[50] He would keep his promise: beginning in June Cremona and Cayley entered into a scientific relationship that was rich in interesting results.

Cremona also maintained a special relationship, in which scientific exchange was intermingled with a profound affinity and friendship, with Thomas Hirst, whom he had met in 1859, when he was still a teacher in a lyceum, and who spent a month in

---

[48]*Per molti anni a Bologna si alzò alla mezzanotte, dopo un brevissimo sonno, per attendere assiduamente a ricerche di scienza fino al sorgere del mattino, dopo di che ristoravasi dormendo ancora qualche poco; ed avendogli il lavoro soverchio date forti emicranie, egli, a correggere i danni dei trascurati esercizi in gioventù, volle già maturo essere nuotatore, alpinista, giuocatore di bigliardo, ciclista, e raggiunse il suo fine. Volle, fortemente volle.* Cf. [40].

[49]Cf. letter no. 9 in [8].

[50]Cf. letter no. 1 in [34].

Bologna from 29 May to 26 June 1864 [35]. During that time the two saw a great deal of each other.

At the end of 1860, he got in touch with Olry Terquem, the editor of the *Nouvelles Annales de Mathématiques*, in which Cremona had published many times, sending him extracts from the *Prolusione*, his opening lecture. This led, in the early months of 1861, to the beginning of ties to the scientific world in France, especially with that presided over by Chasles, and these were consolidated over time. These connections also had political overtones: the ideals expressed in the *Prolusione* resonated with those of the French scientists, who reacted to burning political issues with comments such as this by Prouhet:

> Votre appel au patriotisme de la jeunesse nous a émus. Les savants étrangers seraient bien ingrats s'il ne partageaient pas la joie qui fait naître dans les cœurs italiens la résurrection de l'Italie (*Your appeal to the patriotism of young people has touched us. Foreign scientists would be very ungrateful if they did not share in the joy that is born in Italian hearts over the resurrection of Italy*).[51]

And this by Terquem:

> Garibaldi est un excellent patriote italien. La force est dans la modération et dans une marche opportune. Mazzini est un ennemie de l'Italie plus dangereux que l'Autriche. (*Garibaldi is an outstanding Italian patriot. Strength lies in moderation and an appropriate course. Mazzini is an enemy more dangerous to Italy than Austria*).[52]

As to German mathematicians, Cremona began corresponding to Schröter and Clebsch, again with regards to curves in space. Thus, by the end of 1861 he was no longer unknown in European circles and could, fully self-aware, give free rein to his love for geometry.

In essence, apart from all that we have seen up to now, in comparison to other periods of his life, including those connected to the Politecnico in Milan and the Sapienza in Rome, Cremona's period in Bologna is primarily characterised by his research and the results he achieved.

Stendhal wrote some very beautiful passages about what might lead a poetic spirit to fall in love with mathematics. His description could easily be applied to Cremona:

> My enthusiasm for mathematics might have for its principal base my horror towards hypocrisy ... At the third or fourth lesson we passed to equations of the third degree, and there Gros was entirely original. It seems to me that he transported us straight to the frontier of science, face to face with the doors of science or in front of the veil that had to be lifted ... In those days I was like a great river, like the Rhine above Schaffhausen, where its course is still tranquil, but is about to rush into an immense waterfall. My cascade was a love for mathematics. [39]

---

[51] Cf. letter no. 1 in [26]. Comments on this letter and on the patriotic environment of Cremona may also be found in [32].

[52] Cf. letter no. 2 in [33].

It was during this period, when he above all occupied himself with curves and surfaces, that Cremona began to "lift the veil" that concealed the reality of geometric facts behind an apparent complexity, and found the key that made it possible to trace many results, that were apparently disparate and in any case very diverse, back to a point of unity.

It is in this finally being able to see the order within the apparent chaos that the profound poetry of his mathematics lies, a poetry that, if we believe the testimony that has come down to us, he was often able to communicate to his companions. The discovery of new facts generates a river of feelings and stimuli that he is capable of taming into a sober composedness, but whose richness and violence is evident to all. The "horror towards hypocrisy" that Stendhal speaks of and that we have already encountered in Cremona's life story, as well as the search for an internal consistency that often put him in conflict with the world that surrounded him, filtered into the way he committed himself to the great project, which had first been pursued by Steiner and Chasles, of retracing the theory of algebraic curves to unity.

The study of the second-degree curves, the conics, which dates back to classical Greek mathematics, was followed by the study of the algebraic curves, which can be said to have been begun in a systematic way by Newton, who had included third-degree curves, the cubics, in his famous projective classification of curves generated "by shadows" (*per umbras*).

Chasles' and Steiner's program had then been to classify these curves and identify their general properties, a program that Cremona intended, on the one hand, to complete and, on the other, to extend to curves in space (the "curve gobbe", or "skew curves") and surfaces.

Generally speaking, the instrument that Cremona had at his disposal for the study of geometric objects was that of transformations. Transformations can be used to solve the problems he had outlined in his opening lecture:

– When the properties of a given figure are known, deduce the analogous properties of another figure which is of the same genre but whose construction is more general;
– When some special cases of a certain unknown general property of a figure are known, deduce the general property.

Thus armed, it is possible to "lift the veil" to reveal the theory of plane curves.

Peut donc qui voudra, dans l'état actuel de la science, généraliser et créer en géométrie: le génie n'est plus indispensable pour ajouter une pierre à l'édifice (*This then is what is wanted, in the present state of science, for generalizing and creating in geometry: genius is no longer indispensable for adding a stone to the edifice*). Chasles' epigraph for Cremona's 1861 *Introduzione ad una teoria geometrica delle curve piane*[53] overflows with the enthusiasm of one who has

---

[53] Presented to the Bologna Academy of Sciences on 19 December 1861, this work, which goes far beyond the merely educational intention that lay at its genesis, would have a strong international influence.

discerned a method – that of geometric transformations – which appears decisive for the theory. All that was needed was to add some stones to the edifice by now almost complete; to put the finishing touches on a work already built, not to invent new methods. Only later would Cremona come to recognise that this instrument was not sufficient for a program as vast and ambitious as his own, and he set out to search for more general transformations.

The transformations that he initially had at his disposal were projective, then he gradually drew on the quadratic, and those which led up to the invention of birational transformations: as often happens in mathematics, just when geometry reaches what appears to be its limit, it turns out to be on the threshold of a profound transformation of its methods and aims.

The first part of the *Introduzione* closely follows the outline of the opening lecture: among other topics, it treats anharmonic ratios (cross-ratios) and the projectivities between first-order forms. The definition of projectivity chosen by Cremona, which follows along the lines of contemporary, more accredited texts, says:

> *Two geometrical forms are said to be projective when there is between their elements a relation such that to each element of the first corresponds one and only one given element of the second, and to each element of the second there corresponds one and only one given element of the first.*[54]

As can be seen, this is a definition that requires only a bijective correspondence, without any reference to linearity. This renders the proof of the fact that in a projectivity the cross-ratio is conserved, quite a "daring" one. But if we accept this passage, and thus the cross-ratio is an invariant for the projectivity, then it is easy to prove that, given the images of three elements in a correspondence between two first-order forms (for example, two straight lines), all the images of the other elements are determined in a single way. Let there be two straight lines, r and r', with three points A, B, C on the first, and three points A', B', C' on the second. If we want to determine on r' the image of point X on r by means of a projectivity that transforms A in A', B in B' and C in C', it is sufficient to observe that such an image is the only point X' such that the cross-ratio (A', B', C', X') coincides with the cross-ratio (A, B, C, X).

On one side, this example shows us that the definition given by Cremona is inadequate. (Consider the projectivity above and let D' be the point corresponding to D on r and E' the point corresponding to E. The correspondence which coincides in all points with the previous one except that in D – to which E' is associated – and E – to which D' is associated – is still bijective, but certainly does not conserve the cross-ratio and is therefore not a projectivity).

---

[54] *Due forme geometriche si diranno proiettive quando fra i loro elementi esista tale relazione, che a ciascun elemento della prima corrisponda un solo e determinato elemento della seconda ed a ciascun elemento di questa corrisponda un solo e determinato elemento della prima* [11].

But on the other side, it allows us to grasp one aspect of the progress made in the concept of projectivity over the course of the nineteenth century. It begins with ordinary projection and the fact (known since the time of Euclid) that a projectivity between two straight lines conserves the cross-ratio.

Given that many properties of projections derive from this fact, an attempt was made to extend the theory to all transformations that conserve the cross-ratio. But which transformations were they? Which property of the projections are we to consider? Bijection, chosen by Cremona, turned out to be insufficient. For the entire course of the nineteenth century this problem would give rise to many different extensions of projective geometry.[55]

Just as problematic is the definition of a "tangent to a curve". Having established that the order of a curve is the number of points at which it is intersected by a straight line (naturally counting the multiplicities as well as taking into consideration imaginary points), he expresses it this way: *If a is a position of the generating point, that is, a point of the curve, the straight line A that passes through a and through the successive position of the moving point is the tangent to the curve at that point.*[56] As if it were possible to define the successive position of a point on a continuous curve!

Again typical of the era is the formulation of the "proof" of the theorem regarding the number of points at which two curves, of orders n and n' respectively, intersect (Bézout's theorem):

> At how many points intersect two curves, the orders of which are n, n'? I admit, as an evident principle, that the number of intersections depends solely on the numbers n, n', such that it does not change, when the given curve is substituted by two other loci of the same order. If the curve of order n' is substituted by n' straight lines, these will intersect the curve of order n at nn' points; therefore, two curves, the orders of which are n, n', intersect at nn' points (real, imaginary, distinct, coincident).[57]

This is an audacious, but quite efficient, way of reasoning. The algebraic calculation would be far too complicated: in the case of two curves of orders eight and seven, for example, it would be necessary to verify that the number of solutions to the system formed by the corresponding equations is 56. So the calculation is substituted by a method of reasoning which is unconventional, to say the least: we can substitute one

---

[55] At the time he was writing, the debate on this topic was quite advanced, but Cremona does not appear to have been interested in problems regarding the foundations. Nor does he seem to have been particularly concerned about remaining faithful to the "purist" program that he himself had set out in his opening lecture: for example, he introduces the cross-ratio by means of the metric and not in a synthetic way.

[56] *Se a è una posizione del punto generatore, ossia un punto della curva, la retta A che passa per a e per la successiva posizione del punto mobile è la tangente alla curva in quel punto* [11].

[57] *In quanti punti si segano due curve, gli ordini delle quali siano n, n'? Ammetto, come principio evidente, che il numero delle intersezioni dipenda unicamente dai numeri n, n', talché rimanga invariato, sostituendo alle curve date due altri luoghi dello stesso ordine. Se alla curva d'ordine n' si sostituiscono n' rette, queste incontrano la curva di ordine n in nn' punti; dunque: due curve, i cui ordini siano n, n', si segano in nn' punti (reali, immaginari, distinti, coincidenti)* [11].

of the curves of order n' with a system of n' straight lines, starting from the "evident principle" that the number sought depends only on the order of the curves! In this case the number is easily calculated: each line intersects the curve of order n at n points (by definition of the order), and thus we need only multiply n by n' and the theorem is proved.

It's a shame that the "evident principle" (also called the "continuity principle") is dubious and extremely difficult to prove! But this is not what was important to Cremona. He used the principle as an instrument to overcome the enormous complexity of the problems that the study of curves and surfaces posed for geometers.

This is during the heroic period in which the extremely intricate world of curves and surfaces was about to be dominated, and there was no time to tarry over questions of foundations. That would come later. As André Weil wrote:

> The mathematician who first explores a promising new field is privileged to take a good deal for granted that a critical investigator would feel bound to justify step by step; at times when vast territories are being opened up, nothing could be more harmful to the progress of mathematics than a literal observance of strict standards of rigour. [41]

Cremona's motto was "Forward, later we'll see!" and it is in this spirit that the work must be read.

After having stated the principal properties of curves, including the number of conditions that determine a curve of order n or the maximum number of double points or cusps that they may possess, he extends the definition of projective transformations of the "first-order forms" to each pencil of curves of a given order (we would say, to the linear variety of dimension two).

A pencil of curves of order n is defined as the set of curves obtained as a linear combination of two given curves. The n^2 points at which the curves intersect are the base points of the pencil. Of these points, $\frac{n(n+3)}{2} - 1$ are arbitrary, while the rest are determined. For example, two conics generate a pencil with four base points.

Cremona gives a projective structure to a conic pencil in the following way: let A, B, C, D be the base points of the pencil, and r be variable straight line passing through one of them (D, for example); or, expressed in a different way, let r be a line of the pencil through D. In the conic pencil it determines "the" conic passing through A, B, C and tangent to line r at D and so induces a bijective correspondence between the conic pencil and the pencil of lines. Then he defines the cross-ratio between the four corresponding lines (the cross-ratio does not depend on the choice of point D) and thus he gives a complete projective structure to the conic pencil.

He proceeds in an analogous way for the pencil of curves of order n, thus positioning himself to be able to give the definition of "projective pencils" like that of two pencils that are in bijective correspondence such that the cross-ratio between four elements is conserved.

This allows him to come up with a result that will be widely applied: if two pencils of curves of orders n and m respectively are projective, the points at which they intersect the corresponding elements form a curve of order $n + m$.

Thus, two projective line pencils generate a conic (a special case of which had been noted by Newton), while a projective conic pencil and a projective line pencil generate a cubic.

Let us make only one comment before concluding this part. In his *Introduzione*, Cremona makes a remarkable qualitative leap: the object of study is no longer the line or (by duality) the line pencil, but any linear manifold whatsoever whose elements are abstract objects. The object of geometry is no longer a space of points, but one which is abstract and general. Even if the results set forth in his book are due in large part to Chasles, de Jonquières and Grassmann (all in any case recent, dating to the 1850s), his efforts to systematise and organise the questions fully certainly justify of the book's great international success.

In 1861, after the topic of plane curves, Cremona re-examined and produced a unified vision of skew curves.

Cubic space curves are third-order curves in space (in the sense that a generic plane intersects them at three points), the simplest of those that do not lie entirely in a plane. They are determined by six points and resemble the conics in many respects. For instance, as a conic is generated by two projective line pencils, a skew cubic is similarly generated by two projective point-stars of lines.

The theory of skew cubics was quite recent at the time: Möbius had been the first to treat them in 1827 in his very famous text on barycentric calculus, which was rich in new ideas but which Cremona had only come to know about 2 years earlier, in 1859. Möbius was followed by Seydewitz, Chasles and Schröter. In that frenetic year of 1861, Cremona set himself *le but essentiel de démontrer géométriquement les propriétés énoncées, avec des démonstrations analytiques ou sans démonstrations* (*the essential goal of proving geometrically the properties stated, with analytic proofs or without proofs*), continuing the work of "purifying" the geometric methods that he had begun barely a year earlier.

Finally, he turns his attention to the theory of surfaces, in particular those of the third order. The passage from curves to surfaces is neither obvious nor simple. It requires new techniques and leads to the discovery of unexpected properties. And it was Cremona's pioneering work that paved the way for what would be, for at least the next 70 years, the primary undertaking of the Italian school of algebraic geometry.

The results that are probably Cremona's most famous ones appeared respectively in 1863 and 1865 in the *Memorie* of the Bologna Academy of Sciences: those regarding birational, or Cremona, transformations.

Projective transformations transform lines into lines and can therefore appear insufficient for as ambitious a program as that announced by Cremona in his *Prolusione* (*when the properties of a given figure are known, conclude the analogous properties of another figure which is of the same genre but whose construction is more general*). In a letter of October 1864 to his friend Hirst, Cremona gave a clear vision of the context in which his results were situated:

> *I don't believe that anyone before Steiner and Magnus has spoken of the conic transformation* [i.e., today's quadratic transformation] ... *Magnus alludes to transformations in which*

> to a line corresponds for example a curve of fourth order with three double points and three simple fixed points etc. But he doesn't say anything in this regard.[58]

He, thus, identifies Magnus as the immediate precursor of his own work; he had been an attentive reader of Magnus's works between 1862 and 1864. Shortly after, in December, in another letter to Hirst he transcribed entire passages of Magnus, commenting:

> Magnus had to have realised his error in believing that the most general transformation of one plane figure into another, both of which correspond point-by-point, is of the second order. But he makes no mention of this error.[59]

It was precisely the correction of this error that provided the starting point for his own work.

Cremona had become interested in this problem through a paper by the then very young director of the Observatory of Brera, Giovanni Virginio Schiaparelli [38], an abstract of which had been published earlier, in February 1861. The first draft of that work dated to 1862 [37], and that was the form in which Cremona read it.

Schiaparelli, after first having gone into the method of geometric transformation at length, then examined bijective transformations[60] (in his terminology, first order transformations) of general plane figures. Along with those notes (affinity and projectivity), he finds, using analytic methods, transformations which he calls "conic", which transform lines into conics circumscribed to a fixed triangle, and more generally, curves of order n into curves of order $n + 2$. These transformations also comprised those widely known as inversions.

In the first of the two works, Cremona started with a broader overall vision than that of Schiaparelli, that is, from the observation that:

> ... it is evident that applying to a given figure several conic transformations in succession, from this composition will be created a transformation that will still be first order, although one in which, to the lines of the given figure, there correspond in the transformation, not conics, but curves of a higher order.[61]

---

[58] *Io credo che nessuno prima di Steiner e Magnus abbia parlato della trasformazione ... Magnus fa anche allusione a trasformazioni in cui ad una retta corrisponda p. es. una curva del 4° ord.e con tre punti doppi e tre punti semplici fissi ecc. Ma non dice di più su questo proposito.* Cf. letter no. 18 in [35].

[59] *Magnus deve essersi accorto del suo errore di credere che la più generale trasformazione di una figura piana in un'altra, le quali si corrispondano punto per punto, sia del 2° ordine. Però non fa allusione a questo errore.* Cf. letter no. 21 in [35].

[60] Birational transformations are not actually bijective. They are in a Zariski open set. In what follows, we will use the terminology of Cremona's day.

[61] *... è evidente che applicando ad una data figura più trasformazioni coniche successive, dalla composizione di questa nascerà una trasformazione che sarà ancora del primo ordine, benché in essa alle rette della figura data corrisponderebbero nella trasformata non già coniche, ma curve d'ordine più elevato.* Cf. [12].

This established, he proposes to:

> ... prove directly the possibility of geometric transformations of plane figures in which the lines correspond to curves of any given order whatsoever. Then prove how ... it is possible to project the points of a plane onto a second plane, and thus transform a given figure in the one plane into another figure situated in the other plane.[62]

This is the birth of Cremona transformations.[63]

Cremona carries this undertaking out by first addressing, by means of a purely geometric method, the necessary conditions under which the homaloidal net (that is, the set of curves images of lines) has to satisfy in order to be the image of the net of lines of the domain, and then going on to the question of the existence of a transformation that has the desired characteristics. Here, we will not go into details, but only highlight some aspects of this research that are characteristic of Cremona's procedure.[64]

From the fact that they are the images of lines, it can be deduced that the elements of the homaloidal net of a transformation that transforms lines into curves of order n must have the property that for each pair of points there is one and only one curve of the net that passes through those points. Thus, since there are $\frac{n(n+3)}{2}$ curves of order n, the number of common conditions satisfied by the curves of the net is $\frac{n(n+3)}{2} - 2 = \frac{(n-1)(n+4)}{2}$.

Furthermore, since two lines have one and only one point in common, then likewise, two curves of the net have one and only one point in common; given that two curves of order n have n^2 points in common, all the curves of the net have $n^2 - 1$ common points. Finally, given that a point of multiplicity r common to two curves is equivalent to r^2 their intersections, called x_r the number of points of multiplicity r among those common to all the curves, is evidently $x_1 + 4x_2 + 9x_3 + L + (n-1)^2 x_{n-1} = n^2 - 1$.

Further, since the fact that a curve has a point of multiplicity r is equivalent to $\frac{r(r+1)}{2}$ conditions, there is also the equation: $x_1 + 3x_2 + 6x_3 + L + \frac{n(n-1)}{2} x_{n-1} = \frac{(n-1)(n+4)}{2}$. Basing himself on these two equations, Cremona determines many properties of the net of curves. For example, he proves that for quadratic transformations ($n = 2$) the two equations are reduced to a single $x_1 = 3$, that is, all the curves have three simple points in common, and are thus circumscribed to a triangle, in keeping with what had been established by Magnus and Schiaparelli, while for $n = 3$ there are $x_1 = 4$, $x_2 = 1$, that is, the net of cubics is made up of curves circumscribed to

---

[62] ... mostrare direttamente la possibilità di trasformazioni geometriche di figure piane, nelle quali le rette abbiano per corrispondenti delle curve di un dato ordine qualsivoglia. Poi dimostro come ... si possano projettare i punti di un piano sopra un secondo piano, e così trasformare una figura data in quello, in un'altra figura situata in un altro piano. Cf. [12].

[63] All of the treatises of algebraic geometry from the end of the nineteenth century to the present day discuss, using differing terminologies, Cremona transformations. The classic reference is [28]. An interesting overview using modern terminology while still providing notes on its history is [18].

[64] A faithful account of this part of Cremona's work is given in [22].

a pentagon, one point of which is double (in particular, the curves all have a double point and are therefore rational).

The work goes on to construct geometrically, for each n, transformations of degree n; we will not follow Cremona in this fascinating venture, but will only underline the fact that these are actual constructions, geometric and not algebraic, aimed at demonstrating the power of the purely geometric methods that he had used.

In the second work Cremona had instead to take into account the then very recent article by de Jonquières, who had, among other things, built upon his earlier studies. In this new work, de Jonquières had effectively dealt with one of the possible cases deriving from the solution of the two equations mentioned above, that is, when there are, among the common points of the homaloidal net, one point $(n-1)^{\text{ple}}$ and $2(n-1)$ simple points, that is, the most immediate solution to the equations. Having ascertained that de Jonquières had developed this case in an ample fashion, he set for himself the task of *proving that the same method and the same properties can also be extended to the transformations that correspond to all the other solutions of the two equations that I have mentioned.*[65] Again, without going into the technical details of this work, we will conclude this part with Cremona's own words:

> ... and if the division of the geometric curves is adopted ... in genera, proposed very recently by Mr Clebsch, in relation to the class of Abelian functions on which the curves themselves depend, it is found that our skew curve is of the genus $n-1$.[66]

Cremona's words opened a window to the future. The relationship between Clebsch and Cremona had begun in 1863 and developed into a genuine, deep friendship which would last until Clebsch's death in 1872. From a scientific point of view it was a very original kind of partnership: the two mathematicians developed a closely knit way of comparing methods (that of Clebsch analytical–functional, that of Cremona synthetic). From this comparison and from a careful study of the algebraic-geometric interpretation formulated by Clebsch, Cremona became convinced of the need to master what he called, in a letter to Genocchi written in January 1866, "the Riemannian mysteries". In the meantime, the German school of Clebsch and his student Max Noether demonstrated how powerful the algebraic version of Cremona transformations was for the study of algebraic curves and surfaces.

It would be later in Milan, in close collaboration with Felice Casorati, that Cremona would attempt to develop this point of view to its maximum potential [7], and to follow the developments tied to the new terminology and the new techniques. But in 1873 his move to Rome and the ensuing enormous political and bureaucratic responsibilities would end up frustrating any possibility that the still young Cremona had of making significant scientific contributions in this direction. It should perhaps

---

[65] ... *di mostrare che lo stesso metodo e le stesse proprietà si possono estendere anche alle trasformazioni che corrispondono a tutte le altre soluzioni delle due equazioni che ho accennato* [13].

[66] ... *e se si adotta la divisione delle curve geometriche ... in generi, proposta recentissimamente dal sig. Clebsch, in relazione alla classe delle funzioni abeliane da cui le curve stesse dipendono, si trova che la nostra curva gobba è del genere* $n-1$. [13]

be underlined that he would never lose this conviction, and it would be later fully developed by Eugenio Bertini, Cremona's only student in Bologna who was of a decidedly superior quality.

We do not intend to follow the development of this fruit of Cremona's years in Bologna to its maturity; we need only to remark that he had recognised at once his student's special gifts and had guided his education, taking care that it was complete with regards to analysis as well. This is how he prepared the way for Bertini's arrival in Pisa in a letter to Betti in November 1866:

> He is a young man quite dear to me for his estimable qualities: he is of great intellect and very desirous of learning; if, as I don't doubt, he perseveres, he will achieve something out of the ordinary ... but unfortunately Bertini knows little or nothing about analysis; and he is coming to Pisa precisely to study algebra, calculus, etc. I beg you wholeheartedly to help him and advise him.[67]

Bertini certainly did not disappoint his teacher's hopes for him, and carried to completion the line Cremona had begun to develop. In 1869, he gave a simple and effective proof of the invariance of the genus for Cremona transformations, and in 1877 he classified involutorial Cremona transformations. While Kleiman notes that *Bertini's classification represented a philosophical break with Cremona, who saw his transformations only as a tool for reducing the complexity of given geometric figures, and not as objects of study in their own right*, he also nevertheless confirms that *Bertini's research reflects, by and large, Cremona's influence in its subject and its style. Both geometers employed synthetic methods and analytic methods with equal facility. Both wrote succinct, precise, elegant treatments, which reflect their powerful, penetrating intellects* [29].

Bertini, together with Segre, is the link that connects Cremona's work with that of Castelnuovo, Enriques and Severi.

As it is known, Cremona's period in Bologna concluded with what was perhaps his most prestigious international recognition, the award of the Steiner Prize of the Berlin Academy of Sciences in July 1866, for a detailed study of third-order surfaces [16]. An exhaustive, modern examination of this topic has been provided by Igor Dolgachev [19], and we won't go into it here. We will only underline that, yet again, along with the arguments and the methods, Cremona opened a new window to what would become a basic theme for the Italian school of algebraic geometry. In fact, he wrote:

> Considering that the theory of third order surfaces has as its main foundation the general theory of the surfaces of any order, on which subject there is no known geometric treatise; and that the exhibition of a large number of properties offers no more facility for the cubic

---

[67] *È un giovane a me assai caro per le egregie sue qualità: ha molto ingegno e molto desiderio di imparare; se, come non dubito, egli persevera riuscirà qualcosa di diverso dall'ordinario. ... Ma disgraziatamente il Bertini sa poco o nulla di analisi; e viene a Pisa appunto per studiare algebra, calcolo, ecc. Ti prego caldissimamente di assisterlo e consigliarlo.* Cf. letter no. 38 in [24].

surfaces than it does for surfaces of any order; the author deemed it appropriate to begin with some chapters relative to the latter.[68]

Thus Cremona, during his years in Bologna, can be seen as the leader of a school, in the sense that he is someone who, in addition to achieving significant results in his own research, also indicates a direction for further work. It is worthwhile recalling that when Cremona left Bologna he was only 35 years old!

We conclude this brief overview of Cremona's scientific results by mentioning a work that, although minor, is highly significant (and today widely cited) regarding a special curve, the hypocycloid of three cusps [14].

Cremona had always especially loved this curve, an affection which was well known to his students, who recognised here an aspect of the aesthetics that characterised Cremona's work (Loria, referring to this work, spoke of *la beauté de la méthode géométrique*, the beauty of the geometric method). As it is known, the hypocycloid of three cusps, generated by a point on a circle of a given radius rolling inside a circle whose radius is three times larger, can also be obtained by the envelope of the Simson lines of a triangle (that is, the lines that join the ends of the perpendiculars drawn from a point on the circle circumscribed by the triangle), and features many properties that are frequently unexpected and which link various aspects of geometry and analysis.

This hypocycloid is a curve of fourth-order and third class,[69] and passes through the cyclic points, where it has as a double tangent the line to infinity. Cremona proved that,[70] vice versa, *toute courbe de trosième classe et quatrième ordre, dont la tangente double soit à l'infini et les points de contact sur un cercle, est nécessairement une hypocycloide à trois rebroussements* (*all curves of third class and fourth order whose double tangent is the line at infinity at the cyclic points on the circle, are necessarily hypocycloids with three cusps*). And thus he observes that: *Cette courbe joue donc, parmi les courbes de la troisième classe et du quatrième ordre, le même rôle que le cercle parmi les coniques* (*This curve plays, therefore, among the curves of third class and fourth order, the same role that the circle plays among the conics*).[71] Beginning with this beautiful analogy, Cremona explored the properties of this curve, among which that which is tied to the trisection of the angle by means of a transformation of reciprocal polars.[72]

---

[68]*Attendu que la théorie des surfaces du troisième ordre a son principal fondement dans la théorie générale des surfaces d'ordre quelconque, au sujet de laquelle on ne connaît aucun traité géométrique; et que l'exposition d'un grand nombre de propriétés n'offre pas plus de facilité pour les surfaces cubiques, que pour les surfaces d'ordre quelconque; l'auteur a jugé convenable de commencer par quelques chapitres relatifs à ces dernières.* Cf. [16]. See also [15].

[69]The class of an algebraic curve is the number of its tangents that can be drawn from a generic point.

[70]Many of the theorems proved by Cremona in this work had been (yet again!) stated by Steiner without proof in 1853.

[71]As it is known, the circumference is that conic which is precisely tangent to the line of infinity at the cyclic points.

[72]A brief treatment of this problem is in [25].

Cremona can then justly say that the hypocycloid is a *merveilleuse courbe, douée de propriétés si nombreuses et si élégantes* (*a marvelous curve, whose properties are as numerous as they are elegant*). But the beauty he refers to does not lie in the curve's particular aesthetic beauty (there are other, more "beautiful" star hypocycloids), but rather in the harmony that is revealed in a completely unexpected way by a transparent mathematical reasoning: this is one of those "hidden harmonies" that only the mathematically trained eye can perceive, and which Enriques described so evocatively. The search for hidden harmonies will constitute an unbroken thread in the development of Italian geometry. In this sense as well Cremona was the leader of a school.

References

1. Beltrami, E., Cremona, L. 1878. Domenico Chelini. *Giorn di Mat.* 16: 345.
2. Beltrami, E. 1881. Della vita e delle opere di Domenico Chelini. In *In memoriam Dominici Chelini, Collectanea Mathematica nunc primum edita cura et studio L. Cremona et E. Beltrami*, I-XXXII. Mediolani: Hoepli.
3. Bertini, E. 1917. *Della Vita e delle Opere di L. Cremona*. In *Opere*, III vol, V-XXII. Milano: Hoepli.
4. Brigaglia, A. Di Sieno, S. 2009a. L'opera politica di Luigi Cremona attraverso la sua corrispondenza, Prima Parte. Gli anni dell'entusiasmo e della creatività, La Matematica nella Società e nella Cultura, *Rivista della Unione Matematica Italiana*, (I), II, Dicembre, 353–388.
5. Brigaglia, A. A., Di Sieno, S. 2009b. *La Corrispondenza massonica di Cremona (con G. Carducci e F. Magni)*. Milano: Mimesis.
6. Brigaglia, A. Di Sieno, S. 2010. L'opera politica di Luigi Cremona attraverso la sua corrispondenza, Seconda Parte. Il crollo delle speranze e il lavoro organizzativo, La Matematica nella Società e nella Cultura, *Rivista della Unione Matematica Italiana*, (I), III, Agosto, 137–179.
7. Brigaglia A., Ciliberto C., Pedrini, C. 2004. The Italian school of Algebraic Geometry and Abel's legacy. In *The Legacy of Niels Hendrik Abel*, ed. Laudal, O., Piene, R., 295–348. Berlin: Springer.
8. Carbone, L., Gatto, R., Palladino, F. (eds.). 2001. *L'Epistolario Cremona-Genocchi*. Firenze: Olschki.
9. Cerroni, C., Fenaroli, G. (eds.). 2007. *Il Carteggio Cremona-Tardy (1860–1866)*. Milano: Mimesis.
10. Cremona, L. 1861a. Prolusione al corso di Geometria Superiore letta nell'Università di Bologna nel novembre 1860. *Il Politecnico*, 10, 22–42. In *Opere*, I, 237–253. Milano: Hoepli.
11. Cremona, L. 1861b. Introduzione ad una teoria geometrica delle curve piane. *Mem. dell'Acc. delle Scienze di Bologna*, 12, 305–436. In *Opere*, I, 313–466. Milano: Hoepli.
12. Cremona, L. 1863. Sulle trasformazioni geometriche delle figure piane. *Mem. dell'Acc. delle Scienze di Bologna*, 2, 1863, 621–631. In *Opere*, II, 56–61. Milano: Hoepli.
13. Cremona, L. 1865a. Sulle trasformazioni geometriche delle figure piane, Nota II. *Mem. della R. Acc. Nazionale dei Lincei* 5: 3–35.
14. Cremona, L. 1865b. Sur l'hypocycloïde à trois rebroussements. *J. für Math.* 64: 101–123.
15. Cremona, L. 1867. Preliminari di una teoria geometrica delle superfici. *Mem. dell'Acc. delle Scienze di Bologna*, 6, 91–136, 7, 29–78. In *Opere*, II, 279–387. Milano: Hoepli.
16. Cremona, L. 1868. Mémoire de géométrie pure sur les surfaces du troisième ordre. *J. für Math.* 68, 1–133. In *Opere*, III, 1–121. Milano: Hoepli.
17. Cremona, L. 1870. Lettera in lode del Piani. *Mem. dell'Acc. delle Scienze di Bologna* (3) 1: 40–41.

18. Deserti, J. 2009. Odyssée dans le groupe de Cremona. *SMF-Gazette* 122: 31–44.
19. Dolgachev, I. 2005. Luigi Cremona and Cubic Surfaces. In Luigi Cremona (1830–1903). Convegno di studi matematici. *Istituto Lombardo, Incontri di studio*, 36.
20. Enea, M.R. (ed.) 2009. *Il carteggio Beltrami – Chelini (1863–1873)*. Milano: Mimesis.
21. Enea, M.R., Gatto, R. (eds.). 2009. *Le carte di Domenico Chelini dell'Archivio Generale delle Scuole Pie e la corrispondenza Chelini – Cremona (1863–1878)*. Milano: Mimesis.
22. Galuzzi, M. 1980. Geometria algebrica e logica fra Otto e Novecento. In *Storia d'Italia. Annali* 3, 1001–1105. Torino: Einaudi.
23. Gardner, J.H., Wilson, R.J. 1993. Thomas Archer Hirst – Mathematician Xtravagant. IV. Queenwood, France and Italy. *American Mathematical Monthly* 100: 723–915.
24. Gatto, R. 1996. Lettere di Cremona a Betti. In Menghini, M. (ed.) Per l'Archivio della corrispondenza dei Matematici italiani. La corrispondenza di Luigi Cremona (1830–1903), vol. III, 7–90. Palermo: Quaderni P.RI.ST.EM - Università Bocconi.
25. Ghersi, I. 1988. *Matematica dilettevole e curiosa*. Milano: Hoepli.
26. Giacardi, L. 1994. *Lettere di Eugène Prouhet a Luigi Cremona*. In Menghini, M. (ed.) La corrispondenza di Luigi Cremona (1830–1903), II, 37–47. Roma: Quaderni della Rivista di Storia della Scienza, 3.
27. Gotti, V. 1887. Francesco Magni: Note biografiche. Pavia: Stab. Tip. Succ. Bizzoni.
28. Hudson, H.P. 1927. *Cremona transformations in plane and space*. Cambridge: Cambridge University Press.
29. Kleiman, S. 1998. Bertini and his two fundamental theorems. *Studies in the Hist. of Modern Math., III., Rend. del Circolo Mat. di Palermo*, II, Supplemento, 55, 9–37
30. Loria, G. 1904. Luigi Cremona et son oeuvre mathématique. *Bibliotheca Math.* 3, 5: 125–195.
31. Menghini, M. 1986. Notes on the correspondence between Luigi Cremona and Max Noether. *Historia Mathematica* 13: 341–351.
32. Millán Gasca, A. 2011. Mathematicians and the nation in the second half of century as reflected in the Luigi Cremona correspondence. *Science in Context* 24(1): 43–72.
33. Millán Gasca, A., Nastasi, P. 1994. *Lettere di Olry Terquem a Luigi Cremona*. In Menghini, M. (ed.) La corrispondenza di Luigi Cremona (1830–1903), II, 49–51. Roma: Quaderni della Rivista di Storia della Scienza, 3.
34. Nurzia, L. 1992. *Lettere di Arthur Cayley a Luigi Cremona*. In Millán Gasca, A. (ed.) La corrispondenza di Luigi Cremona (1830–1903), I, 107–128. Roma: Quaderni della Rivista di Storia della Scienza, 1.
35. Nurzia, L. (ed.). 1999. Per l'Archivio della corrispondenza dei Matematici italiani. La corrispondenza di Luigi Cremona (1830–1903), IV. Palermo: Quaderni P.RI.ST.EM. – Università Bocconi.
36. Palladino, N., Mercurio, A.M., Palladino, F. (eds.). 2009. Le corrispondenze epistolari Brioschi-Cremona e Betti-Genocchi. Firenze: Olschki.
37. Schiaparelli, G. 1862. *Sulla trasformazione geometrica delle figure ed in particolare sulla trasformazione iperbolica*. Torino: Stamperie Reali.
38. Schiaparelli, G. 1864. Sulla trasformazione geometrica delle figure ed in particolare sulla trasformazione iperbolica. *Mem. dell'Acc. delle Sc. di Torino* 21: 227–319.
39. Stendhal. 1939. *Life of Henry Brulard*, Catherine Alison Phillips. trans, 249, 256 and 241. New York: A. A. Knopf.
40. Veronese, G. 1903. Commemorazione del socio Luigi Cremona. *Rend. R. Acc. Lincei* (5), 12: 664–678.
41. Weil, A. 1946. *Introduction to Foundation of Algebraic Geometry*. Providence: American Mathematical Society.

Federigo Enriques: The First Years in Bologna

Ciro Ciliberto and Paola Gario

Abstract In this paper, we concentrate on Enriques' first years in Bologna, from 1894 to the turn of the century. More specifically, we consider Enriques' activity relating to his teaching duties in Bologna during those years, showing how the preparation of his Courses stimulated his interest in mathematical and philosophical problems related to the foundations of the discipline.

> *Those who, by examining the intrinsic reason for the development of thought, will be able to rise from history to the knowledge of the laws that rule the body of mathematics, will also be able to draw the most fruitful results from their work. By contrast, the efforts and attempts of those who mould their research on their individual whim rather than on the awareness of cooperating with the natural evolution of thought will remain unfruitful and vain.* (Federigo Enriques, *Conferenze di geometria: fondamenti di una geometria iperspaziale*, Bologna 1895.)

1 Introduction

In this paper, we concentrate on Enriques' first years in Bologna, from 1894 to the turn of the century. More specifically, we consider Enriques' activity relating to his teaching duties in Bologna during those years. In the first section, Sect. 2, we describe the way in which Enriques ended up in Bologna. Then, in Sect. 3 we

---

C. Ciliberto
Dipartimento di Matematica, Università di Roma "Tor Vergata", Via della Ricerca Scientifica, 00133 Roma, Italy
e-mail: cilibert@axp.mat.uniroma2.it

P. Gario (✉)
Dipartimento di Matematica "Federigo Enriques", Università di Milano, Via Cesare Saldini 50, 20133 Milan, Italy
e-mail: paola.gario@unimi.it

move to Enriques' first lectures in projective geometry, showing how the preparation of this course stimulated his interest in mathematical and philosophical problems related to the foundations of the discipline. In Sect. 4, we analyze in some detail the notes for the Lectures in Higher Geometry given by Enriques in the academic year 1894–1895. We show how they reflect Enriques' mathematical knowledge and the basis of his scientific philosophy, which would be developed in the future years and is described in the *Problems of Science* published in 1906.

Finally, in Sect. 5 we move on to the Course on Higher Geometry, devoted to algebraic geometry, which Enriques taught in 1897–1898. We point out how it has to be considered of topical interest for the time, since it touches most of the main developments in algebraic geometry of the previous half century, intertwined with some of Enriques' other research interests at that time and it contains, at the embryonic stage, didactic projects that Enriques would execute years later.

The research carried out by Enriques' during those years in algebraic geometry, and more specifically on algebraic surfaces, is not dealt with systematically but mentioned only in relation to the other topics addressed. The reason is that the experts are already well acquainted with this aspect of Enriques' work and it has already been dealt with by several authors including the writers of the present study.[1]

2 The Route to Bologna

On January 20, 1984, Guido Castelnuovo (1865–1952) warmly welcomed the young Federigo Enriques (1871–1946) at the end of his first lecture at the University of Bologna:

> *Your warm wishes, received just at the end of my lecture, gave me great pleasure, and I am very grateful to you for them. Today I gave my first lecture on Projective Geometry and, soon after lunch, the second (again on Projective Geometry). [...] I will start the course in Descriptive Geometry after the Carnival vacation. I must confess I am pleased with my debut. Today about 70 young people attended my lecture, many of them from the 2nd year, and they welcomed me with a burst of applause. After restoring calm, I started, not without some hesitation, with the words I had prepared beforehand, since I would not have been able to find others in that moment. Soon I felt that I was in command of the quiet and attentive audience; the words came easily to my lips and, when I looked at the students, I perceived that they understood me, and I went on to explain what I was saying with comments and examples.*[2]

---

[1] See, for instance, [18, 19, 35, 52].

[2] [17, 20.1.1894, p. 71]; all translations of the Italian originals have been made by the Authors of the present paper. The episodes from Enriques' biography that we describe in this paragraph, with reference to certain documents, is dealt with in a similar way in the book [75]. We became aware of this only when our article was already in the final stages of revision (C-G).

In this way, with a one year appointment for the Course of Projective and Descriptive Geometry, Enriques started his academic career and his commitment to teaching, which was to be of such importance in shaping his scientific personality.

In the early fall of 1893, Bologna was not exactly the location of choice for the 22-year-old Enriques and the events that led to his appointment there were quite complicated. Indeed, Enriques' original intention was to remain for a further year in Rome, where he had held a post-graduate fellowship for the academic year 1892–1893.[3]

It was in that year that he met Castelnuovo, who became a solid scientific reference point for him as well as a valued friend. However, the answer to Enriques' application for the extension of his fellowship in Rome came late, even though Castelnuovo had received good news about it in October 1893. Meanwhile, the competition in Turin for the Chair in Projective and Descriptive Geometry had closed. Enriques had also applied, but not with the hope of being appointed, since other candidates were older and more experienced than he was.[4]

His intention was simply to have a first positive evaluation of the work he had been doing in the two years following his degree. Segre had spoken to Berzolari, the winner of the competition, proposing that he should take Enriques on "as an assistant to the Chair of Projective and Descriptive Geometry if he did not get the extension of his fellowship" and announcing that he "would delay making any proposals for a few more days, in order to wait for some news about this" [34, from Segre to Castelnuovo, 27.10.1893]. Contrary to all expectations, Enriques' application for the extension of his fellowship was turned down but due to some mistake, this information took too long to reach Turin, arriving when Segre, "taking into account that there seemed to be a strong possibility that Enriques would get the fellowship", had already proposed to the Minister, via the Rector, that "all 5 assistants of the previous year should be reconfirmed".[5]

Having lost the chance of both the assistantship in Turin and the extension of the fellowship in Rome, Enriques went to Pisa to talk to Eugenio Bertini (1846–1933)

---

[3]Enriques graduated from the Scuola Normale Superiore in Pisa under the supervision of Riccardo De Paolis (1854–1892) in 1891.

[4]The results of the competition, with grades on a scale from 1 to 50, were announced by Corrado Segre (1863–1924) to Castelnuovo: "Berzolari 43, Del Re and Pieri 41 (even), Ciani and Enriques 36 (even), Amodeo 33, Gribodo not eligible". [34, from Segre to Castelnuovo, 16.10.1993]; Federico Amodeo (1859–1946), Luigi Berzolari (1863–1949), Edgardo Ciani (1864–1942), Alfonso Del Re (1859–1921), Mario Pieri (1860—1913), Giovanni Gribodo (1846-1924).

[5] [34, from Segre to Castelnuovo, 2.11.1893]. A few days later Segre added "Enriques was wrong not to apply for a position in a technical institute: it is very likely he would got a job there. He should not make a point of avoiding secondary schools. I rather think that he would profit from a period of training there. As for me, I do not feel guilty about what happened. If Berzolari had asked me to propose Enriques rather than Gerbaldi, I would have done so. On the contrary, he left it to me, saying he did not wish to break with tradition, to cause ill feeling, etc. I waited till October 31, the deadline given to me by the University; in the afternoon of that day, after much hesitation, I took my decision"[34, from Segre to Castelnuovo, 5.11.1893]. At that time an initial period spent teaching in secondary schools was quite common in the career of a university professor.

and to Luigi Bianchi (1856–1928) in order to try to "dig out something else" [17, 5.11.1894, p. 34], in particular the possibility of getting a new "Lavagna" postgraduate fellowship at the Scuola Normale Superiore, from which he had already benefited in 1891–1892 in the first year after his graduation. But Ulisse Dini (1845–1918) "was not in favour of this" [34, from Bertini to Castelnuovo, 8.11.1893], since such positions were intended for new graduates. "The only result of my trip to Pisa was that Bertini wrote to Pincherle[6] (as you suggest today); and Volterra[7] also wrote to him for the same reason" [17, 7.11.1894, p. 35].

This is what a rather disappointed Enriques wrote to his friend. It meant that the only possibility to "dig out" was that, with the Chair of Projective and Descriptive Geometry left vacant when Domenico Montesano (1863–1930) moved from Bologna to Naples, there could be "some changes of assistants in Bologna" [17, 5.11.1894, p. 34]. Enriques did not have to wait long for the answer, and he immediately informed Castelnuovo: "Prof. Pincherle has written to Volterra that there are no assistantships available in Bologna. No decision has yet been made about the Projective and Descriptive Geometry Course and he suggests that I should put my name forward, with the warning that it is a temporary position (only for the current year)" [17, 11.11.1893, p. 37]. The prospect of the appointment hinted at by Pincherle was received by Enriques with some trepidation, as can be seen from the sequel to the previous letter:

Dear Castelnuovo, Volterra's message has thrown me into confusion: a multitude of hopes, of fears, of hesitations: when aiming at something that is far away one sees only the good side, but when one is closer to it one often has a feeling of dismay [17, 11.11.1893, p. 38].

The responsibility of preparing two courses and the fear that his teaching duties would deprive him of "a large part of the time for scientific work" prompted him to say that perhaps he would have preferred if "the Faculty did nothing". Nevertheless, in a few days, thanks also to Castelnuovo's encouragement, Enriques developed "a strong desire for Pincherle's proposal to matrialize". There was great uncertainty about the appointment however: another possible candidate was Amilcare Razzaboni (1855–1920), who was already a lecturer (*libero docente*) at the University of Bologna and who enjoyed the "support of other members of the Faculty" [17, 14.11.1893, pp. 38–39]. There was also the possibility that the Faculty would favour a stable solution for cover the empty chair, proposing that the Minister should accept the nomination of Pieri, who had come second in the competition in Turin. This was what actually happened, and Enriques heard about it from Cesare Arzelà (1847–1912) and from Pincherle.[8] At the end of November the Faculty proposed, with 9 votes out of 11, that Pieri "should be appointed extraordinary professor of Projective and Descriptive Geometry, as the successor of prof. Montesano" [4, Adunanza del 18 novembre 1893]. At this point,

---

[6] Salvatore Pincherle (1853–1936).

[7] Vito Volterra (1860–1940).

[8] See, for details, Enriques' letter to Castelnuovo [17, 20.11.1893, pp. 39–41].

Enriques started to nurture the idea of "getting Pieri's position" in Turin, namely the assistantship at the University and the appointment as professor of Projective Geometry at the Military Academy. In order to "speed up the process", Enriques decided to go to Turin, partly with the aim of "getting the chance to see Segre" [17, post card 20.11.1893, p. 39]. He would have preferred this position "rather than the one of a higher level in Bologna" [17, 20.11.1893, p. 41]. Indeed, if Pieri had gone to Bologna and Del Re to Rome, in the event of a "competition in Descriptive Geometry", what chairs would be available in a near future? The two jobs in Turin would give him the possibility of "earning about 3,000 liras" without diverting him from "the pursuit of study". On the other hand it was necessary for him to concentrate "not only on his study" but also to get "a more stable position, for 8 or 10 years", should the "chances of getting a chair" dwindle to a mere "hope for the distant future".[9]

At the beginning of January, Pieri himself informed Enriques that the Minister had rejected the proposal of his nomination as a professor in Bologna and that the Faculty intended to declare the chair vacant and would to proceed to appoint a temporary professor for it. "I believe that they will look for me, since they have almost committed themselves, and I, for the time being, have not taken any steps",[10] Enriques wrote, while making himself available for the coming move to Bologna.

3 First Lectures in Bologna

As Castelnuovo observed several years later, Enriques' scientific personality was moulded through his "profession as a teacher",[11] which was characterized, from his first lectures, by a deeply critical look at the discipline he was dealing with.

Enriques, "especially for Projective Geometry, encountered difficulties at every step":

> Concerning the duality law in S_2 I encountered the difficulty arising from the fact that not all theorems in the plane (e.g. the one on homological triangles) follow from the fundamental properties of the plane.[12] Therefore I deduced the duality law in the plane by observing that

---

[9] He confided to his friend Castelnuovo: "I would certainly be very sorry to renounce the position that you are kindly offering me for next year as your assistant [aiuto] in Rome: you know that, in saying this, I am not trying to flatter you; it is the pleasure of your friendship and the scientific usefulness of the position that appeal to me" [17, 22.11.1893, p. 41].

[10] [17, 4.1.1894, p. 64]. The reason why the Minister rejected Pieri's proposal was that Pieri received, in the competition in which he was a candidate, "a small number of votes and his level was lower or equal to that of the other candidates" [4, Adunanza del 3 gennaio 1894]. The Faculty voted to open a competition and to appoint Razzaboni in the meantime, or, if he did not accept, Enriques. On January 13 the Minister Guido Baccelli (1832–1916) ratified Enriques' appointment.

[11] [32], also in [33, IV, 221].

[12] The remark that Desargues' theorem does not follow from the *position* axioms of plane projective geometry (which deal with points and lines and their intersections, and are common

one can cut a correlative star in S_3. [...] For harmonic groups (which will be the topic of my next lecture) I had to think out an exposition which treats all forms of the first species in a symmetrical way, and this is not usually done. But the greatest difficulties are still to come, namely the concepts of ordered groups on a line (or form) and that of continuity in view of the fundamental theorem of projectivities. For various reasons I do not like to rely on intuitive considerations of mouvement; it seems to me that, in doing this, the doubt remains that, for the conservation of the harmonic groups which is supposed to hold for the correspondence, the assumption of continuity of the correspondence might be added instead of the continuity of the line. [17, 29.1.1894, p. 73].

In order to "fully establish the theories which make it possible to reach the fundamental theorem of projectivities and, at the same time, also to prepare his lectures on descriptive geometry" Enriques used to work "until midnight and into the small hours", nurturing the project of assembling his research "in a little paper on the foundations of Projective Geometry, in which, unlike De Paolis and Pasch",[13] he would establish the theorem "without introducing the notion of equal segments and from a slightly different viewpoint".[14] Indeed, their demonstrations of the fundamental theorem relied on metric notions, "which it would be desirable to avoid if one cares about the purity of the approach", Enriques wrote in his paper,[15] which was published before the summer of 1894. Here, he attacked the foundational problems of projective geometry from a meditated critical point of view and with a strong sense of rigor. He observed that the postulates of *position geometry* did not seem to allow a synthetic treatment of the proof of the fundamental theorem

---

to projective and elementary geometry) is a deep one. It was made, apparently for the first time, together with the same observation for Pappus' theorem, by Hermann Wiener (1857–1939) in a talk at the first meeting of the German Mathematical Society in Halle in September 1891 [117]. Wiener's remark was taken up by David Hilbert (1862–1943), who attended the talk in Halle. While its statement involves only the planar concept, the usual proofs were based on spatial position axioms or on the theory of proportions. This question would be extensively investigated by Hilbert: see Hilbert's Lectures of the period 1891–1902 [53] and his *Grundlagen der Geometrie* [56]. Hilbert would give examples of *non–Desarguesian* and *non–Pappian* geometries. A quite simpler example of non–Desarguesian geometry was provided in 1902 by Forest Ray Moulton (1872–1952). As recognized by Hilbert, the validity of Desargues' theorem is a necessary and sufficient condition in order that can be introduced homogeneous coordinates in the plane with entries in a skew field and Pappus's theorem holds if and only if the skew field is commutative, i.e., it is a field. Later, deep investigations were carried out by Ruth Moufang (1905–1977) (a student of Max Dehn (1878–1952), who in turn had been a student of Hilbert), who discovered the existence of non–Desarguesian planes, in which a weaker version of Desargues' theorem holds, the *little Desargues' theorem*, and this translates in turn into the fact that homogeneous coordinates can be introduced in the plane with entries in an *alternative algebra*. A typical example of such a plane, which is not Desarguesian, is the projective plane over the *octonions* (for more details, see: [110, Sect. 6.8]).

[13] Moritz Pasch (1843–1930).

[14] [17, 19.2.1894, p. 76]. Enriques refers to [40] and to [82]. In Enriques the basic axiomatic approach moved from Pasch's *Vorlesungen*. According to Pasch the axioms correspond to empirical facts but geometry develops in an exclusively deductive way and is therefore independent of the empirical significance of the primitive propositions.

[15] [124] also in [159, vol. I, p. 142].

of projectivities, in particular the part concerning the uniqueness of the projective transformation which sends a given triple of distinct points of a projective line to another assigned triple of points of another line. In order to solve the problem he tried "to establish postulates, derived from the experimental intuition of the space, which appear to be the simplest ones in order to define the object of Projective Geometry".[16] Among them there is a postulate of continuity for the projective line, similar to Dedekind's postulate, expressing an "essential element of our spacial intuition" [129, p. 75], something that other postulates were unable to describe. In Enriques' view, its addition made it possible to overcome the aforementioned gap in the simplest way, with purely synthetic arguments.

Evidence of Enriques' intense activity on this subject can be seen in the lithographed notes of his courses,[17] which he completed the following year, with "a kind of historical introduction" [17, 16.11.1894, p. 149]. His book *Lezioni di geometria proiettiva*[18] was published in 1898.

Enriques' competence in dealing with foundational problems was probably developed through his contacts with the mathematical environment of Turin, which was characterized at that time by the presence of two masters, Segre and Giuseppe Peano(1858–1932) who were particularly sensitive to foundational questions.[19] Enriques had the opportunity to interact with that environment during his visits to Turin, where he used to spend time with Segre, discussing algebraic geometry and

---

[16][124] also in [159, vol. I, p. 142]. Enriques anticipates here one of the main themes of his *critical positivistic* thought and epistemological investigation, namely the concept that the basic principles of sciences are rational elaborations induced by experience and by sensorial perceptions. This would be expressed in [126] (which we will discuss in Sect. 4, see specifically Sect. 4.2) and later in [138] (also in [159, Vol. II, pp. 145–161]) and would be elaborated in an original form in the treatise [141].

[17] See in particular [123].

[18][129]. A second, enlarged edition of the book came out in 1904, and several editions were published over the years. It was translated into German by Teubner in 1903 (with a second edition in 1915), into French by Gauthier–Villars in 1930, into English in 1933 and into Spanish in 1943. Without going into details, it is interesting to mention here the evaluation given by the Committee for the competition for the Chair at the University of Bologna which Enriques won in 1896 (see foonote (32) below): "The lectures in projective geometry are very neat, clear and precise, and they give the right emphasis to the topics treated." By contrast, the evaluation of the descriptive geometry course [122] was unenthusiastic: "In the lectures in descriptive geometry, though author presents a rather extended picture, he does not treat some topics adequately, especially in the practical part" (from the "Relazione della Commissione esaminatrice del concorso alla cattedra di professore straordinario di geometria projettiva e descrittiva vacante presso la R. Università di Bologna", Bollettino Ufficiale del Ministero dell'Istruzione Pubblica, 19 ottobre 1896, 1817–I, 529–535. The quotations are from p. 532).

[19] Only a few years had passed since Peano published the "Arithmetices Principia" [83] and the "Principii di Geometria logicamente esposti" [84], inspired by Pasch's ideas about geometry as a deductive science. In the same year Pieri, at Segre's suggestion, published an Italian edition of K. G. C. Von Staudt's (1798–1867) "Geometrie der Lage" [109], a book which inspired all research in projective geometry from the synthetic point of view. Interesting testimonies about the mathematical environment of Turin can be found in the letters published in [81].

other matters. "Enriques is a pleasant young man. Every night he comes here to see me (and afterwards we meet up with D'Ovidio [...]) and we talk about various things",[20] Segre wrote to Castelnuovo in the fall of 1892, when Enriques spent a few weeks in Turin "in order to make the acquaintance of Segre", before starting his post-graduate fellowship in Roma.[21] As Enriques told Castelnuovo when he brought him the greetings from his friends and former colleagues,[22] in accordance with the local custom he used to meet "informally" with other young people at the Caffè Bergia [17, 9.11.1892, p. 4]. At that time the echoes still reverberated relating to the bitter conflict between Giuseppe Veronese (1854–1917) and Segre on one side and Peano on the other, on the subjects of the definition of higher dimensional spaces and of rigor in mathematics.[23]

Peano and Segre were surrounded by several young scholars who were involved in research on the foundations of mathematics.[24] In his paper, *Su alcuni indirizzi nelle investigazioni geometriche (On some trends in geometric investigations)*,[25] Segre presented his students with the problem of defining "a system of *independent* postulates able to characterize the n-dimensional space, making it possible to deduce from them the representation of its points with coordinates", that is, a system of postulates on which the synthetic theory of higher dimensional projective geometry could be based and from which the possibility of turning to an analytic treatment could also be deduced, as had classically been done for the three-dimensional case.[26]

---

[20] Enrico D'Ovidio (1843–1933); [34, from Segre to Castelnuovo, 16.11.1892].

[21] [17, 6.11.1892, p. 3]. Indeed, in the fellowship application for the academic year 1892–1893, Enriques indicated Turin as the place where he wanted to be, but the Supreme Council (Consiglio Superiore dell'Istruzione Pubblica) decide to send him to Rome: "While I regret not being able to spend next year with Prof. Segre, I am sure that you will give me excellent guidance and I trust that, like Prof. Segre, you will be so kind as to bestow upon me your valuable advice" [34, from Segre to Castelnuovo, 16.11.1892].

[22] Castelnuovo had been D'Ovidio's assistant in Turin from 1887 to 1890.

[23] Segre had a pragmatic attitude on the question of the definition of higher dimensional spaces. From his point of view, it would be relevant to give solid foundations to the synthetic theory of higher dimensional spaces that had played a significant role in his methods of research. In the *Fondamenti di geometria a più dimensioni* [115], Veronese explained his genetic conception of higher dimensional spaces that was harshly criticized by Peano. See: [85, 86, 103, 116].

[24] See [15] for more details.

[25] [102], also in [106, vol. IV, p. 407]. The conflict between Peano and Segre was triggered by Peano's critical reaction in [86] to Segre's positions on mathematical rigor in [102], published in "Rivista di Matematica", the journal created and directed by Peano.

[26] According to Segre, "it had been observed" that with the direct use of coordinates "one does not do true geometry, since the objects considered are analytic objects in their essence", so that "the projective geometry that one constructs in this way is nothing but the algebra of linear transformations". He added that this was intended to be "a distinction, not a reproach! Provided one does some Mathematics!" (see [102] or [106, IV, p. 407]). Indeed, in the previous pages of [102] he had been complaining about the "separation of pure mathematics in Analysis [which used to include algebra] and Geometry" which in Italy young people used to do "in such a distinct way that one may say that very few of them study and cultivate both on an equal basis". According to Segre, this revealed an anachronistic attitude, given the developments that geometry had been

Federico Amodeo (1859–1946) and Gino Fano (1871–1952) responded to Segre's invitation.[27] Shortly afterwards, Pieri also contributed to the topic with a series of papers published between 1894 and 1899.[28]

This is the context in which Enriques elaborated his reflections on the postulates of projective geometry. In Enriques' case, however, problems of a purely logical character, intrinsic to mathematical theory, were combined with more diverse interests such as didactic, philosophical and, more precisely, epistemological considerations, which distinguished his position from that of Amodeo and Fano.[29]

4 The Lectures on Higher Geometry

Some students asked me to give a course on Higher Geometry: I am not against the idea of partly fulfilling their wishes with a series of weekly lectures which however I would start only later (after January) [17, 23.11.1894, p. 151].

This is what Enriques wrote to Castelnuovo in November 1894. Thus, the warm welcome he had received at his "debut", was followed up the year after by the interest of the students in more advanced topics. The lectures on higher geometry were given as a free course. They "would be moulded on a general principle that completes that of Klein's Programm,[30] with the aim of including in this framework various kinds of topics (for instance the geometry on algebraic entities) which are not directly involved in it" [17, 23.11.1894, p. 151]. The first topics he would treat would be:

Higher dimensional spaces and their various interpretations fixing the nature of their elements. Applications to Veronese surface and to the space of lines, the question of linearity

---

going through in Germany thanks to Alfred Clebsch (1833–1872) and Felix Klein (1849–1925) and their schools: "a young person today who wants to cultivate Geometry separating it from Analysis [...] will never be a complete geometer" (Ibidem, p. 394).

[27][2] and [43]. Fano reported that Segre had already pointed out this question during the lectures of his Course on Higher Geometry in 1890–1891. Fano was a student of Segre in Turin, and he graduated in 1892. After graduating, Fano went to Göttingen for a period of study with Klein. Amodeo, who graduated at the University of Naples in 1883, was professor at a technical high school in Turin from December 1890, and therefore he could possibly have attended Segre's lectures.

[28][88–94]. Pieri's interest in foundations of mathematics was probably stimulated by his aforementioned edition of the von Staudt's *Geometrie der Lage*. For Pieri's contributions, see [73].

[29]Concerning this, see an abstract of Enriques' correspondence with Fano of September 1894 about the problem of the independence of postulates, reproduced in [125]. Fano pointed out to Enriques that it was possible to reduce the number of postulates. Enriques objected that this would "certainly be a simplification from a scientific viewpoint, which however would perhaps be inappropriate (as I say in my Note) if one also wants to take into account the practicalities of teaching, or if one wants to follow the route of experimental intuition" (Ibidem or [159, I, p. 161]).

[30]We will come back on Klein's influence on Enriques' later (see Sect. 4.2–4.5).

of systems of curves on algebraic surfaces, varieties and first order differential equations, the null system and Pfaff's equation. The methods of projections and its applications. [17, 15.12.1895, p. 164].

After working during the Christmas vacations to the preparation of the first lectures and their lithographed version, *Conferenze di geometria: fondamenti di una geometria iperspaziale (Lecture Notes)*,[31] Enriques realized that carrying out the whole plan he had in mind was "a utopia, because of the lack of time". He, therefore, wrote to Castelnuovo "perhaps (if I am still here) I will go on next year".[32]

---

[31] [126]. The original plan of the *Lecture Notes* was much broader: this is also shown by the lithographs whose first page shows the still legible indication "Chap. 1", which was later erased.

[32] [17, 11.1.1895, p. 164]. Enriques used to share news about his academic life with Castelnuovo, whom he would turn to for support, advice or help. The Bologna Faculty repeatedly asked in 1894–1895 to the Minister the opening of a competition for the Chair of Projective and Descriptive Geometry. The reply from the Minister Guido Baccelli (1832–1916) arrived only in November 1895: on the one hand he made the statement that "he felt sorry he could not accept the Faculty's proposal to open a competition", on the other he considered it "appropriate to point out to the Faculty that prof. A. Del Re", an extraordinary professor (professore straordinario) at the University of Modena, who had been awarded grades of 41 and 43 over 50, respectively, in competitions for chairs in Turin and Naples, "was aiming at that chair" and he added that if the Faculty decided to hire him in Bologna he would not have "any difficulty in agreeing" ([4, *Adunanza del 30 novembre 1895*]). The Faculty, "taking into account that the Minister definitely refuses to open the competition which has been requested so many times" unanimously resolved that Enriques' temporary job would be reconfirmed also for the current year and, given the fact that non only Del Re, but also Enriques and Pieri were aiming at that chair, requested that the Minister should either appoint a Committee which would decide among the three candidates, or "would leave to the Faculty the opportunity of enlightening his judgement by asking the opinion of particularly competent people". In a subsequent letter, the Minister reconfirmed Enriques' job starting from January 1, 1896: "thus they take two months away from me, but after what was in danger of happening, this is a minor injury", was Enriques' comment when he heard the news [17, 10.1.1896, p. 238]. The Minister also gave the Faculty the chance "to ask the opinion of competent people concerning the choice of the next professor" (see [4, *Adunanza del 7 gennaio 1896*]). The Faculty simply took note of the Minister's reply. The story ended only in 1896. As a consequence of the crisis following the Battle of Adwa, Emanuele Gianturco (1857–1907) succeeded Baccelli in March 1896. Enriques was afraid that the new Minister, who "was disinclined to open the competition for financial reasons", might not respect the obligation he had with the Bologna Faculty: "it would be expedient to let him know that this is an exceptional case, because for three years a competition has been repeatedly asked for, but it has not been accepted by Baccelli, who, meanwhile, has been filling up the Universities with people recommended for preferential treatment" (see [17, 11.3.1896, p. 251]). Gianturco decided to open a competition since, as he made clear to the Dean of the Bologna Faculty, "he did not intend to nominate anybody except by this method" (see [17, 18.3.1896, p. 253]). The Committee was formed by Bertini, Ferdinando Aschieri (1844–1907), Veronese, Montesano and Segre. All the candidates were declared to be eligible, with the following grades: Enriques (44/50), Pieri (42/50), Edgardo Ciani (1864-1942) and Fano (37/50), Amodeo (34/50).

4.1 Segre's Influence

The *Lecture Notes* [126] clearly show the influence of Segre's paper *On some trends in geometric investigations* [102] and Enriques himself acknowledges this.[33] This influence is evident from the purpose itself of the paper, which aimed to stimulate study and research among young people, as well as in the topics addressed, their enunciation and the general considerations about mathematics. Segre's words are echoed by Enriques in his *Lecture Notes*, where the latter agrees with the master's viewpoint about questions concerning purism in mathematics:

> We insisted on the identity between Geometry and Analysis[34] because the formation, in recent times, of pure trends in which one branch of mathematics is developed without involving concepts from other branches (such as the foundation of a pure Analysis with Weierstrass,[35] of a pure Geometry with Steiner, Chasles,[36] Staudt etc.) rise to a harmful, exclusive attitude leading to the temptation to systematically ban any analytic concept from Geometry and vice versa. The intention is not to diminish the importance of such purist trends: one has only to observe the improvement of the mathematical methods obtained in this way. In particular, the synthetic method, which has led to splendid results such as those in the theory of algebraic entities obtained in Italy starting with Cremona[37] would not have reached such a high level without the previous tendency to study Geometry independently from Analysis.[38] However it is not out of place to contrast the aforementioned progress with that determined by a suitable rapprochement between geometric and analytic concepts. And, since we have alluded to the theory of algebraic entities, we might also mention the results thus obtained in the same period by Clebsch, Noether,[39] etc. The tendency to form different pure mathematical trends on a minimal number of concepts, is consistent with a general law of evolution in science.[40] For this reason, in the last few years it has played

---

[33] "As regards the opinions expressed here, we refer to *Segre's* paper «Su alcuni indirizzi delle investigazioni geometriche»[102]; here we can find the exposition of broad views on geometric trends which may turn out to be useful to young people in their researches" (cfr. [126, Sect. 12, p. 45]).

[34] See footnote (26).

[35] Karl Theodor Wilhelm Weierstrass (1815–1897).

[36] Jacob Steiner (1796–1863), Michel Chasles (1793–1880).

[37] Luigi Cremona (1830–1903).

[38] About this, Segre pointed out in [102] (also [106, p. 398]) that the synthetic method "seems to be more penetrating and more enlightening", but the analytic one "in many cases is more powerful, more general or more rigorous" and often a topic "will not be elucidated in every respect unless both methods are used". The choice of the method is also a question of "individual tendencies", so for instance, "Steiner and Cremona would reach the same results with synthetic arguments as those reached by Sylvester [James Joseph Sylvester (1814–1897)] and Clebsch [Alfred Clebsch (1833–1872)] in an analytic way!".

[39] Max Noether (1844–1921).

[40] In [102] research in synthetic geometry had been positioned within the framework of the "purification" process of mathematical disciplines, which together with the "new relevant relations and applications established between them", characterized mathematics in the nineteenth century. This phenomenon was on a parallel with the general law ruling the evolution "of a whatsoever entity, of a living organism, as well of a society or a science" expressed by Spencer [Herbert Spencer (1820–1903)]. "As Spencer says in his general formulation of evolution (*Primi principii*

a determining and very important historical role; today, however, it seems that this role should be considered as pertaining largely to the past.[41] *At the present moment there is a lot to do in order to organize different trends in an organic whole, and, making the most of all the various methods we should, as a rule, direct all our efforts towards the result, which does not belong to only one branch of Mathematics but to Mathematics as a whole, if not to the progress of science in general [126, pp. 44–45]*

Similar words appear in the introduction to the treatise by Enriques and Oscar Chisini (1889–1967),[42] *Lezioni sulla teoria geometrica delle equazioni algebriche (Lectures on geometric theory of equations and algebraic functions)* published between 1915 and 1934.[43]

4.2 Philosophical Reflections on Foundations of Geometry

Segre's vision of the evolution of Mathematics was followed up in greater depth by Enriques, with philosophical arguments and mathematical examples. His treatment, enriched by historical investigations, anticipated some of the research themes which would characterize his more mature scientific personality, such as the *philosophical problem of space*.[44] Indeed, as he wrote to Castelnuovo, in his lectures he "wanted to start with reflections belonging to the philosophy of mathematics, and by developing the concept of abstract Geometry" [17, 13.1.1895, p. 164]:

> The object of Geometry is the study of relations pertaining to the concept of space, as this arises in our mind owing to our sensitivity, in other words as it stems from our intuition [126, p. 2].

The postulates "are deduced by the intuition and, as a whole, they also serve as definitions for the fundamental entities, establishing those mutual relationships which are needed to determine them for geometric purposes". Thus, "in its principles and in its development, Geometry is a subjective science", even though

---

[First Principles]), this induces a transformation of an entity from an undefined, incoherent homogeneity, to a defined coherent heterogeneity"(cfr. [102] or [106, p. 396]).

[41] Segre's words were stronger. "The so-called heroic period of synthetic geometry, in which one not only had to provide new results for science, but also, from Poncelet [Jean-Victor Poncelet (1788-1867)] to Steiner, from Chasles to Staudt, all had to fight to show the usefulness of the geometric method to the analysts who were not inclined to recognize it; that period has gone". Therefore, he added (in a footnote) "it seems that the right time has come for writing its history!" (see [102] or [106, vol. IV, p. 397]).

[42] Chisini obtained a degree in mathematics at the University of Bologna in 1912 under the supervision of Enriques who engaged him as assistant.

[43] "Nowadays, we can no longer presume to comprehend an entire subject at a glance [...] Our age has gone beyond the purism of the analytic and geometric schools, drawing from each the instruments of research; the conciliation of the methods that correspond to Clebsch's eclectic program have led to real and significant progress" [153, p. VIII].

[44] See footnote (16).

"definitions and demonstrations which add new entities to the fundamental ones and to the postulates, are merely logical operations".[45] The question of whether the intuitive space "corresponds to a natural order of things, that is, to a real space for which the postulates express real properties, is a question strictly related to the problem of knowledge, which transcends Geometry as intended in the aforementioned sense, and its development is independent of this question." It, therefore, makes no difference wehether the *problem of knowledge* "is solved in a skeptical (or idealistic) way, as in Kant's[46] theory, or whether one attributes an objective reality to space, as Helmholtz does in the treatise *Die Thatsachen in der Wahrnehmung* (Berlin 1878),[47] discussing the value of postulates as physical truths" [126, pp. 3–4]. From these discussions "nothing can arise that destroys the *logical* value of Geometry, but also nothing that modifies our intuitive conception of space", so that "the *possibility*, but also the *mathematical importance*, of Geometry itself" are independent from them. By contrast, the solution to these problems is not inconsequential from a physical and philosophical viewpoint. Therefore, "the mathematician, as a thinker" cannot neglect it, even though "his main interest is in a broader conception that he can draw from it, thus benefiting the development of mathematics." Exemplary, from this point of view, were the nineteenth century studies about the postulate of parallel lines, which found "their historical foundation in the hypothesis of a physical Geometry that was different from the one that had generally been considered true since Euclid's time" [124, p. 4].

Thus, *subjective geometry* refers to "the concept of intuitive space, given a priori with its postulates which for us are absolute and undeniable". There is, however, also the "possibility of founding various *more general Geometries* in which we disregard some postulate". These geometries arise from the analysis of the postulates according to three criteria: the "physical criterion", the "logical criterion" and the "physical–psychological criterion". This last, in particular, enables us to distinguish "in a complex concept, the various simpler concepts that form it" and the resulting analysis "also follows the physiological origin of intuition through senses". Therefore, when "in conceiving space we take into account only what refers to «sight», we end up by disregarding metric notions and by founding a *projective Geometry* in which we study only the graphic properties of figures". The analyses conducted according to the logical criterion and the physico-psychological criterion are of interest from a mathematical point of view. The logical criterion "is *necessary* for the foundation of a *mathematical* geometry", but "it is not *sufficient* for the mathematical or geometric *relevance* of the subject", that is, to avoid confining mathematics to a "mere syllogistic exercise" [126, pp. 5–6]. This is

---

[45] [126, p. 3]. In the paper on foundations of geometry Enriques had stated that "the experimental origin of geometry should not be forgotten in the search of the hypotheses on which it is based", but at the same time he added that he did not intend "by the way, to introduce, concerning those concepts, anything more than their logical relations". ([124] or [159, vol. I, p. 142]).

[46] Emmanuel Kant (1724–1804).

[47] Hermann von Helmholtz (1821–1894). Enriques refers to [54].

how, in the *Lecture Notes*, Enriques elaborated his view of the connection between the postulates of geometry and cognitive processes and his observations about the foundations of geometry, stating his position more precisely and subtly than in the foundational paper of the previous year.

In order to accomplish the *purification process* of position geometry from improper elements which still contaminated it, Enriques went through a delicate analysis of "the data of the graphic intuition of the line, leaving aside all elements of a metric origin" and from this analysis he derived a system of postulates which were satisfactory for a "logical purpose", as in the paper by Amodeo and Fano, but which, unlike the purpose of the systems described by these authors, had an "intuitive character" [126, p. 5]. The logical analysis of the postulates guided by geometric intuition led him to the analysis of their psychological origin based on sensorial perception, which contributes to the formation of our intuition. As stated above, these themes, which he starts to reflect upon during this first period in Bologna, would keep Enriques busy for many years to come and would be at the basis of the elaboration of his *scientific philosophy*.[48] His interest in this subject would stimulate his commitment to the diffusion of scientific culture in Italy.

In the spring of 1896, Enriques wrote to Castelnuovo saying that "while the mathematical questions are dozing waiting for a better time, for the past few days I have been dealing with a different subject which comes from mathematics only as a pretext: when you hear what it is you will be more horrified than surprised. I am talking about the «philosophical problem of space»" [17, 4.5.1896, p. 260]. And he went on to say that he was savouring "with the greatest pleasure" books about "psychology and logic, physiology and comparative psychology, the critique of knowledge, etc." in order to draw from them "the proof of the concept which is already beginning to germinate in my «Lecture Notes», namely, that the two great directions (metric-analytic and projective-metric) in the treatment of postulates have their foundations in the differentiations of the senses".[49]

---

[48] At the beginning of March, 1900, the Faculty of Sciences of the University of Bologna received the communication that the Superior Council of the Public Instruction had approved the programme of the free Course on "Scientific Philosophy" proposed by Enriques ([4, *Adunanza del 2 marzo 1900*]). On January 22, 1902, Enriques wrote to the Rector begging him "to ask to the Faculties of Philosophy and Letters, of Law and Medicine" that his free course would be "included in their teaching schedules, as a course of general knowledge for their students" ([3, pos. 4-a]). The weekly lectures on scientific philosophy were "attended by a great number of students", as he wrote in a letter to Giovanni Vailati (1863–1909) in June 1902 (quoted from Tina Tomasi in [96, p. 250]), in which he mentions the topics dealt with in the course. The programme will be submitted again to the Superior Council in May 1902 (see [4, Adunanza del 21 maggio 1902]).

[49] [17, 4.5.1896, pp. 260–261]. This passage confutes the myth that Enriques was a *mediocre lettore (bad reader)*, which arises from the following passage of Castelnuovo's obituary [32]: "Federigo Enriques was a bad reader. In the page in front of his eyes he would not see what was written, but what his mind would project onto it". And Carlo Felice Manara adds (see, *Il contributo di Enriques alla matematica italiana*, in [96, p. 39]): "the structure of his mind led him not to read much [...] as a friend rightly says, one could assert that he would not read books but «he would ask others to report on them», then he would draw to his own conclusions". Actually, the whole *Lecture Notes* and the quotations therein, show the opposite. Part of their interest lies in fact in the

Enriques' programme included "the question of the genesis of the concept of space relying on physiological psychology data (especially concerning sight and touch) of Helmholtz, Wundt[50] etc.", the latter considered by Enriques as "the most marvellous philosophical, physiological, psychological, mathematical, etc. mind", who was, in his opinion, little known in Italy where only Spencer's philosophy was "popular" [17, 4.5.1896, p. 261].

The dichotomy between optical and mechanical properties, investigated by the physiological experiments of Helmholtz, and the question posed by Klein in 1890, "in which sense it is psychologically correct to treat projective geometry before metric geometry (*Geometrie der Masses*), and even to consider the former as the basis for the latter?",[51] found a synthesis in Enriques' work. The two great categories into which geometric properties were divided, that is, graphic and metric properties, stem from "two forms of spatial intuition: *graphic intuition* and *metric intuition*" which, for "reasons coming from physiological psychology", connect "in psychogenesis to two different groups of sensations: the *visual sensations* on the one hand, the *tactile sensations and motion sensations* on the other."[52] Indeed, "when we have to verify the graphic properties of a *physical* picture, we rely (mainly) on the sight", whereas in order to "verify metric properties we rely (mainly) on measurement, hence on the touch" [123, pp. 3–4].

The axiomatic presentation of projective geometry given by Pasch was at the basis of the researches on foundations of geometry during the last decade of the nineteenth century. Pasch's *Vorlesungen* were a reference book for Enriques, as we already said, and also for Peano and Hilbert: all of them shared Pasch's view that postulates express the empirical experience.[53] Notwithstanding this, Hilbert set three systems of abstract "*objects*" as the starting point of his axiomatic approach.

---

list of quotations, which can also be considered a list of sources which contributed significantly to Enriques' education. Among them, it is perhaps useful to point out the recent developments in differential geometry thanks to the work of Eugenio Beltrami (1836–1900), Luigi Bianchi (1865–1928), etc., and the great work of Sophus Lie (1842–1899) and Klein on the group theoretical development of geometry and of differential equations. On this subject, see also [24].

[50]Wilhelm Maximilian Wundt (1832 –1920).

[51][65]. The quotation is from [14, p. 250]; see also [15] for Klein's position concerning the role of intuition in the genesis of postulates and his influence on Italian geometers. Klein's question was motivated by his studies at the beginning of the years 1870s according to which non–Euclidean geometries were included in the realm of projective geometry (see [62, 64]). In Klein's papers, using different interpretations of Cayley's *absolute* (Arthur Cayley (1821–1895)), one sees that both, Euclidean and non–Euclidean geometries may be considered as particular cases of the more general projective geometry.

[52]From the "Introduction" of the first edition of the *Lectures of Projective Geometry*, [123].

[53]This is clearly pointed out in Hilbert's lecture notes for his Courses on the foundations of geometry (published in [53]) and by the introduction to "*Grundlagen der Geometrie*", where one can read: "The choice of the axioms and the investigation of their relations to one another is a problem which, since the time of Euclid, has been discussed in numerous excellent memoirs. This problem is tantamount to the logical analysis of our intuition of space." Peano's views on this matter are stressed in [15].

It was with Hilbert that the axiomatic analysis in the modern sense began. Its focus is on the independence and on the coherence of an axiomatic system, aspects almost entirely overlooked by other authors.

Alessandro Padoa (1868–1937), a student of Peano, and Pieri were invited to speak at both International Congresses of Philosophy and Mathematics, held in Paris in 1900. Their lectures concerned foundational topics.[54] But after that date, the Italian mathematicians left the international scene,[55] with the exception of Enriques. His "reflection, elaborated during the decade between 1890 and 1900", had led him "to criticize some problems which are related to the logical and psychological development of scientific knowledge", we can read in the introduction of the *Problemi della Scienza (Problems of Science)*, where the psychological explanation of the postulates of geometry fills a whole chapter.[56] This book was translated into French (in two volumes) in 1907–1913, into German in 1910, into English in 1914, into Russian in 1911 and into Spanish in 1947.

4.3 *A Closer Look at the Contents of the* Lecture Notes

As we have already pointed out, the actual contents of the *Lecture Notes* are reduced compared with respect to Enriques' original plans.[57] One of the reasons for this, was the fact that the students were "less prepared" than Enriques had expected, so that, at the end of January, he had "not yet been able to talk about hyperspaces" [17, 31.1.1895, p. 171]. We give here an account of their contents.

After the first sections, devoted to the nature and meaning of research on foundations, the *Lecture Notes* introduce in Sect. 5 the concept of *"abstract Geometry"*. According to Enriques, the main steps of its evolution consisted in "proceeding by abstraction from some postulates concerning a series of concepts, thus founding a more general trend", in discovering that "such a new trend connects with the one coming from a different series of concepts", in verifying that "from the logical viewpoint the two trends do not differ at all", and in concluding that "one series of theorems translates to another by a simple change of terminology". Therefore "we may agree on considering the basic objects of geometry as abstract entities, related

---

[54] [79,80,95]. Padoa actually attended the congress, while Pieri sent a paper instead. Padoa is said to have been "the first to get all the ideas concerning defined and undefined concepts completely straight" (see [108]).

[55] It is not possible to enter into more details here. As interesting references, see: [15,16,47–49].

[56] [141, Chap. IV, part B]. The secondary literature on Enriques the philosopher is very wide. As interesting references on the general theory of scientific knowledge, see many contributions in the collected papers of the congress held in Livorno in 1981, in Livorno and Rome in 1996, and in Livorno and Paris in 2006, respectively [25,96,97]. For the role played by psychology in Enriques' thought, see [59].

[57] See footnote (31). It is useful to list here the index of the sections of Enriques' *Lecture Notes* [126]:

in a purely logical way and on considering as an *abstract Geometry* the science obtained in this way". Thus Enriques concluded that "the importance we attribute to abstract geometry should not (as one might believe) be opposed to the importance we attribute to intuition: it rather consists in the fact that abstract Geometry may be interpreted in an infinite number of ways as an (intuitive) concrete Geometry, once we fix the nature of its objects: so that, in this way, Geometry can take advantage in its development of an infinite number of different forms of intuition" [126, p. 9].

The next sections are devoted to the relationship between concrete and abstract geometry, considering various examples offered by mathematics at that time. One was the "abstract Geometry of developable surfaces", and more generally of surfaces with constant curvature, as in Beltrami's essay on the interpretation of non-Euclidean geometry ([126, Sect. 6] and [7]). As for the interpretation of "abstract projective Geometry as either the theory of linear systems of surfaces, or of curves, or of involutions of a line",[58] Enriques treats in some detail the geometry of circles in the plane[59] and of the corresponding transformations group, i.e. the group of "transformations by reciprocal vectors", which are also related to interpretations of non–Euclidean geometry. This gives him the opportunity to introduce the concept of groups of transformations, anticipating considerations he would later develop more thoroughly. The "Geometry of third order involutions on a line" (treated in Sect. 9) is related to *invariant theory*, i.e. the study of properties of homogeneous polynomials which are invariant under the action of the group of linear homogeneous invertible substitutions on the variables. This was a very popular subject in the second half of the nineteenth century, with contributes of mathematicians like Siegfried Heinrich Aronhold (1819–1884), Alfredo Capelli (1855–1910), Arthur Cayley (1821–1895),

---

Section 1. The importance of the analysis of foundations of Mathematics; Section 2. How we should intend Geometry; Section 3. How more general geometric trends are formed; Section 4. Compatibility and independence of the postulates of Geometry; Section 5. Abstract Geometry; Section 6. Interpretation of abstract plane Geometry on developable surfaces; Section 7. Interpretation of abstract Projective Geometry as the theory of linear systems of curves and surfaces or of involutions on a line; Section 8. Circle Geometry in the plane; Section 9. The Geometry of involutions of order 3 on a line; Section 10. Interpretation of abstract Projective Geometry as the theory of linear systems of projectivities on a line; Section 11. Hyperspaces; Section 12. Analytic interpretation of abstract Geometry; Section 13. Manifolds; Section 14. Metric determination of a manifold; Section 15. Geodetics on a manifold; Section 16. Angle of two curves on a manifold; Section 17. Manifolds contained in a Euclidean space; Section 18. A mechanical interpretation of metric Geometry of manifolds; Section 19. Historical news and comparisons; Section 20. Postulates of hyperspace Projective Geometry; Section 21. A few basic propositions of Projective Geometry in S_n; Section 22. The fundamental theorem of projectivities and projective coordinates in S_n; Section 23. Imaginary points and analytic manifolds; Section 24. Transformations; Section 25. Groups of transformations; Section 26. The various geometric trends characterized in relation to groups of transformations; Section 27. How two Geometries of S_n, whose groups are contained one inside the other, relate; Section 28. Geometry on a manifold; Section 29: Hesse's transportation principle and Klein's generalization; Section 30. The canonical representation of a Geometry on a manifold; Section 31. The group–theoretical problem of the foundations of Geometry.

[58] Introduced in Sect. 7.

[59] Originally introduced by Sophus Lie in [71], see Sect. 8.

Alfred Clebsch, Paul Albert Gordan (1837–1912), James Joseph Sylvester (1814 –1897).[60] The subject of Sect. 10 is the "interpretation of abstract projective Geometry as the theory of linear systems of projectivities of a line". Enriques shows that "the basic properties of space projective Geometry hold, when we replace «points», «lines», «planes», respectively with «projectivity (over a given line)», «pencils of projectivities», «nets of projectivities»": he suggests that the reader should "refer to the analytic definitions in order to make this immediately clear" and refers him to a paper by Segre [101] for a synthetic treatment. The interpretation of projective geometry of the ordinary space makes clear the advantage to be taken from the concept of "abstract Geometry" and also that there is no reason to "restrict abstract projective Geometry to the three-dimensional space". It was possible therefore "to develop the abstract projective Geometry as that of a space of dimension n (a hyperspace)". For $n > 3$, its interpretation had to be sought "no longer in the intuitive space, but in other varieties of elements, for instance in ∞^n linear systems of plane curves, or of algebraic surfaces of a given (very high) order, or in involutions of nth species (and order $> n$) on a line". Thus the geometric nature of the projective space of dimension n was provided "by a specific interpretation among the aforementioned ones". In this way, Enriques reclaims a geometric nature for the n–dimensional projective space, which is overshadowed by the analytic definition.[61] The introduction of hyperspaces is the object of Sect. 11, and Sect. 12 is devoted to the analytic treatment including the usual Euclidean metric in the affine n-space. It is at this point that Enriques remarks that *abstract Geometry can be identified with Analysis* and hence the ambient in which geometers end up working is the same as the one in which analysts work, developing this argument with the words already quoted above.

From the consideration of analytic geometry as an interpretation of abstract geometry, Enriques proceeds to higher dimensional manifolds, which occupies Sect. 13–18: in Sect. 13 he gives a lucid definition of a manifold of dimension n, warning the reader about pitfalls into which the study of set theory, recently developed by George Cantor (1845 –1918), prevents us from falling: varieties of different dimensions are bijective, and a continuous curve, a *Peano curve*, can even fill up a higher dimensional variety, though there is no bi-continuous map between varieties of different dimensions.[62] In Sect. 14–17, Enriques studies Riemannian metrics on manifolds, including geodesics, measure of angles, embeddings in Euclidean spaces. These pages, devoted to the theory of Bernhard Riemann (1826–1866) and to its impact on mathematics in the second half of the nineteenth

---

[60] Aronhold and Clebsch developed the famous *symbolic calculus*, mentioned by Enriques on p. 31.

[61] To the extent that Segre, as we saw in footnote (26), thought that with the use of coordinates one does not do true geometry.

[62] The so-called *theorem of the invariance of dimension*, by Luitzen Egbertus Brouwer (1881–1966) and Henri Lebesgue (1875–1941), asserting, in modern terms, that two topological spaces have the same dimension if, and only if, they are homeomorphic, had not been proved at the time. Enriques asserts it, but gives no reference.

century, are among the most intense of the lectures, really remarkable for clarity of exposition and richness of quotations. Especially interesting is Sect. 18, in which Enriques shows how Riemannian geometry "can be interpreted as the mechanic of homogeneous systems of material points with r degrees of freedom (related by links independent of time)", including in the framework of differential geometry the classical mechanical theories of Jean-Baptiste d'Alembert (1717–1783) and Joseph–Louis Lagrange (1736–1813).

In Sect. 19, Enriques dwells on drawing comparisons and describing relationships between the theories introduced in the previous sections, looking at them from a historical perspective. In six–seven pages, Enriques skilfully summarizes the main trends of geometric research in the second half of the nineteenth century, and defines relations between them. He also stresses the importance of history in the exposition of mathematical theories, another of the main themes of his more mature thought.[63]

In the last lines of Sect. 19, Enriques introduces his view on the question of the postulates of higher dimensional projective geometry and on basic results like the fundamental theorem of projectivities, and the introduction of imaginary points, treated in Sect. 20–23. We will not dwell on this here, since we have already commented at length on Enriques' general viewpoint in Sect. 4.1 above.

In the final part of the *Lecture Notes*, namely Sect. 24–31, Enriques comes back to the group–theoretic approach to geometry, which he anticipated in Sect. 8. In particular he introduces the reader to Lie's viewpoints.[64] Then he devotes Sect. 26 to analyzing various kinds of geometries in terms of their transformation groups, in

---

[63] We recall the following meaningful passage from the introduction of [153, p. IX]: "The rigid distinction normally made between science and history of science is founded on the concept of the latter as pure literary erudition. In this sense history brings to theory an extrinsic complement of chronological and bibliographical information. But the historical understanding of knowledge has a completely different meaning when it aims to discover what has been acquired from what we already have. This enables us to elucidate the journey of the idea, and to conceive this journey as it extends beyond each stage that is provisionally reached. This progression becomes an integral part of science and it has its place in the exposition of doctrines." For the contributions to history of mathematics, see the full list of Enriques's papers and the list of secondary bibliography quoted in the footnote (100).

[64] Lie's theory was expounded in the recent treatise [72], which Enriques was aware of because of his researches on the positive–finite–dimensional groups of Cremona transformations of the plane and the space (see [118, 119], and especially the joint paper with Fano [128] in which Lie's results are exploited). Enriques and Fano can be considered among the few contemporary Italian algebraic geometers who benefited from Lie's lesson, as witnessed by their contributions to the study of group theoretical questions in algebraic geometry, among them the seminal paper [128], which in fact belongs to this period. In it they classify the positive dimensional Lie subgroup of the Cremona group of the projective three-dimensional space. The two–dimensional case had been considered by Enriques in [118]. The modern treatment of this case by Michel Demazure in [39] is one of the motivations for the introduction of *toric varieties* and *toric geometry*, a flourishing subject in current algebraic geometry. For a modern treatment of the question addressed in [128], see [111–114]. The higher–dimensional case is still wide open. A very interesting question about the Cremona group is proposed by Enriques at the end of Sect. 25. It consists in asking whether the Cremona group of $\mathbb{P}^n$ is simple for $n \geq 2$. Enriques leans towards the affirmative answer, but very recently the problem has been answered negatively in [27] for $n = 2$.

accordance with Klein's viewpoint in the *Erlangen Program*.[65] In Sect. 27 Enriques discusses geometries in relation to the containment properties of their respective transformation groups. In Sect. 28–30 he makes general remarks about the meaning of "Geometry on a manifold" and in which way such a geometry can be realized.

Finally in Sect. 31 Enriques looks for relationships between the group-theoretic approach, the problem of foundations and the psycho-philosophical problem of the meaning of the postulates, a theme which underlies the whole of the *Lecture Notes*. He remarks that group theoretic results like the characterization of the group of motions of a Euclidean or non-Euclidean space in terms of local data "make it possible to characterize the Geometry of the space by means of a hypothesis referring to a limited region of it" and he observes that this "is indeed the philosophical concept which inspired the researches of Helmholtz, who aimed at setting the foundation of physical Geometry on a positive basis, and therefore pointed out the need to restrict the hypothesis to a region which is accessible through experience". Thus, Enriques closes his work with "a remark which shows how the most elevated mathematical questions, such as group theory, relate to the ones concerning foundations and to the most beautiful psychological problems", closing the circle of his ideas and coming back to the first principles from which he had been moving. Hence, "the consideration of the manner in which the various geometric trends ended up organizing themselves from the group–theoretic viewpoint is highly instructive" about the way pure mathematics follows "in its development well determined laws of evolution, though these are not so apparent at first sight as they are in the case of experimental sciences". From this Enriques draws the conclusion we choose as the epigraph of the present paper.

According to the assessment of the Committee for the competition for the chair in Bologna that Enriques won,[66] the *Lecture Notes*, by developing "the foundations of hyperspace geometry, considered in a wide sense, show that Enriques has a deep knowledge of the many theories treated there; but, more than a systematic

---

[65]Enriques was familiar with the *Erlangen Program* [63], which had been translated by Fano with suggestions from Segre, in whose opinion Klein's paper was not "sufficiently well known by the Italian geometers" (see Segre's introductory note in [66]). In the *Trends* Segre stressed the importance of the group theoretic approach, which "solved remarkable geometric problems, putting in a new light foundational questions in geometry". Also Enriques' final teaching examination thesis at the Scuola Normale Superiore of Pisa in 1893 (published in the paper [120]), devoted to the study of surfaces of the three-dimensional space with a positive dimensional group of projective transformations, is within the framework of group–theoretic investigations analogous to those of Klein–Lie. Plane curves with a positive dimensional group of projective transformations, the so-called *anharmonic* or *W–curves*, had been treated by Klein and Lie in the joint paper [70]. Lie treated the analogous case of curves in $\mathbb{P}^3$, in a paper in the Norwegian language, and explained his results in the third volume of his treatise [72] which appeared in the same year as Enriques' paper (for more information see [55]). Fano, too, devoted a number of papers [44–46] to the subject of varieties with a positive dimensional group of projective transformations.

[66]"Relazione della Commissione esaminatrice del concorso alla cattedra di professore straordinario di geometria projettiva e descrittiva vacante presso la R. Università di Bologna", Bollettino Ufficiale del Ministero dell'Istruzione Pubblica, 19 ottobre 1896, 1817–I, 529–535. See footnote (32).

course, they form an admirable collection of the main current concepts and results." Indeed, in not so many pages[67] we find a lucid concentration of the main trends in geometric research in the second half of the nineteenth century, expressed through particularly suitable concepts for a unitary presentation, and it is not by chance that the final pages deal with the group–theoretic viewpoint. Reading this paper is not easy, because of the terse style Enriques decided to adopt. It is however an extremely stimulating reading, in view of the connections outlined between various geometric theories.

4.4 From the Lectures to the Enzyklopädie

The *Lecture Notes* were one of the main vehicles for Enriques' collaboration to the *Enzyklopädie der mathematischen Wissenschaften*. The *Enzyklopädie* project started in 1896. The plan included various contributions of Italian mathematicians on subjects for which they were famous worldwide at the time. "I am very keen to collaborate with you for Meyer's[68] Encyclopedia, whether you want to write each article together, or whether you prefer to split the job between us". This is what Enriques wrote to Castelnuovo in June 1896, suggesting that his friend should accept the subject of algebraic plane curves, because "as for the Geometry on a curve, the subject «algebraic curves» is a unitary theme" [17, 9.6.1896, p. 272]. The plan of the work took definite shape during the same year, so that in February 1897 Enriques received "from Burkhardt[69] the sample articles for the encyclopedia and for the programme." However, there must have been some uncertainty if Enriques wrote to Castelnuovo asking "what are the articles we have to deal with? Algebraic surfaces and algebraic transformations?".[70] A few days later Enriques wrote to Burkhardt to find out who "had been asked to write the article on Foundations of Geometry" [17, 18.2.1897, p. 320]. When he knew that "Burkhardt himself would take charge" of writing the article, Enriques sent him the text of his *Lecture Notes*. Shortly afterwards, Enriques was offered, "with a very kind letter", the job of also writing the article on foundations of geometry: "as you may easily imagine, I am eager to accept this task", he wrote to Castelnuovo [17, 16.3.1897, p. 325]. Indeed, the proposal witnesses to Klein's appreciation for Enriques' *wide* viewpoint in treating the problem of foundations. The original deadline of little more than one year, "June 1898", given him by Burkhardt, was later postponed, and Enriques had time to go deeper into the topic, which appealed so much to his growing interest in the philosophy of science. The plan of the article became more definite when he

---

[67] The original version consists of 136 pages, excluding index and summary.
[68] Friedrich Wilhelm Franz Meyer (1856–1934).
[69] Heinrich Friedrich Burkhardt (1861–1914).
[70] [17, 4.2.1897, p. 314]. See the joint articles [144, 152]. The original plan was modified as the various volumes came out.

had the possibility of looking at Klein's lectures on *Nichteuclidischen Geometrie*,[71] "which only now arrived at the library of the Education Faculty" (Scuola Superiore di Magistero) [17, no date, p. 364].

In the Encyclopedia article, Enriques follows the outline of his *Lecture Notes*. He includes non-Euclidean geometries in the context of foundational research in the second half of nineteenth century,[72] fully sharing Klein's viewpoint. Actually, he had the opportunity of talking "in detail" with Klein about the plan of his article on the occasion of Klein's visit to Bologna in the spring of 1899: "I was glad to see him satisfied, he took very detailed notes of what I was saying", he wrote to Castelnuovo, adding that the topic which "we discussed at greatest length was the one concerning psychological problems in mathematics".[73]

With the aim of going deeper, even from an historical viewpoint, into these matters, Enriques asked his assistant Roberto Bonola (1874–1911) to collect references on non-Euclidean geometries dealt with from various viewpoints, i.e. the elementary one, the group–theoretic–metric–differential one of Riemann–Helmholtz, the projective one of Cayley–Klein.[74] The Encyclopedia article gave Enriques the opportunity to reconsider and go deeper into the subjects treated in the 1894–1895 free course on higher geometry. In particular, the re-reading of Riemann's 1854 memoir [99] "suggested to him a research on the hypotheses necessary to introduce coordinates on a variety" [17, 19.4.1898, p. 366], [130]. Some months later he "encountered the question of whether there is a surface of the ordinary S_3 on which the geometry of the complete hyperbolic or incomplete elliptic plane holds", a question arising from Beltrami's work. He talked about this with Bianchi, who "pointed out to him all the difficulties of the (unsolved) problem, relating to the limits of integrals of partial differential equations".[75]

---

[71][67]. These are notes of a course given in the winter of 1892.

[72]In [17, no date, p. 364], Enriques outlines the contents of the article: "(1) Principles of the ordinary metric Geometry, from Euclid to Legendre [Adrien-Marie Legendre (1752–1833)], Leibnitz [Gottfried Wilhelm von Leibniz (1646–1716)], Bolyai [János Bolyai (1802–1860)], Lobatschewsky [Nikolaj Ivanovič Lobačevskij (1792–1856)], Veronese. (2) Principles of projective Geometry and introduction of the metric. (3) The problem of foundations according to Riemann (ds^2). (4) The group–theoretic problem of Helmholtz–Lie."

[73]The full harmony of views between Enriques and Klein would be stressed by a toast when Enriques visited Göttingen in 1903: "Klein moved me with a toast in a poetic form, in French, in which he tried to find all contact points between him and me" (see [17, 24.10.1903, p. 536]).

[74][11–13]. Bonola got his degree in 1898 with Enriques in Bologna, and he was an assistant of Enriques before becoming a high school teacher. Just before he passed away prematurely, he was nominated professor of Mathematics at the High School of Education (Scuola Superiore di Magistero) in Rome.

[75][17, 28.1.1899, p. 398]. The fact that it is impossible to represent the entire hyperbolic plane on a surface with constant negative curvature was proved by Hilbert in 1901 (see [57]).

4.5 *Towards the* Questions About Elementary Mathematics

By the end of the century Enriques' research on foundations of geometry combined with his interest in elementary mathematics and in problems related to elementary and high school teaching, according to a cultural project which, once again, was strongly inspired by Klein's viewpoint: "I will tell you about a project that I hope to realize with a little effort. Namely, a book devoted to all questions concerning elementary geometry (among which also those problems, treated by Klein, which are not of the 2nd degree), but there are really so many such questions" [17, no date, 1899, p. 419], Enriques wrote to Castelnuovo in the late spring of 1899, stressing that his intention was "not to do it, but to let it be done by young graduates and high school teachers", saving for himself "the possibility of treating some more delicate topic" [17, no date, 1899, p. 419], [134, 135]. This is the way in which the project of the *Questioni riguardanti la geometria elementare (Questions about elementary geometry)*[76] started.

The *Questions*, published in 1900, is a collection of various articles which echo some of the main themes followed by Klein in the 1894 summer course for German school teachers.[77] Castelnuovo contributed with an article [30] in which he included the *segment transporter* introduced by Hilbert in his *Grundlagen der Geometrie*, which in turn were presented in Italy in the same year by Enriques' review [136]. Other collaborators were Ugo Amaldi (1875–1957), Ettore Baroni (1866–1918), Bonola, Benedetto Calò (1869–1917), Alberto Conti (1873–1940), Ermenegildo Daniele (1875–1949), Amedeo Giacomini (b. 1873), Alfredo Guarducci (1871–1944), Giuseppe Vitali (1875–1932).[78]

---

[76][133]. The book was expanded compared with the first edition with the contribution of several authors, coming out in two more editions with the title *Questioni riguardanti le matematiche elementari (Questions about elementary mathematics)* in 1912–1914 and 1924–1927. For Enriques' contributions see [147–150, 154, 155].

[77]Cfr. [68]; the Italian translation [69] has a preface by Gino Loria (1862–1954).

[78] Amaldi graduated in Bologna under Pincherle's supervision, then became Pincherle's assistant. In 1903, he obtained a position as extraordinary professor of Algebra and Geometry at the University of Cagliari. Then he moved to Modena from 1906 to 1918, to Padua from 1919 to 1923, and finally to Rome.

Baroni graduated in Pisa in 1889. First he was Dini's assistant, then he became a high school teacher.

Calò graduated in Pisa in 1893. For two years he was Volterra's assistant of Rational Mechanics in Turin, then he became a high school teacher.

Conti graduated in Pisa in 1895 and became a high school teacher. In 1900, he founded the *Bollettino di Matematiche e di Scienze fisiche e naturali*, a journal intended for elementary school teachers.

Danieli, after graduation in Turin in 1897, became Volterra's assistant. In 1913, he obtained the chair of Rational Mechanics at the University of Catania. Then he moved to Modena and finally to Pisa in 1925.

Giacomini graduated in Pisa in 1896. In the years 1896–1897, he held a "Lavagna" specialist fellowship at the University of Pisa. Then he became a high school teacher in Italy and in Italian schools abroad.

Enriques continued to be interested in problems related to secondary school teaching: the book *Elementi di geometria ad uso delle scuole secondarie superiori (Elements of Geometry, for higher secondary schools)*,[79] written in collaboration with Amaldi in 1903, was published in several versions and new editions continued to come out until very recent times.

5 The *Program* of the Course of Higher Geometry

The opportunity of giving a real course in higher geometry was offered to Enriques in the academic year 1897–1898. As we can see from the *Program* for the course, which Enriques wanted to publish [132] in the section detailing university course programmmes in the *Bollettino di bibliografia e di storia della matematica*, a journal founded by Gino Loria (1862–1954) in 1898.

The subject was completely different from that of the course outlined in the *Lecture Notes*. Indeed, in the 1894–1895 course on higher geometry the emphasis was on themes of general mathematical culture, with a stress on foundations, whereas the present one consisted in an introduction to the main subject of Enriques' mathematical research, namely algebraic geometry.

5.1 The Contents of the Course

The course starts with an introduction, the main contents of which are: generalities about algebraic plane and space curves, linear systems of plane curves, rational and birational transformations, curve singularities and their resolution, with the warning that "given the *direct* analytic definition of higher singularities and the reduction theorems, the projective difficulties arising from more complicated singularities have been systematically left aside".

The second part was devoted to geometry over a curve.[80] Enriques' aim is "an attempt to combine the two ways in which the theory can be developed" with the tendency "to *stress the algebraic content* of definitions and theorems. The

---

Guarducci was a fellow with Enriques at Pisa University. He became a high school teacher. He also devoted his attention to political and administrative matters, becoming the mayor of Prato.

Vitali attended the university in Bologna, starting in 1895–1896. On Enriques' suggestion he moved to Scuola Normale Superiore in Pisa, where he graduated in 1899 under Bianchi's supervision. After about twenty years, during which he taught at high school, in 1922 he obtained the chair of Mathematical Analysis at the University of Modena. He moved to Padua in 1925 and finally to Bologna in 1930.

[79][140]. As an interesting reference see [76].

[80]This central subject was started, soon after Riemann's contributions in [100], by Clebsch and Gordan in [37]. They produced the first algebro–geometric translation, in terms of algebraic plane

experience taught me that this contributes to *clarity*". The central topic here is the concept of *linear series*, which goes back to Brill and Noether. Enriques says that "the *algebraic definition* of linear series by means of rational functions on a curve, turns out to be more *enlightening* than any other, and leads in the *quickest* way to the main theorem on complete series".[81] He then introduces the canonical series in terms of branch points of a rational function, i.e. of a linear series of dimension one. This part ends with the proof of Riemann–Roch's theorem and with its application to the study of curves of low genera and of hyperelliptic curves.

The third and last part of the course was devoted to the birational geometry of plane curves, or, in other terms, to rational surfaces. Enriques starts with plane Cremona transformations, and with the proof that any such transformation is a composition of linear and quadratic transformations.[82] He then introduces the concept of *adjoint systems*[83] and using it, he attacks a series of questions which were crucial not only for plane geometry, but for the general theory of surfaces, in particular the classification up to Cremona transformations of positive dimensional linear systems of plane curves of genus 0 and 1 and of hyperelliptic curves of any genus. Then Enriques states Castelnuovo's theorem on the "rationality of

---

curves, of Riemann's theory of what would later be called *Riemann surfaces*. In the 1870s, Max Noether and Alexander Wilhelm von Brill (1842 –1935), using several geometric ideas introduced by Cremona, made a substantial and extremely successful effort towards a further algebro–geometrization of the theory (see [22, 23]). Brill–Noether's approach was later incorporated by the Italian school and developed in even more geometric terms by Bertini, Castelnuovo and Segre (see [10, 28, 104]), relying on projective geometry of hyperspaces. This gave a great impulse to projective geometry rather than to analytic methods. Later the validity of these methods was confirmed by the contributions of Castelnuovo and Enriques to the classification of surfaces. For an account on the subject, see [18].

[81] The italics in the passages quoted are by C-G. About these passages, two remarks are in order. First, the importance given by Enriques to an algebraic approach, which partly contradicts the myth that Italian algebraic geometers, and he in particular, would only rely on geometric intuition. They relied on algebra as well, but unfortunately their algebraic toolkit was not so large. Secondly, what interpretation should we give to Enriques reference to his *experience*? Rather than from his experience as a teacher, which was certainly modest at the time, Enriques' opinions may derive from a metacognitive analysis of his own learning process.

[82] This is the so-called *Noether–Castelnuovo theorem*. A proof had been given by Noether in [78], and at the time Enriques writes, it was believed to be true. Segre later pointed out in [105] the presence of a serious gap in Noether's proof. Segre's objection affected several results of various authors, including himself, that is, authors who had relied on Noether's argument: among these results was the classification up to Cremona transformations of linear systems of positive dimension of curves of low genera, treated by Enriques in his course. Castelnuovo fixed Segre's objection in [31]. This paper also contained a small gap, fixed by James Waddell Alexander (1888–1971) in [1]. For a modern account, see [26].

[83] The concept of adjoint systems, the use of *subsequent adjoint systems* and the *extinction of adjunction* on rational surfaces is at the core of the surface theory. According to a footnote in [29], also in [33, vol. II, p. 17], the concept was originally introduced by Brill and Noether in 1873 in [22], and later used by Seligmann Kantor (1857–1903) in [60] and by Castelnuovo himself in a series of important papers culminating with [29], in which he solves Lüroth's [Jacob Lüroth (1844–1910)] problem for surfaces, proving that any unirational surface is rational.

plane involution", namely that any unirational surface is rational, and studies some relevant examples, i.e. *Bertini's involutions* [8, 9], Kantor's classification of finite groups of Cremona transformations in the plane up to conjugation,[84] double planes and rational double planes.[85] Next, Enriques treats Noether's theorem[86] on the rationality of surfaces with a linear pencil of rational curves. Finally Enriques presents his recent results on positive dimensional continuous subgroups of the Cremona group of the plane and on surfaces with a positive dimensional group of projective transformations.[87]

5.2 A Few Comments

As we have seen, in the basic courses in projective and descriptive geometry and in the lectures on higher geometry held in 1894–1895, Enriques was very concerned about foundational problems. This concern seems to be absent in his course on advanced topics in algebraic geometry. This course is strongly oriented towards research problems, to the point that Enriques presents here recent results and open problems which would be solved only later. This dichotomy between foundational concern and entering the uncharted waters of open research problems equipped with still not fully structured tools is one of Enriques' characteristic features, which would also characterize his later advanced mathematical treatises.[88]

The fact that Enriques published his course programmme shows the importance he attributed to it, as a possible guide for young students, namely as a list of topics that those who wanted to approach research in algebraic geometry had to know.

It is clear that the *Program* was influenced by Castelnuovo's contributions. At the time Castelnuovo had still not had the opportunity to hold a course on higher geometry.[89] However, essentially, Castelnuovo had left his mark on all the topics

---

[84][60, 61]. This classification has been reconsidered in modern terms only quite recently, and there are several contributions to the subject, so many that we cannot mention all of them here. In [41], which contains all the appropriate references, the Authors complete the classification, though they warn the reader "The large amount of group–theoretical computations needed for the classification of finite subgroups of groups of automorphisms of conic bundles and Del Pezzo surfaces makes us expect some possible gaps in our classification. This seems to be a destiny of enormous classification problems. We hope that our hard work will be useful for the future faultless classification of conjugacy classes of finite subgroups of Cr(2)."

[85]The classification of rational double planes was treated by Castelnuovo and Enriques in the joint paper [137]. The paper contains a gap, which has been fixed only recently (see [6, 26]).

[86][77], which is Noether's Habilitation Thesis. Noether's theorem was later extended by Enriques to surfaces possessing a (not necessarily linear) pencil of rational curves, thus classifying ruled surfaces (see [131]).

[87]See [119, 120], and footnote (64).

[88][153, 156–158], the latter posthumously published.

[89] For Castelnuovo the opportunity presented itself only after Cremona's death (1903). Castelnuovo's notes of his lectures can be downloaded from [34].

touched on by Enriques in previous years and would continue to do so in the future. As we have seen, Segre's influence on Enriques' *Lecture Notes* was strong. Can it be seen in the *Program* too? At that time Segre was giving advanced courses on algebraic geometry in Turin.[90] Segre's course of 1890–1891, on the *Introduction to geometry on simply infinite algebraic entities* had covered topics that can be found in the second part of Enriques' course. These topics had been extensively developed in the two 1894 monographs by Segre and Bertini.[91] In the years between 1894 and 1897, Segre's courses were completely devoted to surface singularities. However, in the summer of 1894, Segre was planning that his next course in higher geometry would focus on algebraic surfaces, but he gave up the idea in the light of the progress being made on the theory relative to this by Castelnuovo and Enriques.[92] At their request, Segre joined in the research, working on the proof of the resolution theorem of the singularities of algebraic surfaces, an important research topic that Enriques mentioned in his course without entering into detail. The difficulties encountered by Segre in approaching the delicate question of singularities of algebraic surfaces were partly responsible for distancing him from the research of Castelnuovo and Enriques on classification, and it was perhaps his long commitment to this field of research that precluded the possibility of a closer collaboration. Also the project for the "great treatise on Higher Geometry" ([34, from Segre to Castelnuovo, 30.12.1896], [50]) which Segre had discussed with Castelnuovo in Turin, as a work that he would have liked to do "willingly together *later*" ([34, from Segre to Castelnuovo, 30.12.1896], [50]) and which had been reproposed to him by Enriques at the end of 1896, never came off. "As expected, Segre would be willing to take up the idea of the treatise again" [17, 30.12.1896, p. 304]. Indeed, Segre was enthusiastic about it because with the addition of "a new, energetic element", the project would become "much easier" to achieve. But he thought that perhaps it would be better "to postpone the undertaking for a few years because, whether we like it or not, the rules demand that he [Enriques] should make the required preparations for his promotion to the Chair". In any case the treatise was to be produced by the three of them and "no one else, because we three fully share the same ideas", Segre wrote ([34, from Segre to Castelnuovo, 30.12.1896], [50]) and, on taking his leave of Castelnuovo, he added that he would have liked to continue to speak about the project. Segre's worries about Enriques' career were not shared by the latter: "since I myself am not concerned about the doubts he expresses regarding the convenience of my participation [...], we can seriously think about transforming the idea into facts" [17, 30.12.1896, p. 304]. The course in higher geometry of 1897–1898 held by Enriques might be seen to be within the perspective of the project of the *great treatise* and the

---

[90] Segre's notes of his lectures can be found in [107].

[91] [10, 104]. The methodological comments in the *Program*, suggest that Enriques may have been inspired by Bertini's *algebro–geometric* approach rather than by the *hyperspatial geometric* approach of Segre.

[92] Fore more details, see [50, 51].

Program could therefore be an outline of the reflections shared with Castelnuovo and Segre in the preceding months.

Enriques course does not directly address the theory of surfaces, whose foundations at the time had been set by Enriques himself. Only a few years had elapsed since the "endless walks through the streets of Rome" during which the young Enriques, after having absorbed "in a short time the achievements of the Italian school on algebraic curves", used to submit his progress in the theory of surfaces to Castelnuovo's supervision every day. A topic which, as Castelnuovo remembers, he "boldly tackled" during his year in Rome, succeeding in founding "the theory of surfaces according to the trends of the Italian school".[93] The results of Enriques' first studies on this subject were collected in the memoir *Ricerche di geometria sulle superficie algebriche* [121] submitted for publication to the Academy of Sciences of Turin by Segre. In his presentation, though a few deficiencies were pointed out, it was stated that the paper "largely deserved to be encouraged" because of the importance of the subject, of the difficulties confronted with by the author and because "besides using the aforementioned geometric tools, in pursuing new results" it went "well beyond the few papers (mainly Noether's)" on the topic [42].

When Enriques, after the first few months in Bologna, during which he was fully absorbed in his theaching duties, went back to his research on surfaces, he noticed a mistake [17, 5.4.1894, p. 88] in the first chapter of his paper, which could have invalidate some results on which Castelnuovo was working. "In order to eliminate from that chapter a certain aura of indeterminacy which certainly does not contribute to its merits" [17, 7.4.1894, p. 88], Enriques started rewriting the chapter. As time went by, he realized that the situation was worse than he had thought and that the work, "like the tale of the long-lasting hardship" [17, 3.10.1894, p. 135] was never ending. "The way to surfaces, like the way to heaven, is strewn with thorns" he would write on October 20, 1894 [17, 20.10.1894, p. 141]. Indeed, the revision was very time-consuming and it ended with the publication of a new memoir [127], in which the whole presentation had been significantly revised.

Probably this material was too new to be presented to students, hence Enriques' decision not to mention it in the *Program*. However, even later, in the 1930s, Enriques would focus his course in *Institutions of High Geometry* in Rome on rational surfaces, rather than on the general theory of surfaces. His lectures were eventually collected in the book *Le superficie razionali (Rational surfaces)* [157], published in 1939 under the name of Fabio Conforto (1909–1954). The racial hatred laws issued by the fascist regime in 1938 forbade Enriques to appear as the author of the book. In the early published copies of the book, Conforto explicitly acknowledged that he had drawn upon the lectures that "Prof. F. Enriques had given over successive years",[94] but Enriques' name disappeared from later copies, replaced by a more generic reference to lectures given by the "Chair" of

---

[93] The quotations here are taken from [32].

[94] [157, p. VII]: see, for example, the copy n. 140, in the Library of the Department of Mathematics, University of Milan.

higher geometry. Enriques' point of view is however well explained in Conforto's foreword:

> The pursuit of such studies[95] is a fruitful exercise, which extends ideas, enhances research methods and cultivates the scholar's taste through the examination of interesting and special cases, making it possible to reach the deeper meaning of the most abstract doctrines. In particular, this is the best preparation for those who wish to confront and understand the genesis of the general theory of algebraic surfaces in relation to their birational transformations [157, p. VIII].

A viewpoint which can be traced back in the Course of Higher Geometry in Bologna of 1897–1898.[96] We can see in the *Program*, in a nutshell, the guiding lines of Enriques' commitment as a teacher of advanced courses in algebraic geometry. In particular, we can see a natural evolution from this course to the monumental treatise *Lectures on geometric theory of equations and algebraic functions* [153] and, as we said, to the *Rational Surfaces* [157].

Just under forty years had gone by between Cremona's famous *Inaugural Lecture* [38] for the first Course in Higher Geometry held at the University of Bologna in 1860 and the publication of Enriques' *Program*. Without entering in details, it is perhaps worthwhile to make a couple of comparative comments between Cremona's and Enriques' courses and their respective motivations.

Cremona's *Inaugural Lecture* is not only and not mainly a scientific manifesto, but a political statement. Cremona starts with bitter criticisms against the old regimes and the level of the universities before unification of Italy:

> In short, until then our university faculties had no Chair from which it was possible to announce the brilliant new scientific discoveries to young Italians. Everyone can see how unseemly it was that State education catered only in small part for the demands of the present conditions of civilization; however, the issue could not be resolved by a foreign government, or by a government under foreign domination, for whom public ignorance was a powerful instrument of sovereignty. Only the national government could carry out this task adequately [38, p. 25].

Indeed, Cremona, who had personally participated in the Italian independence wars, was a strong supporter of the new national State. This State needed a new ruling class, formed by men of culture, in particular scientists, who would raise the poor cultural and social level of the country. It was in this perspective that, in 1860, a number of brilliant young researchers, among them Cremona, had been nominated University professors directly by the Minister of Public Instruction. Cremona's lecture is first of all a *call* for brilliant young people who, having laid down their arms, were recalled to their studies:

---

[95] i.e., the study of rational surfaces.

[96] To support and make even more precise Enriques' point of view, it may be added that it has been remarked that solving the problem of birational classification of all linear systems of plane curves has a *complexity* comparable with the one of classifying all (regular) surfaces ([58]). In other words, by studying rational surfaces one meets all the main problems and technical difficulties of the study of general (regular) surfaces.

> *Young students, you who are preparing to follow this course in modern geometry, do not approach it unless you have a firm resolve to study with tenacity. Without unwavering commitment and constant effort it is impossible to master a science [...]. How fortunate you are, you young people, to witness the resurrection of your homeland in the best years of your life. Awake, arise, to contemplate the new sun that blazes on the horizon!*[97]

These issues had been largely overcome at Enriques' time and Italian mathematics had meanwhile reached a quality that was comparable with that of the most advanced European schools. There were other political issues however, even within the confines of education. For instance, the plague of illiteracy was still rife and a great part of the Italian population was unable to read and write. Nonetheless, unlike Cremona, Enriques showed no political concern. But the *Questions about elementary mathematics*, addressed to secondary school teachers and written together with a considerable number of young collaborators and the *Elements of Geometry, for higher secondary schools*, written with Amaldi, show that the young Enriques was not only interested in studies and advanced research. His interest in the civic progress of the country, as became clear in later years, manifested itself mainly in cultural initiatives. Unlike Cremona, he did not enter the political battlefield directly, but he acted as a reference point for the diffusion of scientific culture in Italy. The seeds of this attitude can be seen in his activity in the first years in Bologna.

Coming to the mathematical content of the courses,[98] for Cremona, higher geometry was projective geometry in the ordinary three dimensional space, and his course was fully dedicated to it. What for Cremona was the top of the mountain, for Enriques was the base, the prerequisite which is not even mentioned, either because it was the object of earlier, more elementary, courses, or because students were simply required to know it. The forty years separating Cremona and Enriques looked like an abyss, the route covered by algebraic geometry in this period was enormous, to the extent that it is even hard to identify it as the same discipline, and great progress was made in only a few years by Castelnuovo and Enriques himself.

6 Conclusions

The thesis which we would like to support with the present contribution is that the first years in Bologna were crucial for shaping Enriques' multiform intellectual personality. In the coming years, his personality matured but did not change and

---

[97][38, pp. 40]. For an account on Cremona's political personality, see [20, 21].

[98]From Cremona's paper we can deduce that the programme of his course included: Basic notions of projective geometry. Projectivities between lines. Cross ratio. Involutions. Fixed points of involutions. Applications to solutions of classical geometric problems, among them Desargues' and Pappus's theorems. Steiner's generations of conics. Pascal's theorem. Duality. Polarity in relation to a conic. Extensions to quadrics in the three-dimensional space. The rational normal cubic. Projectivities and correlations between planes and in the three-dimensional space. Involutions. Homologies. Affine transformations.

would be fully expressed in his contributions to the various fields in which he laid the foundations in the years we have been talking about here.

On the philosophical and epistemological front, the viewpoints in the *Lecture Notes* found a first more mature expression in the paper of 1901 [138] and a full exposition in the *Problems of Science* in 1906. We can also see the roots of Enriques' interest in the history of science, strongly intertwined for him with teaching problems. In the first Bologna period he directed his student Bonola to the history of non-Euclidean geometry whose impressive bibliographic research traced the plot of how it could be inserted in the grandiose story of the geometry of the second half of the nineteenth century. As we showed in Sect. 4.4, the *Lecture Notes* bridged Enriques to the German *Enzyklopädie*, thus strengthening his links with Klein. And, as we saw in Sect. 4.5, it is by the turn of the century that Enriques conceives the idea of the *Question* which marks his involvement in the problems of the secondary teaching which became more explicit in the first decade of the new century.

As for the algebro–geometric side, the early years in Bologna witnessed the improvement of surface theory foundations. Enriques and Castelnuovo, building on these foundations, started to explore the new world they created, classifying interesting classes of surfaces and proving the main results leading to the general classification, established in a paper of 1914 [151]. The collaboration with Castelnuovo on this subject, already extensively documented through their correspondence [17], culminated in the two fundamental joint papers of 1901 and 1906 [139, 142]. International recognition of the work of Castelnuovo and Enriques was validated by the invitation to the two authors to contribute with an appendix [143] to the book *Théorie des fonctions algébriques de deux variables* by Émile Picard (1856–1941) and Georges Simart (1846–1921), and with articles [144, 152] to the German *Enzyklopädie*. And it is just as we finish our story that a new protagonist, Francesco Severi (1879–1961) arrives on the scene. His collaborations[99] and contrasts [5, 36, 98] with Enriques significantly affected many of the developments in Italian algebraic geometry over the following decades.

We would like to conclude our contribution to the biography of Enriques with the words of David Mumford taken from a very recent article in which, with critical insight and attentive observation, he examines Enriques' *visionary* efforts to achieve a demonstration of the so-called "Fundamental Theorem of irregular surfaces"; the questions linked to this theorem were the source of a bitter dispute with Severi in the years 1920–1930.

In short, Enriques was a visionary. And, remarkably, his intuitions never seemed to fail him (unlike those of Severi, whose extrapolations of known theories were sometimes quite wrong). Mathematics needs such people- and perhaps, with string theory, we are again entering another age in which intuitions run ahead of precise theories [74, p. 260].

---

[99]The main result of which was the memoir [145, 146] awarded with the *Bordin Prize* of the French Academy of Sciences in 1906.

References

1. Alexander, J.W. 1916. On factorization of Cremona transformations. *Transactions of the American Mathematical Society* 17: 295–300.
2. Amodeo, F. 1891. Quali possono essere i postulati fondamentali della geometria proiettiva di un S_r. *Atti della R. Accademia delle Scienze di Torino* 26: 741–770.
3. *Fascicolo Federigo Enriques*, Archivio Storico, Università di Bologna.
4. *Verbali delle adunanze della Facoltà di Scienze MM FF NN*, Archivio Storico, Università di Bologna.
5. Babbitt, D., and J. Goodstein. 2011. Federigo Enriques's Quest to Prove the "Completeness Theorem". *Notices of the American Mathematical Society* 58: 240–249.
6. Bayle, L., and A. Beauville. 2000. Birational involutions of $\mathbb{P}^2$, Kodaira's issue. *The Asian Journal of Mathematics* 4: 11–17.
7. Beltrami, E. 1868. Saggio di interpetrazione della geometria non–euclidea. *Giornale di Matematiche* 4: 284–322.
8. Bertini, E. 1877. Ricerche sulle trasformazioni univoche involutorie nel piano. *Annali di Matematica pura ed applicata* (2) 8: 244–286.
9. Bertini, E. 1880. Sulle trasformazioni univoche piane e in particolare sulle involutorie. *Rendiconti del R. Istituto Lombardo di scienze. lettere ed arti* (2) 13: 443–451.
10. Bertini, E. 1894. La geometria delle serie lineari sopra una curva piana secondo il metodo algebrico. *Annali di Matematica pura ed applicata* (2) 22: 1–40.
11. Bonola, R. 1899. Bibliografia sui fondamenti della geometria in relazione alla geometria non euclidea. *Bollettino di bibliografia e storia delle scienze matematiche* 2: 1–10, 33–40, 81–88. Idem for lines: 878,887,891,896,906,911,920,935,971,986,1001, 1008,1011,1014,1016,1030,1041,1082,1094,1096.
12. Bonola, R. 1900. Bibliografia sui fondamenti della geometria in relazione alla geometria non euclidea. *Bollettino di bibliografia e storia delle scienze matematiche* 3: 2–3, 33–41, 70–73.
13. Bonola, R. 1902. Bibliografia sui fondamenti della geometria in relazione alla geometria non euclidea. *Bollettino di bibliografia e storia delle scienze matematiche* 5: 33–41, 65–71.
14. Bottazzini, U. 1994. *I principi della geometria e la filosofia scientifica*. In "Va' pensiero: immagini della matematica nell'Italia dell'ottocento", 241–277. Bologna: Il Mulino.
15. Bottazzini, U. 2001. I geometri italiani e i problemi dei fondamenti (1889–1999). Bollettino dell'Unione Matematica Italiana. *La matematica nella società e nella cultura* IV-A: 281–329.
16. Bottazzini, U. Hilbert e i fondamenti della geometria (1891–1902), "La ricerca logica in Italia. Studi in onore di Corrado Mangione" a cura di E. Ballo e C. Cellucci, Quaderni di Acme 124, Milano 2011, Cisalpino, p. 205–227.
17. Bottazzini, U., Conte, A., and P. Gario. 1996. *Riposte Armonie. Lettere di Federigo Enriques a Guido Castelnuovo*. Torino: Bollati Boringhieri
18. Brigaglia, A., and C. Ciliberto. 1995. *Italian algebraic geometry between the two world wars*. Queen's Papers in Pure and Applied Mathematics. Kingston
19. Brigaglia, A., and C. Ciliberto. 1996. *Enriques e la geometria algebrica italiana*. Lettera Matematica PRISTEM, Dossier19–20: 1–22.
20. Brigaglia, A., and S. Di Sieno. 2009. *L'opera politica di Luigi Cremona attraverso la sua corrispondenza*. La matematica nella società e nella cultura, ed. Unione Matematica Italiana (I) 2, 353–388.
21. Brigaglia, A., and S. Di Sieno. 2010. *L'opera politica di Luigi Cremona attraverso la sua corrispondenza*. La matematica nella società e nella cultura, ed. Unione Matematica Italiana (I)3, 137–179.
22. Brill, A., and M. Noether. 1873. *Ueber die Algebraische Functionen und ihre Anwendungen in der Geometrie*. Nachrichten von der K. Gesellschaft der Wissenschaften und der Georg–August–Universität aus dem Jahre 1873, 116–1323. Göttingen.
23. Brill, A., and Noether, M. 1874. Ueber die Algebraische Functionen und ihre Anwendungen in der Geometrie. *Mathematische Annnanlen* 7: 269–310.

24. Bussotti, P. 2006. *Un mediocre lettore. Le letture e le idee di Federigo Enriques*. Centro Studi Enriques – Quaderni, 5. La Spezia: Edizioni Agorà.
25. Bussotti, P. (ed.). 2008. *Federigo Enriques e la cultura europea*. La Spezia: Edizioni Agorà.
26. Calabri, A. 2002. On rational and ruled double planes. *Annali di Matematica pura ed applicata* (4) 181: 365–387.
27. Cantat, S., and S. Lamy. 2010. *Normal subgroups of the Cremona group*. Math. arXiv: 1007.0895v1
28. Castelnuovo, G. 1889. Ricerche di geometria sulle curve algebriche. *Atti della R. Accademia delle Scienze di Torino* 24: 346–373.
29. Castelnuovo, G. 1894. Sulla razionalità delle involuzioni piane. *Mathematishe Annalen* 44: 125–155.
30. Castelnuovo, G. 1900. Sulla risolubilità dei problemi geometrici cogli strumenti elementari: contributo della geometria analitica. In ed. Enriques, F. *Questioni riguardanti la Geometria elementare, raccolte e coordinate da F. Enriques*. Bologna: Zanichelli; expanded version in two volumes, *Questioni riguardanti le matematiche elementari*, 313–350, 1912–1914; edition in four volumes, 1924–1927.
31. Castelnuovo, G. 1901. Le trasformazioni cremoniane generatrici del gruppo cremoniano nel piano. *Atti della R. Accademia delle Scienze di Torino* 36: 861–874.
32. Castelnuovo, G. 1947. Commemorazione del socio Federigo Enriques. *Rendiconti della R. Accademia dei Lincei* 2: 3–21.
33. Castelnuovo, G. 2002. Opere Matematiche. Memorie e Note, Pubblicate a cura dell'Accademia Nazionale dei Lincei, I; II (2003), III (2004), IV (2007).
34. Fondo Guido Castelnuovo. Lettere e Quaderni delle Lezioni, a cura di P. Gario. Accademia Nazionale dei Lincei, www.lincei.it.
35. Ciliberto, C. 1989. *A few comments on some aspects of the mathematical work of F. Enriques*. In "Geometry and Complex Variables", Proceedings of an international meeting on the occasion of the IX centennial of the University of Bologna, ed. S. Coen. Lecture Notes in Pure and Applied Mathematics, 132: 89–109.
36. Ciliberto, C. 2004. *Enriques e Severi: collaborazioni e contrasti*. In *Enriques e Severi, Matematici a confronto nella cultura del novecento*, ed. O. Pompeo Faracovi, 29–49. La Spezia: Edizioni Agorà.
37. Clebsch, A., and P. Gordan. 1866. *Theorie des Abelschen Funktionen*. Leipzig: Teubner
38. Cremona, L. 1861. Prolusione al corso di geometria superiore letta nell'Università di Bologna nel novembre 1860. *Il Politecnico* 10: 22–42.
39. Demazure, M. 1970. Sous-groupes algébriques de rang maximum du groupe de Cremona. *Annales Scientifiques de l' École Normale Superieure* 4: 507–588.
40. De Paolis, R. 1881. Sui fondamenti della geometria proiettiva. *Memorie della R. Accademia dei Lincei* 9: 489–503.
41. Dolgachev, I., and V. Iskovskikh. Finite Subgroups of the Plane Cremona Group. *Progress in Mathematics* 269: 443–548.
42. D'Ovidio, E., and C. Segre. 1893. Relazione intorno alla Memoria intitolata "Ricerche di geometria sulle superficie algebriche" del Dott. F. Enriques, Seduta del 25 giugno 1893. *Rendiconti della R. Accademia delle Scienze di Torino* 28: 867–868.
43. Fano, G. 1892. Sui postulati fondamentali della geometria proiettiva in uno spazio lineare ad un numero qualunque di dimensioni. *Giornale di Matematiche* 30: 106–132.
44. Fano, G. 1895. Sulle superficie algebriche con infinite trasformazioni proiettive in se stesse. *Rendiconti della R. Accademia Nazionale dei Lincei* (5) 4: 149–156.
45. Fano, G. 1897. Un teorema sulle superficie algebriche con infinite trasformazioni proiettive in sé. *Rendiconti del Circolo Matematico di Palermo* 11: 241–246.
46. Fano, G. 1899. Un teorema sulle varietà algebriche a tre dimensioni con infinite trasformazioni proiettive in sé. *Rendiconti della R. Accademia dei Lincei* (5) 8: 362–365.
47. Freudenthal, H. 1957. Zur Geschichte der Grundlagen der Geometrie. Zugleich eine Besprechung der 8. Aufl. von Hilberts "Grundlagen der Geometrie". *Nieuw Archief voor Wiskunde* (3) 5: 105–142.

48. Freudenthal, H. Die Grundlagen der Geometrie um die Wende der 19. Jahrhunderts. *Mathematisch–Physikalische Senesterberichte N.F.* 7: 2–25.
49. Freudenthal, H. 1962. *The main trends in the foundations of geometry in the 19th century.* In "Logic Methodology and Phylosophy of Science", ed. E. Nagel e.a. Proceedings of 1960 International Congress, Stanford, 613–621.
50. Gario, P. 1991. Singolarità e geometria sopra una superficie nella corrispondenza di C. Segre a G. Castelnuovo. *Archive for History of Exact Sciences* 43: 145–188.
51. Gario, P. 1994. Singolarità e fondamenti della geoemtria sopra una superficie nelle lettere a Castelnuovo. *Rendiconti del Circolo matematico di Palermo, Supplemento* 36: 117–149.
52. Gario, P. 1997. Enriques e le comunità matematiche europee. In *Seminari di Geometria 1996–1997*, 279–306. Bologna: Dipartimento di Matematica, Università di Bologna.
53. Hallett, M., and U. Majer. 2004. *David Hilbert's Lectures on the Foundations of Geometry, 1891–1902*. Berlin: Springer
54. Helmholtz, H. 1879. *Die Thatsachen in der Wahrnehmung*. (Rede gehalten zur Stiftungsfeier der F–W Universität zur Berlin am 3. August 1878). Berlin: A. Hirschwald
55. Hawkins, Th. 1994. Lie Groups and Geometry: the Italian Connection. *Rendiconti del Circolo Matematico di Palermo* (2) Supplemento 36: 185–206.
56. Hilbert, D. 1899. *Grundlagen der Geometrie*. Leipzig: B. G. Teubner
57. Hilbert, D. 1901. Ueber Flächen von Konstanter Gausscher Krümmung. *Transactions of the American Mathematical Society* 2: 87–99.
58. Iitaka, S. 1999. Birational geometry of plane curves. *Tokyo Journal of Mathematics* 22(2): 289–321.
59. Israel, G. 1989. Federigo Enriques: a psychologistic approach for the working mathematician. In *Perspectives on Psychologism*, ed. M.A. Notturno, 426–454. Leiden: Brill.
60. Kantor, S. 1885. Sur une théorie des courbes et des surfaces admettant des correspondances univoques. *Comptes Rendus de l'Academie des Sciences* 100: 343–345.
61. Kantor, S. 1891. Premiers fondements pour une théorie des transformations périodiques univoques. *Atti della R. Accademia delle Scienze fisiche e matematiche di Napoli* (2) 4: 335.
62. Klein, F. 1871. Ueber die sogenannte Nicht–Euklidische Geometrie. *Mathematische Annalen* 4: 573–625.
63. Klein, F. 1872. Vergleichende Betrachtungen über neuere geometrische Forschungen. Erlangen: A. Deichert 1872 (reprinted in Mathematische Annalen 43: 63–100 (1893); italian transl. Klein, F. *Considerazioni comparative su ricerche geometriche recenti*, (transl. ed. G. Fano). Annali di Matematica pura ed applicata (2) 17: 307–343 (1889–1890)).
64. Klein, F. 1873. Ueber die sogenannte Nicht–Euklidische Geometrie, (Zweiter Aufsatz). *Mathematische Annalen* 6: 112–145.
65. Klein, F. 1890. Zur Nicht–Euklidischen Geometrie. *Mathematische Annalen* 37: 544–572.
66. Klein, F. (1889–1890) Considerazioni comparative su ricerche geometriche recenti, (transl. ed. G. Fano). *Annali di Matematica pura ed applicata* (2) 17: 307–343.
67. Klein, F. 1892. Vorlesungen über Nicht–Euklidische Geometrie, I–II, (lithogr.). Leipzig 1892.
68. Klein, F. 1895. Vorträge über ausgewählte Fragen der Elementargeometrie. Leipzig: Teubner
69. Klein, F. 1896. Conferenze sopra alcune questioni di geometria elementare, (transl. F. Giudice). Torino: Rosenberg & Sellier
70. Klein, F., and S. Lie. 1871. Ueber diejenigen ebenen Curven, welche durch ein geschlossenes System von einfach unendlich vielen vertauschbaren Transformationen in sich übergehen. *Mathematische Annalen* 4: 50–84.
71. Lie, S. 1872. Über Komplexe, inbesondere Linien–und Kugelkomplexe, mit anwendungen auf der Theorie der partieller Differentialgleichungen. *Mathematische Annalen* 5: 145–208, 209–256.
72. Lie, S. 1888–1893. *Theorie der Transformationsgruppen*, I–III. Leipzig: Teubner
73. Marchisotto, E.A., and J.T. Smith. 2007. *The Legacy of Mario Pieri in Geometry and Arithmetic*. Boston: Birkäuser
74. Mumford, D. 2011. Intuition and Rigor and Enriques's Quest. *Notices of the Mathematical Society* 58: 250–260.

75. Nastasi, T. 2010. *Federigo Enriques e la civetta di Atena*, Centro Studi Enriques – Collana Linee di Confine. Pisa: Edizioni Plus.
76. Nastasi, P., and E. Rogora. 2007. *Mon cher ami-Illustre professore. Corrispondenza di Ugo Amaldi (1897–1955)*. Roma: Nuova Cultura.
77. Noether, M. 1871. Ueber Flächen welche Scharen rationaler Kurven besitzen. *Mathematische Annalen* 3: 161–227.
78. Noether, M. 1883. Rationale Ausführung der Operationen in der Theorie der algebraischen Functionen. *Mathematische Annalen* 23: 311–358.
79. Padoa, A. 1901. *Essay d'une théorie algébrique des nombres entiers précédé d'une introduction logique à une théorie déductive quelconque*. In Biliothèque du Congrès International de Phylosophie, Paris, 1900, III. Logique et Histoire des Sciences, Coline, Paris, 309–365.
80. Padoa, A. 1902. *Un nouveau système des définitions pour la Géométrie Euclidienne*. C.R. du deuxième Congrés International des Mathématiciens (Paris), 6–12 aout 1900), 353–363. Paris: Gauthier–Villars.
81. Palladino, D., and N. Palladino. 2006. Dalla "Moderna Geometria" alla "Nuova Geometria italiana" viaggiando per Napoli, Torino e dintorni. *Lettere di Sannia, Segre, Peano, Castelnuovo, D'Ovidio, del Pezzo, Pascal ed altri a Federico Amodeo*. Firenze: Olschki.
82. Pasch, M. 1882. *Vorlesungen über neurere geometrie*. Leipzig: Teubner.
83. Peano, G. 1889. Arithmetices Principia nova methodo exposita, Aug. Taurinorum (= *Opere scelte, a cura dell'Unione Matematica Italiana* 2: 20–55).
84. Peano, G. 1889. I Principii di geometria logicamente esposti, Torino (= *Opere scelte, a cura dell'Unione Matematica Italiana* 2: 56–91).
85. Peano, G. 1891. Lettera aperta al prof. G. Veronese. *Rendiconti del Circolo matematico di Palermo* 1: 267–269.
86. Peano, G. 1891. Osservazioni ad un articolo di C. Segre. *Rivista di Matematica* 1: 66–69.
87. Picard, E., and G. Simart. 1906. *Théorie des fonctions algébriques de deux variables*, II. Paris: Gauthiers Villars.
88. Pieri, M. 1894–1895. Sui principii che reggono la geometria di posizione. *Atti della R. Accademia delle Scienze di Torino* 30: 607–641.
89. Pieri, M. 1895–1896. Sui principii che reggono la geometria di posizione. *Atti della R. Accademia delle Scienze di Torino* 31: 381–399, 457–470.
90. Pieri, M. 1896. Un sistema di postulati per la geometria projettiva degli iperspazii. *Rivista di Matematica* 6: 83–90.
91. Pieri, M. 1896–1897. Sugli enti primitivi della geometria projettiva astratta. *Atti della R. Accademia delle Scienze di Torino* 32: 343–351.
92. Pieri, M. 1897–1898. I principii della geometria di posizione composti in sistema logico deduttivo. *Memorie della R. Accademia delle Scienze di Torino* (2) 48: 1–62.
93. Pieri, M. 1898. Nuovo modo di svolgere deduttivamente la geometria projettiva. *Rendiconti del R. Istituto Lombardo di scienze, lettere ed arti* (2) 31: 780–798.
94. Pieri, M. 1898–1899. Della geometria elementare come sistema ipotetico deduttivo. *Memorie della R. Accademia delle Scienze di Torino* (2) 49: 173–222.
95. Pieri, M. 1900. *Sur la géométrie envisagée comme un système purement logique*, Bibliothèque du Congrès International de Philosophie, III, 367–404 (=*Opere sui fondamenti della matematica*, a cura dell'Unione Matematica Italiana, 235–272).
96. Pompeo Faracovi, O. (ed.). 1982. *Federigo Enriques: approssimazione e verità*. Livorno: Belforte.
97. Pompeo Faracovi, O., and F. Speranza (eds.). 1998. *Federigo Enriques: filosofia e storia del pensiero scientifico*. Livorno: Belforte.
98. Pompeo Faracovi, O. (ed.). 2004. *Enriques e Severi, Matematici a confronto nella cultura del novecento*. La Spezia: Agorà.
99. Riemann, B. 1867. Ueber die Hypothesen, welche der Geometrie zu Grunde liegen, Habilitationsschrift, Göttingen 1854. ed. R. Dedekind, Abhandlungen der Königlichen Gesellschaft der Wissenschaften zu Göttingen, 13.

100. Riemann, B. 1857. Theorie der Abel'schen functionen. *Journal für die reine und angewandte Mathematik* 54: 101–155.
101. Segre, C. 1886. Note sur les homographies binaires et leurs faisceaux. *Journal für die reine und angewandte Mathematik* 100: 317–330.
102. Segre, C. 1891. Su alcuni indirizzi nelle investigazioni geometriche. *Rivista di Matematica* 1: 42–66.
103. Segre, C. 1891. Una dichiarazione (risposta a G. Peano). *Rivista di Matematica* 1: 154–156.
104. Segre, C. 1894. Introduzione alla geometria sopra un ente algebrico semplicemente infinito. *Annali di Matematica pura ed applicata* (2) 22: 41–142.
105. Segre, C. 1900–1901. Un'osservazione relativa alla riducibilità delle trasformazioni Cremoniane e dei sistemi lineari di curve piane per mezzo di trasformazioni quadratiche. *Atti della R. Accademia delle Scienze di Torino* 36: 645–651.
106. Segre, C. 1957. *Opere, a cura dell'Unione Matematica Italiana*. Roma: Cremonese, I; II (1958); III (1961); IV (1963).
107. Segre, C. 2002. *I quaderni di Corrado Segre (cd-rom)* ed. L. Giacardi. Dipartimento di Matematica, Università di Torino.
108. Smith, J.T. 2000. *Methods of geometry*. London: Wiley.
109. von Staudt, G.K.C. 1889. *Geometria di posizione*, ed. and transl. M. Pieri. Torino: Bocca (orig. ed. *Geometrie der Lage*, Nuremberg 1847).
110. Stillwell, J. 2005. *The four pillars of geometry*. New York: Springer.
111. Umemura, H. 1980. Sur les sous-groupes algebriques primitifs de groupe de Cremona à trois variables. *Nagoya Mathematical Journal* 79: 47–67.
112. Umemura, H. 1982. Maximal algebraic subgroups of the Cremona group of three variables. *Nagoya Mathematical Journal* 87: 59–78.
113. Umemura, H. 1982. On the maximal connected algebraic subgroups of the Cremona group I. *Nagoya Mathematical Journal* 88: 213–246.
114. Umemura, H. 1985. On the maximal connected algebraic subgroups of the Cremona group II. In *Algebraic groups and related topics*. Advanced Studies in Pure Mathematics, 6, 349–436. North-Holland: Kino–Kuniya.
115. Veronese, G. 1891. *Fondamenti di geometria a più dimensioni e a più specie di unità rettilinee esposti in forma elementare*. Padova: Tipografia del seminario.
116. Veronese, G. 1892. A proposito di una lettera del prof. Peano. *Rendiconti del Circolo matematico di Palermo* 6: 42–47.
117. Wiener, H. 1892. Ueber Grundlagen und Aufbau der Geometrie. *Jahresberichte der Deutschen Mathematiker–Vereinigung* 1: 45–48.

References to Enriques' papers[100]

118. Enriques, F. 1893. Sui gruppi continui di trasformazioni cremoniane nel piano. *Rendiconti della R. Accademia dei Lincei* (5) 2: 468–473.
119. Enriques, F. 1893. Sopra un gruppo continuo di trasformazioni di Jonquiéres nel piano. *Rendiconti della R. Accademia dei Lincei* (5) 2: 532–538.
120. Enriques, F. 1893. Le superficie con infinite trasformazioni proiettive in se stesse. *Atti dell'Istituto Veneto* (7) 4: 1590–1635.
121. Enriques, F. 1893. Ricerche di geometria sulle superficie algebriche. *Memorie della R. Accademia delle Scienze di Torino* (2) 44: 171–232.

---

[100]Some of Enriques' papers quoted here can be downloaded from the web site HTTP: //ENRIQUES.MAT.UNIROMA2.IT/ edited by the Committee for the *National Edition of Federigo Enriques' Collected Papers*. We suggest referring to this site for the full list of Enriques' papers, related detailed citations and secondary bibliography list.

122. Enriques, F. 1894. *Lezioni di geometria descrittiva*, (lithogr.). Bologna.
123. Enriques, F. 1894. *Lezioni di geometria proiettiva*, (lithogr.). Bologna.
124. Enriques, F. 1894. Sui fondamenti della geometria proiettiva. *Rendiconti dell'Istituto Lombardo di scienze, lettere ed arti* (2) 27: 550–567.
125. Enriques, F. 1895. Corrispondenza sui postulati fondamentali della geometria proiettiva, Lettera al dr. G. Fano. *Rendiconti del Circolo Matematico di Palermo* (1) 9: 82–84.
126. Enriques, F. 1894–1895. Conferenze di geometria: fondamenti di una geometria iperspaziale, (lithogr.). Bologna.
127. Enriques, F. 1896. Introduzione alla geometria sopra le superficie algebriche. *Memorie di Matematica e Fisica della Società italiana delle Scienze, detta dei XL* (3) 10: 1–81.
128. Enriques, F., and G. Fano. 1897. Sui gruppi continui di trasformazioni cremoniane dello spazio. *Annali di Matematica pura ed applicata* (2) 26: 59–98.
129. Enriques, F. 1898. *Lezioni di geometria proiettiva*. Bologna: Zanichelli.
130. Enriques, F. 1898. Sulle ipotesi che permettono l'introduzione delle coordinate in una varietà a più dimensioni. *Rendiconti del Circolo Matematico di Palermo* 12: 222–239.
131. Enriques, F. 1898. Sopra le superficie che posseggono un fascio di curve razionali. *Rendiconti della R. Accademia dei Lincei* (5) 7: 344–347.
132. Enriques, F. 1899. Programma di un corso di geometria superiore per l'anno 1897–98. *Bollettino di bibliografia e storia delle scienze matematiche* 2: 76–78.
133. Enriques, F. (ed.). 1900. Questioni riguardanti la Geometria elementare, raccolte e coordinate da F. Enriques. Bologna: Zanichelli; expanded version in two volumes, *Questioni riguardanti le matematiche elementari*. Bologna: Zanichelli, 1912–1914; edition in four volumes, 1924–1927.
134. Enriques, F. 1900. *Sull'importanza scientifica e didattica delle questioni che si riferiscono ai principii della geometria*. In Questioni riguardanti la Geometria elementare, raccolte e coordinate da F. Enriques, 1–31. Bologna: Zanichelli.
135. Enriques, F. 1900. *Sulle equazioni algebriche risolubili per radicali quadratici e sulla costruttibilità dei poligoni regolari*. In Questioni riguardanti la Geometria elementare, raccolte e coordinate da F. Enriques, 353–396. Bologna: Zanichelli.
136. Enriques, F. 1900. D. Hilbert: Grundlagen der Geometrie, (Review). *Bollettino di bibliografia e storia delle scienze matematiche* 3: 3–7.
137. Castelnuovo, G., and F. Enriques. 1900. Sulle condizioni di razionalità dei piani doppi. *Rendiconti del Circolo Matematico di Palermo* 14: 290–302.
138. Enriques, F. 1901. Sulla spiegazione psicologica dei postulati della geometria. *Rivista filosofica, a. III* 4: 171–195.
139. Castelnuovo, G., and F. Enriques. 1901. Sopra alcune questioni fondamentali nella teoria delle superficie algebriche. *Annali di Matematica pura ed applicata* (3) 6: 165–225.
140. Enriques, F., and U. Amaldi. 1903. Elementi di geometria ad uso delle scuole secondarie superiori. Bologna: Zanichelli.
141. Enriques, F. 1906. Problemi della scienza. Bologna: Zanichelli.
142. Castelnuovo, G., and F. Enriques. 1906. Sur les intégrales simples de première espèce d'une surface ou d'une variété algébrique à plusieurs dimensions. *Annales Scientifiques de l'École Normale Supérieure* (3) 23: 337–366.
143. Castelnuovo, G., and F. Enriques. 1906. *Sur quelques résultats nouveaux dans la théorie des surfaces algébriques*. Appendix to [87].
144. Castelnuovo, G., and F. Enriques. 1908. Grundigenschaften der algebrischen Flächen. *Encyklopädie der mathematiscen Wissenschaften, III* 2(6): 635–768.
145. Enriques, F., and F. Severi. 1909. Mémoire sur les surfaces hyperelliptiques. *Acta Mathematica* 32: 283–392
146. Enriques, F., and F. Severi. 1910. Mémoire sur les surfaces hyperelliptiques. *Acta Mathematica* 33: 321–403.
147. Enriques, F. 1912. Sull'insegnamento della geometria razionale. In [Second ed., vol. I] *Questioni riguardanti le matematiche elementari, raccolte e coordinate da F. Enriques*, 19–35. Bologna: Zanichelli.

148. Enriques, F. 1912. I numeri reali. In [Second ed., vol. I] *Questioni riguardanti le matematiche elementari, raccolte e coordinate da F. Enriques*, 365–493. Bologna: Zanichelli.
149. Enriques, F. 1914. Alcune osservazioni generali sui problemi geometrici. In [Second ed., vol. II] *Questioni riguardanti le matematiche elementari, raccolte e coordinate da F. Enriques*, 337–357. Bologna: Zanichelli.
150. Enriques, F. 1914. Massimi e minimi nell'analisi moderna. In [Second ed., vol. II] *Questioni riguardanti la Geometria elementare, raccolte e coordinate da F. Enriques*, 641–798. Bologna: Zanichelli.
151. Enriques, F. 1914. Sulla classificazione delle superficie algebriche e particolarmente sulle superficie di genere lineare $p^{(1)} = 1$. *Rendiconti della R. Accademia dei Lincei* (5) 23: 206–214, 291–297.
152. Castelnuovo, G., and F. Enriques. 1914. Die algebraischen Flächen vom Gesichtpunkte der birationalen Transformationen aus, *Encyklopädie der mathematischen Wissenschaften, III* 2 1(C): 674–768.
153. Enriques, F., and O. Chisini. 1915. *Lezioni sulla teoria geometrica delle equazioni e delle funzioni algebriche*, I; II (1918); III (1924); IV (1934). Bologna: Zanichelli.
154. Enriques, F. 1924. L'evoluzione delle idee geometriche nel pensiero greco: punto, linea, superficie. In [Third Ed., vol. I] *Questioni riguardanti la Geometria elementare, raccolte e coordinate da F. Enriques*, 1–40. Bologna: Zanichelli.
155. Enriques, F. 1925. *Spazio e tempo davanti alla critica moderna*. In [Third Ed., vol. II] *Questioni riguardanti la Geometria elementare, raccolte e coordinate da F. Enriques*, 429–459. Bologna: Zanichelli.
156. Enriques, F., and L. Campedelli. (1934). Sulla classificazione delle superficie algebriche, particolarmente di genere zero. *Rendiconti del Seminario Matematico dell'Università di Roma* (3) 1: 7–190.
157. Enriques, F., and F. Conforto. 1939. *Le superficie razionali*. Bologna: Zanichelli.
158. Enriques, F. 1949. *Le superficie algebriche*. Bologna: Zanichelli.
159. Enriques, F. 1956. *Memorie Scelte di Geometria*, Pubblicate a cura dell'Accademia Nazionale dei Lincei, I (1956); II (1959); III (1966). Zanichelli, Bologna.

Enrico Bompiani: The Years in Bologna

Ciro Ciliberto and Emma Sallent Del Colombo

Abstract In this paper, we consider the first period of scientific and academic life of the Italian mathematician Enrico Bompiani (1889–1975), with special emphasis on the years he spent in Bologna. We put his contributions on projective differential geometry in perspective by illustrating some key features of this discipline, the main contributions of its Italian initiators, and Bompiani's work in the Bologna years.

1 Introduction

The present paper is devoted to the Italian mathematician Enrico Bompiani. He started as a student of Guido Castelnuovo (1865–1952), one of the main characters of the so-called Italian school of algebraic geometry founded by Luigi Cremona (1830–1903).[1] However he soon moved to differential geometry, mainly to the so-called *projective differential geometry* (PDG). Bompiani had a long and intense scientific and academic life. He wrote more than three hundred papers, on different subjects including PDG, Riemannian geometry, algebraic geometry, analysis and science popularization, in a period ranging from 1910 to 1974. PDG is an old field, which in the years of Bompiani's activity saw an enormous and rich production.

---

[1] For general information about the Italian school of algebraic geometry, see [5, 6].

C. Ciliberto (✉)
Dipartimento di Matematica, Università di Roma "Tor Vergata", Via della Ricerca Scientifica, 00133 Roma, Italy
e-mail: cilibert@mat.uniroma2.it

E. Sallent Del Colombo
Departament de Física Fonamental, Universitat de Barcelona, Martí i Franquès, 1, 08028 Barcelona, Spain
e-mail: emma.sallent@ub.edu

Therefore, it would have been impossible here to fully report on the complexity of Bompiani's character and of his discipline, even sticking only to the Italian scientific ambient. So we decided to focus on the first period of his activity, which saw Castelnuovo's presence as advisor and mentor, and very strong scientific influences by Corrado Segre (1863–1924) and Guido Fubini (1879–1943). In particular we concentrated on the few, very productive years Bompiani spent in Bologna in the 1920s.

It would be highly desirable to have a complete historical and mathematical overview of PDG, something which is out of the scopes of this paper. For this reason, several important themes and characters – e.g. the American and German schools, Élie Cartan (1869–1951) – have either been overlooked or remain in the background. As for Bompiani's years in Bologna, we concentrated on his scientific production in PDG, but did not consider either the ambient or the school he created there.[2]

We start our paper in Sect. 2 by casting a glimpse towards the end of our story, namely Castelnuovo's action for hiring Bompiani as a full professor in Rome in the fall of 1927. We then deal with Bompiani's first years as a student in Rome in Sect. 3, with his way to Bologna in Sects. 5 and 6, with a discussion of his contributions to PDG in the Bologna period in Sect. 8. This section is rather technical and may be skipped by a reader who is not interested in mathematical details. The circle closes up in the short Sect. 9, with Bompiani coming back to Rome. In Sect. 10, we give a few informations about Bompiani after Bologna and the developments of PDG. Sections 4 and 7 contain digressions on Segre, Fubini and their main results on PDG. This gives us the opportunity of introducing some mathematical framework necessary to talk about Bompiani's results.[3]

2 The End of Our Story

Dear Pittarelli,[4]

I am sure you will be pleased about the decisions we took in the Faculty meeting. After an endless and often animated discussion it has been decided: (1) by a majority vote (only mathematicians against), to give one of the mathematical chairs to physical Chemistry; (2) by a majority vote, to attribute the remaining chair to a discipline of the first biennium, rather than to higher Analysis, as proposed at the very last moment by Volterra,[5] Severi[6] ... having in mind Fubini or Tonelli.[7] The latter proposal came up suddenly and faded in just a few hours; (3) with a unanimous vote – with only one exception – to chose descriptive

---

[2]For information about Bompiani's scientific influence in Italy and abroad see [91].

[3]More information on Segre, Fubini and on PDG can be found in [5, 6, 87].

[4]Giulio Pittarelli (1852–1934) was a professor in Rome, who was about to retire in that moment. Bompiani will be hired on his chair.

[5]Vito Volterra (1860–1940).

[6]Francesco Severi (1879–1961).

[7]Leonida Tonelli (1885–1946).

Geometry among the disciplines of the first biennium. This should make you particularly satisfied, having in mind the person to be hired; (4) with a completely unanimous vote, to hire Bompiani on this chair. Thus he comes to us with all the honors, according to your and my wishes, and with his greatest satisfaction. It was an epic fight that ended up well [...]
Yours, G. Castelnuovo.[8]

Hiring Bompiani in Rome was not easy, due to various tensions within the group of mathematicians and between mathematicians and other faculty members. Certainly Bompiani owed to Castelnuovo's prestige and influence the possibility of winning resistances, if not oppositions, of people like Volterra and Severi, who, in turn, were supporting very strong candidates like Tonelli and Fubini.[9] This also shows how well reputed was at the time Bompiani, and how close, present and supportive was Castelnuovo towards his former student. A consideration Castelnuovo will always show, till the purge process against Bompiani when Fascism collapsed (see Sect. 10.2). But this will happen long after the end of our story. Let us look now at the beginning.

3 First Years in Rome

Enrico Bompiani was born in Rome on February 12, 1889. His father Arturo was a well reputed physician and two Enrico's brothers followed medical studies. Showing an independent spirit, he chose instead to study Mathematics.[10]

Bompiani was a student of Guido Castelnuovo at the University of Rome. In 1909, Castelnuovo suggested as his master thesis subject a line geometry

---

[8]"Carissimo Pittarelli, Sicuro di farti piacere ti comunico le decisioni prese dalla Facoltà. Dopo una interminabile e spesso vivace discussione si è concluso: (1) a maggioranza (i soli matematici contrari), di dare alla Chimica fisica una delle cattedre di cui godevano i matematici; (2) a maggioranza, di destinare la cattedra residua ad una materia del 1° biennio, anzichè all'Analisi superiore come all'ultimo momento era stato proposto da Volterra, Severi ... pensando a una [soluzione] Fubini o Tonelli, sorta all'improvviso e tramontata in poche ore; (3) all'unanimità, salvo un voto, di scegliere fra le materie del 1° biennio la Geometria descrittiva (e di ciò devi essere particolarmente soddisfatto) in vista della persona che si voleva chiamare; (4) all'unanimità assoluta di chiedere il trasferimento per questa cattedra del Bompiani. Il quale dunque entra fra noi con tutti gli onori, secondo il desiderio tuo e mio, e con suo grande compiacimento. La lotta fu epica ed è terminata bene [...] tuo aff.mo, G. Castelnuovo."
Letter from G. Castelnuovo to G. Pittarelli, November 11, 1927, Università di Roma, Scuola di Geometria. The letter has been reproduced in [80, p. 128] and it has been kindly pointed out to us by Pietro Nastasi, whom we thank also for useful discussion and comments. All translations of the Italian originals have been made by the authors of the present paper. We add in footnotes the Italian original version only for quotation which never appeared in print.

[9]Another, perhaps more realistic, candidate, supported by Tullio Levi-Civita (1873–1941), was Ugo Amaldi (1875–1957). He was at the time professor at the Faculty of Architecture in Rome, a position with which he was not satisfied, so that he aimed at moving to the Faculty of Science. This happened only in 1942, see [83].

[10]For biographical accounts on Bompiani see [72, 81, 98].

interpretation in projective 4-space of Cosserat's[11] results on systems of circles in 3-space. The topic was in the mainstream of the famous *Lie sphere geometry*, inaugurated by Sophus Lie (1842–1899) in his dissertation [79], and was also related to a paper by Castelnuovo [53] on line geometry of $\mathbb{P}^4$. Apparently Bompiani fulfilled his task with success [49, p. 85–86], finding his way on the subject, starting to appreciate both differential and algebro-geometric techniques, i.e. the basis of what was about to be called PDG.

PDG can be defined as the study of properties of analytic subvarieties of a complex projective space $\mathbb{P}^n$, which are invariant under the action of the projective group $\mathrm{PGL}(n+1, \mathbb{C})$. Its purpose is to study local properties, applying it to global classification problems. An elementary example is the local notion of a *flex* of a curve and the theorem that a curve is a line if and only if all its points are flexes.

According to Shiing-Shen Chern (1911–2004), who has been one of the main modern advocates of this discipline, the main problem of PDG

> [...] is to find a complete system of local invariants of a submanifold under the projective group and interpret them geometrically through osculation by simpler geometrical figures. The main difficulty lies in that the projective group is relatively large and invariants can only be reached through high order of osculation [54].

Ernest P. Lane (1886–1969), one of the main characters of the American school of PDG and predecessor of Chern on a chair at the University of Chicago in 1949, adds:

> There was first of all a period of discovery of isolated theorems which were of a projective differential nature but which were not recognized at the time as having this character, because there was then no organized science of projective differential geometry. Later came the initiation of projective differential investigations, and still later the organization of comprehensive theories and the perfection of systematic methods of study [76].

The moment in which Bompiani enters the scene is exactly he *third period* according to Lane's rough historical subdivision. Ernest J. Wilczynski (1876–1932) was one of the initiators of this third period and Lane was a student of Wilczynski.

Bompiani's thesis, entitled "Spazio rigato a quattro dimensioni e spazio cerchiato ordinario", was defended on July 5, 1910. It gave rise to the papers [11, 14, 15]. In particular he introduced in [15] the notion of *index of developability* of a scroll along a ruling, a concept which will lead to the notion of *quasi-asymptotic curves*, which will play a crucial role in future researches.

After Bompiani graduated, his studies suffered constant interruptions due to military duties, which kept him busy for 1 year from 1910 to 1911, then from August 1912 for the whole period of the war of Libia, and finally for the whole first world war. During the first world war he served in the air force and he got a second degree on aeronautical engineering in 1918.

Despite the military duty, in 1912 he managed to participate into the International Congress of Mathematicians in Cambridge, presenting a lucid report [50] on recent

---

[11] Eugène Cosserat (1866–1931).

progresses on PDG, with a very strong emphasis on the role of Italian geometers, but also acknowledging the work made in the previous decade by E. J. Wilczinsky.

Bompiani's results were communicated by Castelnuovo to Corrado Segre, who replied that Bompiani was coming from a «good school» [49, p. 86]. Segre however pointed out he had already published the paper [89] on the subject and he had just submitted for publication (on July 9, 1910, 4 days after Bompiani's thesis defense) another article [90] with results related to Bompiani's.

Analogous studies were at the time undertaken also by Alessandro Terracini (1889–1968), a student of Segre. Though Terracini and Bompiani never wrote any joint paper, they had a very good, long-term scientific and personal relation.

In any event, this was the beginning of a fruitful relationship between Bompiani and geometers in Turin, which was followed years later by collaborations with Eugenio Togliatti (1890–1977), another student of Segre, and especially with G. Fubini [91].

Corrado Segre, another great master of the Italian school of algebraic geometry, was a constant scientific reference point for Bompiani in this period. From the correspondence kept at the *Accademia Nazionale delle Scienze detta dei Quaranta*[12] it is possible to see all Segre's meticulous comments, detailed criticisms and professorial suggestions on Bompiani's papers sent to him for revision and advices in the years around 1913.

Bompiani indicates Segre as the initiator of the Italian school of PDG. This is not completely true. Perhaps the first paper on a PDG subject goes back to Pasquale del Pezzo (1859–1936) [85], who first introduced the crucial notion of *osculating spaces* to a variety. Another important contribution was due to Luigi Berzolari (1863–1949) who studied [7] the PDG of curves in projective spaces. Berzolari followed Georges H. Halphen's (1844–1889) *differential invariants* approach,[13] inspired in turn by Gaspard Monge (1746–1818) who first wrote the *differential equation* satisfied by conics [82] Still referring to Lane's quote on page 146, Monge belongs to the first period of his classification and Halphen and Berzolari to the second.

4 First Digression: Corrado Segre

It is perhaps the right moment for stopping for a while following Bompiani's route, focusing instead on Segre's contributions to PDG. This may shed some light on the reasons why he was considered to be the initiator of the Italian school.

About the birth of Segre's interest in PDG, Lane vindicates it to Wilczynski's influence:

> The distinguished Italian geometer C. Segre (1863–1924) began his geometrical researches at the University of Torino in the early eighties of the nineteenth century. His interest

---

[12]Published in [84].

[13]See Halphen's dissertation [69].

in projective differential geometry is said to have been stimulated by Wilczynski at the Heidelberg Congress of 1904. Beginning with a very significant memoir in 1907 Segre made important contributions to the subject. He was not only interested in the geometry of ordinary space, to which his contributions of the tangents of Segre and the cone of Segre have already been studied in this book, but was a leader in studying the projective differential geometry of hyperspace. Segre gave analytic proofs regularly, but was also an outstanding exponent of the synthetic method, making differential properties even in hyperspace appear intuitive [74]

Julian L. Coolidge (1873–1954), who at the beginning of the century spent a period of study in Torino visiting Segre, confirms, and at the same time complains about a certain separation between Italian and American projective differential geometers:

> [...] the subject of projective differential geometry was that to which he [C. Segre] gave most attention during the later years of his life. He made the acquaintance of Wilczynski at Heidelberg in 1904 and thoroughly appreciated the importance of his work. Yet an outsider cannot help feeling some regret that there still appears a considerable gulf between the Italian doctrine of projective differential geometry, developed by Segre, Bompiani, Fubini and others, and the American contributions of Wilczynski and Gabriel Green,[14] of whom, by the way, no Italian seems to have heard the name.[15]

Terracini adds:

> One should not try to find in Segre unitary analytic methods created with the aim of constructing a whole theory, as it happens, for instance [...] in Wilczynski or in Fubini [...]. In Segre, we do not find anything of this sort: even when the original synthetic approach did not leave a visible trace in its final version, we can perceive that it is well present behind the analytic treatment. This has, in any of its important steps, a geometric meaning [94, p. VI].

It is not completely true that Segre's interest in PDG comes only after he met Wilczynski at the Heidelberg Congress of 1904. His first paper on the subject goes back to 10 years before [88], when he introduced the famous *Wölffing–Mehmke–Segre contact invariant* for two plane curves at a point where they are tangent to each other.

The definition is as follows. Consider two curves C, C' tangent at p, and take their common tangent T. Take a general line L passing close to p, cutting C, C', T in points a, b, c. Take a general point m on L. The limit of the cross ratio (a, b, c, m) when L tends to a general line through p, is independent of the arbitrary choices and is the invariant in question.

This definition was very inspiring, in particular for Bompiani, who studied similar invariants for curves, surfaces and higher dimensional and codimensional varieties, thus creating a PDG of higher order germs which has been cultivated in Italy till the 1960s by followers of Bompiani.

---

[14]Gabriel M. Green (1891–1919).

[15]See [56]. The last statement does not seem to be true, as witnessed by the credit given to Green by Terracini in [93].

In 1907, Segre publishes a fundamental paper [89], which is admittedly motivated by Wilczynski's work, but is also related to the aforementioned paper by Del Pezzo and by purely differential geometric researches by Gaston Darboux (1842–1917).

Taking a wider viewpoint than Wilczynski's, who in general limited his considerations to surfaces in $\mathbb{P}^3$, Segre studies all analytic surfaces Φ in $\mathbb{P}^r$, locally parametrically given as

$$(u_1, u_2) \to [x(u_1, u_2)] = [x_0(u_1, u_2), \ldots, x_r(u_1, u_2)] \tag{1}$$

and *verifying a second order differential equation* or *Laplace equation*[16]

$$A x_{11} + B x_{12} + C x_{22} + D x_1 + E x_2 + F x = 0, \tag{2}$$

where $A, \ldots, F$ are not all identically zero. This means that the second order osculating space to Φ at its general point has dimension smaller then the expected one, which is 5.

Items contained in the paper are

- A surface verifies two independent Laplace equations if and only if it is either developable or it sits in a 3-space.
- Any surface in $\mathbb{P}^4$ verifies at least one Laplace equation.[17]
- A detailed study of the rich projective geometry of surfaces Φ verifying a Laplace equation in $\mathbb{P}^5$, in relation with the behavior of their 4-dimensional tangential varieties.
- A surface has a *double system of conjugate lines* if and only if it verifies a Laplace equation. Accordingly Segre geometrically constructs the *Laplace transform* of the surface. This gives a beautiful interpretation of earlier analytic work of Darboux, relating integration of Laplace equations with Laplace transformations.[18]

Extensions of these results to higher dimensional varieties and higher order differential equations have been produced by followers of Segre, like Bompiani, Terracini, Togliatti.

A paper of 1907 [89] plays a pivotal role in further developments of PDG, since it motivated Fubini's later important work. Given a surface S in $\mathbb{P}^3$ and a smooth point p on it, Segre introduces in a beautiful geometric way three remarkable tangent lines to S at p, later called *Segre tangents*. He compares this triple of tangents with another triple introduced in 1880 by Darboux, the so-called *Darboux tangents* and proves that the Darboux tangents are conjugated to his tangents with respect to the

---

[16] Lower case indices denote differentiation with respect to the corresponding variable.

[17] Riemannian differential geometry of surfaces in 4-space had been studied by Eugenio Elia Levi (1883–1917) in his dissertation [77].

[18] See Sect. 8 for some details, in connection with Bompiani's later work on the subject.

pair of *asymptotic directions* issuing from the point p.[19] We will come back to these constructions in Sect. 8.

In a paper of 1910 [90], Segre studies varieties described by some varying linear subspace. He relates their tangential properties with infinitesimal or, so-called, *focal*[20] properties of the generating space, widely extending the concept of developable surface. In particular he starts the classification of *tangentially defective* varieties, i.e. those varieties of dimension n whose tangential variety has dimension smaller than the expected one, which is $2n$.

This research area has been further pursued by Terracini, Bompiani and others and it is still an open research field.[21]

5 Rome, Pavia and Milan

Back to Bompiani. He appears, since his very beginnings, extremely determined in pursuing his scientific career. He was able to take every single opportunity to produce a new piece of work. His first publication, probably arising from his university studies and interactions with Volterra, was not in geometry but in analysis [10], devoted to Volterra's integro-differential equations.

5.1 Assistant in Rome and Pavia

Bompiani started his career as Castelnuovo's assistant in Rome. In 1911, he was appointed for 2 years. Volterra, who was Dean of the Faculty, wrote two recommendation letters to German colleagues for him,[22] because he wanted to spend a period of study and research in Göttingen. Castelnuovo supported Bompiani's trip abroad dispensing him from teaching duties [4].

Soon after his assistantship in Rome, Bompiani was hired at the University of Pavia, staying there 2 years, from October 16, 1913 to October 30, 1915, as an assistant of Francesco Gerbaldi (1858-1934). In Rome, he had an excellent reputation, the relationship with Castelnuovo being very good under any respect, as witnessed by the following statement that Castelnuovo asked to insert in the report of the Faculty meeting:

> Prof. Castelnuovo communicates to the Faculty that his assistant Dr. Bompiani leaves Rome to Pavia, where a better position has been proposed to him. Prof. Castelnuovo expresses

---

[19]The asymptotic directions at p are the tangent directions at p to the singular curve cut out on the surface by its tangent plane at p.

[20]For a modern approach to the theory of foci following Segre, see [57].

[21]The standard modern references on the subject are [67, 101].

[22]V. Volterra to Emil J. Wiechert (1861–1928) and Eduard Riecke (1845–1915), Roma, May 14, 1913, Fondo Bompiani, Accademia dei XL, Roma.

his regret for loosing the help of a young collaborator who showed, during his 2 years assistantship, excellent qualities both as a scholar and as a teacher. Bompiani, thanks to the diligent accomplishment of his duties and to his special teaching attitudes, largely contributed to the preparation of his students.[23]

By contrast, the years in Pavia were not as satisfactory. Berzolari was there, but there is no apparent trace of any close scientific relation between the two. On the other hand, Bompiani's relationship with Gerbaldi was often tough, just the opposite of the confident and trustful one with Castelnuovo. Gerbaldi was famous for being a solitary, not easy, person. He even wrote down a list of formal duties his assistants had to accomplish, and strict rules concerning their interactions with students. Among them, one forbidding them from teaching "theory", which was reserved to the professor, only "exercises" and "perspective tables" were they allowed to teach and check.[24] In 1914, he became, following Gerbaldi's advice, a *libero docente*.[25]

No surprise that, as soon as he could, Bompiani went back to Rome and to Castelnuovo. This happened in December 1915, when an assistant position for Castelnuovo became vacant, and he appointed Bompiani for it.[26] As we said, for a long period, including his assistantship in Pavia and till October 16, 1919, Bompiani was often on military duty, and even on fields of battle. Nonetheless he was able to partly fulfill his academic duties. After 1919, he went full time back to his assistantship in Rome, where he remained till January 1st, 1923, when he became an *extraordinary professor*.[27]

5.2 Professor in Milan

On October 12, 1922, Castelnuovo writes[28] to Bompiani that Federigo Enriques (1871–1946) let him know, on a strictly confidential base, that Bompiani won the

---

[23]"Il prof. Castelnuovo comunica alla Facoltà che il suo assistente Dott. Bompiani, lascia Roma per Pavia, dove gli fu offerto un posto più elevato. Il Prof. Castelnuovo esprime il rammarico di perdere l'aiuto di un giovane che nei due anni, durante i quali tenne l'assistentato, dimostrò di avere ottime qualità di studioso e di insegnante. Il Bompiani grazie allo zelo con cui adempiva al suo ufficio, grazie alle speciali attitudini didattiche, aveva notevolmente contribuito alla preparazione degli studenti affidati alle sue cure." See [4].

Both Pittarelli and the Faculty Dean joined to Castenuovo's statement.

[24]F. Gerbaldi to E. Bompiani, Pavia, November 20, 1913, in [84, p. 39–42].

[25]See [84, p. 39]. The title of *libero docente* was analogous to the German one of *Privatdozent*, which is used by those who have successfully completed a *Habilitation*. Becoming a *libero docente* in Italy required a national evaluation, and implied the right to give university courses without being supervised by a professor. It was abolished in 1970.

[26]Castelnuovo to the Rector, December 5, 1915, [4].

[27]In the Italian system this is the first step for becoming full or *ordinary* professor, which is a permanent position. The transit from the former position to the latter takes place after 3 years of service and an evaluation by a national committee nominated by the Minister.

[28]G. Castelnuovo to E. Bompiani, October 12, 1922, in [84, p. 25–26].

context for a position of extraordinary professor of projective and analytic geometry at the University of Modena: Bompiani had gotten the first position, with four votes against one, given to Annibale Commessatti (1886–1945). Castelnuovo adds he is very happy about the result, that Bompiani certainly deserved the recognition of a full maturity both from the scientific and didactic viewpoint. Modena is probably not the most desirable destination, Castelnuovo says, but he feels sure that soon better opportunities will come up. Finally he ends by saying: "As for my concern about missing you, this is not the right moment to talk about it." Thus showing his affection for his former student and assistant.

As Castelnuovo predicted, a "better opportunity" than Modena soon came up. Indeed, on November 11, 1922, the Faculty Council of the Higher Technical Institute in Milan resolves to ask the Minister to nominate Bompiani professor of analytic, projective and descriptive geometry. The Faculty's statement contains a flattering evaluation of Bompiani, who "got the best possible recommendations as a teacher and as a scholar devoted to the science of geometry", the "very high qualities of the proposed candidate" having been taken into account.

No doubt that, even for this "better opportunity", Castelnuovo's influence played some role. Indeed, in the Faculty record we read that prof. Mario Dornig (1880–1962), a former student of Castelnuovo, made a clear cut statement of the type: since Bompiani has been trained at Castenuovo's school, he will be for sure a very good teacher for engineers.

On December 18, 1922, Bompiani writes to the Minister, accepting the position in Modena, adding however that, if the Minister and the Higher Council of Public Education would approve the vote of the Higher Technical Institute in Milan, he would be glad to choose this destination. This is in fact what happens, and Bompiani arrives in Milan at the beginning of 1923. However his stay there will be short, only one academic year. A tempting opportunity to go closer to Rome will soon come up, namely the possibiltiy of going to Pisa, a prestigious university [84, p. 119–126]. This will not materialize, but soon another concrete chance will show up, the one of going to Rome, not at the Faculty of Sciences, but at the Architecture School.[29] However this was considered a lower level position.[30] At the same time the possibility of moving to Bologna came up. Probably following Castelnuovo's advices, and taking advantage from the quick evolution of the Bologna solution, Bompiani gave up the Roman chair at Architecture and chose Bologna, where he

---

[29]Bompiani had been teaching there in the two academic years 1920–1921 and 1921–1922 with a temporary, part-time position, before moving to Milan. The Architecture Faculty, in expressing the vote that Bompiani could be hired there with a permanent position as a full professor, made the following statement: "At that time he showed a good scientific competence as well as a wise common sense in his teaching duties, by adapting the contents and the method of his course to the special needs of this School. This surely is, for prof. Bompiani, a specific merit in order to be permanently hired on the chair of Mathematics in this School" [4, Letters to the Minister of the Director of the School, November 3 and November 19, 1923].

[30]This is the chair which in 1924 will be taken by Amaldi, who will be able to move to the Faculty of Science only in 1942 (see [83]).

spent three good and very active years, before going back, as we said, to the Faculty of Sciences in Rome.

6 The Chair in Bologna

Already in 1922, soon after Bompiani won the full-professorship context, Castelnuovo made some moves for giving him a chance of being hired in Bologna, on the chair left vacant by Enriques who, in turn, had just moved to Rome. Probably, he assumed that Bologna, an old and prestigious university, was one of the best possible places for his protégé, waiting for a comeback to Rome. Just a few days after the good news about the context won by Bompiani, Castelnuovo gets in touch with Salvatore Pincherle (1853–1936) who at the time was the Dean of the Faculty of Sciences at the University of Bologna, who answers, on October 21, 1922:

> I got your letter, and I mentioned your idea to my colleagues. Surely the solution you suggest would be very nice, because of the skills of this person, who is well known to everybody, and particularly to me, in a very favorable way; but first of all at the moment the position is not yet available, and moreover we had in mind something slightly different, which I will explain to you in person if, as I believe, I will have to come to Rome within a few days. In any event, it has been good that you made your suggestion, which is certainly worth of being discussed with the greatest favor.[31]

6.1 Reaching Bologna

This possibility concretizes quickly and smoothly. In the meeting of October 23, 1923, the Faculty of Sciences of the University of Bologna proposes to the Minister with an unanimous vote to hire Bompiani on the projective and descriptive geometry chair. The Faculty's enthusiastic statement says:

> Professor Enrico Bompiani, an extraordinary professor at the R. Polythecnical School of Milan, is well known to mathematicians for several and important papers in the geometric area, mainly in the branch of modern projective-differential geometry. Because of his scientific value, which has been also recognized in recent contexts, in one of which he got the first position, and because of his well known teaching skills, the Faculty unanimously proposes to move him on the chair of Projective and Descriptive Geometry in our University.[32]

---

[31] S. Pincherle to G. Castelnuovo, Bologna, October 21, 1922, [84, p. 26–27].

[32] "Il prof. Enrico Bompiani, professore straordinario nel R. Politecnico di Milano, è ben noto ai cultori della matematica per numeroso ed importanti lavori nel campo geometrico, e principalmente nel ramo della moderna geometria proiettivo–differenziale. Per il suo valore scientifico, riconosciuto anche nei recenti concorsi, in uno dei quali è riuscito primo, e per le note sue qualità didattiche, la Facoltà è unanime nel proporre il trasferimento alla cattedra di Geometria proiettiva e descrittiva, presso la nostra Università." See [3].

Despite all, some clash, or perhaps only some misunderstanding, should have come up. Tonelli, who was in Bologna at the time, and was often corresponding with Bompiani, informs him about the Faculty's decision and at the same time urges him to make sure that the bureaucracy in Rome goes as quickly as possible, in order to avoid that "somebody would put an obstacle in our way ...".[33] A few days after,[34] Tonelli is more precise: he is afraid that Enriques may try to scupper the operation and he suggests Bompiani to get in touch with Luigi Bianchi (1865–1928)[35] and at the same time to write to the Minister accepting the chair in Bologna. It is quite unlikely that Enriques, very close to Castelnuovo and with a good opinion of Bompiani, would try to "scupper the operation".[36]

The Minister finally accepted the Bologna Faculty's vote, and Bompiani was nominated professor there on November 16, 1923. A new proposal to him was soon made from Milan again, but now from the recently established Faculty of Sciences. Bompiani was tempted, because the offer was very good also from the economical viewpoint. However, he consulted Castelnuovo who, in a long and interesting letter in which he explains the difficulties of a future call to Rome[37] suggested to avoid moving from one place to another thus showing an inconstant attitude, which could give a pretext for possible detractors of criticizing him. Bompiani will follow Castelnuovo's advice and will spend the next 3 years in Bologna. In any event, it seems to us that all these offers, coming from good and prestigious places, witness the great general esteem enjoyed by Bompiani at the time.

6.2 Visits Abroad

Bompiani showed since the very beginnings of his career,[38] a particular interest in traveling, getting and maintaining contacts abroad. Such a marked attitude will be present also in the years in Bologna.[39]

In July 1924, he was invited to the Swansea University College, to hold a temporary position from January to June 1925. Bompiani was happy with this opportunity, but he had to ask for permission to the University of Bologna. The "Registar" of the University College pointed out the advantages that could come to Swansea from Bompiani's course, and, on the other hand, the appreciation for Italian culture. The Italian Ambassador in London was asked to make moves in

---

[33]L. Tonelli to E. Bompiani, Bologna, October 23, 1923, [84, p. 108].

[34]L. Tonelli to E. Bompiani, Bologna, October 27, 1923, [84, p. 108].

[35]Who was, since September 1st, 1923, a member of the Higher Council of Public Education.

[36]See: F. Enriques to E. Bompiani, Viareggio, July 8, 1923, [84, p. 31].

[37]G. Castelnuovo to E. Bompiani, May 21, 1925, [84, p. 27–29].

[38]Remember the participation to the Cambridge International Congress of Mathematicians in 1912 and his visit to Göttingen after his graduation.

[39]The Fascist regime will strengthen and enforce the opposite attitude, for details see [68].

order that Bompiani could have the required permission. This opportunity however did not concretize, and Bompiani had to give it up [3].

On March 23, 1926, Bompiani asked for a passport to go visit the University of Hamburg, where he had been invited[40] to give a course in May and June on hyperspace geometry. The Rector of Bologna approves and, given that the visit to Hamburg is motivated by "reasons of study and propaganda", asks the Minister of Foreign Affairs to issue the passport for this purposes. This time Bompiani will be allowed to go.

During the Fascist era, it will be definitely clear that some of his travels abroad will have, besides scientific purposes, also political propaganda ones. A clear example of this is given by his participation in the preparation of the Italian pavilion at the Chicago World Exhibition of 1933–1934 ([48] and Sect. 10.1.).

Later on in his career, Bompiani will continue traveling a lot, in Europe, United States and India, giving talks and teaching courses for long periods; he even got the position of "Mellon Professor" at the University of Pittsbourg from 1959 to 1961 [3].

6.3 Becoming an Ordinary Professor

After 3 years of service as an extraordinary professor, Bompiani could ask for the evaluation to become an ordinary professor.[41] The required statement of the Faculty is very positive:

> As for the scientific activity and the effectivity of the courses taught here in the years 1923–1924 e 1924–1925, we, according to conscience, must say that they deserve any possible praise and even more. A conscious, active, hard-working, very competent teacher, possessing an easy and highly communicative attitude, though always being extremely precise and rigorous; he thus obtained affection, respect and the best success from his audience.
>
> For the students of the first 2 years, most of them studying to become engineers, he devoted all his care in order to give to his courses the most suitable contents for the *applications*, always sticking however to a scientifically rigorous standard. He has been giving to applications a space and an interest which are rare to be found in a pure mathematician.
>
> In higher level courses that have been attributed to him, and of which we enclose the interesting programs [...], he treated with a wide, lucid and original mind, topics belonging to higher developments of science. With his passion for the topics he taught, he has been able to inspire, in the best part of his audience, a true love for these disciplines.
>
> As for his scientific activity, it is showed in the clearest way by the number and value of the papers he published in the last 3 years, namely 22 memoirs.[42] His papers belong to areas of mathematics in which he can be considered to be a Master: some are related to

---

[40]Wilhelm Blaschke (1885–1962), the main character of the German school of PDG, was professor there.

[41]See footnote 27.

[42]The full list is contained in [3]. We will report on some of these papers in Sect. 8. To be precise, a good part of these papers where not "memoirs" but short notes, some of them even preliminary notes with no proofs. However, there were also true long "memoirs" among them.

earlier papers, the ones which enabled him to get the full professorship position. Others start and develop new topics, especially in algebraic and differential geometry, treated with new viewpoints.

His papers ensured him the election as a member of the Real Academy of Sciences of the Bologna Institute and gained for him in 1925 the Golden Medal for the best mathematical papers of the Italian Society of Sciences.[43]

As this statement makes clear, Bompiani was an excellent teacher, and his higher level courses were very interesting and very much research oriented. The course in 1923–1924 was devoted to Riemannian geometry and absolute differential calculus. The course of 1924–1925 concerns first order differential equations. Some of his publications in those years[44] are related to the contents of the former course. The course of 1926–1927, his last year in Bologna, was devoted to PDG, and in the audience he had young researchers as well as well established mathematicians, some coming from abroad. In a footnote to one of his paper,[45] Bompiani recalls that he had in his audience Enea Bortolotti (1896–1942), Lane, Ellis Bagley Stouffer (1884–1965) and a few young Italian post-graduates, who will become his scientific followers.[46]

The committee for Bompiani's evaluation was formed by Eugenio Bertini (1846–1933) (Scuola Normale Superiore of Pisa), Gino Loria (1862–1954) (University of Genua) and Federigo Enriques (University of Rome). They met in Rome on January 8, 1926. After having considered with appreciation Bompiani's teaching activity,

---

[43]"In quanto alla operosità didattica e all'efficacia degli insegnamenti impartiti presso questa Facoltà negli anni 1923–1924 e 1924–1925, si deve in coscienza dichiarare che si sono dimostrati superiori ad ogni elogio. Maestro diligente, operoso, infaticabile, di alta dottrina, di facile e persuasiva comunicativa pur conservandosi sempre nel suo dire la massima precisione ed i massimo rigore, ha ottenuto l'affetto, il rispetto ed il massimo profitto da parte della scolaresca. Per quella del primo biennio, in massima parte composta da aspiranti ingegneri, si è preoccupato di dare al suo insegnamento l'indirizzo più confacente alle *applicazioni* pur conservandogli sempre un carattere rigorosamente scientifico, e ha dato alla parte applicativa uno sviluppo ed un interesse rari a trovarsi da parte di un curatore della scienza pura. Nei corsi più elevati, tenuti per incarico e di cui si uniscono gli interessanti programmi [...], egli ha trattato con ampiezza, lucidità e vedute originali, argomenti pertinenti a parti superiori della scienza, ed appassionato per le materie che insegna ne ha saputo infondere l'amore nella parte migliore del suo uditorio. In quanto alla sua operosità scientifica, essa è dimostrata nel modo più luminoso dal numero e dal valore dei suoi lavori pubblicati del triennio ora decorso: sono 22 memorie e note originali oltre a varie recensioni o scritti minori; I suoi lavori si riferiscono a capitoli delle scienze in cui egli si può dire Maestro: alcuni sono continuazione di quelli che gli hanno valsa la cattedra, altri iniziano e sviluppano argomenti nuovi, specie di geometria algebrica e di geometria differenziale trattata secondo moderne vedute. I suoi lavori gli hanno meritato la nomina a Socio della R. Accademia delle Scienza dell'Istituto di Bologna e gli hanno valsa la medaglia d'Oro data, nel 1925 ai migliori lavori matematici della Soc. Italiana delle Scienze detta del XL." See [3].

[44]Like [9, 36].

[45]See [47]: this is a polemical article, in which Bompiani answers some criticisms moved to him by Lane, and vindicates his priority on some concepts, which, according to him, he introduced in that course.

[46]As stated at the beginning of [39], this note and [40] arose form Bompiani's original exposition of Fubini's theory in this course. See Sect. 8.

the committee examined his 22 submitted publications. The committee's report in particular points out:

> [...] the elegant hyperspace representation of the system of plane curves of a given degree with given singularities [25] and the deep studies on subvarieties in Riemann spaces, where one finds, among other things, a simple construction of Ricci–Einstein[47] tensor [22–24,27,33].
> A special mention deserves the important existence criterion for the integral of a first order differential equation, which makes an essential step on the well know Lipschitz[48] condition [9,36].
> But above all one has to comment about the set of papers on classical projective differential geometry, which, in the last few years, after a fundamental memoir by Fubini, attained results which are really important for science. However, the approach of the distinguished mathematician from Turin is purely analytical, and Bompiani deserves the credit for having enlightened its geometric meaning. He thus contributed, in an essential way, to the construction of the new theory. In particular the candidate discovered that Fubini's projective linear element can be interpreted as the main infinitesimal part of a certain cross ratio formed by a fourthuple of points three of which are infinitely near, and he shows it to be independent of the fourth point.[49]

The conclusion is:

> [...] for the whole set of his papers, for the lucidity as well as for the width of his purposes, for the importance of some of his results, for the varieties of tools and the art of the exposition, Bompiani occupies by now an honorable position among geometers of the young Italian school.[50]

The greatest part of Bompiani's papers submitted for the promotion evaluation was on PDG, and most of them had been written during his stay in Bologna. It is, therefore, the case to make some closer comment on them (see Sect. 8). Before that,

---

[47] Albert Einstein (1879–1955), Gregorio Ricci-Curbastro (1853–1925).

[48] Rudolph Otto Sigismund Lipschitz (1832–1903).

[49] "[...] l'elegante rappresentazione iperspaziale del sistema delle curve piane di dato ordine con date singolarità, e gli studi approfonditi sulle varietà immerse in spazi riemanniani, ove s'incontra, fra l'altro, una semplice costruzione del tensore di Ricci–Einstein. Una speciale parola merita l'importante criterio d'esistenza per l'integrale di un'equazione differenziale del primo ordine, che realizza un passo essenziale sulla nota condizione di Lipschitz. Ma soprattutto è da dire del gruppo di lavori attinenti alla Geometria proiettiva–differenziale classica, ha raggiunto negli ultimi anni, e in seguito ad una memoria fondamentale di Fubini, dei risultati veramente importanti per la scienza. Ma la ricerca dell'illustre matematico di Torino è puramente analitica, ed è merito del Bompiani di averne messo in luce il senso geometrico e così di aver contribuito, in una maniera essenziale, al nuovo edificio. In particolare il Nostro ha scoperto che l'elemento lineare proiettivo di Fubini si può interpretare come l'infinitesimo principale di un certo birapporto formato da una quaterna con tre punti infinitamente vicini, questo infinitesimo riuscendo indipendente dal quarto punto." See [3].

[50] "[...] per l'insieme dei suoi lavori, per la lucidità non meno che per l'elevatezza dei propositi, per l'importanza di alcuni risultati per la varietà delle dottrine e per l'arte dell'esposizione, il Bompiani tiene ormai un posto d'onore fra i geometri della giovane scuola italiana." See [3].

The committee proposed to the Minister Bompiani's promotion, which was finally approved by the Faculty on November 27, 1925. He formally became an ordinary professor with effect from January 1st, 1926.

it is the case to stop one more time, as we did for C. Segre, and devote some space (see Sect. 7 below) to Fubini, the main character in PDG of surfaces in $\mathbb{P}^3$. As the committee points out, most of Bompiani's papers of that period have been devoted to Fubini's theory. Thus Fubini, as C. Segre in the 1910s, was in the 1920s Bompiani's main source of inspiration. However, the situation is now quite different: Bompiani, armed with the weapons learned at Castelnuovo's and Segre's school, proceeds to a systematic geometrization of Fubini's analytic approach.

7 Second Digression: Guido Fubini

Fubini was a student of Luigi Bianchi in Pisa. Bianchi was a distinguished differential geometer, with a deep knowledge of algebraic and analytic tools. Fubini's mathematical education was mainly in analysis and classical differential geometry. No better words than Terracini's to explain that:

> [...] there is no doubt that, even in geometry, Fubini was mostly an analyst.

> This is not a minor recognition of his merits, but only a matter of fact: Fubini's work has been extremely important, and he certainly has been driven sometimes by the preliminary knowledge of geometric circumstances that where for him the departure point and his guide. But, in most cases, he discovered new geometric facts with analytic methods. For most geometers, the analytic approach is only a dress which one puts on the body of a geometric property of which one had an a priori intuition: for Fubini, often, it seems to be much more [93].

7.1 Fubini's Idea: Differential Forms

Fubini became interested in projective differential geometry about 1914, when he wrote a first note on the subject [59]. He then undertook the problem of finding invariants determining an analytic surface in the complex projective space $\mathbb{P}^3$, up to projective transformations. In doing this, he certainly had in mind the analogy with classical differential geometry as worked out by Karl Friedrich Gauss (1777–1855). Indeed a surface in the euclidean 3-space is uniquely determined, up to motions, by the two fundamental forms, whose coefficients verify the compatibility conditions expressed by the equations of Codazzi–Mainardi[51] and Gauss–Weingarten.[52]

The problem, as we said already, had been considered in the first decade of 1900 by E. J. Wilczinsky, who attacked it trying to extend to surfaces the theory of *differential invariants* which Halphen had applied to curves in his dissertation.[53]

---

[51] Delfino Codazzi (1824–1873), Gaspare Mainardi (1800–1879).
[52] Julius Weingarten (1836–1910).
[53] See the end of Sect. 3.

Wilczinsky had at his disposal more refined tools, namely Lie's group theoretic viewpoint, which he consistently applied.

Robert Hermann comments:

> Lie would no doubt have been very pleased to see Wilczynski's beautiful book on projective differential geometry [99] (which this subject eventually became), which gave full credit to his ideas as the conceptual foundation of the subject! This theory is one of the simplest and most interesting examples of Lie's differential invariant theory [78].

We cannot dwell on the work made by Wilczynski, who was the founder of a solid American school on the subject, some of whose characters we already mentioned above. In just few words, Wilczynski discovered that a surface in $\mathbb{P}^3$ verifies, and is completely determined by, two Laplace equations of the form (2) (see Sect. 4), which are defined up to a suitable group action. This group is larger than the projective group, inasmuch as it includes also changes of local parameters in a parametric expression of the surface like in (1), and the *lifting* to affine space encoded in the parametric expression.[54] Wilczynski's idea is to come, with suitable manipulations using the group action, to a *canonical form* for the pair of Laplace equations. The canonical form should contain all relevant geometric information, to be revealed by interpreting all functions entering into it.[55]

Fubini's idea is not substantially different. However, the method he devised was based on more agile and effective tools, namely differential forms, rather than differential equations.[56]

As Chern puts it:

> There was ... an "Italian school of projective differential geometry founded by G. Fubini and E. Čech in 1918.
> The American school takes as analytic basis systems of partial differential equations and uses Lie theory to generate invariants, while the Italian school takes the differential forms to be the analytic basis [54].

Fubini completed his approach by 1916, exposing it in two basic papers [60, 61]: at the beginning it was not purely projective, but still entangled with classical Riemannian techniques. It took about 10 years to polish it.

Fubini's theory is exposed in 1926–1927 in a famous book in two volumes [64] and in the later French edition [63], both in collaboration with his Czech student Eduard Čech (1893–1960). The French edition is more agile and open to different viewpoints, like Cartan's ones (see footnote 56). A nice short account of Fubini's theory is contained in his Appendix [62] to Bianchi's monumental book

---

[54] That is, $[x(u_1, u_2)] = [y(u_1, u_2)]$ if and only if $x(u_1, u_2) = f(u_1, u_2) y(u_1, u_2)$ for a suitable never vanishing function $f(u_1, u_2)$ of the parameters.

[55] A quick, elementary introduction to Wilczynski's approach is contained in the expository paper [100], translated in Italian with a short interesting preface by Bompiani in [16].

[56] The reasons why forms are better has been lucidly explained by Ph. Griffiths in [66], and relates to Élie Cartan's work on Lie group, expounded in the book [55]. Unfortunately we cannot dwell on this here.

in four volumes [8]. Fubini starts it with the following words, with which he fully acknowledges Bianchi's heritage:

> The methods used in this book for the study of metric properties of a surface, i.e. properties which are invariants by rigid motions, can be suitably modified in order to be applied also to the study of projective properties. How one can start such a study will be briefly explained in this note.

Fubini's starting point is the concept of Segre and Darboux tangents.[57] According to Bompiani [49, p. 87], it was C. Segre, who was a colleague of Fubini in Turin, who asked him if the distribution of these tangents suffices to characterize a surface up to projective transformations. Fubini answered negatively [61], but, at the same time, he set up the foundations of his theory of differential forms, pointing towards his famous rigidity theorem.

Though Fubini's viewpoint is, in principle, not so distant from Wilczisnsky's, in practice his approach is very different: no differential equations but forms. The connections and the distance between the two methods are explained by Lane:

> The projective differential geometry of surfaces has been studied extensively in the United States by Wilczynski, Green, and others, using the invariants and covariants of a completely integrable system of linear homogeneous partial differential equations.
> In Italy, much progress in projective differential geometry has been made by Fubini, Bompiani, and others, who have approached the subject from the point of view of differential forms and the absolute calculus.
> [...] It is the purpose of this note to show how Fubini's canonical differential equations may be obtained by Wilczynski's method, and to compare Wilczynski's and Fubini's canonical forms for the differential equations of a surface.
> It is evidently desirable that geometers living on different sides of the Atlantic, writing in different languages, and using different analytic apparatus, but interested in the same subject, should be able to exchange ideas freely. It is hoped that this note will to some extent smooth the way for this commerce of ideas by showing how certain equations and formulas obtained in one notation may be written also in the other [73, p. 365].

However Bompiani says:

> One does not take anything out from Wilczynski's substantial merit by saying that, at least at the beginning of his research, Fubini did not know at all Wilczynski's work: this is what he says in his paper on the *Annali di Matematica* of 1916 [61]; i.e. when all computations had been made, Segre informed him of Wilczynski's results.[58]

And Terracini confirms:

> ... I think we can say that Wilczynski's influence on Fubini has been overestimated by Struik,[59] when he outlined Fubini's work with the words "connecting with Wilczynski's work and continuing it".
> No, Fubini, though not unaware of Wilczynski, was not the one who continued his work.

---

[57] See 149 in Sect. 4.

[58] See [49], in which Bompiani acknowledges Fubini's heritage.

[59] Dirk J. Struik (1894–2000).

And even Lane's statement in Wilczynski's obituary [75], though more generic and vague, that his influence was particularly strong in Italy and Czechoslovakia, goes, in my opinion, well beyond an objective evaluation of reality [93].

7.1.1 Fubini's First Two Fundamental Forms and the First Rigidity Theorem

A surface S in $\mathbb{P}^3$ is parametrically given as

$$(u_1, u_2) \to [y(u_1, u_2)] = [y_1(u_1, u_2), \ldots, y_4(u_1, u_2)].$$

Assume S is *non-developable*.[60] Fubini defines two homogeneous differential forms

$$F_2(u_1, u_2, du_1, du_2), \quad F_3(u_1, u_2, du_1, du_2),$$

where:

- F_2 is quadratic in du_1, du_2 and vanishes along the two distinct *asymptotic directions*[61]: this is the analogue of Gauss' second fundamental form;
- F_3 is cubic in du_1, du_2 and vanishes along the Darboux tangent directions; F_3 is obtained from the cubic part of the local expression of y in such a way that F_3 is *apolar* to F_2, i.e. the *Hessian* of F_3 is F_2.

The forms F_2 and F_3 are defined up to a factor, hence only their ratio

$$\omega = \frac{F_3}{F_2}$$

is well defined: this is called the *linear projective element*; it plays the same role as the metric volume form in Riemannian geometry.

Fubini's first rigidity theorem says that two surfaces have the same linear projective element if and only if they are *projectively applicable*. Projective applicability is the analogue of the notion of applicability in Riemannian geometry: a bianalytic map $f : S \to S'$ is a *projective application* if it is linear up to the the second infinitesimal neighborhood of corresponding points, i.e. if it maps curves on S through a point p with osculating planes in p passing through a line, in curves through $p' = f(p)$ with the same property. In other words, f has to map any curve with a flex in p to a curve with a flex in p'.

---

[60] A surface is *developable* if its tangent plane at a general point is tangent along a whole curve, which turns out to be a line: they are either cones or the locus of tangent lines to a curve.

[61] Since S is not developable, the tangent plane to S at a general point p cuts out on S a curve with a *node* at p, i.e. an ordinary double point with two distinct tangent lines, whose directions are the *asymptotic tangent directions* to S at p.

As we said, Fubini's first rigidity theorem answers in the negative to Segre's question whether the Darboux distribution of tangent lines determines the surface up to projective transformations. However it shows that the question does make sense! Indeed the equation

$$F_3 = 0$$

characterizes the quadrics: this is a beautiful extension to any dimension of Monge's differential equation of conics, mentioned at the end of Sect. 3

7.1.2 Fubini's Third Fundamental Form and the Main Rigidity Theorem

If
$$F_2 = a_{11}du_1^2 + 2a_{12}du_1du_2 + a_{22}du_2^2$$

Fubini considers the point

$$Y = \frac{1}{2}(A_{11}y_{11} + 2A_{12}y_{12} + A_{22}y_{22})$$

which does not sit on the tangent plane to the surface at y[62]; the line joining y with Y plays the role of the *normal line* to the surface in the classical case. It is called the *projective normal line*, a notion independently introduced by G. Green ([65] and [76, p. 87]). The asymptotic tangent lines plus the projective normal line, plus suitable points on them, define a *movable frame* à la Darboux.[63] Differentiating one gets:

$$y_{rs} = p_{rs}y + \lambda_{rs}y_1 + \mu_{rs}y_2 + \rho_{rs}Y$$

and Fubini defines the *third fundamental form* as

$$P = p_{11}du_1^2 + 2p_{12}du_1du_2 + p_{22}du_2^2.$$

Fubini's main rigidity theorem says that a surface is uniquely determined, up to projective transformations, by the three fundamental forms F_2, F_3, P, which can be suitably put into a *normal form*. The integrability conditions give the analogues of Codazzi–Mainardi and Gauss–Weingarten equations in the PDG setting.

---

[62] A_{ij} is the cofactor of a_{ij} in the metrics $A = (a_{ij})_{1 \leq i,j \leq 2}$.

[63] Jean Gaston Darboux (1842–1917) has been one of the great masters in analysis and differential geometry in the second half of nineteenth century. He wrote the monumental treatise [58] in which he made a systematic use of the method of *moving frames*. This method was introduced by Martin Bartels (1769–1836), published by his student Carl Eduard Senff (1810–1848) in 1831, and independently by Jean Frenet (1816–1900) for curves in his thesis in 1847. Later is was fully developed by Élie Cartan in a wider and more abstract setting.

8 Some Glimpses to Bompiani's Work in the Bologna Period

In his long life, Bompiani wrote about 360 papers. Reporting on them would be impossible in the present paper. Even in the relatively short period he spent in Bologna, his production, of about 30 paper, is too wide to be reported on here in detail. Together with some deep and substantial contributions, Bompiani shows a certain fragmentary attitude, which leads him sometimes towards high level exercises more than stepping on new, difficult uncharted research areas. Moreover he seems to prefer to remain confined in the PDG techniques, that he mastered, rather than looking for general important problems to be solved. Castelnuovo himself was aware of this attitude. In a letter of 1914, in which one finds also comments and advices about Bompiani's moves between Pavia and Rome and his contacts with Bianchi, Castelnuovo says:

> Rather than devoting yourself to a *research method*, it is useful to aim for relevant questions and try to solve them, whatever route leads you to the result.[64]

On the other hand, Terracini remembers that Fubini himself used to complain about the dispersion of the discipline, which likely was not a Bompiani's exclusive feature:

> Fubini didn't have any particular predilection for certain aspects of projective geometry, properly speaking.
> Perhaps this attitude was not unrelated with the circumstance of having seen projective geometry lose itself somewhat in the thousands rivulets of the investigation of particular figures – which he shunned, since he always had general facts and properties in mind.
> Fubini had no love for geometric configurations in which the minute properties came into play.
> In regard to this, I remember that more than once, in projective differential geometry itself, seeing the geometric entities associated to a neighborhood of a point on a curve or surface multiply themselves, he expressed his heartfelt fear that so was created which he called, perhaps a bit too derogatorily, a second geometry of the triangle [93].

All this given, we decided to concentrate on what we think are his most significant contributions to the geometrization of Fubini's theory, which, as we said in Sect. 6.3, has been his main scientific occupation in those years. In doing this, we will explain some of the concepts outlined before.

Let us start with the paper [28].[65] Bompiani studies here the properties of bianalytic maps between two surfaces in 3-space. He introduces a Cremona transformation of degree 3 from the dual of the stars of lines through corresponding points of the two surfaces.[66] He shows that this Cremona transformation has in general 3 base points: one is the tangent plane and the other two are not. The

---

[64]G. Castelnuovo to E. Bompiani, Roma, November 13, 1914 in [84, p. 21–22].

[65]Of which [26] is a preliminary note. According to Bompiani the results here come from a collaboration with Bianchi and Castelnuovo.

[66]He studies also some degeneracy cases: this is done quickly and may be would deserve more attention.

intersection of the latter two planes is a line, so that one has a distribution of lines, which he calls the *axes* of the correspondence. Given such a distribution, a 2-dimensional system of curves arises: the curves of the system passing through a given point have osculating planes containing the axe through that point. This is what Bompiani calls an *axial system*. This concept, which will come back often in future papers by Bompiani and others,[67] is studied here, with applications to metric and conformal geometry.

One of the central points in Fubini's theory is the concept of projective normal line to a surface. Fubini's definiton, as we said in Sect. 7.1.2, is analytic, but, how to interpret and construct it geometrically? This was one of Bompiani's main objectives, shared with other geometers at the time, including Terracini and Lane. This is considered in the short, preliminary note [31]. More detailed papers will follow in the course of the years.

Let us look at [29, 30],[68] which give a good example of what Bompiani meant by "geometrization" of Fubini's theory. Bompiani starts from the concept of *Darboux curves* which (locally) integrate the distribution of *Darboux tangents*. Take a general point p of a surface S in projective 3-space and consider the linear system of quadrics osculating S at p, i.e. cutting S in a curve with a point of multiplicity at least 3 in p. Taking into account that there is one single quadric cutting S in a curve with a 4-tuple point at p, i.e. the double tangent plane, the triples of tangent lines to these curves at p form a g_3^2. It possesses three triple points, corresponding to the three flexes of the cubic curve related to the g_3^2 which Bompiani claims (but does not prove) to be in general nodal. These are the Darboux directions. The *Segre directions* are conjugated to the Darboux lines with respect to the asymptotic directions, and give rise to the *Segre curves* on S.

Given two 1-dimensional systems of curves, the intersections of the osculating planes to the curves of the two systems at a general point of S give rise to a congruence of axes and therefore to an axial system which includes the two starting systems. Bompiani proves that the axial system determined by two systems of Darboux curves contains a system of Segre curves. Then he gives other geometric properties relating Segre and Darboux curves. Moreover he interprets some projective invariants of the surface, e.g. *Fubini's canonical plane* and *Fubini's canonical curves* and he proves that the canonical curves are indeterminate if and only if the Darboux lines are geodesics with respect to Fubini's first quadratic fundamental form. Bompiani shows that these surfaces are *applicable* onto the surface $xyz = 1$. This is the *projective analogue of a sphere*, because all its projective normal lines pass through a point.

Next, Bompiani considers the *Lie quadric*, i.e. the osculating quadric Q_p in p to the scroll of tangent lines to the asymptotics of one system along the asymptotics of

---

[67] See, e.g., [32,35]. In the latter paper, Bompiani alludes to extensions and recent work by Blaschke on the subject.

[68] These are the first and the second part of a single paper.

the other system.[69] The quadric Q_p is also the limit of the quadric of lines meeting the tangents to three asymptotic curves on S of the same system, one passing through p, the other two passing through two points p', p'', tending to p on the asymptotic curve through p of the other system.

The Lie quadric Q_p varies (with p) in a 2-dimensional system, which envelops a surface, one of whose branches is S. It has three more branches, meeting Q_p in three more points p_1, p_2, p_3 which are vertices of a quadrilater contained in Q_p, the so called *Demoulin*[70] *quadrilateral*. The diagonals of the quadrilateral are polar to each other with respect to Q_p. The line through p meeting these two lines is the so called *Wilczynski line*. The remaining line, also called a Wilczynski line, sits in the tangent plane to S at p but does not pass through p.

Then, Bompiani passes to the *Moutard*[71] *quadric* relative to a tangent direction t in p: this is the locus of all conics which osculate the plane sections of S through p with tangent direction t. Consider the g_3^1 in the $\mathbb{P}^1$ of tangent directions containing the two triple asymptotic directions. For each divisor $D = t_1 + t_2 + t_3$ of this g_3^1, consider the three Moutard quadrics corresponding to t_1, t_2, t_3. They intersect (off the asymtpotic tangents) at a point q_D. Consider the line r_D joining p and q_D. When D varies in the g_3^1, r_D describes a quadric cone. Bompiani studies the case in which this cone is reducible: then it consists of the planes through the projective normal and either one of the asymptotic lines. This gives a neat geometric description of the projective normal line in this case. Then Bompiani relates surfaces with indeterminate canonical curves studied in the first part of the note with the geometry of this cone.

More geometric constructions of the same type can be found in other papers of this period, e.g. in [37, 39, 41, 44].

First, a classical construction which goes back to Pierre–Simon Laplace (1749–1827), explained by Darboux in Chap. II of the second volume of is treatise [58], and translated in a projective-differential language first by C. Segre [89, Sect. 17] and then by Bompiani.[72] Given a surface S in projective 3-space, consider the ruled threefold $V(S)$ on the *Klein quadric* $Q = \mathbb{G}(1, 3)$ in $\mathbb{P}^5$ described by all tangent lines to S. In modern terms, this is the projective bundle associated to the vector bundle Ω_S^1 of holomorphic 1-forms on S. The 2-dimensional system of lines of $V(S)$ has two foci on each general line, which can be interpreted by saying that it is locally siting in a 3-space. The foci lie on the asymptotic lines and describe two surfaces Φ, Φ^*, which verify a Laplace equation of the form (2). The tangent

---

[69]The construction is symmetric, i.e. it does not change by exchanging the two systems of asymptotics.

[70]Alphonse Demoulin (1869–1947).

[71]Théodore–Florentin Moutard (1827–1901).

[72]In [12, 41]. These papers have been written as an answer to two papers [95, 96] by Gerhard Thomsen (1899–1934), devoted to a similar approach, but using analytic methods. Bompiani vindicates his idea, already appeared in [12], and claims his treatment to be "simpler and therefore more fruitful".

lines to the u_i curves, for $1 \leq i \leq 2$, on either Φ or Φ^* form a 2-dimensional system of lines which has again two foci on each general line. These foci describe further surfaces, which are called *Laplace transforms* of Φ and Φ^*, and so on. The first step of the Laplace transformation of Φ and Φ^* in one direction is the other surface Φ^* and Φ. Hence, the Laplace transformation of Φ and Φ^* give rise to only one (for each of them) new surface Φ_1 and Φ_1^*.[73] Bompiani proves that the tangent planes to these surfaces cut the Klein quadric Q along conics corresponding to different rulings of the same quadrics in $\mathbb{P}^3$ which osculate (in a certain sense) the original surface S. These are nothing else than the Lie quadrics. The main result in [37] is the reconstruction of Fubini's first two differential forms in terms of cross ratios related to Φ and Φ^*. An idea which goes back to Segre's definition of the Wölffing–Mehmke–Segre invariant.[74] Here Bompiani gives a geometric definition of two differential forms, which he calls *elementary* forms, from which the first two Fubini's differential forms (which Bompiani calls *normal forms*) can be deduced. For instance, the first form is obtained in this way. Take a point O on the surface S and a parametric representation $x = x(u, v)$ of S around O in which the asymtotic curves are the coordinate curves. Take a curve C through O on S, with direction corresponding to du/dv. Take O' on C different from O. Then take the asymptotic quadrilateral with opposite vertices at O, O' and let $O_1' O_2'$ the other two vertices. Let T_1 be the point where the line $O'O_1'$ intersects the tangent plane to S at O. Let M be a general fixed point of the same line and take the limit of the cross ratio $(O_1' O' T_1 M)$, when O' tends to O. This is independent on M and is the value of the differential form at du/dv. Replacing O_1' with O_2' one gets another differential form.

Finally, in [44] Bompiani presents another ingenuous geometric construction and interpretation of the most important PDG invariants of a surface in 3-space. One considers the Lie quadric Q_p at a point p of the surface S. The pencil defined by Q_p and by the tangent plane counted twice is the so called *Darboux pencil* $\mathcal{D}_p$. Bompiani sorts out a projectively well determined quadric in it, different from Q_p and the double tangent plane. This gives, via the cross ratio, a well determined projective coordinate in $\mathcal{D}_p$. Take a quadric in $\mathcal{D}_p$, as p varies, defined by a constant coordinate θ. This is a 2-dimensional system of quadrics $\mathcal{Q}_\theta$, which envelopes a surface Q_θ. The contact points of Q_θ with the general surface of $\mathcal{Q}_\theta$ consists of 4 points: in the case of Q_p these are the vertices of the Demoulin quadrilateral. The curve described by these 4 points when θ varies has a singular point at p and one of its tangents is Fubini's projective normal. This seems to be one of the most efficient geometric interpretation of this concept.

---

[73] The Laplace transformation had been devised in order to try to solve, by simple integrations, a Laplace equation. The result is that this is possible if the Laplace transformation process *degenerates* after finitely many steps, producing a curve (or even a point), rather that a surface.

[74] See (4). This idea comes up again in [39, 40]. These are the first and second part of one and the same paper. They look preliminary notes, since they contain little or no proofs.

Fubini was apparently enthusiastic about Bompiani's work:

> Bravo, bravissimo! I like your result a lot and your method looks very promising to me. This correspondence between V_2 of S_3 and of S_5 is certainly very interesting. Who knows what happens in S_5 when one starts from two V_2 of S_3 which are W-transformed one of the other?[75]
>
> Your results look really interesting and remarkable to me.[76]

Fubini's favour can be measured from the fact that he let Bompiani write one of the four appendices to his book [64].[77] This is an expository paper in which Bompiani lists, with no proof, the results he has being obtaining in the previous years, from 1923 to 1926 on the subject.

Also [34] is mostly expository: reading it is pleasant, since Bompiani gives here, in the style of Fubini's appendix to Bianchi's treatise, a very nice and lucid account of the development of Fubini's projective theory of surfaces in projective 3-space.

He starts with a quick historical excursus, locating in Monge the beginnings of PDG. This gives him the opportunity of describing what he calls the "anatomy" of surfaces, with the introduction of the main concepts we have been discussing above. Among them Fubini's projective normal, of which Bompiani gives a "not so brilliant" – his words – geometric construction (too long to be reported on here).

Then he passes "from anatomy to biology", trying to develop a theory from the projective invariants introduced above. Here he compares the classical Riemannian case with the PDG viewpoint, showing how Darboux's *moving frame* and Gauss' *differential forms* approaches mirror in Wilczynski's and Fubini's ones. Concentrating on Fubini's theory he points out his geometric obscurity, promising suitable geometric interpretations. For example he shows that the projective linear element measures, in a suitable sense, the distance of a curve having a given direction from the tangent plane with respect to the metric determined by the corresponding Lie quadric.

Finally, Bompiani passes "from anatomy and biology of our individual to sociology", meaning that he intends to talk about *classification results* of classes of surfaces verifying particular interesting conditions. Among which he mentions surfaces with undetermined canonical curves.[78]

From Bompiani's papers on Fubini's theory in the period we are considering, their strong and weak points are apparent. It is certainly impressive the remarkable quantity of work he had been doing on the subject; his equally remarkable computational ability; his lively geometric intuition corroborated by a good basic knowledge of algebraic geometry, learned for sure at Castelnuovo's school. At the same time one is impressed by the fragmentation of his research in this period: he does not

---

[75] G. Fubini to E. Bompiani, n. 1, from Turin, March 22, 1926, in [84]. Fubini probably refers here to the papers on Laplace transformations.

[76] G. Fubini to E. Bompiani, n. 2, from Turin, March 30, 1926, [84].

[77] That is, [38]. The other three appendices are [92, 97].

[78] Studied in [29, 30].

attempt either to construct a theory or to proceed on the way opened by Fubini investigating tempting higher dimensional/codimensional problems analogous to the ones Fubini solved for surfaces in 3-space.[79] He seems to be satisfied with the plethora of diverse geometric interpretations he gives. They are nice, and show how Bompiani mastered projective geometry. On the other hand they also show how much of indeterminacy there is in this field and it is not clear where all this ends up.

We cannot end this overview without mentioning a couple of more papers, which, though related to the above, have, in our view, some extra interest, and we will try to explain why.

In [42, 45], Bompiani continues the study of the geometry of surfaces verifying Laplace equations. In the former paper, he associates to a Laplace equation two second order symmetric differential forms on the surface which completely determine the equation. He gives some geometric interpretation for them, but does not relate them to Fubini's ones. Then, following ideas by Lie, he studies equations having *infinitesimal transformations* fixing them. In the latter paper, Bompiani uses interesting, but complicated, projective constructions related to the second order neighborhood of a point on a surface verifying a Laplace equation. His aim is to find suitable local coordinates, sort of Frenet frames in the present situation, in the osculating $\mathbb{P}^4$ to the surface, which have an invariant meaning with respect to change of local coordinates on the surface. Everything related to these coordinates is also an invariant of the surface and therefore of the Laplace equation. The interest in these papers resides in the fact that Bompiani tries here to extend to higher codimensional situations Fubini's approach. Proudly enough, Bompiani claims in the introduction: "In this way, from a geometric viewpoint, the projective theory of surfaces Φ with a double conjugate system, has essentially been completed". We think this is not the case. The results are not outstanding, and moreover parts of these papers are (for Bompiani's good expository standards) unusually obscure. However, continuing in this direction would have been, and would still be, interesting.

Finally, paper [43] despite its brevity has the ambition of indicating some guidelines for the study of PDG of varieties in higher projective spaces, relating it to recent contributions by É. Cartan on *varieties with a projective connection* [52], giving for them a strong projective geometric interpretation. This is also the object of a paper by Bortolotti [51], published in the same volume of the Rendiconti Lincei, in which the author goes even deeper into this analysis, showing how various notions of Riemannian, affine, projective connections introduced in the previous years by Ludwig Berwald (1883–1942), É. Cartan, Jan Shouten (1883–1971), Hermann Weyl (1885–1955), all fall into a unique pattern, which relates to the notions of Bompiani's *quasi-asymptotic curves*[80] and *axial systems* discussed above.[81]

---

[79]With one little exception, we will mention in while.

[80]See [13] and later [15, 17–19].

[81]By the way, the concept of quasi-asymptotic curve will play an essential role in a beautiful geometric characterization of *Veronese varieties* given in [20, 21], which still deserves a better understanding and a modern treatment.

Bompiani shows that, to give a projective connection à la Cartan is the same as giving for each point of a variety, a cone of a suitable dimension inside the osculating spaces of a given order (two for ordinary connections). The geodetics of the connection are those curves whose osculating spaces (of the appropriate dimension, in general the osculating planes) meet the cone along generators. He shows how this idea applies in the projective setting, for surfaces and for ordinary connections (i.e., for osculating 2-spaces) under the regularity assumption that the surface does not verify a Laplace equation. Then he shows how this specializes to the metric situation.

It looks like as if these papers of Bompiani and Bortolotti try to vindicate the originality and power, as well as the priority, of their projective geometric ideas. A defensive attitude, which shows however how close were the ideas and tools of Italian projective differential geometers to the new trends which were developing in those years, or about to be developed.[82]

9 Back to Rome

Finally the circle closes up and, as we know, thanks to Castelnuovo, Bompiani will succeed going back to the Faculty of Science of the University of Rome. In the meeting of November 9, 1927, Castelnuovo proposes to hire Bompiani with the following motivation, which will be unanimously approved:

> Enrico Bompiani, ordinary professor of projective and descriptive Geometry at the University of Bologna, developed in the last 15 years, with a strong commitment and strenuous work, a coherent research project on various topics of metric and projective differential Geometry, following Monge's viewpoint, the founder of descriptive geometry. Bompiani's value is universally well known, his teaching ability makes clear why our faculty decided to propose him for the vacant chair of descriptive Geometry, which is a part of the discipline taught by him in Bologna.[83]

Answering to the Rector's letter, Bompiani writes:

> Thank You for the communication concerning my new position and for the welcome You give me in this glorious University. I feel proud to come back, as a professor, in the

---

[82]The concept of a *connection in a vector bundle*, which subsumes all previous ones, will be introduced only in 1950 by Jean-Louis Koszul.

[83]"Enrico Bompiani, professore stabile di Geometria proiettiva e descrittiva nella R. Università di Bologna, va svolgendo da quindici anni, con altro impegno e assiduo lavoro, un piano organico di ricerche in piani svariati della Geometri differenziale metrica e proiettiva, nell'indirizzo segnato da Monge, il fondatore della geometria descrittiva. Il valore scientifico del Bompiani universalmente riconosciuto, le sue eminenti qualità didattiche spiegano perché la nostra Facoltà abbia pensato al suo nome per coprire la cattedra vacante di Geometria descrittiva, che fa parte della materia di cui il Bompiani è titolare a Bologna." See [3].

University in which I have been a student and where I started my career as a teacher. I count on my strength in order to fulfill the task which trustfully my Colleagues devolve upon me.[84]

10 Final Comments

10.1 Bompiani in the Following Years

Coming back to Rome opens a new season in Bompiani's academic career, in which he will be involved in organizational duties at a national and international level. He will intensively pursue this, together with mathematics, for all the rest of his life.

In 1927, the National Committee for Mathematics was created inside the Italian National Reserach Council (CNR), founded in 1923 with Volterra as a president. The National Committee will have the responsibility of coordinating all research projects and funding in mathematics. Its creation takes place in the framework of a reorganization of CNR promoted by the Fascist government, which will lead to the substitution of Volterra with Guglielmo Marconi (1874–1937).

Bompiani was, since the beginning, nominated secretary of the National Committee for Mathematics, with Bianchi as president. Bianchi's health conditions were very poor, and he was willing to accept the responsibility in view of Bompiani's effective collaboration.[85] Bianchi died in 1928, and the next president was Gaetano Scorza (1876–1939), who was assisted by a Presidency Executive Committee formed by Berzolari, Bompiani and Mauro Picone (1885–1977).[86]

Bompiani was also a member of the Scientific Committee of the Italian Mathematical Union (UMI) since its foundation in 1923.[87] He was vice-president of UMI from 1938 to 1949, then president from 1949 to 1952 and finally honorary president from 1952 to his death. From 1951 to 1956 he was general secretary of the International Mathematical Union and he was one of the Italian delegates to the General Assembly for its new foundation in 1951.

---

[84]"La ringrazio della comunicazione che mi riguarda e del benvenuto ch'Ella mi porge in questo glorioso Ateneo. È per me un orgoglio rientrare come professore nella Università che mi ha accolto studente e all'inizio della mia carriera didattica. Confido nelle mie forze per assolvere il compito cui mi chiama la fiducia dei Colleghi. Mi è particolarmente gradito porgerLe i miei ossequi in attesa di riverirLa personalmente appena le operazioni relative al cambiamento di sede me lo permetteranno." E. Bompiani to the Rector of the University of Rome, December 1, 1927, see [4].

[85]L. Bianchi to E. Bompiani, May 2, 1928, Fondo Bompiani, 2. CNR, Comitato matematico italiano, Accademia dei XL.

[86]To understand the role played by the Committee in the organization of mathematics in Italy and specifically by Bompiani and Picone, see [68, p. 183–214].

[87]With Volterra's dismissal as a president of CNR, the UMI lost its function as the Committee for Mathematics of CNR, see [68, p. 184].

Bompiani's relationship with Bologna remained solid due to the presence of his former students there. This is also witnessed by the fact that the University of Bologna awarded him with an honorary degree on June 23, 1966.[88]

10.2 Bompiani and the Fascist Regime

Bompiani's relationship with the Fascist regime lies outside the scope of this paper. He became a member of the Fascist Party only in July 1930. It is impossible here to make a serious analysis about the reasons and nature of his involvement with Fascism. However, it is worth to point out a couple of significant episodes.

In 1939, Bompiani was involved in the organization of the Exhibition of the Italian Civilization. For this he wrote a document,[89] in which there is no mention of contributions of Italian Jewish mathematicians.[90] Bompiani was also one of the members of the Scientific Committee of UMI who, after the hatred racial laws, signed the shameful declaration[91] containing an absurd statement about the arian character of Italian mathematics. Besides any political consideration, the meanness and ungratefulness of Bompiani with respect to his masters and mentors is absolutely breathtaking. It is likely that this was among the reasons why Castelnuovo did not support him on the occasion of the purge process at the end of the Fascist regime.[92] This purge regarded both the Accademia dei Lincei (Castelnuovo was a member of the jury) and the University. The accusation regarding the University position mainly concerned propaganda activities supporting the Fascist regime abroad. As an example, Bompiani's involvement in the XI Annual Conference of the Entr'aide Universitaire Internationale in Brno in 1932 was much criticized [84, p. 127–137].

---

[88] Bompiani received the same honor from the University of Groningen (The Netherlands) in 1964 and from the University of Iaşi (Romania) in 1970.

[89] "Contributi italiani alla matematica", [Archivio Centrale dello Stato, Busta 1025].

[90] By contrast, in a talk of 1931 given in the United States, with essentially the same title "Italian contributions to modern mathematics" (see [46]), these names abundantly appear. The omissions in the 1939 document are constructed in a sneaky, ambiguous way. First of all, only dead scholars are mentioned. This could have justified neglecting people like Castelnuovo, Enriques, Fubini, Volterra, but not Corrado Segre who died in 1924. On the other hand also important non-Jewish mathematicians close to Bompiani, like Berzolari and Scorza, are neglected, which could justify neglecting whomever else.

[91] Of December 10, 1939; see "Bollettino UMI", S II, 1 (1939), 89–90.

[92] Bompiani complains about this in the letter: E. Bompiani to L. Berzolari, Roma, April 2, 1946. Archivio Istituto Storico Italiano per il Medioevo, Fondo R. Morghen. Raffaello Morghen (1896–1983) was General Director of the National Academy of Lincei. Among his papers there is a section regarding Castelnuovo's activity concerning purge processes at the Academy in the period 1944–1946. This has been discovered by Maurizio Mattaliano, the responsible person for the Archive of the Istituto per le Applicazioni del Calcolo (private communication by P. Nastasi).

As a matter of fact, the purge only resulted in an official blame, which in fact had no practical effect.[93] There are however documents, whose relevance should be analyzed, attesting Bompiani's involvement in the organization of courses when the University was closed for the German occupation, in saving from destruction the mathematical library of the University of Rome, in helping young people prosecuted from the Nazis [4, 98].

10.3 Final Comments on Projective Differential Geometry

PDG is a vast area of mathematics, where a great variety of tools (ranging from differential geometry, to Lie groups, analysis, topology, algebraic geometry) has been displayed in the course of the last century or so. As vast as it is, it is at the same time a rather dispersive area. A number of serious, difficult and fascinating problems, are mixed with more extemporaneous topics. So far no leading trend emerged driving the research, as it happened, for instance, for algebraic geometry, where birational classification of varieties is the main objective. This caused sometimes a certain distrust about the discipline and its scholars. So that PDG has been considered to be marginal with respect to other important disciplines. To the point that even from an historical viewpoint, it is still missing a coherent reconstruction of its development, and a systematic analysis of the main contributions and of its main characters. It should not be believed, however, either that PDG is less important because of this, or that it is a dead subject. We address the reader to the beautiful paper by Griffiths and Harris [67] to get a flavor of the importance of PDG, of the open problems and of a modern way of looking at the subject.[94] In particular, the reader will find there interesting comments on Fubini's work and suggestions on how to try to extend it to higher dimension and codimension.

11 Conclusion

As we said in the Introduction, this article had a limited and well defined scope. Precisely the one of introducing Bompiani's personality, following him in a first segment of his career, with stress on his activity on PDG in the years spent in Bologna. From our presentation the following main features are apparent. First, Bompiani's resolute and ambitious personality, with a strong inclination and passion for scientific and didactic work. Second, Castelnuovo's constant presence in his life as an advisor and mentor. Third, Segre's and Fubini's heavy scientific influence. Finally, Bompiani's determination to pursue his academic career, having as one of

---

[93]For more information, see, e.g., [68, 71].

[94]Other recent books on the subject are [1, 2, 70, 101]. See also the interesting thesis [86].

his main objectives to become full professor in Rome. A position which immediately gives him a large visibility and an active influence on scientific policy at a national and even international level. This is the result of the scientific and academic connections he was able to construct in the course of the years with people we have been mentioning in this paper. We hope that our contribution showed the interest in pursuing further research on Bompiani, on the ambient in which he lived and worked, on his scientific environment and on PDG.

Acknowledgements The first author is a member of G.N.S.A.G.A. of INdAM; the second author is a member of HAR2010-17461 and of 2009-SGR-417.

References

1. Akivis, Maks, A. and Vladislav V. Goldberg. 1993. *Projective Differential Geometry of Submanifolds*. Amsterdam: Elsevier.
2. Akivis, Maks, A. and Vladislav V. Goldberg. 2004. *Differential Geometry of Varieties with Degenerate Gauss Maps*. CMS Books in Mathematics, 18. London: Springer.
3. Archivio Centrale dello Stato. 1922–1964. Enrico Bompiani. Fascicolo personale. M.P.I. Professori Universitari B. 65.
4. Archivio Storico Università di Roma. 1913–1975, Roma. Facoltà di Scienze: Fascicolo personale E. Bompiani.
5. Aldo Brigaglia and Ciro Ciliberto. 1995. Italian algebraic geometry between the two world wars. *Queen's Papers in Pure and Applied Math.*, 100. Canada: Queen's University, Kingston.
6. Brigaglia, Aldo and Ciro Ciliberto. 1998. *La geometria algebrica italiana tra le due guerre mondiali*. In: La Matematica Italiana italiana dopo l'unità, Gli anni tra le due guerre mondiali, S. Di Sieno, A. Guerraggio, P. Nastasi (eds.) Milano: Marcos y Marcos.
7. Berzolari, Luigi. 1897. Sugli invarianti differenziali proiettivi delle curve di un iperspazio. *Annali di Matematica Pura e Applicata* 26 (1): 1–58.
8. Bianchi, Luigi. 1927. *Lezioni di geometria differenziale*. 3a ed. 4 vol. Bologna: Zanichelli.
9. Bompiani, Enrico. 1924–1925. Sull'andamento degli integrali di un'equazione di Riccati. *Memorie dell'Accademia delle Scienze di Bologna* 8 (2): 61–65.
10. Bompiani, Enrico. 1910. Sopra le funzioni permutabili. *Atti della Reale Accademia dei Lincei: Rendiconti* (5) 19: 101–104.
11. Bompiani, Enrico. 1912. Sopra una trasformazione classica di Sophus Lie. *Atti della Reale Accademia dei Lincei: Rendiconti* (5) 21: 607–704.
12. Bompiani, Enrico. 1912. Sull'equazione di Laplace. *Rendiconti del Circolo Matematico di Palermo* 34: 383–407.
13. Bompiani, Enrico. 1913. Alcune estensioni dei teoremi di Meusnier e di Eulero. *Atti della Reale Accademia delle Scienze di Torino* 48: 393–410.
14. Bompiani, Enrico. 1913–1914. Contributo allo studio dei sistemi lineari di rette nello spazio a 4 dimensioni. *Atti del Reale Istituto Veneto di Scienze, Lettere e Arti* 73: 579–616.
15. Bompiani, Enrico. 1914. Alcune proprietà proiettivo-differenziali dei sistemi di rette negli iperspazi. *Rendiconti del Circolo Matematico di Palermo* 37: 305–331.
16. Bompiani, Enrico. 1914. E.J. Wilzinski. Alcune vedute generali della geometria moderna (traduzione e note). *Bollettino di Bibliografia e storia delle scienze matematiche* 16: 97–109.
17. Bompiani, Enrico. 1914. Sullo spazio d'immersione di superficie possedenti dati sistemi di curve. *Rendiconti del Reale Istituto Lombardo di Scienze e Lettere*. 47: 177–192..

18. Bompiani, Enrico. 1916. Analisi metrica delle quasi-asintotiche sulle superficie degli iperspazi. *Atti della Reale Accademia dei Lincei: Rendiconti* 5(25): 493–497; 576–578.
19. Bompiani, Enrico. 1919. Sur les courbes quasi-asymptotiques des surfaces dans un espace quelconque. *Comptes Rendus de l'Académie des Sciences* 168: 755–757.
20. Bompiani, Enrico. 1921. Proprietà differenziale caratteristica delle superficie che rappresentano la totalità delle curve algebriche di dato ordine. *Atti della Reale Accademia dei Lincei: Rendiconti* (5) 30: 248–251.
21. Bompiani, Enrico. 1921. Proprietà differenziali caratteristiche di enti geometrici. *Atti della Reale Accademia dei Lincei: Memorie* (5) 13: 451–474.
22. Bompiani, Enrico. 1921. Studi sugli spazi curvi: del parallelismo di una varietà qualunque, parte i. *Atti del Reale Istituto Veneto di Scienze, Lettere e Arti* 80: 355–386.
23. Bompiani, Enrico. 1921. Studi sugli spazi curvi: del parallelismo di una varietà qualunque, parte ii. *Atti del Reale Istituto Veneto di Scienze, Lettere e Arti* 80: 839–859.
24. Bompiani, Enrico. 1921. Studi sugli spazi curvi: la seconda forma fondamentale di una V_m in una V_n. *Rendiconti del Circolo Matematico di Palermo* 80: 1113–1145.
25. Bompiani, Enrico. 1922. Sulla rappresentazione iperspaziale delle curve piane. *Atti della Reale Accademia dei Lincei: Rendiconti* (5) 31: 471–474.
26. Bompiani, Enrico. 1923. Corrispondenza puntuale fra due superficie e rappresentazione conforme. *Atti della Reale Accademia dei Lincei: Rendiconti* (5) 32: 376–380.
27. Bompiani, Enrico. 1923. Spazi riemanniani luoghi di varietà totalmente geodetiche. *Rend. della R. Acc. Nazionale dei Lincei* (5) 32(2): 14–15.
28. Bompiani, Enrico. 1923–1924. Proprietà generali della rappresentazione puntuale fra due superficie. *Annali di Matematica Pura e Applicata* (4) 1: 259–284.
29. Bompiani, Enrico. 1924. Contributi alla geometria proiettivo-differenziale di una superficie. *Bollettino della Unione Matematica Italiana* (1) 3: 49–56.
30. Bompiani, Enrico. 1924. Contributi alla geometria proiettivo-differenziale di una superficie. *Bollettino della Unione Matematica Italiana* (1) 3: 97–100.
31. Bompiani, Enrico. 1924. Nozioni di geometria proiettivo–differenziale relativa ad una superficie dello spazio ordinario. *Atti della Reale Accademia dei Lincei: Rendiconti* (5) 3(1): 85–90.
32. Bompiani, Enrico. 1924. Sistemi coniugati e alcuni sistemi assiali di linee sopra una superficie dello spazio ordinario. *Bollettino della Unione Matematica Italiana* 3: 10–16.
33. Bompiani, Enrico. 1924. Spazi riemanniani luoghi di varietà totalmente geodetiche. *Rendiconti del Seminario Matematico dell'Università di Roma* 48: 121–134.
34. Bompiani, Enrico. 1924. Sulla geometria proiettivo-differenziale delle superficie. *Rend. del Sem. Mat. dell'Univ. di Roma* 2(2): 121–134.
35. Bompiani, Enrico. 1925. Sulla corrispondenza puntuale fra due superficie a punti planari. *Bollettino della Unione Matematica Italiana* (1) 4: 195–199.
36. Bompiani, Enrico. 1925. Un teorema di confronto e un teorema di unicitá per l'equazione differenziale $y' = f(x, y)$. *Atti della Reale Accademia dei Lincei: Rendiconti* (6) 1: 298–302..
37. Bompiani, Enrico. 1926. Ancora sulla geometria delle superficie considerate nello spazio rigato. *Atti della Reale Accademia dei Lincei: Rendiconti* 6(4): 263–367.
38. Bompiani, Enrico. 1926. I fondamenti geometrici della teoria proiettiva delle curve e delle superficie. In G. Fubini e E. Čech, *Geometria proiettiva differenziale*, 671–727. Bologna: Zanichelli.
39. Bompiani, Enrico. 1926. Le forme elementari e la teoria proiettiva dalle superficie. *Bollettino della Unione Matematica Italiana* (1) 5: 167–1730.
40. Bompiani, Enrico. 1926. Le forme elementari e la teoria proiettiva dalle superficie. *Bollettino della Unione Matematica Italiana* (1) 5: 209–214.
41. Bompiani, Enrico. 1926. Sulla geometria delle superficie considerate nello spazio rigato. *Atti della Reale Accademia dei Lincei: Rendiconti* (6) 3: 395–400.
42. Bompiani, Enrico. 1926. Sulla geometria dell'equazione di Laplace. *Atti della Reale Accademia dei Lincei: Rendiconti* (6) 5: 143–149.

43. Bompiani, Enrico. 1927. Alcune idee generali per lo studio differenziale delle varietà. *Atti della Reale Accademia dei Lincei: Rendiconti* (6) 5: 383–389.
44. Bompiani, Enrico. 1927. Fascio di quadriche di Darboux e normale proiettiva in un punto di una superficie. *Atti della Reale Accademia dei Lincei: Rendiconti* (6) 6: 187–190.
45. Bompiani, Enrico. 1927. Ricerche analitiche e geometriche sull'equazione di laplace. *Atti della Reale Accademia dei Lincei: Rendiconti* (6) 5: 84–90.
46. Bompiani, Enrico. 1931. Italian contributions to modern mathematics. *The American Mathematical Monthly* 38(2): 83–95.
47. Bompiani, Enrico. 1931. Sulle superficie integrali di due o piu equazioni lineari a derivate parziali del 3 ordine. *Annali di Matematica Pura e Applicata* (4) 9: 287–306.
48. Bompiani, Enrico. 1934. L'italia all'esposizione di Chicago. *Credere, Istituto fascista di cultura di Bologna* 4–5: 3–12.
49. Bompiani, Enrico. 1966. Dopo cinquant'anni dall'inizio della geometria proiettiva differenziale secondo G. Fubini. *Rendiconti del Seminario Matematico di Torino* 25: 83–106.
50. Bompiani, Enrico. 1967. Recenti progressi nella geometria proiettiva differenziali degli iperspazi. In *Proceedings of the Fifth International Congress of Mathematicians: Cambridge, 22–28 August 1912*. Nendeln: Kraus Reprint.
51. Bortolotti, Enea. 1927. Sistemi assiali e connessioni nelle V_m. *Atti della Reale Accademia dei Lincei: Rendiconti* (6) 55: 390–395.
52. Castelnuovo, Guido. Élie, Cartan. 1924. Sur les variétés a connexion projective. *Bull. Soc. Math.* 52: 205–241.
53. Guido, Castelnuovo. 1891. Ricerche di geometria della retta nello spazio a quattro dimensioni. *Atti del Reale Istituto Veneto di Scienze, Lettere e Arti.* (7) 2: 855–901.
54. Chern, Shiing-Shen, S.-Y. Cheng, G. Tian, and P. (Peter) Li. 1992. *A Mathematician and his Mathematical Work: Selected Papers of S.S. Chern*. World Scientific series in 20th century mathematics ; v. 4. Singapore: World Scientific.
55. Cartan, Élie and Jean Leray. 1937. *La théorie des groupes finis et continus et la géométrie différentielle, traitées par la méthode du repère mobile*. Cahiers scientifiques; fasc. 18. Paris: Gauthier-Villars.
56. Coolidge, Julian Lowell. 1992. Corrado Segre. *Bulletin of the American Mathematical Society* 33 (3): 352–357.
57. Ciliberto, Ciro and Edoardo Sernesi. 1992. Singularities of the theta divisor and congruences of planes. *Journal of Algebraic Geometry* 1: 231–250.
58. Darboux, Gaston 1914–1925. *Leçons sur la théorie générale des surfaces et les applications gómétriques du calcul infinitésimal*. Cours de géométrie de la Faculté des Sciences. Paris: Gauthier-Villars, 2e. éd.
59. Fubini, Guido. 1914. Definizione proiettivo–differenziale di una superficie. *Atti della Reale Accademia delle Scienze di Torino* 48: 542–558.
60. Fubini, Guido. 1916. Applicabilità proiettiva di due superficie. *Rendiconti del Circolo Matematico di Palermo* 41: 135–162.
61. Fubini, Guido. 1916. Invarianti proiettivo–differenziali di curve tracciate su una superficie e definizione proiettivo–differenziale di una superficie. *Annali di Matematica Pura e Applicata* (3) 25: 229–252.
62. Fubini, Guido. 1927. Introduzione alla geometria differenziale di una superficie. In *Luigi Bianchi*. Lezioni di geometria differenziale, 3a ediz., 812–833. Bologna: Zanichelli.
63. Fubini, Guido and Eduard Čech. 1931. *Introduction a la géométrie projective différentielle des surfaces*. Paris: Gauthier-Villars.
64. Fubini, Guido and Eduard Čech. 1926–1927. *Geometria proiettiva differenziale*. Bologna: N. Zanichelli.
65. Green, Gabriel. Abstract. 1916. *Bulletin of the American Mathematical Society* 23: 73.
66. Griffiths, Phillip. 1974. On Cartan's method of Lie groups and moving frames as applied to uniqueness and existence questions in differential geometry. *Duke Math. J.* 41(4): 775–814.
67. Griffiths, Phillip and Joe Harris. 1979. Algebraic geometry and local differential geometry. *Ann. Scient. Ec. Norm. Sup.* 12: 355–432.

68. Guerraggio, Angelo and Pietro Nastasi. 2005. *Italian Mathematics between the two World Wars*. Basel: Birkhäuser.
69. Halphen, Georges Henri. 1878. *Sur les invariants differentiels*. Paris: Gauthier-Villars.
70. Ivey Thomas A. and Joseph M. Landsberg. 2003. *Cartan for beginners: differential geometry via moving frames and exterior differential systems*. Graduate studies in mathematics, 61. Providence: American Mathematical Society.
71. Israel, Giorgio and Pietro Nastasi. 1998. *Scienza e razza nell'Italia fascista*. Bologna: Il Mulino.
72. Israel, Giorgio. 1988. *Bompiani Enrico, Dizionario Biografico degli Italiani, Primo supplemento*, XXXIV: 471–473. Roma: Istituto della Enciclopedia Italiana.
73. Lane, Ernest P. 1926. Wilczynski's and Fubini's Canonical System of Differential Equations. equations. *Bulletin of the American Mathematical Society* 32 (4): 365–373.
74. Lane, Ernest P. 1932. *Projective Differential Geometry of Curves and Surfaces*. Chicago: The University of Chicago Press.
75. Lane, Ernest P. 1934. Biographical Memoir of Ernest Julius Wilczynski (1876–1932) *Nat. Acad. of Sci.* 16 (6): 295–327.
76. Lane, Ernest P. 1942. *A Treatise on Projective Differential Geometry*. Chicago: The University of Chicago Press.
77. Levi, Eugenio Elia. 1908. Saggio sulla teoria delle superficie immerse in un iperspazio. *Annali della Scuola Normale Superiore di Pisa* 10: 1–99.
78. Lie, Sophus and Robert Hermann. 1884. *Sophus Lie's 1884 Differential Invariant Paper*. Lie groups; 3. Brookline, MA: Math Sci Press, 1976. In part a translation of "Über Differentialinvarianten", Mathematische Annalen 24: 573—578.
79. Lie, Sophus. 1872. Ueber Complexe, insbesondere Linien- und Kugel-Complexe, mit Anwendung auf die Theorie partieller Differential-Gleichungen. *Mathematische Annalen* 5: 145–208, 209–256.
80. Lisio, Carlo de. 2005. *Giulio Pittarelli e i matematici molisani dell'Ottocento*. Ferrazzano: Enne.
81. Martinelli, Enzo. 1975. Enrico Bompiani. *Annali di Matematica Pura e Applicata* 107(1): 1–3..
82. Monge, Gaspard. 1810. Sur les équations différentielles des courbes du second degré. *Correspondance sur l'École Polytechnique* 2: 51–54.
83. Nastasi, Pietro and Enrico Rogora. 2007. *Mon cher ami. Illustre professore. Corrispondenza di Ugo Amaldi (1897–1953)*. Roma: Nuova Cultura.
84. Paoloni, Giovanni. 1991. *Il fondo "Enrico Bompiani"*. Quaderni P.RI.ST.EM. N. 2. Per l'archivio della corrispondenza dei matematici italiani. Milano: Bocconi.
85. Pezzo, Pasquale del. 1886. Sugli spazii tangenti ad una superficie o ad una varietà immersa in uno spazio a più dimensioni. *Rendiconti della Reale Accademia delle Scienze Fisiche e Matematiche di Napoli* 25: 176–180.
86. Pirio, Luc. 2004. *Équations fonctionelles abéliennes et géométrie des tissus*. PhD thesis. France: Université de Paris VI.
87. Pizzocchero, Livio. 1998. Geometria differenziale. In *La matematica italiana dopo l'unità, Gli anni tra le due guerre mondiali*, S. Di Sieno, A. Guerraggio, P. Nastasi ed., Geometria differenziale, 321–379. Milano: Marcos y Marcos.
88. Segre, Corrado. 1897. Su alcuni punti singolari delle curve algebriche e sulla linea parabolica di una superficie. *Atti della Reale Accademia dei Lincei: Rendiconti* (5) 6: 168–175.
89. Segre, Corrado. 1906–1907. Su una classe di superficie degli iperspazi, legata con le equazioni lineari alle derivate parziali di 2° ordine. *Atti della Reale Accademia delle Scienze di Torino* 42: 1047–1079.
90. Segre, Corrado. 1910. Preliminari di una teoria delle varietà luoghi di spazi. *Rendiconti del Circolo Matematico di Palermo* 30: 87–121.
91. Segre, Beniamino. 1978. Enrico Bompiani (1889–1975). In *E. Bompiani Opere scelte*, 1–17. Bologna: Unione Matematica Italiana.

92. Terracini, Alessandro. 1927. *Esposizione di alcuni risultati di geometria proiettiva differenziale negli iperspazi and Sulle superficie aventi un sistema, o entrambi, di asintotiche in complessi lineari*. In G. Fubini e E. Čech, *Geometria proiettivo-differenziale*: 729–769 and 770–782. Bologna: Zanichelli.
93. Terracini, Alessandro. 1949–1950. Guido Fubini e la geometria proiettiva differenziale. *Seminario Matematico dell'Università e Politecnico di Torino.* 9: 97–123.
94. Terracini, Alessandro. 1958. *Prefazione al vol. II delle Opere di Corrado Segre*, V–XVIII. Roma: Cremonese.
95. Thomsen, Gerhard. 1924. Über eine gemeinsame Behandlungsweise verschiedener Differentialgeometrien. *Mathematische Zeitschrift* 21 (1) : 254–285.
96. Thomsen, Gerhard. 1925. Über eine liniengeometrische Behandlungsweise der projektiven Flächentheorie und die projektive Geometrie der Systeme von Flächen zweiter Ordnung. *Abhandlungen aus dem Mathematischen Seminar der Universität Hamburg* 4 (1): 232–266.
97. Tzitzeica, M. Gheorge. 1927. *Appendice al trattato di Geometria proiettivo differenziale di G. Fubini e E. Čech*, chapter Sur la déformation de certaines surfaces tétraédales, les surfaces S et les réseaux R, 663–668. Bologna: Zanichelli.
98. Vaccaro, Giuseppe. 1975. Enrico Bompiani (1889–1975). *Bollettino della Unione Matematica Italiana* (VI) 12: 1–36.
99. Wilczynski, Ernest J. 1905. *Projective Differential Geometry of Curves and Ruled Surfaces.* New York: Chelsea Publishing Co.
100. Wilczynski, Ernest J. 1913. Some General Aspects of Modern Geometry. *Bulletin of the American Mathematical Society* 19: 331–342.
101. Zak, Fyodor. 1993. *Tangents and secants of algebraic varieties. Translations of Mathematical Monographs. 127.* American Mathematical Society.

Dario Graffi in a Complex Historical Period

Mauro Fabrizio

Abstract This essay examines a selection of the scientific works of Dario Graffi and frame them within the historical period of his life. The selection is restricted to a some papers, essentially those with the more outstanding results, which though are enough to show why Dario Graffi has been an eminent mathematical physicist of the last century. The essay ends with a biography.

Dario Graffi was an leading mathematical physicist. But what is mathematical physics? If we think of two eminent personalities, such as Enrico Betti and Eugenio Beltrami, considered by many among the founders of mathematical physics in the modern sense, we see that we cannot label them simply as mathematical physicists, but more generally as distinguished mathematicians who have also worked on mathematical problems applied to physical or engineering issues; if we look at their academic life, we remark that they held only at the end of their career a chair of Rational Mechanics or *Mathematical Physics*, whereas they previously held chairs of Analysis, Geometry or Algebra. In order to better characterize mathematical physics, and with a view to those who work in this area, we see that some people feel close to theoretical physics; however one can easily argue that mathematical physics is not theoretical physics, nor is Fluid-dynamics, Electromagnetism, Solid mechanics, Thermodynamics and even cannot be associated with partial or ordinary differential equations, or with differential geometry, algebraic structures or whatever. We can say that mathematical physics addresses issues which not only border but also often deeply intersect with all these disciplines. But if mathematical physics is not anything like that, then what is it? You have probably already realized that I will not fall in danger to provide a definition. But to understand what we intend as *Mathematical Physics*, I will follow the researches of Dario Graffi.

M. Fabrizio (✉)
Department of Mathematics, University of Bologna, Italy
e-mail: fabrizio@dm.unibo.it

For this analysis, I decided to limit them to only four research themes, which I consider most important and which can be associated with the following titles:

1. Problems connected with the study of materials with *fading memory* in continuum mechanics and electromagnetism
2. Uniqueness and reciprocity theorems
3. Study of nonlinear differential problems
4. Problem solving, such as Luxembourg effect and the boundary condition of Graffi–Schelkunoff

Regarding the first issue, we can now say that Graffi was the main continuator of the work of Volterra, in the study of materials with fading memory, which were then called *Phénomènes héréditaires*. In an early work (when he was 23-years old), entitled *Sui problemi della eredità lineare*, Graffi reaches one of its best-known results. The problem stems from the study of restrictions that should be imposed on the parameters of constitutive equations for materials with memory. Consider the linear viscoelasticity described by the relationship between the stress tensor $\mathbf{T}$ and the infinitesimal strain tensor $\mathbf{E} = \frac{1}{2}(\nabla \mathbf{u} + (\nabla \mathbf{u})^{\mathrm{T}})$

$$\mathbf{T}(t) = \mathbf{G}_0 \mathbf{E}(t) + \int_0^\infty \mathbf{G}'(s) \mathbf{E}^t(s) \mathrm{d}s, \tag{1}$$

where $\mathbf{G}_0$ and $\mathbf{G}'(s)$ are symmetric tensors of fourth order, while $\mathbf{E}^t(s) = \mathbf{E}(t-s)$ is the history of $\mathbf{E}$ at time t. Moreover, $\mathbf{G}_0$ is supposed to be a positive definite tensor, while no other hypothesis is given on $\mathbf{G}'(\cdot)$.

In this paper, Graffi remarks that the function $\mathbf{G}'(\cdot)$ cannot be taken arbitrarily, because the work on each closed loop of length d must be positive. So the inequality

$$\oint_0^d \mathbf{T}(t) \cdot \dot{\mathbf{E}}(t) \mathrm{d}t > 0, \tag{2}$$

must be satisfied. We can now prove that this restriction is a consequence of the *Second Law of Thermodynamics*, when isothermal processes and closed cycles are considered. By the inequality (2), Graffi proves that the kernel $\mathbf{G}'(\cdot)$ must verify the relationship

$$\mathbf{G}'(\omega) = \int_0^\infty \mathbf{G}'(s) \sin \omega s \ \mathrm{d}s < 0, \text{ for all } \omega > 0, \tag{3}$$

currently called *Graffi's inequality*.

This restriction is crucial for the asymptotic stability of the dynamic problem and also for the uniqueness of the quasi-static problem, as proved by Fichera in 1988. Finally, (3) also ensures the invertibility of (1). From this result of Graffi, we can begin to understand the role of *mathematical physics*, which is directed towards the study of the compatibility of mathematical models with the constraints resulting

Fig. 1 Paris 1958. Graffi with his daughters and Renato Nardini

from general physical properties such as invariance, objectivity, and the principles of thermodynamics which hold for any particular context of physical models. In fact, Graffi's results about viscoelasticity, can be taken over to other phenomena occurring in electromagnetism, fluid dynamics, heat conduction. As a final remark, we can say that although this is not the definition of mathematical physics, it is certainly one of the topics that characterize it.

Let us now turn to another important topic investigated by Graffi in several researches and well-summarized in the work: *L'espressione dell'energia libera in viscoelasticità*, Ann. Mat. Pure Appl. (4) 98 (1974). Here, again for a viscoelastic material, Graffi studies the properties of a functional, now known in the literature as Graffi–Volterra free energy, defined as

$$\psi(\mathbf{E}^t) = \frac{1}{2}\mathbf{G}_\infty \nabla \mathbf{u}(t) \cdot \nabla \mathbf{u}(t) - \frac{1}{2}\int_0^\infty \mathbf{G}'(s)(\nabla \mathbf{u}^t(s) - \nabla \mathbf{u}(t)) \cdot (\nabla \mathbf{u}^t(s) - \nabla \mathbf{u}(t))ds \quad (4)$$

where $\mathbf{G}_\infty = \lim_{s \to \infty} \mathbf{G}(s)$.

He proposes the following conditions to characterize the *free energy* (7):

a. the functional $\psi(\mathbf{E}^t)$ is positive definite and on each process holds the inequality

$$\dot\psi(\mathbf{E}^t) \leq \mathbf{T}(t) \cdot \dot{\mathbf{E}}(t), \quad (5)$$

b.
$$\mathbf{T}(\mathbf{E}^t) = \partial_{\nabla u} \psi(\mathbf{E}^t), \quad (6)$$

c. for all histories such that $\lim_{s \to 0^+} \mathbf{E}^t(s) = \mathbf{E}(t)$, we have

$$\psi(\mathbf{E}^t) \geq \psi(\mathbf{E}^t_c), \quad (7)$$

where $\mathbf{E}^t_c$ is the constant history defined by $\mathbf{E}^t_c(s) = \mathbf{E}(t)$, for all $s \in [0, \infty)$.

Moreover, Graffi shows that to make $\psi(\mathbf{E}^t)$ a free energy, the kernel $\mathbf{G}'(s)$ has to be a function of $L^1(0, \infty)$ and to satisfy

$$\mathbf{G}'(s) < 0, \qquad \mathbf{G}''(s) \geq 0, \quad s \in [0, \infty). \tag{8}$$

The first condition follows from positive definiteness of the free energy (4), while the latter is necessary when the condition (5) is imposed. The time derivative, of $\psi(\mathbf{E}^t)$ gives

$$\dot\psi(\mathbf{E}^t) = \mathbf{T}(t) \cdot \dot{\mathbf{E}}(t) - \frac{1}{2} \int_0^\infty \mathbf{G}''(s)(\nabla \mathbf{u}^t(s) - \nabla \mathbf{u}(t)) \cdot (\nabla \mathbf{u}^t(s) - \nabla \mathbf{u}(t)) \mathrm{d}s, \tag{9}$$

which shows that inequality (5) is satisfied only when the restriction (8)$_2$ holds. This result is very important, because besides having a definite interest in thermodynamics, it has significant implications from a mathematical point of view. In fact, the expression (4), together with the restriction (8) allows the definition of a natural topology on the history space and the proof of a theorem on the domain of dependence. Moreover, it implies that the operator associated with the integro-differential equation (9), can be represented as a strongly continuous semigroup. Finally, it should be noted that the functional (4) is called Graffi–Volterra free energy, because it was proposed by Volterra in a work of 1928 though he considered it as a potential energy only. The properties (5) and (7) were proved by Graffi in the above mentioned works and ascribe to the expression (5), which shows that inequality (7) is satisfied only when the restriction (8)$_2$ holds. In this regard, I would also like to well on an issue that may belong to the historical questions associated with the early decades of the 1900. As is known, Volterra is considered the initiator of functional analysis by his research on what he called *fonctions de lignes*, which were to be called functionals. Volterra was criticized by Hadamard and others, for having introduced in the field of functional analysis the definition of continuity, differentiability in a non-satisfactory form to the theory of the variational calculus. Subsequently, by an appropriate use of Cantor set theory, M. Fréchet settled this issue with a new and satisfactory definition of limit, continuity and differentiability. However, in spite of this, in his survey paper, Graffi shows how the expression (4) obtained from Volterra in 1928 with his techniques and tools, allows the definition of a natural topology on the history space, through the property of the Hilbert space associated with the definition (4). But especially, when we set this problem over the spatial domain $\Omega \subset R^3$, we have that the Graffi–Volterra free energy defined by

$$\int_\Omega \psi(\nabla \mathbf{u}^t(x)) \mathrm{d}x = \frac{1}{2} \bigg(\int_\Omega \mathbf{G}_\infty(x) \nabla \mathbf{u}(x,t) \cdot \nabla \mathbf{u}(x,t)$$

$$- \int_0^\infty \mathbf{G}'(x,s)(\nabla \mathbf{u}^t(x,s) - \nabla \mathbf{u}(x,t))$$

$$\times (\nabla \mathbf{u}^t(x,s) - \nabla \mathbf{u}(x,t)) \mathrm{d}s \bigg) \mathrm{d}x,$$

provides the typical structure of a standard Sobolev space with weight. In other words, by this work of Graffi we can see that in the research of Volterra in 1928, there were already the seeds of what will be some of the basic elements of functional analysis. And also, this is a nice example of how physics can constrain or better guide the development of mathematics.

The second theme: The reciprocity and uniqueness theorems.

Dario Graffi in a work of 1939 titled: *Theorem of reciprocity in time-dependent phenomena*, Ann. Mat., s. 4, 18, 173–200, began the study of reciprocity theorems for dynamic problems for various physical problems, such as elasticity, electromagnetism, propagation of heat and also fluid dynamics. In this research, he begins by using the Laplace transform but then he is able to obtain his results merely by the convolution product. Despite the formal simplicity, this research gives rise to a brilliant result and, furthermore, is very important *per se*. The crucial idea is that the algebra of convolution product can be treated as the product of the usual algebra, if we look at problems with opposite time directions. The importance also follows by the remark that this result is a dynamic extension of the Betti static theorem, and is obtained after more than 67 years of research on this topic. For this reason, in the major treaties on these issues, this result is called Graffi's reciprocity theorem (see, for example M. Gurtin, Handbuck der Physik, Mechanics of Solids II, vol. VI / 2, Springer, 1972).

I would note that this theorem is based on the variational formulation of the dynamic problem. That is, by the language of mathematical physics, on the *virtual power* principle. In fact in the variational setting, as appears from the work of Graffi, two solutions that correspond to two different data, the second appear to be the test function (or virtual solution) of the first and vice versa. So for the validity of this theorem the definition of weak or variational solution is to use. In addition, for an Hermitian operator the duality of the problem allows to establish and to build the natural spaces of solutions and data. In fact, if A is the Hermitian operator of the problem $Au = f$, then by the reciprocity theorem we have

$$(u_1, f_2) = (u_2, f_1), \tag{10}$$

which is equivalent to the identity

$$(u_1, Au_2) = (Au_1, u_2), \tag{11}$$

from which we can say that the space of solutions will be identified with the function set satisfying the following restriction

$$(Au, u) < \infty,$$

while the data space will be given by the set of the functions f such that

$$(A^{-1} f, f) < \infty. \tag{12}$$

For a dynamic problem, the internal product coincides with the convolution product. These properties of duality and the variational approach allow us to the study of existence theorems for differential problems.

The variational formulation (or the principle of virtual power) is also closely linked with the uniqueness theorems studied by Graffi. A particularly important and famous theorem is the one for the Navier Stokes equations for a compressible fluid, made in 1953 and published in the Journal of Rational Mechanics and Analysis. It holds on a unbounded domain with appropriate boundary conditions on the velocity at infinity. This result aroused great interest and it is mentioned in the monograph of James Serrin, Mathematical Principles of Classical Mechanics Fluids. Handbuck der Physik, VIII / 1, Springer, 1959. It has also been taken over by the Serrin school, by Neapolitan group of Salvatore Rionero, and have involved other foreign researchers. As noted above, the uniqueness theorem as well as the reciprocity theorem, still need of a variational or weak form. This setting allows you to easily extend the Graffi theorem originally formulated for strong solutions to the space of weak solutions. It is well known that the variational formulation of the motion laws, which relies on the Principle of *virtual power*, did not have the right and duty location by the Anglo-Saxon school. We must instead say that it has always been well understood and assessed by the French and Italian schools. The theoretical and numerical results on the good position for the differential problems, makes us understand how the laws of motion are more consistently expressed through the Principle of virtual power, instead of Newton's second law. This is another reflection that in a clear and well motivated form, only the mathematical physics can make. This was well known to Graffi, who observed, with subtle wit, as the Truesdell school had rightly emphasized the word *rational*, but Graffi did not understand why Truesdell had ignored the laws of mechanics by means of the Principle of virtual power.

Another research field, where Graffi showed his mathematical ability and physical sensitivity, has been the non-linear mechanics which is now more often called dynamic systems. It is perhaps pertinent to note, that the name is born from non-linear problems of classical mechanics, which then were common to other

Fig. 2 Kiev 1962. Graffi with Luigi Caprioli

disciplines and other natural systems. He was the first in Italy to deal systematically with these issues. The first paper *Sul modello di Rocard per le oscillazioni di rilassamento*. Atti Acc. Sc. Mediche Naturali Ferrara (2) **21–22,** is dated 1938, but thirty other papers will follow on this topic, where he obtained several significant results, largely connected with the research that the Russian school was pursuing. In particular, he was able to apply the methods of Krylov–Bogolyubov–Mitropolsky to study the existence of forced oscillations related to nonlinear differential systems with periodic coefficients. These works were widely appreciated in the international literature. In fact, Solomon Lefschetz did publish perhaps the most important of Graffi's papers on this subject, Forced Oscillations for Several non-linear circuits, (1951), on Annals of Mathematics. Also he reported the results of Graffi in his important book, Differential Equations: Geometric Theory. These studies were also recorded by Lamberto Cesari and Jack Hale in the United States and Yuri Mitropolsky and his school in Kiev. In his works of the 1940s, Graffi faces the study of the Liénard equation

$$\ddot{y} + g(y)\dot{y} + \varphi(y) = 0,$$

where he obtained the existence of periodic solutions whose amplitude does not decay at infinity. In other works, for a system of two nonlinear equations, he determines a lower limit for the period of the solutions. While for solutions with forcing term of the same Liénard equation, he gets an important bound to the amplitude of the periodic solutions, later called Graffi's bound. Moreover, further researches was directed towards the determination of the error of approximate solutions of the generalized Van der Pol equation

$$\ddot{y} + \dot{G}(y) + \left[\alpha\omega^2 + m \sin \omega_0 t\right] y = 0.$$

Finally, we must remember the vast amount of investigations on these issues which Graffi promoted and directed with his students in Bologna and other universities.

In the years 1935–1936 Graffi published five papers on Rendiconti dell'Istituto Lombardo di Scienze e Lettere, which earned him a major prize of the Accademia dei Lincei. In this study of more than 80 pages, entitled *Sopra alcuni fenomeni ereditari dell'elettrologia*, Graffi provides a systematic and comprehensive modeling of electromagnetic phenomena by a *hereditary theory*, which is crucial for interpreting the broad class of phenomena, showing effects of dispersion of electromagnetic waves. In fact, Graffi notes that the phenomenon of dispersion can be interpreted only by a model, where the constitutive equations between the electromagnetic fields are with fading memory. In these notes, among other things, Graffi proposes for the first time a new hereditary model for the electric current **J** given by

$$\mathbf{J}(\mathbf{E}^t) = \int_0^\infty \sigma(s)\mathbf{E}^t(s)\mathrm{d}s, \tag{13}$$

where the kernel $\sigma \in L^1(0, \infty)$ is the coefficient of conductivity, while $\mathbf{E}^t(\cdot)$ is the history of electrical field.

In the following year (1937), on Rendiconti del Seminario Matematico dell'Università di Padova, he publishes one of the most important work, entitled *Una teoria ereditaria dell'effetto Lussemburgo*, in which extends the (13) to the following new functional

$$\mathbf{J}(\mathbf{E}^t) = \int_0^\infty \sigma(s)\mathbf{E}^t(s)ds + \int_0^\infty \gamma(s_1, s_2, s_3)\mathbf{E}^t(s_1)\mathbf{E}^t(s_2)\mathbf{E}^t(s_3)ds_1 ds_2 ds_3,$$

where the coefficient $\gamma s_1, s_2, s_3)$ is a fourth-order tensor. Through this nonlinear model with memory, he manages to describe and justify the *Luxembourg effect*, only four years after the first observations of this phenomenon. As noted by Graffi, the Luxembourg effect is due to the contemporary action of hereditary and non-linear effects. The study of this problem has implied the apparent resolution of technical problems. But also the need to describe the phenomenon through the modeling of new materials, such as ionized gases. This certainly has been possible thanks to the genius of Graffi that understands how hereditary effects in electromagnetism are crucial in a large set of phenomena. So again, in this work, Graffi plays well the role of mathematical-physicist, because he provides a broad interpretation of the phenomenon using a new model that not only solves the problem under study, but contributes to a better understanding of the theory and to extend its validity.

This research allows me to propose a reflection. The equation (13) relates the electric current $\mathbf{J}$, with the electric field $\mathbf{E}$. When there are no hereditary effects, this relationship reduces to the Joule law

$$\mathbf{J}(t) = \sigma \mathbf{E}(t). \tag{14}$$

However, when in (13) we put $\sigma(s) = \sigma_0 \exp(-\beta s)$, then this constitutive equation is reduced to rate-type relationship

$$\dot{\mathbf{J}}(t) = \sigma \mathbf{E}(t) - \beta \mathbf{J}(t). \tag{15}$$

It is known from reflections on the mesoscopic models that the behaviour of electrical conduction is very similar to that of heat conduction. So the ideas or theories applicable to the electrical conductivity can be transmitted to the thermal models. So if the analogue of the Joule's law (14) is the relation of Fourier

$$\mathbf{q}(t) = -k\nabla\theta(t),$$

then the equation connected with (15) will be the relation of Cattaneo–Maxwell

$$\gamma\dot{\mathbf{q}}(t) = k\nabla\theta(t) - \mathbf{q}(t). \tag{16}$$

So this means that in the paper of Graffi published in 1935, already appears an interpretation of the problems related to the conductivity through a relationship with fading memory. After 10 years, Cattaneo will use a similar idea to achieve the equation (16). This example is useful to understand what many people say. That is, the researches of Dario Graffi in electromagnetism, though very important, have not had the success they deserved.

Another important application developed by Graffi on these topics of electromagnetism is the boundary condition of Graffi–Schelkunoff introduced to describe a boundary consisting of a good conductor, but not perfect. The first work on this subject was presented in 1958 on Atti Acc. Sc. Ist. Bologna Serie XI, V, 88–94, entitled *Sulle condizioni al contorno approssimate in elettromagnetismo*. This research was repeated in other works and by other researchers. A good summary appears in the monograph of the Cime Course *Onde superficiali*. Cremonese. Roma 1961. In this work, Graffi generalizes the usual boundary condition

$$\mathbf{E} \times \mathbf{n} = \mathbf{0}, \text{ on } \partial\Omega,$$

which describes a domain Ω bounded by a perfect conductor. If the edge is made by a good conductor, then for the harmonic fields Graffi proposes the condition

$$\mathbf{E}_\tau(x, \omega) = \lambda(x, \omega)\mathbf{H}(\mathbf{x},\omega) \times \mathbf{n}(x), \text{ on } \partial\Omega, \tag{17}$$

where $\lambda(x, \omega)$ is a complex scalar coefficient, which depends on the material making up the border, called *surface impedance*.

Graffi observed that the boundary condition (17) holds only for harmonic fields, and he proposed an extension to arbitrary fields using again a relationship with fading memory. His suggestion was correct and important, but unfortunately this has been achieved only three years after his death.

Finally, I want to conclude this partial review of the scientific work of Dario Graffi, studying his work of 1955, *Il teorema di unicità per fluidi incompressibili, perfetti, eterogenei*. Rev. Union. Mat. Argentina, 17, 73–77, in which he proposes a model for the study of a pollutant in a mixture of compressible fluids. For this problem, Graffi provides an idea that, as it has been correctly observed, "it is capable of describing a practical situation." This research was taken up by several authors and it is a significant anticipation of the famous Cahn Hilliard equation proposed in a paper of 1958.

Dario Graffi lived in a time of great change and innovation relative to science and culture, as well as to social and political life. Indeed, he began his academic career in the 1920s, a decade extremely important for research in various physical disciplines, but especially in General relativity and Quantum mechanics. A revolution begins in those years that will touch not only the laws of physics, but also cause a major change in contents even on the epistemological foundations of physics. However, this world in Bologna is well represented by the photo in Fig. 3, where I think, we can say that it still does not perceive the change that will invest the whole environment of Physics.

Fig. 3 Bologna 1925. Graffi on the day of graduation outside the Institute of Physics. Quirino Maiorana is sitting in the middle

Later in the 1950s and 1960s, Dario Graffi witnessed a remarkable expansion of Mathematics (mainly Analysis, Geometry, and Algebra) also under the impact of the axiomatic approach of the Bourbaki school. His attitude towards research however did not change. Having in mind his character and his personality, I would interpret his attitude in the sense that he understood that we have to continue to maintain our interest toward applied mathematics, motivated and powered by real problems arising in physics and engineering. He understood how any new theory of physics or mathematics cannot erase the many practical problems that science and technology continue to propose. For this reason, in the research he showed a way of working completely oriented to practical problems. He made use of classical mathematics, at a level which is sufficient to describe the phenomena under study. Graffi was aware that sophisticated mathematical tools, could not by themselves represent a real step towards the solution of problems. Motivated by the successes of the Russian school, both in mathematics and physics, many of the subsequent comments and critics to the formal setting of the Bourbaki school were in favour of his view. Apparently, this period was not easy for Dario Graffi in the attempt to keep an important role of mathematical physics in the academic community, despite the strong and important advances of Mathematics and Physics. Yet his personality, his charisma, his talent for research allowed him to maintain the prestige of this discipline. Nevertheless, in this country in the fifties and even more in the sixties mathematical physics was considered by many people within the mathematical community, and within the physical one as well, an old-fashioned discipline in need of a deep change. For this I wonder if in those years Dario Graffi faced the question of what the Mathematical Physics was and how it could be characterized. To my view the possible answer was determined by his well-known pragmatism, his vision of research and his impatience with the formal patterns. So it seems more likely that he did not devote much attention to the characterization of mathematical physics, which he probably considered an absolutely futile exercise. Rather, he thought about how to renew the discipline in operational terms. Equally, however, I think that if someone wants to investigate the root and the development of this discipline would be facing an ill-posed problem in that there is no uniqueness theorem, and even existence is conditioned by the environment that generates it.

I cannot end this remembrance of Dario Graffi, without recalling the clear figure of a kind and generous man. He was kindly available towards anyone who required his opinion or, more often, his help in the preparation of a paper. It was surprising to observe the efforts that he was doing to provide answers and solve problems relative to questions from even poorly-known researchers.

As a teacher, he transmitted to the students his discipline, never in academic or pedantic form, but instead in a vivid manner. He avoided the academic behaviors and also why his lessons were brilliant, sliding. During his lessons, he showed his acumen and his ability in the scientific method. Even during examinations, he was trying to establish with students a basic understanding about the content of questions, always showing great generosity on the more technical aspects.

Equipped with a remarkable capacity for work, he was engaged in research activities with continuity and passion until the last days. Besides, supported by a strong and constant Christian faith, he has always maintained a great serenity, together with an extraordinary lucidity.

He possessed a special ability to capture the essence of scientific reasoning, immediately identifying the fundamental problem with its difficulties. He was especially able to communicate his science, never with presumption, but always with great simplicity. He was equipped with a vital energy that allowed him to dominate the situation and to project with confidence and hope toward the things of life.

For his scientific stature, and personality of a Maestro, Dario Graffi deserves to be located in the prestigious Italian school of Mathematical Physics together with the glorious names of Eugenio Beltrami, Enrico Betti, Vito Volterra, Tullio Levi-Civita.

After more than 20 years after his death, we feel that his remembrance is renewed strong in our memory and we discover with joy that our memory is not fading.

Biography of Dario Graffi[*]

Dario Graffi was born in Rovigo on January 10, 1905, the son of Michele, a yarn wholesale trader, and of Amalia Tedeschi. He attended the physical-mathematical section of the *Istituto Tecnico* high school of his hometown but got his degree in 1921 in Bologna, where his family had moved the previous year. After enrolling in the local university, he graduated in physics in July 1925. Among his teachers at the Alma Mater he remembered with admiration mathematicians Pietro Burgatti, Federigo Enriques, Leonida Tonelli and physicist Quirino Maiorana. Among his classmates, he became a close friend of Bruno Rossi (1903–1988), who was to be a reknown physicist, and of Giuseppe Evangelisti (1903–1984), who later became an illustrious teacher in hydraulic engineering.

---

[*]From the preface of the volume "Dario Graffi – Opere Scelte – C.N.R. (National Research Council) Rome, 1999. Translated by Carlo Alberto Bosello".

Right after his graduation he was appointed *assistente* to the chair of Technical Physics at the *Scuola di Applicazione di Ingegneria* (*Facoltà di Ingegneria*, i.e. Division of Engineering, from 1982 onwards). The chair was held by Emanuele Foà (1885–1949), who came from the famed school of gas dynamics of the *Politecnico* (Institute of Tecnology) of Turin. Between Dario Graffi and Foà a strong bond of esteem and friendship would develop, made all the more solid by the hardships the latter had to endure because of the racial laws. To the same period dates another of Graffi's life lasting bonds: that with Giulio Supino (1898–1987) who too was *assistente* at the *Scuola di Applicazione* and who later became one of the most outstanding personalities in hydraulic engineering in our country. In 1927, Graffi graduated in mathematics too, under the supervision of Pietro Burgatti. He also worked as a supply teacher in high schools, something the *assistenti* were allowed to do at the time, though unfortunately not any more nowadays. During one of these appointments, at the "Augusto Righi" *Liceo Scientifico* (Scientific High School) in Bologna, he had Lamberto Cesari (1910–1991) as a pupil: another friendship was born which would last for sixty years.

His academic career was quick. Having earned the title of *Libero docente* in Mathematical Physics in 1930, for several years he was appointed to teach the course of Electric Oscillations for students of physics in Bologna; in 1935, having left the *Facoltà di Ingegneria*, he took up a post as teacher of the course of Rational Mechanics at the University of Cagliari as well as teaching Mathematics and Physics at the "Dettori" *Liceo classico* (Classical high school). The following year he ended up ranking first among the three candidates selected for the chair of Rational Mechanics at the University of Messina. In December 1936, he was appointed *Professore straordinario* of the same subject by the University of Turin, where he remained two years. He always recalled those years with pleasure because of the scientific exchanges he had with his collegues in that famed university, with particular reference to Guido Fubini (1879–1943, professor at the *Politecnico*), Francesco Tricomi (1897–1978), and Enrico Persico (1900–1973). In October 1938, he succeeded Pietro Burgatti, who had died a few moths earlier, as chair of Rational Mechanics at the University of Bologna, a post he would hold continuously for 37 years until the official end of his teaching career on October 31, 1975. Immediately after his retirement on November 1st, 1980, he was appointed *Professore Emerito* and kept performing his research work serenely until his death, which occurred on December 12, 1990 after a short illness, at the age of 86.

During the 52 years between his return to Bologna and his demise, Dario Graffi devoted himself incessantly and tirelessly to his job as professor: teaching, research, training younger scholars, institutional appointments. He thus ended up becoming one of the most eminent figures both in the university of Bologna and within mathematical physics in general in this country. His teaching, at all levels, was wide and diversified. Beside the course of Rational Mechanics, which he held until 1960 and was aimed at the students of the old two-year course schedule common at first to students of mathematics, physics, and engineering, then to those of mathematics and physics and finally to those of mathematics only, since 1949 he continuously taught the course of Mathematical Physics (renamed Fundamentals of

Mathematical Physics since 1960) to students of mathematics and physics. During and after the war he then taught the course of General Physics II to students of mathematics and physics and of Technical Physics to students of engineering and later on the courses of Electromagnetic Waves and of Mathematical Methods of Physics to students of physics. It was not uncommon for his courses of Mathematical Physics or Electromagnetic Waves (the latter having also being held for several years at *Scuola Superiore di Telefonia e Telegrafia* of the Ministry of Mail and Telephones) to be attended by young scholars already pursuing their own academic career. One particularly shining example of such a scholar was Ercole De Castro (1928–1984), founder of the school of electronic engineering of the university of Bologna; he attended the courses of Dario Graffi from 1954 to 1956 and always proclaimed himself to be a pupil of Graffi's, resulting in still another, very fruitful human and scientific relationship. The honorary degree in electronic engineering that the *Facoltà di Ingegneria* of Bologna conferred to Dario Graffi in 1983, the only such instance where the recipient was a mathematician, represented the best acknowledgment of the influence of his scientific teaching on the engineering sciences.

Among the different aspects of his activity as a professor, though he never eschewed his academic and institutional duties, scholarship was the one closer to the deep nature of Dario Graffi. By far his prevailing interest lay in scientific research: both in his own and in directing that of his pupils. Being scientifically open minded and bright, having an easy approach to other people and remarkable communication skills, he constantly found himself acting as thesis advisor for a relevant number of students (overall more than 500 theses, of which at least sixty became published papers) and he naturally attracted many bright young people toward his research field, namely classical mathematical physics in a very broad sense. Thus he originated one among the most important schools in the country. Among his direct pupils eight (Giuseppe Aymerich, Renato Nardini, Luigi Caprioli, Marziano Marziani, Marialuisa Pini de Socio, Carlo Banfi, Italo Ferrari, Mauro Fabrizio) would hold a chair and three (Marisa Merri Manarini, Giovanna Pettini, Franca Franchi) would become *professori associati*. Even more significantly, many other scholars in mathematical physics from several universities in and out of the country worked in more or less direct contact with him and were profoundly influenced by him. This will be recalled below in an albeit brief description of his scientific activity, which was long and very intense: in 65 years of uninterrupted work he published more than 180 original papers, 3 textbooks and 2 treatises.

While Pietro Burgatti (1868–1938), professor of Rational Mechanics in Bologna from 1908 to 1938, did not, strictly speaking, have Dario Graffi as a pupil, (for instance, Graffi never acted as his teaching assistant), it was him who decisively steered Dario Graffi toward the field of mathematical physics. Burgatti was a high ranking scholar. He had trained in the school of Eugenio Beltrami (1835–1900), for whom he was for some years the teaching assistant in the last decade of the nineteenth century, and he left important contributions both to analytical mechanics and to hyperbolic partial differential equations, like extending Riemann's integration method to higher order equations (a result to be found in the classical treatise by Courant–Hilbert). The topics of mathematical physics that Burgatti cultivated found fertile ground in Dario Graffi also thanks to the latter excellent background in mathematics as well as in physics. In fact, he was among the last to attend the "glorious physics and mathematics division" of the Technical High Schools (abolished by the Gentile school reform bill of 1923) where the majority of the scientists and engineers of the preceding generations had been trained. This type of high school of very modern conception, which consisted of four years instead of the usual five, allowed its graduates to enter the two-year initial program (the so called *biennio*) of the *Facoltà di Science* and to make the most of it. So did Dario Graffi; anybody who knew him was astonished by his excellent command of all the subjects treated in the *biennio*, and anybody in our line of work knows very well what such command means. Later, often after his graduation, he followed the courses of Projective Geometry and of Differential Geometry held by Federigo Enriques and Enrico Bompiani and, above all, those of Calculus and of Higher Analysis held by Leonida Tonelli. Dario Graffi's research activity could thus deal from the start with a wide array of issues, all of them very demanding technically. As the years passed he added other themes, but never wholly abandoned those of his youth; at most he updated them. His first research on adiabatic invariants (of fundamental importance in the Bohr–Sommerfeld quantum theory) date to the 1920s, and was taken up again in the 1960s (when the issue became topical again with the appearance of KAM theory). In the same period, he did his first research on hereditary phenomena, including electromagnetism, where he immediately applied the variational techniques he had learned from Tonelli; on fluid mechanics, particularly concerning viscous fluids, a topic to which he was led by Foà, especially concerning uniqueness of solutions theorems; and finally his first research on electromagnetic phenomena in continuous media, more specifically on electromagnetic wave propagation. Therefore, right from the start Dario Graffi dealt with all the main topics into which his production can be subdivided.

Dario Graffi never saw himself as an author of treatises; nevertheless, he wrote several. The one about Mechanics, the first edition of which appeared in 1938 was repeatedly updated in the following years (in the end with the help of Renato Nardini, and of Carlo Banfi for the exercises), was by far his main didactic work in the sense of having been conceived as a textbook for students. His book about the mathematical theory of electromagnetism, a collection of his lectures of mathematical physics of the early 1950s, in spite of being considered a masterpiece by many, reached a small audience, perhaps also because it was never published in print but only by lithography; his collected lectures on the fundamentals of mathematical physics suffered of a similarly low circulation. This latter book, contrary to what had happened to the textbook on mechanics, was not kept updated because Graffi often liked to change the content of his courses aimed at senior students. The book on mechanics was a standard reference in many universities in the country and still enjoys a high circulation today. Contrary to what many others did, Graffi was little inspired by the great classic on the subject, namely the *Lezioni di meccanica razionale* by Tullio Levi-Civita and Ugo Amaldi, especially in his treatment of some crucial topics, particularly in statics (for instance, the principle of virtual works). In order to find an equally illustrious predecessor one needs to look (via his pupil Burgatti) at Eugenio Beltrami, who was professor of Rational Mechanics in Bologna from 1866 to 1873. Beside the already quoted *Nonlinear partial differential equations in mathematical physics*, Pitman, London 1978, Graffi wrote only another book, published in 1962: the volume "Electromagnetic waves" in the series of mathematical monographs of the CNR (National Research Council), the same series where, for example, the book on partial differential equations by F. G. Tricomi was to be found. This series did not achieve a lot of success, by reasons to be explained presently.

The value of Dario Graffi's scientific activity was acknowledged at an early stage (already in 1933, aged 28, he came very close to being ranked among the three winners of a selection for professors of Rational Mechanics, with Mario Picone voting for him) and acknowledgments multiplied with time. He won the *Accademia d'Italia* prize for research in mathematics in 1942 and the *Presidente della Repubblica* prize given by the *Accademia dei Lincei* in 1965; the *Accademia Nazionale dei Lincei* also appointed him Corresponding Member in 1953, National Member in 1967 and Academic Secretary in 1975; he was *Accademico Benedettino* of the *Accademia delle Scienze dell'Istituto di Bologna* in 1942, being its president from 1973 to 1976; he was a member of the *Accademia delle Scienze di Torino*, of the *Istituto Lombardo*, and of many other societies. As Fullbright Fellow he spent the fall of 1950 at the Institute of Advanced Studies in Princeton, where he had close contacts with Solomon Lefschetz; in 1967/1968 he was appointed *Professeur Associé* at the *Faculté des Sciences de Paris*, at that time still part of the Sorbonne before the latter got fragmented into 13 independent universities. He was in the editorial board of many important journals, including international ones: for example, the Archive for Rational Mechanics and Analysis. He was scientific coordinator for many courses organized by *CIME* (International Mathematical Summer Center), of whose board of directors he was a member for many years.

Dario Graffi never sought to play any official administrative role but neither he refused to when offered, because it considered it to be part of his duties as professor. Thus he was dean of the *Facoltà di Scienze Matematiche, Fisiche e Naturali* (Division of mathematical, physical and natural sciences) of the University of Bologna (1960–1965), member of the mathematics committee of the *CNR* (1958–1966), Administrator and member of the Scientific Committee of the Italian Mathematical Union (1952–1964), President of the Scientific Council of *GNFM* (National Group for Mathematical Physics) (1970–1980). Never having sought such positions, he always left them with no regrets at the end of his term; besides he believed, like many others of his generation, that scientists, once past their full maturity, should leave the managing of their organizations to subsequent generations. He was thus able to spend the last part of his life serenely, working, listening to young people and guiding them in their research; he also found time to indulge in the pleasure of conversation in scientific as well as in other subjects (especially history), a pleasure he shared with colleagues and lifetime friends; among such friends, having already mentioned a few, one would find Gaetano Fichera (1922–1996) and, above all, Gianfranco Cimmino (1908–1989), famous scholar in the field of mathematical analysis, a colleague of his in Bologna since 1940, whose broad and deep knowledge both inside and outside mathematics Graffi admired and with whom he shared his great love of the *Divina Commedia*.

In the late 1950s, Italy too was being swept by the bourbakist wind coming from France. In its most widely accepted interpretation, it took strict observance of a faultless formalism as the criterion of belonging to the mathematical community, while the Italian school of mathematical physics had always considered mathematical rigour more as a matter of substance than of form. Moreover, another wind had started blowing from the quarters of theoretical physics. Thus, around the end of 1960s Italian mathematical physics found itself in a situation of isolation, if not in an outright crisis. While being less welcome within mathematics, for essentially formal reasons, it found itself the target of hegemonic attempts on the part of theoretical physics. It also witnessed the shrinking of its traditional interdisciplinary role between mathematics, physics and engineering (essential at the times of the old *biennio* which remained in common between the different curricula up to 1960) due to the progressive fragmentation of training in those three subjects. Dario Graffi saw clearly that this crisis could not be ascribed to scientific or structural causes and that it could be easily overcome by letting some (his words) "fresh air" get into the field. By that he meant that it was necessary to cope with the (still valid) traditional topics of Italian mathematical physics by means of the new technical tools that had been emerging in analysis and geometry; moreover, that other topics should be dealt with as well, such as kinetic theory, statistical mechanics and others, that in Italy had been treated in different circles. He actively operated toward this goal, first of all by encouraging his pupils to walk new paths and also, inclined as he was to appreciate bright young people even if they worked in areas far removed from his own, by favoring the recruiting of the best experts of the new areas. The present florid condition of research in mathematical physics in our country shows, ten years after his demise, how far sighted Dario Graffi was.

During the last years of his official academic activity Dario Graffi not only had to face a crisis in his own field, luckily turning out to be a temporary one, but, like all the university professors of his generation, a time of epoch-making changes. The strong, turbulent and unexpected growth of the country during the 1950s had triggered a big upheaval within academia as well, culminating in the turmoils of 1968 and following years. The generation to which Dario Graffi belonged, and this holds both for politicians and academicians, who not uncommonly were one and the same, was unable to notice at the right time the needs entailed by the pouring into higher education by ever larger masses of students. Therefore, they did not take the appropriate legislative steps. Maybe they could not have: that generation had actually trouble understanding even the origin of the crisis, having grown up in a country where access to higher education had remained severely limited for a century, during which the structure of society had not even changed much. It is always easy to judge retrospectively. Anyway, the storm of 1968, that swept away even the timid reform proposals put forward till then, caught university professors totally unprepared. Dario Graffi was no exception, but he faced it with firmness and dignity and personally he never had to endure anything. But watching the principles in which he believed and for which he had lived being publicly reviled not just by pseudo-revolutionary students but also (his words) by "colleagues looking for idle popularity" and seeing illustrious scientists, colleagues and friends (such as Bruno Finzi (1899–1974) then *Rettore* of the *Politecnico* of Milan), being subjected to physical aggression in the acquiescing silence (bordering on approval) of the then Minister, were wounds from which he never fully recovered. He knew well that effort, commitment to work, seriousness, spirit of sacrifice are absolute values; only through them should people gifted with individual innate abilities earn the achievement of posts of great responsibility. He knew equally well that in those values every civilized society, not just the progress of science, is grounded. All his life he remained faithful to such values, with a lack of pretentious display reflecting the depth of his beliefs.

Pietro Burgatti and His Studies on Mechanics

Paolo Freguglia and Sandro Graffi

Abstract A brief reconstruction is attempted of the scientific figure of Pietro Burgatti (1868–1938) by giving some detail on his more significant contributions to mathematics: general vector calculus, hyperbolic partial differential equations, analytical mechanics. An attempt is also made to put Burgatti's contributions into the perspective of Italian mathematics of his times.

Pietro Burgatti was born in Cento (Ferrara) in 1868. He was commissioned as an artillery officer after graduating from the Military Academy in Turin in 1888; later he studied Mathematics in the University of Rome under Eugenio Beltrami and Valentino Cerruti. In Rome he was also for some years Assistant to the Chair of Mathematical Physics held by his teacher, Eugenio Beltrami. Appointed Professor of Rational Mechanics in the University of Messina in 1904, four years later, in 1908, he moved to the same chair at the University of Bologna, left vacant by Ernesto Cesaro who had died in an accident just before taking office. In Bologna, he stayed until his death (1938). He served for several years as Dean of the Science Faculty of this University, and taught also in the University of Ferrara. He was elected in the Accademia dei Lincei in 1923. He was among the founding members of Unione Matematica Italiana, and served also two terms as vice-president. His studies in Mechanics were very significant for the development of the investigations in this subject, in particular in Italy. His treatise (1914) on Mechanics had an important role in the formation of mathematicians, physicists and engineers. Since a large and significant part of the scientific activity of Burgatti dealt with *general vector calculus* [1] we begin by a short reminder on the situation of the treatises on

---

[1] Calcolo vettoriale generale.

P. Freguglia (✉) · S. Graffi
Dipartimento di Matematica, Università di Bologna, Bologna, Italy
e-mail: fregugli@univaq.it; graffi@dm.unibo.it

Mechanics published in Italy around the turn between the XIX and the twentieth century, mainly as far as the vector calculus is concerned.

1 Vector Calculus in Mechanics Treatises in Italy in the Early Twentieth Century

The mathematical study of the rational mechanics requires advanced mathematical techniques of calculus and differential geometry. The Italian contribution to this field has been very significant. For instance, on one hand we have the investigations on the geometrical or vector calculus (G. Peano, C. Burali-Forti, R.Marcolongo, vector homographies); on the other hand we have the investigations on the absolute differential or tensor calculus (T. Levi-Civita, G. Ricci-Curbastro, tensors of n-th order). Let us now shortly review the Italian treatises of rational mechanics. The turn of the century was the time of birth of Einstein's relativity theory (1905); in Italy we find many articles and books in favour of this theory (i.e. R. Marcolongo's book *Relatività* [21]), as well as against it (i.e. the treatise by C.Burali-Forti and T.Boggio *Espaces courbes, critique de la relativité* (1924)); moreover, many debates took part among physicists and mathematicians on its mechanical foundations (i.e. comment about Poincaré-Andrade controversy, see (1901)), etc. Of course, the treatises have a didactic and formative role; hence the explained topics must be well-established. Therefore, relativity is excluded from treatises intended primarily as textbooks. We have for instance the books by Gian Antonio Maggi (1896, 1912), by Roberto Marcolongo (1905), by Pietro Burgatti (1914; second printing 1919) and the *Lezioni di meccanica razionale* by Tullio Levi-Civita and Ugo Amaldi (final edition: 1926–1930). If we consider Maggi's treatise *Principi della teoria matematica del movimento dei corpi*, we find an expository structure which does not utilize, at least in the first edition, the synthetic vector calculus apparatus. Given the vector notion, Maggi proceeds in a completely Cartesian way. On the other hand, if we examine some treatises coming from the other side of the Alps, like the *Cours de Mécanique à l'usage des candidats à l'École Polytechnique* (1903) by Ch. Michel, there is not even a shadow of the synthetic vector calculus. In fact, the use of the Cartesian co-ordinates was traditional, as already well illustrated in *Cours de Mécanique pour l'École Polytechnique (1861)* by M. Sturm. Other foreign mechanics treatises in this period are for instance L. Lecornu (1914), P. Appell (1921, 1926), P. Painlevé (1930, 1936), A. Föppl (1933), M. Planck (1933). The contents of Maggi's book consists of two parts: Kinematics and Dynamics. The latter has two sections: general laws of the motion, determination of the motion. In the Preliminaries, the author introduces the notion of vector (and other notions such as torque, vector tern, etc.) according to the Cartesian method.

2 The Vector Calculus of Burgatti and of Burali–Forti–Marcolongo

Even if in a somewhat rhetorical style, Burgatti motivates very clearly his preference for the vector calculus (of which he gives an ample presentation in the first chapter of Lezioni di meccanica razionale (1914))[2]

The Cartesian method, which is sometimes excellent and sometimes unavoidable in the applications, is not very flexible and is often cumbersome in dealing with the general development of physical theories; it hides the concepts among labyrinths of calculations [...]. This serious inconvenient can be avoided by means of the vector calculus, which economizes and enlightens the thought [...], in particular according to the recent sharp and elegant version worked out by Professors Burali-Forti and Marcolongo.

The main feature of the Burali–Forti–Marcolongo vector calculus is the independence from any Cartesian system. According to Peano and Burali-Forti, the co-ordinates method represents a numerical intermediation for the study of geometrical objects and of their properties, while geometric calculus achieves absoluteness and conciseness, and the approach through it in order to study geometrical problems is immediate and direct. However, this calculus does not exclude the use of co-ordinates (On this matter, see e.g. [13, 15, 16]).

In his 1888 volume Peano presents Grassmann ideas in an original way: as we have already said, he gives an Euclidean interpretation to the fundamental Grassmannian notions, by limiting his considerations on the nature of the system of unities, and he goes not beyond the three dimensions. At first, he introduces the notion of *geometrical formation* in the following way:

$$\sum_{i=1}^{n} m_i \tau_i^q$$

where m_i are real numbers and τ_i^q are q hedrons, with $1 \leq q \leq 4$. These are Peano's interpretations of Grassmanns unities. That is, we have for the system of unities the following possibilities:

1 − hedron	point
2 − hedron	segment
3 − hedron	triangle
4 − hedron	tetrahedron

---

[2]See P. Burgatti [6], p. VIII: *Il metodo cartesiano, talvolta eccellente, talaltra indispensabile nelle applicazioni, è poco agile e più spesso ingombrante per lo sviluppo generale delle teorie fisiche; maschera i concetti tra i labirinti dei suoi calcoli [...]. A togliere sì grave inconveniente provvede, economizzando e rischiarando il pensiero, il calcolo vettoriale [...] nella forma recentemente elaborata con fine acume e con bell'arte dai Professori Burali-Forti e Marcolongo.*

With Peano, we call the geometrical formations of the first kind (or degree) those which have points (and only points) as system of unities, of the second kind if the unities are segments, of the third kind if the unities are triangles, and finally of the fourth kind if the unities are tetrahedrons. We will in general denote a geometrical formation by F_q, where q expresses the kind of formation considered. Even if the possibility of geometrical formations with $q \geq 5$ is not considered by Peano, it should represent an easy and natural generalization. Veronese, who was a contemporary of Peano, should have done it. But Peano is bound to a traditional vision of geometry. Between two geometrical formations we can establish the operation of algebraic addition, which complies with the rules of the algebra of polynomials. But conceptually the more important operation is the alternating product, which is introduced by Peano[3] (and by Burali-Forti[4]) in the following way:

If we have two geometrical formations in $3D$, F_r and F_s, the alternating product is a product which complies with the rules of the algebra of polynomials, but without changing the order of the letters which denote points. If $r + s \leq 4$ the product is called *progressive*[5] and expresses the geometrical operation of projection. If $r + s > 4$ the product is called *regressive*[6] and represents the geometrical operation of section. In the plane case, $2D$, in the above definition we must replace the conditions on $r + s$ with $r + s \leq 3$ and $r + s > 3$, respectively. The alternating product is not commutative.

For instance, a segment is represented by the product AB, if A and B are two points which determine the segment. But a segment can be represented by the formation of the first kind BA. We also have that: $AB = -BA$ and therefore it follows that $AA = 0$.

In Chap. IX (first pages) of his 1888 treatise Peano introduces the definition of linear system. In today's language, he establishes the notion of vector space (on $\mathbb{R}$). And he introduces the notion of linear transformation as well. Still in the context of the Grassmann-Peano line of thought, Burali Forti and Roberto Marcolongo developed their studies about the vector calculus. Marcolongo was not, in strict sense, a student of Peano, but had a systematic scientific collaboration with Burali-Forti. We must consider the following important works of these two mathematicians:

– *Elementi di calcolo vettoriale con numerose applicazioni alla geometria, alla meccanica e alla fisica-matematica* (1st edition 1909, second enriched ed. 1921)
– *Analyse vectorielle générale, I. Transformations lineaires* (ed. 1912),
– II. *Applications la Mécanique et la Physique* (ed. 1913).

---

[3] see G. Peano (*Calcolo geometrico secondo l'Ausdehnunglehre di H.Grassmann, preceduto dalle operazioni della logica deduttiva* Fratelli Bocca Editori, Torino 1888), pp. 110–111.
[4] See [2], pp. 5–6.
[5] See G. Peano (*Op.cit.* 1888), p. 30.
[6] See G. Peano (*Op.cit.* 1888), p. 107.

In the 1909 book, they present vector calculus as a structure of vector space with the operations of scalar product and vector product. This is the *minimum system*, while we have the *general system* when the geometrical formations are introduced and we use the alternated product (see Burali-Forti Elementi di calcolo vettoriale, in *Enciclopedia delle Matematiche Elementari* [vol. II, part II]). In the case of the minimum system, they introduce the notions of gradient, rotor, divergence. The Hamilton quaternions are presented through the techniques of the minimum system in the edition of 1920. The sixth chapter of both editions is devoted to applications to electrodynamics and to Maxwell equations and the Lorentz transformations. Let us now examine how, according to Burali-Forti and Marcolongo, the general system contains the minimum system, i.e. from the general system we deduce the minimum system. In fact, Burgatti presents and applies just the minimum system in his Rational mechanics treatise of 1914. To this end they introduce the notions of bi-vector and of three-vector (inherited from Peano and Grassmann). The first is seen as an alternated product of two vectors, in th following way:

$$\hat{u} = v'w' = (B - A)(C - A).$$

A bi-vector is a geometrical object determined by three elements: a modulus, a plane position and an orientation. mod$\hat{u}$ is the area of the parallelogram determined by v' and w'. These two vectors also determine a plane position and the direction depends on the property $v'w' = -w'v'$. We define *index of a bi-vector* $\hat{u}$, and we denote it by $|\hat{u}$, a vector such that: 1. mod$\hat{u}$ = mod$|\hat{u}$; 2. if $\hat{u} \neq 0$ the direction of $|\hat{u}$ is normal to the plane position of $\hat{u}$. 3. the orientation of $|\hat{u}$ is such that the number $O\hat{u}(|\hat{u})$ is either positive or zero.

If v' is a vector, we define *index of a vector* v', and we will thus denote it $|v'$, the bi-vector $\hat{u}$ which has as index $|v'$.

A three-vector is, in its turn, a alternating product of three vectors. If we consider three vectors u', v', w', we can show that the product Ou', v', w' is a real number (which represents the affine volume of the oriented parallelepiped constructed through the vectors u', v', w'). In particular, we put $\hat{U} = O\mathbf{ijk}$, where $\mathbf{i, j, k}$ determine the usual unitary orthogonal right system. $\hat{U}$ is called unitary three-vector. Hence, we can establish the following definitions:

Scalar product (or inner product) definition: *If u' and v' are two vectors, we will call scalar product (or inner product) the operation:*

$$u' \times v' = \frac{u'(|v')}{\Omega}.$$

We can show that

$$u' \times v' = \text{mod}\hat{u}' \cdot \text{mod}\hat{v}' \cos(u', v').$$

Vector product (or outer product) definition: *If u' and v' are two vectors, we will call vector product (or outer product) the operation*:

$$u' \wedge v' = |(u'v').$$

To sum up, from the general system of geometrical calculus we have obtained the minimum system.

The theory of homographies is well presented and analyzed in the treatise *Analyse vectorielle générale* [5]. Burali-Forti and Marcolongo define a homography as a linear operator which transforms vectors into vectors.

Certainly, from a foundational point of view also, the Grassmann-Peano approach shows its relevance. But nevertheless, as we have seen, the goal of the geometrical calculus, or the vector calculus or the homographies goes beyond a philosophical justification of the bases of the geometry. In fact, the applications to mathematical physics and to the geometry of transformations are very important and, notwithstanding the matrix techniques, can be proposed even today . Of course, suitable adaptations could be required. Analogously, *mutatis mutandis*, as it happened in the reappraisal of the quaternion techniques. In the Preface to *Analyse vectorielle générale* Burali-Forti and Marcolongo[7] explain once more that their calculus is not, in itself, just a tachygraphy in comparison to the techniques of vector calculus which are introduced by means of Cartesian co-ordinates. In fact, the geometrical synthetic calculus is, according to our mathematicians, intrinsic, or absolute, or autonomous . In our opinion, what is important is the absoluteness, because this calculus presents concepts and procedures which disregard the particular Cartesian co-ordinate system. We can say to have an intrinsic invariance. Burgatti makes this calculus in his studies on mechanics and on differential geometry. Another remarkable application is Signorini's treatment on nonlinear elasticity (see e.g. [A. Signorini, *Trasformazioni termoelastiche finite*, Annali di Matematica Pura e Applicata **22**, 33–143 (1943)]). Concerning this, it is very interesting to read the second and third volumes (edited by Burgatti) of the work Analisi vettoriale generale e applicazioni. A work which contains contributions written together with C. Burali-Forti, T.Boggio and R. Marcolongo.[8]

3 Burgatti's Treatise on Mechanics

Pietro Burgatti publishes his treatise *Lezioni di Meccanica Razionale* (540 printed pages) in 1914 (publisher: Nicola Zanichelli). In the second edition (1919), he dedicates the book to his students fallen in the field of honour in World War I. The contents of this treatise covers the subject of mechanics as usually taught in Italian universities at that time, including the basic elements of continuum mechanics. In this aspect, he is closer to Marcolongo [20] than to Levi-Civita and

---

[7] See C. Burali-Forti, R. Marcolongo (1912, 1913), p. IX (foonote 4).
[8] See P. Burgatti, *Opere scelte* [8].

Amaldi [18, 19]. As already mentioned, the vector calculus is expounded as an Introduction. The first part, divided into five sections, deals with the Kinematics, in particular relative kinematics and rigid kinematics. The second part, divided in four chapters, is devoted to Statics; of course he starts by introducing the notion of force. Then he analyses the role of the constraints, and the equilibrium conditions. In this connection he states the principle of virtual work according to the original formulation of Lagrange, namely as a Statics principle: a necessary and sufficient condition for the equilibrium involving only the virtual work of the active forces. The part on Statics is concluded by the friction and and the applications to wire mechanics (wire equilibrium). The Dynamics is explained in third part, consisting in seven chapters. There we find, among other things, D'Alembert's principle, Lagrange's and Hamilton's equations. It is interesting to remark that analytical mechanics is presented in dealing with the subject which represents both its origin and its main application, namely celestial mechanics. He gives a brief sketch on canonical perturbation theory, also in the secular case, and concludes the chapter with the following prophetic words, in view of today's KAM theory: *Thus, the theory reproduces perfectly the observations; but, moreover, gives life to them in our thought; it shows to the eyes of our mind, as if we were an immortal audience, year after year, the mechanical evolution of the solar system. Whatever the future will be of this perturbation theory, it will remain one of the highest creations of human mind.*[9] The fourth and final part of the book covers the mechanics of the deformable bodies, with special emphasis on linear elasticity and fluids. Burgatti then concludes the book with some details about the historical development of mechanics.

It may be remarked that the formulation of the virtual work principle in the original one of Lagrange is at variance with that given in the treatises of Marcolongo [20] and Levi-Civita–Amaldi [18, 19], who prefer the formulation in terms of the constraint reactions, valid also off equilibrium (adopted also in the treatise of B.Finzi [14]. On this point see also [22]). It is natural to conjecture that Burgatti may have had this point of view from his teacher Beltrami, who never wrote a treatise on Mechanics (nor on any other subject, for that matter) even though he taught it in several universities (Bologna, Pavia, Roma). So far, however, we were unable to find any support for the validity of this conjecture.

4 Some Other Relevant Contributions of Burgatti to Mechanics and Analysis

We do not have the possibility, in this short contribution, to fully analyze the scientific legacy of Pietro Burgatti, and therefore we limit ourselves to a short description of some among his more lasting results beyond the general vector

---

[9]Così la teoria traduce perfettamente i fatti osservati; ma, per dippiù, dà loro vita nel nostro pensiero: facendoci assistere con gli occhi della mente, quali spettatori immortali, anno dopo anno, alle vicende meccaniche del sistema solare. Qualunque sia l'avvenire di questa teoria, essa rimarrà una delle più alte creazioni dell'ingegno umano.

calculus recalled above. As we have seen, even though the center of gravity of the scientific activity of Burgatti was Mathematical Physics, he worked in differential geometry and analysis as well, according to the italian tradition of the time, when the boundaries of the different sectors of mathematics were less sharply defined than they are today.

As a matter of fact, one of the most relevant scientific contributions of Burgatti is doubtless purely analytic, namely the generalization to equations (in two variables) of order higher than two of the classic integration method of Riemann for second order hyperbolic equations, condensed in the famous Riemann representation formula. In the simplest case (we follow here the presentation of [12], §V.1), Riemann's method deals with the Cauchy problem for the hyperbolic second order equation

$$u_{yy} - u_{xx} + au_x + bu_y + cu = f, \qquad (1)$$

where $c(x, y)$ and $f(x, y)$ are given smooth functions and the initial datum is a curve C nowhere tangent to the characteristic curves $x + y = $ const, $x - y = $ const. Aim of the formula is to represent the value $u(P)$ of the solution, at a point $P = (\xi, \eta)$ inside the domain of influence, in terms of the initial data, namely $u|_C$ and its outgoing derivatives also evaluated on C, so that also $u_x = p$ and $u_y = q$ are known on C. Then the formula is

$$u(P) = \frac{1}{2}[u(A)R(A;\xi,\eta) + u(B)R(B;\xi,\eta)]$$

$$+ \frac{1}{2}\int_A^B [Ru_x - (2bR - R_x)u]\,dx$$

$$+ \frac{1}{2}\int_A^B [Ru_y - (2aR - R_y)u]\,dy + \iint_G Rf\,dxdy,$$

where

- A and B are points on the noncharacteristic curve C;
- G is the bounded plane region whose boundary is formed by the arc C and the two segments along the characteristic lines emanating from P and their intersection points with C;
-

$$R(x,\eta;\xi,\eta) := \exp \int_\xi^x b(\lambda,\eta)\,d\lambda,$$

$$R(\xi,y;\xi,\eta) := \exp \int_\eta^y a(\lambda,\xi)\,d\lambda.$$

In 1906, Burgatti [9] first obtained the generalization of this formula to the case of an equation of order n, under the same assumptions of linearity and two independent variables of Riemann. This result is explicitly referred to in the classic treatise of Richard Courant on partial differential equations.

Two other lasting contributions of Burgatti deal with analytical mechanics, and are both included in the famous treatise of Whittaker [24], in addition to [18]. The first one concerns one of the most classic problems of that discipline, initiated by Liouville: given an Hamiltonian system with n degrees of freedom, formulate conditions on the potentials under which the corresponding Hamilton–Jacobi equation can be separated into n ordinary first-order differential equations. These ODE can be immediately integrated and therefore the original system is integrable. This problems admits a differential geometric formulation, as first remarked by Stäckel in his famous criterion[23]; subsequently, a criterion easier to verify in terms of the derivatives of the original Hamiltonian was worked out by Levi-Civita [17]. This is the starting point of Burgatti's investigation [10]. To describe that contribution,[10] we just quote *verbatim*[11]

LIOUVILLE *was the first (in 1946) to look for the* HAMILTON-JACOBI *equations integrable through simple quadratures, corresponding to mechanical problems in two degrees of freedom. He found one very important form, which is now studied in the best treatises of mechanics and differential geometry.*

Forty five years later Professor STÄCKEL *[23] formulated the problem in the precise and general terms written in the title of this note and found, in n-variables, and integrability case, which is the immediate generalization of that of* LIOUVILLE.

---

[10]Entitled *Determination of the Hamilton–Jacobi equations integrable through separation of the variables* = Determinazione dell'equazioni di Hamilton–Jacobi integrabili mediante la separazione delle variabili.

[11]LIOUVILLE fu il primo (nel 1946) a cercare l'equazioni di HAMILTON-JACOBI, corrispondenti ai problemi della meccanica con due gradi di libertà, integrabili mediante semplici quadrature. Egli ne trovò una di forma molto importante, che si trova ora studiata nei migliori trattati di meccanica e geometria differenziale.

Quarantcinque anni dopo il Prof. STÄCKEL pose il problema nei termini precisi e generali scritti nel titolo di questa Nota, e trovò per le n variabili, un caso d'integrabilità, che è l'immediata generalizzazione di quello del LIOUVILLE. Qualche anno prima, studiando l'equazioni con due sole variabili indipendenti, egli aveva risoluto completamente il problema, determinando due nuovi casi di integrabilità, oltre a quello del LIOUVILLE, e mostrando che non ne esistono altri.

Più tardi, nel 1904, il prof.LEVI-CIVITA determinò tutte le condizioni necessarie e sufficienti, affinché un'equazione di HAMILTON–JACOBI sia integrabile per separazione delle variabili; lo studio delle quali condurrebbe senza dubbio alla soluzione completa del problema, vinta che fosse la ripugnanza alla fatica di calcoli assai lunghi e laboriosi.

Applicando il metodo in certe ipotesi restrittive, egli pervenne con una elegante analisi a un nuovo caso di integrabilità che è la generalizzazione alle n variabili di uno dei casi trovati dallo STÄCKEL per due variabili. Poco dopo, nel 1906, il prof.DALL'ACQUA [...] nel caso di tre variabili e trovò quattro espressioni per l'energia cinetica: le sole che possano dar luogo ad equazioni di HAMILTON–JACOBI integrabili per separazione delle variabili.

In questo scritto io ho ripreso il problema in tutta la sua generalità, e abbandonati i metodi finora usati, guidato più dall'intuizione che da una logica rigorosa, son pervenuto alla determinazione di $n + 1$ equazioni di HAMILTON-JACOBI integrabili mediante la separazione delle variabili.

Resterebbe da dimostrare che quell'equazioni sono le sole rispondenti al problema, sulla qual cosa io non ho dubbio alcuno; ma non ho trovato finora una dimostrazione soddisfacente.

Some years before, studying the equations with only two independent variables, he had completely solved the problem. determining two new integrability cases, beyond that of LIOUVILLE, *and shown that further ones do not exist.*

Later, in 1904, Professor LEVI-CIVITA[17], *determined all necessary and sufficient conditions in order that a* HAMILTON-JACOBI *equation be integrable by separation of variables; the study of these conditions would doubtless lead to the complete solution of the problem, once the reluctance is overcome to the labor of long and painstaking computations.*

Applying the method under certain restricitve assumptions, he came through an elegant analysis to a new integrability case, which is the generalization to the n variables of one of the cases found by STÄCKEL *in two variables.*

A short time later, in 1906, Professor DALL'ACQUA *[...] in the case of three variables found four expressions for the kinetic energy: the only ones giving rise to* HAMILTON-JACOBI *equations integrable by separation of variables.*

In this paper, I came back to the problem in all its generality; left aside all methods emplyed so far, and led more by intuition than by rigorous logic, I reached the determination of n + 1 HAMILTON-JACOBI *equations integrable by separation of variables.*

It remains to be proved that these equations are the only ones corresponding to the problem; on the validity of this statement I have no doubts; however, I did not find so far a convincing proof.

The conditions written by Burgatti are too long and cumbersome to be reproduced here; we limit ourselves to remark that his conclusion was correct, but the full proof had to wait 96 years. The reader is referred to the paper of S. Benenti [1] for a complete clarification of the matter.

The second contribution we want to recall concerns another famous problem of analytical mechanics, the motion of rigid body with a fixed point O (different from the center of mass) under the action of gravity, when the inertia ellipsoid through it is an ellipsoid of rotation and the center of mass lies on the rotation axis. That is, denoting A, B, C the principal moments of inertia, and by ξ, η, ζ the coordinates of the center of gravity with respect to the mobile frame defined by principal inertia axes, then $A = B \neq C$, and $\xi = \eta = 0$. This system admits three first integrals, namely energy, angular momentum and projection of the angular velocity along the top axis. Their explicit expression as a function of the components (p, q, r) of the angular velocity ω and of the direction cosines (a, b, c) of the vertical with the principal inertia directions is

$$\begin{cases} r = r_0 \\ Apa + Aqb + Crc = k \\ A(p^2 + q^2) = 2Mg\zeta\gamma + h. \end{cases}$$

These integrals are polynomials in the variables $p, q, r; a, b, c$), and this entails that the motion can be integrated by quadratures in terms of hyperelliptic functions. In 1888, Sofia Kowalewsajia discovered that when the two equal principal moments of inertia are twice the third, $A = B = 2C$, the system a fourth polynomial first integrals: the energy and the projection of the angular momentum along the symmetry axis. A natural question was to determine whether or not this discovery depended critically on the above ratios among the principal moments of inertia. The question was therefore the persistence of a fourth polynomial integral under the more general assumption $A = B = kC$ for $k \neq 1, 2$. Burgatti [11] proved that this is not the case, actually showing the more general result of the nonexistence of algebraic integrals other than the three known.

5 Some Conclusions

Pietro Burgatti represents a significant figure in the great tradition of the italian Mathematical Physics blossomed right after the unification of the country 150 years ago. It may be remarked that much in the same way as the founding fathers of that school, Enrico Betti (1823–1892) and Eugenio Beltrami (1835–1900), and their most prominent followers in the subsequent generations, Vito Volterra (1860–1940) and Tullio Levi-Civita (1873–1941), he is remembered maybe more for his contributions to pure mathematics than to mathematical physics. In particular, he inherited from his teacher Beltrami the powerful view of the differential geometry in his approach to mathematical research.

The fact that Burgatti made lasting constributions to several different fields of mathematics (his list of publications of 114 entries is almost evenly split among partial differential equations, mechanics and differential geometry) is not surprising. Indeed, he lived and worked at the turn between the XIX and the twentieth century, the Golden Age of italian mathematics, and therefore he operated in an environment of high scientific level where the diversified interests were the rule. To give an idea of that environment, it is enough to recall that when at age 40 he joined the Faculty in Bologna, one of the seven major Universities of the country at the time, his senior colleagues were mathematicians of world renown: the analysts Cesare Arzelà (1853–1912) and Salvatore Pincherle (1852–1936), and the algebraic geometer Federigo Enriques (1871–1946). In the course of his tenure as Dean he had as younger colleagues other mathematicians of comparable standing: Leonida Tonelli (1885–1946), Beppo Levi (1873–1951), Beniamino Segre (1903–1977) and Luigi Fantappiè (1901–1967).

Neither Burgatti nor his closely associated colleagues of his generation, Marcolongo and Burali-Forti, established a school comparable to that of the slightly younger Levi-Civita. This may due to the fact that their most original technical contribution, the general vector calculus and the vector homographies, was rapidly superseded by the theory of exterior differential forms which in a way includes also the multilinear algebra and tensor calculus. Actually, only one scholar of international reputation can be considered a student of Burgatti, namely Dario Graffi (1905–1990), while no comparable students of Burali-Forti and Marcolongo are known.

References

1. Benenti, S. 1997. *Intrinsic characterization of the variable separation in the Hamilton-Jacobi equation*, J.Math.Phys. **38**, 6570–6590.
2. Burali-Forti, C. 1926. *Geometria analitico-proiettiva*. Casa Editrice G.B. Petrini: Torino.
3. Burali-Forti, C., and T. Boggio. 1924. *Espaces courbes. Critique de la relativité*, Sten Editrice: Torino.
4. Burali-Forti, C., and R. Marcolongo. 1909. *Elementi di calcolo vettoriale con numerose applicazioni alla geometria, alla meccanica e alla fisica-matematica*. Nicola Zanichelli: Bologna; (seconda edizione 1921).
5. Burali-Forti, C., and R. Marcolongo. 1912, 1913. *Analyse vectorielle générale*. (I et II vol.), Mattei & C. Éditeurs: Pavia.
6. Burgatti, P. 1919. *Lezioni di meccanica razionale*. Nicola Zanichelli Editore: Bologna.
7. Burgatti, P., T. Boggio, and C. Burali-Forti. 1930. *Analisi vettoriale generale e applicazioni*. v. II (Geometria differenziale) and v.III (Teoria matematica dellelasticit, a cura di P. Burgatti), Nicola Zanichelli: Bologna.
8. Burgatti, P. 1951. *Opere scelte* (pubblicate sotto gli auspici delle universit di Bologna e di Ferrara, dellAccademia delle Scienze di Bologna e dellUnione Matematica Italiana). Nicola Zanichelli: Bologna.
9. Burgatti, P. 1906. *Sull'estensione del metodo di integrazione di Riemann alle equazioni lineari di ordine n in due variabli indipendenti*. Rend. Lincei Serie V. 15: 602–609.
10. Burgatti, P. 1911. *Determinazione dell'equazioni di Hamilton-Jacobi integrabili mediante la separazione delle variabili*. Rend. Lincei Serie V. 20: 108–11.
11. Burgatti, P. 1906. *Metodo di integrazione di Riemann alle equazioni lineari di ordine n in due variabli indipendenti*. Rend. Lincei Serie V. 15: 602–609.
12. Courant, R., and D. Hilbert. 1989. *Methods of mathematical physics*, Vol. II, *Partial differential equations*. by R. Courant, Wiley Classics Editions: New York.
13. Dall'Aglio, L. 2004. *Un case study nell'accettazione delle teorie matematiche. Sviluppo e diffusione del calcolo differenziale assoluto in epoca prerelativistica*. Bollettino di Storia delle Scienze Matematiche 2: 9–65.
14. Finzi, B. 1946. *Meccanica razionale*. (2 vol.), Zanichelli: Bologna.
15. Freguglia, P. 2006. *Geometria e numeri. Storia, teoria elementare ed applicazioni del calcolo geometrico*. Bollati Boringhieri: Torino.
16. Freguglia, P. 1992. *Dalle equipollenze ai sistemi lineari. Il contributo italiano al calcolo geometrico*. (con appendici di S. Briccoli Bati e G. Canepa), Quattroventi ed., Urbino.
17. Levi-Civita, T. 1904. *Sull'integrazione dell'equazione d Hamilton-Jacobi per separazione delle variabili*. Math. Annalen 59: 383–394.
18. Levi-Civita, T., and U. Amaldi. 1926–1930. *Lezioni di Meccanica Razionale*. (3 voll.), Nicola Zanichelli Editore: Bologna.
19. Levi-Civita, T., and U. Amaldi. 1938. *Compendio di Meccanica razionale*. Nicola Zanichelli Editore: Bologna.
20. Marcolongo, R. 1905. *Meccanica razionale*. (2 voll.), Manuali Hoepli: Milano.
21. Marcolongo, R. 1921. *Relatività*. Casa Editrice Giuseppe Principato: Messina.
22. Nastasi, P., and R. Tazzioli. 2006. *Tullio Levi-Civita*. (Lettera matematica PRISTEM 57, 58), Springer: Milano.
23. Stäckel, P. 1893. *Habilitationschrift*. Halle a.d.Saale 1891; = Mathematische Annalen 42: 537–563; (Lettera matematica PRISTEM 57, 58), Springer: Milano 2006.
24. Whittaker, E. 1904. *A treatise on the analytical dynamics of particles & rigid bodies*. Cambridge University Press: Cambridge (fourth edition: 1947).

Federigo Enriques (1871–1946) and the Training of Mathematics Teachers in Italy

Livia Giacardi

> What has the greatest influence on the effectiveness of teaching, more than differences in methods or the guidelines of programs, is the skilfulness of the teachers: their mentality, their ability to communicate, the passion they bring to the subjects they teach, the breadth of interests that make it possible for them to put themselves in the students' place and feel with them.
>
> Federigo Enriques[1]

Abstract This essay will illustrate Federico Enriques' vast, multifaceted efforts to improve the preparation of mathematics teachers, situating them in their historic context and within the framework of the cultural project that formed the basis of his whole scientific output. The first part of the essay is dedicated to a brief presentation of the principal steps in the history of Italy's *Scuole di Magistero* (teacher training schools), with reference to the most significant legislative measures, to the contribution of teachers' associations, and to debates among mathematicians. The second part will show how Enriques' cultural project for the creation of a scientific *humanitas*, which was rooted in the philosophy and history of science, developed gradually during his years in Bologna, and how this was reflected in

---

This research was carried out as part of the Project *PRIN 2009, Scuole matematiche e identità nazionale nell'età moderna e contemporanea*, Unit of the University of Torino.

[1] See [53, p. 188]: "Più che le differenze dei metodi o le indicazioni dei programmi influisce sull'efficacia dell'insegnamento il valore degli insegnanti: la loro mentalità, la comunicativa, la passione che portano nelle cose insegnate, la larghezza degli interessi che li fa capaci di mettersi al posto degli allievi e di sentire con essi."

L. Giacardi (✉)
Department of Mathematics, University of Torino, Turin, Italy
e-mail: livia.giacardi@unito.it

his vision of mathematics teaching. The influence of Felix Klein will also be highlighted. The third part examines Enriques' involvement in teacher training and the various strategies he adopted, and frames his initiatives and methodological assumptions within his cultural program. Finally, three appendices containing previously unpublished letters and documents conclude the essay.

1 Introduction

Immediately following the constitution of the Kingdom of Italy and the establishment of an educational system at a national level, the Italian political class, which included many high-level mathematicians, understood the importance of having corps of adequately trained teachers in order to guarantee the formation of the future ruling class of the new nation. The problem was urgent, given the fact that at the time people without a degree were permitted to teach. It was only in 1906 that the legislation was approved regarding the legal status of teachers, making it official that only those who had won a competition could teach, and that a degree was required for admission to the competition [GU 1906, 106, p. 2085]; it was not until 1914 [GU 1914, 174, pp. 4086–4101] that mathematics teachers were placed on an equal footing with teachers of Italian, rectifying the inequality that had existed in the system since the Casati legislation (1859). Among those who were personally involved in the effort to improve Italian schools and in training teachers for various levels of schools were many members of the well-known Italian school of algebraic geometry. Many factors led them to embrace this commitment, although for each of these mathematicians the various factors had different overtones. First of all, it was not incidental that geometry is the discipline that best makes it possible to bring into focus the problems of methods that are inherent in mathematics teaching, and to clarify the delicate relationship between *formation* and *information*, which has always played a particular role in education. Secondly, of indubitable importance was the influence of Felix Klein, a mathematician who was not only active in advanced research, but was also involved in the reform of secondary and university mathematics teaching in Germany, and who had been president of the International Commission on Mathematical Instruction since 1908. Klein's influence (see [71, 75, 117]), which can be seen in the trends and methods of research – it is sufficient to think of the number of young mathematicians who gathered around him to perfect their scientific training – also affected the way that mathematics teaching and curriculum reform was conceived. One sign of this is the translation into Italian of Klein's *Erlanger Programm*,[2] as well as some of his other writings which were more specifically concerned with didactics, in particular his 1895

---

[2] See [86]. The translation was by Gino Fano at the suggestion of Corrado Segre.

Vorträge über ausgewählte Fragen der Elementargeometrie translated by Francesco Giudice with the title *Conferenze sopra alcune questioni di geometria elementare*,[3] and the 1895 lecture entitled *Über Arithmetisirung der Mathematik*, translated by Salvatore Pincherle with the title *Sullo spirito aritmetico nella matematica* [89]. Another important reason why many Italian algebraic geometers were committed to teacher training is that they belonged to a school that shared a well-defined program for research and a common vision of mathematics and its teaching, a strong school that was at the forefront of the international scene at the turn of the century. Finally, political and social motivations cannot be overlooked: the mathematicians working in the period immediately following Italian unity were animated by the strong spirit of the Risorgimento, and believed that the creation of national identity depended on an efficient school system and adequate preparation of an educated ruling class.

1.1 The Scuole di Magistero

The *Scuole di Magistero*, or teacher training institutes, were established by Minister of Public Instruction Ruggero Bonghi in 1875 to respond to the need to train future teachers and thus guarantee a higher level of secondary schools, and they survived with successive modifications until 1920, when they were abolished by the minister Benedetto Croce. Their history was especially troubled, as shown by the great number of decrees that concerned them.[4]

The initial purpose of the Scuole di Magistero was fundamentally ambiguous, emphasising both research and professional teacher training as can be seen from articles 32 and 33 of the Royal Decree of 2 November 1875 (R. Bonghi):

> The program of the Scuole di Magistero consists, in addition to the studies required for the corresponding degree, in special exercises aimed at instilling in the students an aptitude for research and the original exposition of that discipline that they wish to profess ... [and students] will take a course on the limits and methods of teaching of the sciences instituted by the Minister ...[5]

We need only consider the fact, for example, that Francesco Faà di Bruno, who was responsible for teaching mathematics in the Scuole di Magistero of the

---

[3] See [88]. It was Gino Loria who was behind the translation; see the letter from Loria to Klein dated Genova, 22 July 1895, SUB Göttingen, *F. Klein* 10.

[4] All of the legislative measures cited here can be consulted in the section *Teacher Training* of [76].

[5] See [GU 1875, 255, p. 6835]: "Il corso delle scuole di Magistero consiste, oltre che negli studi richiesti per la corrispondente laurea, in esercitazioni speciali dirette a produrre negli studenti l'attitudine alla ricerca e alla esposizione originale e propria di quella disciplina che vogliono professare ... Di più egli seguirà un corso sui limiti e sui metodi dell'insegnamento delle scienze instituito dal Ministro."

University of Torino, in the year 1882–1883 dealt only with the theory of elliptic functions.

In order to clarify the aims of the Scuole di Magistero and to improve their effectiveness, a commission was created in 1885 by the *Consiglio Superiore della Pubblica Istruzione* (High Council for Public Instruction). Its members were Luigi Cremona, Eugenio Beltrami and Sebastiano Richiardi. In its report, the commission placed particular emphasis on the "practical preparation for secondary teaching," which had up to that time been neglected in favour of a specifically scientific preparation, and insisted on the need for practical training dedicated to the study of the foundations of mathematics and critical analysis of the methods.[6] These suggestions were reflected in article 2 of the Royal Decree of 30 December 1888 (P. Boselli), which states that "The Scuola di Magistero is aimed at the practical training for secondary teaching" and underlines the importance of a preparation for teaching by means of practical training that "consists in the examination of the postulates of science, in written works, and in lessons by students on subjects chosen by them at the suggestion of their professor and with his approval. The discussion of the didactic rules to be applied to the aforementioned subject in secondary teaching will be included."[7] Further, in article 4, a certain emphasis was given to mathematics by assigning to it four years of courses, while only two years were assigned to other scientific disciplines.

The nature of the lessons were further defined by the historian and statesman Pasquale Villari in the Royal Decree of 29 November 1891, which underlined from the beginning that the primary aim of the courses (of a minimum of two years) was to "render the students expert in the art of teaching the different disciplines" in the various kinds of secondary schools. In particular, article 6 regarded lectures of a didactic nature:

> In these, the professor should: 1. set forth the method to be used in Secondary Schools for teaching the subject assigned to him, assessing its extents and limits; 2. make the students perform appropriate practical exercises that serve to accustom them to applying the method being taught. Among these practical exercises there are also actual lessons given in the Scuole di Magistero, and, when possible, in a Secondary School as well; 3. present and examine the best textbooks for Secondary Schools.[8]

---

[6] See *Sull'istruzione secondaria classica. Notizie e documenti presentati al Parlamento Nazionale dal Ministro della Pubblica Istruzione Paolo Boselli* (Roma, Sinimberghi, 1889), pp. 266–269 and 273–274.

[7] See [GU 1889, p. 219]: "La Scuola di Magistero ha per fine la preparazione pratica all'insegnamento secondario" ... "consistono nell'esame di postulati della scienza, in lavori scritti e in lezioni degli studenti sopra soggetti scelti da loro con approvazione del professore, indicati da questo. Vi sarà compresa la discussione delle regole didattiche da applicarsi alle suddette materie nell'insegnamento secondario."

[8] See [GU 1892, p. 80]: "In esse il professore dovrà quindi: 1. esporre il metodo da seguirsi nelle Scuole secondarie per l'insegnamento della materia a lui affidata, determinandone l'estensione ed i limiti, 2. fare eseguire agli alunni opportune esercitazioni che valgano ad abituarli alla applicazione del metodo insegnato. Fra queste esercitazioni vi sono anche saggi di lezioni date nelle Scuole di

It also recommended holding lectures on education in general, to be given by only those who had acquired long years of practical experience in secondary teaching.

In many cases, the Scuole di Magistero were completely inadequate for reliably addressing the problem of teacher training. There were many reasons for this: above all, the professors who taught there were the same ones who taught institutional courses at university, and because these had, with rare exceptions, no experience in secondary teaching, they were unprepared to address questions about pedagogy and method. Furthermore, supporting structures (libraries, laboratories, etc.) and teaching materials were practically nonexistent, the number of assigned course hours was inadequate, and there was scant funding.

There are various testimonies to these shortcomings by both pedagogists and mathematicians. For example, Saverio De Dominicis, professor of pedagogy at the University of Pavia, wrote in 1882:

> The Scuola di Magistero does not exist in many faculties, although these create professors; where it exists it simply provides an illusion, because it has no distinct purpose; it is always, even the blind can see it, incomplete. ...The Scuola di Magistero should come after the specialised studies in this or that faculty; ...it should be, not the Scuola di Magistero of one faculty or another, but the Scuola di Magistero for secondary teaching. ...This would be a serious school: a school that would oblige various professors to ponder the problems of pedagogy ...It is the Scuola di Magistero, not the faculty, that can create good teachers: the faculty has always created and will continue to create erudite young people, but erudite young people are not professors.[9]

In his report to the Senate on the reorganisation of university teaching, Luigi Cremona, at that time a professor at the University of Rome, wrote:

> The great interest that the State has in the formation of qualified teachers demands that these be trained in only a few centres ...under the guidance of men who are not only scientists but also masters of the art of teaching. No country has an abundance of such men, and we fewer than others, because for a long time no one seems to have cared about this. It is precisely in this that the regulations of 1875 were mistaken, relying too much on the pedagogical-didactic training of our professors.[10]

---

magistero, e, quando si possa, anche in una Scuola secondaria; 3. far conoscere ed esaminare i migliori libri di testo per le Scuole secondarie."

[9]See [18, pp. 184–185]: "La scuola di magistero dunque in molte facoltà, che pure creano professori, manca; dove trovasi è una semplice illusione, perché non ha scopo a sé e distinto; sempre, anche ad essere ciechi, è incompleta. ...La scuola di magistero ...dovrebbe essere non la scuola di magistero di questa o di quella facoltà, ma la scuola di magistero per l'insegnamento secondario ...scuola seria sarebbe questa; scuola che obbligherebbe professori vari a ponderare i problemi pedagogici ...È la scuola di magistero, non la facoltà, che può fare de' bravi insegnanti: la facoltà ha fatto e farà sempre de' giovani dotti, ma i giovani dotti non sono i professori."

[10]See [16, p. 85]: "L'alto interesse che lo Stato ha per la formazione di valenti maestri, esige che questi siano educati soltanto in pochi centri ... sotto la direzione d'uomini che non siano solo scienziati, ma anche maestri nelle arti educative. D'uomini siffatti non v'à abbondanza in alcun paese, e in casa nostra meno che altrove, perché da gran tempo pare che nessuno se n'occupi In ciò appunto errarono i regolamenti del 1875, facendo soverchia fidanza con la scienza pedagogico-didattica de' nostri professori."

The lack of interaction between the university world and secondary teachers was a further reason why the Scuole di Magistero were inadequate. This chasm was highlighted by Gino Fano in 1894, in an article written upon his return from a year of professional development spent in Göttingen with Klein. Fano pointed out the initiatives promoted by Klein to address this problem:

> ...each year during the Easter holidays the secondary school teachers are invited to convene, those of the eastern provinces in Berlin, those in the western provinces in Göttingen; and there they stay for about fifteen days, in contact with university teachers. Lectures and lessons make it possible on one hand for the numerous participants to stay up-to-date with the many, many advancements that are continually being made by science, while on the other, the university professors as well have a way to understand fully the needs and desires of the secondary school teachers.[11]

From the very beginning, the problem of the professional training of teachers was one of the most hotly debated topics in the *Associazione Mathesis*, an association of teachers of mathematics founded in Torino in 1895–1896 by Rodolfo Bettazzi, Aurelio Lugli and Francesco Giudice with the aim of "improving the school and the training of teachers, from the points of view of science and didactics."[12] From the association's very first congress, held in Torino in 1898, the Mathesis Association sponsored an enquiry among the members regarding the theme "Modifications to be introduced in the regulations of university mathematical studies, intended to produce good secondary teachers." The theme was taken up again in subsequent meetings and congresses, and different proposals were formulated.[13]

Within the multiplicity of presentations, it is possible to identify two main lines of thought. Some, such as Salvatore Pincherle, proposed the separation of curricula, and more precisely, the institution, after the first two years of university, of a special course leading to a degree in education (*laurea didattica*) to be attended by all those who intended to pursue a career in secondary teaching; this was to be distinct from the degree in pure mathematics, which was instead to be sought by those who intended to pursue a career in research. According to Pincherle, the future teachers should, in a biennial course dedicated to mathematical methodology, inspect and analyse in depth all of the chapters of elementary mathematics [108, p. 86]. This proposal was shared by Guido Castelnuovo, as well as by Enriques, as we shall see in greater depth in a moment. What Pincherle intended can be clearly seen in the classes he held at the Scuole di Magistero of the University of Bologna from 1899–1900 to 1920–1921. All of his annual reports show the importance he attached to

---

[11] See [67, pp. 181–182]: "ogni anno nelle vacanze Pasquali gli insegnanti delle scuole secondarie sono invitati a riunirsi, quelli delle province orientali a Berlino, quelli delle province occidentali a Gottinga; e lì rimangono circa quindici giorni, a contatto degli insegnanti universitari. Conferenze e lezioni permettono da un lato ai numerosi convenuti di tenersi al corrente dei tanti e tanti progressi che la scienza va continuamente facendo, mentre d'altra parte anche gli insegnanti di Università hanno modo di rendersi conto esatto dei bisogni e dei desideri dei primi."

[12] See "Statuto dell'Associazione," *Bollettino dell'Associazione Mathesis*, 1, 1896–1897, in *Periodico di Matematica*, 1896, p. 161.

[13] See [72, 101], and the section *Mathesis' Congresses* in [76].

elementary mathematics from an advanced standpoint, and his growing interest in questions regarding the principles and foundations of mathematics, in all probability a result of time spent with Enriques, who at that time was teaching in Bologna.[14]

Others, such as Alessandro Padoa, supported by Gino Loria and Giuseppe Peano, disapproved of the separation into two different curricula, and instead believed that it was urgent to strengthen the Scuole di Magistero. In particular, they proposed instituting, in addition to an obligatory period of practice teaching in a secondary school, a two-year course of Mathematical Methodology in place of the didactic lectures in the Scuole di Magistero, which would make it possible to address not only topics of arithmetic, algebra and geometry useful for the future teacher, but also include an examination of teaching methods, an analysis of school textbooks, as well as show the educational usefulness of the history of mathematics and mathematical games [96]. Loria and Padoa wrote that the history of mathematics should permeate the entire program, aiming above all to reconstruct the various phases of development of each theory, as well as to render the subject "less arid and more attractive" [96, pp. 4, 6]:

> The new university course we are suggesting would serve, in our opinion, to fill the deplorable abyss that separates university teaching from secondary teaching today, ... which F. Klein has recently referred to as "a system of double forgetting" [n.b. *doppelte Diskontinuität*]: the university student's forgetting what he studied in secondary school, and the secondary school teacher's forgetting all that he studied while he was at university.[15]

References to Klein emerge in all the Mathesis congresses, and are an index of the influence he exerted in the Italian debates, an influence that can be also perceived at the base of the project for the *Enciclopedia delle matematiche elementari* presented by Luigi Berzolari and Roberto Bonola during the congress in Padua in 1909. Intended for mathematics teachers as well as the students of the Scuole di Magistero, the encyclopaedia was aimed at addressing elementary mathematics from an advanced standpoint as well as contain suitable remarks regarding the history of mathematics and questions of education.[16]

When the Minister Croce abolished the Scuole di Magistero with the Royal Decree of 8 October 1920 [BUMPI 1920, p. 2064], some of the most vigorous opposition came from the Mathesis Association and the two members of the Italian school of geometry, Loria and Fano. In a lecture to the Liguria section of the Mathesis Association, Loria expressed indignation for this "sudden and violent measure," saying that the Scuole di Magistero represented "a bridge, the only

---

[14] ASUB, *Scuole di Magistero* (pos. 53/b), busta 3 (1880–1921).

[15] See [96, pp. 3–4]: "Il nuovo corso universitario da noi suggerito servirebbe, a parer nostro, a colmare il deplorevole abisso che oggi separa l'insegnamento universitario dall'insegnamento secondario, la cui esistenza venne segnalata da uno di noi sin dal 1898 e che F. Klein ha recentemente designato come 'sistema del duplice oblio': oblio da parte dello studente universitario di quanto studiò nelle scuole secondarie, oblio dell'insegnante secondario di tutto quello che lo occupò mentre trovavasi all'università."

[16] For more about this, see the article by Erika Luciano in this present volume.

one that exists between upper-level and middle-level teaching." He criticised the identification of scientific training with educational training, the lack of interest in questions of methodology, and the fact that "future teachers were not put in front of school students in the way the future health worker is put in contact with human suffering" [95, p. 163]. During the 1921 Mathesis congress in Naples, Fano formulated an item for the agenda in which he asked for "the reinstatement of the Scuole di Magistero for mathematics, in a broader and more comprehensive form than the previous one." Convinced that *"knowing more* than what you teach is worthless, if this *more* does not make you know better what should be taught," he energetically proposed the establishment of courses of Elementary Mathematics from an Advanced Standpoint, with an emphasis on the historical, critical, methodological, and didactical aspects, citing the lessons of Corrado Segre and Enriques as examples. He also invited the faculties to accept as dissertations for degree theses in complementary mathematics (*matematiche complementari*), that is concerning those sectors of mathematics more strictly connected to elementary mathematics, and urged his colleagues to establish, without awaiting ministerial decrees, practice teaching programs in secondary schools for the future teachers [68, pp. 103, 109].

The proposals were accepted at least in part by the Minister for Public Instruction Orso Mario Corbino, who in 1921 established "combined" degrees (*lauree miste*) in physical and mathematical sciences [BUMPI 1922, p. 22] aimed at qualifying young people to teach scientific subjects in secondary schools, and in 1922 instituted a course in complementary mathematics, accompanied by didactic and methodological exercises [BUMPI 1922, p. 349].

2 The Emergence of Enriques' Cultural Project and the Project's Effects on Mathematics Education

2.1 The Teaching of Projective Geometry in Bologna

After earning his degree at the Scuola Normale di Pisa in 1891, in 1892 Enriques obtained a Lavagna scholarship and although he had hoped to study in Torino with Corrado Segre,[17] he was sent to the University of Rome. This was where the extraordinary friendship with Guido Castelnuovo was born, a friendship that would last for the rest of his life and lead to the publication of the well-known works on algebraic surfaces.[18] However, the young Enriques went to Torino anyway in

---

[17]See Enriques to Castelnuovo, s.l. 6 November 1892, in [8, p. 3].
[18]For Enriques' contribution to algebraic geometry, and related bibliography see the paper by P. Gario and C. Ciliberto in this present volume.

November 1892, staying there for some weeks and then returning there a year later, in November 1893, at the end of a year spent in Rome perfecting his studies, in hopes of becoming an assistant to Luigi Berzolari and thus being able to work with Segre.[19] The months in Torino between November 1893 and January 1894 were very intense for his scientific research, and stimulating for his reflections on the foundations of geometry. It should be recalled that in 1889 Segre had been behind Mario Pieri's translation of K. G. Staudt's *Geometrie der Lage*, as well as Fano's 1890 translation of Klein's *Erlanger Programm*, and had urged Fano and Federico Amodeo to study the foundations of projective geometry. In contrast to Segre, who left physical or philosophical aspects of the problem aside,[20] Enriques was attracted by these very aspects, and explicitly said as much in a 1894 paper on the foundations of projective geometry:

> The route followed by them [Fano and Amodeo] is quite different from that we intend to take, especially in that, while the two esteemed authors propose to establish an arbitrary system of hypotheses that is capable of defining a linear space to which the results of ordinary geometry be applicable, here we will seek to establish the postulates deduced from experimental intuition of the space that appear to be the simplest for defining the object of projective geometry.[21]

In a note he added, "It only seems to us that geometry's experimental origins must not be forgotten in the search for the hypothesis on which it is founded."[22]

Enriques' interest in problems connected to mathematics teaching was born in close connection to his philosophical, historical, and interdisciplinary interests, and in particular, to research into the foundations of geometry, stimulated by the course in projective geometry at the University of Bologna, which, thanks to a series of lucky events, he was assigned to teach on 16 January 1894.[23] A month after the course began, Enriques wrote to Castelnuovo about the difficulty of reconciling the need for rigour with that of intuition in his lessons:

> ...whether I already have or will yet sin in aiming too high in the course, depends on the fact that I don't yet have an adequate idea of the difficulty that young people run up against. I can only realise it during the lesson, as I explain, when by then the order of the topics to cover is fixed and the notes written; but I believe I compensate for the difficulties of some

---

[19]See the letters from Enriques to Castelnuovo in [8, pp. 39 e 44].

[20]See for example [118, p. 61].

[21]See [21, p. 551]: "L'indirizzo da essi *[Fano e Amodeo]* seguito è alquanto diverso da quello a cui noi intendiamo attenerci, specialmente per ciò che, mentre i due egregi autori si propongono di stabilire un qualunque sistema di ipotesi capace di definire uno spazio lineare al quale siano applicabili i risultati dell'ordinaria geometria, noi cerchiamo qui di stabilire i postulati desunti dall'intuizione sperimentale dello spazio che si presentano più semplici per definire l'oggetto della geometria proiettiva."

[22]See [21, p. 551]: "Ci sembra soltanto che l'origine sperimentale della geometria non debba essere dimenticata nella ricerca delle ipotesi su cui essa è fondata." See also [1, pp. 391–401].

[23]For the reconstruction of the events based also on archival documents, see [105, pp. 72–84].

points by insisting on them vigorously, since it is precisely on those that it occurs to me to be more energetic in my explanation.[24]

The first steps in Enriques' path towards the research on the psychological and physiological origin of the postulates of geometry are found in the lessons in higher geometry that he gave at the invitation of several students in the year 1894–1895,[25] and which were collected in the lithograph printed that same year, entitled *Conferenze di geometria tenute nella R. Università di Bologna. Fondamenti di una geometria iperspaziale*. The discussion is preceded by an introduction, four aspects of which we want to underline here, because they will be examined in depth and clarified by Enriques in his later works. First of all, Enriques affirms the importance of the history of mathematics. Second, he reflects on the importance of the study of the foundations, whose value derives from the fact that "in mathematics every step forward has drawn attention back to an analysis of the foundations and, vice versa, such an analysis has often resulted in new and important concepts that made it possible to extend known results to a more general area" [22, p. 1].[26] The third aspect is that he underlines the importance of comparing mathematics with other sciences, because that is the only way to ascertain the true significance of the scientific importance of mathematical research, in keeping with the conviction that science is "an organic whole" [22, p. 2]. The fourth and final aspect is that he addresses the problem of what geometry is in order to arrive, at the end of the introduction, at an explanation of what is meant by abstract geometry, thus tying in with the guiding thread that Segre had provided for algebraic geometry:

> Bearing in mind from the very beginning the extension that we want to give to the results obtained in a given field by applying them to other fields, it avails us to consider the fundamental elements of Geometry as objects of an abstract nature connected by purely

---

[24] See Enriques to Castelnuovo, Bologna (?) 22 February 1894 [8, p. 77]: "...se io ho peccato o peccherò di troppa elevatezza del corso, ciò dipende da che non ho ancora un'idea adeguata delle difficoltà che incontrano i giovani. Io me ne accorgo soltanto nella lezione, spiegando, quando ormai è fissato l'ordine delle cose da svolgere e sono scritti gli appunti; ma credo di compensare le difficoltà di alcuni punti con una vibrata insistenza, poiché su quelli appunto mi avviene di animarmi di più nell'esposizione."

[25] See Enriques to Castelnuovo, 23 November 1894 [8, p. 151]: "Some young people have asked me to give a class in Higher Geometry: I am not against the idea of satisfying them in part with a series of weekly lectures that I will, however, only begin later (after January). Just in case, I'll tell you my plan for these: they would be concerned with a general principle that completes that of Klein (Programm) in order to encompass ideas about various other kinds of research ... which for now escapes it, at least directly" ("Alcuni giovani mi chiedono che faccia un corso di G*[eometria]* Sup*[eriore]*: non sono alieno dall'idea di contentarli in parte con un seguito di conferenze settimanali che però comincerei solo più tardi (dopo Gennaio). Ti dirò in caso il piano di queste: s'informerebbero ad un principio generale che completa quello di Klein (Programm) per far rientrare in quell'ordine d'idee vari altri tipi di ricerche ... che ad esso sfuggono, almeno direttamente").

[26] See [22, p. 1]: "...nella Matematica ogni passo avanti ha richiamato l'attenzione all'analisi dei fondamenti, e viceversa da una tale analisi sono scaturiti spesso concetti nuovi ed importanti che hanno permesso di estendere i risultati noti ad un campo più generale."

logical relations, and in this sense conceive the science founded as an *Abstract Geometry*. This way of thinking, to which we are naturally led by the previous observations, on the other hand makes no difference in the mathematical development of Geometry.

The importance that we attribute to Abstract Geometry is not (as may be believed) in opposition to the importance attributed to intuition: rather, it lies in the fact that *Abstract Geometry can be interpreted in infinite ways as a concrete (intuitive) Geometry by fixing the nature of its elements: so in that way Geometry can draw assistance in its development from infinite divers forms of intuition.*[27]

However, Enriques – as Castelnuovo observed[28] – also affirmed that in analysing the genesis of the postulates of geometry it is useful to take into account the psychological criteria, and to conduct an investigation of the sensations and experiences that led to the formulation of those postulates. This kind of investigation would carry Enriques to study German physiological psychology in 1896, and would have a systematic presentation in his 1906 publication *Problemi della scienza*.

The correspondence between Enriques and Castelnuovo makes it possible to retrace his steps. In January 1896, he began to study biology;[29] in February, he undertook the study of the physiology of cells;[30] in May that same year he gathered information on the studies in psychology and physiology of Hermann von Helmholtz, Ewald Hering, Ernst Mach, and above all the German psychologist and physiologist Wilhelm Wundt. He tried unsuccessfully to involve his friend in the discussion:

> While the mathematical questions are sleeping until a better day, I have been occupied for several days with a high question that only takes its pretext from mathematics ... It is the 'philosophical problem of space'. Books of psychology and logic, of physiology and of comparative psychology, of critique of knowledge, etc., all cross my desk, where I savour them with sensuous delight in the attempt to extract the essences that concern my problem. ... Since included in my program is the question of the genesis of the concept of space on the basis of physiological psychology (especially from the eye and the sense of touch) of Helm[h]oltz, Wundt, etc.[31]

---

[27]See [22, pp. 9–10]: "Tenendo di mira fin da principio la estensione che vogliam dare ai risultati ottenuti in un dato campo applicandoli ad altri campi, ci converrà considerare gli elementi fondamentali della Geometria come enti di natura astratta legati da relazioni puramente logiche e concepire in questo senso la scienza fondata come una Geometria astratta. Tale modo di considerare a cui si è naturalmente condotti dalle precedenti osservazioni, è d'altronde indifferente nello sviluppo matematico della Geometria.
L'importanza che attribuiamo alla Geometria astratta non è (come si potrebbe credere) da contrapporsi all'importanza attribuita all'intuizione: essa sta invece nel fatto che la Geometria astratta si può interpretare in infiniti modi come una Geometria concreta (intuitiva) fissando la natura dei suoi elementi: sicché in tal modo la Geometria può trarre aiuto nel suo sviluppo da infinite forme diverse d'intuizione."

[28]See [14, p. 7] and the letter of Enriques to Castelnuovo dated 4 May 1896 [8, p. 261].

[29]See Enriques to Castelnuovo, Firenze 19 January 1896 [8, p. 237].

[30]See Enriques a Castelnuovo, Firenze 9 February 1896 [8, p. 246].

[31]See Enriques to Castelnuovo, s. l., 4 May 1896 [8, pp. 260–261]: "Mentre le questioni matematiche sonnecchiano fino al miglior tempo, io mi sto occupando da più giorni di un'alta

Four days later he observed:

> I draw the elements of physiology of the sensations from Wundt, who reproduces and corrects the experiments of his predecessors, and especially of Helm[h]oltz. In many points his ideas correspond to mine, but, for example, his observation that "the idea of the straight line comes from the sense of touch and from the sensation of muscular motion because the mechanical conditions of the organism favour rectilinear motion of the muscles" does not seem to me to be correct. ... Instead, the notion of the straight line comes directly from the eye, like all other graphic notions of shape. Likewise, it is strange that W[undt] admits that the notion of "distance" comes (also) from the eye, while the experiments he cites prove the opposite, that is, that "the eye is never capable of perceiving the equality of two distances that are not equally situated."[32]

The theme of the psychogenesis of geometrical properties is also mentioned in the introduction to Enriques' *Lezioni di Geometria proiettiva*, which came out in 1898 and were the fruit of four years' experience in teaching at the university.[33] Here, along with the problem of the scientific presentation of the subject,[34] he also addresses that of the educational presentation, as evidenced in the dense correspondence with Castelnuovo, and as Enriques himself writes in the preface:

> Having resolved the problem as far as the scientific aspect was concerned, it was still necessary to articulate the form of the exposition and carry it out more completely in its details, in order to make it acceptable from an educational point of view. It seems to me that, during the past three years, the lessons that I am now publishing in print have come ever closer to this educational end. In them I have sought to reconcile the needs of the logical mind with the advantages and the attractions that intuition confers on studies of geometry, ... observations of an intuitive nature ... appear in any case to illuminate some of the more abstruse concepts or explanations, and in some places can even take the place of the rigorous procedure of proof to the advantage of didactics.[35]

---

questione che dalla matematica prende solo il pretesto ... Si tratta del "problema filosofico dello spazio". Libri di psicologia e di logica, di fisiologia e di psicologia comparata, di critica della conoscenza ecc. passano sul mio tavolino dove li assaporo con voluttà tentando di estrarne il succo per ciò che concerne il mio problema. ... Giacché vi è nel mio programma la questione della genesi dei concetti di spazio sopra i dati della psicologia fisiologica (specie dell'occhio e del tatto) di Helm[h]oltz, di Wundt ecc."

[32] See Enriques to Castelnuovo, s. l., 8 May 1896, [8, pp. 264–265]: "Io traggo gli elementi di fisiologia delle sensazioni dal Wundt che riproduce e corregge le esperienze dei predecessori e specialmente di Helm*[h]*oltz. In molti punti le sue idee collimano con le mie, ma non mi sembra p*[er]* e*[sempio]* accettabile la sua osservazione che 'l'idea di retta viene dal tatto e dalla sensazione di movimento muscolare perché le condizioni meccaniche dell'organismo favoriscono il moto rettilineo dei muscoli'. ... Invece la nozione di retta proviene direttamente dall'occhio, come ogni altra nozione grafica di forma. Similmente è strano che il W*[undt]* ammetta che la nozione di 'distanza' proviene (anche) dall'occhio mentre le esperienze che cita provano il contrario, e cioè che 'l'occhio non sa mai apprezzare l'uguaglianza di due distanze se non sono ugualmente poste'."

[33] See [24, pp. 3–4]; see also [23, pp. 4–5].

[34] For more on this, see [1, §§ 6 and 7].

[35] See [24, pp. V–VI]: "... risoluto il problema sotto l'aspetto scientifico, occorreva ancora elaborare la forma della trattazione e svolgerla più compiutamente nei suoi dettagli, in guisa da renderla accettabile nella scuola. A questo scopo didattico mi sembra si sieno venute avvicinando, durante i tre anni scorsi, le lezioni che ora pubblico per le stampe. Nelle quali ho cercato di contemperare le

Among the educational instruments used by Enriques the history of mathematics had already a place: he inserted an appendix[36] about history at the end of the book, in order to show his students the genesis of the fundamental concepts of projective geometry and to make clear how

> the various branches of pure and applied Mathematics interweave and connect to each other in unexpected ways; and [how] the ideas, which arise from elementary practical problems, seem to require long process of thought in order to mature, in the highest regions of theory, before they can descend and bear fruit in the field of daily activity.[37]

Enriques' letters to Castelnuovo, the lecture notes, and the class registers for these years show what a tight mix Enriques' activities were of the study of foundations, the history of mathematics, and the needs of education. These can be summarised in the following points:

- The refusal to resort to artifices in the proofs. He wrote to Castelnuovo: "I am disposed for educational reasons even to the greatest compromises in order not to oblige one to introduce artifices. For me any proof that is not remembered once understood is artificial. Such proofs do not illuminate, and the students prefer them precisely because there is nothing substantial in them to understand: thus I hold them to be educationally futile: we might as just as well give the student only the statement".[38]

---

esigenze dello spirito logico coi vantaggi e colle attrattive che l'intuizione conferisce agli studi geometrici ... osservazioni di carattere intuitivo ... compariscono tuttavia a lumeggiare alcuni concetti o ragionamenti più astrusi, ed in taluni punti possono anzi sostituire con vantaggio didattico il procedimento rigoroso della dimostrazione." See also the comment by Segre [119, p. 11]: "In his book he presents his research, taking due account of the needs of education. Some intuitive observations lead to stating certain postulates with precision, chosen so that from them it is possible to deduce, not only with rigour, but also with simplicity, all of the fundamental propositions of the geometry of position. This condition of simplicity is essential for teaching purposes. It would not be satisfied by an author who wanted to break each postulate into its most minute parts, discarding those that could be deduced logically from others, and demonstrating the logical independence of the remaining propositions; the result would be a bad *educational* job even if it were excellent from a scientific point of view" ("Nel suo libro egli si serve di quelle sue ricerche, tenendo il debito conto delle esigenze didattiche. Alcune osservazioni intuitive conducono ad enunciare con precisione certi postulati, scelti per modo che da essi si possan poi trarre, non solo con rigore, ma anche con semplicità, tutte le proposizioni fondamentali della geometria di posizione. Questa condizione della semplicità è essenziale per la scuola. Ad essa non soddisfarebbe, e quindi riuscirebbe ad una cattiva opera didattica, se anche ottima dal punto di vista scientifico, chi volesse scindere ogni postulato nelle sue più minute parti, scartando quelle che si possono dedurre logicamente dalle altre, e dimostrando l'indipendenza logica delle proposizioni rimaste").

[36]See [24, pp. 358–371].

[37]See [24, p. 371]: "... i vari rami della Matematica pura ed applicata si annodano e si collegano fra loro per vie inaspettate; e le idee, che traggono origine da elementari problemi della pratica, sembra debbano maturarsi per lunga elaborazione di pensiero, nelle regioni più alte della teoria, prima che possano discendere feconde nel campo di attività della vita."

[38]See Enriques to Castelnuovo s. l, 24 November 1895 [8, p. 224]: "Io sono disposto per ragioni didattiche a quelle maggiori transazioni che non obbligano a introdurre artifici. È per me artificiosa

- The importance of using intuition.
- The digressions into higher mathematics; Oscar Chisini, for examples, recalled that Enriques had a habit of amplifying his lessons in elementary projective geometry with frequent digressions into advanced geometry, topology, logic and economy [15, p. 119].
- The use of the history of mathematics as a tool for understanding the genesis of the concepts presented.[39]
- A unified vision of science and culture.

The open course in the philosophy of science that Enriques taught in 1902–1903 is emblematic, because the program interweaves scientific, philosophical and educational aspects of the subject.[40] The correspondence with Giovanni Vailati shows that although the program for the course proposed by Enriques was not approved at first, he did not give up, and asked the High Council for Public Instruction to decide whether or not philosophy of science could be part of the open courses of the faculty of science at the University of Bologna. The answer was positive, under the condition that not too much space was given to philosophy. The letters also show that the topics addressed in the course were the objects of six lectures given by Enriques in March 1902 at the *Univérsité Nouvelle* in Brussels.[41] Significant is the fact that Enriques expressly asked the rector for and was granted permission to open his course entitled *Filosofia scientifica*, "Scientific Philosophy," to students in the Faculty of Philosophy and Letters and that of Law as well as to students of mathematics. Moreover, all of the 366 lire that he was paid for that course was spent in buying books for the library of the Scuola di Magistero where Pincherle was teaching at the time.[42]

Enriques' cultural project, of which his vision of mathematics teaching was part, was beginning to take shape, and a few years later, it would lead Enriques to formulate his proposal for a reform of the university, as an expression of a unified vision of knowledge.

---

ogni dimostrazione che capita una volta non si ricorda senz'altro. ... Siffatte dimostrazioni non illuminano, e gli studenti le preferiscono appunto perché non vi è in esse nulla di sostanziale da capire: quindi io le ritengo inutili didatticamente: tanto varrebbe dare gli studenti il solo enunciato."

[39] The history of mathematics, among other things, is also found in the courses in higher analysis that Enriques was assigned to teach at the University of Bologna; for example, in the course of 1917–1918 no fewer than fifteen lessons were dedicated to it; see ASUB, *Enriques prof. Federigo. Fascicolo personale.*

[40] See Appendix 1.

[41] See the letters of Enriques to Vailati dated 11 November 1901, 1 January 1902, 24 January 1902 [92, pp. 570–571, 575–576].

[42] See Enriques to Castelnuovo s. l, 31 October 1902 [8, p. 503].

2.2 Klein's Influence

In the evolution of Enriques' cultural project and his vision of mathematics teaching, along with his experience in teaching which we have described above, an important role was played by the influence of Klein,[43] to whom Enriques refers often, and to whom Enriques reserved a special place in the section devoted to teaching in the 1934 entry entitled *Matematica* that he wrote in the *Enciclopedia Italiana*. Here, we will briefly mention some of the characteristics of Klein's vision that were taken up and reinterpreted by Enriques.

For Klein, theoretical research had to be very strictly connected to experimental research:

> From the point of view of pure mathematical science I should lay particular stress on the *heuristic value* of the applied sciences as an aid to discovering new truths in mathematics. ...Such separation [between abstract mathematical science and its scientific and technical applications] could only be deplored; for it would necessarily be followed by shallowness on the side of the applied sciences, and by isolation on the part of pure mathematics [87, pp. 46, 50].

He classifies geometry as one of the applied sciences, and he affirms that the mathematical treatment of any applied science "substitutes exact axioms for the approximate results of experience, and deduces from these axioms the rigid mathematical conclusions" [87, p. 47].[44] Klein also shows a refusal of the axiomatic point of view and a conviction that progress in science originates from the combined use of intuition and logic:

> The science of mathematics may be compared to a tree thrusting its roots deeper and deeper into the earth and freely spreading out its shady branches to the air. Are we to consider the roots or the branches as its essential part? Botanists tell us that the question is badly framed, and that the life of the organism depends on the mutual action of its different parts [90, pp. 248–249].

As far as intuition is concerned, Klein distinguishes between *naïve intuition* and *refined intuition* and highlights the fact that naïve intuition is important in the discovery phase of a theory (as an example Klein cites the genesis of differential and integral calculus) and at the time when its foundations are being established, refined intuition (shown, for example, in Euclid's *Elements*) intervenes in the elaboration of data furnished by naïve intuition, and in the rigorous logical development of the theory itself: "The naïve intuition is not exact, while the refined intuition is not properly intuition at all, but arises through the logical development from axioms considered as perfectly exact" [87, pp. 42]. In the article "The Arithmetizing of Mathematics," Klein further hypothesises that the clarification of the relationship

---

[43]See, for example, [106] and [83]. See also the four letters from Enriques to Klein (SUB Göttingen, *F. Klein* 4A, 8, 34 and 51) and one letter from Klein to Enriques (SUB Göttingen, *F. Klein* 51).

[44]See also [91, II vol., pp. 201–202].

between the intuitive process and the logical process may be achieved through physiology and experimental psychology [90, p. 247], a theme he discussed with Enriques during his second visit to Italy in 1899. On that occasion Enriques wrote to Castelnuovo:

> ...I passed two splendid days with Klein; the first in Florence where (except for a two-hours visit to the Institute of Geography) I had him all to myself, and the second in Bologna where I was again able to talk with him at length.... Saturday during the visit to the galleries, I told him in detail about the outline for my article on the Foundations of Geometry, and I was very pleased to see that he was satisfied. He took very detailed notes about what I told him. ...But the problem we discussed at greatest length was that regarding the psychological issues relating to mathematics. Yesterday morning, as he took leave of me, he said, "We must take up our conversation on these subjects again, which I will not forget."[45]

In fact, Klein had invited him to write a chapter on the foundations of geometry for the *Encyklopädie der mathematischen Wissenschaften*. This was the principal theme discussed during Enriques' stay in Göttingen in 1903:

> As far as my conversation with Klein goes, you already know how interesting it was. In addition to talking about the foundations of geometry, we discussed educational issues at length, and in just a few hours I learned a great deal from him about a lot of things I knew nothing about – specifically about the way in which mathematics teaching is developing in England and Germany.[46]

Enriques would make many of Klein's pedagogical assumptions his own. These can essentially be summarised as follows. First, he desired to bridge the gap between secondary and higher education. In particular, he proposed transferring the teaching of analytic geometry and, above all, of differential and integral calculus, to the middle school level, even in those schools which did not specialise in the sciences. The concept of function would pervade the whole mathematics curriculum: the famous expression "functional thinking" (*funktionales Denken*) was adopted as a slogan for his reform program. Furthermore, he favoured a genetic teaching method, that is, one that takes account of the origins and evolution of the subject, and

---

[45] See Enriques to Castelnuovo, s. l, 28 March 1899 [8, p. 404]: "...ho passato col Klein due giornate bellissime: la prima a Firenze ove (tranne due ore di visite all'Istituto geografico) me lo sono goduto interamente, e la 2ª a Bologna dove pure ho conferito lungamente con lui ...Sabato durante la visita alle gallerie, gli ho esposto dettagliatamente il programma del mio Art*[icolo]* sui Fondamenti della Geometria, e sono stato lieto di vederlo soddisfatto. Egli ha preso note assai minute su ciò che gli ho esposto. ...Ma il soggetto di cui abbiamo discorso più lungamente è quello che si riferisce ai problemi psicologici matematici. Ieri mattina congedandosi da me, mi ha detto: riprenderemo la nostra conversazione su questi argomenti, che non dimenticherò." Klein had already been in Italy the first time in 1878, and on the occasion of this second visit he stopped in Florence, Bologna, Roma and Padova, meeting amongst others, Enriques, Castelnuovo, Cremona, Veronese and Fano.

[46] See Enriques to Castelnuovo, Brussels 24 October 1903 [8, p. 536]: "Quanto alla conversazione di Klein sai già quanto era interessante; oltre che delle questioni sui principii abbiamo discorso molto di questioni didattiche e da lui solo in poche ore ho imparato tante cose interessanti, di cui non avevo mai avuto notizia, sullo sviluppo dell'istruzione matematica in Inghilterra e in Germania."

he believed that teachers should capture the interest and attention of their pupils by presenting the subject in an intuitive manner. He stressed the importance of showing the applications of algebra to geometry and vice versa. He suggested highlighting the applications of mathematics to all the natural sciences. He believed in looking at the subject from a historical perspective. In addition, he argued that more space should be dedicated to the "mathematics of approximation" (*Approximationsmathematik*), that is, "the exact mathematics of approximate relations." Lastly, he firmly believed that it was crucial that elementary mathematics viewed from an advanced standpoint play a key role in teacher training.

It was thanks to Klein's intervention that a German translation of Enriques' *Lezioni di geometria proiettiva* was published in 1903. In his introduction to this book, Klein expresses particular appreciation for Enriques' treatment of the subject, which "is always intuitive, but thoroughly rigorous," and underlines the impact of this kind of research on didactics, writing:

> Italian researchers are also well ahead of us from a practical point of view. They have by no means disdained exploring the educational consequences of their investigations. The high quality textbooks for secondary schools which came out from this exploration could be made available to a broader audience through good translations. And it would seem particularly desirable in Germany when we consider that our own textbooks are completely out of touch with active research.[47]

As Enriques would write twenty years later in his review of Klein's *Gesammelte mathematische Abhandlungen*, it was precisely the "tendency to consider the objects to be studied in the light of visual intuition" [*Periodico di Matematiche,* (4), 3, p. 55] that brought Klein and the Italian geometers so close together intellectually.

Klein's example, and in particular that of the 1895 *Vorträge über Ausgewählte Fragen der Elementargeometrie*, inspired Enriques to begin to collaborate with his friends and followers on a series of monographs on elementary geometry from an advanced standpoint for the students of the Scuole di Magistero. In the spring of 1899, he wrote to Castelnuovo:

> Now I shall tell you about a project that I hope to turn into a reality without a great deal of effort. It would be a book dedicated to all the questions that concern elementary geometry (included in which are also the problems not of the second-degree which are dealt with by Klein, but there are very many questions). I do not propose to do it myself, but to have it done by students newly graduated and by secondary school teachers, reserving for myself, or for any other mathematician who wants to take it on, the treatment of some of the more delicate arguments.[48]

---

[47]F. Klein, "Zur Einführung," in [27, p. *III*]: "Aber die italienischen Forscher sind längst nach praktischer Seite weitergegangen: sie haben es nicht verschmäht aus ihren Forschungen pädagogische Folgerungen zu ziehen. Die sehr bemerkenswerten Lehrbücher für Hoch- und Mittelschulen, welche solcherweise entstanden sind, können den weiten Kreisen, für die sie Interesse haben, nur durch geeignete Übersetzungen zugänglich gemacht werden. Und daß dies geschieht, erscheint gerade in Deutschland um so erwünschter, als unsere Lehrbüchliteratur den Kontakt mit der vorwärts drängenden Forschung gar zu sehr verloren hat."

[48]See Enriques to Castelnuovo, undated [8, p. 419]: "Ora vengo a parlarti di un progetto, che spero di attuare con poca fatica. Si tratta di un libro dedicato a tutte le questioni che interessano

The idea of a collective work aimed at teacher training in Italy was actually not completely new. Cremona had already thought of it when in 1865–1868 he edited the translation of Richard Baltzer's *Elemente der Mathematik*. In fact, Cremona, as he wrote to Genocchi, considered that book too difficult to be used as a manual for secondary schools, while he believed it could be quite useful for teachers.[49]

Among the friends and colleagues that Enriques involved in the realisation of his project were Ugo Amaldi, Ettore Baroni, Roberto Bonola, Benedetto Calò, Castelnuovo, Alberto Conti, Ermenegildo Daniele, Amedeo Giacomini, Alfredo Guarducci and Giuseppe Vitali.[50] In 1900 the volume entitled *Questioni riguardanti la geometria elementare* was published.[51] It was a work specifically aimed at teacher training: although for Enriques Euclidean geometry remained "the most effective tool for educating the mind, the most consistent with geometric reality," he, like Klein, nevertheless believed that the teaching of geometry could "take advantage of the progress made, in the field of the elements as well, by a more mature criticism and recent developments in higher mathematics," and that "the teacher entrusted with secondary school education must possess a much broader knowledge of such progress so that his work is inspired by much larger perspective" [25, p. *II*].[52] The topics treated were congruence, equivalence, the parallel theory, problems that could or could not be solved with straightedge and compass, the constructibility of regular polygons. Enriques' own contribution to the volume regarded algebraic equations

---

la G*[eometria]* elementare (fra queste vi sono anche quei problemi non di 2*°* gr*[ad]*o trattati dal Klein, ma le questioni sono moltissime). Mi propongo non di farlo, ma di farlo fare a giovani laureati e ad insegnanti delle scuole secondarie, serbando a me, o a qualche altro matematico che volesse occuparsene, la trattazione di qualche argomento più delicato." See also the introduction to *Questioni riguardanti la geometria elementare* [25, p. *VII*] where Enriques writes: "Such questions were recently addressed in a series of lectures by Mr Klein, to whom we are indebted for the idea behind this collection;' ("Tali questioni sono state svolte recentemente in una serie di conferenze del signor Klein, alla quale dobbiamo in parte l'idea di questa raccolta").

[49] See the letter of L. Cremona to A. Genocchi, Milano, 6 November 1867 [10, p. 110].

[50] Many of these were teachers in secondary schools: Ettore Baroni (1866–1918) taught from 1901 at the Liceo E. Q. Visconti in Rome; Benedetto Calò (1869–1917) taught from 1900 at the Istituto tecnico in Naples; Alberto Conti (1873–1940) taught in secondary schools in Florence, and in 1900 founded the *Bollettino di Matematica*, a journal mainly for teachers; Amedeo Giacomini (1873–1948) was from Pisa; Alfredo Guarducci was professor of mathematics at the Liceo classico in Prato.

[51] See [25]. The German edition, entitled *Fragen der Elementargeometrie* (vol. *II*, Leipzig: Teubner, 1907; vol. *I*, Leipzig: Teubner 1910) contained an additional article by Giovanni Vailati on the theory of proportions. The reviews by M. Grossmann and H. Fehr appeared in *L'Enseignement mathématique* (vol. 11 (1909): p. 322; vol. 13 (1911): pp. 427–428) underlined the high level of research and teaching in geometry in Italy.

[52] "... possa avvantaggiarsi dei progressi portati, anche nel campo degli elementi, da una critica più matura e dagli sviluppi recenti delle alte Matematiche"; '... di tali progressi debbano possedere una cognizione assai larga gli insegnanti cui la scuola secondaria è affidata, affinché l'opera loro possa ispirarsi a più larghe vedute' [25, p. *II*].

and the constructibility of regular polygons,[53] but he also prefaced the anthology with an essay on the scientific and educational importance of the questions that refer to the principles of geometry,[54] which merits discussion because provides us with a clear picture of how Enriques conceived mathematics teaching and the training of mathematics teachers.

The essay is divided into two parts. In the first he outlines his vision of geometry as an "experimental science" just like physics (p. 5), and then he once again addresses the concept of abstract geometry, already presented in his 1894–1895 *Conferenze di geometria*, making evident its merits, but also cautioning:

> Abstract Geometry can be variously interpreted and thus draw new aid from various forms of intuition. But where, in contrast, it is desired leave aside any consideration of possible ways of interpreting it, and construct an edifice that is purely logical, on the basis of criteria that are exclusively logical, there is a danger of falling into a void. ...It should not be forgotten that this science is a science of facts, physical or intuitive, however we want to consider them. The logical formalism must be conceived, not as an end to achieve, but as a means aimed to use and increase the faculty of intuition. The results themselves, logically established, no matter how far-reaching, must still not be considered as mature achievements until they can be in some way comprehended intuitively. But *in the principles the intuitive evidence must shine brightly.*[55]

Enriques situates geometry in a central position in mathematics because he considers it the most fertile terrain for reconciling abstract formal procedures with experimental procedures, as he will say more clearly in his 1906 *Problemi della Scienza*, which is an organic formulation of the ideas born when he was teaching in Bologna, as we have seen; here he shows his refusal of dogmatic Kantism and his divergence from Poincaré's conventionalism.[56]

He then addresses the problem of the psychological acquisition of fundamental concepts of geometry and, on the basis of his study of physiological psychology that he had been pursuing for a number of years, he states that the three branches into which Geometry is divided, that is, topology, metric geometry, and projective geometry, appear to be connected to three orders of sensations: respectively, to

---

[53] See F. Enriques, "Sulle equazioni algebriche risolubili per radicali quadratici e sulla costruibilità dei poligoni regolari" in [25, pp. 353–396].

[54] See F. Enriques, "Sull'importanza scientifica e didattica delle questioni che si riferiscono ai principii della Geometria" in [25, pp. 1–31].

[55] See [25, p. 12]: "La Geometria astratta può ricevere varie interpretazioni e trarre così nuovi aiuti da varie forme di intuizione. Ma ove, all'opposto, si voglia prescindere affatto da ogni maniera d'interpretarla, costruendo un edifizio puramente logico, in base a criterii esclusivamente logici, si corre il pericolo di cadere nel vuoto. ... non bisogna dimenticare che tale scienza è scienza di fatti, fisici o intuitivi, che vogliano considerarsi. Il formalismo logico deve essere concepito, non come un fine da raggiungere, ma come un mezzo atto a svolgere e ad avanzare le facoltà intuitive. Gli stessi risultati più lontani, logicamente stabiliti, non debbono ancora considerarsi come un acquisto maturo, fino a che non possano essere in qualche modo intuitivamente compresi. Ma nei principii l'evidenza intuitiva deve risplendere luminosa."

[56] See [28, Chapter *IV*]. See also the article by G. Lolli in this present volume, and [105, pp. 87–127].

general tactile-muscular sensations, special tactile sensations (like the hand that allows man to measure objects) and those of vision (p. 19); The detailed explanation will be provided by Enriques in his 1901 article entitled "Sulla spiegazione psicologica dei postulati della geometria," and in the 1906 *Problemi della scienza*.[57]

In the second part of the essay Enriques examines educational questions in light of the reflections set out in the first part. He addresses himself directly to mathematics teachers, exhorting them to enter more deeply into the "philosophical spirit of [their] science: that spirit of relation that coordinates everything into a synthesis, and makes the great light of the general idea shine on the humble details!" (p. 23).[58] The teachers he is addressing are above all those at the gymnasiums-lyceums and those teaching physics and mathematics at the technical schools, but he says in a note (p. 24) that the teachers at the Normal Schools, which specialised in primary schoolteacher training, could also benefit from his reflections.

Here, we will underline only a few salient points to which Enriques returns more than once. First of all, the object of secondary teaching is not merely to provide useful notions, but to train the mind to reason, and to foster the spirit of initiative in young people; the teacher should be familiar with critical analyses and philosophical investigations, and although these should not enter into the practice of teaching, because the students are not capable of appreciating them, they should nevertheless enlighten their lessons; artificial technical developments and abstruse problems are to be avoided. Regarding geometry, Enriques observes:

> ...it seems to us that the essential goal of teaching is achieved *when we are able to make it understood how the logical development of Geometry rests on an empirical basis, destroying the strange illusion that the postulates founded on immediate experience appear to have a degree of certainty that is inferior to theorems, even though* [the theorems] *depend on the* [postulates].[59]

He then specifies what the method of teaching should be, a method which will later be called rational-inductive: the teachers should begin with a series of observations, and then on the basis of these present the fundamental concepts as "ideal representations of objects of reality" and state the postulates "as expressions of elementary facts." From these they will then deduce the theorems, beginning with the simplest and going on to consider the most complex. The rigorous proof of the theorems can be followed by experimental verification. Enriques then invites the teachers to keep empirical facts and logical facts well separated, and remarks that "a new datum of intuition, which has been neglected in the premise, should never insinuate itself in a hidden way in the reasoning of the proof" (p. 29). This, he says,

---

[57] See [28, pp. 177–187]; see also [1, pp. 401–406].

[58] See [25, p. 23]: "...lo spirito filosofico della vostra scienza; quello spirito di relazione che tutto coordina in una sintesi, e fa brillare sugli umili particolari la grande luce dell'idea generale!"

[59] See [25, p. 28]: "...ci pare che il fine essenziale dell'insegnamento sia raggiunto, se si riesce a far comprendere come lo sviluppo logico della Geometria riposi sopra una base empirica, distruggendo la strana illusione per cui i postulati fondati sopra un esperimento immediato, sembrano quasi avere un grado di certezza inferiore ai teoremi, che pur da quelli dipendono."

"is the only, important, even necessary, condition for rigour," while "for rigour it is not at all important to seek the independence of the postulates, and indeed, in terms of education it is preferable to draw a greater number of evident principles from the observation" (pp. 29, 30).[60]

On the basis of Enriques' reflections over time, gathered in the *Problemi della scienza*, this essay was later fleshed out and split into two chapters ("Sull'importanza filosofica delle questioni che si riferiscono ai principii della Geometria" and "Sull'insegnamento della Geometria razionale") in the second, enlarged edition of the *Questioni*, which appeared in 1912 with the title *Questioni riguardanti le matematiche elementari*. This second edition extended to questions of arithmetic and algebra, and also featured new collaborators: his disciple Oscar Chisini, and three capable secondary school teachers (at that time), Duilio Gigli, Alessandro Padoa and Umberto Scarpis.

2.3 Epistemological Assumptions at the Basis of Enriques' Vision of Mathematics Teaching

From the research projects carried out from 1896 to 1906 emerge a rather clear picture of Enriques' particular vision of mathematics and its influence on teaching. There also emerges a very precise cultural program in which active research in the field of algebraic geometry and philosophical, psychological and historical reflections are all closely intertwined. Enriques' aim was to communicate to his intended audience – scientists, philosophers, and educators – his vision of a scientific *humanitas* in which the boundaries between disciplines were overcome and the abyss between science and philosophy was bridged. The history of science constituted the path of first choice for achieving this end, or at least it was the tool used by Enriques, as we have seen, in his university teaching from the very first years, and had over time gradually become the most important one in the various initiatives aimed at teacher training.

Broad, rich, and sometimes contradictory, it is impossible within the limits of this paper to outline the epistemological vision on which all of Enriques' scientific work was founded, so I will confine myself to indicating the most important factors which inspired his idea of mathematics education.

---

[60]See [25, p. 30]: "...per il rigore non importa affatto cercare l'indipendenza dei postulati, ed anzi didatticamente è preferibile trarre dall'osservazione un maggior numero di principi evidenti."

2.3.1 A Genetic and Dynamic Vision of the Scientific Process and the Significance of Error

First of all, Enriques held a dynamic and genetic view of the scientific process, which he described as one

> ...at once inductive and deductive, which ascends from specific observations to abstract concepts, only to descend again to practical experience. It is a process of continuous development, which establishes a generative relationship between theories and perceives in their succession only an approximation of the truth.[61]

Science was therefore not conceived by Enriques as a closed system of definitive propositions. He writes:

> ...if the truth is only one step towards truth, the value of science would consist in moving forward rather than in stopping at a terminus reached provisionally. The facts, laws, theories will become meaningful not so much as a finished and static system, as in their reciprocal concatenation and their development.[62]

In such a vision of science, errors become valuable as well, because in the dynamic process of science truth and error are constantly mixed: "every error always contains a partial truth that must be kept, just as every truth contains a partial error to be corrected."[63]

According to Enriques, the error/gap (which is found when there is a missing link in the deduction that leads to a true statement) and the error proper (when a false proposition is stated as true) are errors that are almost necessarily encountered in the psychological acquisition of a theory, and are often reflected in the historical development of science. They do not appertain to either the faculty of logic or to the faculty of intuition, but are introduced "at the delicate moment of their juncture," that is, when the abstract concepts are developed from the objects effectively perceived [55, pp. 64–65]. The correction of errors leads to scientific progress, and from this derives their heuristic value.[64]

This vision is necessarily reflected in mathematical education. Enriques in fact criticised the tendency to present a mathematical theory in a strictly deductive

---

[61] See [34, p. 132]: "...processo induttivo e deduttivo, che dalle osservazioni particolari sale ai concetti generali ed astratti per ridiscendere all'esperienze di fatto, processo di sviluppo continuo, che pone fra le teorie un rapporto generativo e scorge nel loro succedersi un'approssimazione alla verità."

[62] See [52, p. 3]: "...se la verità è solo un passo verso la verità il valore della scienza consisterà piuttosto nel camminare che nel fermarsi ad un termine provvisoriamente raggiunto. I fatti, le leggi, le teorie riceveranno il loro senso non tanto come sistema compiuto e statico, quanto nella loro reciproca concatenazione e nel loro sviluppo."

[63] See [33, p. 417]: "...ogni errore contiene sempre una verità parziale da mantenere, come ogni verità un errore parziale da correggere."

[64] See also Enriques' criticism of the theory of error proposed by Croce, according to which "error is the product of practical motives that deter the spirit from contemplation of the truth. Thus error is to be corrected with thrashing" ("l'errore è il prodotto di motivi pratici che distolgono lo spirito dalla contemplazione della verità. Dunque l'errore si corregge con le bastonate") [33, p. 417].

manner at school, as in this way it appears something closed and already perfect, leaving no room for further discovery. Instead, teachers should approach problems with a number of different methods, paying attention to the errors which have allowed science to move forward, and indicating open questions and new fields of discovery.

On the other hand, the good teacher must also take into account the errors of his students and quickly learn "to distinguish the significant errors from those that are not actually errors – rather gratuitous statements by insolent [learners] who try to guess – where no effort is made to think."[65] Since, in Enriques' opinion, "errors proper" represent "natural steps along the way of thought in search of truth," the teacher must attribute an educational value to them:

> ...they are educational experiences that he pursues, encouraging the student to discover for himself the difficulties that impede right judgment, and thus also to err in order to learn to correct himself. Every kind of possible errors is also a kind of opportunity for learning.[66]

2.3.2 Inductive Aspects of Scientific Research and the Dialectic Between Intuition and Rigour

These views on science are connected to Enriques' conception of the nature of mathematical research – typical of the Italian school of algebraic geometry – as something aiming above all at discovery and particularly emphasising the inductive aspects and intuition:

> The main thing is *to discover*. ...A posteriori it will always be possible to give a proof, [which,] translating the intuition of the discoverer into logical terms, will provide *everyone* with the means to recognise and verify the truth.[67]

Much has been written on the working method of the Italian geometers, and about Enriques in particular, so here I will limit myself to underlining by means of a quotation the importance that he attached to intuition in scientific research:

> The faculty which comes into play in the construction of science and which thus expresses the actual power of the mathematical spirit is intuition. ...There are in any case different forms of intuition. The first is the intuition or imagination of what can be seen. ...But there is another form of intuition that is more abstract, that – for example – which makes it possible for the geometer to see into higher dimensional space with the eyes of the mind.

---

[65] See [52, p. 14]: "...a distinguere gli errori significativi da quelli, che non sono propriamente errori – affermazioni gratuite di sfacciati che cercano d'indovinare – dove manca lo sforzo del pensiero."

[66] See [52, p. 14]: "...sono esperienze didattiche che egli persegue, incoraggiando l'allievo a scoprire da sé la difficoltà che si oppone al retto giudizio, e perciò anche ad errare per imparare a correggersi. Tante specie di errori possibili sono altrettante occasioni di apprendere."

[67] See [63, *II*, p. 307, 318]: "La cosa essenziale è di regola scoprire ...a posteriori si riesce sempre a darne una dimostrazione ...*[che]* traducendo l'intuizione dello scopritore in termini logici, vuol dare a tutti il mezzo di riconoscere ed appurare la verità."

And there is also a sense of formal analogies which, in the work of many analysts, takes the place of the visual representation of things. ... [I]ntuition protracts and surpasses itself in the unifying power of reason, which is not something exclusive to the mathematician, but – in every field of science and application – marks the greatest reaches of the spirit.[68]

This belief is naturally reflected in the style of teaching, which should, according to Enriques, take into account the inductive as well as the rational aspect of theories. Logic and intuition are not two distinct faculties of the human intellect; rather, they represent two inextricable aspects of the same process. Teachers should therefore find the right balance between the two. The important thing is to distinguish clearly between empirical observation and intuition on the one hand, and logic on the other. On this subject, Enriques distinguishes between what he calls "small scale logic," the refined and almost microscopically accurate analysis of thought, and "large scale logic," which considers the organic connections in science. He maintained that teaching should above all take "large scale logic" into account, gradually preparing young people to develop a more refined and rigorous approach. He writes:

> It is of no use to develop with impeccable deduction the series of theorems of Euclidean geometry, if the teacher does not go back to contemplate the edifice constructed, inviting the students to distinguish the truly significant geometric properties from those which are valuable only as links in the chain.[69]

At the first level of teaching it is convenient to keep to a method which appeals to intuition and calls for active work on the part of the students:

> a logical education (indeed the most appropriate one for minds little disposed to abstraction) is also comprised in the exercise of intuition, when this is put to the test by making the students *work*. Thus, for example, the construction of a geometric figure requires not only the attitude of passively seeing a model ... but also the capacity to shape a possible model, on which are imposed, a priori, certain conditions: and this kind of constructive activity which orders the data of observations and past experience, is not pure imagination ... but rather true logical activity.[70]

---

[68] See [53, pp. 173–174]: "La facoltà che viene in opera nella costruzione della scienza e che esprime perciò il reale potere dello spirito matematico è l'intuizione. ... Vi sono del resto più forme d'intuizione. La prima è l'intuizione o immaginazione del visivo. ... Ma c'è poi un'altra forma d'intuizione più astratta, quella – per esempio – che consente al geometra di vedere con gli occhi dello spirito negli spazi a più dimensioni. E c'è ancora un senso delle analogie formali che, presso molti analisti rimpiazza la rappresentazione visiva delle cose. ... *[L']* intuizione stessa si prolunga e si supera nel potere unificatore della ragione che non è qualcosa di esclusivo del matematico, ma – in ogni campo della scienza e della pratica – contrassegna la maggiore altezza dello spirito."

[69] See [38, p. 10]: "Non giova sviluppare con impeccabile deduzione la serie dei teoremi della geometria euclidea, se non si ritorni a contemplare l'edificio costruito, invitando i discepoli a distinguere le proprietà geometriche veramente significative da quelle che hanno valore soltanto come anelli della catena."

[70] See [38, p. 8]: "Un'educazione logica (anzi la più appropriata alle menti poco disposte ad astrarre) è pur contenuta nell'esercizio dell'intuizione, quando questa venga messa alla prova facendo lavorare il discepolo. Così, per esempio la costruzione di una figura geometrica, importa – non solo – l'attitudine a vedere passivamente un modello ... ma anzi la capacità di foggiare ... un

With regard to the fact that many Italian teachers resisted the introduction of methods that were more intuitive and empirical, lamenting that a certain incompleteness and a non-rigorous way of reasoning is inherent in these, Enriques observed with a touch of humour:

> Resisting the ideas that... relate to the eye, the ear, the sense of touch, and seeing in sensations, not the doors to knowledge, but only occasions for sinful errors, this strange chastity of mathematical logicians brings to mind Plotinus and those Christian ascetics of the Middle Ages who were ashamed of having a body.[71]

Teaching how to reason abstractly without recourse to intuition must be done gradually, so that the student is able to grasp its importance. Enriques suggested, for example, beginning by presenting some proofs *ad absurdum* from which, he wrote, logic "draws its historical origins," adding:

> Only at the end of a course in geometry, looking at the system of science, is it useful to explain the logical structure, pointing out the significance of the primitive concepts and the postulates which must come at the beginning of a written treatise ... but not in a lively lesson, in which those principles should be left aside, informing the student that [the principles] contain only a precise recapitulation of things already known, and they will be introduced along the way as need arises.[72]

Enriques also believed it was counterproductive educationally to persist in proving everything that is intuitively evident because of the danger of depriving intuition of its value and leading the student to doubt the importance of reasoning. Further, a good teacher should not overindulge in the search for generality:

> ...a too abstract form of the statement can obscure the true meaning of the theorem, concealing its origins, and – in the second place – awakens in the young scholars the allurement of easy, purely formal generalisations.[73]

---

modello possibile, cui s'impongono, a priori, talune condizioni: ed una tale attività costruttiva che ordina i dati di osservazioni ed esperienze passate, non è pura fantasia ... bensì vera attività logica."

[71] See [53, p. 145]: "Respingere le idee che hanno ... rapporto con l'occhio, o con l'orecchio, o col tatto, vedendo nelle sensazioni non le porte della conoscenza, ma soltanto l'occasione di errori peccaminosi, questo strano pudore dei logici matematici ci richiama alla memoria Plotino e quegli asceti cristiani del Medio Evo che si vergognavano di avere un corpo."

[72] See [38, p. 11]: "Solamente al termine d'un corso di geometria, riguardando al sistema della scienza, gioverà spiegarne l'organismo logico, rilevando il significato dei concetti primitivi e dei postulati, coi quali si deve cominciare un trattato scritto ... ma non la lezione viva, che lascia dietro di sé quei principii, avvertendo il discepolo che contengono soltanto una ricapitolazione precisa di cose note, da richiamare di mano in mano che se ne presenti il bisogno"; see also [36] and [49].

[73] See [63, I, p. XI]: "la forma troppo astratta dell'enunciato riesce ad oscurare il vero significato del teorema nascondendone le origini, ed – in secondo luogo – crea nei giovani studiosi la lusinga delle facili generalizzazioni, puramente formali"; see also [53, p. 153].

2.3.3 Science as a "Conquest and Activity of the Spirit" and Unified Vision of Culture

For Enriques, science is the "conquest and activity of the spirit, which ... merges in the unity of the spirit with the ideas, feelings and aspirations which find expression across all the different aspects of culture" [54, p. 130].[74] In this, Enriques thus ran counter to Croce and Gentile, the leading proponents of Italian neo-idealism, who tended to devalue science, recognising in it only a practical function and a role that was completely instrumental, and separating it from the world of philosophy and culture.[75] He was aware of the grave danger that cultural isolation poses to science, so he continually emphasised the importance of "cultivating one's own field of study as a segment of the greater body of science!."[76] He held that:

> The end that should be sought today is a scientific education that allows a person working in any given field to understand how the object of his own research is subordinated to more general problems. ... Nothing is as dangerous as enclosing oneself in a circle from which everything that does not agree with the results of limited experience is banished according to rigorous logic.[77]

Furthermore, for Enriques the fact that science does not have goals that are purely utilitarian does not imply a separation between pure and applied science, but means that scientific research is valuable in itself, and does not necessarily have to aim at applications. Like Klein, he believed it was useful and necessary to maintain close ties between abstract science and applied sciences because pure sciences offer instruments that are needed for the purposes of applied science, and in their turn, applied sciences perform functions that are essential for stimulating the development of theoretical sciences, as history makes amply clear.[78]

Such a unitary vision of culture found expression in Enriques' constant efforts to bridge the gap between mathematics and other scientific and scholarly fields, such as physics, biology, psychology, physiology, philosophy, and history. Only by overcoming narrow specialisation could the sciences, and especially mathematics, realise their true humanistic and educational value.

From this viewpoint derive some of the fundamental tenets of Enriques' idea of education: the importance of establishing links between the various parts of mathematics and between mathematics and the other intellectual activities, because

---

[74]See [54, p. 130]: "conquista e attività dello spirito ... *[che]* si fonde nell'unità dello spirito colle idee, coi sentimenti, colle aspirazioni che si esprimono nei vari aspetti della cultura."

[75]See for example, [83, 110, 112].

[76]See [35, p. 35]: "...coltivare il proprio ramo di studii come un frammento della scienza generale!."

[77]See [28, pp. 3–4]: "...il fine a cui oggi si deve tendere è un'educazione scientifica, la quale faccia meglio comprendere a colui che lavora in un campo qualsiasi come l'oggetto della propria ricerca venga subordinato a problemi più generali. ... Nulla è così pericoloso come il rinchiudersi in un cerchio, donde si bandisca con una logica rigorosa ciò che non si accorda coi resultati di un'esperienza ristretta!."

[78]See [43, p. 4].

these are simply different moments within a single cognitive process; the need for active teaching; the importance of training teachers who are capable of transmitting a vivacious knowledge to students' minds, "like sparks from one fire ignite other fires" [38, p. 15];[79] and finally, his firm belief in the educational and cultural value of mathematics.

According to Enriques, the duty of the teacher consists in communicating to the student "his need for knowledge before allowing him to possess it," and attaining anew that knowledge along with the student, with the joy of discovery; this kind of teaching is certain more difficult, but is much more effective for the student as well as more gratifying for the teacher [46, p. 68]. He, therefore, proposes that teachers adopt the method that Socrates used with his students, which consists in conversing with them, acting "a little ignorant" and, through dialogue and a guided search, leading them to a personal discovery of mathematical truth:

> The greatest advantage of this method is, in my opinion, its sincerity, because the postulate of ignorance is infinitely closer to the truth than the presupposition of knowledge already certain in the mind of the student, which the pedantic lesson starts off with [38, p. 14].[80]

It is only through personal conquest that the student can arrive at the true comprehension of mathematics. Enriques writes:

> Teaching should not be a gift from a teacher to a person who comes to hear his perfectly prepared lessons ... but rather it should be an aid given to the person who wants to learn by himself or is, at any rate, disposed not merely to absorb passively, but to attain to knowledge, as if it were a discovery or a product of his own spirit.[81]

To stimulate the students to active participation, the teacher must not limit himself to repeating mechanically the old lessons he himself learned when was a student, but must show himself capable of offering a clearer and broader point of view born from a mastery of higher mathematics:

> ... there is no gap or schism between elementary and higher mathematics, because the latter is a development of the former, as a tree develops from a seedling. And as by studying the tree we discover new aspects of the seedling, and understand characteristics whose meaning had escaped our understanding, so the development of mathematical problems will throw light on the elementary theories in which they have their roots.[82]

---

[79] See [38, p. 15]: "... come scintilla di fuoco ad accendere altro fuoco."

[80] See [38, p. 14]: "Il più grande vantaggio di questo metodo è, a mio avviso, la sincerità, perché il postulato dell'ignoranza è infinitamente più vicino al vero che la presupposizione di conoscenze già sicure nella mente dell'allievo, da cui muove la lezione cattedratica."

[81] See [38, p. 6]: "... l'insegnamento non può essere un regalo che il maestro faccia a qualcuno che viene ad ascoltare le sue ben tornite lezioni (che, se sta disattento, merita di essere rimproverato per la sua ingratitudine!); ma è piuttosto un aiuto a chi voglia imparare da sé e però sia disposto, anziché a ricevere passivamente, a conquistare il sapere, come una scoperta o un prodotto del proprio spirito."

[82] See [38, pp. 15–16]: "... non vi è iato o scissura fra matematiche elementari e matematiche superiori, perché queste si sviluppano da quelle, al pari dell'albero dalla tenera pianticina. E come, riguardando l'albero, potremo scoprire nella pianticina nuovi aspetti o comprendere caratteri di

Further, in order for mathematics teaching to improve the faculty of logic, the teacher must be able to coordinate the various aspects of mathematics and relate them to each other:

> ...we are pleased to see recognised today the rights of education, on the condition that this fact leads the teachers ...to account for the psychology of the students and the usefulness of reconciling mathematical doctrines, too separated by purist concerns, which the history of science shows to be related.[83]

2.3.4 The History of Science

There are three methods that belong to Enriques' "positive gnoseology": historical, for retracing the genesis and development of scientific theories; psychological, for studying the formation of concepts; and scientific, which "consists in the direct critical examination of Science, regarding science itself as a fact to be explained."[84] Of the three, the historical method, which is also closely connected to the dynamic vision of science, was to assume an increasingly important role for Enriques.[85] In his own words:

> A dynamic vision of science leads us naturally into the territory of history. The rigid distinction that is usually made between science and history of science is founded on the concept of this [history] as pure literary erudition; ...But a very different meaning is obtained by the historical comprehension of scientific knowledge that aims at ...clarifying the progress of an idea. ...Such a history becomes an integral part of science.[86]

History is in fact intended as a science in itself:

> The history of science ...must be constructed thanks to the scientific reasoning which is useful for coordinating and evaluating the traditions, the testimonies, the sources, investigating first the *possibility* in order to infer the *reality*. In this manner the antithesis of

---

cui ci era sfuggito il significato, così anche lo sviluppo dei problemi matematici recherà luce sulle dottrine elementari in cui essi profondano le loro radici."

[83]See [39, p. 123]: "...non ci dispiace di vedere riconosciuti oggi i diritti della didattica, a condizione che l'indirizzo così affermato conduca gli insegnanti ...a rendersi conto della psicologia degli alunni e dell'utilità di ravvicinare dottrine matematiche, troppo separate da preoccupazioni puristiche, di cui la storia della scienza è atta a metter in luce la parentela."

[84]See [28, p. 78]: "...consiste nell'esame critico diretto della Scienza, riguardata essa stessa come il fatto da spiegare."

[85]See for example [113], especially the essays by G. Israel, M. Galuzzi and P. Freguglia; see also [105, pp. 150–173 and pp. 186–226].

[86]See [63, *I*, p. *XI*]: "Una visione dinamica della scienza porta naturalmente sul terreno della storia. La rigida distinzione che si fa di consueto fra scienza e storia della scienza, è fondata sul concetto di questa come pura erudizione letteraria; ...Ma assai diverso significato ha la comprensione storica del sapere che mira a ...chiarire il cammino dell'idea ...Una tale storia diviene parte integrante della scienza."

science and history is reconciled into a collaboration regarding the concrete progress of our knowledge.[87]

It should be emphasised that the kind of historiography that Enriques proposed required an in-depth knowledge of scientific theories, including their technical aspects, and this couldn't help but render it unpalatable for pure historians and philosophers.

Furthermore, history also offers the cultural legitimisation of the function of mathematics, and thus for Enriques has a central educational role in both teacher training as well as in teaching proper. He rues the fact that too often,

> ... mathematics has been studied as an organism in itself, looking at the abstract formulation achieved after centuries of development, rather than at the profound historical reasons. Therefore the concrete problems that confer interest on the theories are forgotten, and the facts by then long since acquired are no longer visible behind the formula or the development of the reasoning, but only the concatenation into which we have artificially restrained them.[88]

For this reason, according to Enriques, future teachers should study the origins of each theory, together with its relationships and developments, not some static formulation;[89] they should be familiar with the work of ancient mathematicians, analysing the ways they addressed problems and the methods used to solve them, in order to better understand the more general and complex developments in modern science. Young people too should be "educated in the masterpieces of the masters" by means of readings of significant passages from their works during class:

> For developing culture that is serious and effective, it is necessary that [the students] be put in touch with the great thinkers, and thus set on the path to knowing the historic genesis of scientific ideas. The poets develop their knowledge in the company of poets, merchants in the company of merchants, philosophers in the company of philosophers. For that particular philosophy which is science, it is also time to turn from the textbooks and anthologies to the sources.[90]

---

[87]See [53, p. 166]: "La storia della scienza ... deve essere costruita mercé il ragionamento scientifico che vale a coordinare e valutare le tradizioni, le testimonianze, le fonti, indagando prima la possibilità per inferire la realtà. In tal guisa l'antitesi scienza-storia si risolve in una collaborazione per riguardo al progresso concreto del nostro sapere."

[88]See [29, p. 71]: "... le matematiche sieno state studiate come un organismo a sé, riguardandone piuttosto la sistemazione astratta conseguita dopo uno sviluppo secolare, che non l'intima ragione storica. Si dimenticano per tal modo i problemi concreti che conferiscono interesse alle teorie, e sotto la formula o lo sviluppo del ragionamento non si vedono più i fatti ormai da lungo tempo acquisiti, ma soltanto la concatenazione in cui noi artificialmente li abbiamo stretti."

[89]See [38, p. 16].

[90]See [64, p. 11]: "Per una cultura seria e veramente fattiva è necessario che questi vengano messi a contatto coi grandi pensatori, e avviati così a conoscere la genesi storica delle idee scientifiche. I poeti sviluppano la loro coscienza in compagnia dei poeti, i mercanti in compagnia dei mercanti, i filosofi dei filosofi. Anche per quella filosofia che è la scienza è tempo di volgersi dai manuali e dalle compilazioni alle fonti."

He also writes:

> The school is not a place in which individual imagination can do what it likes in attempting arbitrary experiments, indeed, the more it aims at grasping the spirits and voices of the society around it, the more it is nourished by the tradition in which it is rooted: not by preserving outdated forms and repeating dead words, but reconnecting ... past culture to the present, in striving towards the future. And as in school, so too in science. Also for science there is no real progress if new generations do not frame their vision of problems within the continuity of scientific thought, honing their skills in the study of the great models.[91]

The history of science, furthermore, can also constitute an important auxiliary tool for education in making it possible to better understand certain concepts or properties. Here, I will only cite by way of example the use that Enriques himself suggested of Pythagorean figurate numbers to facilitate comprehension of some arithmetic properties in one of his texts for middle schools:

> If the student is to participate in an active way in this study, he cannot be given definitions and rules without explanations, like gifts rained down from above, which he would not be able to use. ... The history of science comes to our aid here, showing us how arithmetical truths were recognised by the Pythagoreans by means of the geometric models of numbers, which are the figurate numbers: square and rectangular numbers, triangular numbers, etc.[92]

In discussing Enriques' cultural project, mention must be made of another brilliant exponent of the scientific movement in Italy at the beginning of the twentieth century, Giovanni Vailati, who shared the idea of promoting a scientific *humanitas* and who even proposed creating a unified front of all Italian scientists, especially including Enriques, Volterra and Peano, to fight against the separation of science and philosophy (see [81]). His premature death, and the fact of having underestimated the evident differences in the various methodological and epistemological approaches to mathematics, led to the failure of this project. It is emblematic, for example, that Enriques and Vailati were never able to reach an effective agreement about the nature of logical and philosophical research, even though their ideas regarding the role of philosophy and history within science

---

[91] See [45, p. 8]: "La scuola non è un campo in cui la fantasia individuale abbia a sbizzarrirsi tentando esperimenti arbitrarii, anzi tanto più è atta ad accogliere gli spiriti e le voci della società circostante, quanto più si alimenti della tradizione in cui anche questa prolunga le sue radici: non già serbando viete forme e ripetendone la morta parola, ma riattaccando ... il passato al presente della cultura, in uno sforzo verso l'avvenire. E come la scuola la scienza. Anche per questa non vi ha un vero progresso, dove le nuove generazioni non attingano alla continuità del pensiero scientifico la visione dei problemi, facendosi valenti nello studio dei grandi modelli." Enriques had certainly absorbed and interiorised into his own vision of science the teaching of his own mentors Segre and Beltrami. From Beltrami he had received the conviction that study of the history of mathematics can assume "the interest and value of scientific research" and he quotes verbatim [64, p. 11] the invitation to young people to study "the masterpieces of the great masters." See [77].

[92] See [50, pp. IX–XI]: "Se l'allievo deve partecipare in modo attivo a questo studio, non si può dargli definizioni e regole senza spiegazione, come doni piovuti dal cielo, di cui poi quegli che riceve il dono non saprebbe servirsi. ... La storia della scienza viene qui in soccorso, mostrandoci come le verità aritmetiche siano state riconosciute dai Pitagorici mediante modelli geometrici dei numeri, quali sono i numeri figurati: numeri quadrati e rettangolari, numeri triangolari, ecc."

were quite similar, and they also shared many pedagogical assumptions regarding mathematics teaching. Their correspondence sheds a great deal of light on this, as historical studies have shown.[93]

3 The Battle for a Scientific *Humanitas*: Strategies and Teacher Training

Enriques used several strategies to make his vision of a scientific *humanitas* clear to and accepted by mathematicians, philosophers and teachers. They were aimed in many directions – cultural, institutional and editorial – in addition to the channel of university courses, where research and the history of science were intertwined in a significant way. This is made evident by the registers of lessons given in both Bologna[94] and Rome.[95]

In particular, his efforts and commitment to the training of teachers, and thus more generally to the improvement of mathematics education in secondary schools, are truly remarkable.

3.1 The Textbooks for Secondary Schools

In 1903, Enriques inaugurated a long and successful series of textbooks for secondary schools in collaboration with Ugo Amaldi. This was the year which saw the publication of the very well known textbook *Elementi di geometria* [57], successive editions of which were published up to 1992,[96] and various adaptations released for schools of different levels and specialties: middle and high schools, classical and technical schools, normal (*normali*) schools for primary school-teachers training, and *scuole complementari*. The historical catalogue of Zanichelli, the famous Bologna publisher who brought out all the textbooks Enriques wrote for schools, show that before the Gentile reform (1923) eight different kinds of textbooks were published, while seventeen were published after the reform with

---

[93] See [92, pp. 559–602; 1, pp. 406–411].
[94] See ASUB, Enriques prof. Federigo. *Fascicolo personale* and Appendix 1.
[95] See ASUR, Facoltà di Scienze Matematiche Fisiche e Naturali, *Libretti delle lezioni*. See also [105, Appendice 4, Le lezioni di Storia delle scienze a Roma].
[96] See also [62], with a preface by G. Israel [84].

various later editions.[97] Most of these are textbooks for geometry, but there are also textbooks for algebra, trigonometry and calculus.

It is worthwhile to describe briefly at least two of these in order to show how Enriques' vision of mathematics teaching was translated into practice. The edition of the 1903 geometry textbook had been carefully prepared from the scientific and methodological points of view with the preliminary publication of *Questioni riguardanti la geometria elementare* (1900), which is often referred to in the notes. Amaldi, who that same year had been appointed professor of algebra and analytic geometry at the University of Cagliari, made good use of the studies undertaken for the *Questioni* on basic concepts of geometry and the equivalence theory, but the methodological vision which underpins the book is, without a doubt, that of Enriques. The preface opens with a clear indication of the method its two authors will follow:

> An elementary geometry textbook must satisfy two sets of needs: the scientific and the didactic. A mistaken idea of scientific rigour leads some mathematicians to believe that the ideal of the science of geometry consists in a systematic exclusion of intuition. According to this premise one would arrive at an abstruse treatment of the elements which would be inaccessible to a beginner and irreconcilable with the educational purpose of geometry. Geometry is a science of observation and reasoning. It should educate young people in both of these faculties. Scientific rigour, as we understand it, has a formative value because it accustoms students to distinguishing between the activity of one faculty and that of the other.[98]

---

[97] Before the Gentile reform: *Elementi di geometria* [*scuole normali*], *Elementi di geometria* [*scuole secondarie superiori*], *Elementi di geometria elementare* [*ginnasi superiori*], *Elementi di geometria* [*scuole tecniche*], *Nozioni di geometria* [*ginnasi inferiori*], *Nozioni di geometria* [*scuole complementari*], *Geometria elementare* [*scuole secondarie superiori*], *Nozioni di matematica* [*licei moderni*]. After the Gentile reform: *Elementi di geometria* [*edizione ridotta*], *Elementi di geometria* [*scuole complementari*], *Elementi di geometria* [2 vols. *scuole secondarie superiori*], *Elementi di geometria con esercizi* [*istituti tecnici*], *Elementi di geometria con esercizi* [*edizione ridotta*], *Geometria elementare* [*scuole secondarie superiori*], *Geometria elementare con esercizi* [*edizione ridotta*], *Nozioni di geometria* [*ginnasi inferiori*], *Nozioni intuitive di geometria* [*istituti magistrali inferiori*], *Nozioni di geometria* [*scuola media*], *Algebra elementare* [*ginnasi superiori e corso inferiore degli istituti tecnici*], *Algebra elementare* [*licei classici*], *Algebra elementare* [*corso ordinario degli istituti tecnici*], *Algebra elementare* [*primo biennio dei licei scientifici*], *Complementi di algebra e nozioni di analisi* [*secondo biennio del liceo scientifico*], *Elementi di Algebra* [*scuole medie superiori*], *Elementi di trigonometria piana* [*licei*]. See *Le Edizioni Zanichelli 1859–1939* (Bologna: Zanichelli, 1984) and the *Catalogo storico* on the Zanichelli website (http://www.catalogo.zanichelli.it/Page/t01?siteLang=IT&idp=24).

[98] See [57, p. 1]: "Un trattato elementare di geometria deve soddisfare a due ordini di esigenze: scientifiche e didattiche. Un falso concetto del rigore scientifico, fa ritenere a taluno che l'ideale della scienza geometrica consiste nel bandire sistematicamente l'osservazione intuitiva, onde si sarebbe condotto ad una trattazione astrusa degli elementi, inaccessibili ai principianti, ed inconciliabili collo scopo educativo della geometria. La geometria è scienza d'osservazione e di ragionamento; essa deve educare nei giovani queste due facoltà. Il rigore scientifico come lo intendiamo noi, serve allo scopo educativo, insegnando a discernere l'esercizio dell'una dall'esercizio dell'altra."

The subject is presented using a "rational-inductive" method, with the aim of avoiding the shortcomings typical of Euclidean-style exposition, which by "presenting propositions which are analysed at length in their logical connections and coordinated in a deductive system, hides the process of discovery under a rigidly dogmatic framework" [35, p. 24].[99] The procedure is as follows: beginning with a series of observations, the authors enunciate certain postulates from which the theorems that depend on them are developed by logical reasoning; from these theorems, they then continually return to observations or intuitive explanations. In this case as well Enriques acknowledged Klein's influence; in fact, he wrote to him:

> I am sending you a copy of the 2nd edition of my *Elementi di geometria*. In the explanation of a method which, while remaining rational, lays emphasis on the inductive aspects, you will recognise the influence of your own ideas and our conversations in Göttingen.[100]

Among the textbooks that Klein would mention in his essay on geometry teaching in Italy, *Der Unterricht in Italien*, he refers to the *Enriques–Amaldi*, which he praises for having taken didactic requirements into consideration, thus reconciling logical rigor and intuition [91, *II*, pp. 245–250]. Similar praise is found in the long, in-depth review of the textbook written by Vailati, who goes so far as to observe that some of the theorems whose proofs lead to conclusions which for the student are no less evident than the postulates they use, could have been stated in the form of a postulate, because the student has to learn *"as soon as possible* to see in the process of demonstrating a means to go from the known to the unknown."[101] In contrast, Beppo Levi was not in agreement with the "philosophical" part; he believed that too much emphasis had been placed on observation and experience in the explanation of geometric concepts, and he was equally unenthusiastic about the approach to the theory of congruence, which was developed in part by following Hilbert's formulation.[102] In fact, Enriques and Amaldi had assumed the notion of congruence as a primitive for segments and angles, and used movement, intended as a "physical operation," to explain its meaning and check its first properties. They then define it case by case for the more complex figures as they arise. Particular attention is given to the constructions and use of the instruments for making them in order to achieve the aim of "stirring up in young people the spirit of geometrical research" [57, p. 5]. The textbook is supplemented with some 600

---

[99]See [35, p. 24]: "...presentando coordinati in un sistema deduttivo dei resultati lungamente analizzati nei loro rapporti, nasconde sotto la forma dommatica il cammino della scoperta."

[100]See Enriques to Klein, 10 January 1905, SUB Göttingen, *F. Klein* 34: "Le invio una copia della 2a ediz.e dei miei Elementi di geometria. Nell'avviamento ad un metodo che, pur essendo razionale, accentua il carattere induttivo, Ella potrà riconoscere una influenza delle sue idee e delle conversazioni di Gottinga." See also Appendix 2.

[101]See [124, p. 24]: "...il più presto possibile a vedere nel processo di dimostrazione un mezzo per passare dal noto all'ignoto."

[102]See the letter of B. Levi to U. Amaldi, Piacenza, 19 October 1902 [103, pp. 28–31]. For details regarding technical aspects of the textbook, and for a comparison with textbooks of the time, see [73, pp. CXV-CXIX]. See also [84, 125] and [122].

exercises, between problems to solve and propositions to prove. In later editions, the manual was gradually refined and simplified, especially in the parts regarding equivalence and proportions; above all, the texts were enriched with numerous notes about history of mathematics.[103] In some cases, as mentioned earlier, the history of mathematics is also used in order to facilitate understanding of certain concepts. Moreover, in the textbooks for middle schools and normal schools, frequent use is made of experiments with folded or cut paper, sand, or small models.[104]

Another textbook which became a classic was the two-volume *Nozioni di matematica ad uso dei licei moderni* (1914–1915), written with Amaldi for use in the modern secondary school instituted by the Minister of Education Luigi Credaro in 1911. The mathematics programs, formulated by Guido Castelnuovo, introduced the concepts of function, derivative and integral, and gave greater emphasis to numeric approximation.[105] The *Nozioni di matematica* opens with a chapter on approximate measures and irrational numbers, discusses the calculation of areas and volumes from an elementary point of view, establishing connections between geometry and algebra, introduces the concept of function with ample use of grid paper, presents the elementary functions, and trigonometry with particular attention to practical problems, and introduces the concepts of limit, derivative and integral. With respect to the manuals of geometry, this one reveals other characteristic aspects of Enriques' vision of mathematics education. First of all, the various theories are seen as parts of a single organism, and thus the authors try to re-establish the unity of mathematics, making evident the connections between the various branches, especially algebra and geometry in keeping with their historic development; they "abolish the boundary" (vol. 1, p. *III*) between elementary and higher mathematics and between mathematics and the other sciences, from which are drawn problems, exercises and examples, especially to illustrate the concept of function. The history of mathematics makes its appearance in some digressions intended to show how science had moved forward, as in the note on the history of *pi* from the Egyptians to Lindemann (vol. 1, pp. 35–36). It is also used as a means of approach to certain concepts; for example, to calculate the volumes of the pyramids, of the cone, and the sphere, the authors "set forth in an elementary fashion the classical procedure of integration used by the precursors of infinitesimal calculus, which goes back to Archimedes" (vol 1, p. *V*).

---

[103] See, for example, [59] and [60].

[104] See, for example, [61, pp. 64, 78, 88–89, 96–97]. The importance of developing the students' faculty of intuition "with *drawing*, with *cutting*, and with *folding paper*, with the *construction of solid models*" starting in the very first grades of teaching is explicitly stated in [65].

[105] See [75, pp. 6–8], and the website [76].

3.2 The Initiatives of the First Decades of the Twentieth Century

The first two decades of the twentieth century were extremely busy for Enriques. He took an active part in the congresses of the Italian National Federation of Middle School Teachers (the *Federazione Nazionale Insegnanti Scuola Media*, or FNISM) beginning with the first one in Florence in 1902; at the fifth one in Bologna in 1906 he spoke on the topic of teacher training. From 1912 to 1915, he was president of the Italian National Association of University Professors, and presented a project for a university reform. From 1908 to 1920, he was one of the Italian delegates, together with Guido Castelnuovo and Giovanni Vailati, to the International Commission on Mathematical Instruction (ICMI) under Klein's presidency.[106] In 1906, Enriques was one of the founders of the reorganised Zanichelli publishing house, with which he collaborated not only by publishing his own works but also by soliciting publications by esteemed scientists (see [66]). In that same year, he founded the Italian Philosophical Society, and was its president until 1913; in that capacity he organised and presided over the fourth international congress of philosophy, which took place in Bologna in 1911, and provoked the well-known, harsh criticism of Croce and Gentile.[107]

The idea of bringing together philosophy and mathematics was not the product of a extemporaneous improvisation on the part of Enriques; rather, it was the primary concept underlying an entire intellectual movement in Europe, one that spread above all in France, and which found an expression in the first four international congresses of philosophy.[108] As we shall see, Enriques declared more than once that this movement should influence the ordering of schools and universities. For example, it is emblematic that Enriques wanted to organise an international meeting of philosophy in Paris in 1914 to coincide precisely with the congress of the International Commission on Mathematical Instruction (1–4 April 1914).[109]

In 1907, together with Eugenio Rignano, he founded the *Rivista di Scienza* (*Scientia* from 1910 on), "international organ of scientific synthesis," aimed at fighting the excessive specialisation in the field of science and putting an end to the hegemony of literary and historic studies (see [93]).

Here, we will mainly focus on the initiatives directly aimed at improving mathematics teaching and teacher training.

Enriques' position on institutional ways of providing an adequate scientific and educational training for teachers emerges from the report prepared on the occasion of the fifth congress of FNISM in 1906 [29]. In a rather long introduction, he presented his vision of scientific teaching, and his idea of a philosophical university based on the German model, which makes possible "the free and full development

---

[106] See [74], and L. Giacardi, *Timeline* 1908–1910, in the website [70].

[107] See for example [80, 94, 110, 112, 120, and 105, pp. 139–150].

[108] See for example, [114].

[109] See the letters of Enriques to Xavier Léon in [115, pp. 311–315].

of all the elective affinities among the various branches of knowledge" (p. 73). He then suggested the establishment of a *pedagogical degree* in addition to the *scientific degree*: the first two years of study would be dedicated to acquiring basic knowledge of the discipline, and by the end of that time, a distinction would be made between those who intended to dedicate themselves to research and those who wanted to teach. For the future teachers, the next two years would take place in the Scuole di Magistero and would be aimed at providing professional training by means of "(1) courses on those parts of science that aim at a more profound understanding of the elements, (2) lectures on concrete questions of pedagogy that interest the various areas of teaching, particular in relation to the analysis of the textbooks, (3) exercises comprising practice teaching, partly in the university and partly in secondary schools, drawing, and experimental technique" (p. 78).[110] He further expressed his hope that those called to teach in the Scuole di Magistero would include all the professors of the scientific faculties and the best of the secondary school teachers; he also proposed that the selection of teachers be based on the results of a competition comprising both written and oral exams in order to make evident the candidates' attitudes towards science and education.

Enriques' proposals, as he himself emphasised at the beginning of his presentation, were directly related to his project for university reform,[111] which had grown out of the ascertainment of the defects of the Italian university system. Above all he criticised the lack of interaction between the various faculties, the excessive fragmentation, and the separation of disciplines with programs that were obligatory and too heavy:

> Heaven help you if you pass from one laboratory to another, interrupt the process to meditate or study, or worse still, to attempt research that goes beyond the limits set in the definition of the chair!
> The rash one who dares set foot in new territory, investigating the relationship between two *different* disciplines, knows well the fate that awaits him.[112]

In addition, the tendency of each professor to defend his own discipline favoured the pre-eminence of already consolidated areas of research over those which were interdisciplinary or unexplored, with serious repercussions for research, teaching and the work world:

---

[110] See [29, p. 78]: "(1) corsi su quelle parti della scienza che si riattaccano ad una più profonda visione degli elementi, (2) conferenze sulle questioni di pedagogia concreta che interessano i varii rami d'insegnamento, particolarmente in rapporto colla critica dei testi (3) esercitazioni comprendenti il tirocinio parte nell'università e parte in una scuola secondaria, il disegno e la tecnica sperimentale."

[111] See the two articles [31, 32] and Enriques' paper in *Atti dell'Assemblea della Associazione nazionale fra i professori universitari*, Torino, 1911, pp. 122–141; rpt. in [121, pp. 91–132].

[112] See [121, p. 99]: "Guai a passare da un laboratorio ad un altro, a interrompere la produzione per meditare o studiare, o peggio ancora per tentare ricerche che oltrapassino i limiti stabiliti nella definizione delle cattedre! Il temerario che si sarà avventurato sopra un terreno nuovo, indagando i rapporti fra due discipline diverse, sa bene quale sorte l'attenda."

Now all of these deficiencies and difficulties are directly reflected in middle school teaching ... The exaggerations of rigour – in the form of minutia and senseless pedantry – in schools of mathematics, the empiricism of physics teaching ..., the morphological erudition that suffocates the natural sciences ..., all these defects – so often lamented – are in direct correlation with the conditions of the university training of middle school teachers.[113]

He also predicted that future workers would be "devoid of initiative, ... ready at any moment to take refuge in the excuse of procedure and the observance of form."[114]

Enriques, in accordance with his strategy, explained his point of view to the philosophers of the first philosophical congress in Milan in 1906, and later that same year to the middle school teachers in Bologna, to the mathematicians and scientists in his 1908 article in the *Rivista di Scienza*, and finally to university professors in 1911. The solution he proposed was that of conjoining in a single faculty of philosophy all of the theoretical disciplines: mathematics, physics, physiology, history, law, economy, etc. He also proposed the institution of "special schools of Application" which were to group together professional teaching aimed at a specific career, the polytechnical schools for engineers and the polyclinical schools for physicians, and the Scuole di Magistero for the training of teachers. With more specific regard to the programs, the courses and the examinations, Enriques believed that it was necessary to "reduce the science to be learned to a minimum" (in [121, p. 97]); to give the students the freedom to choose the courses to attend within a given number established by the faculty, which, however, would be responsible for guaranteeing the reliability and coherence of the courses; and to introduce a different way of testing what knowledge had been acquired and the capacity for putting it to use.

To sum up, Enriques wrote:

The reform of the Italian university

(1) Must correspond to the synthesis required by renewed philosophical consciousness and practical life, as opposed to the scientific-educational particularism of the previous era
(2) Must give new life to the spirit of initiative of our universities, promoting their free differentiation
(3) Must sanction the principle of the freedom of study and, emancipating young people from the weight of formal erudition, prepare them for professions and for life through a more active exertion of their faculties.[115]

---

[113]See [121, p. 94]: "Ora tutte queste deficienze ed angustie si rispecchiano direttamente nell'insegnamento medio, ... Le esagerazioni del rigore – sotto forma di minuzie e di pedanterie senza scopo – nelle scuole di Matematica, l'empirismo dell'insegnamento fisico ..., l'erudizione morfologica che soffoca i corsi di scienze naturali ..., tutti questi difetti – spesso lamentati – sono in correlazione diretta colle condizioni della preparazione universitaria dei docenti delle scuole medie."

[114]See [121, p. 96]: "... fiacchi, ... pronti a rifugiarsi ogni momento nelle scuse della procedura e nell'osservanza della forma."

[115]See [121, p. 114]: "La riforma dell'Università italiana 1) deve corrispondere alle esigenze sintetiche della rinnovata coscienza filosofica e della vita pratica, avverso il particolarismo scientifico-didattico dell'epoca precedente; 2) deve ravvivare lo spirito d'iniziativa dei nostri

From the same need to combat excessive specialisation was born the *Rivista di Scienza*, which Enriques co-directed with Rignano until 1915, and then again from 1930 to 1938, when he had to quit because of the racial laws. The vision underlying the *Rivista* was that of a scientific philosophy which, "free of direct ties to traditional systems, arises to promote the coordination of work of science, the criticism of its methods and theories, and to assert a broader appreciation of its problems."[116] It was precisely for this reason that from the very beginning the *Rivista* had an international dimension: it came out in two editions, one Italian, the other foreign, which was distributed by prestigious publishers, William & Norgate in England, F. Alcan in France, and W. Englemann in Germany. Moreover, thanks to his personal prestige, Enriques was able to count on the collaboration of well known scholars – mathematicians, physicists, chemists, geologists, historians of science, sociologists, linguists, economists – including, just to name a few, Einstein, Mach, Michelson, Ostwald, Picard, Russell, and Volterra. Enriques himself, from 1907 to 1938, wrote twenty-three articles and critical notes, sixty-three reviews and twenty-five surveys of journals. His imprint is particularly noticeable in the early years, and it is no coincidence that in addition to aspects of history, philosophy and methodology, attention was also given to aspects of education: in 1907, articles appeared by G. Castelnuovo, J. Tannery, T. Bonnesen; in 1908 were published Enriques' own articles on university reforms [31, 32] and his review of the book by A. Galletti and G. Salvemini entitled *La riforma della scuola media* (1908); between 1913 and 1915 there were three reviews by G. Scorza of works aimed at mathematics teaching in secondary schools; and in 1915 there was another article by Enriques on the art of writing a mathematics treatise.

3.3 Enriques' Mathesis Presidency and Direction of the Periodico di Matematiche

In 1919, Enriques was nominated president of the *Mathesis* association, a position he held until 1932. Then, since in 1921 the *Periodico di Matematiche* had gone back to being the association's publishing venue, he assumed its direction together with Giulio Lazzeri. The imprint of the fourth series, which began with the 1921 volume, is exquisitely Enriques', starting with the title – *Periodico di Matematiche. Storia-Didattica-Filosofia* – and from the introductory sentence that appears on the inside of the front cover of each issue of the journal:

Atenei, promuovendone la libera differenziazione; 3) deve sancire il principio della libertà degli studi ed, emancipando i giovani dal peso di un'erudizione formale, prepararli alle professioni ed alla vita con un esercizio più attivo delle loro facoltà."

[116]See ["Preface"], *Rivista di Scienza*, 1, 1907, pp. 1–3, at p. 2: "...libera da legami diretti con i sistemi tradizionali, sorge appunto a promuovere la coordinazione del lavoro, la critica dei metodi e delle teorie, e ad affermare un apprezzamento più largo dei problemi della Scienza."

The *Periodico* publishes above all articles regarding elementary mathematics in a broad sense, and others that tend towards a wider comprehension of the spirit of mathematics. It also contains reports on movements in mathematics abroad, notes on bibliographies and treatises, miscellany (problems, games, paradoxes, etc.) as well as news of a professional nature, and finally, the Proceedings of the Italian Matematical Society "Mathesis."[117]

According to Enriques' project, the *Periodico* was intended to disseminate the idea of mathematics as an integral part of the philosophical culture, an idea he had always supported, as well as to fill the gap that existed in scientific education at that time in Italy. In the letter to the readers that opened the 1921 issue, he presented an actual working program for the journal, which was at the same time a working program for teachers. The cardinal points are: teachers should study the science that they are teaching in depth from various points of view, so as to master it from new and higher points of view, and thus make evident the connections between elementary mathematics and higher mathematics; use the history of the science seeking to attain, not so much erudite knowledge as a dynamic consideration of concepts and theories, through which students can recognise the unity of thought; bring out the relationships between mathematics and the other sciences, and physics in particular, in order to offer a broader vision of science and of the aims and meanings of the many different kinds of research [40, pp. 3–4].

This open letter was followed by his famous article, "Insegnamento dinamico" [38], which is almost a manifesto of Enriques' working program, and of his particular vision of mathematics education: active teaching, Socratic method, learning as discovery, the right balance between intuition and logic, the importance of error, the historic view of problems, the connections between mathematics and physics, elementary mathematics from an advanced standpoint, and the educational value of mathematics.[118] A look through the issues shows above all an increase in the number of articles about physics and history of physics (mostly written by Enrico Persico, Umberto Forti and Enrico Fermi), and those dealing with history of mathematics and science in general: the principal collaborators are the mathematics historians Ettore Bortolotti, Gino Loria and Amedeo Agostini, but there are also contributions by Ugo Cassina, Giulio Vivanti, Alpinolo Natucci (a secondary school teacher in Pisa), Emilio Artom (a secondary school teacher in Torino), and Maria Teresa Zapelloni, among others. Noteworthy are the articles written by Oscar Chisini, which clearly show Enriques' influence. Enriques had made Chisini editorial secretary of the *Periodico* in 1921, and it was Chisini who, after Enriques' death, succeeded him as director. Chisini's articles mostly concern the elementary aspects of mathematics which show the connections with the recent progress in mathematics, making

---

[117]"Il Periodico pubblica soprattutto articoli riguardanti le matematiche elementari intese in senso lato, ed altri tendenti ad una più vasta comprensione dello spirito matematico. Esso contiene inoltre relazioni del movimento matematico straniero, note di bibliografia e di trattatistica, varietà (problemi, giuochi, paradossi, etc.) nonché notizie di carattere professionale, ed infine gli Atti della Società Italiana di matematiche 'Mathesis'."

[118]See the new edition [56] accompanied by essays by F. Ghione and M. Moretti published by the Centro Studi Enriques.

reference to history and to educational aspects. Enriques himself wrote no fewer than twenty-seven articles and brief notes, and thirty-four reviews, most of which regarded history of science or mathematics teaching.[119]

The desire to open up to other sciences is also evident in the new charter for the Mathesis association, which, on 7 May 1922, welcomed teacher of physics into its ranks, and led the society to assume a new name: *Società italiana di scienze fisiche e matematiche "Mathesis"*. Under the leadership of Enriques, the number of members in 1920 grew from 775 to 895; by 1924 there were more than 1, 200. During his presidency, the society organised six national congresses (Trieste, 1919; Naples, 1921; Leghorn, 1923; Milan, 1925; Florence, 1929; Milan, 1931). The congresses of 1929 and 1931 were organised in collaboration with the Italian Society for the Progress of the Sciences (SIPS),[120] directed at that time by phycisist and geologist Gian Alberto Blanc. The SIPS had also the aim of contrasting excessive specialisation and stimulating interdisciplinary dialogue, but its project was complementary to that of Enriques because it was primarily addressed to the world of technology and industry.

The inaugural lectures that Enriques gave at the congresses were all aimed at upholding the educational and cultural value of mathematics and the sciences.[121] The problems that he had to grapple with were not simple, not least because they were contingent on historical and political situations, but a strong point of Enriques was his constant attention to the opinions of the teachers and the various local sections of the association, as emerges for example from the unpublished correspondence with Giacomo Furlani, president of the Trieste section.[122] In particular, after the first World War, it was necessary to solve the delicate problem of how to harmonise the mathematics programs of the provinces of Trento and Trieste, recently annexed from Austria, with those of the Kingdom of Italy.[123] After the advent of Fascism, it was necessary to address the problems related to the Gentile reform: the devaluation of mathematics and of sciences in general, the reduction of the number of teaching hours, the combination of mathematics and physics, and teacher training.

---

[119]On the historiography of mathematics in Italy see [20].

[120]See "Congresso della Società Italiana *Mathesis*," *Periodico di matematiche*, (4) 11, 1931, pp. 322–325. See also *Atti della Società Italiana per il Progresso delle Scienze, Firenze 18–25 Settembre 1929*. Roma 1930, and *Atti della Società Italiana per il Progresso delle Scienze, Milano 12–18 Settembre 1931*. Roma 1932. Enriques had already interacted with the SIPS at the beginning of the twentieth century; see [105, pp. 134–139].

[121]Enriques gave the following inaugural lectures: Trieste, 1919: *Il valore delle Matematiche nella Filosofia italica* [37]; Naples, 1921: *Evoluzione del concetto della Scienza nei pensatori matematici* [42]; Livorno, 1923: *Il significato umanistico della scienza nella cultura nazionale* [43]; Milan, 1925: *L'essenza della matematica* (see [*Periodico di Matematiche*, (4) 1, 1925, p. 378]; Florence, 1929: *La geometria non-euclidea e i presupposti filosofici della Teoria della Relatività* [47].

[122]See Appendix 3.

[123]See Giacardi, L., ed. 2006. Da Casati a Gentile. *Momenti di storia dell'insegnamento secondario della matematica in Italia*. Lugano: Lumières Internationales, pp. 47–53 and [127].

In 1923, in the space of a single year, Giovanni Gentile, minister for education, put into effect a complete and systematic reform of the Italian scholastic system in keeping with neo-idealist philosophy. Secondary education was divided into two branches: classical-humanistic and technical-scientific. The classical-humanistic branch was intended to train the ruling class and was considered overwhelmingly superior to the scientific-technical one, which, moreover, made access possible to only a limited number of university degrees. The principles of Fascism and neo-idealist ideology were opposed to the widespread diffusion of scientific culture and, above all, to its interaction with other cultural sectors. Humanistic disciplines were to form the main cultural axis of national life and, in particular, of education; it was symptomatic that even the courses of history of science introduced into the scientific high schools were taught by philosophers.[124] In addition, Gentile, who identified knowledge with knowing how to teach, paid no attention at all to professional training of teachers. This point of view was, of course, opposed to the scientific *humanitas* to which Enriques aspired. As president of the Mathesis Association, he engaged in intense negotiation with Gentile, both before and after the law on secondary education was enacted, in the hope of avoiding the devaluation of science teaching. However, the pleas of the Mathesis fell on deaf ears.[125] Unlike Vito Volterra and Guido Castelnuovo, who were in absolute opposition to the Gentile Reform, Enriques assumed and maintained a conciliatory position. In fact, he agreed with Gentile on many points: he was convinced that among the various kinds of secondary schools, those which best performed the function of education were the *ginnasi-licei* (classical schools); he conceived of knowledge as a personal conquest; he was in agreement with the need to fight encyclopaedism and he considered education to be the free and unfettered development of inner energy. Moreover, he did not want to renounce his idea of the fusion of scientific knowledge and humanistic idealism which was the basis of the cultural program he had dedicated his whole life to: the creation of a scientific *humanitas* which would express and make manifest the universality of human reason.[126]

Enriques' position emerges clearly from his correspondence with Gentile [80], as well as from the report on the reform which he prepared for the ICMI in 1929 [48]. His account appears less critical than might be expected: he limited himself to pointing out the reduction in the number of the hours devoted to mathematics, and the unsolved problem of teacher training. Instead, he gave ample space to the flourishing of new textbooks, citing the manuals which he himself had written with Amaldi and two series, one directed by Roberto Marcolongo and Onorato Nicoletti, the other by Francesco Severi. He also presented his many initiatives aimed at teacher training: in addition to the *Questioni riguardanti le matematiche elementari*, of which the third edition had just been published, he cited the school

---

[124] See "Atti della Società Italiana di Scienze fisiche e matematiche "Mathesis". Relazione del Congresso di Milano," *Periodico di matematiche*, (4) 5, 1925, pp. 374–383, at p. 383.
[125] This is discussed in greater detail in Giacardi, L., ed. 2006. Da Casati a Gentile. cit., pp. 54-63.
[126] See [43, p. 4]. See also [80, 83, 94, 112].

for specialisation in history of the sciences annexed to the *Istituto Nazionale per la Storia delle Scienze*, which he had created after the Gentile reform and the book series on the history of mathematics, *Per la storia e la filosofia delle matematiche*, which he had begun in 1925 and was expressly intended for teacher training. These are precisely the initiatives we will examine in the next section.

3.4 The Roman Initiatives and Teacher Training

In December 1920 Alberto Tonelli died; Tonelli had held the chair in algebraic analysis at the University of Rome. Many were interested in succeeding him, including Enriques and Severi. In the end, it was Severi who prevailed; Enriques' being called to Rome was only made possible thanks to Castelnuovo's having given up the chair in higher geometry, as has been recently shown.[127] In fact, Enriques, who had been called for a temporary position at the University of Rome in 1921–1922 to "teach lessons in mathematics for the [Scuola di] Magistero," and in 1922–1923 to teach the then newly established course in complementary mathematics, had not even been successful in obtaining a transfer to the chair of complementary mathematics.[128]

His lessons and related practical exercises in complementary mathematics of that year are the translation into practice of Enriques' way of conceiving teacher training: the history of mathematics is interwoven with the mathematical theory, elementary mathematics are linked to higher mathematics; mention is made of the theory of relativity, and a comparative examination is proposed of textbooks.[129] With regard to this course he wrote to Gentile:

> I should add that the difference between this course and the other two in advanced mathematics given during our second biennium (higher analysis and advanced geometry) is this: that here come into play precise arguments – such as the problems of the trisection of the angle and the squaring of the circle, etc. – which we believe the teacher needs to know about, and which cannot be dealt with in courses in higher analysis and advanced geometry, the only ones in our university which are aimed at pure mathematics! ... Further, by means of these problems that are closer to elementary mathematics and which have a history that is twenty centuries old, we aim to reach young people with a vocation for teaching, who ... must be protected from the risk of becoming mechanical propagators of a culture that they have received from outside and is truly foreign to their spirit: this is a conclusion to

---

[127] See [105, Appendice 2, *Il trasferimento di Enriques a Roma*].

[128] See the documents of 17 February 1922, 4 September 1922 in ASUB, *Enriques prof. Federigo. Fascicolo personale*, and the document of 30 December 1922 in ASUR, *Enriques, Federigo. Fascicolo personale*.

[129] See the *Libretto delle lezioni di Matematiche complementari* and the *Libretto delle esercitazioni di Matematiche complementari* of Enriques, 1922–1923, ASUR *Facoltà di Scienze Matematiche Fisiche e Naturali. Libretti delle lezioni*. See also Appendix 1.

which you have arrived by means of metaphysical premises, but to which I have also – as far as my powers allow – contributed with the actions of my life.[130]

History of mathematics was also used, sometimes quite extensively, in the lessons in higher geometry,[131] and its centrality in Enriques' program is clearly shown by the many initiatives that went hand in hand with his leadership of the Mathesis association during this period. Effectively the campaign Enriques was conducting amounted to a genuine battle aimed at projecting an image of science both as a unified whole and as an integral part of culture.

In 1923, he founded the *Istituto nazionale per la Storia delle Scienze fisiche e matematiche*, with the aim of giving an impetus to studies in the history of the physical and mathematical sciences, and in particular, to promote:

> the collection, in some of the most suitable centres, of books and documents that are necessary for the pursuit of serious and wide-reaching research projects; the diffusion of research ..., the arrangement and publication of unpublished manuscripts ..., the publication ... of works either classic or representative of some special interest.[132]

In connection with the Rome Institute, the following year Enriques founded the *Scuola universitaria per la Storia delle scienze*, annexed to the University of Rome, whose threefold aim was to provide incentives for historical research, train future teachers, and promote the consolidation of the idea of scientific *humanitas*. In 1924–1925, Enriques taught a course on the history of scientific concepts, while Giovanni Vacca taught one on history of mathematics; the next year Enriques and Vacca taught the same courses again, broadening and enriching them with new material, while Aldo Mieli taught history of chemistry, Federico Raffaele gave lectures on the evolution of cellular theory, Silvestro Baglioni taught history of medicine and Roberto Almagià taught history of geography. In the following years,

---

[130] See the letter of Enriques to Gentile, Rome, 23 December 1922 [80, pp. 149–150]: "Aggiungo che la differenza fra questo corso e gli altri due di matematiche superiori del nostro secondo biennio (analisi superiore e geometria superiore) è questa: che qui entrano argomenti precisi – come i problemi della trisezione dell'angolo o della quadratura del cerchio, ecc. – intorno a cui si ritiene che l'insegnante debba essere informato, ed a cui non si può costringere i corsi di analisi superiore e di geometria superiore, i soli che mirino presso di noi alla pura scienza matematica! ... Inoltre attraverso quei problemi che toccano più da vicino le matematiche elementari e che hanno una storia venti volte secolare, si mira soprattutto ai giovani chiamati all'insegnamento, i quali ... debbono essere preservati dal pericolo di diventare ripetitori meccanici di una cultura ricevuta dal di fuori e però estranea veramente al loro spirito: che è una tesi a cui Ella giunge da premesse metafisiche, ma a cui io ho pur dato da parte mia – nella misura delle mie forze – il contributo dell'azione della mia vita."

[131] See [105, Appendice 4].

[132] See "Istituto Nazionale per la Storia delle Scienze fisiche e matematiche," *Periodico di matematiche*, (4), 3, 1923, pp. 149–153, at p. 151: "... la raccolta, in alcuni centri più adatti, dei libri o dei documenti che occorrono per proseguire serie e larghe ricerche; la divulgazione delle ricerche ...; l'ordinamento e la pubblicazione di manoscritti inediti ...; la pubblicazione ... di opere classiche o rappresentanti qualche speciale interesse." For the later fusion of the *Istituto* with Aldo Mieli's *Federazione nazionale fra le Società, gli Enti, gli Insegnanti ed i Cultori di Storia della Scienze*, see [105, pp. 160–163].

the courses were almost all continued, and in 1934–1935 were added courses in history of astronomy taught by Pio Emanuelli and history of biology taught by Giuseppe Montalenti; in that same year, Ettore Carruccio and Attilio Frajese joined the school as volunteer assistants.[133] In order to bolster and consolidate the school, in a lecture at the Accademia dei Lincei in 1938 Enriques asked for the institution of a chair in history of mathematics:

> A minister who is a philosopher ... had the merit of understanding the educational and didactical value of the history of science, and to introduce its teaching in several orders of Italian middle schools [but there being no] adequate preparation of teachers, his reform could not be carried out seriously. But the idea remains, and more than the idea, the incumbent duty to translate it into action [54, p. 134].[134]

Enriques' proposal was thwarted by Bortolotti and by Severi.[135] In any case, in that same year the the racial laws excluded Enriques from teaching. In 1938–1939 the course in history of mathematics was taught by Fabio Conforto, who had helped Enriques with this course the previous year and was then collaborating with him on the treatise *Le superfici razionali* (1939). Baglioni taught a course in the history of discoveries in biology and physiology, and Adalberto Pazzini taught history of medicine.[136] In February 1939,[137] Severi was named director of the School, and this marked the end of an important period for the history of science in Italy.

Among the initiatives collateral to the School, two deserve special mention. The first was the book series created in 1925 entitled *Per la storia e la filosofia delle matematiche*, the second was the *Settimana della Scuola di Storia delle scienze* organised by Enriques and his collaborators in 1935 in Rome.

The idea for the book series had been suggested to him "from practical experience in the Scuola di Magistero" [45, p. 7]; primarily intended for a readership of educators, it also aimed at students and educated people in general. Twelve volumes were published from 1925 to 1938:[138] a look at the titles shows that

---

[133]For more about Enriques' courses, see [105, "Appendice 4"]; for the others see ASUR, *Facoltà di Scienze Matematiche Fisiche e Naturali. Libretti delle lezioni*, 1924–1941.

[134]See [54, p. 134]: "Un ministro filosofo ... ha avuto il merito di comprendere il valore educativo e didattico della storia della scienza e d'introdurne l'insegnamento in alcuni ordini della scuola media italiana [ma, mancando] un'adeguata preparazione degli insegnanti, la sua riforma non ha potuto essere ancora seriamente attuata. Ma l'idea rimane; più che l'idea il dovere incombente di tradurla in atto." See also the letter of Enriques to Gentile dated 20 December 1924 in [80, pp. 151–153].

[135]On Enriques' relation with Severi and with Fascism, see for example [102] and the essays of E. Vesentini, C. Ciliberto, A. Brigaglia and S. Linguerri in [111].

[136]See ASUR *Facoltà di Scienze Matematiche Fisiche e Naturali. Libretti delle lezioni*, 1938–1939.

[137]See the letters of Pietro De Francisci to Francesco Severi, Rome 16 February 1939 and 13 January 1943, ASUR *Personale docente. Severi, Francesco.*

[138]1) F. Enriques (ed.), *Gli Elementi d'Euclide e la critica antica e moderna*, (Libri I–IV) (Rome: Alberto Stock, 1925); 2) J. L. Heiberg, *Matematiche, scienze naturali e medicina nell'antichità classica*, Gino Castelnuovo, trans. (Rome: Alberto Stock, 1924); 3) F. Enriques and U. Forti (eds), *I. Newton: Principii di Filosofia naturale, teoria della gravitazione* (Rome: Alberto Stock,

Enriques particularly favoured translations with commentaries, often accompanied with historical notes, of works by important authors of the past (Euclid, Archimedes, Bombelli, Galileo, Newton, Dedekind, etc.) which might be of relevance to mathematics teaching. Collaborators on the book series included colleagues, followers, students and friends working in various areas: Ettore Bortolotti, Guido Castelnuovo, Umberto Forti (professor of mathematics in secondary schools and historian of science), Amedeo Agostini, Oscar Zariski, Enrico Rufini (teacher at *Liceo Tasso* in Rome), Ettore Carruccio, Attilio Frajese, Maria Teresa Zapelloni, Gino Castelnuovo (son of Guido Castelnuovo, at that time a student in the school for engineering in Rome), Maria Lombardini (from the geophysics observatory at Rocca di Papa), Guido Rietti, Ruth Struik (wife of Dirk Struik). The first volume was dedicated to the first four books of Euclid's *Elements*, a text which, in Enriques' opinion, all teachers should know. The second was Gino Castelnuovo's Italian translation of the Danish Johan Ludvig Heiberg's treatise on mathematics, natural science and medicine in antiquity. Heiberg embodied Enriques' ideal of the historian of science: a philologist with profound knowledge of the sources, but capable of "hiding all burdensome erudition," thus writing inspired, panoramic works capable of shedding light on the relationships between science of the past with contemporary and later culture, and ready to collaborate with scholars in other fields. In Enriques' words:

> ...it is well known that Heiberg worked, especially in the history of mathematics, alongside the great geometer Zeuthen, and Zeuthen alongside Heiberg, with a communion of spirit that constitutes a splendid example of collaboration between scholars differently trained, and thus to the benefit of both, and above all fortunately for our knowledge.[139]

The importance that Enriques attributed to the history of science in teacher training is twofold: it not only helps to understand the genesis of the ideas and problems, but also makes it possible to participate in scientific research:

> The training of mathematics teachers who are capable of carrying out their educational responsibilities requires, generally speaking, that they understand science not only in its

---

1925); 4) E. Rufini, *Il Metodo di Archimede e le origini dell'analisi infinitesimale nell'antichità* (Rome: Alberto Stock, 1926); 5) O. Zariski (ed.), *Riccardo Dedekind: Essenza e significato dei numeri. Continuità e numeri irrazionali* (Rome: Alberto Stock, 1926); 6) M. Lombardini (ed.), A. C. Clairaut: *La teoria della forma della terra dedotta dai principi dell'idrostatica* (Bologna: Zanichelli, 1928); 7) E. Bortolotti (ed.), *L' Algebra, opera di Rafael Bombelli da Bologna, Libri IV e V comprendenti "La Parte geometrica" inedita tratta dal manoscritto B. 1569, Biblioteca dell'Archiginnasio di Bologna* (Bologna: Zanichelli, 1929); 8) F. Enriques (ed.), *Gli Elementi d'Euclide e la critica antica e moderna*, (Libri $V - IX$) (Bologna: Zanichelli, 1930); 9) U. Forti, *Introduzione storica alla lettura del "Dialogo sui massimi sistemi di Galileo Galilei"* (Bologna: Zanichelli, 1931); 10) F. Enriques (ed.), *Gli Elementi d'Euclide e la critica antica e moderna*, (Libro X) (Bologna: Zanichelli, 1932); 11) F. Enriques (ed.), *Gli Elementi d'Euclide e la critica antica e moderna*, (Libri $XI - XIII$) (Bologna: Zanichelli, 1936); 12) G. Castelnuovo, *Le origini del calcolo infinitesimale nell'era moderna* (Bologna: Zanichelli, 1938).

[139]See [82, pp. 6–7]: "...è ben noto come Heiberg abbia lavorato, particolarmente nella storia delle matematiche, accanto al grande geometra Zeuthen, e Zeuthen accanto ad Heiberg, con una comunione di spiriti che costituisce un esempio splendido di collaborazione fra studiosi diversamente educati, e così con profitto di entrambi e soprattutto per fortuna del nostro sapere."

static aspect, but also in its developing state; and thus that the scholar learn from history to reflect on the genesis of the ideas, and on the other hand, take an active interest in research.[140]

The second initiative, the *Settimana della Scuola di Storia delle Scienze* (Rome, 15–22 April 1935) organised by Enriques and the teachers of the School (Almagià, Baglioni, Montalenti and Vacca), deserves mention because it documents Enriques' aperture to other countries in oppostion to all forms of nationalistic isolation and distortion in the field of history of science. The participants included Castelnuovo, Bompiani and Giuseppe Armellini and twenty-six members from London's Unity History School as well as scholars from other European countries, including the Belgian Paul Libois, who would draw various aspects of his own vision of mathematics teaching from Enriques,[141] and the French historian Hélène Metzger,[142] who shared Enriques' unitary concept of science. The topics addressed went from philosophy to the history of physics, astronomy, biology and technology, and the debate was lively, as can be seen from the detailed summary of the week's activity written by Metzger and published in Aldo Mieli's journal *Archeion* [99].

During the same period Enriques also participated in the meetings (Paris, Vienna, Berlin) and congresses (Heidelberg, 1927; Barcelona, 1929; Paris, 1933; Budapest, 1934; Zurich, 1938) of the *Institut International de Coopération Intellectuelle* (IICI) inaugurated on January 1925 in Paris,[143] in addition to various other international congresses of philosophy, history of philosophy and philosophy of science: it was no coincidence that Enriques remained in contact with the IICI, whose aim was to promote international cultural exchange between scientists, researchers, teachers, artists and other intellectuals. He also directed two sections of the book series *Actualités scientifiques et industrielles* published by Hermann in Paris: "Philosophie et histoire de la pensée scientifique"[144] and "Histoire de la pensée scientifique."[145]

---

[140]See [53, p. 190]: "La formazione di docenti di matematiche, che siano all'altezza dei loro compiti didattici, richiede, in genere, che la scienza sia da loro appresa non soltanto nell'aspetto statico, ma anche nel suo divenire. E quindi che lo studioso apprenda dalla storia a riflettere sulla genesi delle idee, e d'altro lato partecipi all'interesse per la ricerca."

[141]See [98].

[142]See [104].

[143]See ASUR, *Enriques, Federigo. Fascicolo personale* and [97].

[144]F. Enriques, *Signification de l'histoire de la pensée scientifique* (1934); G. Castelnuovo, *La probabilité dans les différentes branches de la science* (1937); F. Gonseth, *Qu'est-ce que la logique?* (1937); H. Metzger, *Attraction universelle et religion naturelle chez quelques commentateurs anglais de Newton. Première partie, Introduction philosophique* (1938); H. Metzger, *Attraction universelle et religion naturelle chez quelques commentateurs anglais de Newton. Deuxième partie, Newton, Bentley, Whiston, Toland* (1938); H. Metzger, *Attraction universelle et religion naturelle chez quelques commentateurs anglais de Newton. Troisième partie, Clarke, Cheyne, Derham, Baxter, Priestley* (1938); F. Enriques, *La théorie de la connaissance scientifique de Kant à nos jours* (1938); F. Enriques, *Causalité et déterminisme dans la philosophie et l'histoire des sciences* (1941).

Between 1934 and 1939 eight volumes were published in the first series, with the collaboration of Hélène Metzger, Ferdinand Gonseth and Guido Castelnuovo, and six in the second series, written in collaboration with de Santillana; these were developed on the bases of the book they had published together in 1932 entitled *Storia del pensiero scientifico. Il mondo antico* (Milan, Treves). The first volume to appear was *Signification de l'histoire de la pensée scientifique* in 1934, and in the opening chapter, titled "La science et son histoire" Enriques once again presented his dynamic vision of science and the conviction that it is precisely from history that science draws its meaning.

In 1937, together with de Santillana, he published the *Compendio di storia del pensiero scientifico dall'antichità ai tempi moderni* (Bologna, Zanichelli), which aimed at filling the gap in the teaching of philosophy and history in secondary schools. Although some parts now appear dated, Paolo Casini has written that "the two authors' political commitment and their efforts to overcome the impasse of then current trends in textbook writing' is above all evident in the "brief sections concerning nineteenth century, positivism, pragmatism and neo-idealism" [11, pp. XIV, XV].[146] The following year was published the volume *Le matematiche nella storia e nella cultura* [53], aimed primarily at students in secondary school and the first two years of university. As the title indicates, the objective was to show the significance and place of mathematics in the context of other sciences and in its relations with technology, art, history and philosophy in order to reconstruct the unity of thought in the face of increasing specialisation. A few dense pages dedicated to mathematics teaching (pp. 184–191) gave Enriques the chance to reaffirm the educational and cultural value of mathematics and the importance of having adequately prepared teachers.

It is abundantly clear that history played an increasingly central role in the struggle for a scientific *humanitas*, and that the teachers were a very important channel for Enriques. This can also be seen in the third edition of the *Questioni riguardanti le matematiche elementari* (1924–1927), which was republished in a reorganised form and enriched with new material[147] drawn principally from the courses Enriques had taught at the University of Rome in the previous two years.

---

[145]*Les Ioniens et la nature des choses* (1936); *Le problèmes de la matière: Pythagoriciens et Eléates* (1936); *Les derniers "Physiologues" de la Grèce* (1936); *Le problème de la connaissance. Empirisme et rationalisme grecs* (1937); *Platon et Aristote* (1937); *Mathématiques et astronomie de la période hellénique* (1939).

[146]See [11, pp. XIV, XV]: "…l'impegno politico dei due autori e il loro sforzo di superare le impasses della manualistica corrente'" … "nei rapidi scorci concernenti il diciannovesimo secolo, il positivismo, il pragmatismo e il neoidealismo."

[147]Enriques himself wrote six articles: "L'evoluzione delle idee geometriche nel pensiero greco: punto, linea e superficie" [III ed., I.1 pp. 1–40]; "I numeri reali" [III ed., I.1 pp. 231–389]; "Spazio e tempo davanti alla critica moderna" [III ed. I.2 pp. 429–459]; "Sulle equazioni algebriche risolubili per radicali quadratici e sulla costruibilità dei poligoni regolari" [III ed., II, pp. 263–305]; "Alcune osservazioni generali sui problemi geometrici" [III ed., II pp. 575–596]; "Massimi e minimi nell'Analisi moderna" [III ed., III pp. 311–471].

New collaborators flanked the original ones, including Enrico Bompiani, Alfredo Sabbatini e Vittoria Notari Cuzzer, Enriques' assistant first in Bologna and then in Rome, and his collaborator on questions of didactics in the *Periodico di matematiche* as well. In the preface Enriques affirmed that the aim was that of "giving scientific theory a basis in history," at the very moment when various circumstances threatened "to diminish science and mathematical culture ... precisely among those whose highest duty it is to diffuse them in the schools." The work is thus addressed to teachers, to the students of the course of complementary mathematics, and to those who were preparing for the state examinations, but Enriques underlined that another aim was that of "opening the fruitful field of historical investigation to a greater number of scholars."[148]

In his 1931 preface to the index of the first ten years of the second series of the *Periodico di matematiche* that he himself had inaugurated, Enriques underlined with pride the role played by the journal in teacher training:

> No other journal of this sort, in no other country in the world, has been able to realise a program that is as lofty and attuned to the exigencies of education and culture of teachers of middle schools.[149]

Once again, Enriques highlighted the effectiveness of using advanced mathematics to improve comprehension of elementary questions, the importance of criticism of the principles which avoids logical subtleties and makes evident the philosophical meanings of the problems, and the use of history of mathematics to cultivate in teacher the idea of "becoming" in science.

Enriques' aspiration of diffusing his unitary vision of science and avoiding the cultural isolation of mathematics also lay at the basis of his collaboration with the *Enciclopedia Italiana*, and was joined to the need for a wider dissemination. As he expressed it, addressing members of the Mathesis Association:

> Nor should this work of dissemination and propaganda appear superfluous ... And, even if the need for propaganda distracts us for a time from other useful work, we must not regret it, because, by strengthening our scientific faith and recreating the need [of science] in the society around us, we prepare younger and more daring energies for scientific progress.[150]

In 1925 Gentile, with the financial support of Giovanni Treccani, relaunched an earlier project of the *Società Italiana per il Progresso delle Scienze* for a national

---

[148] See [44, vol. I, Prefazione]: "dare alla teoria scientifica una base storica," "diverse circostanze minacciano oggi di menomare la scienza e la cultura matematica ... nella schiera di coloro che hanno l'alto compito di diffonderla nella scuola," "illuminare la ricerca più elevata e aprire anche il campo fruttifero dell'investigazione storica ad un più vasto numero di studiosi."

[149] See "Indice generale Serie IV – Volumi I a X – Anni MCMXX-MCMXXX", *Periodico di matematiche*, (4) 11, 1931, pp. 3–21: "Nessuna rivista dello stesso genere, in nessun paese del mondo, ha saputo realizzare un programma così alto e intonato alle esigenze formative e culturali dei docenti delle scuole medie."

[150] See [43, p. 3]: "Né quest'opera di divulgazione e di propaganda deve apparirvi superflua ... E, se anche la necessità della propaganda ci distragga per alcun tempo da altro utile lavoro, non dobbiamo rammaricarcene, perché, rinfrancando in noi la fede scientifica e ricreandone il bisogno nella società circostante, prepariamo pure al progresso della scienza più giovani e balde energie."

encyclopaedia. After Volterra refused to collaborate, Enriques joined the enterprise, and enthusiastically accepted the direction of the scientific part, with the help of Fermi and Amaldi. The significance of his contribution and his relationship with Gentile have been the subjects of recent studies (see [5–7]). Here we will only underline the fact that it is above all the mathematical entries (see [3]) that reflect Enriques' cultural project and his vision of mathematics, where the theoretical results are seen in connection to the problems from which they originated, even outside of the specific disciplinary field, and are closely related to aspects of history, epistemology and, of more specific interest to us here, education. This is already evident in the guidelines sent by Enriques to the collaborators, the salient points of which are: address both mathematicians and non-mathematicians; present fundamental problems, shedding light on their scientific and philosophical significance and the mutual connections; use history to illustrate the development of ideas and to connect mathematics with other aspects of culture; avoid too minute technical details; develop the elementary questions in greater depth with respect to the more advanced questions because they are of interest to a wider public; reduce symbolism to a minimum.[151] Emblematic of this point of view is the entry "Matematica" [51], where, among others, there is a paragraph specifically dedicated to teaching: the approach to the topic is historical, and Enriques manages to reaffirm the educational value of mathematics, which

> reveals itself not only in the elevation and strengthening of those minds which, by means of classical instruction, want to prepare themselves for more advanced studies, but also in the early grades of education of children and the working classes.[152]

Here, Enriques also makes explicit reference to the renowned pedagogists Pestalozzi and Fröbel,[153] attributing to them above all the merit of having introduced mathematics into the education of children as an important element for their intellectual development. Instead, in spite of evident points of contact, no reference is made to Adolphe Ferrière, the father of the "école active," nor to Ovide Decroly, both of whom were well known in Italy.

In all, from 1925 to 1935 Enriques wrote a total of thirty-eight entries for the *Enciclopedia*.[154]

All of these activities were brought to a sudden halt after the enactment of the racial laws in 1938. Nevertheless, Enriques continued his involvement as far as possible, writing in the *Periodico* under a pseudonym, and giving courses in

---

[151] See "Norme per la collaborazione dei matematici all'Enciclopedia Italiana," *Periodico di matematiche*, (4) 6, 1926, pp. 46–47.

[152] See [51, p. 553]: "...si palesa non soltanto nell'elevamento e nel potenziamento delle intelligenze che, attraverso l'istruzione classica, vogliono abilitarsi ai più alti studi, bensì anche nei primi gradi di educazione dell'infanzia e delle classi popolari."

[153] Enriques had already cited the two pedagogists in [25, p. 26] and would cite them again in [53, p. 185].

[154] See the list in [6, pp. 129–130].

geometry and history of mathematics in the so-called clandestine university in Rome. This university had been organised by Castelnuovo beginning in 1941 and continued under his direction until 1943 to offer courses to enable Jewish students who had been banned from the Italian university to take the examinations at the *Institut Technique Supérieur* in Fribourg (see [12]). One of his students recalled:

> The course that [Enriques] gave in the history of mathematics was a memorable event, which drew not only students of engineering. This handsome old man, this fascinating gentleman ... spoke with the soft and direct voice of the great persuaders. He guided the listeners to the limpid comprehension of complex relations, to the identification of connections never dreamed of.[155]

After the Liberation in 1944, as Castelnuovo wrote, Enriques "resumed teaching, but his body was by then worn out, and He no longer had the strength to take up fighting positions" [14, p. 12],[156] but he never abandoned his interest in educational questions. He was one of the supporters of the *Instituto Romano di Cultura Matematica* founded at the beginning of 1945 by Tullio Viola and Emma Castelnuovo to foster discussion of educational problems and teacher training, and he gave two lectures on topics that were dear to him: the significance of mathematics in general culture, and the significance of mathematics for physics (see [107]). As Emma Castelnuovo recalled, he also organised meetings in his own home for students and teachers, with the aim of improving teaching of geometry in secondary schools:

> ... in addition to these meetings [of the Roman Institute for Mathematical Culture] of about a hundred people, there were also small meetings at the home of the mathematician Enriques, we were about eight or ten at most. Enriques had proposed to study the books ... of 1700 − 1800 of elementary geometry in order to have an idea of how the school courses could be modified by moving away from Euclid.[157]

It was in one of these meetings that Emma learned about the 1741 *Eléments de Géométrie* by Alexis Clairaut, which led her to change her way of teaching by introducing the active method of teaching intuitive geometry: "In a single stroke I change," she writes, "the class changes in my hands."[158]

In those years, Enriques also influenced Carleton Washburne, who had been sent to Italy by the United States in the summer of 1943 to eliminate all traces

---

[155] See [19, p. 96]: "Il corso che tenne di storia delle matematiche fu un memorabile avvenimento, che richiamò non soltanto gli studenti d'ingegneria. Il bel vecchio, l'affascinante signore ... parlava con la voce piana e diritta dei grandi persuasori. Conduceva gli ascoltatori alla comprensione limpida di relazioni complesse, all'individuazione di nessi mai sospettati."

[156] See [14, p. 124]: "... riprese l'insegnamento, ma l'organismo era ormai stanco ed Egli non sentiva più la forza di assumere posti di combattimento."

[157] See [13, p. 25]: "... oltre a queste riunioni di circa cento persone [dell'Istituto Romano di Cultura matematica] c'erano riunioni in piccolo, a casa del matematico Enriques, eravamo 8,10 al massimo. Enriques aveva proposto di studiare dei libri ... del 1700–1800, di geometria elementare per avere un'idea di come si poteva forse modificare il corso allontanandosi da Euclide."

[158] See [13, p. 26]: "Di colpo cambio. ... la classe mi cambia fra le mani."

of Fascist propaganda from the schools and to begin the process of democratising the country. A well known pedagogist who had created the "Winnetka School" and a supporter of the active method of teaching, Washburne was the director of the Allied Forces Education Review Board of the Allied Control Commission in Italy which, with the help of a subcommittee of Italian experts, produced new programs for elementary and secondary schools and for the *Istituti magistrali* (primary schoolteacher training schools).[159] The methodologies that inspired the mathematics programs of the Allied Commission reflect Enriques' influence: the new programs stressed the importance of a teaching that is intuitive-dynamic in close connection with the historic process, and invited teachers to pay greater attention to the psychological needs of the students.

Enriques died suddenly in Rome on 14 June 1946. Up to the end he was involved in teacher training, which he believed to be the crucial element for the formation of good schools and one of the channels for achieving his cultural project. In his own words:

> These ideas were defended by us, even with battles, in the social science area of scientific institutions and in the ordering of studies; and we have not given up hope that they are about to leave some fertile seeds.[160]

Acknowledgements I am very grateful to Aldo Brigaglia, Philippe Nabonnand, Pietro Nastasi, Ornella Pompeo Faracovi and Gabriele Lolli for reading the text and making suggestions. I also want to express my gratitude to Paolo Bussotti, Cristina Chersoni (*Archivio storico dell'Università di Bologna*), Gian Paolo Brizzi and Carla Onesti (*Archivio storico dell'Università di Roma*), Helmut Rohlfing (*Niedersächsische Staats-und Universitätsbibliothek Göttingen*), Sandra Linguerri, Emma Sallent and Gert Schubring for their help. My most sincere thanks go to Kim Williams, who with intelligence and professionalism translated the text. Thank you also to the personnel of the *Biblioteca "Giuseppe Peano"* of the Department of Mathematics of the University of Torino for invaluable help in bibliographic research.

---

[159] See [126]. See also the Decree of 9 February 1945 [BUMPI I 1945.1, pp. 253–313]; the *Piano di Studi per gli Istituti magistrali superiori 1944–1945*. Roma: Signorelli, 1945; and Commissione Alleata in Italia (Sotto-Commissione dell'Educazione), *La politica e la legislazione scolastica in Italia dal 1922 al 1943 con cenni introduttivi sui periodi precedenti e una parte conclusiva sul periodo postfascista* (Milan: Garzanti, 1947), in particular pp. 382–386.

[160] See [41, p. 287]: "Queste idee sono state sostenute da noi, anche con battaglie, nel campo sociale delle istituzioni scientifiche e dell'ordinamento degli studi; e non abbiamo perduto la speranza che esse sieno per lasciare qualche seme fruttifero."

Appendix 1: *Class Registers* [161]

1. ASUB, *Enriques prof. Federigo. Registri delle lezioni: Corso libero di Filosofia delle Scienze. Programma per l'anno 1902–1903*
Introduzione – Rapporti della Filosofia colle scienze fisico-matematiche da una parte e colle scienze biologiche dall'altra
I problemi filosofici attinenti ai principii della geometria
Questioni pedagogiche che ne dipendono
I problemi filosofici attinenti ai principii della meccanica
Questioni pedagogiche
Maggio 1902
Federigo Enriques

2. ASUR, Facoltà di Scienze Matematiche Fisiche e Naturali, *Libretti delle lezioni: Libretto delle lezioni di Matematica Complementare dettate dal Sig. Prof. Enriques Federigo nell'anno scolastico 1922–1923*
16.11.1922 Gli Elementi di Euclide
18.11.1922 Sulle origini della geometria greca: i pitagorici
21.11.1922 Critica eleatica
23.11.1922 Segue: origini dell'analisi infinitesimale
25.11.1922 Def.^{i} assiomi e postulati in Euclide
28.11.1922 Concetti primitivi e post.^{i} nella geom. moderna
30.11.1922 Analisi di Pasch dei primi post.^{i} della geom. piana
02.12.1922 Segue
05.12.1922 I numeri naturali
07.12.1922 I numeri fratti e negativi
09.12.1922 Non fatta per chiusura dell'Università
12.12.1922 Numeri irrazionali: potenza del continuo
14.12.1922 Numeri non archimedei
16.12.1922 Varie forme del post. della continuità
19.12.1922 Applicazioni elem.^{i} del post. di continuità
21.12.1922 Segue: intersez.^{i} di rette e circoli
11.01.1923 Sviluppo storico della geom. proiettiva
13.01.1923 Teoria fondamentale della pr.
16.01.1923 Eq.^{ne} funzionale di Darboux
18.01.1923 Omografie piane punti uniti
20.01.1923 Om.^{e} particolari metriche del piano
23.01.1923 Omogr.^{e} nello spazio
25.01.1923 Rappresentazione delle quadriche
27.01.1923 Class.^{e} p.^{ti} uniti omogr.^{e} spaziali
30.01.1923 Cubica gobba e om.^{e} con p.^{ti} un.^{i} multipli

---

[161] All transcriptions are by L. Giacardi unless otherwise noted.

01.02.1923 Movimenti dello spazio: traiettorie dei [...] eliche.
03.02.1923 Sup. con ∞^2 movimenti. L'immaginario: introduzione
06.02.1923 Eq. di gr. n e rad. complesse
10.02.1923 L'imm.° e la teoria delle funz., condiz.^{e} di monogeneità
13.02.1923 Segue: integrali e teor. di Cauchy
15.02.1923 Sviluppi in serie di potenze
17.02.1923 Principio di cont. e immaginario in geometria
20.02.1923 Segue: l'imm.° nella teoria delle coniche
22.02.1923 Discussione sul sistema delle coniche omofocali
24.02.1923 Linee di lunghezza nulla
29.02.1923 Applic.^{i}: teor. Beltrami trasf. conformi nello spazio
01.03.1923 Principio di dualità e di trasporto logico
03.03.1923 Introd.^{e} coor. proiettive. Trasf.^{i} che mutano sfere in sfere
06.03.1923 Trasf.^{e} quadratica
08.03.1923 Post. d'Euclide sulle parallele
10.03.1923 Teor. di Saccheri-Legendre ecc.
13.03.1923 Principii di Geom. non euclidea
15.03.1923 Interpretaz. su sup. a curv. cost. neg. e in p.^{no} rispetto a conica
17.03.1923 Cerchi, ipercicli, oricicli in geom. non-euclidea
20.03.1923 Non fatta per funerale prof. Semeraro
22.03.1923 Sul valore fisico della geom. non euclidea
24.03.1923 Segue: cenni sulla teoria della relatività
12.04.1923 Geometria ellittica
14.04.1923 Principio di corr. sulle rette
17.04.1923 Involuzioni [...] teor. di Lüroth.
19.04.1923 Il birapporto e l'inv. ass. F
24.04.1923 Gruppi finiti di proiett.: analisi di Klein
26.04.1923 Segue
28.04.1923 Non fatta perché aula occupata per libera docenza
01.05.1923 Poliedri regolari
03.05.1923 Le coniche e le prop. focali elementarmente
05.05.1923 Non fatta per esami scritti di cultura per lauree miste
08.05.1923 Sezioni circolari del cono quadrico
12.05.1923 Critica comparativa dei testi di geometria
15.05.1923 Segue: criteri di ug.^{a} dei Δ
17.05.1923 Segue: equivalenza
19.05.1923 Non fatta per motivi personali
22.05.1923 Segue: teor. di Pitagora
24.05.1923 Schiarimenti riassuntivi
26.05.1923 Segue
29.05.1923 Segue
09.06.1923 Schiarimenti
14.05.1923 Schiarimenti
Visto: Il Preside

3. ASUR, Facoltà di Scienze Matematiche Fisiche e Naturali, *Libretti delle lezioni*: *Libretto delle lezioni di Esercitazioni di Matematiche Complementari dettate dal Sig. Prof. Enriques Federigo nell'anno scolastico 1922–1923*

21.11.1922 Indicazioni bibliografiche
23.11.1922 Proporzioni
25.11.1922 Uguaglianza dei triangoli
28.11.1922 Eq.[i] di 2° gr.° in Euclide
30.11.1922 Volume della sfera
2.12.1922 Segue
5.12.1922 Costruzione del triang.° date le mediane e le altezze
7.12.1922 Impossibilità di costruire in generale il triang. date le bisettrici
9.12.1922 Non fatta per chiusura dell'Un[à].
11.12.1922 Segue: discussione dei problemi di 2° grado
14.12.1922 Sviluppi in serie delle funzioni
16.12.1922 Segue: campo di convergenza
19.12.1922 Caduta dei gravi
21.12.1922 Frazioni continue
11.01.1923 Irraz.[i] quadratici e fraz.[i] cont.[e] periodiche
13.01.1923 Duplicazione del cubo
16.01.1923 Volume del tetraedro
18.01.1923 Geometria del compasso
20.011923 Costruzioni del tr. eq. e del pentagono regolare
23.01.1923 Numeri primi
25.01.1923 Poligoni equivalenti
27.01.1923 Costr.[i] 1° gr. con riga
30.01.1923 Probl. 1° grado con riga e cerchio fisso
01.02.1923 Analisi indeterminata di 1° gr.
03.02.1923 Geom. del compasso
06.02.1923 Trasf.[i] per raggi vettori reciproci
10.02.1923 Geom. sferica
13.02.1923 Segue: dualità
15.02.1923 Segue: confronto colla geom. euclidea
17.02.1923 Trigonometria sferica
20.02.1923 Sist.[i] di eq.[i] di 1° grado
22.02.1923 Angoli nel cerchio
24.02.1923 Logaritmi
27.02.1923 Prima lez.[ne] sulle equazioni
01.03.1923 Segue
03.03.1923 Interpolazioni
06.03.1923 Sist.[a] d'eq.[i] di 1° e 2° gr.
08.03.1923 Interpretazione di geom. analitica
10.03.1923 Frazioni decimali periodiche
13.03.1922 Eq.[ne] di 4° grado come resultante di due eq.[i] di 2° gr. con 2 incognite
15.03.1923 Eq.[ne] di 3° grado
17.03.1923 Segue: soluz.[ne] geometrica. Progressioni
20.03.1923 Non fatta per funerale Prof. Semeraro

27.03.1923 Sistemi di eq.^{i} di 2° grado risolubili con l'eq. di 2° gr.
24.03.1923 Ciclometria
12.04.1923 Massimi e minimi in algebra
14.04.1923 Segue
17.04.1923 Teoria elementare degli isoperimetri
19.04.1923 Sul resultante di due eq.^{i}
24.04.1923 Massimi e minimi in analisi
26.041923 Segue
28.04.1923 Segue: massimi e minimi di funz.^{i} di 2 e più var.^{i}
01.05.1923 Decimali illimitati
03.05.1923 Eq.^{i} alg.^{e} e funz.^{i} simm.^{e} delle radici
05.05.1923 Non fatta per lauree miste
08.05.1923 Funzioni simmetriche delle rad. d'un'eq.^{ne} nel piano complesso
12.05.1923 Numeri negativi
15.051923 Segue
17.05.1923 Numeri frazionari
19.05.1923 Non fatta per motivi personali
22.05.1923 Segue: num.^{i} frazionari, teorie sintetiche
24.05.1923 Esercitazioni riassuntive
26.05.1923 Segue
Visto: Il Preside

Appendix 2: *Letters to Felix Klein*

1. *SUB Göttingen. F. Klein 34, F. Enriques to F. Klein, Bologna 10 January 1905*
Illustre Sig. Professore
In risposta alla sua lettera Le ho spedito stamani i programmi ufficiali dei nostri Licei, Ginnasii e Istituti tecnici. Quello per i Licei e Ginnasii subisce in questi giorni un rinnovamento, ma il nuovo testo non è stato ancora pubblicato a parte. Quando vedrà la luce glie ne manderò una copia. La modificazione introdotta è assai profonda perché si tratta di rendere possibile la scelta agli studenti fra il Greco e la Matematica dal primo anno di Liceo in su.
Insieme ai programmi anzidetti Le invio una copia della 2ª ediz.^{e} // dei miei Elementi di Geometria. Nell'avviamento ad un metodo che, pur essendo razionale, accentua il carattere induttivo, Ella potrà riconoscere una influenza delle sue idee e delle conversazioni di Gottinga.
Aspetto dal Fleischer comunicazione delle osservazioni intorno al mio art. per l' Enciclopedia, osservazioni che terrò nel massimo conto.
Coi migliori e più distinti saluti, mi abbia per
Suo dev.mo
Federigo Enriques
Bologna 10/1/1905

2. SUB Göttingen. F. Klein 51, F. Enriques to F. Klein, Bologna 19 July 1920

Bologna viale Gozzadini 9: 19 Luglio 1920

Caro ed illustre professore,

col prossimo anno mi propongo di riprendere la pubblicazione di un Periodico di Matematiche diretto agli insegnanti secondari, a cui vorrei dare nuova vita, valendomene per promuovere la cultura dei detti insegnanti, specie col richiamare la connessione fra i campi più elevati delle matematiche e gli elementi, nonché dando sviluppo alle questioni storiche. Non ho bisogno di spiegare a Lei l'interesse ed anche la difficoltà di una tale impresa, che risponde proprio ad una delle vedute che Lei stesso ha fatto brillantemente valere con tanti modi diversi di operosità. Ma non Le dispiaccia che, ricordando appunto il Suo interesse per tali questioni, io venga a chiederle il dono della Sua collaborazione, ed il Suo prezioso consiglio.

Se Ella può collaborare alla nuova Rivista con un suo proprio scritto (che procureremo di volgere, nel miglior modo in italiano, e che – per tale motivo – vorrei pregare fosse scritto, possibilmente a macchina, o almeno con caratteri latini molto leggibili) questo sarà effettivamente un grosso regalo per i nostri lettori.

Oltre a ciò io Le sarei pur grato di additarmi questioni che, a suo avviso meriterebbero di attrarre l'attenzione del Periodico, ed anche il nome di qualche collaboratore che ritenga specialmente adatto, per un tale lavoro.

Ringraziandola in ogni caso per qualsiasi contributo che Ella voglia recare al disegnato Periodico mi abbia, coi migliori devoti saluti suo Federigo Enriques

2. SUB Göttingen. F. Klein 4A, F. Enriques to F. Klein, Bologna 18 January 1921

Bologna viale Gozzadini 9: 18 Gennaio 1921

Caro ed illustre professore,

ebbi a suo tempo la Sua lettera e La ringrazio dell'appoggio indiretto che Ella promette al Periodico, di cui ho ora il piacere di inviarle in omaggio la prima copia.

Per quello che concerne le condizioni dello spirito pubblico italiano verso la Germania, e segnatamente nel mondo della cultura, credo di poterle affermare che la grande maggioranza è favorevole alla migliore ripresa dei rapporti; il tempo vincerà le riluttanze di coloro che credono di essere ancora in stato di guerra ammesso come fatto, e non concesso secondo il mio sentimento, che la guerra delle nazioni debba estendersi al campo intellettuale!).

Inviandole ora, come ho detto, la prima copia – che ha appena veduto la luce – del periodico di matematiche, vorrei chiederle in pari tempo, se il nostro programma non le sembri tale che il Periodico stesso meriti di essere diffuso anche in Germania, presso le biblioteche delle scuole di magistero o quelle che sono alla portata degli insegnanti. Ma se così è, io non mi nascondo tuttavia la difficoltà che a questa diffusione crea l'altezza dei cambi. L'editore Zanichelli, tenuto conto del prezzo di trasporto ecc., cede la rivista all'estero per fr. (francesi) 20 all'anno, e questa somma, che probabilmente non varrà a compensare le spese della pubblicazione, riesce ora un po' alta se deve esser pagata // in marchi. In considerazione di ciò, ho ottenuto dal detto editore Zanichelli il consenso ad una proposta, che – se potrà attuarsi – avrà un carattere simpatico, come quella che tende a facilitare la ripresa degli scambi intellettuali dei nostri paesi. Il pagamento degli abbonamenti al Periodico di matematiche potrà essere fatto in libri (da scegliere dall'editore Zanichelli). Siccome ora i libri tedeschi, venduti all'estero a prezzi assai più alti che all'interno, vengono – per noi – a costare molto cari, sicché non trovano più quel largo mercato che ebbero in passato, e che – nell'interesse della cultura – dovrebbero nuovamente acquistare, sarebbe questo un mezzo assai atto a promuovere ciò che si ha in vista. Non è escluso poi che la cosa possa estendersi dal Periodico, anche ai libri italiani in cambio dei quali si accetterebbe sempre, come pagamento, libri anziché denaro.

Io sottopongo questa proposta all'editore Teubner. Ma se essa potesse avere il Lei un patrocinatore, e se Lei stesso credesse di raccomandare il Periodico di Matematiche a biblioteche ecc., la cosa avrebbe grande probabilità di riuscire.

Ringraziandola intanto La prego gradire l'espressione dei miei sentimenti devoti, Suo
Federigo Enriques[162]

3. *SUB Göttingen. F. Klein 8, F. Enriques to F. Klein, Bologna 1March 1921*

Bologna 1 marzo 1921

Caro e illustre Collega,
ebbi la gentilissima Sua e – soltanto ieri, dati i soliti ritardi postali – il libro inviato dall'editore Springer, che mi è giunto molto gradito. Mi è caro avere così sott'occhio la raccolta dei suoi lavori, iniziata con questo primo volume. Da parte mia molto volentieri ne parlerò in qualche rivista, ma sono in dubbio se io sia il recensore più adatto per scriverne nel Circolo di Palermo con quella diffusione che ivi è desiderabile, e specialmente per ciò che riguarda gli ultimi lavori sulla fisicomatematica che a Lei giustamente preme di vedere messi in rapporto colle antiche ricerche di geometria non euclidea. Nel caso dunque che veda la cosa riuscire meno facile per me (e dati i molteplici impegni che mi tolgono di dedicarvi eventualmente tutto il tempo necessario) resta inteso che io stesso cercherò chi si occupi della cosa, e frattanto – come ho detto – mi procurerò il piacere di fare un cenno più breve del libro sopra qualche altra rivista. In ogni modo temo che la pubblicazione nel Circolo non potrà essere tanto sollecita, perché le condizioni della stampa da noi sono ora difficili, e le riviste matematiche sovraccariche d'impegni; ma in proposito scriverò alla redazione.

---

[162]The following manuscript note by Klein appears in the margin of the letter: "Hrn. [Herrn] Koll.[egen] Krazer zur fr.[eundlichen] Kenntnisnahme, mit der Bitte um spätere Rücksendung. Die Sache fügt sich sehr gut in unsere allgemeinen Austauschpläne ein!", that is: "To my colleague Krazer for his kind consideration, with the request of a subsequent return. The matter fits perfectly into our general plans for an exchange!". The note means that Klein forwarded Enriques' letter, dated Bologna 18 January 1921, to Adolf Krazer (1858–1926) on 25 January 1921. This is made clear by Klein's note in the third line: "G.(öttingen) 25 I 21," which is the date he sent off Enriques' letter to Krazer. I thank Helmut Rohlfing for help in interpretating the note.

L'editore Zanichelli è in massima disposto all'accordo per organizzare uno scambio di libri tedeschi ed italiani, specie di matematica: attendiamo perciò le proposte del Geh. Krager (sic).
Mi abbia intanto cordialmente e devotamente Suo.
F. Enriques

4. *SUB Göttingen. F. Klein 8, F. Enriques to F. Klein, Bologna 25 June 1923*

Roma 25 giugno 1923

Caro e illustre Professore,
Le esprimo i più vivi ringraziamenti per l'invio del 3° Volume delle Sue Opere, che porta tanti concetti e risultati interessanti nel campo delle funzioni algebriche intese nel più vasto senso, e a cui conferisce mirabile unità di pensiero e singolare pregio l'insieme delle Sue note e spiegazioni.

Naturalmente anche di questo Volume, come dei due primi, sarà dato un cenno nel Periodico di Matematiche.

Mi è grata l'occasione per ricordarmi a Lei col sentimento di devota e reverente amicizia.
F. Enriques

Appendix 3: *Correspondence with Giacomo Furlani President of the Trieste Section of Mathesis Association*[163]

BDMIUT. Fondo Mathesis, Serie II "Carteggio"

1. *F. Enriques to G. Furlani, Bologna 8 July 1920*

Bologna 8 luglio 1920

Caro prof. Furlani,
ebbi la gentile cartolina che Ella mi ha inviato con Amaldi, e Le son grato del buon ricordo.

Ricevo ora i verbali e le relazioni della Sezione della Mathesis; e prima di tutto rivolgo un saluto e un ringraziamento ai colleghi del cessato consiglio direttivo della detta Sezione per l'opera da loro prestata, ed un cordiale saluto ai nuovi eletti: mentre mi compiaccio, in special modo, che Lei – caro amico e solerte presidente – abbia a continuare nell'ufficio tenuto per il bene della nostra società e della scuola.

Passo ora a rispondere ad alcune sue domande.

---

[163] My most sincere thanks to Luciana Zuccheri who provided me with copies of the letters. See the catalogue of the *Fondo Mathesis* of Trieste in *Animi divisi. Vicende dell'insegnamento della matematica nella Venezia Giulia dal 1918 al 1923*, L. Zuccheri, V. Zudini, eds. Trieste: EUT, pp. 47–71. The correspondence is entirely transcribed in [122, pp. 354–379].

1) Ho comunicato al Franchi (ditta Zanichelli) la sua lettera circa il testo del Battelli; ma egli dice di non poter sobbarcarsi al rifacimento da loro desiderato. Egli ritien dubbio che la cosa possa convenirgli in massima (poiché ha avuto l'avviso in contrario da fisici); ma – in ogni caso – si trova nella impossibilità pratica di realizzare la cosa, per quest'anno.
2) Quanto alle Nozioni di matematiche, compilate dall'Amaldi e da me, la ditta Zanichelli ha già risposto a scuole che gliene han domandato, di avere a disposizione le copie richieste. Ma, poiché par di comprendere che vi sarà una richiesta più larga, provvede ad una immediata ristampa.

Però a proposito di libri, mi consenta una preghiera. Se loro, nella Sezione vogliono discutere dei libri di testo, questa discussione non potrà che illuminare le questioni didattiche; ma non mi par compito della nostra società, o delle sue Sezioni, di raccomandare o meno l'adozione di dati libri di testo. So bene che voi avete fatto e fareste questo nel modo più elevato, per il puro interesse della scuola; pure non si può dimenticare che a tali questioni si legano interessi personali, sicché – almeno negli interessati – potrebbe ingenerarsi qualche malumore. Se, domani, una discussione di questo genere si porta in una Sezione dove si trova qualche autore di libro di testo, la cosa viene ad assumere un aspetto imbarazzante.

Aggiunga che, nel concetto italiano – secondo la tradizione – i proff. considerano che la scelta dei libri deve esser fatta con scelta individuale libera, e però non amano di vedere raccomandazioni aventi l'aspetto di inviti, più o meno officiosi.

Per tali motivi, La prego consentirmi di non pubblicare nel Bollettino l'elenco dei libri raccomandati: tanto più che, fra questi ce n'è anche uno mio.

Mi abbia, intanto, coi più cordiali saluti, Suo aff.mo
F. Enriques

2. *F. Enriques to G. Furlani, Torino 9 January 1920*

Torino, 9 Gennaio 1921

Carissimo Furlani,
ricevo qui la Sua lettera e mi compiaccio per l'acquisto dei nuovi soci. Ora Le raccomando il *Periodico*: è proprio necessario che la gran maggioranza dei soci si abbonino! Anche in questo caso l'editore ha preventivato una perdita.

Ha ricevuto il primo numero, che impressione ne ha riportato?

Può fare un pò di propaganda?

Ho scritto all'ufficio per le nuove province del Ministero, nel senso convenuto per la visita a Trieste e Trento (studio de visu delle scuole e anche conferenze): ho atteso per far ciò la rielezione del Cons. dir. in nome del quale è fatta la domanda. Lei può ora appoggiare la cosa presso gli amici dell'ufficio?

Saluti cordiali suo
F. Enriques

3. G. Furlani to F. Enriques, Trieste 23 January 1921

Trieste, 23 gennaio 1921

Carissimo e stimatissimo professore,
in risposta alla Sua gradita del 9 corr. confermo il ricevimento del periodico. Incaricandosi dell'incasso dei canoni pro 1921 sarà facile al C. D. della sezione di fare abbonati per il periodico.

Questo fu accolto qui con simpatia. Con piacere Le comunico impressioni e idee raccolte. Io lessi con vivo compiacimento il Suo articolo, dove ravviso delle direttive nell'insegnamento che sole possono contribuire alla diffusione estensiva della cultura matematica. Mentre per la maggior parte dei lettori delle nuove province sono interessanti gli articoli riguardanti le matematiche antiche o la critica dei fondamenti, mi pare che per quelli delle vecchie province sarebbero interessanti la trattazione di argomenti che si riferiscono all'insegnamento nelle scuole medie di capitoli della matematica moderna. Io eccitai più di uno dei miei colleghi a dare la propria collaborazione. Un desiderio espresso da qualcuno, che non mi pare di facile attuazione è questo: che fossero pubblicati degli articoli sintetici sugli ultimi progressi fatti dalla scienza matematica nei vari rami compilati da maggiori cultori specialisti in quei rami. Forse delle recensioni di pubblicazioni recenti fatte con queste intenzioni gioverebbero a questo bisogno.

O' raccomandato tosto in via amichevole a Roma quella domanda avanzata dalla Direz. della Mathesis

Vivissimi saluti, anche dai miei colleghi e famigliari

[Giacomo Furlani]

4. F. Enriques to G. Furlani, Roma 10 March

Roma 10 Marzo
Park Hotel

Caro Furlani,
ringrazio cordialmente Lei e il prof. Cermeli!

Per quel che riguarda la traduzione dei lavori del Boscovich, mi consulterò con Chisini nello spazio che avremo disponibile nel Periodico, ma temo che – al solito – non ne abbiamo molto; certo sarebbe interessante pubblicare lavori di questo genere, ma essi andrebbero a scapito di ciò che già dobbiamo, e non possiamo fornire nella misura in cui vorremmo ai lettori.

Però io sto pensando al modo di dar vita ad un *istituto per la storia delle scienze matematiche*; se la cosa (che è stata interrotta dalla crisi bancaria) mi riesce (e al momento solleciterò anche il vostro appoggio!) potremo pubblicare appunto, // e in primo luogo memorie e trattati di classici, specie italiani.

Saluti cordiali
dal suo
F. Enriques

5. G. Furlani to F. Enriques, Trieste 4 June 1923

Trieste, 4.VI.23

Chiarissimo prof Enriques,

Le notizie contenute nell'appello e implicite nel memoriale per il Ministero che sono stati inviati dalla Presidenza della "Mathesis" alle sezioni sono tali da suscitare un vivo allarme e richiedere la più pronta e vivace azione. Io avevo deciso quindi di convocare i soci della mia sezione e provocare tale azione. Senonché quasi contemporaneamente mi occorse di // leggere qua e là che il decreto sulla riforma era già bello e fatto e vi era fissato anche la distribuzione della materia. In tal caso piuttosto che fare una azione affrettata varrebbe meglio prepararla in modo più opportuno e fondarla su dati più precisi.

Considerate le difficoltà di radunare i soci, per non sciupar tempo in un'azione poco utile, mentre ò già provvisto a suscitar l'interessamento dei colleghi col diffondere le informazioni avute, vorrei ulteriori chiarimenti // per convocare la sezione. In particolare vorrei sapere se e quando sia stato pubblicato quel decreto oppure che cosa si conosca in particolare di preciso da cui partire per un'azione.

La ringrazio per la gent. cartolina.

[Giacomo Furlani]

6. F. Enriques to G. Furlani, Viareggio 10 August [1923]

Viareggio, Pens. Margherita al [...]
20 Agosto

Caro Furlani,

sarebbe bene che tutte le Sezioni fossero rappresentate al Congresso di Livorno (25 – 27 Sett) ove si discuteranno i problemi sollevati dalla riforma e si prenderanno accordi d' importanza pratica, anche per le disposizioni transitorie ecc. Verrà qualcuno di voi? Faccia propaganda e procuri // che i probabili congressisti inviino fin d'ora l'adesione al prof. Lazzeri (via Indipendenza 15)

Cordiali saluti

suo

F. Enriques

References

Abbreviations

ASUB: Archivio Storico dell'Università di Bologna
ASUR: Archivio Storico dell'Università di Roma
BUMPI: Bollettino Ufficiale del Ministero della Pubblica Istruzione
BDMIUT: Biblioteca del Dipartimento di Matematica e Informatica dell'Università di Trieste

GU: Gazzetta Ufficiale
SUB Göttingen: Niedersächsische Staats-und Universitätsbibliothek, Göttingen.

1. Avellone, M., A. Brigaglia and C. Zappulla. 2002. "The Foundations of Projective Geometry in Italy from De Paolis to Pieri." *Archive for History of Exact Sciences* 56, 5: 363–425.
2. Berzolari, L. and R. Bonola, 1909. "Sopra una Enciclopedia di Matematiche Elementari". *Atti del II° congresso della "Mathesis" Società Italiana di Matematica*, Allegato E, 1–5. Padova: Premiata Società Coperativa.
3. Biggiogero, G. 1933. "Le Matematiche nell'Enciclopedia Italiana". *Periodico di matematiche* 4, 13: 69–73.
4. Brigaglia, A. and C. Ciliberto. 1995. *Italian Algebraic Geometry between the two World Wars*. Queen's Papers in Pure and Applied Mathematics, vol. 100. Kingston, Ontario: Queen's University.
5. Bolondi, G. 1998. "Federigo Enriques e la sezione di Matematica dell'Enciclopedia italiana". pp. 117–159 in *Federigo Enriques. Filosofia e storia del pensiero scientifico*, O. Pompeo Faracovi and F. Speranza, eds. Livorno: Belforte.
6. Bolondi, G. 2002. "Periodico di Matematiche e Enciclopedia Italiana: tracce di un intreccio". pp. 121–133 in *La Mathesis. La prima metà del Novecento nella "Società Italiana di Scienze matematiche e fisiche"*. Note di Matematica, Storia, Cultura 5, G. Bolondi, ed. Milano: Springer.
7. Bolondi, G. 2004. "Enriques, Severi, l'Enciclopedia italiana e le istituzioni culturali". pp. 79–106 in *Enriques e Severi, matematici a confronto nella cultura del Novecento*, O. Pompeo Faracovi, ed. Sarzana: Agorà.
8. Bottazzini, U., A. Conte and P. Gario. 1996. *Riposte armonie. Lettere di Federigo Enriques a Guido Castelnuovo*. Torino: Boringhieri.
9. Campedelli, L. 1973. "Un cinquantenario: Federigo Enriques nell'insegnamento". pp. 75–90 in *Atti del convegno internazionale sul tema: storia, pedagogia e filosofia della scienza*. Rome: Accademia Nazionale dei Lincei.
10. Carbone, L., R. Gatto and F. Palladino 2001. *L'epistolario Cremona-Genocchi (1860–1886). La costituzione di una nuova figura di matematico nell'Italia unificata*. Firenze: Olschki.
11. Casini, P. 1973. "Premessa alla ristampa anastatica". pp. V–XVI in F. Enriques and G. De Santillana, *Compendio di storia del pensiero scientifico dall'antichità fino ai tempi moderni*. Bologna: Zanichelli.
12. Castelnuovo, E. 2001. "L'Università clandestina a Roma: anni 1941–1942 e 1942–1943". *Bollettino della Unione Matematica Italiana. La Matematica nella Società e nella Cultura* 8, 4 − A: 63–77.
13. Castelnuovo, E. 2007. *Insegnare matematica. Lectio magistralis (Roma, 15 marzo 2007)*. E. Peres and S. Serafini, ed. Rome: Iacobelli, 2008. See also: http://files.splinder.com/716025ca91a7d007d2f2b2551216ae2b.pdf. (last accessed 05 April 2011).
14. Castelnuovo, G. 1947. "Commemorazione del socio Federigo Enriques". *Atti della Accademia Nazionale dei Lincei. Rendiconti Cl. Sci. Fis. Mat. Nat* 8, 2: 3–21.
15. Chisini, O. 1947. "Accanto a Federigo Enriques". *Periodico di matematiche* 4, 25, 2: 117–123.
16. Cremona, L. 1885. Relazione, 15.3.1885. *Atti parlamentari. Senato del Regno* (N. 100–A). Rome: Forzoni, Tip. del Senato, pp. 1–90.
17. Dedò, M. 1982. "Federigo Enriques e la matematica elementare". pp. 251–263 in *Federigo Enriques. Approssimazione e verità* Pompeo Faracovi, O., ed. Livorno: Belforte Editore.
18. De Dominicis, S.F. 1882. "Le nostre università e le nostre scuole secondarie". *Rivista di filosofia scientifica* 2: 166–185.
19. Della Seta, F. 1996. *L'incendio del Tevere*. Udine: Paolo Gaspari editore.
20. Di Sieno, S. and M. Galuzzi. 1995. "La storia della matematica in Italia tra le due guerre mondiali ed il "Periodico di Matematiche", pp. 25–68 in *Aspetti della matematica italiana del Novecento*, L. Carbone and A. Guerraggio, eds. Naples: La Città del Sole.

21. Enriques, F. 1894. "Sui fondamenti della geometria proiettiva". *Rendiconti. R. Istituto Lombardo di Scienze e Lettere* 2, 27: 550–567 (rpt. *Memorie, I*, 141–157).
22. Enriques, F. 1894–1895. *Conferenze di geometria tenute nella R. Università di Bologna. Fondamenti di una geometria iperspaziale*. Lithograph. Bologna.
23. Enriques, F. 1895–1896. *Lezioni di Geometria proiettiva*. Lithograph. Bologna: R. Università.
24. Enriques, F. 1898. *Lezioni di Geometria proiettiva*. Bologna: Zanichelli.
25. Enriques, F. 1900. *Questioni riguardanti la geometria elementare*. Bologna: Zanichelli.
26. Enriques, F. 1901. "Sulla spiegazione psicologica dei postulati della geometria". *Rivista di Filosofia* 4: 171–195 (rpt. *Memorie, II*, 145–161).
27. Enriques, F. 1903. *Vorlesungen über projektive Geometrie*. Introduction (*Zur Einfürung*) by F. Klein. H. Fleisher, trans. Leipzig: Teubner. (2nd. ed. 1915).
28. Enriques, F. 1906. *Problemi della Scienza*. Bologna: Zanichelli.
29. Enriques, F. 1907. "Sulla preparazione degli insegnanti di Scienze". pp. 69–78 in *Quinto Congresso nazionale degli insegnanti delle scuole medie. Bologna, 25-26-27-28 settembre 1906, Atti*. Pistoia: Tip. Sinibuldiana.
30. Enriques, F. 1907–1910. *Fragen der Elementargeometrie*, 2 vols. Leipzig: Teubner. (German translation of *Questioni riguardanti la geometria elementare*; vol. *II*, 1907; vol. *I*, 1910).
31. Enriques, F. 1908a. "L'Università italiana. Critica degli ordinamenti in vigore". *Rivista di Scienza* 3: 133–147.
32. Enriques, F. 1908b. "La riforma dell'Università italiana". *Rivista di Scienza* 3: 362–372.
33. Enriques, F. 1911. "Esiste un sistema filosofico di Benedetto Croce?" *Rassegna contemporanea*, 4.6: 405–418.
34. Enriques, F. 1912a. *Scienza e razionalismo*. Bologna: Zanichelli.
35. Enriques, F. 1912b. "Sull'insegnamento della geometria razionale". pp. 19–35 in *Questioni riguardanti le matematiche elementari*. Bologna: Zanichelli.
36. Enriques, F. 1919. "Sul procedimento di riduzione all'assurdo". *Bollettino della Mathesis* 11: 6–14.
37. Enriques, F. 1920. "Il valore delle Matematiche nella Filosofia italica". *Bollettino della Mathesis* 12: 4–7.
38. Enriques, F. 1921a. "Insegnamento dinamico". *Periodico di Matematiche* 4, 1: 6–16.
39. Enriques, F. 1921b. (Review of) "C. Ciamberlini, *Saggi di didattica matematica* (Milano, Paravia, 1921)". *Periodico di Matematiche* 4, 1: 122–124.
40. Enriques, F. 1921c. "Ai lettori". *Periodico di matematiche*, 4, 1: 1–5.
41. Enriques, F. 1922a. *Per la storia della logica*. Bologna: Zanichelli.
42. Enriques, F. 1922b. "Evoluzione del concetto della Scienza nei pensatori matematici". *Periodico di Matematiche* 4, 2: 90–92.
43. Enriques, F. 1924. "Il significato umanistico della scienza nella cultura nazionale". *Periodico di Matematiche* 4, 4: 1–6.
44. Enriques, F. 1924–1927. *Questioni riguardanti le matematiche elementari. Raccolte e coordinate da Federigo Enriques*. Bologna: Zanichelli (Anastatic rpt. Bologna: Zanichelli 1983).
45. Enriques, F. 1925. *Gli Elementi d'Euclide e la critica antica e moderna (Libri I-IV)*. Rome: Alberto Stock.
46. Enriques, F. 1928. "La riforma Gentile e l'insegnamento della Matematica e della Fisica nella Scuola media". *Periodico di Matematiche* 4, 8: 68–73.
47. Enriques, F. 1929a. "La geometria non-euclidea e i presupposti filosofici della Teoria della Relatività". pp. 411–413 in *Atti della Società Italiana per il Progresso delle Scienze, Firenze, 18-25 Settembre 1929*, vol. *I*. Rome: S.I.P.S.
48. Enriques, F. 1929b. "Italia. Les modifications essentielles de l'enseignement mathématique dans les principaux pays depuis 1910". *L'Enseignement mathématique* 28: 13–18.
49. Enriques, F. 1930. "Assurdo". *Enciclopedia Italiana di Scienze, Lettere ed Arti*. Rome: Istituto della Enciclopedia Italiana, vol. 5, pp. 70–71.
50. Enriques, F. 1934a. "Prefazione" in Enriques, A. *Aritmetica ad uso delle scuole medie inferiori*. Bologna: Zanichelli

Enriques, F. 1934a. *Aritmetica ad uso delle scuole medie inferiori*. Bologna: Zanichelli.
51. Enriques, F. 1934b. "Matematica". *Enciclopedia Italiana di Scienze, Lettere ed Arti*. Rome: Istituto della Enciclopedia Italiana, vol. 22, pp. 547–554.
52. Enriques, F. 1936. *Il significato della storia del pensiero scientifico*. Bologna: Zanichelli.
53. Enriques, F. 1938a. *Le matematiche nella storia e nella cultura*. Bologna: Zanichelli.
54. Enriques, F. 1938b. "L'importanza della storia del pensiero scientifico nella cultura nazionale". *Scientia* 63: 125–134.
55. Enriques, F. [Adriano Giovannini, pseudonym]. 1942. "L'errore nelle matematiche". *Periodico di Matematiche* 4, 22: 57–65.
56. Enriques, F. 2003. *Insegnamento dinamico*. (accompanied by essays by F. Ghione and M. Moretti) Centro Studi Enriques. La Spezia: Agorà
57. Enriques, F. and U. Amaldi. 1903. *Elementi di geometria ad uso delle scuole secondarie superiori*. Bologna: Zanichelli.
58. Enriques, F. and U. Amaldi. 1914–1915. *Nozioni di matematica ad uso dei licei moderni*, 2 vols. Bologna: Zanichelli.
59. Enriques, F. and U. Amaldi. 1925. *Geometria elementare per le scuole secondarie superiori*, vol. 1. Bologna: Zanichelli.
60. Enriques, F. and U. Amaldi. 1926. *Geometria elementare per le scuole secondarie superiori*, vol. 2 Bologna: Zanichelli.
61. Enriques, F. and U. Amaldi. 1933. *Nozioni di geometria ad uso dei ginnasi inferiori*, 3rd ed. Bologna: Zanichelli.
62. Enriques, F. and U. Amaldi. 1992. *Elementi di geometria*. Pordenone: Studio Tesi.
63. Enriques, F. and O. Chisini. 1915–1934. *Lezioni sulla teoria geometrica delle equazioni e delle funzioni algebriche*, 4 vols. Bologna: Zanichelli.
64. Enriques, F. and U. Forti, eds. 1925. *I. Newton. Principii di Filosofia naturale, teoria della gravitazione*. Rome: Alberto Stock.
65. Enriques, F., A. Severi and A. Conti. 1903. "Estensione e limiti dell'insegnamento della matematica, in ciascuno dei due gradi, inferiore e superiore, delle Scuole Medie". *Il Bollettino di Matematica*, a. 2, n. 3-4: 50–56.
66. Fabietti, U. 1982. "Enriques, l'editoria e la Zanichelli". pp. 265–272 in *Federigo Enriques. Approssimazione e verità*, O. Pompeo Faracovi, ed. Livorno: Belforte.
67. Fano, G. 1894. "Sull'insegnamento della matematica nelle Università tedesche e in particolare nell'Università di Gottinga". *Rivista di matematica*, 4: 170–187.
68. Fano, G. 1922. "Le scuole di Magistero". *Periodico di matematiche*, s. 4, 2: 102–110.
69. Fehr, H. 1911. (Review of) *Fragen der Elementargeometrie*, vol. *I*. *L'Enseignement mathématique L'Enseignement mathématique* 13: 427–428.
70. Furinghetti, F. and L. Giacardi, eds. 2008–2011. The History of ICMI. http://www.icmihistory.unito.it/index.php.
71. Gario, P. 2006a. "Quali corsi per la formazione del docente di matematica? L'opera di Klein e la sua influenza in Italia". *Bollettino della Unione Matematica Italiana. La Matematica nella società e nella cultura* (8), *IX* − *A*: 131–141.
72. Gario P. 2006b, "Quali corsi per la formazione del docente di matematica? I congressi dei professori di matematica". *Bollettino della Unione Matematica Italiana. La Matematica nella società e nella cultura* (8), *IX* − *A*: 483–497.
73. Giacardi, L. 2003. "I manuali per l'insegnamento della geometria elementare in Italia fra Otto e Novecento". pp. XCVII–CXXIII in *TESEO, Tipografi e editori scolastico-educativi dell'Ottocento*. Milan: Editrice Bibliografica.
74. Giacardi, L. 2009. "The Italian contribution to the International Commission on Mathematical Instruction from its founding to the 1950s". pp. 47–64 in *Dig where you stand. Proceedings of the Conference on On-going Research in the History of Mathematics Education* (Garðabær, Iceland, June 21–23, 2009), K. Bjarnadóttir, F. Furinghetti, and G. Schubring, eds. Reykjavik: University of Iceland.
75. Giacardi, L. 2010. "The Italian School of Algebraic Geometry and Mathematics Teaching in Secondary Schools. Methodological Approaches, Institutional and Publishing Initiatives". International Journal for the History of Mathematics Education 5, 1: 1–19.

76. Giacardi, L., ed. 2005–2011. *Documents for the History of Mathematics Teaching in Italy*. http: //www.subalpinamathesis.unito.it/storiains/uk/documents.php
77. Giacardi, L. and R. Tazzioli. 2011. "Saggio introduttivo". In "Pel lustro della Scienza italiana e pel progresso dell'alto insegnamento". *Le lettere di Beltrami a Betti, Tardy e Gherardi*. In press.
78. Giacardi, L. 1999 (2001). "Matematica e humanitas scientifica Il progetto di rinnovamento della scuola di Giovanni Vailati". *Bollettino della Unione Matematica Italiana. La Matematica nella società e nella cultura* (8), 3 — A: 317–352.
79. Grossmann, M. 1909. (Review of) *Fragen der Elementargeometrie*, vol. *II*. *L'Enseignement mathématique* 11: 322.
80. Guerraggio, A. and P. Nastasi. 1993. *Gentile e i matematici italiani. Lettere 1907–1943*. Torino: Bollati Boringhieri.
81. Guerraggio, A. 1987. "Il pensiero matematico di Giovanni Vailati". pp. V–XXVII in Vol. *II* of G. Vailati, Scritti. 3 vols., M. Quaranta, ed. Sala Bolognese (Bo): Arnaldo Forni Editore.
82. Heiberg, J. L. 1924. *Matematiche, scienze naturali e medicina nell'antichità classica*, Gino Castelnuovo, trans. Rome: Alberto Stock.
83. Israel, G. 1984. "Le due vie della matematica italiana contemporanea". pp. 253–287 in *La ristrutturazione delle scienze tra le due guerre mondiali*. Vol. *I* : *L'Europa*. G. Battimelli, M. De Maria and A. Rossi, eds. Rome: La Goliardica.
84. Israel, G. 1992. "F. Enriques e il ruolo dell'intuizione nella geometria e nel suo insegnamento" pp. IX–XXI (Prefazione) in Enriques, F. and U. Amaldi, *Elementi di geometria*. Pordenone: Studio Tesi.
85. Israel, G., Nurzia L. 1989. "Fundamental trends and conflicts in Italian mathematics beetween the two world wars". *Archives internationales d'histoire des sciences*, 39: 111–143.
86. Klein, F. 1890. "Considerazioni comparative intorno a ricerche geometriche recenti", Gino Fano, trans. *Annali di matematica pura ed applicata*, 2, 17: 307–343.
87. Klein, F. 1894. "On the mathematical character of space-intuition and the relation of pure mathematics to the applied sciences (1893)". pp. 41–50 in *Lectures on mathematics delivered from Aug. 28 to Sept.9, 1893 . . . at Northwestern University Evanston, Ill. by F. Klein, reported by A. Ziwet*. New York, Macmillan and C.
88. Klein, F. 1896a. *Conferenze sopra alcune questioni di geometria elementare*, Francesco Giudice, trans. Torino: Rosenberg & Sellier.
89. Klein, F. 1896b. "Sullo spirito aritmetico nella matematica". *Rendiconti del Circolo matematico di Palermo*, 10: 107–117.
90. Klein, F. 1896c. "The Arithmetizing of Mathematics". Isabel Maddison, trans. *Bulletin of the American Mathematical Society* 2, 8: 241–249.
91. Klein, F. 1925–1933. *Elementarmathematik vom höheren Standpunkte aus*, I *Arithmetik, Algebra, Analysis*, II *Geometrie*, III *PräzisionsundApproximationsmathematik*. Berlin: Springer, 1925-1933 (1st ed. 1908–1909).
92. Lanaro, G., ed. 1971. *Giovanni Vailati, Epistolario 1891–1909*. Torino: Einaudi.
93. Linguerri, S. 2005. *La grande festa della scienza. Eugenio Rignano e Federigo Enriques. Lettere*. Milan: Franco Angeli.
94. Lombardo Radice, L. 1973. "Il confronto filosofico di Federigo Enriques con il neoidealismo". pp. 295–306 in *Atti del convegno internazionale sul tema storia, pedagogia e filosofia della scienza*. Rome: Accademia Nazionale dei Lincei.
95. Loria, G. 1921. "La preparazione degli insegnanti medi di matematica". *Periodico di Matematiche* 4, 1: 149–164.
96. Loria, G. and A. Padoa. 1909. "Preparazione degli insegnanti di matematica per le scuole medie". *Atti del II° congresso della "Mathesis" Società Italiana di Matematica*, Allegato A: 1–10. Padova: Premiata Società Coperativa.
97. Mayoux, J.-J. 1946. *L'Institut International de Coopération Intellectuelle, 1925–1946*. Paris: Edition de l'IICI.
98. Menghini, M. 1998. "Klein, Enriques, Libois: variations on the concept of invariant". *L'Educazione matematica* 5, 3: 159–181.

99. Metzger, H. 1935. "La Settimana della Scuola di Storia delle Scienze a Roma". *Archeion* 17: 203–212.
100. Moretti, M. 2003. "Insegnamento dinamico". Appunti sull'opera scolastica di Federigo Enriques (1900–1923). pp. 15–91 in Enriques, F. 2003. *Insegnamento dinamico*. Centro Studi Enriques. La Spezia: Agorà.
101. Nastasi, P. 2002. *"La Mathesis e il problema della formazione degli insegnanti"*. pp. 59–119 In *La Mathesis. La prima metà del Novecento nella "Società Italiana di Scienze matematiche e fisiche"*. Note di Matematica, Storia, Cultura 5, G. Bolondi, ed. Milano: Springer.
102. Nastasi, P. 2004. "Considerazioni tumultuarie su Federigo Enriques". pp. 79–204 in *Intorno a Enriques. Cinque conferenze*, L. M. Scarantino, ed. La Spezia: Agorà.
103. Nastasi, P. and E. Rogora. 2007. *Mon cher ami-Illustre professore. Corrispondenza di Ugo Amaldi (1897–1955)*. Rome: Nuova Cultura.
104. Nastasi, T. 2008. "Hélène Metzger: un'allieva ideale". pp. 121–162 in *Enriques e la cultura europea*, P. Bussotti, ed. Lugano: Lumières Internationales.
105. Nastasi, T. 2011. *Federigo Enriques e la civetta di Atena*. Centro Studi Enriques. Pisa: Edizioni Plus.
106. Nurzia, L. 1979. "Relazioni tra le concezioni geometriche di Federigo Enriques e la matematica intuizionista tedesca". *Physis* 21: 157–193.
107. Perna, A. 1950. "L'azione dell'Istituto Romano di Cultura Matematica a favore degli insegnanti secondari e neolaureati". *Archimede* 2: 36–40.
108. Pincherle, S. 1906. "Sulle riforme scolastiche da compiersi e in particolare su quelle relative all'insegnamento della matematica". *Bollettino di Matematica* 5, 5–6: 83–87.
109. Pompeo Faracovi, O., ed. 1982. *Federigo Enriques. Approssimazione e verità*. Livorno: Belforte.
110. Pompeo Faracovi, O. 1984. *Il caso Enriques: tradizione nazionale e cultura scientifica*. Livorno: Belforte.
111. Pompeo Faracovi, O., ed. 2004. *Enriques e Severi matematici a confronto nella cultura del Novecento*. Centro Studi Enriques. La Spezia: Agorà Edizioni.
112. Pompeo Faracovi, O. 2006. "Enriques, Gentile e la matematica". pp. 305–321 in *Da Casati a Gentile. Momenti di storia dell'insegnamento secondario della matematica in Italia*, L. Giacardi, ed. Centro Studi Enriques. La Spezia: Agorà Edizioni.
113. Pompeo Faracovi, O. and F. Speranza, eds. 1998. *Federigo Enriques. Filosofia e storia del pensiero scientifico*. Livorno: Belforte.
114. Pompeo Faracovi, O. 2010. "La filosofia scientifica tra Francia e Italia nel primo Novecento: un carteggio fra Enriques e Léon". pp. 95–116 in *Geometria, intuizione, esperienza* P. Bussotti, ed. Pisa: Edizioni Plus University Press.
115. Quilici, L. and R. Ragghianti 1989. "Il carteggio Xavier Léon: corrispondenti italiani. Con un'appendice di lettere di Georges Sorel". *Giornale critico della filosofia italiana* 6, 9: 295–368.
116. Scarantino, L. 2001. "Federigo Enriques e l'Istituto Internazionale di Cooperazione Intellettuale". pp. 45–52 In *Federigo Enriques, Matematiche e Filosofia. Lettere inedite. Bibliografia degli scritti* Pompeo Faracovi, O. and L. Scarantino, eds. Livorno: Belforte.
117. Schubring, G. 1989. "Pure and Applied Mathematics in Divergent Institutional Settings in Germany: the Role and Impact of Felix Klein". pp. 170–220 in *The History of Modern Mathematics*, D. Rowe, J. McCleary, eds. London: Academic Press, Vol. *II*.
118. Segre, C. 1891. "Su alcuni indirizzi nelle investigazioni geometriche. Osservazioni dirette ai miei studenti". *Rivista di Matematica* 1: 42–66. (Rpt. *Opere*, vol. *IV*, pp. 387–412.)
119. Segre, C. 1898. (Review of) "Lezioni di Geometria projettiva, Bologna, Zanichelli, 1898". *Bollettino di Bibliografia e Storia delle Scienze Matematiche* 1: 11–15.
120. Simili, R., ed. 1989. *Federigo Enriques. Filosofo e Scienziato*. Bologna: Cappelli.
121. Simili, R., ed. 2000. *Per la scienza. Scritti editi e inediti*. Napoli: Bibliopolis.
122. Tealdi, A. 2010. *Federigo Enriques e l'impegno nella scuola*. Thesis directed by L. Giacardi, University of Torino.

123. Tomasi, T. 1982. *La questione educativa nell'opera di Enriques.* pp. 223–250 In *Federigo Enriques. Approssimazione e verità* Pompeo Faracovi, O., ed. Livorno: Belforte Editore.
124. Vailati, G. 1904. (Review of) "F. Enriques e U. Amaldi. Elementi di Geometria ad uso delle Scuole secondarie superiori". *Bollettino di bibliografia e storia delle scienze matematiche* 7: 16–24.
125. Viola, T. 1963. "Didactique sans Euclide et pédagogie euclidienne". *L'Enseignement Mathématique* 9: 5–27.
126. Washburne, C. 1970. "La riorganizzazione dell'istruzione in Italia". *Scuola e Città*, 6–7: 273–277.
127. Zuccheri, L. and V. Zudini. 2010. "Discovering our history: A historical investigation into mathematics education". *International Journal for the History of Mathematics Education* 5, 1: 75–87.

Beppo Levi and Quantum Mechanics

Sandro Graffi

Abstract Beppo Levi, known as a pure mathematician because of his fundamental contributions to analysis and algebraic geometry, was also deeply interested in the development of theoretical physics of his times. His contributions to quantum mechanics are here reviewed.

On November 26, 1933, Beppo Levi gave at the monthly meeting of the Accademia delle Scienze dell'Istituto di Bologna the required yearly talk in his capacity of *Accademico Benedettino*. The subject he chose was a possible interpretation of wave mechanics. The written contribution appears as a paper entitled *Sulle equazioni della meccanica ondulatoria*.[1]

An attempt to understand its contents necessarily requires to recall some basic facts about wave mechanics.

1 Wave Mechanics and the Schrödinger Equation

Quantum mechanics can be formulated abstractly in Hilbert spaces.[2] Wave mechanics is its concrete representation most commonly used. It involves the *wave function*, which describes the amplitude of DeBroglie's matter wave associated to

---

[1] *On the equations of wave mechanics*. It appeared on the Memoirs of this Accademia, Serie IX, Tomo I, Sezione Fisiche e Matematiche, 1933–1934.

[2] The most classical references are the books of Dirac [1] in the language of the physicists, and of Von Neumann [6] in the language of the mathematicians.

S. Graffi (✉)
Dipartimento di Matematica, Università di Bologna, Italy
e-mail: graffi@dm.unibo.it

any particle, and depends on its classical coordinates, energy, time. The knowledge of the wave function determines all properties of the quantum system.

The *fundamental equation* of wave mechanics is the Schrödinger equation (published in January 1926), which is indeed the partial differential equation satisfied by the wave function.[3] His formal derivation went essentially as follows.

1.1 "Derivation" of the Schrödinger Equation

The starting point is represented by the fundamental idea of matter wave of L. DeBroglie (1923), based on relativistic invariance. According to Einstein's explanation of the photoelectric effect (1905), a particle of zero mass and energy $E = h\nu$ has to be associated to an electromagnetic wave of frequency ν. h is the Planck constant. On the other hand, energy E and momentum $\vec{p}$ of a particle form the components of the four-vector $(E/c, \vec{p})$ in Minkowski's space. Since $E/c = h\nu/c = h/\lambda$, relativistic invariance suggests, and this was DeBroglie's idea, the relation $|\vec{p}| = h/\lambda$, that is: a wave must be associated to any free particle of momentum $\vec{p}$; this is DeBroglie's *matter wave*, and its length should be

$$\lambda = \frac{h}{|\vec{p}|}.$$

h = Planck's constant, $\vec{p} = m\vec{v}$, $\vec{v}$ velocity of the particle.

Natural question: how to find the amplitude ψ of the matter wave?
Schrödinger's reply: exploit the analogy with optics.

The equation determining the amplitude ψ of a stationary light wave propagating through an optical medium of variabile refraction index $n(x, y, z)$ is indeed

$$\Delta\psi + n^2\psi = 0. \qquad (1)$$

Let us assume, with Schrödinger:

The particle moves along the line under the action of a conservative force of potential V.

Denoting q the coordinate of the particle, and E its energy, the energy conservation theorem $T + V = E$, where $T = \dfrac{m}{2}\dot{q}^2$ is the kinetic energy, can be written as

$$\frac{p^2}{2m} + V(q) = E.$$

---

[3] A standard reference for the origins of quantum mechanics is [7].

This yields
$$|p| = \sqrt{2m[E - V(q)]}.$$
Now recall that the refraction index is the reciprocal of the propagation speed of the wave in the medium, or, equivalently
$$n = \frac{2\pi}{\lambda}.$$
Hence, by DeBroglie's formula
$$n = \frac{2\pi |p|}{h} = \frac{2\pi \sqrt{2m(E-V)}}{h},$$
whence
$$\frac{d^2\psi}{dq^2} + \frac{4\pi^2 2m(E-V)}{h^2}\psi = 0,$$
and finally
$$-\frac{\hbar^2}{2m}\frac{d^2\psi}{dq^2} + V\psi = E\psi \qquad \hbar := \frac{h}{2\pi},$$
which is the one-dimensional *stationary* Schrödinger equation.

In three dimensions the same argument applies, up to the trivial variant of replacing p by $|\vec{p}|$, so that DeBroglie's wavelength is $\lambda = \frac{\hbar}{|\vec{p}|}$, and the conservation of energy yields
$$|\vec{p}| = \sqrt{2m[E - V(\vec{q})]}.$$
The three-dimensional Schrödinger equation is, therefore,
$$-\frac{\hbar^2}{2m}\Delta\psi(\vec{q}) + V\psi(\vec{q}) = E\psi(\vec{q}).$$
Formally, this last equation can be written as a spectral problem:
$$H\psi = E\psi, \qquad H = -\frac{\hbar^2}{2m}\Delta + V, \tag{2}$$
which becomes mathematically well posed provided H can be realized as a self-adjoint operator in the Hilbert space $L^2(\mathbb{R}^3)$.

Schrödinger was able to solve explicitly the above spectral problem in the following two cases

1. $V = \omega^2 q^2/2$ (harmonic oscillator of frequency ω);
2. $V = -\frac{Ze}{|\vec{q}|}$ (hydrogen-like atom of atomic number Z and electron charge e).

In both cases, the eigenvalues coincide with the energy levels experimentally observed. This result provided a very satisfactory description of the quantization phenomenon, namely the identification between quantized energy values and eigenvalues of a linear partial differential operator acting on $L^2(\mathbb{R}^3)$, the Schrödinger operator (2).[4]

This very simple recovery of the quantization phenomenon made wave mechanics immediately supersede the earlier but much more cumbersome formulation of quantum mechanics, the matrix mechanics of Heisenberg and Heisenberg, Born e Jordan (1925; see [7]).

1.2 What About Classical Mechanics?

A key question is obviously the relation of all this with classical mechanics. Here we limit ourselves to the following elementary, formal remarks: since $\vec{p} = m\vec{\dot{q}}$, the classical Hamiltonian $\mathcal{H} := T + V$ has the expression

$$\mathcal{H}(\vec{p},\vec{q}) = \frac{1}{2m}\vec{p}^{\,2} + V(\vec{q}).$$

Schrödinger remarked that his equation can be obtained from the conservation of energy theorem $H(\vec{p},\vec{q}) = E$ performing the formal substitution

$$\vec{p} \mapsto -i\hbar\nabla_q, \quad \vec{q} \mapsto \vec{q},$$

and letting the formal operator acting on $\psi(q)$.

This formal relation between the Schrödinger equation and the classical Hamiltonian tells nothing on the relation between wave mechanics and the classical trajectories. This relation is in general unclear, and in any case an *approximate* one.

Anyway, quantum mechanics should reduce to classical mechanics as $\hbar \to 0$. The standard way to study this transition is through the WKB method.[5] The essence of the method is as follows: one looks for solutions of the Schrödinger equation in the form

$$\psi(E,q) = \varphi(E,q)e^{iS(E,q)/\hbar},$$

---

[4]Standard references for the inclusion of Schrödinger operators in the framework of linear operators in Hilbert spaces, and in particular spectral and scattering theories, are [4] and [5].

[5]The initials stand for G. Wentzel, H.A. Kramers and L. Brillouin, who introduced in wave mechanics the well known Liouville–Green method to obtain approximate solutions of linear ODE with a small parameter.

where the *amplitude* $\varphi(E,q)$ and the *phase* $S(E,q)$ have to be determined. Inserting this form into the Schrödinger equation one obtains

$$(\nabla_q S(E,q))^2 + V(q) + \hbar R = E \qquad (3)$$

$$R := \left(\hbar \frac{\Delta \varphi}{\varphi} + 2 \frac{\nabla \varphi \cdot \nabla S}{\varphi} + \Delta S \right).$$

For $\hbar = 0$ (3) reduces to the classical Hamilton–Jacobi equation generated by $\mathcal{H}$. This is the starting point of a recusrive procedure which formally yields the wave function ψ as a power series in $\hbar$ with coefficients depending on q through the solution $S(E,q)$ of the Hamilton–Jacobi equation and its derivatives. As is well known, the mathematical implementation of this procedure is an important direction in mathematical research, closely related to microlocal analysis. A classical reference is [2].

2 The Contribution of Beppo Levi

Beppo Levi's approach to mathematical research lies well inside the italian tradition of Enrico Betti, Eugenio Beltrami, Vito Volterra, etc No mathematical subject was foreign to him; moreover, mathematics included much of what today would be considered mostly theoretical physics. Again within the italian tradition, as well as, for example, Tullio Levi-Civita, he was closely following also the development of the new quantum mechanics.
Actually, he wrote two monographies on the new mechanics.
The first is:

B. Levi, *Nuove teorie della Meccanica quantistica*,[6] Annuario Scientifico e Industriale – Anno LXIII (1926), Ed. Fratelli Treves, 1928

and the second:

B. Levi, *Fondamenti della logica e fondamenti della Meccanica quantistica*,[7] ibidem, Anno LXIV (1927), Ed. Fratelli Treves, 1929

To the best of my knowledge, those monographies are the first expositions of quantum mechanics ever published in Italy. The *Lezioni di meccanica ondulatoria*[8]

---

[6]*New theories of Quantum mechanics.*
[7]*Foundations of logic and foundations of quantum mechanics.*
[8]*Lectures on wave mechanics.*

by Enrico Persico,[9] based on the notes taken by Giulio Racah and Bruno Rossi[10] from his lectures at the University of Florence (known as "il Vangelo," the Gospel, in italian physics community of the time), appeared only in 1929.

The 1933 paper under examination here is therefore the product of a line of thought concerning wave mechanics going back to its very first formulation. It is interesting to begin with the reflections of Beppo Levi concerning the "derivation" of the Schrödinger equation.

I quote *verbatim*:

[..il..] fatto, a prima vista straordinario, che concordanze notevoli siansi potute ottenere tra l'esperienza fisica e una teoria più vicina alla divinazione che alla deduzione.[11]

This remark defines very clearly the purpose of Levi's investigation (again I quote *verbatim*):

introdurre un procedimento matematico non formale per riconciliare l'aspetto ondulatorio e l'aspetto corpuscolare della materia.[12]

Now as well as then, it is the main interpretation problem posed by quantum mechanics. Beppo Levi approach is along the following line of thought:

Cercare delle equazioni del moto deterministiche per il corpuscolo quantistico tali che vi si possa associare univocamente un moto di propagazione ondosa che soddisfi l'equazione di Schrödinger.[13]

which is made more precise in the following words:

Nel caso presente si tratta di associare alle equazioni differenziali ordinarie, che caratterizzano lo studio del moto secondo la meccanica classica, delle equazioni alle derivate parziali relative al comportamento di funzioni diffuse nello spazio, cui si possa attribuire l'apparenza ondosa.[14]

---

[9]Enrico Persico (Rome 1900–Rome 1969), Professor of Theoretical Physics in Florence (1927–1930), Turin (1930–1947), and in Rome since 1947.

[10]Giulio Racah (Florence 1909–Florence 1965) and Bruno Rossi (Venice 1905–Boston 1993) became world famous physicists. As well as Beppo Levi, both had to emigrate in 1938 because of the racial legislation. Racah became Professor at the Hebrew University in Jerusalem, and Rossi at the MIT in Boston.

[11]*[..a..] fact, extraordinary at first sight, that remarkable agreements could be obtained between physical experiments and a theory closer to divination than to deduction.*

[12] *Introduce a non-formal mathematical procedure aimed at reconcile wave and corpuscolar aspects of the matter.*

[13]*Look for deterministic equation of motion for the quantum corpuscle such that it is possible to associate with them a wave propagation motion satisfying Schrödinger's equation.*

[14]*In the present case we deal with associating to the ordinary differential equations, which characterize the study of the motion according to classical mechanics, some partial differential equations relative to the behaviour of functions smeared in space which admit the wave appearance.*

The basic idea for a similar construction is as follows:

Pare che a ciò si presti una inversione del punto di vista di Hugoniot e Hadamard nello studio dei moti ondosi di un mezzo continuo.[15]

In other words, the Hugoniot–Hadamard theory shows that in the framework of the hyperbolic partial differential equations describing wave motions in a continuous medium there is a natural system of ordinary differential equations: the bicharacteristic one, whose solutions are the rays, always orthogonal to the propagating wave fronts.

The inversion of this point of view amounts, as we will recall shortly, to the following procedure: one tries to immerse the Schrödinger equation into a class of hyperbolic equations, whose bicharacteristics should coincide with the classical motions.

If successful, this procedure would represent by far the simplest explanation of the wave-particle duality.

Beppo Levi proceeds in the following way (with little concern for mathematical rigour in this instance): let

$$\mathcal{H} = T + V,$$

the Hamiltonian function describing the particle motion *in classical mechanics*, depending on the canonical coordinates (p, q). He writes the corresponding equations of motion under the Jacobi form of the differential quotients:[16]

$$dt : dq_i : dp_i = 1 : \frac{\partial \mathcal{H}}{\partial p_i} : -\frac{\partial \mathcal{H}}{\partial q_i}. \qquad (4)$$

Let us again quote *verbatim*:

Introduciamo una nuova coordinata Q proporzionale all'azione, per modo che questo sistema differenziale può ancora scriversi[17]

$$dt : dq_i : dp_i : dQ = 1 : \frac{\partial \mathcal{H}}{\partial p_i} : -\frac{\partial \mathcal{H}}{\partial q_i} : 2T. \qquad (5)$$

As a matter of fact, one has by definition : $Q = \int 2T \, dt$.

---

[15]*To this purpose an inversion seems to be suited of the Hugoniot and Hadamard point of view in the study of wave motion in a continuous medium.*

[16]The notation used today is obviously

$$\frac{dq_i}{dt} = \frac{\partial \mathcal{H}}{\partial p_i}, \qquad \frac{dp_i}{dt} = -\frac{\partial \mathcal{H}}{\partial q_i}.$$

[17]*Let us introduce a new coordinate Q proportional to the action, in such a way that this differential system can be written also as.*

Here, Levi pauses to state his first conclusion:

Abbiamo così scritto il sistema delle linee caratteristiche dell'equazione alle derivate parziali del primo ordine nella funzione incognita Q delle variabili q_i, t [18]

$$F(q_i, p_i, t, \tau) = \mathcal{H} + \tau = 0, \quad (6)$$

dove si interpreti [19]

$$p_i = \frac{\partial Q}{\partial q_i}, \quad \tau = \frac{\partial Q}{\partial t}.$$

It is important to remark that (6) is the Hamilton–Jacobi generated by the Hamiltonian $\mathcal{H}$.

Then he goes on:

Si indichi allora con $\mathcal{T}$ l'operatore che si ottiene da T sostituendovi a p_i il simbolo operatorio $\frac{\partial}{\partial q_i}$, con Ψ un operatore differenziale del primo ordine rispetto alle variabili Q, t, q_i, arbitrario, [20]

$$\Psi = \Psi\left(\frac{\partial}{\partial Q}, \frac{\partial}{\partial t}, \frac{\partial}{\partial q_i}\right),$$

con $f(Q, t, q_i)$ una funzione arbitraria; sia infine Φ una funzione incognita delle medesime variabili. L'equazione alle derivate parziali del secondo ordine: [21]

$$\left(\mathcal{T} + \frac{\partial^2}{\partial Q \partial t} + V\frac{\partial^2}{\partial Q^2} + \Psi + f\right)\Phi = 0. \quad (7)$$

(remark the hyperbolicity!) *ha per equazione delle caratteristiche (6) e quindi per sistema differenziale delle bicaratteristiche le (5). Le (4) sono allora le equazioni differenziali delle projezioni nello spazio-tempo di queste bicaratteristiche. Poichè le (4) definiscono d'altronde le trajettorie possibili (e cioè non ancora determinate tramite le condizioni iniziali) del moto caratterizzato dalla funzione Hamiltoniana $\mathcal{H}$, si conclude che la connessione fra meccanica classica e meccanica*

---

[18] *We have thus written the system of the characteristic curves of the first order partial differential equation in the unknown function Q.*

[19] *under the interpretation.*

[20] *Denote then by $\mathcal{T}$ the operator obtained from T replacing p_i by the operator symbol $\frac{\partial}{\partial q_i}$, Ψ being an arbitrary first order differential operator with respect to the variables Q, t, q_i.*

[21] *the function $f(Q, t, q_i)$ being also arbitrary; let finally Φ be an unknown function of these variables. The second order partial differential equation.*

ondulatoria che colla equazione di Schrödinger era soltanto approssimata, e di approssimazione non valutabile, [recall (3)] *diventa, con la equazione (7), totalmente esatta.*[22]

Remark: so far Levi has constructed a completely general relation between the mechanical evolution of the particle (i.e., Hamilton's equations) and a wave propagation law, the semilinear hyperbolic equation (7). Still no link is established with the Schrödinger equation.

To include the Schrödinger equation, Levi's argument goes as follows: he selects Ψ as a linear homogeneous operator, with coefficients indipendent of Q, and assumes $f = f(q,t)$. He then looks for solutions in the form

$$\Phi = \varphi(q)e^{aQ+bt}, \quad a,b \quad \text{costants}. \tag{8}$$

One then finds

$$(\mathcal{T} + \Psi_0 + a^2 V + ab + f + af_1 + bf_2)\varphi(q) = 0, \tag{9}$$

where it has been set

$$\Psi := \Psi_0 + f_1 \frac{\partial}{\partial Q} + f_2 \frac{\partial}{\partial t}.$$

Let us quote *verbatim* once more:
Ponendo[23]:

$$b:a = \lambda, \quad \frac{1}{a^2}(f + af_1 + bf_2) = F,$$

questa equazione (9) diviene[24]

$$\left(\frac{1}{a^2}\mathcal{T} + \frac{1}{a^2}\Psi_0 + (V + F + \lambda)\right)\varphi(q) = 0, \tag{10}$$

---

[22] *has (6) as equation of the characteristics and hence (5) as differential system of the bicharacteristics. The equations (4) are then the differential equations for the space-time projections of these bicharacteristics.*

Since on the other hand the equations (4) define the <u>possible trajectories</u> (that is, still undetermined through the initial conditions) of the motion characterized by the Hamiltonian function H, the conclusion is that the connection between classical and wave mechanics, which as only approximate in the Schrödinger equation, with degree of approximation impossible to estimate, becomes exact by means of equation (7).

[23] *Setting.*

[24] *This equation becomes.*

e l'operatore $\frac{1}{a^2}\mathcal{T}$ si ottiene da T sostituendovi alle variabili p_i gli operatori differenziali $\frac{1}{a}\frac{\partial}{\partial q_i}$. Prendendo infine[25]

$$a = \frac{2\pi i}{h}, \qquad F = \Psi_0 = 0,$$

(10) diviene esattamente l'equazione di Schrödinger.[26]

Levi then proves that $a \in \mathbb{R}$, $b \in \mathbb{R}$ or $ia \in \mathbb{R}$, $ib \in \mathbb{R}$. This entails that λ in (10), which is eigenvalue of the Schrödinger equation if $F = \Psi = 0$, is real as it should be.

The program seems therefore to be completed.

However, Beppo Levi himself immediately raises the relevant question:

How much substance there is beyond these formal considerations?

Let us quote once more:

Non si potrebbe con questo affermare che l'equazione (7) non sia altro che una generalizzazione dell'equazione di Schrödinger dalla quale si passi a questa mediante una scelta conveniente di costanti e funzioni indeterminate (scelta che potrebbe essere anche quella consigliata dall'esperienza fisica); inquantoché è essenziale in (7), per la sua correlazione a (4), la presenza della variabile Q che in (10) scompare soltanto a causa della posizione (8), legame indispensabile fra (10), (7), (4).[27]

After some considerations on how to generalize his argument to the N-body case and to non-Euclidean metrics, Levi concludes:

Il significato fisico delle funzioni Φ e φ, come della f e dell'operatore lineare Ψ resta ancora del tutto sospeso; ebbene ciò è avvenuto <u>e avviene tuttora per la funzione d'onda di Schrödinger</u>; ma la presente discussione mostra che il fissare il significato fisico di queste funzioni od operatori è veramente la parte essenziale del problema; e mi pare renda conto del fatto, a prima vista straordinario, che concordanze notevoli siansi potute ottenere tra l'esperienza fisica e una teoria più vicina alla divinazione che alla deduzione.[28]

---

[25] *and the operator $\frac{1}{a^2}\mathcal{T}$ is obtained from T replacing the the variables p_i by the differential operators $\frac{1}{a}\frac{\partial}{\partial q_i}$. Taking finally.*

[26] *(10) becomes exactly Schrödinger's equation.*

[27] *It cannot be stated that (7) is nothing else than a generalization of the Schrödinger equation from which one goes to it through a convenient choice of constants and undetermined functions (choice which could also be the one suggested by physical experience); this is because it is essential in (7), by its correlation to (4), the occurrence of the variable Q which in (10) disappears only because of the position (8), unavaoidable link between (10), (7), (4).*

[28] *The physical meaning of the functions Φ and φ, as well as of f and of the linear operator Ψ, remains so far completely undecided; now this has happened <u>still happens for the wave function</u>*

A first important remark on the above considerations: Beppo Levi *does not accept* the statistical interpretation of the wave function introduced by Max Born in July 1926, and subsequently by Wolfgang Pauli in December in today's wording (see [7]; Beppo Levi describes this interpretation right after its appearance, in the Annuario of 1927 quoted above). This attitude is in complete agreement with his attempt at a purely wave formulation of quantum mechanics.

Let us now examine more closely the crucial point, namely the physical interpretation.

The whole construction is based on the possibility of treating Q and q as *independent variables.*

I believe that this independence cannot hold in this context.

Levi himself remarks indeed that the action Q must satisfy the Hamilton–Jacobi equation (8).

This equation defines Q as a function of q, q_0 and t:

$$Q = Q(q, q_0; t)$$

where q_0 describes the set of the initial conditions, and q the set of the corresponding configurations reached at time t. Once the action Q is known, the phase space trajectories generated by the Hamiltonian $\mathcal{H}$ are obtained (Jacobi's theorem) putting:

$$p_i = \frac{\partial Q}{\partial q_i}, \qquad p_{0,i} = -\frac{\partial Q}{\partial q_{0,i}}.$$

Inverting the second equation for q_i, and inserting in the first:

$$p_i = p_i(p_{0,i}, q_{0,i}; t), \qquad q_i = q_i(p_{0,i}, q_{0,i}; t).$$

If Q depends on q, setting always $F = \Psi_0 = 0$, $a = 2\pi i/h$, the position (8), once the computation is performed, reproduces exactly the (3) with $S = Q$, $E = -\lambda$.

Thus the connection between classical and wave mechanics becomes again an approximate one.

This consideration might account for the fact that the paper does not seem to have ever been noticed.

On the other hand, it has been otherwise determined that the propagation of DeBroglie's matter waves and the wave propagation in continuous media are different phenomena: the last one is indeed hyperbolic, while the first one is not.

---

of Schrödinger; but the present discussion shows that the determination of the physical meaning of these functions and operators actually is the essential part of the problem, and it seems to me that it accounts for the fact, extraordinary at first sight, that remarkable agreements could be obtained between physical experiments and a theory closer to divination than to deduction.

This is a manifestation of that aspect of quantum mechanics which, although experimentially verified beyond any possible doubt, is most repugnant to our intuition: the wave-particle duality.

It is mostly to account for this phenomenon that the standard formalism of quantum mechanics has been developed (Dirac and von Neumann). It is admittedly obscure and abstract, up to the point that according to many people it is impossible to understand (a famous quotation of Feynman).

To conclude, it seems to me that Beppo Levi is maybe the first of a long list of famous scientists who tried, with alternating success, to reconcile mathematically the wave aspect of quantum mechanics with the corpuscolar one. To mention only a few of them (for a review, see e.g. [3]) R. Furth, Wilhelm Weizel and more recently Edward Nelson advocated the stochastic interpretation, Erwin Madelung the fluid mechanics analogy, Louis DeBroglie the hidden variable theory, David Bohm the Bohmian particle. The idea of Beppo Levi looks very natural and mathematically elegant, but at a closer examination does not appear to really provide a viable solution to the problem.

References

1. Dirac, P.A.M. 1930. *Quantum mechanics*, Oxford University Press: Oxford (London /Melbourne).
2. Fedoryuk, M.V., and V.P. Maslov. 1981. *Semiclassical approximations in quantum mechanics*. D. Reidel Kluwer Publishing, Dordercht.
3. Jackiw, R., and D. Kleppner. 2000. *One hundred years of quantum physics*. Science (2 August 2000) 289: 893–924.
4. Kato, T. 1966–1976. *Perturbation theory of linear operators*. Springer: Berlin.
5. Reed, M., and B. Simon. 1972–1978. *Methods of modern mathematical physics*. Voll. I-IV, Academic Press: New York.
6. Von Neumann, J. 1955. *Mathematical foundations of quantum mechanics*. Princeton University Press: Princeton, NJ.
7. Van der Waerden, B.L. 1966. *Sources of quantum mechanics*. Dover, New York.

Leonida Tonelli: A Biography

Angelo Guerraggio and Pietro Nastasi

Abstract This paper aims at going through the work by Leonida Tonelli, with a focus on the results he got in the fields of Real Analysis and Calculus of Variations, and put it within the socio-political context of his time. One of the most outstanding Italian analysts in the first half of twentieth century, first in Bologna and then in Pisa and Rome, Leonida Tonelli's path combined elements of different, conflictual in some moment, relationships which featured at that time Italian mathematics and which linked the latter with Giovanni Gentile, the Fascism and the after-war "new" Italy.

1 Introduction

Leonida Tonelli is a central figure for those who intend to deal with the developments of Analysis in Italy and the dynamics of the scientific institutions in the first half of the twentieth century. He is a renowned mathematician, not only at national level. In addition to the general acknowledgements for his scientific career and the results he achieved in the Calculus of Variations, in Real Analysis and in the study of trigonometric series, there are also the international acknowledgements for his speeches at the International Congresses of Toronto [1924], Bologna [1928]

A. Guerraggio (✉)
Università Bocconi, Via Sarfatti, 25, 20136, Milano
e-mail: angelo.guerraggio@unibocconi.it

P. Nastasi
Università di Palermo, Via F. Sacchetti 7, 00137, Roma
e-mail: pgnastasi@libero.it

and Zurich [1932],[1] his affiliation with the "Moscow Mathematical Society" and the "Calcutta Mathematical Society," the publication of numerous articles for the *Bulletin of the American Mathematical Society*, the *Mathematische Annalen*, the *Acta Mathematica*, etc. These are all events that moreover refer to a period in which Italian Mathematics, for a number of reasons that we will also analyse in this essay, certainly did not shine in terms of intensity or interest in developing international relationships.[2]

The sources that we used to reconstruct Tonelli's figure and scientific career were mainly the biographies written on the occasion of certain academic commemorations.[3] This material was ordered, reviewed and analyzed in depth with further historical analyses that relied on the consultation of collections of letters and conversations with other mathematicians, who had been his students or knew him in some way. The collections of letters used were those of Tullio Levi-Civita and Vito Volterra, both preserved at the Accademia dei Lincei; those of Giovanni Vacca, which we were able to look through thanks to the courtesy of his son, Roberto; and the work of Giovanni Gentile, also in Rome, conserved at the Foundation named after him and housed in the Faculty of Letters and Philosophy.[4]

2 From Bologna to Bologna

Leonida Tonelli was born in Gallipoli (in the province of Lecce) on 19 April 1885. He spent his childhood in various cities: Gemona, Pavia and Pesaro, where he attended the technical Institute (the only high school with specialization in classical studies, which allowed enrolment in the college preparation program in Mathematics).We do not have significant testimonies from his early years, only one minor episode that nonetheless marks the beginning of his social and political undertakings.The last years of the century were particularly lively in Italy, in terms of social conflicts, with numerous demonstrations of popular discontent and acts of repression by the army and police. Precisely in Pesaro, Tonelli, who back then

---

[1] At the congress of Toronto, Tonelli held a speech on the Calculus of variations, a theme that he also discussed at the Congress of Zurich, while the plenary *lecture* of Bologna was dedicated to the "Italian contribution to the theory of functions of real variables." A second *plenary lecture* was foreseen at the Congress of Cambridge (Mass.) in September 1940 but, as known, that congress was postponed due to the breaking out of the Second World War.

[2] For an overall view on the Italian Mathematics in the period between the two wars, see [15].

[3] For the bibliography and an overall description of Tonelli's scientific production, see [27] S. Cinquini, "Della vita e delle Opere di Leonida Tonelli" (in [32, I]) ; L. Cesari, L'opera di Leonida Tonelli e la sua influenza nel pensiero scientifico del secolo (in [3, pages 41–73]); A. Faedo, Leonida Tonelli e la scuola matematica pisana (in [4, pages 21–41]); E. De Giorgi, Il calcolo delle variazioni origini antiche e prospettive furure (in [4, pages 59–68]); A. Guerraggio, L'Analisi, in [9, pages 1–158], particularly, about Tonelli and his "school," pages 13–43).

[4] The letters of mathematicians to G. Gentile are published in [14].

was still a student, gave a speech about the unrest, during a "conference" in a small socialist-leaning circle.

In 1902, Tonelli enrolled at the University of Bologna to attend the courses of Mathematics.[5] In Bologna, his socialist militancy grew and he partook in founding the "Gruppo Socialista Universitario." Later he joined the "Syndicalist group," for which he wrote a pamphlet entitled "Le nostre premesse."

The degree earned in the summer of 1907 was preceded by only a few weeks prior (in July) by an apparently minor episode that nonetheless bore important consequences: Leonida contracted a serious form of typhus during a walk in the hills on Bologna. His thesis was discussed with Cesare Arzelà (1847–1912) and published the following year in *Annali di Matematica*, with title: "I polinomi d'approssimazione di Tchebychev." The topic was chosen by Tonelli himself and then it was obviously approved by Arzelà who put him in contact with Hilbert and Fréchet. These men had already dealt with some aspects of the topic. Tonelli joined the research – typical of the period straddling the nineteenth and twentieth centuries – that aimed to generalize known results to N-dimensional spaces, for the functions of a variable. He succeeded in extending the procedures used by Tchebychev to the functions of two variables to approximate a function through assigned degree polynomials; in this case, however, unlike the functions of one variable, the uniqueness of the approximating polynomial was lost. Immediately after graduating, Tonelli became the assistant of Salvatore Pincherle (1853–1936) for Algebra and Analytical Geometry. In 1910, Pincherle, despite the suboptimal relationship he had with Arzelá, agreed to have Arzelá (who was ill) replaced by Tonelli in his lectures of Infinitesimal Analysis and High Analysis. During the winter of 1910–1911, Tonelli was also a Mathematics substitute teacher at the High School and in just four years he garnered full acknowledgment for his scientific skills. In 1910, he became a lecturer in Infinitesimal Analysis and the following year he was given the Chair of Infinitesimal Analysis at the University of Parma.

At that time, Tonelli had already published 16 works that already contained what would become the main themes of his research. Besides the predominant contents of algebraic problems there were also Notes on the theory of analytical functions, clearly inspired by Pincherle. More important in the Notes were his contributions of Real Analysis that are still remembered today in association with his name. After drawing his first inspirations from the problems of the relationships between derivation and integration,[6] in his Note "Sulla rettificazione delle curve" published in 1908 in the *Atti dell'Accademia delle Scienze* di Torino, Tonelli proved that the only condition required for the length of a curve to be given by the usual integral, is that the same curve must be expressed through absolutely continuous functions. The realm in which Arzelà had been exploring was significantly expanded by the now systematic reference to Lebesgue's integral. In 1909, he wrote the

---

[5]In reality, Tonelli registered to the first year of engineering, prior to finally choose the mathematical studies.

[6]Continuous references are obviously made to the work of Volterra, Arzelà and Vitali.

Note "Sull'integrazione per parti" whose first Lemma contains the classic "Fubini–Tonelli theorem," demonstrating that for a non-negative function of two variables, measurable and having subsequent integrals, there is a double integral and its value is independent of the integration order.

During those same years, Tonelli started his great adventure in the Calculus of Variations. In 1911, he published the Memory "Sui massimi e minimi assoluti del Calcolo delle variazioni" on the Rendiconti del Circolo Matematico di Palermo in which he was not yet using the expression "direct methods," but he applied them in practice, doing resuming the attempts he had already made in this direction for Dirichlet's problem and "reasoning directly on the integrals in question." The classic methods, in his search for a complete analogy with the ordinary optimization of real functions, were not able to overcome a series of criticalities: the reduction of the problems concerning the Calculus of Variations to problems of differential equations (and not vice versa) with the consequent difficulty of the calculus and prior to that, of the existence of the solution of a boundary problem, the strong limitation applied to the functional class by the consideration of differential equations, the privilege given to relative extrema, the search for sufficient conditions. The procedure followed by Tonelli is based on the concept of semi-continuity (and on compactness conditions). Though the functional $J[y] = \int_a^b (x, y, y') \, dx$, that characterises the simplest problem of Calculus of variations, poses serious problems in terms of continuity, it is semi-continuous in the uniform topology (in conditions of "regularity" not particularly restrictive). And this is sufficient, also in view of certain hypotheses, to guarantee the existence of an equation curve $y = y(x)$ that makes it minimum. Indeed, once verified that within the functional class taken into consideration, it is $\inf J = j > -\infty$, it is possible to construct[7] a minimizing sequence of equation curves $y = y_n(x)$ such that $J[y_n]$ tends to j and from this, through compactness theorems, extract a sequence $\{y_{n_k}\}$ converging to a function $\bar{y}$ for which $j = J[\bar{y}]$ is obtained.

Tonelli was removed from the chair of Parma because the Superior Council of Public Education annulled the acts of the competition during the meeting of 17 June 1912. Its Commission (made up of Pincherle, Pascal, Torelli, Vivanti and Bagnera) had used the test of teaching abilities, generally optional, as decisive "after having verified that the [Commission] was not able to reach a decision through the discussion of the scientific publications."[8] Among Tonelli's students, however, it was widely believed that the intervention of the Superior Council was suggested by Mauro Picone (1885–1977). Picone had filed an appeal against Luciano Orlando

---

[7] See B.U.M.I., (2), a. II (1940), no. 2, pages 142–146.

[8] It must be remembered that among the applicants of Parma competition, there was also Giuseppe Vitali in addition to Tonelli (and Orlando and Picone), who was not awarded a chair until 1922. His unfortunate academic events (and not proper, for the Italian mathematical community) are analysed by L. Pepe, Una biografia di Giuseppe Vitali, (in [22, pages 1–33]). Tonelli, together with Fubini, Levi-Civita, Pincherle and Torelli, was part of the Examination Board that assigned the chair of mathematical Analysis to Vitali, awarded by the University of Modena, after the resignation of Sannia, who was declared first winner.

(1887–1915), claiming that the latter had stayed at the same hotel in Rome where the Commissioner Giuseppe Bagnera had stayed. Bagnera was also from Sicily and was Picone's teacher in Messina.[9] One thing is certain: when the competition was reopened, Picone was the third winner (in the tern) while Orlando was excluded from the tern. Tonelli then applied for the competition for the Chair of Algebraic Analysis at the University of Cagliari; he won and moved to the island. His transfer to Parma took place with a one-year delay when the previous competition was called again in 1917 and he was again declared the winner. In Parma, he also taught Mechanics upon assignment, becoming a full professor.[10]

The First World War is another circumstance that helps us understand the figure of Tonelli, as well as the climate that fed nationalism and patriotism in Italy in that period. Tonelli was initially declared unfit for military service and exempted from taking part in the war; he underwent a surgery to change the previous opinion and to be sent to the front.[11] When he returned unharmed from the war, he went back to teaching and research.

His first post saw him involved in a harsh (and brief) polemic with Attilio Vergerio (1877–1937) who had presented, through Carlo Somigliana, a Note entitled "Sulla derivazione per serie" in 1916, when Tonelli was enrolled in the army. He made reference to previous theorems of Fubini and Tonelli on the term to term derivation of a series of functions $u_n(x)$, to demonstrate a result similar to that of Tonelli, with slightly more restrictive hypotheses, but with a definitely simpler procedure compared to the "fairly laborious geometric calculations" of Tonelli. Reference was mainly made to his statement, contained in the Note "Successioni di curve e derivazione per serie"(1916), to demonstrate with a counter-example, the incorrectness of the presumption to eliminate the hypothesis of the convergence of the derivative series, because it was considered superfluous. The response of Tonelli is contained in a Note, presented to the *Lincei* by Pincherle, where he evidences a gross mistake made by Vergerio, who incorrectly cited the theorem demonstrated by Tonelli.

After the years 1911–1912, some ten years later between 1921 and 1922 Tonelli's career was boosted even further. The first volume of *Fondamenti di calcolo delle variazioni* was published in 1921 by Zanichelli in Bologna, which described in an organized and fairly definitive manner, that which turned out to be results that were no longer singular, but gave life to an actual theory. The first volume was mostly introductory: after a historic outline (in reality, an interesting chapter on the problems and evolution of the methods concerning the Calculus of Variations),

---

[9] See A. Faedo, Leonida Tonelli and the mathematic school of Pisa, cit., page 22.

[10] The Commission that promotes him is the same one as before, with the replacement of Castelnuovo by Pascal.

[11] In the Archive of Fondazione Gentile (herein after AFG), annexed to the letter sent to Gentile on 8 April 1938, there is a memo that summarises the military position of Tonelli. Another "military curricula" is published (Appendix 2) in ([4, pages 90–91]). We only add that Tonelli obtains two promotions (as lieutenant and captain) very close to each other, for exceptional merits, and two distinctions: the "War merit cross" and the "Bronze medal".

Tonelli points out definitions and properties of functions of bounded variation, absolutely continuous and measurable, and Lebesgue's integral that introduces the concept of pseudo-interval and quasi-continuous function. The volume ends dealing with the key concept of semi-continuity, referring to the functional object of the "simplest problem" of the Calculus of Variations (in parametric and ordinary form). At the same time Tonelli published some fundamental Memories on the topic, among which "Delle funzioni di linee" [*Rendiconti Accademia Lincei*, 1914], "La semicontinuità nel calcolo delle variazioni" [*Rendiconti Circolo Matematico di Palermo*, 1920], "Criteri per l'esistenza della soluzione in problemi di calcolo delle variazioni" [*Annali Matematica pura e applicata*, 1921]. The second volume of Fondamenti was published in 1923 with almost definitive results on the existence of solutions for the "simplest problem" and for the isoperimetric problem, which are presented separately in this volume (in the parametric and ordinary case) always in one dimension, that is when the unknown function depends on one single variable. In the history of the Calculus of Variations Tonelli's name is mainly associated with the existence theorems.

During the spring of 1922, Tonelli took the chair of Advanced Analysis at the University of Bologna, and on 1 August 1922, he was appointed Correspondent Member of the Accademia dei Lincei.[12] He was never elected National Member for the reasons included in the Annals of the Academy: "Prof. Tonelli, already proposed as National Affiliate in 1935, did not obtain the relative appointment for political reasons." In fact, his political ideas mentioned in the brief outline of his young days in Pesaro, had taken root during the early days of his time in Bologna. It was there that he became a member of the "Circolo di Cultura" when it was being founded and he promoted his ideas and his admiration of Antonio Labriola, Enrico Leone (one of the main figures of the revolutionary syndicalism in the early decades of the century), the socialist ideals and later the anti-fascist movement. In this context, the most relevant episode was his adhesion to the "Manifesto Croce."[13]

---

[12] In a letter sent by Pincherle to Volterra (29.7.1922), the mathematician from Bologna comments Tonelli's election: "I am very pleased of the Tonelli's election at the Lincean Academy, and Im convinced that the Academy could have not chosen better. His book was particularly successful abroad, and I think the 2nd volume will be even more successful than the first."

[13] On 21 April 1925 the Italian press published the "Manifesto degli intellettuali del fascismo," inspired by Giovanni Gentile, and tending to define in general lines, a doctrine of Fascism towards culture. Only two mathematicians signed the Gentile Manifesto: the statistician Corrado Gini, and Salvatore Pincherle, one of Tonelli's teachers! The following May 1st, *Il Mondo* published the counter-manifesto of Benedetto Croce, entitled: "Una risposta di scrittori, professori e pubblicisti italiani al manifesto degli intellettuali fascisti," that criticized the relations between Fascism and culture. Tonelli was the only mathematician who initially signed the "Croce Manifesto." Later on, on May 10th and 22nd, *Il Mondo* published other two lists of signers. These lists included the mathematicians Ernesto and Mario Pascal, Vito Volterra, Giusepe Bagnera, Guido Castelnuovo, Beppo Levi, Tullio Levi-Civita, Alessandro Padoa, Giulio Pittarelli and Francesco Severi. For a complete discussion on the "two manifestos" see E. R. Papa, Storia di due manifesti. *Il fascismo e la cultura italiana*, Feltrinelli, Milan, 1958. For the mathematical aspect, see [15].

Prior to addressing this topic, it must be mentioned for the sake of completeness, that his transfer to Bologna took place in 1922, after the failed attempt of Volterra to call Tonelli to Rome, to occupy the vacant post of the deceased homonymous teacher, Alberto Tonelli (1849–1921), teacher of Infinitesimal Calculus since 1879. The intent of Volterra and Castelnuovo to strengthen and improve Roman mathematics had already led to some transfers to the capital in the immediate post-war period: Tullio Levi-Civita in 1919 and Federigo Enriques in 1921, initially as ordered. The plan was to complete the picture by calling an analyst. As we learn in a letter from Tonelli to Giovanni Vacca (26/2/1921), Volterra chose Tonelli, who declared his was available after realizing that Salvatore Pincherle[14] was not interested. Unfortunately Tonelli – who had already come across Picone during the contest in Parma – on this occasion crossed with Severi's inspirations to go to Rome and take up the chair of Analysis. Severi was supported by Levi-Civita, with whom he had established, a relationship based on esteem and friendship since the times of their stay in Padua. This support was decisive in Severi's triumph over Tonelli. Severi used his call to Rome as an opportunity to attain positions with increasing power, during the years and in line with the Fascistization of the Italy. But that outcome was not foreseeable beforehand; nor can any objections be made on a scientific level, since Severi was another top-notch mathematician.

3 The Transfer to Pisa

During mid-1920s, Fascism, after overcoming the turmoil created by the murder of Matteotti, focused on searching for alliances in high culture. During the fascistization of Italian culture, Tonelli was always indicated by the extremist groups of the regime, as one of the main figures of the anti-Fascism since his name was among the first who had signed the *Manifesto Croce*. It is in this context that he decided to move to Pisa to administer the renovation of the *Scuola Normale Superiore di Pisa*, strongly promoted by Giovanni Gentile. But before reaching 1930, we must dwell a bit longer on the period in Bologna that, besides maintaining his research at the same levels as previous decades, Tonelli was notably involved in teaching.

During those years in Bologna, Tonelli dealt with the problem of the squaring of areas in Cartesian form $z = z(x, y)$. One of his main Memories is "Sulla quadratura delle superficie" published in 1926 in the *Rendiconti* of the *Accademia dei Lincei*. In order to demonstrate a condition necessary and sufficient for the area to be finite and for its calculation, he extended to the functions of two variables the concepts of bounded variation function and absolutely continuous function. Some important Notes on trigonometric series and the double series of Fourier date back to this period, with results which are better described in the volume *Serie trigonometriche*, published by Zanichelli in 1928.

---

[14] The letter is included in [20].

The prestige attained was confirmed by the acknowledgements and prizes awarded him during those years by the *Società dei XL* and by the *Accademia dei Lincei*. In Bologna, Silvio Cinquini (1906–1998), among others, met Tonelli as a teacher, and then who followed him as his assistant in Pisa. He most likely became the interpreter and most faithful follower of his researches; then there was Antonio Mambriani (1898–1989), whose education involved an equally important role played by Pincherle. In 1928, Pincherle left the chair of Analysis and Theory of the functions upon retirement and this circumstance ignited a lively debate in the Faculty of Sciences between Tonelli and the group of chemists led by Mario Betti (1875–1942). The mathematics group had two vacant chairs: the chair of infinitesimal Analysis left by Pincherle and the chair of Projective Geometry left vacant by Bompiani who in 1927 had been transferred to Rome (once again frustrating Volterra's desire to call Tonelli for the teaching post). The group of chemists wanted to obtain one of these chairs at all costs, to institute a chair of Physics-Chemistry. Tonelli was in the minority and Betti took advantage of the climate in favour of Chemistry that reigned in Italy for, privileged by the regime for national defence and mainly for industrial development, The only chair spared was that of the Theory of Functions assigned to Beppo Levi, who relocated from Parma,[15] while Tonelli was transferred to the chair of Infinitesimal Analysis.

The transfer of Tonelli to Pisa (to the University and to Scuola Normale Superiore) took place in the academic year of 1930–1931. The decision was made by Giovanni Gentile, the erstwhile royal commissioner of the Scuola Normale Superiore di Pisa [SNSP] and determined supporter of the project to create a research centre in Pisa of international calibre and prestige. We can follow the different phases of the negotiation and agreement in the letters exchanged between Tonelli and Gentile over the period of a few weeks, starting in April 1930. There were difficulties, linked to the political image of the mathematician, as explained by Gentile in a letter dated June 10th:

> Dear Colleague,
> I'm very sorry not to have written you since you had the courtesy to take in consideration my proposal about the position in Pisa, but due to the fact that I was far away from that location and the consequent difficulty of promptly handling the necessary negotiations with the professors of the Faculty and with the authorities, together with the seriousness of the problems back then (...) caused by the long delay in following up the proposal I had made to you. I wish to mention you that a small minority of the professors of the Faculty were somewhat hesitant and undecided due to the character of certain political demonstrations in which you were involved: the usual vileness against which I strongly fought in these last few years; to put an end to them, I called the Prime Minister, since I would have been disappointed if you had not been called to Pisa due to the unanimous vote of the Faculty. The Prime Minister wrote recently a letter to the Prefect of Pisa that was communicated to the rector and naturally, eliminates any doubt.[16]

---

[15]On the event, see [17].

[16]"Egregio Collega,
sono dispiacentissimo di non averLe scritto dacché Ella ebbe la cortesia di prendere in considerazione la mia proposta della Sua chiamata a Pisa, ma la mia lontananza da quella sede e la

The political difficulties are clearly due to the adhesion of Tonelli to the *Croce Manifesto*. We do not have other testimonies of his opposition to Fascism during the second half of the 1920s or of his political commitment which was perhaps diminishing. On the other hand, Tonelli, still in 1930, continued not to be registered in the Fascist Party and paid the price of malicious attacks and insinuations against him in the local press of Bologna. Another consequence of his political persuasion was that, after 1926, he was no longer called to be part of the commission for the competitions to get teaching posts. In 1932–1933, he took part for the last time (with Guido Fubini and Francesco Severi) in a commission for the position of lecturer. In view of this situation, Gentile was determined to create a free "science citadel," or less bound to political events. The same attitude was expressed by Gentile towards the *Enciclopedia* and, in Tonelli's case, he found the support of Mussolini. The testimonies of the students clarify (perhaps with some additional "touches")," Gentile's reference to the meeting with the Prime Minister: Mussolini asked only whether Tonelli knew Math (well) and when he was given certainty of Tonelli's expertise, he wrote the relative instructions to the Prefect of Pisa.[17]

Gentile also took actions to deal with the economic requests brought forward by Tonelli in the first letter. This attitude may seem astonishing for an Italian university environment such as mathematics, scarcely characterised by self-employed professionals. In any case, it cannot be seen as separate from Tonelli's determination in taking up the new appointment that soon thereafter become a specific project. Both factors concur to portray a figure who was aware of his value and willing to take up a heavy work load, also in terms of the fundamental organization of research, which he considered vital.

We shall end this paragraph by underscoring that in July 1930, Tonelli still harboured serious doubts about his transfer, though it had been decided by the Faculty of Sciences of Pisa on June 20th. Confirmation was given in a letter dated 13 July 1930, sent by Beppo Levi to Vitali[18]:

---

conseguente difficoltà di condurre sollecitamente le necessarie trattative coi professori della Facoltà e con le autorità, insieme con la gravità dei problemi che erano in corso (...) sono state causa di questo lungo ritardo con cui m'è dato di riprendere la proposta che già ebbi il piacere di farLe. A mia scusa mi permetto di accennarLe che in una piccola minoranza dei professori della Facoltà era sorta una certa esitazione e titubanza pel carattere di talune manifestazioni politiche che di Lei si ricordano: le solite miserie, contro le quali io vengo combattendo energicamente in questi ultimi anni; e a troncare le quali poiché troppo mi sarebbe dispiaciuto che la Sua chiamata a Pisa non dovesse aver luogo per voto unanime della Facoltà ho creduto opportuno far intervenire lo stesso Capo del Governo. Il quale ha scritto recentemente al Prefetto di Pisa una lettera che è stata comunicata al rettore e che dissipa, naturalmente, ogni dubbio."

[17] Alessandro Faedo, in his "Leonida Tonelli e la scuola matematica pisana" [4, page 26], wrote: "Gentile (...) convinced that Tonelli was the most suitable person for the School, visited Mussolini directly to talk about the problem. Mussolini asked him what Tonelli was teaching and after discovering that he was a mathematician, he answered: «What does mathematics have to do with politics?»and ordered the Minister of National Education to sign the transfer decree."

[18] See *Lettere a Giuseppe Vitali* (by M.T. Borgato and L. Pepe), Tecnoprint, Bologna, 1984, pages 511–512.

The faculty of Pisa is actually insisting so that T. [Tonelli] accepts the transfer. T. is still totally undecided because he does not want to leave Bologna. Nonetheless, I wish to discuss with you the possibility that he accepts. I would be very pleased in this case to have you here with me to form the analytical group. And since many delicate issues are currently influencing the decisions of the Faculty and mine, I would be pleased if you could tell me as soon as possible, if there is a possibility that you will accept the proposal. (...)[19]

4 Tonelli in Pisa

Tonelli's determination was immediately in line with that of Gentile, and from this meeting and the consequent implementation of Tonelli's projects, Pisa was able to achieve its fortunes and modern prestige, as the centre of mathematical research. At the University Tonelli was teaching Infinitesimal Analysis and High Analysis, and he also became director of the Mathematics Institute. From his teachings in Pisa, we get the vivid image that Sandro Faedo (1913–2001) had left us[20] :

> Tonelli was holding courses of Infinitesimal Analysis and High analysis at the University. His lectures were marvellous (...).I had joined the Scuola Normale Superiore as a student of Physics, attracted by the news of the discoveries of the school of Fermi; after hearing Tonelli's lectures, I felt a strong attraction for Mathematical Analysis because of the method for solving problems, the harmony and rigor of the reasoning. Therefore, upon my own decision, trying to understand and follow my inclinations, I left the degree course in Physics and enrolled in the third year of Mathematics, convinced that this was my path. (...)
> He was also holding two seminars at the Scuola Normale, for beginner and more advanced students, in addition to a course in which he described his method to introduce Lebesgue's integral.

---

[19]"La facoltà di Pisa sta effettivamente facendo insistenze col T.[Tonelli] perché egli accetti il trasferimento. Il T. non per nulla deciso al riguardo, perché non lascia volentieri Bologna. Nondimeno io voglio parlarti della eventualità che egli accettasse. Io sarei molto contento in tal caso di averti qui con me a formare il gruppo analitico. E poiché molte cose delicate influiscono in questo momento sulle direttive che può avere la Facoltà e io stesso, avrei piacere che tu mi dicessi prontamente, sia pure in via del tutto ipotetica, se tu accetteresti la proposta. (...)"

[20]See A. Faedo, op. cit., page 27: "Tonelli teneva all'Università corsi di Analisi infinitesimale e di Analisi superiore. Le sue lezioni erano meravigliose (...). Io ero entrato alla Scuola Normale Superiore come studente in Fisica, attratto dagli echi che mi erano giunti delle scoperte della scuola di Fermi; udite le lezioni di Tonelli, sentii una forte attrazione per l'Analisi Matematica, per il modo di affrontare i problemi, per l'armonia e il rigore dei ragionamenti. Così in modo autonomo cercando di capire e seguire le mie inclinazioni, spontaneamente lasciai il corso di laurea in Fisica e mi iscrissi al terzo anno di Matematica, convinto che quella fosse la mia strada. (...) Alla Scuola Normale teneva inoltre due seminari, per gli studenti di primo livello e per i più anziani, oltre a un corso in cui espose il suo metodo per introdurre l'integrale del Lebesgue. Dai seminari dei primi anni egli ci conduceva gradualmente ai problemi aperti della Matematica, stimolandoci a prepararci seriamente e a iniziare con impegno quell'attività così appassionante che avevamo spontaneamente scelto per la nostra vita. Anche dei seminari pielementari ricordo alcuni episodi salienti, come quelli in cui Renzo Cisbani, allora studente del secondo anno, ci espose il libro appena uscito di W. Sierpinski sui numeri transfiniti, aprendoci orizzonti di insospettata bellezza."

> From the seminars of the first years, he was gradually leading us to the unsolved problems of Mathematics, stimulating us to achieve a serious education and start this interesting activity that we had autonomously chosen for our life with a real sense of commitment, I remember a few main episodes even during the most basic seminars, like those in which Renzo Cisbani, then a second-year student, disclosed the recently published book by W. Sierpinski on transfinite numbers, opening horizons of unexpected beauty.

In his role as Director of the mathematics institute, Tonelli pursued as his first objective, the renewal of the top positions, by freeing new positions in order to find a suitable placement for the young students who were graduating from his school[21]: A. Del Chiaro, G. Dantoni, S. Petralia, L. Cesari, R. Cisbani, A. Pedrini, D. Dainelli, E. Pizzetti, G. Zappa, C. Santacroce, G. Ottaviani. It must be remembered that in those years, accesses to a research and university career in Italy was extremely competitive and this compelled very promising mathematicians to teach in secondary schools for a fairly long period of time, in order to prepare for the contest to become a university professor. The foundation of the *Istituto Nazionale per le Applicazioni del Calcolo (INAC)* by Mauro Picone was the first one to break this rule at the beginning of the 1930s, and to create a certain number of positions as researchers and calculators for young graduates. Lamberto Cesari (1910–1990) and Antonio Pedrini (born in 1914) found employment at the INAC in 1937. Tonelli was nonetheless able to also exploit the "pastoral visits" realized periodically by his former student at the Scuola Normale, Enriques, in Pisa. The purpose of his visits was to find new talent and assess the great possibilities offered – thanks to the contribution of Guido Castelnuovo and Francesco Paolo Cantelli from the Faculty of Statistics, Demographics and actuarial Sciences in Rome. There positions would be given to A. De Chiaro, R. Cisbani, E. Pizzetti, C. Santacroce, G. Ottaviani and Guido Zappa later recuperated by Severi for "pure" Mathematics and establishing himself as an algebraist under the guidance of Scorza).

The direction of the *Annals* is another point of strength of Tonelli's renewal project. We already mentioned the scarce regularity with which the volumes of the first series were published, and they only presented abstracts from his undergraduate thesis. Just by browsing the table of contents of the first volumes of the *Annali della Scuola Normale Superiore di Pisa. Scienze Fisiche e Matematiche*, under Tonelli's direction, it is possible to notice immediately the remarkable qualitative leap of the magazine. Once again, prestige and fame – this time, that of the Annali – were achieved (and reinforced) with the help of Tonelli.

The direction of the *Annali*, the relaunching of the *Normale* and the Mathematics Institute of the University clearly show the weight, in terms of work and responsibilities, carried by Tonelli.[22] In addition to this, were his usual teaching assignments which included, in addition to the jobs at the University, his presence in the *Normale* as the supervisor of two series of seminars and conferences (the first involved more teaching and the second was more focused on preparing for

---

[21] See A. Faedo, op. cit., pages 28–29.

[22] For a few years, Tonelli was also vice-president of *Mathesis*.

research). Therefore, it should come as no surprise that the figure that stands out in the Thirties is different from the figure described in his Bologna period. The "turning point" in Pisa can be summarized in four points. We already discussed the first (a significant increase of the work load). The second, which is related to the first, concerns his health conditions. We already mentioned the episode of the serious illness that Tonelli contracted in the final months of his university studies, for a distressing red light indicating that his body was not extremely strong. In 1932 the signs of his physical decline became more serious and frequent. In 1938, Tonelli had both kidneys severely impaired and died in 1946, due to renal tuberculosis (though his discretion and that of his family about the illness, make this detail very likely but not certain). The third element that characterised the 1930s was the definitive and widely acknowledged fame attained by Tonelli. Though in Bologna he was not unknown and his prestige was not unfounded, it was certainly in Pisa where he established himself as a public figure. If it is true that Tonelli relaunched Pisa (in mathematical terms), there may also have been a sort of reverse beneficial reaction. Tonelli moved to Bologna when he was still relatively young and was definitely influenced (and suffered) by the presence major mathematicians, of the likes of Pincherle, and Ettore Bortolotti. His relations with the latter were not optimal, as it can be inferred in his correspondence with Volterra. On the contrary, when Tonelli moved to Pisa he was a famous mathematician, to the point that even Gentile and Mussolini "went out of their ways" to help him. Here, we have a public figure at the height of his glory and career as a scientific organizer. Therefore, the fourth point comes as no surprise: his research could not keep up with the pace of the previous decades. It was not so much the frequency of his publications, which had diminished – but here is a snapshot showing the difficulties encountered by Tonelli in confirming the quality of his research at that time.

The *Fondamenti* had substantially run out of the application of direct methods to the simplest problem of the Calculus of Variations and to the isoperimetric calculus in the one dimensional case. Starting from this publication, Tonelli's studies focused on the possible generalization of the method and results to the multi-dimensional case, with regards to the Calculus of Variations. Important contributions were not lacking, but they did not reach the finality and depth shown for $n = 1$. Today, we know that the use – even the refined use that Tonelli was accustomed to – of the instruments provided by Real Analysis, is insufficient for dealing with the new and more complex problem that requires the knowledge and practice of abstract functional spaces that Tonelli did not have, at least to the same extent that he mastered Classical Analysis. In Zurich, at the International Congress of mathematicians in 1932, Tonelli presented an announcement "On the calculus of variations" with the aim to "handle the issue of the absolute minimum for the double integral:

$$I_D[z] = \iint_D f(x, y, z, p, q) \, dx dy$$

(where $p = z'_x, q = z'_y$), with the same direct method that I developed, in my *Fondamenti di Calcolo delle Variazioni*, for curvilinear integrals, a method based on the concept of semi-continuity." The results pre-announced in this direction were scarcely followed up in the future works of Tonelli and this seemed to , indicate that the originality (and disposable time) he once had, was now less and up to the challenge of the new tasks, while the role and prestige he had attained would require new results of a comparable level.

5 The Relations with Severi and Picone

Among Tonelli's students, we already mentioned Silvio Cinquini and Lamberto Cesari (in addition to Antonio Mambriani from Bologna). Then there was Basilio Manià (1909–1939), who is deemed one of the most brilliant students of the "school." Born in Fiume, Manià adhered to Fascism, driven by strong national feelings and went to fight the war in Spain as a volunteer, also to prove to himself, that a disabled leg did not prevent him from living a "normal life." Manià tragically committed suicide due to a troubled love affair, a few hours after having written the Note "Sopra una questione di compatibilit nel metodo variazionale" Tonelli published it in the Annali in Pisa, preceded by a brief, touching introduction to remember him.

The generation after Tonelli's students (in reality of only a few years younger) consisted of the already mentioned Lamberto Cesari, Sandro Faedo, and Emilio Baiada (who was given the chair in Palermo in 1952) and Landolino Giuliano, who would become a professor at the Naval Academy of Livorno.

Within the group, Sandro Faedo was one of the main students who established relations with Federico Enriques, the Roman group, and Mauro Picone. Faedo graduated with Tonelli in 1936 and initially he was given a position as a substitute teacher of Geometry. During the winter of 1936–1937 he met Enriques who arrived in Pisa with the alleged and purpose of the meeting the students of the *Normale*, but he really wanted to find among Tonelli's students an assistant who could take the place of Campedelli, who had become a professor. Enriques' choice fell on Faedo, who moved to Rome in March 1937 after overcoming certain perplexities associated with the relocation and to a discipline itself (Geometry) in which he had not earned a degree. The following year Faedo substituted Enriques (who also had Conforto as senior assistant) in the teaching of his courses for a brief period during his trip to Scotland for an honorary degree. It was also Faedo (who in the meantime had established a scientific collaboration with Enriques, documented by some Notes written in this period), who would inform Enriques on the happenings at the university during the years of the racial discrimination measures, when Enriques was forbidden to attend the Mathematics Institute. In Rome Faedo once again had the opportunity to frequent Tonelli who in the late 1930s accepted to move to the capital, leaving his teaching position in Pisa, but not the *Normale*.[23] We have two

---

[23]In Rome, Tonelli was also member for one year, of Istituto di Alta Matematica founded by Severi in 1939. On Severi's Institute, see G. Roghi, Materiale per una storia dell'Istituto Nazionale di Alta

letters regarding this episode: the first is dated 3 November 1938, addressed to Gentile: it mentioned the allegiance to the Fascist party (which we will discuss later): "Please also do not forget to solve the issue with my card. If you can solve it, I assure you that you will be very pleased with what I will do"; the second letter, written to Volterra's wife on 4 February 1939, seems to imply that Tonelli was embarrassed to be transferred to Rome by within a series of "operations" destined to cover the empty spots left by Jewish mathematicians who had been expelled from the Universities: "my transfer to Rome takes place now in a period in which it no longer has the meaning it would have had when your husband was in my Faculty in Rome. Now, as he knows, things have completely changed."

For a few years he would later return to Pisa in 1942 Tonelli spent the early part of the week in Rome, where he arrived on Monday and left on Wednesday. Faedo helped him with the small issues related to a new residence and their relations became less "professional." Outside the strictly academic environment, Tonelli appeared to be less "pedantic" and ready to show his humanity. For example, he helped a former student who had been hospitalized in Rome due to a typhus seizure that turned out to be fatal.[24]

In Rome Faedo, developed relationships with Severi and Picone and, unlike what happened with Enriques, for whom Tonelli had always shown great reciprocal esteem and friendship, there were a few problems. The relations with Severi in particular, became extremely tense in 1940 in view of the Mussolini Prize,[25] on which Tonelli vented with Gentile (letter of 12 April 1940): "when I told you about the Mussolini Prize , you said that Severi really liked me. And indeed, the other day he had me fired from the Academy. Our friend has an aching pain in his heart: the Scuola Normale. But I have no intention to take it away from him. I'll tell you

---

Matematica dal 1939 al 2003, *Bollettino della Unione Matematica Italiana*, (8), 8-A (December 2005/2), monographic issue.

[24] This is Renzo Cisbani. The episode is told in A. Faedo, Leonida Tonelli e la scuola matematica pisana, cit., page 29.

[25] The *Mussolini Prize* awarded by *Accademia d'Italia* and financed by *Corriere della Sera* for the best work or series of works selected among those published or elaborated in the last ten years, concerning moral and historic disciplines, sciences, literature, arts, was conferred by the Academy without any competition, every year (starting from 1931). Until 1939, it was constituted by four prizes, one for each class, of £50,000 each. For the selection, the four classes were forming an examining board each, with the addition of three members, one for each of the other classes. An academic speaker was in charge to report to the general assembly, which had the final vote. The academic council, during the meeting of 19 July 1938, established that, since 1939, the prize would have been only one (of £200,000) assigned to the four classes in shift, as indicated by the prize regulation, a "high and solemn acknowledgement to an enlightened activity carried out at the benefit of fine culture, for increasing historical and philosophical studies, scientific research, and literary and artistic creation." Here is the list of the people awarded of the Class of physical, mathematical and natural sciences during the period 1931–1938: 1931: F. De Filippi, physiologist; 1932: A. Castellani, pathologist; 1933: O.M. Corbino, physicist; 1934: A. Garbasso, physicist (in memoriam); 1935: M. Panetti, aeronautic engineer; 1936: Giulio Chiarugi, embryologist; 1937: G.B. Bonino, chemist; 1938: F. Rasetti, physicist. As it can be seen, the net prevalence of physical–chemical and biomedical disciplines is obvious.

how things turn out when I see you." In 1940 the Class of Physical, Mathematical, and Natural Sciences of *Accademia d'Italia* was awarded the Mussolini Prize and held two preliminary discussions on the topic during the meetings of 16 December 1939 and 19 January 1940[26] (the contents of which were never recorded in the minutes). Despite the reticence, we know that there were self-candidatures and candidatures by other academicians, among which the powerful candidature of Nicola Pende, already a signatory of the notorious *Manifesto della razza* of July 1938 and recent author of the *Trattato di biotipologia umana* (1939). Pende's candidature was proposed by the academician Lucio d'Ambra, who wrote a letter about it to the Dean of the Academy, Luigi Federzoni.[27] The decisive meeting was no one held on 21 March 1940, in which the Class President, the Turinese electro-technician Giancarlo Vallauri, opened the discussion on the specific topic of the Prize, excluding a priori the candidates who had directly proposed their candidatures or had been recommended by the academicians of other classes.[28] The first to speak was Severi, whose speech was recorded in the minutes:

> The Academician Severi illustrated the work of Prof. Tonelli, insisting on the opportunity to take into special consideration this time, the experts of mathematical disciplines that until then had not benefited from any of the previous Mussolini Prizes.
> He deemed appropriate to also propose the candidatures of prof. Picone, director of Istituto di Calcolo del Consiglio nazionale delle Ricerche, a talented mathematicians distinguished for his studies and researches on technical applications. Academicians Crocco, Giordani, Orestano and Guidi supported this proposal, remembering the importance of the contributions that, thanks to Prof. Picone, the Istituto Nazionale di Calcolo had made in past years to solve many problems of great national interest. They also applauded his work as a teacher through which beyond the field of study noted by Severi he was able to encourage and direct the work of many students, many of whom ended up becoming his students.
> After a prolonged discussion, the Academician Severi, supported by his colleagues Giorgi and Somigliana, ended the speech, stating that the candidature of Prof. Tonelli should be preferred.[29]

---

[26] See Minutes in the Archive of the *Accademia dei Lincei, Fondo Accademia d'Italia* (from 5 December 1929 to 9 April 1943).

[27] The letter in the Archive of the *Accademia dei Lincei, Fondo dell'Accademia d'Italia*, title VII, envelope 100, folder no. 164 and 165 ("Mussolini Prize 1940").

[28] In his report on Pende, entrusted to the Milanese doctor Pietro Rondoni, it is read that "many statements are not supported by experimental data and some of them are inaccurate."

[29] "L'Accademico Severi illustra l'opera del prof. Tonelli, insistendo sull'opportunitL di prendere questa volta in speciale considerazione i cultori delle discipline matematiche, che finora non hanno beneficiato di nessuna delle precedenti assegnazioni del premio Mussolini.
Ritiene suo dovere di avanzare anche la candidatura del prof. Picone, direttore dell'Istituto di Calcolo del Consiglio nazionale delle Ricerche, valoroso matematico, distintosi per gli studi e le ricerche che hanno attinenza con le applicazioni tecniche. Si associano a questa proposta gli Accademici Crocco, Giordani, Orestano e Guidi, ricordando la importanza dei contributi che, per merito del prof. Picone, l'Istituto Nazionale di Calcolo ha dato in questi ultimi anni alla soluzione di molti problemi di grande interesse nazionale ed esaltando la sua opera di maestro, che anche al di là di quel campo di studi ricordato dall'Eccellenza Severi ha saputo animare e dirigere il lavoro di numerosi allievi, molti dei quali sono pervenuti alla cattedra.

Severi was thus able to naïvely hinder both candidatures that he himself had proposed: in an environment hardly favourable to mathematicians, only an unambiguous and strongly motivated candidature would have been able to convince the "non-believers" to assign the prize to a representative of the "mathematical disciplines" [30] The great variety of proposals immediately brought forward after Severi's speech, prove that the environment was ready to welcome the unexpected "slip." Among these proposals, the one of Vallauri stands out inasmuch as it "describes the activity of Prof. Maiorana, and elegant and original researcher" and points out his discoveries "of new effects and in particular his contribution to the first radio-telephone transmissions." The proposal was supported by Academicians Pession, Lo Surdo, Orestano and Giorgi, who also emphasized the extraordinary organizational skills showed by Maiorana, as founder of the *Istituto per le Comunicazioni Elettriche*. At this point Somigliana invited his colleagues to "concentrate their votes on Maiorana and Tonelli" which, in his opinion, were the ones that would prevail over all the other candidatures. The voting result revealed a landslide victory for Maiorana with 10 votes against 4 for Tonelli and 5 scattered votes. As we will see later, Tonelli won the Mussolini Prize only in 1944, thanks to the strong "remedial" pressure of Gentile, but in a critical situation, that would suggest a polemic detachment from the spreading of the news that he used the prize to sponsor a scholarship for a young student of the *Normale*. His response to Severi's attitude was immediate. Feeling "betrayed," Tonelli did not take part in the 2nd Congress of Unione Matematica Italiana (held in Bologna on 4–6 April 1940), though the program referred to him as the Chairman of the Analysis session and was looking forward to his general report on the "Scuola Normale di Pisa and mathematical studies."

Nonetheless, between Tonelli and Mauro Picone there was unbearable friction, the latter already fully engaged in this career during the contest of 1911. In the years that followed, Picone founded INAC, one of the main touchstones of Italian Mathematics during that twenty-year period in terms of research activity and innovations introduced in the organizational sphere. The position of absolute importance attained by Picone was accompanied by his ready and deeply felt allegiance to Fascism, from which he drew questionable identities between Applied Mathematics and "Fascist" Mathematics (in a polite but obstinate polemic with Severi, who had also converted to Fascism but continued to believe in the profound values of "pure" mathematics).

The relations between Tonelli and Picone always remained tense, also for objective reasons. The former represented one of the latest experts of the great

---

Dopo lunga discussione l'Accademico Severi, al quale si associano i colleghi Giorgi e Somigliana, conclude affermando che debba essere preferita la candidatura del prof. Tonelli."

[30]It would have been sufficient for example, that Severi would have omitted the name of Picone and repeated the positive opinion on Tonelli, that he gave when he was transferred to Rome (in the Historic Archive of the University of Rome, Minutes of the Faculty of Sciences, meeting of 27 December 1938–XVII).

Italian Mathematics at the end of the Nineteen century; analyst (and only analyst), he expanded the high levels of tradition, also with regards to the organization of the research and education of the students. He did not understand the versatility of Picone's research or at least who was very sceptical and began his insinuations starting from the famous shooting boards of the colleague; neither had he fairly evaluated the value of the school founded by Picone, and also brought forward critics and reservations towards Renato Caccioppoli. He hardly accepted the abstract layout, more ready to acknowledge the new results of the Functional Analysis that Picone tends to give to his works, rejecting firmly at didactical level. In this respect, Tonelli accused Picone to be the "most general" professor who starts from an artificial generality to reach levels that did not require said generalizations.

The polemic developed (after various "academic disagreements," due to negative results or failures in the competitions of some students), following the publication of *Appunti di Analisi Superiore* (Rondinella, Naples, 1940), for which Picone asked a review to Tonelli, through Faedo. And the "sweet and sour" evaluation given by the latter, does not hide in conclusion, the reasons for the disagreement,[31] when he brought forward some critics on the contents of the single chapters. Picones response is contained in a long Note entitled "Sull'integrazione delle funzioni," published on the *Rendiconti di Matematica e delle sue applicazioni* (1942), that starts by reconstructing the steps that give "a structure of utmost transparency, that I would judge as final" to Lebesgue's theory of integration. The focus on Tonelli's review is limited to a note at the bottom of the page, where Picone explicitly declares not to agree with the evaluation of the Colleague that was denying any theoretical and didactical utility to the procedure followed by him to construct the integration theory. For Picone, the "criticized" procedure was, on the contrary, extremely useful in providing a significant example of a generalization, presented also in a constructive way, to a functional class that mainly concerns physical applications. The most aggressive act of the polemic is constituted by the following reply from Tonelli ("Sull'integrazione delle funzioni," in *Annali della SNSP*, (1942) having a sarcastic tone, and addressing the contradictions in Picone's article, between the content of the text and the content in the note. In the text Picone "strongly supported" with "juvenile enthusiasm," the thesis supported by Tonelli, according to which it is possible and appropriate to define Lebesgue's integral through one single limit step, starting from that of Mengoli-Cauchy defined on continuous functions. The note seemed to be "in contrast to all his work." The different opinions "do not surprise me, and nor do they bother me. Whoever viewed the Lectures on Infinitesimal Analysis of Picone knows that it is not always easy to agree with him." Obviously, the attack was not overlooked, and Picone contacted Gentile directly (11.6.1943):

> In the Annali of the Scuola Normale, which I am very fond, and which I have always defended, and recently strenuously so against attempts to diminish it; in those Annali, that by the will of Dini and Bianchi, published my thesis and my certification doctorate, in the last issue, my prestige as a researcher and teacher was attacked, also in front of my students

---

[31] See L. Tonelli, review on [21].

of the 1st two-year course of engineering, who are not able to realize in full, the foolishness of that attempt.
I request that you publish in the same Annali the typewritten note herein, in which I respond to the observations and insults of that letter, as it is my duty and incontestable right.[32]
Unfortunately said publication damages the reputation of Italian abroad, and this is what truly pains me most sharply![33]

The polemic was therefore extremely harsh, though the correspondence[34] between the two was never suspended, except during the period 1943–1945 because of the war. Then Tonelli died. It seems sadly ironic that not even his death was able to end the dispute. Picone, appointed to hold a series of seminars at Normale, chose as the topic Lebesgue's integral which is what had triggered the debate.

6 Return to Pisa

His stay in Rome did not last long. The strain (during wartime) of a double teaching load and less than -exhilarating relationships in the Roman environment, soon drove Tonelli back to Pisa. The negotiations had become laborious due to the' specific economic demands s made by Tonelli, along with acknowledgment of the importance of his possible return to Pisa. Those assurances were received late. The first letter to Gentile on the topic was dated 12 November 1940:

> As you well understand, my resignation from the chair of Rome imposes various sacrifices on my family and me; these, at least, should be partially compensated. After all, I have been teaching in Pisa for over ten years and never had the smallest acknowledgment in all these years for my work to promote mathematical studies in Pisa. In saying this, it is not my intent to mention the amusing conferral of honors of Cherubino, to which I am totally indifferent. By now, I have completely adapted to the new times and I aim only to achieve concrete things.
> I would also like to add that the most significant demonstration of approval for my work was given to me here in Pisa, by the Faculty of Sciences, presided by our fellow colleague Gigi

---

[32] It is useless to say that Picones Note was not published.

[33] "Negli Annali della Scuola normale, della quale sono affezionatissimo figlio, che ho sempre, ed anche recentemente, strenuamente difeso nei commessi tentativi di diminuirla, in quegli Annali che, per volere del Dini e del Bianchi, pubblicarono la mia tesi di laurea e la mia tesi di abilitazione, si è, nell'ultimo fascicolo, attentato al mio prestigio di studioso e di insegnante, anche di fronte ai miei scolari del I biennio d'ingegneria i quali non possono essere in grado di rendersi conto di tutta la stoltezza di quel tentativo.
Io Vi chiedo, Eccellenza, la pubblicazione negli stessi Annali, della qui acclusa nota dattiloscritta, con la quale rispondo, come è mio imperioso dovere e inoppugnabile sacrosanto diritto, alle osservazioni ed alle contumelie di quello scritto.
Purtroppo tali pubblicazioni recano, comunque, danno alla reputazione all'estero dell'universit e della scienza italiane, ed è in ci che io provo il mio più pungente dolore!"

[34] It is conserved at the historic Archive of IAC.

Puccianti; this faculty, upon the announcement of my transfer to Rome, spoke no words of regret on my behalf.[35]

The letter from the Rector of the University of Pisa, Mr. Biggini, arrived only on 13 June 1942, and he accepted the last conditions proposed by Tonelli.

The final step, equally important, in Tonelli's career took place at the beginning of the academic year 1943–1944, when he became director of the *Normale*. The mathematician felt the need to contact his friend Gentile with words of sincere gratitude for his help in improving the for his help in improving the fortunes of the *Scuola Normale* (18.11.1943):

> I became director of the Normale, despite knowing it would be difficult to maintain the School at the great level to which you had raised it. I will try to draw inspiration from your work and put all my efforts into it, hoping that luck will be on my side. (...)
> I confide in the fact that I will always be able to rely on you for advice and help; and I even dare request that you maintain your name as director of the Annali della Scuola for the literature class. Please dont say no!

The backstage intrigues of this event were indirectly told by [26, page 4]:

> [Russo] was appointed [as the Director of SNSP], after the end of the Fascist regime, on 1 September 1943. This appointment only lasted a few days because on September 13th the last director [Russo] of the School was forced to leave Pisa due to particular circumstances. On October 20th, in his seaside home in Versilia, the reluctant director was visited unexpectedly by Prof. Carlo Alberto Biggini, the new Minister of the Italian republic of Salò, accompanied by his principal private secretary Umberto Biscottini. Biggini used caring and profound words in his attempt to convince him to resume his appointment as Rector of the University of Pisa and Director of Scuola Normale. Upon the refusal of the recalcitrant director, the self-styled minister who was a vain but not evil youngster, rather than threaten him, warned him by saying that his arrest warrant was ready. On that occasion (...) I was forced to emigrate (...)
> Two close friends of mine (...) were the ones who first brought me the news about the University of Pisa and the Scuola Normale Superiore and the role of director temporarily filled by prof. Leonida Tonelli who was objectively respected for his high authority as a scientist on matters such as mathematics which by nature were a political and he was validly helped in his defence of the School by his assistant and student Dr. Landolino Giuliano. I was glad to learn that Tonelli and Giuliano had succeeded in putting the brakes

---

[35]"Come tu comprendi benissimo, la rinunzia alla cattedra di Roma impone a me ed alla mia famiglia vari sacrifizi; ed è necessario che essi vengano, almeno in parte compensati. D'altronde, io, che insegno a Pisa da oltre dieci anni, non ho mai avuto in tutto questo lungo periodo nemmeno il più piccolo riconoscimento per l'opera che vado svolgendo in favore degli studi matematici pisani. Non intendo con ciò di accennare alla sollazzevole distinzione del Cherubino ad onorificenze, che mi lascerebbero completamente indifferente. Io mi sono ormai pienamente adeguato ai nuovi tempi e miro soltanto al sodo.

Voglio anche aggiungere che la più significativa dimostrazione di consenso all'opera mia mi è stata data qui a Pisa dalla Facoltà di Scienze, presieduta dal nostro Gigi Puccianti; la quale Facoltà, all'annunzio del mio trasferimento a Roma, non ha creduto di rivolgermi neppure una parola di rincrescimento."

on the petulance and invasiveness of the Germans who wanted to requisition the School for their military needs and that normal life continued as usual, though with a haggard family. Youth preferred not to be drafted and they carried out their courageous partisan struggle elsewhere.[36]

In Russo's reconstruction, there is not the slight trace of any accusation of Tonelli's collaborationism with the German invaders. Nonetheless, as for the majority of university professors, right after the Liberation and in the climate of those months, Tonelli's activity was investigated by the local screening commission. Tonelli complained repeatedly about this fact, attributing the submission to an implicit negative opinion of him, held by Russo, who had returned to direct the University of Pisa and was boasting a different and "more militant" concept of anti-fascism. As can be read in the report of the screening commission,[37] the process ended, but in any event with the complete acquittal of Tonelli (26 October 1944):

> Prof. LEONIDA TONELLI – Of long-held and constant anti-fascist ideas; among the first signers of the Croce Manifesto. In November 1943, in the absence of the Director of the Scuola Normale, Superiore Prof. Luigi Russo, in hiding after his arrest warrant, Tonelli was appointed director by the Republican Government, a position he held with decorum and dignity. In April 1944, he was awarded 200,000 Liras from the Accademia d'Italia for his outstanding scientific work, and he gave the money to the School he was directing in the form of a scholarship for mathematicians. Prof. Tonelli worked hard for the anti-fascist organization and is currently the deputy mayor of the city of Pisa.

The trust given to Tonelli was therefore full and unconditional: the Direction of the *Normale* was the (successful) attempt to protect the ideal and concrete heritage of the great school of Pisa, and nor could the anti-fascist ideas of Tonelli be doubted,

---

[36]"[Russo] ebbe la nomina [di direttore della SNSP], con il crollo del regime fascista, il 1° settembre del 1943. Nomina che ebbe effetto per pochi giorni, poiché il 13 settembre il postremo direttore [Russo] della Scuola fu costretto, per circostanze speciali, ad allontanarsi dalla città di Pisa. Il 20 ottobre, in una sua casa marina in Versilia, il direttore renitente ebbe la visita inaspettata del prof. Carlo Alberto Biggini, nuovo ministro della repubblica italiana di Salò, accompagnato dal suo capo-gabinetto Umberto Biscottini, il quale con sue parolette affettuose ed ornate tentò di persuaderlo a rientrare in carica come rettore dell'Università di Pisa e come direttore della Scuola Normale. E al rifiuto animoso oppostogli dal recalcitrante direttore, il sedicente ministro, che era un giovane vano ma non cattivo, più che per minacciare, per mettere sull'avviso, gli comunicò che era pronto il suo mandato d'arresto. In quell'occasione (...) io mi vidi costretto a rapidamente emigrare (...)

Furono (...) due bravi amici che mi portarono le prime notizie dell'Università di Pisa e della Scuola Normale Superiore, e della direzione assunta interinalmente dal prof. Leonida Tonelli, che era obiettivamente rispettato per la sua alta autorità di scienziato in materie come le matematiche di loro natura apolitiche, ed era validamente coadiuvato nella difesa della Scuola dal suo assistente e scolaro dott. Landolino Giuliano: seppi con vivo piacere che il Tonelli e il Giuliano erano riusciti a tenere a freno la petulanza e l'invadenza dei tedeschi, che avrebbero voluto requisire la Scuola per i loro bisogni militari, e che la vita normalistica continuava, sia pure in sparutissima famiglia, a svolgersi. I giovani preferivano sottrarsi al reclutamento forzato, e svolgevano altrove la loro animosa opera di partigiani."

[37]The Commission was constituted by Augusto Mancini, Vincenzo Rossi, Lorenzo Mossa, Enrico Pistolesi, Giovan Battista Funaioli, Alberto Chiarugi, and presided by the deputy rector Russo.

witnessed also by his support to the partisans in the safe home of Asciano.[38] As a consequence of his complete acquittal, on 11 December 1944, the Prefect of Pisa appointed Tonelli "member of the preliminary Commission for the screening of State staff, in substitution of Prof. Italo Simon, who resigned" [4, Appendix 4, pages 92–93]. The outcome of this event, based on testimonies and documents, appears substantially correct. There is also a slight shadow represented by the *Mussolini Prize* in 1944, contained in the report of the screening Commission, perhaps upon Russo's suggestion.

The director of this last acknowledgement was once again Gentile, who nonetheless was unable to attend the award ceremony of the Prize, since he was killed on 15 April 1944 in front of the gate of his Florentine home. In Florence, after September 8th and the constitution of the Salò Republic, *Accademia d'Italia* was also relocated. Gentile had been appointed President on 20 November 1943, and it was located in Palazzo Serristori on December 1st. The philosopher asked and obtained from the Duce, authorization to proceed and award the prizes without the contest, including the most prestigious one the *Mussolini Prize*, which the *Corriere della Sera* continued to finance. The choice made by the Academic Council which met in Florence on 10 February 1944, was to entrust the proposal to a Commission that could discuss it by correspondence and leave the final decision to the Council. Since the designation for 1944 was entrusted to the Science Class, Vallauri president of the Science Class and vice-president of the Academy proposed submitting the name of Francesco Pentimalli to the colleagues that could be found, which had already been proposed in 1940. Gentile, for precaution, asked Biggini if there were political objections against the Neapolitan cancer specialist. The response arrived extremely late, and specified "that there are no obstacles to awarding the prize to Prof. Francesco Pentimalli who, during the period of Badoglio, had been revoked from the appointment of President of the Istituto del Cancro in Naples, and after the enemy occupation of the city, he was apparently dismissed from his teaching post." In the meantime, another proposal arrived. We know this from a confidential letter sent by Vallauri to Gentile on 23 March 1944, in which the Turinese electrotechnician wrote:

> Once back in Turin, I will proceed to carry out our decision concerning the Mussolini prize. I wrote to De Blasi mentioning the difficulties of selecting an expert of biological sciences, and asked his opinion in this regard.

---

[38] In addition to the oral testimony given by his daughter in the early 1990, see the testimonies of Augusto Mancini and Annibale Mazzola in [2, page 129 and page 133, respectively]. But mainly see the following passage of a letter written by Tonelli (of 1 September 1945) to the young Enrico Magenes (1923–2010), in response to a letter of the latter, in which he notified his Teacher, about his return after 18 months of detention in Nazi lagers: "Dear Magenes, I was very pleased about your letter. After having seen here and in my home, the ferocity of the Germans, I was very worried for you. (...) My family suffered many material damages; but it is safe. But we had terrible blows to our moral; and when we came back to our rural house, from which the Germans chased us away last summer, we found out horrible things, and sought to give a holy burial to the youngsters slaughtered in our garden."

On the other hand, our class has awarded so far: 4 physicists, 2 physicians, 1 chemist, 1 engineer, 1 explorer and no mathematician. Therefore, it would be fair to propose a mathematician, instead of another researcher.

During the meeting of 4 June 1943, the last one held by our class, we had the opportunity, after the death of Marcolongo, to examine in depth, during a non-confidential exchange of ideas, our "mathematical horizon." We all agreed in recognizing the superiority of Tonelli over the others (Berzolari, Bompiani, Picone, Comessatti, Sansone, Signorini, Krall, Fantappiè).

Therefore, I deem that the choice of Tonelli would be appropriate. I wrote the enclosed letters to colleagues Giorgi and Severi, and would be pleased, if you approve the content, to mail these letters to them in the safest and fastest way. I wrote another similar letter to the other mathematician of the class, Somigliana, who resides in Casanova. Please consider whether or not it is appropriate for you to talk to Severi, regardless of my letter, but on the same topic.

Should the replies of the three mathematicians be positive, we will express our objections to the biological sciences, which we will postpone to another contest, and we can ask, if you agree, again by letter, the approval of the other academicians of the class or at least those who are involved in sciences similar to mathematics. In any case, I am afraid that due to the postal difficulties (when will you receive this letter?), they won't be ready by April 21st.[39]

There is no time to lose because the usual date for the awarding of the prize is April 21st, "the birthdate of Rome." Gentile did not make it – he was buried on April 18th in Santa Croce – but the date of April 21st was not postponed. After obtaining the consent on the name of Tonelli from the Executive Council of the Academy – though there was some disagreement (for example Somigliana, who proposed the name of Picone, in case Tonelli was appointed academician[40]) – Gentile wrote promptly to

---

[39]"Rientrato a Torino, mi appresto a provvedere a quanto fu con te concordato riguardo al premio Mussolini. Ho scritto a De Blasi accennandogli alle difficoltà che si oppongono alla scelta di un cultore delle scienze biologiche e chiedendo il suo parere in merito.

D'altro canto la nostra classe ha finora premiato: 4 fisici, 2 medici, 1 chimico, 1 ingegnere, 1 esploratore e nessun matematico. Sembrami quindi assai più giusto pensare ad un matematico, anzi che ad un altro medico.

Nell'adunanza del 4 giugno '43, l'ultima tenuta dalla nostra classe avemmo occasione, in seguito alla scomparsa del Marcolongo, di esaminare a fondo, in uno scambio di idee affatto confidenziale, il nostro "orizzonte matematico". Fummo tutti d'accordo nel riconoscere la preminenza del Tonelli sugli altri (Berzolari, Bompiani, Picone, Comessatti, Sansone, Signorini, Krall, Fantappiè).

Ritengo pertanto, che la scelta del Tonelli sarebbe particolarmente felice. Ho scritto in tal senso ai colleghi Giorgi e Severi le lettere che ti accludo, pregandoti, se ne approvi il contenuto, di disporre che esse siano inoltrate col mezzo più sicuro e sollecito. Analoga lettera scrivo all'altro matematico della classe, Somigliana, che risiede a Casanova. Lascio a te di giudicare, in base a quanto ti accennai, se non sarebbe opportuna una tua parola al Severi, indipendente dalla mia lettera, ma nel medesimo senso.

Se le risposte dei tre matematici fossero favorevoli, lasceremmo cadere la riserva circa le scienze biologiche, rimandandole ad altro turno e potremmo, se tu lo credessi opportuno, chiedere, sempre per lettera, l'approvazione degli altri accademici della classe o almeno di quelli che coltivano scienze affini alla matematica. Temo comunque che, a causa delle difficoltà postali (quando ti giungerĹ questa mia?), non potranno esser pronti per il 21 aprile."

[40]In the acceptance letter of 30 May 1944, he wrote indeed: "I agree on the opportunity to award a mathematician (...). I think that Tonelli well deserves the prize, even if his production is not completely of my liking, due to a certain heaviness and scarcity of elegance that renders the subject,

Severi (on April 5th) notifying him that the other members approved this choice and asking him to "send him, with the quickest medium," his "brief report on Tonelli that might be needed as justification for the prize." There is no trace of the detailed letter sent to Severi or of his reply in the documentation preserved. But certainly, even if the letter was sent (to Arezzo, where he found refuge after September 8th), Severi did not reply, and nor did he send any report to Gentile. He was upset with Gentile for having transformed Vico's commemoration the previous March into a political compromising manifestation, and had no intention to assist him in further academic practices.

We could end here our narration except that the following events force us to go back to the last period of Tonelli's life. Pisa was freed in early September 1944 and Luigi Russo returned about ten days later, as told by him [in Simoncelli 1994, p. 180], to resume his appointment of Director at Normale and Rector of the University. His return marked the removal of Tonelli from that School that he had defended from the pushiness of the German soldiers. His disappointment must have been significant, because he mentions it to his friend from Bologna, Alfeo Liporesi who stated, in his letter to Tonelli's wife: "His last letter vaguely mentioned his sadness and disappointments after the liberation." Most likely, the sadness was linked to the sensation perceived by Tonelli of a negative judgement created by connivances with the authorities of Northern Italy. This sadness drove Tonelli to write an incredible letter to the Ministry of Education on 10 September 1945, in which he asks for the annulment of his transfer to Pisa in 1942, so that he can return to Rome:

> it is with great sadness that I found out I had been completely discarded. This sadness does not allow me to continue by work as a Teacher in Pisa, a job for which my longed-for acknowledgement was denied through the appointment of prof. Russo as director of Normale.

The disagreement between the two directors of the Normale interlaced with the other disagreement that we mentioned, with Picone, according to the logic that "the enemies of my enemies are my friends." The occasion, a few months after the death of Tonelli, is represented by the outcome of the competition of Analysis for the University of Genoa, that saw the loss of Sandro Faedo. The tension created with the "school" of Pisa and Tonelli's students is remarkable. Rector Russo consults Picone directly:

> I criticized this concept of "school," that here in Pisa, they want to assert: the "school of Pisa" is intended as the small churches of our Sicilian towns: where there was S. Antonio of

---

an artistic science. But I also remember the circumstance that during the last awarding, voting took place between Majorana and Tonelli, designated since then, as worthy of the prize. (...) In addition to Tonelli for the appointment of Academician, I think that we could think about another mathematician worthy of the Mussolini Prize, Mr. Picone, who is not at the same scientific level of Tonelli, but he has the merit of the foundation of the Istituto del Calcolo, which acquired a high level of importance and utility thanks to him. And I think that the Mussolini Prize could also be awarded for non scientific merits."

Padova, we cannot put the statue of S. Francesco or S. Antonio abate (...) In one of my long held opinions, the "school," the "local tradition" is always defended by mediocre people, the powerless, who cannot protect the school and the studies, but only their weakness. Should I say the same for the several men like Danieli,[41] Lazzarino,[42] Cecioni, Giuliano, who defended the local tradition? I dont know, because I'm not a mathematician.

Please give this letter to Guido Castelnuovo, who I proclaimed today as my mentor and guide in a press release, due to his superior intelligence and seniority of years, he can advise me on these mathematical issues. I have been troubled by mathematics for three years. I'm sure you saw the generous words I wrote for Landolino Giuliano and Leonida Tonelli[43]: I did not mention that he obtained the position from the government of the Republic of Salò, that he was awarded the prize of 200 thousand liras from the dying Accademia d'Italia, and that he pretended to give the 200,000 liras to the School to which he never gave anything.

Until now, I gave given proof of having religious patience, of which I thought I was not capable. Sooner or later I shall explode and write some of my lively notes on Belfagor, to report to the public opinion, all these twists of certain men that lower the level of our Universities more and more.

Be patient my dear Picone, and give me the patience of my sovereign master Guido Castelnuovo. Please always be there for me with your care, advice and help: I will not bring forward any polemic, for the love of peace and unfortunate Italy, which in addition to the seven politics, also the small academics sects are blooming.[44]

Russo not only limits himself to this. He grasped the opportunity of Tonelli's honours by the University of Pisa on 12 March 1947, in view of the first anniversary of his death, to point out the detachment of *Normale* and his personal detachment

---

[41] Pietro Ermenegildo Daniele (1875–1949).

[42] Orazio Lazzarino (1880–1963).

[43] Reference is made to his opening lecture on *La Scuola Normale Superiore (1944–1946)*, held in occasion of the inauguration of the year of the school, on 15 December 1946, that we have already mentioned.

[44] "Ho criticato questo concetto di «scuola», che qui a Pisa vorrebbero far valere: la «scuola pisana»è intesa come le chiesette dei nostri paesi siciliani: dove c'è stato S. Antonio di Padova, non ci può entrare la statua di S. Francesco o di S. Antonio abate, quello del porco. (...) Mi confermo in una mia vecchia opinione, che la «scuola», la «tradizione locale»è difesa sempre dai mediocri, dagli impotenti, i quali in tal modo non proteggono la scuola e gli studi, ma soltanto la loro debolezza. Che debba ripetere lo stesso per i vari Danieli, i vari Lazzarino, i vari Cecioni, i vari Giuliano, che difendono la tradizione locale? Io non lo so, perché non sono matematico.
Ti prego di far leggere questa mia lettera a Guido Castelnuovo, che io in una prolusione oggi a stampa ho proclamato mio mentore e guida, perch per l'altezza dell'ingegno e per la casta superiorità degli anni può consigliarmi in queste faccende della matematica. Sono tre anni che tribolo per la matematica. Avrai visto le generose parole che io scritto per Landolino Giuliano e per Leonida Tonelli: ho passato sotto silenzio che egli aveva avuto la direzione dal governo della Repubblica di Salò, che aveva riscosso il premio di 200 mila lire della allora moribonda Accademia d'Italia, che aveva finto l'attribuzione delle 200 mila alla Scuola a cui non ha dato mai un bel nulla.
Io finora ho dato prova di una pazienza religiosa, di cui non mi credevo capace. Una volta o l"altra, scoppio e scrivo qualcuna delle mie note vivaci su Belfagor, per denunciare all'opinione pubblica tutto questo rigiro di alcuni omuncoli, che abbassano sempre più il livello delle nostre Università.
Abbia pazienza, mio caro Picone, e impetra sul mio capo la pazienza del maestro sovrano Guido Castelnuovo. Soccorretemi sempre del vostro affetto, del vostro lume, dei vostri aiuti: io terrò imbrigliata la mia musa polemica, per amore della pace e di questa sventurata Italia, in cui fioriscono oltre che le sette politiche, anche le piccolissime sette accademiche."

from the celebrations. Then, he prevented the Minister of the Public Education Gonella from taking part in the ceremony, despite it was announced with a telegram and advertised on the local press. In Tonelli's personal folder stored at the Central Archive of the State, there is the following letter (of 5 March 1947) written by Russo to the general director of Higher Education, Giuseppe Petrocchi:

> Dear Petrocchi,
> from the local newspapers, I learned that the minister Gonella should come to Pisa for the anniversary of the death of Prof. Tonelli. I would have been pleased if you could be a guest of Scuola Normale on this occasion and I had already prepared an invitation letter that I enclose herein, for your information, when the Prefect told me that it would have been difficult for the Minister to come here.
> I take this opportunity to inform you about a few very confidential facts. You know that Tonelli during the period of the republic of Salò, was appointed director of Scuola Normale; you also know that he obtained the Mussolini Prize of 200,000 liras from Accademia d'Italia in April 1944 for mathematics; and in conclusion, you know that Tonelli donated this money to the School prior to the arrival of the allies, but he never followed up with his commitment, and nor is it likely that his heir will do so.
> We know all of this in Pisa: and stand in this atmosphere an official ceremony, in which the Minister would participate would be inappropriate. Aside from the fact that similar ceremonies never had famous mathematicians of the past such as Betti and Bianchi, of whom there is not even a marble plaque at the university.[45]
> I tell you this confidentially, so the Minister realizes that this is not the best occasion should he decide to come to Pisa.[46]

The detachment of Tonelli from the Italian academic world is therefore marked by a belated polemic in which personal aspirations and envies mix to different conceptions of politics and anti-fascism. Russo represents a democratic and militant Italy, but in spite of this, he did not hesitate to seek the support of Mauro Picone, who was supporting the other party until a few months before. Even Tonelli's parable is educational. His initial anti-fascism cannot be doubted. Then there was the choice

---

[45] Reference is made to the commemorative tombstone located in the courtyard of La "Sapienza" palace in Pisa, that states: "L'Università ricorda fra i suoi Maestri più insigni Leonida Tonelli, matematico, che dischiuse nuove vie alla ricerca – modello di civiche virtù in pace e in guerra."

[46] "Caro Petrocchi, dai giornali locali si è appreso che il ministro Gonella dovrebbe venire a Pisa per le onoranze al prof. Tonelli, deceduto un anno fa. Io avrei avuto piacere che potesse essere ospite della Scuola Normale in questa occasione e avevo già preparato una lettera di invito che Le accludo, per conoscenza, quando dal Prefetto mi è stato fatto capire che sarebbe stato difficile che il Ministro venisse qui.
Io mi permetto farLe presente alcuni fatti di carattere riservatissimo. Lei sa che il Tonelli, durante il periodo della repubblica di Salò, fu nominato direttore della Scuola Normale; Lei sa anche che dall'Accademia d'Italia di allora egli ottenne nell'aprile 1944 il premio Mussolini di £200.000 per le matematiche; e infine Lei sa troppo bene che il Tonelli ha donato questi denari alla Scuola poco prima dell'arrivo degli alleati, ma poi non ha più mantenuto il suo impegno, né è probabile che i suoi eredi vogliano mantenerlo.
Tutto questo a Pisa si sa: e c'è un ambiente che troverebbe inopportuna una cerimonia ufficiale, cui partecipasse anche il Ministro. A parte il fatto che simili cerimonie non ebbero illustri matematici del passato, quali il Betti e il Bianchi, di cui non esiste all'Università neppure un ricordo marmoreo.
Io Le dico tutto questo, in via privata, perch il Ministro qualora avesse deciso di venire a Pisa, si renda conto che non è questa l'occasione più propizia."

of the work and defence of the position attained as the most suitable instrument for the pursuit of certain values, even to the detriment of certain formal compromises with the regime. In 1934, Tonelli asked for the membership card in the National Fascist Party. Personal envies and grudges in the "extreme" wing of the party, which feared being polluted by non-fascist elements since the beginning, thus delaying the positive procedure of the request; Tonelli registered in the P.N.F. only in 1940 (with a creative backdating on 3 March 1925, therefore at the same time of the signature of *Manifesto Croce*[47]). The acceptance to direct the *Normale* is a manifestation of the same logic, in reality much less compromising and more understandable. The case of the *Mussolini Prize* is different, which Tonelli accepted at first, as compensation for the one he had missed in 1940. And only a month later, perhaps upon the suggestion of more canny political councillors, his will to devolve the amount in favour of a scholarship was advertised, but he was not able to pursue this due to lack of time. A sad outcome for a democratic generation and mentality that wanted to find a compromise with the Fascism a few hours prior to its fall.

References

1. Agostini, A. 1935. Matematici, Fisici, Direttori e Professori della Scuola Normale Superiore di Pisa, in *Annuario Scuola Norm. Sup. Pisa, vol. I*. 285–301.
2. AA. VV. 1952. *Leonida Tonelli. In memoriam*. Arti Grafiche Tornar; Pisa; (it is the statement of the anniversary of the 1st year of death).
3. AA. VV. 1986. *Convegno celebrativo del centenario della nascita di Mauro Picone e di Leonida Tonelli*. (Rome, 6–9 May 1985), Accad. Naz. Lincei: Rome.
4. AA. VV. 1998. *Leonida Tonelli e la matematica nella cultura italiana del '900* (Atti del Convegno omonimo, Pisa, Aula magna storica dell'Università degli studi, 1 March 1996), Scuola Normale Superiore (Quaderni della Direzione), Pisa.
5. Bernardini, G. 1969. *La Scuola Normale Superiore e la vita culturale del paese*. Accad. Naz. Lincei: Rome; (collection: "Problemi attuali di Scienza e di Cultura, book no. 118).
6. Cantimori, D. 1963. Conferenza celebrativa del 150° anniversario della fondazione della Scuola Norm. Sup., In *Annuario Scuola Norm. Sup. Pisa*. vol. V: 31–38.
7. Cinquini, S. 1950. Leonida Tonelli, *Annali Scuola Norm. Sup. Pisa*. cl. Sci., (2) t. 15, no. 1–4, 1–37.
8. Cugiani, M. 1963. Commemorazione di G. Ricci, *Rendiconti del Seminario Matematico e Fisico di Milano*. vol. XLIII, 6–23.
9. Di Sieno, S., and A. Guerraggio. 1998. e Nastasi P. (a cura di), *La matematica italiana dopo l'Unità. Gli anni tra le due guerre mondiali*. Marcos y Marcos: Milan.
10. Ferrarotto, M. 1977. *L'accademia d'Italia. Intellettuali e potere durante il fascismo*. Liguori: Naples.

---

[47] Again in 1939, in view of Tonelli's requests to adhere to the international committee for the honours to the French mathematician Émile Borel (1871–1956) and hold a relation with the international Congress of mathematicians foreseen in 1940 in Cambridge (Mass.) and then cancelled, the ministerial practices labelled him as signer of *Croce Manifesto*, not registered to PNF (see folder "Tonelli Leonida di Gaspare" in ACS, Min. Interni, Dir. Gen. P.S., Aff. Gen. and Ris., cat. A 1, a. 1939, envelope 79).

11. Fichera, G. 1963. L'analyse fonctionnelle en Italie, *Cahiers d'histoire mondiale*, no. 2, vol. VII: 407–417.
12. Fichera, G. 1979. Il contributo italiano alla teoria matematica dell'elasticità, *Rendiconti del Circolo Matematico di Palermo*, S. II. t. XXVIII, 5–26.
13. Fichera, G. 1999. L'analisi matematica in Italia fra le due guerre, *Rend. Mat. Acc. Lincei*, (9) vol. 10: 279–312; (ora in Id., *Opere storiche, biografiche, divulgative* (by L. Carbone, P.E. Ricci, C. Sbordone, D. Trigiante), Giannini, Naples, 2002, 409–442).
14. Guerraggio, A., and P. Nastasi. 1993. *Gentile e i matematici italiani. Lettere 1907–1943*, Boringhieri: Turin.
15. Guerraggio, A., and P. Nastasi. 2005. *Italian Mathematics Between the Two World Wars*, Birkhäuser: Basel.
16. Isnenghi, M. 1979. *Intellettuali militanti e intellettuali funzionari. Appunti sulla cultura fascista*. Einaudi: Turin.
17. Karachalios, A. 2001. I chimici di fronte al fascismo. Il caso di Giovanni Battista Bonino (1899–1985), Istituto Gramsci Siciliano: Palermo; pages 71–74.
18. Marino, G.C. 1983. *L'autarchia della cultura. Intellettuali e fascismo negli anni '30*. Editori Riuniti: Rome.
19. Mercuri, L. 1989. *L'epurazione in Italia, 1943–1948*. L'Arciere: Bergamo.
20. Nastasi, P., and A. Scimone (ed.), 1995. Lettere a Giovanni Vacca, "Quaderni Pristem," no. 5, Palermo.
21. Picone, M. 1941. "Appunti di Analisi superiore," *Boll. Un. Mat. It.*, (2) 3: 402–405.
22. Pepe, L. 1984. *Giuseppe Vitali, Opere sull'Analisi Reale e Complessa*. Carteggio, Tecnoprint: Bologna.
23. Pucci, C. L'Unione Matematica Italiana dal 1922 al 1944: documenti e riflessioni, In *Symposia Mathematica*. vol. XXVII, Academic Press: London; 1986 pages 187–212.
24. Racinaro, R. 1975. Intellettuali e fascismo, *Critica Marxista*. a. XIII, no. 1, pages 177–214.
25. Ricci, G. 1948. La Scuola matematica pisana dal 1848 al 1948, Conferenza letta in occasione del III Congresso dell'Unione Matematica Italiana (Pisa, 23–26 September 1948).
26. Russo, L. 1946. La Scuola Normale Superiore durante gli ultimi tre anni, In *Annali della Scuola Norm. Sup. di Pisa*. (Lettere, Storia e filosofia), (2), vol. XV, 1–18.
27. Sansone, G. 1948. L'opera scientifica di Leonida Tonelli, *Atti Accademia Nazionale dei Lincei*. a. CCCXLV, s. VIII, Rendiconti Cl. Sc. F., Mat. e Nat., Vol. IV, 1st semester, folder 5, 594–624.
28. Sansone, G. 1977. *Algebristi, Analisti, Geometri differenzialisti, Meccanici e Fisici-Matematici ex normalisti del periodo 1860–1929*. Scuola Norm. Sup.: Pisa.
29. Sansone, G. 1977. *Geometri algebristi ex normalisti del periodo 1860–1929*. Scuola Norm. Sup.: Pisa.
30. Simoncelli, P. 1994. Cantimori, *Gentile e la Normale di Pisa. Profili e documenti*. Franco Angeli: Milan.
31. Simoncelli, P. 2009. *L'epurazione antifascista all'Accademia dei Lincei. Cronache di una controversa "ricostituzione"*. Le Lettere: Florence.
32. Tonelli, L. 1960–1963. *Opere Scelte*. 4 voll., Cremonese: Rome.
33. Turi, G. 1980. *Il fascismo e il consenso degli intellettuali*. Il Mulino: Bologna.
34. Turi, G. 1995. *Giovanni Gentile. Una biografia*. Giunti: Florence.

Bruno Pini and the Parabolic Harnack Inequality: The Dawning of Parabolic Potential Theory

Ermanno Lanconelli

Abstract In this article, we describe the pioneering work of Bruno Pini toward the modern Potential Analysis of second order parabolic Partial Differential Equations. We mainly focus on the *parabolic Harnack inequality*, discovered by Pini in 1954, jointly, and independently, with Jacques Hadamard. Pini made of this inequality one the crucial tools in his construction of a *generalized Wiener-type solution* to the Dirichlet problem for the Heat equation. To this end, he also used an average operator on the level sets of the Heat kernel, characterizing *caloric* and *sub-caloric* functions, in analogy with the classical Gauss–Koebe, Blaschke–Privaloff and Saks Theorems for harmonic and sub-harmonic functions. He also proved a Riesz-type representation Theorem for subcaloric functions. To complete his research design, Pini established the notion of *Heat capacity*, and proved a *Wiener-type criterion* for regularity of the boundary points in the generalized Dirichlet problem for the Heat equation.

1 Bruno Pini: A Short Biography

On February 26, 1918 Bruno Pini was born in Poggiorusco, a small town not far from Mantova, where his family moved during the last year of the Big War. He died in Forlì on November 24, 2007. His long life was entirely devoted to Mathematics, and teaching.

Bruno Pini studied at the University of Bologna, where he became a student of Gianfranco Cimmino, and received his Laurea Degree in Mathematics on December 1941. After the second World War, Pini became assistant professor at the Istituto Matematico "Salvatore Pincherle" in Bologna, and in 1953, after winning

---

E. Lanconelli (✉)
Dipartimento di Matematica, Università di Bologna, Piazza di Porta San Donato, 5, 40126 Bologna, Italy
e-mail: lanconel@dm.unibo.it

a national competition, he moved to the University of Cagliari as full professor of Mathematical Analysis. In 1956, Pini moved to the University of Modena, and finally, in 1958, he returned to Bologna, on one of the two chairs of Mathematical Analysis of that University. In Bologna, Pini started his major activity as a Master and mentor of young researchers, ended in 1993 when he retired from teaching. He supervised a large number of students who then became full professors at the University of Bologna.

He was member of the "Accademia Nazionale dei Lincei," of the "Accademia delle Scienze dell'Istituto di Bologna" and of the "Accademia Nazionale di Scienze Lettere e Arti di Modena." In 1978, he received the Gold Medal of the "Accademia detta dei Quaranta."

After his retirement, Pini was named Professor Emeritus of the University of Bologna.

The scientific activity of Bruno Pini mainly focused on the Theory of Partial Differential Equations. He published over seventy scientific papers, and a series of Monographs containing an impressive amount of detailed and deep theoretical results in that vast theory, together with a rich description of many related applied topics.

2 The Parabolic Harnack Inequality

The Harnack inequality for harmonic functions is the ancestor of a family of inequalities playing a key role in the general theory of second order Partial Differential Equations (PDE). It was originally formulated for harmonic functions in the plane by Carl Gustav Axel von Harnack and first published in the book [8] in 1887. In its modern version, currently used in the theory of partial differential equations, the Harnack inequality for harmonic functions reads as follows.

Let $N \geq 2$, and let $\Omega \subset \mathbb{R}^N$ be an open set. Then, there exists a positive constant C, only depending on the dimension N, such that

$$\sup_{B_r} u \leq C \inf_{B_r} u, \qquad (1)$$

for every nonnegative harmonic functions $u : \Omega \to \mathbb{R}$, and for every Euclidean ball B_r whose concentrical ball B_{2r} is contained in Ω.

Before proceeding, we want to recall that a function $u : \Omega \to \mathbb{R}$ is called *harmonic* if its second order derivatives are continuous, and it satisfies the Laplace equation

$$\Delta u = 0 \quad \text{in } \Omega, \qquad \Delta := \sum_{j=1}^{N} \frac{\partial^2}{\partial x_j^2}.$$

Inequality (1) can be easily proved by using the Poisson formula, giving the solution of the Dirichlet problem for Δ on the Euclidean ball $B_r(x_0)$ with continuous boundary data φ:

$$u(x) = \frac{r^2 - |x - x_0|^2}{\omega_n r} \int_{\partial B_r(x_0)} \frac{\varphi(y)}{|x - y|^N} \, d\sigma(y), \quad \omega_n = \text{area of } \partial B_r(x_0). \quad (2)$$

In spite of its almost trivial proof, the Harnack inequality (1) has deep and powerful consequences. Two of the most important of them are the Harnack's second convergence theorem and the strong maximum principle:

(1) *Let (u_n) be a sequence of harmonic functions in a connected open set $\Omega \subset \mathbb{R}^N$. If (u_n) is monotone increasing and there exists a point $x_0 \in \Omega$ such that $(u_n(x_0))$ is convergent, then (u_n) is uniformly convergent on every compact subset of Ω.* (Harnack's Second Convergence Theorem)
(2) *Let u be a harmonic function in a connected open set $\Omega \subset \mathbb{R}^N$. If there exists a point $x_0 \in \Omega$ such that $u(x_0) = \max_\Omega u$, then $u \equiv u(x_0)$ in Ω.* (Strong Maximum Principle)

We would also like to mention the Harnack's *first* convergence theorem, which can be stated as follows.

Let (u_n) be a sequence of harmonic functions in a bounded open set Ω with smooth boundary. If u_n is continuous up to $\partial \Omega$ and the sequence of the restrictions $(u_n|_{\partial\Omega})$ is uniformly convergent, then, on every compact subset of Ω, (u_n) is uniformly convergent, together with all the sequences of its derivatives. Moreover $u := \lim_{n \to \infty} u_n$ is harmonic in Ω.

As a consequence, in the statement of Harnack's Second Convergence Theorem, the limit function of the sequence (u_n) is harmonic.

The Harnack's inequality (1), and the *unique solvability of the Dirichlet problems for the Euclidean balls* (given by the Maximum Principle, and the Poisson formula (2)), are the main basic ingredients toward the development of an exhaustive Potential Analysis of the Laplace equation. The depth and completeness of these explorations, consequences of so few assumptions, lead to the birth of the Potential Theory in abstract Harmonic Spaces, mainly performed by Marcel Brelot and his school. The results of Brelot's theory, however, only apply to Elliptic, or degenerate Elliptic, PDEs. In particular, they are not applicable to the case of the Heat equations and, more generally, of parabolic partial differential equations.

The breakthrough toward a parabolic potential theory was the discovery of a kind of Harnack inequality for nonnegative solutions to the Heat equation in $\mathbb{R}^2$, $Hu := (\partial_x^2 - \partial_t)u = 0$.

It was not obvious what the analog of (1) for caloric functions (i.e. solutions to the Heat equations) could be. Indeed, consider the Heat kernel, that is the fundamental solution of H:

$$\Gamma(x, t) = \frac{1}{\sqrt{4\pi t}} \exp\left(-\frac{x^2}{4t}\right), \text{ if } t > 0, \quad \Gamma(x, t) = 0 \text{ if } x \neq 0 \text{ and } t \leq 0. \quad (3)$$

This function is nonnegative and caloric in $\mathbb{R}^2 \setminus \{(0,0)\}$, but there is no positive constant C satisfying

$$\sup_B \Gamma \leq C \inf_B \Gamma, \quad \text{with} \quad B = \left\{(x-1)^2 + t^2 < \frac{1}{4}\right\}, \tag{4}$$

since the left-hand side is strictly positive while the right-hand one is zero.

In 1954, almost seventy years after the appearance of C.G. Axel Harnack's inequality, the right analogy between harmonic and caloric functions was found, independently, by Bruno Pini and Jacques Hadamard. Roughly speaking, this analogy can be illustrated as follows.

First, remark that the classical Harnack inequality (1) says:

Suppose we are given a non-negative harmonic function u in a region containing an Euclidean ball $B_{2r}(x_0) = \{|x - x_0| < 2r\}$. Then, if $u(x_0) \geq 1$, at every point of the ball $B_r(x_0)$ the function u is larger than a positive constant C independent of u, of the given region and of the ball $B_{2r}(x_0)$.

Analogously, suppose we have a heat conducting body, containing a certain amount of heat, in absence of a heat source. Assume the body contains a ball $B_{2r}(x_0)$ and that its temperature on $\{x_0\}$ at a time t_0 is larger than 1. Then, *at the time $t_0 + r^2$ the body's temperature on every point of the ball $B_r(x_0)$ is larger than a positive constant C independent of the temperature, of the given body, and of the ball $B_{2r}(x_0)$*.

It has to be noticed that *we really have to wait some time* for the ratio between the temperature on two different points of a body, to be bounded from below by an absolute positive constant. Indeed, consider the sequence of nonnegative caloric functions $\left(u_n(x,t) := \Gamma\left(x, t + \frac{1}{n}\right)\right)$ in the region $\{(x,t) : x > 0\}$. For every fixed $\theta \in]0, 1[$ the quotient

$$\frac{u_n(1+\theta, 0)}{u_n(1, 0)} = \exp\left(-\frac{n}{4}(\theta^2 + 2\theta)\right),$$

goes to zero as n goes to infinity. We also have to remark that the "waiting time" has to be proportional to r^2, if $2r$ is the radius of the ball contained in the given body. This comes from the *homogeneity* of the Heat operator with respect to the anisotropic dilations

$$\delta_\lambda(x,t) = (\lambda x, \lambda^2 t). \tag{5}$$

Let us now come to the original version of the *parabolic Harnack inequality*. In the paper [17], dated 1954, Bruno Pini proved the following inequalities:

$$0 \leq u(x,y) \leq K u\left(\frac{1}{2}, \frac{1}{4}\right) \quad \forall \, (x,y) \in R(\delta), \tag{6}$$

where u is a continuous function in the rectangle

$$R = \{0 \le x \le 1\} \times \left\{0 \le t \le \frac{1}{4}\right\},$$

caloric in the interior of R and nonnegative on the *parabolic boundary*

$$\partial_p R = \{(x,0) : 0 \le x \le 1\} \cup \left\{(0,t) : 0 \le y \le \frac{1}{4}\right\} \cup \left\{(1,t) : 0 \le t \le \frac{1}{4}\right\}.$$

Moreover, $0 < \delta < \frac{1}{8}$,

$$R(\delta) = \{\delta \le x \le 1-\delta\} \times \left\{\delta \le y \le \frac{1}{4} - \delta\right\},$$

and $K > 0$ only depends on δ.

To prove inequality (6) Pini started from the Green representation formula for u, and used the explicit expression of the Green function for the rectangle R. This expression was well known at that time: Pini gave as a reference for it the Doetsch's monograph [4], Chap. 20. With the right intuition of what the parabolic Harnack inequality could be, it was not difficult for Pini to prove (6) by a simple direct computation. Surprisingly enough, quite exactly the same computations, and at the same time, were independently performed by Jacques Hadamard, who published his version of the parabolic Harnack inequality in [7].

In his work [17], Pini extended inequality (6) to nonnegative caloric functions on what he called *domini normali*, by using a kind of covering argument allowing to obtain the parabolic Harnack inequality in the following modern form. Let $\Omega \subset \mathbb{R}^2$ be a connected open set and let $z_0 \in \Omega$. Denote by Ω_{z_0} the subset of the point $z \in \Omega$, which can be reached from z_0 along piecewise smooth paths with decreasing time component. Then, *if u is a nonnegative caloric function in Ω and K is a compact set contained in the interior of Ω_{z_0}*, one has

$$\sup_K u \le C u(z_0), \qquad (7)$$

where C is a positive constant only depending on K and Ω.

In the modern literature, it is usually stressed a so called *scale invariant* form of inequality (7), i.e. an inequality with constant C independent of the domain of the involved caloric function. This is the parabolic analog of inequality (1), and it comes from (7) and the invariance of the Heat equation with respect to the anisotropic dilations (5). Let $z_0 = (x_0, t_0) \in \Omega$ and let $r > 0$ be such that the cylinder

$$Q_{2r}(z_0) := \{|x - x_0| < 2r\} \times \{t_0 - 4r^2 < t \le t_0\},$$

is contained in Ω. Finally, let us define

$$Q_r^-(z_0) := \{|x - x_0)| < r\} \times \{t_0 - 2r^2 < t < t_0 - r^2\}.$$

Then *there exists a positive constant C, independent of z_0, Ω and r such that*

$$\sup_{Q_r^-(z_0)} u \le C\, u(z_0), \tag{8}$$

for every nonnegative caloric function u in Ω.

As for classic harmonic functions, inequality (7) (or equivalently, its scale invariant version (8)), has important and deep consequences, such as the analog of the Harnack second convergence theorem, and the strong maximum principle. Here are the precise statements of these results:

(i) Let (u_n) be a sequence of caloric functions in a connected open set $\Omega \subset \mathbb{R}^2$. If (u_n) is monotone increasing and there exists a point $z_0 \in \Omega$ such that $(u_n(z_0))$ is convergent, then (u_n) is uniformly convergent on every compact subset of Ω_{z_0}.
(ii) Let Ω be a connected open subset of $\mathbb{R}^2$ and let $u : \Omega \to \mathbb{R}$ be a caloric function. If there exists a point $z_0 \in \Omega$ such that $u(z_0) = \max_\Omega u$, then $u \equiv u(z_0)$ in Ω_{z_0}

The original Harnack-type second convergence Theorem for caloric functions was stated by Pini in [17], page 429. The Strong Maximum Principle in (2) had been proved one year before by Louis Nirenberg in [14], with an argument resembling the one introduced by Hopf in the proof of his Strong Maximum Principle for Elliptic second order PDE's.

It has to be remarked that also the analog for caloric functions of the Harnack's first convergence Theorem was already known at the time of Hadamard's and of Pini's works. These authors were aware of it: Pini gave for granted the convergence to a caloric function of the sequences of caloric functions uniformly convergent on every compact subset of a given region. Hadamard quoted the paper by Sven Taeklind [23], where this theorem is explicitly stated on page 13.

Some page pictures of the original papers by Pini and Hadamard can be found in the Appendix.

Additional remarks are now in order.

- The extension of the parabolic Harnack inequality to the Heat equation in spatial dimension $N \ge 2$,

$$Hu = 0, \quad H := \Delta - \partial_t = \sum_{j=1}^{N} \partial_{x_j}^2 - \partial_t,$$

appeared in 1955. It is due to Oscar Montaldo, and it was published in [12].

- In 1955, Pini extended his parabolic Harnack inequality to equations with variable coefficients of the kind

$$\frac{\partial^2}{\partial x^2}u - \frac{\partial}{\partial t}u + a\frac{\partial}{\partial x}u + b = 0,$$

where a and b are Hölder continuous functions. For this extension, appeared in [21], Pini used an approach not requiring the explicit knowledge of the Green function. As he wrote, *la presente trattazione può facilmente imitarsi per ottenere l'estensione alla più generale equazione parabolica lineare omogenea in quante si vogliano variabili* (the present treatment can be easily imitated to obtain the extension to the general linear homogeneous parabolic equation in any dimensions). However, such a general extension only appeared in literature in 1964, due to Jürgen Moser. As a matter of fact, in his work [13], Moser proved the Harnack inequality for general parabolic operators in divergence form with merely bounded measurable coefficients. His proof, based on a today celebrated iteration procedure, opened the way to the extension of the Harnack inequality to nonlinear elliptic and parabolic equations. For a recent survey on these and other developments we refer the reader to the paper [9] by Moritz Kassmann.
- We want to explicitly mention one of the most important and unexpected developments of the researches on the parabolic Harnack inequality. About twenty years ago, A. Grygor'yan and L. Saloff-Coste discovered that nonnegative caloric functions on a complete and noncompact Riemannian manifold M satisfy a scale invariant Harnack inequality if and only if the geodesic balls verify the so called *doubling property* and support a *Poincaré-type inequality* (caloric function on M are the solution to the equation $\Delta_M u - \partial_t u = 0$, being Δ_M the Laplace–Beltrami operator on M). For a precise statement of this and of related results, we refer the reader to the Saloff-Coste monograph [22].

3 The Pini's Wiener-Type Solutions for the Heat Equation

Pini's motivation for the parabolic Harnack inequality was a research program accomplished in a series of papers dated 1954 and 1955. That program was aimed at introducing a Wiener-type solution of the first boundary value problem for the Heat equation, by adapting ideas, devices and methods of the classical Potential Theory for the Laplace equation.

3.1 The Mean Value Property for Caloric Functions

A crucial tool in Pini's investigations was a Mean Value formula characterizing the caloric function that he had proved three years earlier, in [15]. Extending to the solutions of the Heat equation the classical Gauss Mean Value Theorem for harmonic functions, in [15] Pini showed that a smooth function $u : D \to \mathbb{R}$ solves the equation $\mathrm{H}u = 0$ if and only if

$$u(z) = u(x,t) = m_r(u.z) := \int_{\partial \Omega_r(z)} u(\zeta) K(z-\zeta) d\sigma(\zeta), \quad K := \frac{|\nabla_x \Gamma|^2}{|\nabla_{x,t} \Gamma|} \quad (9)$$

for every $z \in D$ and $r > 0$ such that the *Heat ball*

$$\Omega_r(z) := \left\{ \zeta : \Gamma(z-\zeta) > \frac{1}{r} \right\}, \quad (10)$$

has the closure contained in D. Here Γ denotes the fundamental solution of the Heat equation

$$\Gamma(x,t) = \frac{1}{\sqrt{4\pi t}} \exp\left(-\frac{|x|^2}{4t}\right) \text{ if } t > 0, \quad \Gamma(x,t) = 0 \text{ if } t \le 0.$$

Then, the *Heat sphere* $\partial \Omega_r(z_0)$ is a level set of the fundamental solution of the adjoint operator $\mathrm{H}^* = \partial_x^2 + \partial_t$.

In [15], the caloric functions are also characterized in terms of the "area average"

$$\mathrm{M}_r(u,z) := \frac{2}{r^2} \int_0^r \rho\, m_\rho(u,z)\, d\rho.$$

Pini's caloric average operators were extended to every spatial dimension $N \ge 2$ by O. Montaldo [12]. Later on, W. Fulks in [6] and N.A. Watson in [24] gave nice explicit expressions of Pini-Montaldo's operators in dimension $N \ge 1$.

3.2 Subcaloric and Supercaloric Functions

The first step of Pini's program toward the caloric Wiener-type solutions, was performed in the paper [16]. The notions of subcaloric and supercaloric functions (*funzioni subvalenti* and *funzioni prevalenti*, in Pini's language) were introduced there. These are the analog for the Heat equation of the classical subharmonic and superharmonic functions.

A continuous function $u : \Omega \to \mathbb{R}$ is called *subcaloric* (*supercaloric*) if for every open rectangle $R \subset \Omega$ the solution v of the boundary value problem

$$\mathrm{H}v = 0 \text{ in } R, \quad v = u \text{ on } \partial_p R,$$

satisfies the inequality $v \ge u$ ($v \le u$) in R (actually, instead of the rectangles, Pini considers the slightly more general family of piecewise regular *domini normali*). In [16], subcaloric functions are characterized in terms of submean property with respect to the operators m_r and M_r and also in terms of the monotonicity with respect to r of the caloric averages. Pini also extended to this new setting the classical Blaschke and Privaloff Theorems. As a byproduct, he obtained the Heat analog of the classical Koebe Theorem for harmonic functions: a continuous function satisfying the caloric Mean Value property on every caloric ball contained in its domain, actually is smooth and solves the Heat equation.

Again in 1954 Pini wrote the papers [18] and [19]. In the first one, after extending the notion of sub-caloric function, weakening the continuity assumption of the involved functions, only assuming their uppersemicontinuity, it is proved the following Theorem.

Let $u : \Omega \to [-\infty, \infty[$ *be a subcaloric function in the open subset* Ω *of* $\mathbb{R}^2$, *and let* Ω_1 *be a bounded open region with closure contained in* Ω. *Then there exists a nonnegative measure* μ *supported in* $\overline{\Omega_1}$, *and a caloric function* h *in* Ω_1 *such that* $u = \Gamma * \mu + h$ *in* Ω_1, *being* Γ *the fundamental solution of the Heat equation.*

In the paper [19], Pini extended to his subcaloric functions Saks Theorem on classical subharmonic functions.

We would like to mention that several of these last Pini's notions and results were extended thirty years later to the higher dimension by Watson in [24]. In his paper Watson called *temperatures* and *subtemperatures* what we call caloric and subcaloric functions, respectively, and made a crucial use of the caloric average operator we have mentioned above.

3.3 The Caloric Wiener-Type Solutions

The core of Pini's research program was established in the paper [17]. There he proved the parabolic Harnack inequality and constructed his generalized Wiener-type solution of the first boundary value problem for the Heat equation on *domini normali*. Pini called *domino normale* a subset D of $\mathbb{R}^2$ of the kind

$$D = \{(x,t) : a < x < b, \ \chi_1(t) < x < \chi_2(t)\}.$$

where χ_1 and χ_2 are continuous functions. The *parabolic boundary* of D is

$$\partial_p := \partial D \setminus \{(x,b) : \chi_1(b) < x < \chi_2(b)\}.$$

Now, let $f : \partial_p \to \mathbb{R}$ be a continuous function, restriction to $\partial_p D$ of a polynomial function F which is subcaloric in a neighborhood of the closure of D. Let (D_n) be a sequence of *piecewise smooth domini normali* exhausting D and such that $\overline{D_n} \subset D_{n+1}$. Then Pini defined a sequence of functions (u_n) in D by setting $u_1 = F$ and

$$u_{n+1} = F \text{ in } D \setminus D_n, \quad u_{n+1} = v_n \text{ in } D_n,$$

being v_n the solution of the boundary value problem

$$Hv_n = 0 \text{ in } D_n, \quad v_n = F \text{ on } \partial_p D_n.$$

This solution v_n does exist, because of the piecewise regularity of the boundary of D_n. Then, as Pini proved, u_n is subcaloric, and $u_n \leq u_{n+1}$ for every n. An easy application of the parabolic Harnack inequality allowed to recognize the uniformly convergence of (u_n) on every compact subset of D to a function u_f^D, which is caloric in D thanks to the Harnack-type first convergence Theorem. The function u_f^D is what Pini called the *generalized Wiener solution of the first boundary value problem*

$$Hu = 0 \text{ in } D, \quad u = f \text{ on } \partial_p D.$$

It is independent of the choice of the sequence (D_n) exhausting D. To complete the construction in the case of f merely continuous, Pini chose two sequences of polynomial functions (F_n) and (G_n), subcaloric in a neighborhood of the closure of D, such that $(F_n - G_n)$ is uniformly convergent to f on the parabolic boundary or D. The corresponding sequence $(u_n - v_n := u_{F_n|\partial_p D}^D - u_{G_n|\partial_p D}^D)$, converges to a caloric function u_f^D, independent of the choice of the sequences (F_n) and (G_n), the generalized Wiener solution to the first boundary value problem with boundary data f. Once this construction was completed, Pini faced the problem of the boundary behavior of u_f^D. He called H-regular a point $z_0 \in \partial_p D$ satisfying

$$\lim_{z \to z_0} u_f^D(z) = f(z_0), \quad \text{for every contiuou function } f : \partial_p D \to \mathbb{R},$$

and after giving a suitable definition of *H- barrier function*, Pini extended the classical Bouligand Theorem by showing that z_0 is H-regular if and only if there exists a H-barrier function at z_0 for D.

It was easy to show the H-regularity for the bottom points of the parabolic boundary $\partial_p D$. The problem of giving geometrical criteria for the regularity of the "lateral" boundary points was left open in [17].

3.4 Caloric Capacity, Heat Equilibrium Potential, Wiener-Type Test

In looking for geometric H-regularity criteria, Pini introduced in [20] the notion of caloric capacity for compact paths $x \mapsto \gamma(x)$ proceeding as follows. Chosen a strip $S := \mathbb{R} \times]a, b[$ containing γ, he constructed a generalized solution to the problem:

$$Hu = 0 \text{ in } S \setminus \gamma, \quad u = 0 \text{ in } \mathbb{R} \times \{a\} \text{ and at infinity}, \quad u = 1 \text{ on } \gamma.$$

The solution, represented as a Heat potential of a positive measure μ supported on γ, was called *potenziale conduttore parabolico* of γ by Pini, and the total mass of μ *capacità parabolica* of γ (parabolic equilibrium potential and parabolic capacity, respectively). He immediately extended to the Heat equation the classical de La Vallée–Poussin Theorem: *a point $z_0 = (x_0, t_0)$ of the lateral parabolic boundary of a dominio normale D is H-regular if and only if the parabolic equilibrium potential of $\partial_p D \cap \{t_0 - r \le t \le t_0\}$ is equal to one for every $r > 0$.*

The notions of Heat capacity and of Heat equilibrium potential were extended to any compact subset of $\mathbb{R}^N \times \mathbb{R}$ in the 1973 paper [10], by using several results from abstract capacity and potential theory, at that time well settled in literature.

Pini's work on the Heat equation ended with a Wiener-type criterion for a boundary point to be regular for the Heat equation. This criterion, contained in [20], was formulated for domini normali of $\mathbb{R}^2$ with continuous boundary. However, as was proved more than 30 years later, it holds for general domains of $\mathbb{R}^N \times \mathbb{R}$. Here is a modern version.

Let D be a bounded domain of $\mathbb{R}^N \times \mathbb{R}$ and let $z_0 \in \partial D$. For every $k \in \mathbb{N}$ define

$$D'_k(z_0) := \{z \in \mathbb{R}^n \times \mathbb{R} \setminus \Omega : 2^k \le \Gamma(z_0 - z) \le 2^{k+1}\}. \tag{11}$$

Then z_0 is H-regular if and only if

$$\sum_{k=1}^{\infty} 2^k \, \mathrm{cap}\,(D'_k) = \infty,$$

where Γ and cap denote the Heat fundamental solution and the Heat capacity, respectively.

The necessity part of this theorem was proved in [10]. The hard part, the sufficiency, was proved almost ten years later by L.C. Evans and R.F. Gariepy [5]. It should be noticed that in 1969, E.M. Landis had proved a necessary and sufficient condition for H-regularity in terms of series of caloric potentials of suitable rings modeled like the level sets of Γ [11].

If D is a dominio normale of $\mathbb{R}^2$, $z_0 = (x_0, t_0)$ is a point of its lateral boundary, and the parabolic boundary close to z_0 and under t_0, is contained in a suitable paraboloid of the kind $\{|x - x_0|^2 \le M(t_0 - t)\}$, Evans–Gariepy's Theorem gives back exactly what Pini proved twenty-seven years earlier.

4 Final Remarks: The Abstract Parabolic Potential Theory

In these concluding remarks, we would like to briefly hint to some general developments subsequent to, and partially anticipated by, Pini's works. We already mentioned the inadequacy of Brelot's Abstract Potential Theory to study parabolic PDE's. The desire to remove this obstacle was the motivation for the birth of more general potential theories. Here we refer the reader to the 1966 Lecture Notes [1] by Heinz Bauer, one of the major contributors, together with Joseph Leo Doob, to these new theories. An abstract Harmonic Space is a topological space $\mathbb{X}$ endowed with a linear sheaf $\mathbb{H}$ of continuous functions (the solutions of a given partial differential equations, in the PDE's applications of the theory). A function u is called $\mathbb{H}$-harmonic in an open set Ω if $u \in \mathbb{H}(\Omega)$. The open set Ω is called $\mathbb{H}$-regular if for every continuous function f on the boundary of Ω there exists only one function $H_f^\Omega \in \mathbb{H}(\Omega)$, nonnegative if f is nonnegative, satisfying $\lim_{x \to y} H_f^\Omega(x) = f(y)$ for every $y \in \partial\Omega$. The family of the regular open sets is assumed to be a basis for the topology of $\mathbb{X}$ (for the classical harmonic space related to the Laplacian this *Axiom* is satisfied thanks to the Poisson Integral formula, while in the case of the Heat equation the family of the parabolic regions $\{|x - x_0|^2 < t_0 - t < r^2\}$ forms a basis of H-regular open sets).

An upper semi continuous function u is defined to be $\mathbb{H}$-subharmonic in a domain Ω if it is finite in a dense subset of Ω and, for every regular open set $V \subset \overline{V} \subset \Omega$, and every continuous function $f \geq u$ on the boundary of V, one has $H_f^V \geq u$ in V. The family of the continuous $\mathbb{H}$-subharmonic functions is assumed to separate the points of $\mathbb{X}$.

A crucial assumption, inspired by the typical form of the parabolic Harnack inequality, is the Doob Convergence Axiom: *An increasing sequence of $\mathbb{H}$-harmonic functions in an open set Ω, which is convergent in a dense subset of Ω, actually converges everywhere in Ω to a function u $\mathbb{H}$-harmonic in Ω.*

Obviously, due to the classical and the parabolic Harnack inequalities, this Axiom is satisfied both for the Laplace and the Heat-harmonic spaces. We stress that in Brelot's Axiomatic Potential Theory, the convergence of the sequences, in the Convergence Axiom, is a-priori assumed only on a point on every connected component of Ω.

Starting from the previous Axioms it has been developed a Potential Theory that parallels the classical one, which is also applicable to parabolic equations and, more generally, to wide classes of linear second order PDE's with nonnegative characteristic form (see e.g. the paper by Jean Michel Bony [2], and the recent monograph [3]) .

Appendix

In this appendix we display some original pages of the papers by Pini and by Hadamard, with courtesy of, respectively:

Professor Francesco Baldassarri , Managing Editor of the Rendiconti del Seminario Matematico dell'Università di Padova,

Professor Pasquale Vetro, Manager Editor, and Springer Verlag, Publisher, of the Rendiconti del Circolo Matematico di Palermo

SULLA SOLUZIONE GENERALIZZATA DI WIENER PER IL PRIMO PROBLEMA DI VALORI AL CONTORNO NEL CASO PARABOLICO

Nota (*) *di* Bruno Pini (*a Cagliari*)

Siano γ_1 e γ_2 due archi di equazioni $x = \chi_1(y)$, $x = \chi_2(y)$, $a \leq y \leq b$, con $\chi_1(y)$ e $\chi_2(y)$ continue e $\chi_1(y) < \chi_2(y)$ per $a < y \leq b$; indichiamo con C_1 e C_2 i segmenti $\chi_1(a) \leq x \leq \chi_2(a)$, $y = a$ e $\chi_1(b) \leq x \leq \chi_2(b)$, $y = b$, non escludendo che C_1 possa ridursi a un punto; indichiamo poi con D il dominio limitato di frontiera $\gamma_1 + \gamma_2 + C_1 + C_2$ e poniamo $S = \gamma_1 + C_1 + \gamma_2$, che brevemente diremo un *contorno parabolico*. Com'è noto, esiste ed è unica la soluzione del problema

(1) $\begin{cases} \mathfrak{L}(u) = u_{xx} - u_y = 0 & \text{in } D - S \\ u = f & \text{su } S \end{cases}$

se $f(P)$ è una funzione continua del punto P variabile su S e se $\chi_1(y)$ e $\chi_2(y)$ sono funzioni lipschitziane su $a \leq y \leq b$ (o anche soltanto hölderiane d'ordine $> 1/2$). Chiameremo *normale* un dominio D per cui γ_1 e γ_2 soddisfano tale condizione. È ben noto che non per ogni dominio T dello spazio euclideo ad n dimensioni ($n \geq 2$), esiste una funzione armonica in $T - \mathfrak{F}T$ la quale assuma in senso ordinario nei punti di $\mathfrak{F}T$ assegnati arbitrari valori continui; la presenza su $\mathfrak{F}T$ di certi punti *irregolari* (tipica ad esempio la *spina* di Lebesgue nel caso di $n = 3$) esclude l'esistenza della soluzione ordinaria del problema di Dirichlet per l'equazione di Laplace.

Nel caso parabolico è banale che se il dominio D, anzichè essere del tipo indicato all'inizio, è ad esempio quello che ha per frontiera due segmenti di caratteristica C_1 e C_2, un arco del tipo γ_1 e un arco di equazione $x = \chi_2(y)$ ($> \chi_1(y)$) per $a \leq y \leq c$ ($< b$)

(*) Pervenuta in Redazione il 9 agosto 1954.

Fig. 1 Original first page of Pini's article [17]

$(\chi_1(c) <) \chi_2(c) - \delta \leq x \leq \chi_2(c)$ per $y = c$ ($\delta > 0$), $x = \chi_3(y)$ ($> \chi_1(y)$) per $c \leq y \leq b$ con $\chi_3(c) = \chi_2(c) - \delta$, non si possono prefissare ad arbitrio i valori di f su $\chi_2(c) - \delta \leq x \leq \chi_2(c)$, $y = c$.

Ma anche supponendo che le funzioni $\chi_i(y)$ siano continue, possono presentarsi punti irregolari [1]). Tuttavia è possibile anche nel caso parabolico parlare di *soluzione generalizzata* del tipo di WIENER [2]) e di *barriera* per i punti del contorno parabolico, capace di discriminare i punti regolari dai punti irregolari, essendo i primi, a differenza dei secondi, tali che la soluzione generalizzata prende in essi nel senso ordinario i valori assegnati. Per far ciò è sufficiente l'uso di una formola di media, di un analogo del cosidetto secondo teorema di HARNACK e della nozione di funzione $\mathcal{L}$-prevalente ed $\mathcal{L}$-subvalente [3]), le quali sono delle super ed infer-funzioni relativamente alle soluzioni di $\mathcal{L}(u) = 0$; in possesso di ciò, basta ricalcare ragionamenti noti nel caso del problema di DIRICHLET. Noi svolgeremo per completezza anche quest'ultima parte seguendo l'esposizione di KELLOGG [4]), e ci riserviamo di tornare in altro luogo sull'argomento.

Vogliamo però espressamente rilevare che, anziché usare il procedimento delle successioni, si può giungere allo stesso risultato col procedimento di mediazione di ZAREMBA, applicato a una formola di media caratteristica delle soluzioni di $\mathcal{L}(u) = 0$, ciò che permette di giungere a un risultato in tutto simile a quello di LEBESGUE-PERKINS [5]).

1. · Vogliamo qui indicare un esempio di punto la cui irregolarità, sebbene deducibile dalle condizioni di irregolarità

---

[1]) Un primo studio sui punti regolari e irregolari di un contorno parabolico è stato fatto da I. PETROWSKY, *Zur ersten Randwertaufgabe der Wärmeleitungsgleichung*, Compositio Mathematica, I (1935).

[2]) N. WIENER, *Certain notions in potential theory*, Journal of Math. and Phys. Massachussetts Inst. of Technology, 1924.

[3]) B. PINI, *Maggioranti e minoranti delle soluzioni delle equazioni paraboliche*, Annali di Mat. pura ed appl., s. 4, XXXVII (1954).

[4]) O. D. KELLOGG, *Foundations of potential theory*, Berlin 1929.

[5]) H. LEBESGUE, *Sur le problème de Dirichlet*, C. R. Paris, 154 (1912) ed F. W. PERKINS, *Sur la résolution du problème de Dirichlet par des médiations réitérées*, C. R. Paris, 184 (1927).

Fig. 2 Original second page of Pini's article [17]

EXTENSION À L'ÉQUATION DE LA CHALEUR
D'UN THÉORÈME DE A. HARNACK

par J. Hadamard (Paris)

Deux théorèmes de A. Harnack sont bien connus dans la théorie des fonctions harmoniques.

I. *Une série de fonctions régulièrement harmoniques dans une aire D et continues sur son contour S qui converge uniformément sur S converge uniformément dans D ainsi que toutes ses séries dérivées. Elle représente une fonction harmonique dans D.*

II. *Une série de fonctions régulièrement harmoniques et positives qui converge en un point déterminé quelconque a intérieur à D converge dans tout l'intérieur de D, et cela uniformément dans tout domaine intérieur D', ainsi que toutes ses dérivées et représente par conséquent une fonction harmonique.*

Le premier de ces deux théorèmes s'étend aisément ([1]) aux fonctions « paraboliques », c'est à dire aux solutions de l'équation de la chaleur

$$(E) \qquad \frac{\partial^2 u}{\partial x^2} - \frac{\partial u}{\partial y} = 0.$$

Je ne sais si une semblable extension, également utile à la théorie, a été obtenue pour le théorème II. Par une analogie évidente avec ce qui se passe en théorie des fonctions harmoniques, toute la question est d'obtenir une limite inférieure de la fonction de Green $G(a, M)$ ainsi que de l'une de ses dérivées intérieures en se donnant, d'autre part, une limite inférieure de la distance du point a au contour.

([1]) Voir p. ex. S. Täcklind, *Actes de la Soc. Roy. Sc. Upsal*, t. X_4, 1936.

Fig. 3 Original first page of Hadamard's article [7]

References

1. Bauer, H. 1966. *Harmonische Raüme und ihre Potentialtheorie.* Lectures Notes in Mathematics, 22 Springer: New York.
2. Bony, J.M. 1969. *Principe de maximum, inegalité de Harnack et unicité du problème de Cauchy.* Ann. Inst. Fourier (Grenoble) 19: 277–304.
3. Bonfiglioli, A., E. Lanconelli, and F. Uguzzoni. 2007. *Stratified Lie Gruops and Potential Theory for their Sub-Laplacians.* Springer Monograph in Mathematics. Springer: New York.
4. Doetsch, G. 1937. *Theorie und anwendung der Laplace-Trasfromation.* Berlin: New York.
5. Evans, L.C., and R.F. Gariepy. 1982. *Wiener criterion for the heat equation.* Archive for Rational Mechanics and Analysis, 78: 293–314.
6. Fulks, W. 1966. *A mean value theorem for the heat equation.* Proc. Amer. Math. Soc. 17: 6–11.
7. Hadamard, J. 1954. *Estension à l'équation de la chaleur d'un théorème de A. Harnack.* Rendiconti del Circolo Matematico di Palermo, Serie II 3: 337–346.
8. Ahel Harnack, C.G. 1887. *Die Grundlagen der Theorie des logarithmischen Potentiales und der eindeutigen Potentialfunktion in der Ebene*, Teubner, Leipzig, Germany.
9. Kassmann, M. 2007. *Harnack inequalities : an introduction.* Boundary Value Problem, Article ID 81415, 21 pages.
10. Lanconelli, E. 1973. *Sul problema di Dirichlet per l'equazione del calore*, Annali di Matematica Pura ed Applicata, Serie IV, 97: 83–114.
11. Landis, E.M. 1969. *Necessary and sufficient conditions for regularity of a boundary point in the Dirichlet problem for the heat equation.* Dokl. Akad. Nauk, SSSR 185: 517–520; Soviet Math. Dokl. 10: 1969. 380–384.

12. Montaldo, O. 1955. *Sul primo problema di valori al contorno per l'equazione del calore.* Rendiconti del Seminario della Facoltà di Scienze, Università di Cagliari, 25: 1–14.
13. Moser, J. 1964. *A Harnack inequality for parabolic differential equations.* Comm. Pure and Appl. Math. 17: 101–134.
14. Nirenberg, L. 1953. *A strong maximum principle for parabolic equations*, Comm. Pure and Appl. Math, 6: 167–177.
15. Pini, B. 1951. *Sulle equazioni a derivate parziali lineari del secondo ordine in due variabili, di tipo parabolico.* Annali di Matematica Pura ed Applicata, Serie IV 32: 179–204.
16. Pini, B. 1954. *Maggioranti e minoranti delle soluzioni delle equazioni paraboliche*, Annali di Matematica Pura ed Applicata, Serie IV, 37: 249–264.
17. Pini, B. 1954. *Sulla soluzione generalizzata di Wiener per il primo problema di valori al contorno nel caso parabolico*, Rendiconti del Seminario Matematico dell'Università di Padova 23: 422–434.
18. Pini, B. 1954. *Estensione al caso parabolico di un teorema di F. Riesz relativo alle funzioni subarmoniche.* Rivista Matematica della Universitá di Parma 5: 269–280.
19. Pini, B. 1954. *Su un integrale analogo al potenziale logaritimico* , Bollettino dell'Unione Matematica Italiana, Serie III 9: 244–250.
20. Pini, B. 1955. *Sulla regolarità e irregolarità della frontiera per il primo problema di valori al contorno relativo all'equazione del calore*, Annali di Matematica Pura ed Applicata, Serie IV, 40: 69–88.
21. Pini, B. 1955. *Sul primo problema di valori al contorno per le equazioni parabilche lineari.* Rivista Matematica dell'Univeresità di Parma 6: 2i5–237.
22. Saloff-Coste, L. 2002. *Aspect of Sobolev-type inequalities.* London Mathematical Society, Lectures note Series 289: Cambridge University Press.
23. Taeklind, S. 1936. *Sur le classes quasianalitiques des solutions des équations aux dérivées partielle du type parabolique.* Nova Acta Regiae Societatis Scientiarum Upsaliensis, Ser. IV 3:3 3–57.
24. Watson, N.A. 1973. *A theory of subtemperatures in several variables.* Proc. Lond. Math. Soc. (3) 26: 385–417.

Federigo Enriques as a Philosopher of Science

Gabriele Lolli

Abstract We discuss Enriques' production and activity in the field of the philosophy of science. Enriques' involvement in philosophy was stronger in the first years after graduation; the same years in which he became famous as a mathematician for his work in algebraic geometry are the years he formed the views he would maintain for his whole life. Enriques stressed the necessity of a psychological foundation of the scientific concepts, and gave an example of such an analysis with regard to the geometrical postulates in 1901. His general proposal is contained in the book *Problems of science* of 1906, which was translated in the USA in 1914; in the book he defended the cognitive value of science against the irrationalistic movements of the time. His position became known as a sort of critical anti-positivism, with an insistence on the attention to the historical growth of theories; he was later recognized by the leaders of neo-positivism as one of their founding fathers, not to his complete agreement. After 1906, Enriques became a prominent figure in Italian and european philosophy, and was engaged in fierce battles with the Italian idealistic philosophers. He did not appreciate the logical work of Giuseppe Peano, claiming that his formalism didn't have the right psychological basis. Later, he dedicated himself and promoted the study of the history of science. The 1922 book *Per la storia della logica* is the second, and last, important philosophical work of Enriques, where he gave the best description and justification of the modern axiomatic method as rooted in the development of nineteenth century mathematics.

Federigo Enriques states in the preface to [8], the book that gave him the *status* of a philosopher, that his philosophical interests kept growing from 1890 to 1900;

G. Lolli (✉)
Scuola Normale Superiore di Pisa, Piazza dei Cavalieri 7, 56100 Pisa, Italy
e-mail: gabriele.lolli@sns.it

then when he was sure of himself he began to go public.[1] Since he graduated in 1891, this means that for the whole period in which he worked with Guido Castelnuovo in Rome and Corrado Segre in Turin on algebraic surfaces, becoming a worldwide acclaimed geometer, he continued to reflect on philosophical matters. This is confirmed by the letters he exchanged with Castelnuovo.[2] He was appointed professor in Bologna in 1896, where he remained till 1922, when he moved to Rome.

He was by then considered an important philosopher, not only in Italy. His book of 1906 had been translated in the USA as [9], besides french and german translations; Enriques had become president of the SFI, the italian philosophical society, and organized the Bologna international congress of philosophy in 1911; he had been co-founder in 1907 of the *Rivista di scienza*, later *Scientia*, among whose contributors in the earlier years were Freud, Mach, Poincaré, Russell, Einstein and others.

Apart from his scientific prestige however, as a philosopher he was respected abroad more as a charismatic figure, for his defense of rationality, than as a scholar. In Italy not even that, outside the scientific circles, since the acrimonious and disparaging critique of the idealistic philosopher Benedetto Croce had succeeded in isolating him.[3]

Later in 1935 he spoke at the Paris international congress of scientific philosophy which launched the neopositvistic movement ([13]), and Otto Neurath invited him to contribute to the *Encyclopaedia of Unified Science* with an essay on the history of science.

Enriques was there representing, as director of the journal *Scientia*, the only italian group in the international panorama of scientific philosophy, and had been invited along with the Centre de Synthèse in France, the Vienna Circle, the Berlin group, anglo-saxon and scandinavian thinkers in order that all European nations in which there was some activity in the philosophy of science would be involved. Enriques was all in favour of the neopositivistic antimetaphysical stance, but he had reservations on the insistence on the logical analysis of language. *Scientia* had however published papers by Schlick, Frank, Carnap, Neurath, and Enriques himself had reviewed two articles by Frank and Carnap in [14]. He was inserted in the genealogical tree of the neopositivists, but behind the scenes they found it difficult to cope with his old style rationalism.

The difficulty for philosophers of other languages and cultures to understand Enriques' theses and positions is made clear by the review of Charles D. Broad, of [9]; although the book is presented as containing interesting topics ("there are some excellent remarks on the nature of definition"), the review is interspersed with expressions of regret and complaint: "[the book] deals with so many difficult and important subjects that the argument is obscure through its condensation [...] the

---

[1] Probably referring to [7].

[2] In 1896, he confessed to Castelnuovo that he wasn't putting his heart and soul into mathematics, but into the philosophical problem of space ([16, p. 260]).

[3] See [4] and references therein to [5].

style is very heavy [...] the book is also disfigured by an immense number of notes of exclamation [...] I am not perfectly sure that I understand this [...] Again I cannot see precisely what Prof. Enriques' special argument about the actual infinite is supposed to prove," and so on ([1]).

1 Psychology and Geometry

In 1901, as the result of years of reflections "of mathematical, psychological and philosophical order,"[4] Enriques laid down his project, which was to give a psychological genetic explication of the postulates of geometry. He viewed himself as belonging to a tradition comprising Herbart, Gauss, Lobacevskij, the psychologists Bain, Taine, Delboeuf, Lotze. According to Enriques, the physio-psychological experiments of Helmholtz and Wundt for showing the emergence of the representation of space by means of sensations and associations fell short of explaining the sense of necessity of the mathematical intuition of space; what was missing was a link of the postulates to the logical structure of thought.[5]

Physio-psychological investigations had to be integrated by a pure psychological research in order to show how concepts are formed from sensations, and to discover the necessary conditions of their genesis, dependent on the psychic structure.

The psychic structure is given by the laws of association, the logical laws and the sense of temporal order, the distinction of "before" and "after," prior to the sense of duration. Enriques was convinced that these capabilities were expression of not yet known biological laws, not the product of cultural inheritance ([7, p. 81]).

For Enriques "the three groups of representations tied to the basic concepts of the theory of the continuum (analysis situs), of differential geometry and of projective geometry are immediately connected to three groups of sensations: respectively, the tactile-muscular ones, those of touch and those of vision."[6]

For example a line as viewed by the eyes either passes through the center of vision, and then it appears as a point, or not, and it is seen as a line; to relate these two representations it is necessary that the line be determined by any two of its points, not only as a postulate, but from the physiological determination. Looking from A at point B, the segment AB is defined as the set of points whose image falls in $B' = B$; the same set is viewed from outside as a segment and from B as a point, hence first of all $AB = BA$. If C is on AB, then $AC = AB$ for the visual

---

[4]See the article of Livia Giacardi in this volume for more on these early years of Enriques' philosophical apprenticeship.
[5]Enriques was a great admirer of W. Wundt and his *Logik*, whose influence on Enriques is deep and transparent. For Wundt, logic was the study of "those laws of thinking active in scientific knowledge" ([22, vol. 1, p. 1]); symbolic logic was a discipline apart, called "Logistik."
[6]All translations from Italian are by the author, but for those from [8] which are taken from the american translation [9].

ray, and the same for another point D; in conclusion, $AB = CD$, that is the line is determined by any two of its points.

The postulates of the line are the density and the continuity axioms. To perceive the density already requires more than sensations: the concept of a line represents all possible successions of points; when two points are too near to appear distinct to the senses, the thought uses a correspondence with another line in which the two point are further off, so that in between a new point can be inserted.

Dedekind's continuity postulate does not seem derivable from representations of the line as succession of points; according to Enriques it could be based on two superimposed concepts, that of a corpuscle and that of the infinite divisibility of the line.

Such were the reflections of [7].

2 The Problems of Science

The general aim of the book [8] was to oppose both the irrationalistic tendency characterised by the *ignorabimus* of Du Bois Reymond and the positivistic philosophy, proposing a synthetic view of knowledge. There is in the book a "relatively pragmatistic element," probably due to the influence of Giovanni Vailati, the philosopher whose untimely death in 1909 swept away the only italian interlocutor of Enriques; such an element, which Enriques labelled "critical positivism," is nothing however compared to James' pragmatism and Bergson's spiritualism, the thinkers who in the following years originated a new anti-intellectualistic movement, as Josiah Royce remarks in his preface to [9, p. x].

The book begins with "an analysis of what constitutes reality," and a defense, or foundation of objective knowledge. Enriques tries to distinguish "the positive content of science on the one hand, and its subjective aspects on the other."

Objective knowledge has different subjective versions if a prevision of the right answer is obtained by means of different images; the resort to some image is necessary, so pure objective knowledge does not exist. In the mean time there is also an objective element in subiectivity, in that subiectivity is able to influence the previsions.

Knowledge is based on sensations, but not only on these, or not precisely on these, rather on the correspondence of sensations to voluntary acts.

"There are fixed groupings, independent of us, amongst our actual or supposed volitions on the one hand, and the sensations produced by them on the other hand [...] These groupings correspond to what we call 'the real'. The real gets defined, in this way, as *an invariant in the correspondence between volitions and sensations."* [9, p. 65].

In analogous way, a scientific fact is a relation of succession between facts conceived as invariant. For example, "[the scientific fact] teaches us that we can experience the sensations that testify to the heating up of the plate of metal, whenever we perform the acts which produce the blow of the hammer against it"

(p. 67). A scientific fact is actually an abstraction, with elimination of data to obtain a "type of a series of possible facts," which type Enriques calls a concept.

A theory is the subjective aspect of the knowledge of a scientific fact, and precisely "a concept or system of concepts [...] from which by deduction we can subordinate [other] supposed facts."

There arise two classes of problems: "(1) the problems connected with the logical transformation of concepts, regarded both as a psychological development and as an instrument of knowledge; and (2) those problems which refer to the significance and to the acquisition of the more general concepts of space, time, force, motion and so on." The latter are treated in the second part of the book.

Logic is discussed at length in the third chapter, covering more than one hundred pages (pp. 153–259) and we dwell on it since it is the focus of Enriques' reflection, as confirmed by his second important publication [11].

As usual, Enriques inserts his own considerations in a historical perspective. The thinkers of the Middle Ages were mistaken in believing that studying only formal rules they could research the truth. The study of reasoning has been enlarged in the modern times to include the treatment of empirical data, giving origin to what is now called inductive logic. Enriques is not interested in the latter.

Formal logic is only an instrument, it does not change the data of knowledge, but it is essential because, as Jevons remarked, it transforms hypotheses inaccessible to experience into hypotheses which are equivalent but can be verified and possibly refuted by experience. Formal logic is important, the question is whether it is possible.

For a foundation of the possibility of a formal logic Enriques discusses two ways. The first one is that pursued so far, and it consists in isolating verbal schemes corresponding to ideal forms, and in eliminating from them all ambiguities. For example, the word "some" in common talk frequently means "some but not all," but this meaning has been eliminated from its scientific use. This path leads to symbolic logic, perfectioned in a tradition in which Enriques includes Leibniz, Lambert, Boole, Schröder, Peano, Frege.

The second way consists in a study of the actual functioning of thought; psychological logic is an empirical science, falling under psychology. This is why Enriques didn't mention it in relating the range of his studies in [7]. Vailati told Vacca in 1901 [21, p. 188] that Enriques was studying logic, but he didn't understand what kind of logic it was, certainly not mathematical logic.

Definitions and deductions are psychological operations; their combination is the logical process. Logical processes are those mental processes in which some conditions of consistency are voluntarily satisfied.

In order to give definitions, one must suppose that some objects are given in thought, capable of entering in some associations satisfying a sort of invariance. Psychological operations in fact are associations and dissociations, which permit to build new objects.

From the association of objects, either simultaneously or in succession, we acquire the operations of conceiving classes and series. Comparing classes gives rise to the concept of correspondence.

Dissociations on the contrary give the inversion of series, the disjunction, the intersection and the abstraction, that is an element of a class substitutable to any other element.

Among the propositional laws for example the associative and commutative ones come from the associativity and commutativity of the psychological operation of making a heap.

The invariance of objects with respect to operations is expressed by the principles of identity, non contradiction and excluded middle; this last means that for two objects it is always possible a judgement of identity or diversity. It is based on the fundamental intuition of before and after. Such intuition are not representable by formulae; one should not write $a = a$.

Enriques' description of logical operations is a genetic one; he admits that the working of the logical process can be represented by symbolic logic as a system of actual relations, thus conferring a reality independent from time to the objects of thought. One would expect in the end a convergence of the two ways of founding formal logic, but it is not so, as we will see.

The logical relations among objects built in the thought are a-priori compatible. As to arithmetic, consistency derives from the fact that acts of thought can be repeated in unlimited way. Enriques calls however this statement a psychological postulate. So arithmetic is part of pure logic.

When one reasons, one assumes as evident that the reality of the premises entails that of the conclusions; the transformation of concepts determined by subjective laws reflects or becomes a transformation of real relations. In order that this be possible there must be a relation between the invariants of experience and those of thought. One should only put a word of caution: applicability of logic assumes that under the conditions of invariance expressed by the logical principles, the systems of things approximately satisfy the properties expressed by the axioms.

In the light of such postulate, it is curious to remember the dislike Enriques had for Russell; he reproached him the idea that logical relations (as discussed in [20]) express the most general relations holding among entities of every possible world; thus they wouldn't relate to the analysis of thought but would be truths of a metaphysical universe. Hence that Russellian curse of the "the totally irrelevant notion of a mind." To characterize his own position in opposition to Russell's, Enriques resorted to Boole, for whom logic was the set of laws guiding mental processes, which can only have a representation in the static image of a formalism.

It is not clear whether the approximation in the correspondence between the invariants of the thought and the invariants of reality can be related to a fundamental character of definitions. According to Enriques it is impossible to establish the real meaning of a concept through a definition; definitions, at least implicit ones, do not give an exact determination but throw only a partial light on the real sense of concepts. They have to be completed with a concrete interpretation through suitable rules of correspondence.

This idea is related to the axiomatic method, which is discussed at greater length in [11]. At the end of the chapter on logic Enriques summarizes a few neurological findings then known on the brain functioning, expressing the hope that from advances in this field his own approach could be confirmed in the future.

3 The History of Logic

In [11], Enriques enlarges the analysis of logic, and makes clear what he meant in *The Problems* by saying that the possibility of formal logic follows from the development of mathematics. The subtitle of [11] is "The principles and order of the science in the conception of mathematical thinkers." The order, or ordering (it. *ordinamento*) of the science is an obsolete word, but central in Enriques' vision.[7] The problems of the ordering of the theories concern the definitions, the axioms, the place they have in the body of the theory, the criteria of their choice, how to judge their aceptability. These problems should be discussed in the study of logic.

With his analysis, Enriques qualifies as one of the most sensible and clear minded interpreter of the modern formal axiomatic method.

The order of sciences for the Greeks displayed a naive realism, the necessary character of the principles, no theory of definitions, a concept of deduction based on the meaning of the terms.

At the beginning of the modern age a central figure is that of Leibniz. Enriques underlines some of his contributions to the reform of logic: the idea that a concept is not tied to the real but to the realm of the possible; the principle of sufficient reason, from which the existent is singled out among the many possibles (in Leibniz's words: *Ens quod distincte concipi potest. Existens quod distincte percipi potest*[8]). Great importance is given by Enriques, as was by Peano, to Leibniz's analysis of a concept into simple ideas; the regress must lead to purely logical truths, or identities.

Notwithstanding the break with the ancient tradition, the development of logic up to the nineteenth century has not changed the traditional concept of the order of demonstrative sciences. This task had to be accomplished by mathematicians: several intellectual movements with different origins concurred in the same reform. Enriques mentions projective and non Euclidean geometries, Riemann, Beltrami's models, the formal algebra and logic in Great Britain, the arithmetization of analysis, the new trend in physics towards the building of models.

But only through the recent critique of the principles of geometry "the mathematical thinkers acquire a full conscience of a revolution prepared for centuries."

The final phase for Enriques begins with Gergonne, his notion of implicit definition and the duality principle as a principle of substitutivity of concepts. It comes to a full blooming with Plücker and von Staudt: with Plücker's coordinates a unified treatment of correlated entities invites to translate one into the other different forms of intuition.

---

[7] It has been rendered as "structure" in the american translation [12]. There have been also french and german editions of the book.

[8] An entity is what can be distinctly conceived; the existent is what can be distinctly perceived.

Here, Enriques gives a lyrical paean of the axiomatic method: "Nothing is more fecund than the multiplication of our intuitive powers enhanced by this method: it is almost as to our bodily eyes, with whom we examine a figure under a certain perspective, a thousand spiritual eyes be added that allow us to contemplate several different transfigurations, while the unity of the object is resplendent under the enriched reason."[9]

The notion of abstract theory is thus so presented: some primitive concepts A, B, C are given; a postulate states a certain relation $f(A, B, C)$ among them; when we ask whether the relation is true or false for some new interpretation, the translation makes no sense if the relation f depends on the intuitive meanings of A, B, C. Hence, the formal nature of mathematics.

When two systems A and B are both possible intepretations of the same abstract theory, if to discover the consequences we look at the objects given with A we must be careful that they do not depend on particular intuitions that fail for B.

Quoting Pasch [19, p. 82]: "in order that geometry could become a true deductive science, it was necessary that the derivations should be independent from the sense of the geometrical concepts, as they were independent from the figures. In the course of a deduction it is permissible and sometimes it can be useful to bear in mind the meaning of the geometrical concepts involved, but it is not necessary; when it becomes necessary, it is a sign of a shortcoming of deductions, or of the insufficiency of the assumptions of the proof."

Enriques mentions Pasch's work as an example of the new order of the geometric theory; he mentions also Peano and Hilbert, and in general he says that almost each mathematician of his generation had to re-discover by himself the sense of the logical form.

4 Logical Controversies

In [10], a review of [2], Enriques uses strong words of irony and contempt against Peano's ideography, disregarding the concliatory attempts of Burali-Forti to come to a compromise in the ensuing exchange [3].

The controversy is a sad episode in the history of the italian mathematical community; there are clearly political undertones; Enriques couldn't suffer Burali-Forti coarse nationalism. He ridicules the contorted definition of vectors depending on the concept of volume, and many other minor points. Some of his remarks are correct: the unavoidability of the choice principle in the proof of equivalence of the

---

[9]Enriques has had this idea of the multiple interpretations as the bonus of the axiomatic method since the beginning of his reflection: "The importance that we attribute to Abstract Geometry is not [...] in opposition to the importance attributed to intuition: rather, it lies in the fact that *Abstract Geometry can be interpreted in infinite ways as a concrete (intuitive) Geometry by fixing the nature of its elements:* so in that way Geometry can draw assistance in its development from infinite different forms of intuition ([6, pp. 9–10]).

definitions of finiteness, given by Burali-Forti by means of his partiton principle; the necessity of an additional axiom for arithmetic, to the effect that 0 is the only element without a predecessor, if induction is substituted by well ordering[10]; the disadvantage of characteriz ing in a unique way the equality instead of considering equivalence relations in different fields. He too, however, is sometimes confused, for example with reference to the distinction between an individual and its singleton, which comes up in his main argument and which he does not understand.

Enriques' critique focuses on the the following idea: there is a fatal shortcoming of Peano's logic in the impossibilty to explain the following paradox: Peter was an Apostle, the Apostles were twelve, so Peter was twelve. Peano's way out consists in a distinction between membership and inclusion, but this move is contrary to use of "is" in speech. Enriques is strongly opposed to it; his own solution of the paradox is that the middle term "Apostle" has a double *status*, appearing once as a class and once as an abstract concept.

In the item "Mathematical Logic" written for the italian encyclopaedia Treccani by Beppo Levi ([18]), an unsigned appendix bearing the title "Meaning of logic," and apparently written by Enriques, makes a comparison between Peano and Enriques: "G. Peano and his followers see in the logical symbolism an ideography well suited to the exposition of 'deductive and mathematical sciences'; from the positive development of these sciences they argue a sort of experimental revelation of the logical schemes of reasoning, which logic is meant to analyse. No further question on the meaning of reasoning, hence on logic, is considered by this school. To Enriques, logic is the study of the operations of the exact thought and their laws, without reference to anything outside the mind"; for Enriques the problem of the application of logic to the knowledge of reality is devolved to epistemology. There follows a short presentation of Enr iques' ideas on the logical process, and finally Enriques' critique of Peano's distinction between membership and inclusion is repeated: "a classroom conceived as a union of individuals is something different from the abstract concept of the classroom: the union of students A, B, C, ... gives the school; from this by abstraction one gets the student (of that school) [...] Ordinary language makes here a distinction that it is impossible to translate in Peano's symbolism, and as a consequence he has to distinguish instead two meanings in the copula of ordinary language [...]."

5 Conclusion

So in the end Enriques didn't believe what he nevertheless had recognized, that the regimentation of language realized by symbolic logic had been an adequate answer to the task of designing the exact tools of reasoning. His refusal to use simbolic

---

[10]M. Pieri had rightly assumed it, but Burali-Forti in formalizing the axioms apparently didn't notice its role. Enriques suspected an error, giving as a counterexample $0, \ldots, n, \ldots \omega, \omega + 1, \ldots$; in the end Burali-Forti took notice of Pieri's axiom.

languages to express the formal character of mathematics was probably due to a positive vision of the axiomatic method: the true value of the abstract conception of mathematics did not reside for him in the elimination of all meanings, but in the possibility to find the same content in different systems of images, each of which adds to the economy of knowledge.

However, Enriques' philosophical ideas were formed at the end of the nineteenth century and didn't evolve. He did not confront seriously with the principal new trends in philosophy and in logic, in particular with logical positivism and the new mathematical logic born of Hilbert's work.

References

1. Broad, C.D. 1914. "Problems of Science". *Mind* 24: 94–8. review of [9].
2. Burali-Forti, C. 1919. *Logica matematica* Hoepli, Milano, 2nd ed.
3. Burali-Forti, C., and F. Enriques. 1921. "Polemica Logico-matematica". *Periodico di Matematiche* 4: 2, 354–65.
4. Ciliberto, M. "Scienza, filosofia e politica: Federigo Enriques e il neoidealismo italiano". in 17: [pp. 131–66].
5. Croce, B. 1981. *Lettere a Giovanni Gentile (1896–1924)* (A. Croce ed.), Mondadori: Milano.
6. Enriques, F. 1895. *Conferenze di geometria tenute nella R. Università di Bologna. Fondamenti di una geometria iperspaziale*, Lithograph: Bologna.
7. Enriques, F. 1901. "Sulla spiegazione psicologica dei postulati della geometria". *Rivista di Filosofia* 4: reprinted in 15: [pp. 71–94].
8. Enriques, F. 1906. *I problemi della scienza*. Zanichelli, Bologna, American Transl. as [9].
9. Enriques, F. 1914. *Problems of Science*. The Open Court Pub. Co., Chicago.
10. Enriques, F. 1921. "Noterelle di Logica matematica". *Periodico di Matematiche* 4: 1, 233–44.
11. Enriques, F. 1922. *Per la storia della logica. I principi e l'ordine della scienza nel concetto dei pensatori matematici*, Zanichelli: Bologna.
12. Enriques, F. 1929. *The Historic Development of Logic: The Principles and Structure of Science in the Conception of Mathematical Thinkers*, Holt: New York.
13. Enriques, F. 1936. "Philosophie scientifiques". *Actes du Congrès international de philosophie scientifiques*, Hermann: Paris.
14. Enriques, F. 1935. Reviews of R. Carnap and Ph. Frank. *Scientia*, 57: 69–71 and 227–229.
15. Enriques, F. 1958. *Natura, ragione e storia*. (L. Lombardo-Radice ed.), Edizioni Scientifiche Einaudi, Torino.
16. Enriques, F. 1996. *Riposte armonie. Lettere di Federigo Enriques a Guido Castelnuovo*. (U. Bottazzini, A. Conte, P. Gario eds.), Torino: Einaudi.
17. Faracovi, O.P. (ed.) 1982. *Federigo Enriques. Approssimazione e verità*, Belforte Editore Librario: Livorno.
18. Levi, B. 1936. "Logica matematica". *Enciclopedia italiana*, Treccani: Roma, vol. xxi.
19. Pasch, M. 1882. *Vorlesungen über neuere Geometrie*, Teubner: Leipzig.
20. Russell, B. 1901. "Sur la logique des relations avec des appplications à la théorie des séries", *Rivista di Matematica* 7: 115–36 and 137–148.
21. Vailati, G. 1971. *Epistolario (1891–1909)* (G. Lanaro ed.), Einaudi: Torino.
22. Wundt, W. 1880–83. *Logik*, Enke: Stuttgart, 3.

The *Enciclopedia delle Matematiche elementari* and the Contributions of Bolognese Mathematicians

Erika Luciano

Abstract It may seem strange, in a volume dedicated to Bolognese mathematicians, to come across a paper about an initiative that, as is well-known, has its roots in Pavia, namely, the *Enciclopedia delle Matematiche Elementari*. Yet the mathematical community of Bologna was in fact involved in this editorial enterprise and several of its authors, including E. Bortolotti, A. Comessatti, S. Pincherle, F. Sibirani, B. Segre and G. Vitali, together with its conceiver R. Bonola and one of its directors G. Vivanti, were either from Bologna or had links with the University there. In the present paper, after summarising the volume's somewhat turbulent publication history, we will focus on the involvement of Bolognese mathematicians and, in particular, on the scientific dialogue which took place between L. Berzolari, G. Vivanti and G. Vitali regarding the coordination of *Enciclopedia* chapters on analysis.

1 The Complicated Circumstances of the Publication

It was at the meeting of the Lombardy Section of the teachers' association Mathesis[1] held in Pavia on 23 May 1909 with Luigi Berzolari as chair, that one of its members, Roberto Bonola,[2] first put forward the idea of publishing an *Enciclopedia delle*

This research was performed within the Project *PRIN 2009 Scuole matematiche e identità nazionale nell'età moderna e contemporanea*, Unit of Turin University.

[1] The Lombardy section had been just founded in Milan the previous month, on 8 April 1909.

[2] Bonola had studied mathematics at the University of Bologna. Assistant to E. Bertini, at Pavia, and teacher at the nearby Scuola Normale, he was a brilliant scholar in non-Euclidean geometry and, a

E. Luciano (✉)
Department of Mathematics, University of Turin, via Carlo Alberto 10, 10123 Turin, Italy
e-mail: erika.luciano@unito.it

Matematiche Elementari under the direction of Mathesis.[3] Bonola noted that Italian mathematical literature to date had not yet produced a comprehensive guide to the mathematics taught in middle and secondary schools.[4] Such an encyclopaedia would provide a point of reference for teachers, students attending the University and the *Scuole di Magistero*, teachers in training, and all those interested in elementary mathematics and wishing to update and improve their teaching methods. Such a tool, he went on, should be developed with a view to:

> spreading knowledge of the exact sciences amongst the many teachers who, due to the posts they hold, live far from intellectual hubs, from libraries, from modern scientific movements. It should, ultimately, forge a kind of intellectual network between all those interested in the advancement of elementary mathematics, and by making valuable information and interesting news about recent developments more widely available, promote research in fields which are neither sterile nor lacking in interest.[5]

From the very beginning, a general overview of the *Enciclopedia* and its aims were well defined. According to Bonola's proposal, the *Enciclopedia* was to contain all the contents of syllabi taught in secondary schools, with those developed in the physics–mathematics section of technical institutes being considered to be particularly exemplary. In addition, it was to incorporate those "complementary" subjects aimed at facilitating a reconsideration of various points of elementary mathematics from an advanced standpoint and which could be used to elaborate exercises for use in the classroom or as homework. Considering the fact that introductory elements of infinitesimal calculus could be usefully added to the curricula of high schools (*Licei*) and technical institutes (*Istituti tecnici*), Bonola suggested that space also be made for theories which had so far been considered as pertaining to the first two years of university, such as infinitesimal analysis and the theories of sets and of analytic functions. Finally, in order to broaden the scope of the references provided, he expressed his desire that the *Enciclopedia* also include chapters on applied mathematics (physics, mechanics) and on history. With reference to the style of exposition, Bonola observed that:

> although they should not eschew the critical and philosophical aspects of certain subjects, entries should be of a primarily descriptive character, avoiding bias towards specific *Schools* or methodologies. For each topic discussed, the most noteworthy relevant propositions should be assembled in a systematic way. Each theory expounded should be accompanied

---

few months before his sudden death on 16 May 1911, obtained a chair at the Scuola superiore di magistero femminile in Rome.

[3]See [58, 10]. The report is also published in [11]. See also *Bollettino di Bibliografia e Storia delle Scienze Matematiche* (G. Loria), XI, 1909, p. 91 and XIII, 1911, p. 44–45.

[4]In what follows the expression "middle school", *scuola media*, is used to refer to levels of education between elementary and university.

[5][10, p. 41]: ...*a diffondere la coltura delle scienze esatte fra i numerosi insegnanti che, per ragioni d'ufficio, risiedono lontani dai centri di studio, dalle Biblioteche, dal movimento scientifico moderno; dovrebbe, infine, formare una specie di legame intellettuale fra quanti s'interessano del progresso delle matematiche elementari e, porgendo a tutti dati preziosi, notizie interessanti, promuovere ricerche in campi non sterili, né privi di qualsiasi interesse.*

by indications as to the various critical modes which can be used to develop a rational and comprehensive treatment of it. As regards the fundamental theorems, especially those which comprise difficulties of concept or development, *Enciclopedia* entries should provide brief proofs. Otherwise, only the simple statement should be provided, together with extensive historical and bibliographical references, which will make it possible for interested readers to find the works that deal with the relevant subjects in detail.[6]

Bonola's proposals, which would be largely adopted during the execution phase of the project, met with unanimous approval. F. Severi, then president of the Mathesis, took it upon himself to involve the association's central office and to "express the Lombardy section's wish that the future committee responsible for bringing the work into being be based in Pavia" [58, p. 36].

The preparations got off to a quick start and, in May of the same year, the executive council of Mathesis reached a unanimous agreement establishing the editorial board for the *Enciclopedia*, which consisted in a group of university and secondary school teachers, all of whom were in Pavia: L. Berzolari, F. Gerbaldi, G. Vivanti, R. Bonola and E. Veneroni.[7] The editorial board was given absolute authority and autonomy in choosing contributors and in deciding on the content and form of the articles.

The board's first task was to find an Italian publishing house willing to underwrite the costs of the *Enciclopedia*, the Mathesis itself lacking the financial resources to fund publication and having therefore simply authorised its publication "under the auspices of Mathesis". In two meetings held in July 1909, the members of the editorial committee refined their intentions in preparation for the Mathesis Congress to be held in Padua the following September. On that occasion, Berzolari and Bonola would make another report, in which they illustrated a preliminary outline of the activities managed up to that time. An addition was made to the original plan, namely, the chapter *Questioni pedagogiche*. The objectives, orientation and contents of the work remained unchanged, although it was now specified that the *Enciclopedia* would be published with a view "not only to saving readers' time and effort in seeking out accurate and reliable information on elementary mathematics, but also and above all to promoting mathematical culture".[8] In the meantime, a three-volume structure had been outlined for the *Enciclopedia*, the first treating

---

[6][10, p. 41]: ...*pur non evitando di toccare in certe quistioni il lato critico e filosofico, dovrebbe essere nettamente espositivo, senza pregiudizi di metodo o scuola. Intorno a ciascun argomento vorrebbe raccolte sistematicamente le proposizioni più notevoli che vi si riferiscono, e per ciascuna teoria indicati i vari modi secondo cui si può darle una trattazione razionale e completa. Delle proposizioni fondamentali, segnatamente di quelle che racchiudono difficoltà di concetto o di sviluppo, vorrebbe che l'«Enciclopedia» porgesse rapide dimostrazioni; delle altre il solo enunciato, insieme a larghe citazioni storiche e bibliografiche, che permettano all'interessato di risalire facilmente alle opere che ne trattano esplicitamente.*

[7][58, p. 48]. The editorial board changed several times during the years. In 1923 it was composed of L. Berzolari, F. Gerbaldi, G. Vivanti, D. Gigli and R. Serini.

[8][6, p. 2]: ...*con l'intendimento non solo di risparmiare tempo e fatica a chi desidera notizie precise e sicure su qualche argomento elementare, ma con la mira principale di diffondere la cultura delle matematiche.*

Analisi, the second devoted to *Geometria* and the third to the *Applicazioni, Storia della matematica* and *Questioni didattiche*. In order to speed up publication, with the aim of releasing the entire work within a maximum of three years, the board decided to bring out the *Enciclopedia* in a series of issues, in imitation of the German *Encyclopäedie der Mathematische Wissenshaften*.

At the Padua congress, Berzolari and Bonola were already able to present an index and abstract for each individual volume, close to the definitive one and sufficiently detailed. The volumes were to consist in 12, 17 and 9 chapters, respectively. They also explained to the members of the Mathesis attending the Congress the considerable problems involved in selecting the contributors and the publisher. With regard to the contributors, they announced that they had contacted a set of some forty scholars, both university and secondary school teachers, so as to entrust the various chapters "to academics who have already carried out specialised research on the subjects at hand and will therefore be able to carry out the work required of them rapidly and effectively" [6, p. 4]. The financial side of the project came up for considerable discussion at the congress. Indeed, the entire session of 23 September was dedicated to this matter, and some noteworthy differences of opinion amongst G. Loria, G. Lazzeri, R. Bonola, F. Severi, A. Padoa, G. Castelnuovo, A. Conti, R. Bettazzi and E. Lenzi emerged with regard to the layout, the format, the payments due to contributors, etc. For example, Berzolari and Bonola suggested opening a subscription amongst the member of the Mathesis, so that each member committed himself to purchasing one copy of the entire encyclopedia at a pre-established price payable in instalments over a 3-year period. Volterra and Castelnuovo, in contrast, gave assurances of their willingness to present a request for a publication subsidy to the Ministries of Education and of Agriculture.[9]

Work on the *Enciclopedia* resumed in Pavia in November 1909 and during the first months of 1910 the Committee began distributing assignments for specific entries.[10] In May 1910, for example, Mario Pieri, Berzolari's former assistant at Turin, was assigned to write and edit the chapter on *Logica Matematica*, "including the most recent departures from the Peano system (Russell, etc.)."[11] At the same time, Bonola and Vivanti were sounding out F. Amodeo, A. Perna and G. Vacca as possible authors for the chapters on *Geometria del piano e dello spazio, Teoria della misura*[12] and *Storia della matematica*,[13] respectively. Berzolari would discuss the selection and substitutions of contributors with his colleagues on many occasions over the years to come. Thus, for example, in April 1931 he wrote to Peano:

---

[9] In the end, Loria's proposal to offer the same discount to Mathesis members as that offered to booksellers was unanimously approved.

[10] See *Giornale di matematiche ad uso degli studenti delle Università italiane* (Battaglini), 49 (3, 2), 1911, p. 171.

[11] L. Berzolari to M. Pieri, 29 May 1910, in [3, p. 7–8].

[12] G. Vivanti to F. Amodeo, 4 August 1910, in [51, p. 477].

[13] G. Vivanti to G. Vacca, 11 May 1910, in [45, p. 185].

I am very grateful to you for your words and for the interest you have shown in our *Enciclopedia*, which is a source of infinite perplexities for me. I agree with you that it would be an excellent choice to assign the article on *Calcolo approssimato* to Maccaferri. There has been some discussion with Cassina. But if nothing is achieved in that direction, I will surely turn to Maccaferri.[14]

Actually, the Commission

> sought the collaboration of the greatest scientific minds of our country, and had the satisfaction of receiving positive responses from almost all the universitary and secondary teachers contacted. On the rare occasions on which, for various reasons, our invitations have not been accepted, we have been able to find alternative candidates quickly.[15]

The publication ran smoothly through the years immediately before the first World War, the "happiest years" [9, p. 396] of the academic careers of Berzolari and Vivanti. However, difficulties soon began to emerge. Bonola, Calò and Veneroni all died young, and even though Berzolari was president of Mathesis from 1915 to 1918, World War I and the post-war crisis paralysed the activities of the association. The editorial board of the *Enciclopedia*

> for a long period was reduced to a state of inertia, and was unable to respond to the many enquiries from Italian and foreign mathematicians as to how its endeavours were proceeding. Only when the most acute crisis had been overcome and when it seemed possible that things might gradually return to normal was there a revival of hope among the board members that their still unrealised initiative might be brought to fruition. Negotiations with the publisher were resumed and the contract signed, manuscripts were sent back to their authors for correction and updating, and replacements were found for those contributors who had passed away during the long interval.[16]

F. Enriques' presidency, starting in 1919, saw a certain shift in the matters discussed within Mathesis, although attention continued to be given to the *Enciclopedia di Berzolari*, as it had come to be called, at the various congresses of the association. After having shelved the idea of a serial publication, the first part of the first volume

---

[14]L. Berzolari to G. Peano, Pavia, 23 April 1931, Library of Cuneo, Peano Archives, c.p. N. 100299: *Ti sono molto grato di quanto mi scrivi e dell'interesse che mostri per la nostra Enciclopedia, la quale mi procura un'infinità di grattacapi. Sono d'accordo con te che l'articolo del Calcolo approssimato sarebbe al prof. Macc[aferri] ottimamente affidato. In passato ci fu qualche parola col prof. Cassina. Ma, se non si farà nulla con quest'ultimo, mi rivolgerò senz'altro al Maccaferri.*

[15][7, I¹, p. VIII]: ... *cercò di mettere a contributo le migliori energie scientifiche del nostro paese, ed ebbe la soddisfazione di vedere accolte le sue richieste di collaborazione da quasi tutti i professori universitari e medi ai quali si era rivolta, e di poter prontamente sostituire quei pochissimi che, per vari motivi, non avevano accettato l'invito.*

[16][7, I¹, p. VIII–IX]: ... *ridotta per lunghi anni all'inerzia, e all'impossibilità di dare qualsiasi risposta ai molti che dall'Italia e dall'estero chiedevano notizie dello stato de' suoi lavori. Fu soltanto quando le crisi più acute furono superate e apparve possibile un lento ritorno alla normalità, che rinacque nella Commissione la speranza di poter riprendere l'opera interrotta sul nascere. Si riallacciarono e si condussero a termine le pratiche con l'editore; si rinviarono i manoscritti agli Autori per la revisione e l'aggiornamento; e si sostituirono i collaboratori scomparsi nel lungo intervallo.*

was ready to go to print in December 1923 [5, p. 47], but did not actually come out until June 1929. It was published by Ulrico Hoepli in Milan, was enthusiastically presented the following November by Berzolari at the Unione Matematica Italiana and at the Istituto Lombardo,[17] and was sent off to numerous colleagues.[18] The second part of the first volume appeared between the end of 1931 and the beginning of 1932.[19] However, a series of sudden deaths which struck the editorial board led

---

[17] See *Bollettino dell'UMI*, s. 1, 8, 1929, p. 171 and *Rendiconti dell'Istituto Lombardo*, s. 2, 62, 1929, p. 802, meeting of 21 November 1929.

[18] See L. Berzolari to G. Peano, 16 November 1929, Library of Cuneo, Peano Archives, c.p. N. 103011: *Carissimo Peano avrai ricevuto da Hoepli, o riceverai presto, la 1ª parte del 1° volume dell'Encicl. delle matem.^{e} elementari. Un'altra copia ho fatto spedire, in omaggio a codesta R. Accademia delle Scienze. Ti sarei grato se volessi presentarla tu stesso, con qualche "buona parola". Spero che l'opera, malgrado i difetti che potrà avere, farà del bene alla scuola. Questo sarà l'unico compenso che potrò avere per le incredibili fatiche durate nel raccogliere i vari articoli* (Dearest Peano, you will by now have received, or will receive soon by Hoepli, the 1st part of the 1st volume of the *Encicl. delle matem.^{e} elementari*. I have had another complimentary copy sent off to the R[oyal] Academy of Sciences. I would be grateful if you would present it to them yourself, putting in a "good word". I hope that this work will do some good for the School, despite its defects. That would be the only compensation that I receive for the incredible strain I've been put under while assembling the various articles). As requested, Peano made a formal presentation of the work during the meeting of the Turin Academy of Sciences of 1 December 1929 (see Turin Academy of Sciences Archives, Cat. 3ª, *Adunanze di Classe e Verbali, Classe I, Mazzo 38, Verbali originali della Classe di Scienze Fisiche e Matematiche*, 1920–1933 p. n.n.): *Il socio Peano, per incarico del prof. Berzolari, socio corrispondente dell'Accademia, presenta il primo volume dell'opera Enciclopedia di Matematiche Elementari, a cura di L. Berzolari, G. Vivanti, e D. Gigli, con le seguenti parole: "La matematica elementare è la base della matematica superiore, ed è spesso fine a se stessa. Da alcuni anni, in Italia ed all'estero, si pubblicarono numerosi libri che esaminarono sotto l'aspetto storico e critico i fondamenti della matematica, rilevando definizioni viziose, dimostrazioni illusorie. Le prime 80 pagine del libro costituiscono la Logica del prof. Padoa. Essa tratta le questioni di Logica generale, specialmente quelle che si riferiscono alla matematica, per mezzo d'un calcolo simile al calcolo algebrico. È una chiara ed esauriente esposizione storica e critica di quanto fu fatto finora e contiene vari perfezionamenti dell'autore. Seguono le trattazioni dei Professori Gigli, Bortolotti, Cipolla, Finzi, Tacchella, dell'Aritmetica, Teoria dei numeri, Logaritmi e Calcolo meccanico. Questo libro è di massima importanza per tutti gli insegnanti di matematica, di ogni grado"* (Academy member Peano, at the request of prof. Berzolari, honorary member of the Academy, presented the first volume of the *Enciclopedia di Matematiche Elementari*, edited by L. Berzolari, G. Vivanti, and D. Gigli, with the following words: "Elementary mathematics are the basis for higher mathematics, and are frequently as well an end in themselves. Over the last few years, in Italy and abroad, a number of books have been published that examine the foundations of mathematics from historical and critical points of view. The first 80 pages of this volume consist in an exposition of *Logic* by prof. Padoa. It considers logical issues in general, and particularly those issues that refer to mathematics, through a kind of calculus analogous to algebraic calculus. This is a clear and comprehensive historical and critical overview of research in Logic lead till now and it also provides various particular refinements on the part of the author. This is followed by the articles of Professors Gigli, Bortolotti, Cipolla, Finzi, Tacchella: *Aritmetica, Teoria dei numeri, Logaritmi* and *Calcolo meccanico*. This book is of the greatest value to all teachers of mathematics, at all levels.").

[19] Berzolari presented it as forthcoming at the meeting of the Istituto Lombardo on 5 November, 1931, *Rendiconti dell'Istituto Lombardo*, s. 2, 64, 1931, p. 1091. The second part of the first volume was also presented by Peano to the Turin Academy of Sciences, at the meeting of 15

to a further suspension of activities until September 1936, the date of publication of the second volume. D. Gigli, one of the editors in chief, died unexpectedly, leaving his article on the *Teoria della misura* incomplete, to be finished by L. Brusotti, on the basis of his deceased colleague's notes. G. Biggiogero found herself in a similar position, when she was called to replace V. Retali in writing the chapter on the *Geometria del triangolo*. P. Benedetti also passed away, leaving his essay the *Fondamenti di Geometria* in proofs.

The Second World War once again brought the publication of the *Enciclopedia* to a standstill. The first section of the third volume, the last to be published under the direction of Berzolari, appeared in 1949. The second was approved and sent to press in the same year by Berzolari's collaborator, Brusotti. It opened with Berzolari's obituary and a warm tribute by Carlo Hoepli to the deceased *Maestro* "on the brink of the end of his 20-year effort dedicated to the direction and publication' of the *Enciclopedia*, which 'is and will remain the most apt and lasting monument to the memory of this great mathematician and teacher" [39, p. n.n].

In 1950, the publication process finally reached its conclusion, with the printing of the third part of the third volume of the *Enciclopedia delle Matematiche Elementari*, dedicated in its entirety to statistics, developed by C. Gini and G. Pompilj. By the end, in its complete form the *Enciclopedia* stood at seven volumes and 63 chapters, for a total of over 4,300 pages.

It occupies a unique place in the panorama of Italian publishing,[20] differing in contents, methodology and style of exposition from the other encyclopaedic works of the same period (F. Enriques' *Collectanea* and *Questioni riguardanti le matematiche elementari*, E. Pascal's *Repertorio di Matematiche superiori*, G. Peano's *Formulario di matematica* and M. Cipolla's *La matematica elementare nei suoi fondamenti*). It met with almost universal approval[21] both in Italy and abroad. The

---

November 1931 (see Turin Academy of Sciences Archives, Cat. 3ᵃ, *Adunanze di Classe e Verbali*, Classe I, Mazzo 38, *Verbali originali della Classe di Scienze Fisiche e Matematiche*, 1920–1933 p. n.n.): *Il socio Peano presenta il Vol. I, Parte II, dell'Enciclopedia delle matematiche elementari, donato all'Accademia dagli autori Berzolari, Vivanti, Gigli, coadiuvati da altri illustri matematici. Già due anni or sono presentò la parte prima di detto volume. Il socio Peano illustra lo scopo e l'utilità di tale Enciclopedia, veramente notevole per chiarezza e rigore* (Academy member Peano presented Vol. I, Part II, of the *Enciclopedia delle matematiche elementari*, donated to the Academy by the authors Berzolari, Vivanti and Gigli, in collaboration with other eminent mathematicians. Two years or so ago he presented the first part of the same volume. Peano explains the aim and the usefulness of this *Enciclopedia*, truly remarkable for its clarity and rigour).

[20]The same might be said upon comparison with similar encyclopaedias from abroad, such as the German and French versions of the *Encyclopädie der Mathematische Wissenschaften*, F. Klein's volumes of *Elementarmathematik vom höheren Standpunkte aus* or H. Weber and J. Wellstein's *Encyklopädie der elementar Mathematik*.

[21]Enriques is an exception here. He criticised the planning out of the first two volumes, arguing that, in their effort to meet the demands of eclecticism, the aspiration to seek out unity in diversity has been neglected, resulting in an overly fragmented overview which levelled and blurred distinctions between theories which actually varied in outlook and value. His criticisms mainly referred to the treatment of the foundations of mathematics and of didactic considerations, on which latter point he wrote [28, p. 113]: *Mi auguro che queste osservazioni sieno tenute in conto dai redattori dei volumi*

repeated requests received by Berzolari to provide a German edition, published at the same time as the original Italian one, the excellent reviews in the top scholarly journals in Italy, France, Germany and the United States [2,4,8,15,19–22,24,26,28, 30–36,40,41,43,44,46–50,54,56,59,66], and the many anastatic reprints over the years all bear witness to the appreciation of a collection of essays which preserves its indisputable mathematical, historical and cultural value even today.

2 Contents and Contributors

The first historiographical question which the *Enciclopedia delle Matematiche Elementari* presents to us is the attribution of a precise meaning to the term "elementary" in its title. The term is particularly rich in nuances and made frequent appearances in analogous series between the end of the nineteenth and the beginning of the twentieth centuries. Indeed, as E. Bompiani writes:

> When used with reference to geometry (or to mathematics in general), the term "elementary" can assume at least three different meanings. Firstly, it expresses simplicity and straightforwardness, delimiting subjects that are the object of a qualitative middle-school teaching. Secondly, it refers to the elements, to the foundations or bases, and provides an opportunity for critical analysis to enter more deeply into the concepts which underlie common intuition and consequently to broaden its field of research. The third refers to the elementary nature of the tools employed to attain to properties which might be more easily arrived at with other methods. Of these three aspects, the first is necessary, the second is of the greatest interest, because it goes beyond the first, giving it a firm foundation and indicating new lines of development; the third often has to do with technical skills.[22]

Without a doubt, the *Enciclopedia* reflects all of these different connotations, while there are at the same time noticeable differences from one article to the

---

in preparazione, e in ispecie di quello che concernerà le questioni didattiche: l'Enciclopedia deve fornire al lettore un'informazione quanto è possibile obiettiva dei varii indirizzi e criterii scientifici, logici e didattici: i redattori non hanno il compito di risolvere, secondo un proprio criterio questioni disputate, che sono passibili soltanto d'un giudizio storico di là del presente (I hope that the above observations will be borne in mind by the editors of the forthcoming volumes, and in particular by those dealing with didactic considerations. The *Enciclopedia* should provide readers with information as objectively detached as possible from the various scientific, logical and didactic schools and criteria. It is not the task of the editors to resolve, with their own criteria, controversies which will only be truly surmounted with the benefit of hindsight).

[22][8, p. 156]: *La qualifica di «elementare» data alla geometria (o alla matematica in genere) può assumere almeno tre significati diversi: il primo è quello di semplice, piano, e vale a delimitare qualitativamente la materia che è oggetto dell'insegnamento medio; il secondo è quello di relativo agli elementi, ai fondamenti ed offre all'analisi critica l'opportunità di approfondire le nozioni che soggiacciono all'intuizione comune e di allargare conseguentemente il campo della ricerca; il terzo si riferisce alla elementarietà dei mezzi impiegati per raggiungere proprietà che più rapidamente si troverebbero con altri mezzi. Di questi tre aspetti il primo è necessario; il secondo è il più interessante perché supera il primo dandogli sicuro fondamento e indicando nuove vie di sviluppo; il terzo sa spesso d'artificio.*

other. Thus, while in some sections the treatment seem overly elementary and rather too close in style to a common textbook, in others, to the contrary, it seems too abstract and "high-minded" with regard to the work's aims and the readers targeted. The various authors adopt different solutions to provide a bridge between "elementary" and "higher" mathematics. B. Colombo, for example, seeks to confine "higher matters" to the very end of his article *Sistemi lineari di cerchi e di sfere*, presenting them in the guise of an appendix. In his chapter on *Trasformazioni geometriche elementari*, which are elegantly outlined in accordance with the *Erlangen program*, U. Cassina adopts a different strategy. Here, "higher matters" are dealt with in specifically reserved paragraphs, which are indicated with asterisks and frequently include the results of original research by Cassina himself.

The *Enciclopedia* articles can be divided into four general groups. First, there are those which illustrate theories or methods fundamental for the practice of teaching, such as the discussion of problems and the *Questioni didattiche*, edited by R. Marcolongo and L. Brusotti, respectively. Next, there are those which address particular topics in elementary mathematics from a higher standpoint. Then, we find those chapters which aim at providing a historical framework for given scientific theories. The last typology consists in those articles providing specific aids for teaching, such as Cipolla's essay *Matematica ricreativa*, useful for finding exercises and problems which can make lessons more lively and interesting. In most cases, the choice of themes for entries echoes those included in the syllabi issued by the Italian Ministry for Education for the curricula to be taught at the secondary level of various kinds of schools. They thus reflect school reforms between 1909 and 1950 and, at the same time, economic development during that period. For example, in T. Boggio and F. Giaccardi's treatment of financial and actuarial mathematics it is possible to trace the effects of the improvement of social and industrial conditions after the second World War as well as the changes to the ministerial syllabi for technical and trade colleges. Similarly, C. Gini's chapter on *Metodologia statistica* was added after school reforms in the late 1940s and in order to meet the new demands of industry regarding the production planes. However, the chapters in the *Enciclopedia* are much more than mere summaries of traditional courses[23] or syntheses of treatises.[24] Indeed, the *Enciclopedia* includes a number of valuable and comprehensive articles on little-known and very circumscribed topics, such as G. Biggiogero's *Geometria del tetraedro* and Biggiogero and V. Retali's *Geometria del triangolo*. Moreover, even when they are treating aspects of mainstream didactic tradition, the authors seldom neglect to outline generalisations, new developments,

---

[23] Among the more traditional chapters, we can mention Brusotti's article on *Poligoni e Poliedri*, Artom's on *Proprietà elementari delle figure del piano e dello spazio* and Agostini's on *I Problemi geometrici elementari e i Problemi classici*, which, despite their conventional approach, would nonetheless prove useful to teachers in need of rapid and ready guidelines.

[24] Here, we might refer to the articles by C. Burali-Forti, G. Loria, G. Fano and U. Cassina. This last, in particular, closely follows the author's 1928 volume *Calcolo numerico con numerosi esempi e note storiche originali*, Bologna, Zanichelli.

applications and updatings of those theories, also taking into account research in progress at the time of writing.[25] For example, after providing a rather traditional exposition of polyhedrons, linked to the teaching of the theory of polygons at the middle school level, Brusotti goes on to append some interesting considerations on topology. Although the treatments primarily emphasize theoretical aspects, there is no lack of references to the applications of the various themes expounded. Thus, A. Comessatti, after illustrating the general aims and principles of descriptive geometry, dwells on its applications in photography and aerial photography.

The majority of the chapters in the *Enciclopedia* seek to give an account of the various scientific and methodological approaches, without affiliating themselves with any single author or approach and without privileging one kind of School over another.[26] They, thus, fulfil the requirements of the editorial board, whose task it was to supervise the balanced development and the *generale affiatamento*, or overall harmony, of the various sections. There are a few exceptions, the most blatant of which is without a doubt the article *Logica*, entrusted to A. Padoa.[27] The choice to include an entry on logic in an *Enciclopedia delle matematiche elementari* in itself bears witness to the breadth of Berzolari's outlook of mathematics. Despite the general acknowledgment of the contributions of Peano's School in this field, that a section on logic should serve as an introduction to the field of elementary mathematics was by no means a given. Even though the circumstance is important for its cultural meaning, Padoa limited himself to presenting a new system of propositions of ideographical logic based on three primitive concepts. He displayed a rather blinkered vision in neglecting to contextualize the relationship of symbolic to classical logic and to other contemporary systems. Indeed, his chapter was met

---

[25] E. Togliatti, for example, includes recent results by O. Chisini in his chapter on *Massimi e minimi*.

[26] Notwithstanding this, in the paragraphs on *Progressioni* and *Logaritmi*, A. Finzi paid no heed whatsoever to the contributions of the Peano School, such as those of T. Boggio, E. Viglezio, R. Frisone, A. Borio etc. which, instead, are expounded in detail in Cassina's chapter on numerical calculus.

[27] A similar evaluation might be attached at Benedetti's article entitled *Fondamenti di Geometria*, which Enriques strongly criticizes, arguing that [28, p. 113]: ...*l'esposizione dell'argomento per l'Enciclopedia delle matematiche elementari non aveva da conformarsi a vedute subiettive, anzi doveva seguire un ordine storico* (The treatment of this subject for the *Enciclopedia delle matematiche elementari* must not comply with subjective points of view. It should rather follow the historical order of developments). Enriques' criticism was not without foundation. Indeed, as a kind of supplement to Benedetti's article, an appendix on *Fondamenti della Geometria*, edited by G. Giorgi is inserted in the *Enciclopedia*. Here, the various aspects of foundational problems are discussed from an abstract, experimental and psychological point of view, also taking into account contributions to the field given by the logic and the advanced mathematics.

with a great deal of perplexity and criticism, from those both inside[28] and outside Peano's circle. Enriques' words were particularly harsh:

> But I cannot refrain from making some specific reflections on the article on Logic. First of all, there is the title of the article: simply "Logic", rather than "mathematical logic," "symbolic", "pasigraphic", or "ideographic"! We completely agree with the choice of this title, which implies the concept that there are not two different kinds of logic, one of which is worked by mathematicians, or by those mathematicians who use symbolic languages, while the other is cultivated by other scholars (philosophers, scientists in general or non-symbolist mathematicians). There is only one "Logic" and we recognise that, *for Padoa*, this Logic is given its highest expression in the symbolic treatment and in the system to which Padoa adheres. The full explanation of this view, and a comparison with other approaches, would be very interesting, also for those who do not share Padoa's point of view. Thus, what we really deplore is the fact that the author cloistered himself within a particular School and practically within his own current system.[29]

In other cases, instead, there are shortcomings in the coordination of related articles. Thus, for example, Gigli's chapter on *Aritmetica generale* makes no cross-references to the article on *Logica*, even where the two overlap (for example, in expounding on the notion of finite classes), nor does it take into complete account recent developments in the study of the foundations of arithmetic, providing rather instead vague definition of the concept of number.[30]

Contributors were repeatedly asked to take special care with the historical and bibliographical apparatus, whose breadth and precision was considered essential both for those wishing to carry out further research by going back to the sources and for those in need of a concise overview on the various theories.[31]

---

[28] See G. Vacca to G. Peano, 24 November 1929, Turin, Peano Archives, *Vacca correspondence*, c.p.: *Non ho ancora visto l'Enciclopedia del prof. Berzolari, se non di sfuggita. Ho avuto l'impressione che la Logica del Padoa sia troppo lunga, e forse anche, per quanto me ne ha detto il Padoa stesso, non corrispondente in tutto al mio modo di vedere* (I have not as yet had more than a fleeting glance at prof. Berzolari's *Enciclopedia*. My first impression was that Padoa's *Logica* was too long and perhaps also, from what Padoa himself has told me, rather dissonant with my own ideas on the subject). See also [19, p. LV–LVI].

[29] [26, p. 40–41]: *Ma non posso astenermi da alcune considerazioni particolari sull'articolo relativo alla logica. Anzitutto sul titolo dell'articolo: "Logica", anziché "Logica matematica" o "simbolica" o "pasigrafia" o "ideografia"! Approviamo toto corde la scelta di codesto titolo che implica il concetto non esservi due diverse logiche, ad una delle quali lavorano i matematici o quei matematici che usano linguaggi simbolici, mentre l'altra sarebbe coltivata dagli altri (filosofi, o cultori delle scienze in genere o anche matematici non simbolisti). C'è una sola logica, e si capisce anche che, per il Padoa, questa tocchi la più alta espressione nella trattazione simbolica e nel suo proprio sistema. La spiegazione d'una tale veduta, col confronto delle idee altrui, sarebbe riuscita interessantissima anche e specialmente per chi ne dissenta. Perciò deploriamo tanto più che l'autore si sia chiuso in una scuola particolare e quasi nel suo sistema attuale.*

[30] Similarly, Artom's article on *Proprietà elementari* ... makes only passing references to that of Benedetti on the foundations of geometry. By contrast, Gigli's *Teoria della misura* is outstanding in its constant cross-referencing to other articles.

[31] Of particular value are the "many thousands" of bibliographical citations, fruit of extensive and careful research by Berzolari himself who, in order to simplify this task, over the years had gradually created and enriched a miscellanea of 3,176 booklets and off-prints. The *Miscellanea*

The comprehensiveness of bibliographical data, which also include references to manuscripts, has, in general,[32] been recognised as one of the *Encicopedia*'s greatest merits.

With its 45 authors and 8 board members, the *Enciclopedia delle Matematiche Elementari* was one of the most "choral" and interactive forms of collaboration of Italy's mathematical community. Its true team-leader was Berzolari. A pupil of Pincherle, Berzolari benefited from his early experience in secondary school teaching in Pavia (1885–1887) and Vigevano (1888). This was followed by a period of university teaching in Turin (1893–1899), where he made contact with many scholars involved in the Schools of Segre, Peano and Volterra, before winning, in 1899, a chair in Algebraic analysis and Analytical geometry in Pavia, where he would remain until his retirement. Berzolari's research in the field of higher geometry, and, in particular, differential-projective geometry, is well-known, but it represents "a small part of Berzolari's services to mathematics" [9, p. 400]. What made him such an excellent candidate for the role of editorial director of the *Enciclopedia* was his involvement in "a particular form of academic endeavour" which began around 1906 and continued from then on "intensifying, increasingly absorbing almost all his time free from academic tasks".[33] This "particular form

---

Berzolari was donated to the Collegio Ghislieri and the University of Pavia, where it is still treasured today, together with a general miscellany of 1,000 booklets and with the Casorati miscellany, donated by G.A. Maggi, which consists in 2001 off-prints.

[32] See by contrast [20, p. LXVIII]: ... *estesissime, fin troppo a mio avviso (in un'opera di cultura generale, dedicata specialmente ai neo-laureati e professori di scuole medie), perché la sovrabbondanza straordinaria ... finisce con lo smarrire più che coll'aiutare il lettore novizio. E la consueta ricchezza di indicazioni bibliografiche fa poi maggiormente risaltare le lacune, talvolta non trascurabili* (They are immensely thorough (almost too thorough for a work of general education, aimed at new graduates and middle school teachers), because the extraordinary abundance ends up baffling rather than helping the novice reader. And the wealth of bibliographical data usually provided tends to throw light on shortcomings which are often far from negligible).

[33] [14, p. XXVII]: On this point, Bompiani writes [9, p. 402–403]: ... *ché anzi essa è di gran lunga superata, sia per l'ampiezza dei campi considerati sia per mole di lavoro, dalla organizzazione in vaste opere enciclopediche del sapere raggiunto. Questa attività richiede anzitutto il sacrificio del piacere e dell'orgoglio della ricerca scientifica personale per porre la propria dottrina a servizio e a profitto di altri ricercatori; e non dà neppure la soddisfazione che prova l'A. di un libro nel legare il proprio nome ad un'esposizione originale della materia al cui sviluppo abbia contribuito. Ma oltre a questa mortificazione della propria personalità come ricercatore e come autore, essa richiede una completa conoscenza della materia da raccogliere, chiari criteri ordinativi secondo cui raggruppare i prodotti delle varie fantasie individuali, senso di equilibrio onde contemperare le necessità di un'esatta e minuta documentazione bibliografica con l'esigenza di porre in rilievo le idee direttrici, vigilanza continua delle proprie preferenze, obiettività e serenità di giudizio* (That activity is indeed exceeded, in terms of both the immense range of fields considered and the sheer amount of work, by the activity of organisation of the knowledge obtained into vast encyclopaedic works. This activity requires sacrificing the pleasure and pride of individual scientific research in order to place one's own learning at the service and profit of other scholars. Nor does it give the satisfaction which the author of a book feels in seeing his name connected with an original treatment of a subject to which he himself has made some contributions. But aside from this mortification of one's own identity as researcher and author, this activity requires a complete

of academic endeavour" was his involvement in the main encyclopaedic projects of the period: the *Encyclopädie der Mathematischen Wissenschaften*, for which he prepared three articles on algebraic curves in the plane and in space (with K. Rohn), E. Pascal's *Repertorium der höheren Mathematik*, for which he was responsible for ten essays, and the *Enciclopedia italiana Treccani*, to which he contributed his masterful 49-column entry on *Algebra*. All of these experiences, undertaken in a patriotic spirit with the intention of performing a "great national service" [14, p. XXVII], predated his involvement with the *Enciclopedia delle Matematiche Elementari* and represented a kind of apprenticeship, perfecting his "mastery in fully satisfying both the aims of an entry of an encyclopedia: the need for a broad, all-encompassing perspective, and that for detailed bibliographical information. He was able to conciliate these requirements, which appear at first glance antithetical, into an harmonic and effective framework".[34] It was Berzolari who, with his sensitivity to the intellectual needs of teachers, his erudition and his organisational capacities and tenacity, outlined the *modus operandi* for the *Enciclopedia*'s preparation, established the editorial plan, selected most of the contributors, wrote various articles of his own and continued to correct manuscripts and proofs until the very final days of his life, suggesting changes, integrations, links among the articles, etc. Again, it was Berzolari who strictly dictated the form, extension, content and style of the work, thus maintaining its unity of vision, together with the coordination and balance of the various parts. It is no coincidence that many of the contributors had had some form of specific contact with Berzolari: Brusotti was his student at secondary school, then his assistant and colleague at Pavia; Biggiogero had also worked as his assistant; Boggio and Severi had been his students in Turin; R. Serini, E. Daniele and F. Sibirani served as professors when the Mathematical Institute in Pavia was under his direction; A. Palatini and L. Gabba were his colleagues at Pavia; Loria his fellow student, and so forth with others. This notwithstanding, the selection of contributors was remarkable in its breadth of perspective. The *Enciclopedia* included the involvement of scholars representative of the various *Schools* of mathematics present in Italy, even those whose outlooks and attitudes were diametrically opposed, and involved the universities of Turin, Pavia, Milan, Bologna, Genoa, Naples, Padua, Rome, Pisa, Modena and Trieste.

Duilio Gigli (1878–1933) and Giulio Vivanti (1859–1949) would share the direction of the *Enciclopedia delle Matematiche Elementari* with Berzolari. Gigli, Berzolari's son in law, after having been his student at Pavia's *Liceo* and then his assistant at the University there for the chair of Infinitesimal calculus, taught and

---

knowledge of the subjects that must be illustrated, clear regulating criteria according to which the products of many individual personalities are to be grouped, and an ability to find the right balance between the need to provide exact bibliographical information and the need to highlight the guiding principles. And finally this activity requires a constant supervision of one's own preferences, objectivity and serenity of judgement).

[34][14, p. XXVII]: ...*maestria nel soddisfare pienamente ad entrambi gli scopi di un articolo di Enciclopedia: una larga veduta d'assieme ed una minuta informazione bibliografica, conciliando tali esigenze a prima vista antitetiche in un quadro armonico ed efficace.*

served as deputy head teacher of the secondary school in Pavia. He contributed three articles to the *Enciclopedia* and dedicated "all the brief moments of respite from his arduous school commitments to the scrupulous correction of bibliographical references, ensuring the standardisation of notations and symbols and compiling the indices of authors and periodicals" [65, p. 256].

From a scientific point of view, Giulio Vivanti's role in the publishing project was more significant. After graduating in Civil engineering from the University in Turin (1881), Vivanti went on to study Mathematics at the University of Bologna, where he was a student of C. Arzelà and S. Pincherle, graduating on 30 June 1883. He obtained his habilitation (*libera docenza*) in Infinitesimal calculus on 13 May 1892. During this period, Vivanti taught a course in Bologna on algebraic numbers, of which a precise documentary record remains. On 23 October 1892 he sent Peano a note from Mantua entitled *Sull'uso della rappresentazione geometrica nella teoria aritmetica dei numeri complessi* for publication in the *Rivista di Matematica*, of which Peano was the editor-in-chief. Vivanti introduced the note as a digest of his lessons held on this topic:

> While teaching a course on the general theory of algebraic numbers, I came to observe that the theory of complex whole numbers assumes a fairly clear and intuitive form thanks to the constant recourse to geometrical representation. Here I outline the part of my lessons that refers to the above mentioned theme, in the hope that it might be of some interest to readers of the *Rivista* from a didactic point of view. For brevity's sake, I will omit all that reflects the extension of elementary operations to complex numbers, an extension to be carried out in conformity with Hankel's *principle of the permanence of formal rules*.[35]

The treatment begins with the definitions of the system T of whole complex numbers and the definitions of sum, difference, product and division, of which the primary properties are enounced. Vivanti then proceeds to outline an original geometric interpretation of the product of numbers of the system T. Geometric interpretations, visualized in the form of diagrams, are used for the introduction of the concept of lowest remainder of a number, concluding with the study of the ideals of the T system and of various theorems on congruence classes, including that of Wilson.

Vivanti moved to Pavia in 1893, where he held a chair at the *Scuola Normale* annexed to the Faculty of Sciences, remaining there until 1895, when he took over the chair in Infinitesimal calculus at the University of Messina, winning a full professorship in 1901. He returned to Pavia in 1907, as full professor of Infinitesimal calculus. He would hold this chair uninterruptedly, together with that in Higher analysis, until 1924, when he transferred to the University of Milan. There he taught

---

[35][60, p. 167]: *Nell'occasione d'un corso sulla teoria generale dei numeri algebrici tenuto nell'Università di Bologna, ebbi ad osservare che la teoria dei numeri interi complessi prende una forma assai chiara ed intuitiva mediante l'uso costante della rappresentazione geometrica. Espongo qui quella parte delle mie lezioni che si riferisce all'accennato argomento, nella lusinga che essa possa riuscire di qualche interesse ai lettori della Rivista dal punto di vista didattico. Ometterò per brevità tutto quanto riflette l'estensione ai numeri complessi delle operazioni elementari, estensione la quale deve farsi conformemente al* principio della permanenza delle leggi formali *di Hankel.*

Analysis (algebraic, infinitesimal and higher) and was dean of the faculty until his retirement in 1934. The author of some 200 publications,[36] dedicated prevalently to the theory of analytic functions, Vivanti, like Berzolari, also had to his credit some prior experiences of collaboration on specialised encyclopedias, having been one of the editors of Peano's *Formulario* and of the *Enciclopedia Treccani*, for which he prepared the entries *F. Brioschi*, *Equazioni* and *Calcolo infinitesimale*.

Thanks to his mastery of modern languages, he was solidly inserted in the international mathematical community, above all in that of Germany, where his indefatigable activity as reviewer for the *Jahrbuch über die Fortschriften der Mathematik* was well-known (he published a total of 1,740 reviews between 1884 and 1938).

Vivanti was, moreover, sensitive to didactic issues. The clearest testimony to his teaching skills lies in his 19 university and six secondary school textbooks, together with the clearly-explained booklets published by Hoepli on analytic functions, on integral equations, and on polyhedric and modular functions, some of which were translated into German and favourably received in Italy and abroad.

Given his training and his academic activities, it was natural that Vivanti's collaboration on the *Enciclopedia delle Matematiche Elementari* was primarily focused on analytic slope. He contributed two articles on this matter and oversaw their coordination with other related entries. Member of the Pavia section of Mathesis, and later, starting from 1925, President of the Milan section, he was nonetheless only fully involved in the first volume of the *Enciclopedia*. Subsequently, his role became increasingly marginal and eventually ceased altogether, partly as a result of his blindness, and partly because of the racial laws, which led to his marginalisation from the academic world.

3 The Contribution of Bolognese Mathematicians

In the first decades of the twentieth century, Bologna was a well-known hub for the debates on pedagogical issues. Here, the Mathesis had its headquarters from 1900 to 1902, during the presidency of G. Frattini, and then again, from 1919 to 1932, under the direction of F. Enriques. The Federazione Nazionale degli Insegnanti di Scuola Media, for a long time directed by Vitali, placed here its first congress in 1902. The presence of publishing houses such as Zanichelli, sensitive to didactic issues and committed to the publication of textbooks, and the fact that the *Periodico di Matematiche* had its publishing seat in Bologna were components that together contributed to create fertile ground for the promotion of interactions between the world of secondary schools and that of the university, whose members were called

---

[36]Today Vivanti is principally remembered for the theorem which carries his name, which states that for a function represented by a series of real nonnegative coefficients, the intersection of the circumference of convergence with the real-positive semi-axis is a singular point. On Vivanti's scientific output see [23, p. 184–205].

to face themselves on methods and program contents, on the initial training of teachers and their continuing education after graduation. It is not surprising, then, that in the "geographic distribution" of the members of the team of the *Enciclopedia*, the contribution of Bolognese community is prominent. This involvement assumed the widest possible range of forms and titles and enabled the scholars involved to develop or reinforce their network of professional relations with colleagues throughout Italy and, above all, in Pavia. The contributors with ties to Bologna range from those – such as Vivanti – who had carried out here the entirety or a part of their studies at the University, to those – such as Bortolotti, B. Segre, Sibirani, Vitali and Pincherle – who taught at the University of this town during the period in which they were involved in the *Enciclopedia*, to those who were linked to this project in a more indirect and tangential way, such as Enriques, in his role as Mathesis president, as driving force behind a similar editorial enterprise – the *Questioni riguardanti le Matematiche elementari* – as well as a critic and reviewer of the *Enciclopedia delle matematiche elementari* itself.

The occasions of contact between the editorial board of the *Enciclopedia* and the Bolognese community passed presumably through Pincherle, to whom both Berzolari and Vivanti had ties. Pincherle, the "Nestor of Italian mathematicians", had taught Vivanti at university, and Berzolari – who would succeed him as president of the Unione Matematica Italiana – in secondary school. In addition to this trait d'union, we should however also recall the large number of former students and colleagues of Berzolari and Vivanti from their periods in Turin and Pavia who had become fixtures to a greater or lesser extent in the Bolognese milieu. B. Levi and F. Sibirani all fell into this category, as did U. Amaldi and F. Enriques, who were also in their turn involved in the *Enciclopedia*, albeit in a less formal way.[37]

From a historiographical point of view, the analysis of the participation of a *local* community in a such *national* and polyphonic enterprise as the *Enciclopedia* implies:

- examining whether and to what extent the chapters reflect the authors' training received at the University of Bologna and/or their teaching practices there;
- reconstructing, through recourse to relevant documentary sources, the reasons that justify the choice of these authors and the desire of the editorial board to involve them in the *Enciclopedia*'s team; and finally,
- establishing the existence of any common ground between these mathematicians, beyond the obvious geographical location, evaluating the presence and nature of

---

[37] It was not by coincidence that, on the occasion of celebrations of Berzolari's 50 years as a teacher, in October 1935, many of those who participated came from a Bolognese background. Together with the mathematicians mentioned above, there were also G. Charrier, E. Lodi, M. Manarini, G.B. Zecca, and messages arrived from Pincherle's family, from the University, from the Unione Matematica Italiana, from the Istituto superiore di scienze economiche e commerciali and from the Accademia delle Scienze dell'Istituto di Bologna, of which Berzolari was a corresponding member (*socio corrispondente*).

any direct relations or interactions between them over the period in which they were working on the *Enciclopedia*.

Due to a plethora of internal and external factors, it is impossible to answer all these questions. The often quotidian and informal nature of the relationships which develop among a group of colleagues and scholars working in the same place, the lack of archival materials and the component of chance that mark the ties to Bologna of some of these mathematicians all impede us from being exhaustive. For example, it is impossible, on the basis of our extant historical knowledge, to establish whether and in what way Agostini's years in Bologna, first as a student in Mathematics and then as assistant and lecturer in free courses of History of mathematics, influenced him as he was preparing his chapter on *Problemi geometrici elementari e problemi classici*. Similarly, we do not fully understand the reasons which led E. Artom, who had spent two years in Bologna as Enriques' assistant, to request a transfer immediately after the First World War to become secretary of the editorial board of the *Enciclopedia delle matematiche elementari*.

The picture is clearer for the six full professors linked to Bolognese milieu of whom whole entries were commissioned: G. Vitali, who prepared the article on *Limiti, serie, frazioni continue e prodotti infiniti* (I, 2, p. 391–439), G. Vivanti, who wrote the *Elementi di analisi infinitesimale* (I, 2, p. 441–547) and the *Teoria degli aggregati* (I, 2, p. 549–563); S. Pincherle who was responsible for *Le funzioni analitiche da un punto di vista elementare* (I, 2, p. 565–597), B. Segre who prepared the entry on *Geometria analitica* (II, 2, p. 141–249), F. Sibirani who outlined the *Calcolo delle probabilità* (III, 2, p. 193–244) and finally E. Bortolotti, of whom was commissioned the *Storia della matematica elementare* (III, 2, p. 539–750).

Without attempting to reconstruct the genesis and stages of development for each of these articles, we will seek to identify their distinguishing features and their relationship to the other sections of the *Enciclopedia*, indicating, at the same time, the plausible reasons, both scientific and not, why Bolognese mathematicians were chosen as their authors.

The commission for the chapter on *Calcolo delle probabilità* went to Filippo Sibirani who was, together with Bortolotti, the most "authentically" Bolognese of the team involved in the *Enciclopedia*. Born in S. Agata in 1880, Sibirani had graduated in the Emilian chief town in 1902, remaining there as assistant in Infinitesimal calculus and Rational mechanics until 1907 and then returning in 1929 as dean of the Istituto Superiore di Scienze Economiche e Commerciali. A former colleague of Berzolari in Pavia, where he taught from 1915 and 1922 and again in the 1926–1927 academic year, Sibirani's credentials as a mathematician were already firmly established by the time of his involvement with the *Enciclopedia*, as was his reputation as a scholar with superb organisational capacities, as his activity as *Commissario prefettizio* (Prefectorial Commissioner) of the Unione Matematica Italiana in 1945-46 attests. He could boast a wealth of significant findings on ordering functions and many years of teaching experience both at the secondary school and the university level, which particularly qualified him to be the author of an article for an *Enciclopedia* especially geared towards teachers.

Sibirani's chapter, which was quite similar to that which he would contribute to M. Villa's *Repertorio di Matematiche* (1951), developed a theme which had been inserted in ministerial programmes for secondary schools thanks to the insistence of G. Vailati and, above all, G. Castelnuovo. The treatment begins with the notion of an event's probability, which is considered from a historical perspective, moving from the classical and statistical definitions given by P.S. de Laplace and L. Cournot respectively, to the critiques by H. Poincaré and G. Castelnuovo to the axiomatic definition of F.P. Cantelli and to the abstract theory of R. Von Mises. There follow the fundamental theorems (of total probability, of repeated trials, of Tchebychef, Bernoulli, Bayes, Poisson etc.) and, in conclusion, the applications of Gauss's theory on errors of observation and the method of least squares. Sibirani's primary emphasis is on the contributions of the Italian School to the probability theory[38] from the eighteenth to the twentieth century, and he makes repeated references to the results of Castelnuovo, Cantelli, Tedeschi and De Finetti.[39] He does not neglect to place his own article in relation to those on financial and actuarial mathematics by T. Boggio and F. Giaccardi. The insights into the applicative relapses of the calculus of probability provided by Sibirani would be further enriched by F. Severi and F. Conforto who, in the third volume of the *Enciclopedia*, would illustrate the significance that this field was acquiring in the experimental sciences:

> especially in relation to physics during the great crisis that brought us from a decidedly deterministic perspective to those more recent conceptions, whereby the universe is thought to be governed by laws of an indeterministic and therefore probabilistic character (quantum physics, atomic physics).[40]

The editorial board's choice regarding the commissioning of the chapter on *Storia della matematica elementare* appears to be "natural" from a scientific point of view. In the by no means crowded ranks of historians of mathematics at that time, Ettore Bortolotti, who taught Analytical geometry in Bologna from 1919 to 1936, was one of the most illustrious figures. It is therefore no surprise that, after Giovanni Vacca's refusal, the choice fell on Bortolotti. With a wealth of publications to his name, focusing above all on the history of algebra, Bortolotti also fully shared Berzolari's patriotic fervour and, although sometimes prone to lapse into nationalistic tones, seemed the scholar most qualified to highlight the contribution of Italian mathematicians to the history of elementary mathematics. His article, indeed, stands as a true monograph, accompanying in the *Enciclopedia* that of M. Gliozzi on the *Storia del pensiero fisico*. After providing an overview

---

[38] The contributions of Italians are also emphasised in Berzolari's articles, although without nationalistic asides.

[39] He includes some suggestive reflections on the subjective theory of probability by B. De Finetti, who had been student of Vivanti at the University of Milan. These thus make their first appearance in an *Enciclopedia* aimed at teachers.

[40] [57, p. 809]: ... *accompagnando in ispecie la fisica nella grande crisi che ha portato dalle vedute decisamente deterministiche alle più recenti concezioni, nelle quali si pensa l'universo retto da leggi di tipo indeterministico e quindi probabilistico (teoria dei quanti, fisica atomica).*

of mathematics in ancient civilizations (Chinese, Egyptian, Sumerian), Bortolotti, rightly, dwells at some length on the description of Greek mathematics, from its "formative" period (sixth to fourth centuries B.C.) to the classic and Alexandrian periods. He then provides a detailed examination of the problems of Renaissance mathematics and arrives at the first origins of infinitesimal calculus. There is no lack of references to the history of mathematics beyond Europe – in China, India and the Islamic world, in particular. His treatment of the subject is systematic, harmonious and engaging, echoing the author's own research and thoughtfully selecting only general methods and ideas, without getting lost in superfluous details or erudite excursus. The precision with which the author refers back to the sources, of which he often also provides significant excerpts, is exceptional. However, various canons of historiographical inquiry, now outdated but typical of that time, crop up, such as the use of rather imprecisely-defined categories such as *School* and *Maestro*, the tendency to search for precursors and to make claims about the contribution of Italian mathematical Schools compared to those in other states.

More closely related to the author's didactic experiences than to his research activity is B. Segre's chapter on *Geometria analitica*. In fact, at the time of his involvement in the *Enciclopedia delle matematiche elementari*, and prior to the exile imposed on him as a consequence of the racial laws, he had already taught courses on this topic both in Turin, as assistant from 1925–1926, and in Bologna, where he had been awarded a chair in Analytic and Descriptive geometry in 1931. He too had been invited to collaborate on the *Enciclopedia Treccani*, for which he prepared the entry on *Coordinate*, the commission for which had first been offered to Berzolari, but refused by him.

Although there is no lack of interactions between these authors, all of whom were colleagues working together at the University of Bologna in the early 1930s, the contributions of this community show themselves to be particularly close-knit with regard to analysis, the sector with regard to which it had the most opportunities for dialogue.

The second part of the first volume of the *Enciclopedia delle matematiche elementari* is devoted to analysis, both in its "classical" part and in those themes called at that time "complimentary" (i.e., the theory of analytic functions and set theory). This section was coordinated by Berzolari, who was, however, flanked by Vivanti somewhat more strongly than elsewhere.

Even though it had always been a fundamental discipline in the curricula and one of the most important courses for the training of not only pure but applied mathematicians as well (engineers, scientists, naturalists, etc.), infinitesimal analysis was inserted into the syllabi for Italian secondary schools only after 1909,[41] thanks

---

[41] The history of the insertion of the first elements of analysis into Italian secondary school syllabi is fairly complex. As early as 1867, L. Cremona had recommended introducing the concept of function and the method of limits. In effect, the first elements of infinitesimal calculus made their appearance in algebra textbooks written for the physics-mathematics section of *Istituti Tecnici* in around 1871. In 1887, however, Italian school syllabi were noticeably behind the times, especially in comparison to those of Prussia where, thanks to F. Klein's interventions, differential calculus had

to the efforts of Vailati and Castelnuovo, who had, in turn, been strongly influenced by Klein's opinions in this regard. It is therefore natural that Berzolari and Vivanti should seek to bring out the best of the calculus section of the *Enciclopedia*, assigning it to an individual who was prepared to take into account the different methodological approaches which had emerged over the time. Between 1880 and the first two decades of the 1900s, Italy had witnessed fierce debates between mathematicians of contrasting tendencies concerning calculus, with regard to both the choice of research themes and its teaching. The role of logic, set theory and foundational criticism as theories preparatory to calculus, the relationship with numerical and physical applications, and the connection between the courses on Infinitesimal calculus and those on Algebraic and Higher analysis are just a few of the themes which came up for discussion amongst renowned scholars such as G. Peano, E. Cesàro, C. Arzelà, S. Pincherle, G. Fubini, L. Tonelli and F. G. Tricomi. This, then, is the reason why, for the treatment of such a delicate topic, the editorial board of the *Enciclopedia* selected two scholars known in the academic world for their balanced outlook: Vitali and Vivanti, both of whom had deliberately remained on the sidelines, so to speak, of all those polemics.

By the beginning of the 1930s, both could boast experience in the teaching of calculus which spanned decades. Vitali had taught at high schools in Voghera and Genoa for more than 20 years before being given the chair left vacant by Tonelli at the University of Bologna in 1930. Vivanti had won general approval as a teacher of Differential and Integral calculus in all three universities where he had taught during his career: Messina, Pavia and Milan.

Both Vitali and Vivanti adopted a traditional content structure. Vitali offered a schematic development of the main properties of the infinite algorithms of sequences of real numbers and the properties of limits of numerical sequences and series. The paragraphs on the convergence criteria for series with positive terms and double series are particularly meticulous, as are those on continuous fractions and infinite products, which were completed by Cassina in his article on *Calcolo numerico*. Vivanti instead expounds the elements of infinitesimal calculus, stretching so far as to make a brief reference to the calculus of finite differences. A distinguishing feature of his exposition, as had already been the case in his university textbooks, is the historical notes provided, fruit of his own research activity in the sector.

The genesis and the various rewritings of these entries of the *Enciclopedia* can be well documented, thanks to various letters by Berzolari and Vivanti, collected in the *Fondo Vitali* conserved at the Archives of the Unione Matematica Italiana and as yet only published in part.

---

been part of the syllabus since 1882. Under the influence of Klein, Vailati introduced the concept of differentiation (1909–1910) in all three kinds of *Liceo* and, at the *Liceo scientifico*, added that of integration. This was the object of lively debates. In 1911, Castelnuovo released the new syllabi for the *Liceo moderno*, which included the concepts of function, derivative and integral. Soon after, these concepts appeared in textbooks by F. Enriques and U. Amaldi (1914) and by S. Catania (1914). See [37, p. 5, 13, 17–26, 29, 35–37, 43–47, 51–52].

Vivanti and Vitali began corresponding in 1904, when Vivanti, who was still teaching at Messina at the time, asked Vitali for an extract of his note *Sopra le serie di funzioni analitiche*, which he intended to review for the *Jahrbuch*.[42] It therefore seems highly probable that the fact that Vitali was called on to contribute to the *Enciclopedia* from the first decade of the twentieth century resulted from his relationship with the Mantovan mathematician. We need, instead, to look ahead to January 1928 to trace Vitali's first contact with Berzolari, as a consequence of the latter's need to inform him of the economic conditions which the editorial board of the *Enciclopedia* had managed to "wrench" from the Hoepli publishing house:

> For the publication of the first volume of the Enciclopedia delle mat.^{e} elem.^{i} I had contacted Zanichelli, as well as Hoepli. However, Zanichelli rejected the proposal, as they already had many other commitments. Our only option, then, is Hoepli, who offers us the following conditions: *that each author should receive 10 per cent of the cover price, payable in three instalments: one third upon publication of the volume, another after the first thousand copies sold (the first print run will be of 1,500 copies) and the last when the first print run has sold out. Each author will have 25 free off-prints of his article.* I do not like the fact that payment is in three instalments, but Hoepli is unshakeable on this point. The publisher also wishes to have a written declaration of acceptance of the aforementioned conditions from each author. If you accept, please send the declaration (referring to the conditions) to me and I will send it on to Hoepli together with those of the other authors.[43]

Vitali's involvement in the *Enciclopedia* project strengthened his ties to Berzolari, as shown by another letter, written a year later, in which Berzolari thanked Vitali for the gift of his important monograph, *Geometria nello spazio Hilbertiano*, about which he expressed a fairly positive opinion.[44]

In spring 1930, the dialogue between the two was further confirmed in consideration of Vitali's work on the chapter on *Limiti*, the manuscript version of which had already been submitted to Berzolari in the previous months. On 4 February, the Pavese mathematician contacted Vitali, indicating his desire that the first draft be modified and partially extended, in order to also take into account the complex field:

> Dear Professor, I must trouble you once more in order to ask you a favour. It would be good if in one of the articles in our *Enciclopedia* there appeared at least the most elementary notions pertaining to limits and series in the complex field. It seems to me that the most

---

[42] See G. Vivanti to G. Vitali, Messina, 16 October 1904, in [12, p. 447–448].

[43] L. Berzolari to G. Vitali, Pavia, 20 January 1928, Bologna, UMI Archives, *Vitali correspondence*, c.p.: *Per la pubblicazione del 1° volume dell'*Enciclopedia delle mat.^{e} elem.^{i} *mi sono rivolto, oltre che all'Hoepli, allo Zanichelli, ma questi ha rifiutato, avendo già molti altri impegni. Non resta dunque che Hoepli. Il quale pone le condizioni seguenti. Dare ad ogni autore il 10 per cento sul prezzo di copertina, pagando in tre volte: un terzo appena uscito il volume, un altro terzo dopo vendute mille copie (l'edizione sarà di 1500 copie) e l'ultimo terzo ad edizione esaurita. Ogni autore avrà 25 estratti gratuiti del suo articolo. La ripartizione in tre rate non mi piace ma l'Hoepli è irremovibile. Egli desidera pure avere da ogni autore la dichiarazione scritta dell'accettazione delle dette condizioni. Se Ella le accetta, La prego di mandare a me la dichiarazione (contenente le condizioni) ed io le trasmetterei all'Hoepli con le altre degli altri autori.*

[44] L. Berzolari to G. Vitali, Pavia, 27 December 1929, in [12, p. 505–506].

fitting place for this would be in your article. Would you mind adding a few pages dealing with these subjects?[45]

This request met with a warm response from Vitali, who immediately declared himself willing to modify his original manuscript. On 10 March, Berzolari thanked him "profoundly for your cordiality in showing yourself so to be so willing to help me in the best interests of the *Enciclopedia*"; at the same time, he also put forward another request:

> Your proposal to modify the current text of your article here and there so as to embrace the complex field to the greatest possible extent seems an excellent one to me. However, I must appeal to your good will again in requesting one further small addition. The last article in the first volume is that by Pincherle (*Le funzioni analitiche da un punto di vista elementare*). It is therefore necessary that there should be some treatment of function series, and in particular of power series, circle of convergence, etc., in one of the articles that precedes it. Vivanti's entry (*Elementi di analisi infinitesimale*) should already touch on these matters, but only with limited reference to the field of real numbers. Nor would it be possible, without perverting the nature of that article and without rendering it over-long, to incorporate there the treatment of the complex field. As a consequence, I would ask you to leave your own article in its current form, with the exception of the additions which you yourself have suggested, and to add a brief chapter, in which the properties of limits, of function series and, in particular, of power series are extended to the field of complex numbers. It would be good if you could also draw forth the definitions of power and logarithm, which are deduced in Gigli's article on the basis of the permanence principle. To minimise overlaps with Vivanti's article, you could get in direct touch with him, to whom I will write myself, in order to inform him about this question.[46]

Berzolari, then, invited Vitali to get in touch with Vivanti and, at the same time, urged Vivanti to agree with his colleague upon the subdivision of topics which the two entries might both cover so as to minimise overlaps and gaps. Vitali's chapter was thus revised once more, this time by Vivanti who, in addition to noting, like Berzolari, some lacunae with regard to complex sequences and series, also pointed out the lack of a section on the theory of function series, which, in his article on infinitesimal calculus, he had assumed had been treated elsewhere. Thus he expressed his willingness to add these missing sections – to his mind essential in such a didactic enterprise – himself, while leaving the decision as to where it should best be placed within the *Enciclopedia* up to Vitali.[47]

---

[45]L. Berzolari to G. Vitali, Pavia, 4 February 1930, Bologna, UMI Archives, *Vitali correspondence*, c.p.: *Caro Professore, debbo nuovamente disturbarla per pregarla d'un favore. Sarebbe bene che in qualche articolo della nostra* Enciclopedia *comparissero almeno le cose più elementari relative ai limiti e alle serie nel campo complesso. E mi pare che il posto più adatto sia il Suo articolo. Le spiacerebbe aggiungere ad esso poche pagine, dove quegli argomenti fossero esposti?*

[46]Translation of the letter by L. Berzolari to G. Vitali, Pavia, 10 March 1930, in [12, p. 507–508].

[47]G. Vivanti to G. Vitali, Milan, 17 March 1930, in [12, p. 508–509]: *Bisognerà poi dire qualche cosa delle serie di funzioni, e in particolare delle serie di potenze a variabili reali o complesse (la cui teoria occorre per l'articolo del Pincherle); questo capitolo dovrà trovar posto, mi sembra, nello stesso articolo in cui si parlerà dei limiti delle funzioni. Attendo di sapere quello che Lei propone di fare a questo riguardo* (You should then say something about function series and, in particular, about power series with real or complex variables (whose theory is required for

After this exchange of letters, the two directors of the *Enciclopedia* agreed to contact Vitali, requesting that he fill the gaps indicated. This new request, too, was complied with, and Vitali provided the additions, receiving thanks once more from Berzolari for his promptness.[48]

By 1 April 1930, Vivanti had already examined the second version of the article on the *Limiti*, which he now found to be excellent, and was going to make the necessary adjustments to his own chapter.[49] By September, the division of the contents had become definitive, thanks to the immediate understanding between Vitali and Vivanti. The former would concentrate on the numerical aspect, the latter on the functional ones:

> My esteemed colleague, I am in full agreement with what you have written to me. It is arranged that you will be concerned with the numerical part and I with the functional. Since you tell me that you yourself will write to prof. Berzolari, I see no use in me writing as well to inform him of the agreement reached. I have started reading your fine essay, but I have been sidetracked by other occupations. I hope to get back to it as soon as possible.[50]

In October 1931, a few months before Vitali's sudden death, his correspondence with the editorial board of the *Enciclopedia* came to a close, with the board expressing their recognition of the excellent quality of his work and of his proactive collaboration.[51] In addition to the *Elementi di Analisi infinitesimale*, Vivanti also

---

Pincherle's article). This chapter should, to my mind, be placed in the same article in which the limits of functions are treated. I look forward to hearing what you propose to do on this point).

[48] L. Berzolari to G. Vitali, Pavia, 25 March 1930, Bologna, UMI Archives, *Vitali correspondence*, c.p.: *Carissimo Professore, oggi Le ho spedito – in franchigia, a mezzo dell'Università – le bozze del Suo articolo. Non so come ringraziarla della Sua cortesia e della buona volontà che mi dimostra, e soltanto sono dolente che tutto ciò aumenti il Suo lavoro e contribuisca ad accrescere la Sua stanchezza. Non lavori troppo, e si prenda un po' di riposo!* (My dear Professor, today I have sent off the drafts of your article, with postage paid via the University. I cannot thank you enough for your courtesy and for the goodwill that you have shown me. I am only sorry to have added to your workload and increased your fatigue. Do not work too hard and get some rest!).

[49] G. Vivanti to G. Vitali, Milan, 1 April 1930, Bologna, UMI Archives, *Vitali correspondence*, c.p.: *Carissimo Collega, La ringrazio di quanto mi scrive. La nota sul* Limite *va benissimo. Da parte mia sto preparando le modificazioni e le aggiunte al mio articolo, in modo che alla fine nulla abbia a mancare* (My dear Colleague, thank you for your letter. The note on the *Limite* is absolutely fine. At my end, I am preparing the modifications and additions to my article, so that nothing will be lacking in the end).

[50] G. Vivanti to G. Vitali, Milan, 23 September 1930, Bologna, UMI Archives, *Vitali correspondence*, Bologna, c.p.: *Egregio Collega, sta bene quanto Ella mi scrive; resta fissato che Ella si occuperà della parte numerica, mentre io provvederò alla parte funzionale. Poiché Ella mi dice che scriverà al prof. Berzolari, credo inutile informarlo io pure degli accordi presi. Ho incominciato a leggere il Suo bel trattato, ma poi altre occupazioni sono venute ad interrompermi. Spero riprendere al più presto.* The correspondence between the two includes another missive from Milan, dated 31 December 1928.

[51] L. Berzolari to G. Vitali, Pavia, 23 October 1931, Bologna, UMI Archives, *Vitali correspondence*, Bologna, c. 1r: *Caro Professore, con la presente Le mando un vaglia della Banca d'Italia (N. 67188) di lire 340,95 come compenso per il Suo articolo inserito nell'*Encicl. di matem.[e] elem.[i]*. Poiché le copie in vendita sono 1500 e ognuna costa lire 82 ad ogni rata di pagamento*

prepared the *Enciclopedia* entry on the *Rapporti fra la teoria degli aggregati e la matematica elementare*, thus confirming his reputation as a *Maestro* in this field, a reputation which he had begun to earn in the last decade of the nineteenth century, starting from the publication of a *Notice historique sur la théorie des ensembles* in G. Eneström's *Bibliotheca Mathematica* [Vivanti 1892, VI, p. 9–25]. That had been followed by his involvement in Peano's *Formulario di Matematica*, an encyclopedic treatise written in ideographic language, for the first edition of which Vivanti had written the chapter on the *Teoria degli aggregati*.[52] Referring back to the works of C. Burali-Forti, here Vivanti focused primarily on the theory of transfinite numbers and order types, on well-ordered sets, on the genetic definition of real numbers and on the *classi ordinate secondo n dimensioni*, to which Peano had alluded at in his *Lezioni di Analisi infinitesimale* (Torino, Candeletti, 1893, vol. 2, p. 1–30).[53] He completed the exposition with a *Lista bibliografica* on the theory of sets, which appeared in the *Formulario*,[54] as an article in the *Rivista di Matematica* directed by Peano [61, p. 189–192], and finally, after having been updated to include the

---

ciascuna pagina viene compensata con lire 7.45. Il Suo articolo, compreso l'indice è di 46 pagine; quindi le sono dovute lire 342,70. Detraendo lire 1,75 per la spedizione, rimane la detta somma di lire 340,95. Per mia tranquillità, La prego di farmi avere un cenno di ricevuta. Con i saluti più cordiali e con vivi ringraziamenti per il valido contributo dato alla nostra Enciclopedia *mi creda Suo aff.mo L. Berzolari* (Dear Professor, please find enclosed a bank order from the Banca d'Italia (N. 67188) for 340.95 lire, in payment for your article published in the *Encicl. di matem.^{e} elem.^{i}*. Since 1,500 copies are on sale and the cost of each is 82 lire, 7.45 lire will be paid per page per instalment. Your article, including the index, is 46 pages long, so you are due 342.70 lire. Subtracting 1.75 lire for postage, the sum due stands at 340.95 lire. For my peace of mind, I would ask you to send me notice of receipt. With my kindest regards and heartfelt thanks for your valuable contribution to our *Enciclopedia*, I am truly, affectionately yours, L. Berzolari). See also L. Berzolari to Vitali, Pavia, January 1939, Bologna, UMI Archives, *Vitali correspondence*, Bologna, c. 1r: *Chiar.mo Professore, della 2^a parte del I volume dell'*Enciclopedia delle Matematiche Elementari *si sono sinora vendute 1360 (delle 1500) copie, e l'editore Hoepli mi ha mandato il compenso da distribuire ai vari autori (limitatamente alle copie vendute). Poiché l'articolo redatto dal compianto suo fratello è di 46 pagine, gli spettano lire 589,70. Le mando perciò un vaglia della Banca d'Italia N. 0.010.090. per la somma di lire 587,70, avendo trattenuto due lire per le spese postali. Con distinti saluti e auguri mi abbia suo dev.mo L. Berzolari. P.S. Le sarò grato di un cenno di ricevuta. Veramente la spesa fu di 1.75 quindi le accludo un francobollo da 0.26. Risponde 25 gennaio 1939.* (Most esteemed Professor, so far 1,360 (of the 1,500) copies of the 2nd part of the first volume of the *Enciclopedia delle Matematiche Elementari* have been sold, and the publisher Hoepli has sent me the payment to distribute to the various authors (limited to the number of copies sold). Since the article by your late lamented brother is 46 pages long, he would be due 589.70 lire. I therefore enclose a bank order from the Banca d'Italia N. 0.010.090 for the sum of 587.70 lire, having detracted 2 lire to cover postal expenses. With sincere best wishes, I remain yours truly, L. Berzolari. P.S. I woud be grateful if you could send me notice of receipt. Postal expenses in fact stood at 1.75, so I have attached a 0.26 lire stamp. Answer 25 January 1939).

[52]G. Vivanti, *Teoria degli aggregati*, *Chap. VI* in [52, p. 65–70].

[53]Instead no appearance is made in the *Formulario* of those "new denominations... whose introduction would not have been advantageous in the exposition with the logical symbols" [62, p. 135]. Amongst the absent concepts, we can cite, for example, the notions of closed, perfect, isolated, separate and concentrated sets.

[54]Vivanti in [63, p. 71–74].

period 1893–1899, in Eneström's *Bibliotheca Mathematica* [64, p. 160–165]. Here, he not only quoted the papers of G. Ascoli, G. Loria, R. Bettazzi, C. Arzelà, R. De Paolis, L. Milesi and F. Giudice, together with his own works, in which Cantor's theory of infinite sets had been assimilated and employed in analytical and geometric contexts, but he also mentioned those university textbooks in which these themes were already partially treated, such as U. Dini's *Fondamenti per la teorica delle funzioni di variabili reali*, G. Veronese's *Fondamenti di geometria*, and Peano's *Arithmetices Principia* and *Lezioni di Analisi*. Although carefully examined and annotated by Peano,[55] Vivanti's chapter did not appear in later editions of the *Formulario*, as did not that by G. Fano on algebraic numbers. However, Vivanti translated another famous article by Cantor, which again appeared in the *Rivista di Matematica* [16], thus contributing, together with Peano and F. Gerbaldi [17, 18] to the spread of Cantorian set theory in Italy. Vivanti had also entered into direct epistolary correspondence with Cantor[56] and on several occasions between 1891 and 1894 he had taken part in the thorny discussions on the use of actual infinitesimals in mathematics, adopting original critical positions.

The sum of these activities, at once scientific and didactic, placed him in an ideal position to fashion a suggestive chapter on set theory for the *Enciclopedia delle Matematiche Elementari*, which met with the special praise of Enriques for its remarkable sobriety:

> And we would ascribe to the illustrious authors the merit of having contained these developments – always clearly and precisely – within the appropriate limits for a book aimed at teachers in our middle schools.[57]

Adopting a completely different style from that used in the *Formulario*, Vivanti not only completely dropped here the logical and ideographic notations, but also shed a somewhat different light on the various issues dealt with. Foregoing any forays into the historical or foundational aspects of set theory, he instead dwelt, in a deliberately elementary manner, only on those aspects which were most useful from a teaching perspective. He, therefore, placed his primary emphasis on the differences between the properties of finite and infinite sets, the concepts of cardinal and ordinal number, the definitions of bijective and continuous correspondences, and the notion of dimension. In doing so, he relinquished opportunities to make even passing reference to the expositions of set theory which he himself had prepared for university teaching and assembled in his textbooks on infinitesimal calculus.

Indeed, to find in the *Enciclopedia delle Matematiche Elementari* a broader reference to the links between the set theory, the critique of principles, the theory of real functions and the functional analysis, one must consult F. Severi and

---

[55] See [53, p. 65–69].

[56] See G. Cantor to G. Vivanti, 3 December 1885, 6 November 1886, 30 January 1888, 2 April 1888, in [42, p. 250–251, 269–270, 300–301, 302–306].

[57] [27, p. 125]. ... *Ed ascriviamo a merito degli illustri autori di avere contenuto questi sviluppi - sempre chiari e precisi - nei limiti che si convengono ad un libro rivolto agli insegnanti delle nostre scuole medie.*

F. Conforto's article on the *Caratteri e indirizzi della matematica moderna*, written in a time of fully Bourbakist *Weltanschauung*. There, completing the propaedeutic exposition laid out by Vivanti, set theory is presented as a possible basis for mathematics and a demonstration given of how, from this starting point and by adding specific postulates, one can attain to a *matematica generale* or *assiomatica* or *astratta*

> in which various theories and concepts which were previously developed autonomously and independently from one another fit together in a unitary frame, thus obtaining an economy of thought which, given the ever-growing field of knowledge and mathematical developments, is increasingly necessary, and managing at the same time to grasp the common logical weave of a group of theories, which historically often grew out of intuitive data.[58]

In the same article, Severi and Conforto also illustrated how set theory becomes modern algebra, at the moment in which operations having all or some of the properties of the four fundamental arithmetic operations are defined as abstractly as possible for the sets on which one is working.[59] The calculus section of the *Enciclopedia* closes with an overview on analytic functions. Here again, the choice of author was a "foregone conclusion", with the commission going to Pincherle, who had been involved in the field since his youth, when he had been invited to give a series of lessons on the general theory of analytic and elliptic functions in Pavia, expounded on the basis of the notes which he himself had taken attending Weierstrass' lessons in Berlin [1, p. 4–5]. Over the course of the years spent as full professor of the chair of Calculus in Bologna, he had had many occasions to expound upon this theory, and lithographs of his *Lezioni sulle funzioni analitiche* had circulated widely in Italy, meeting with large praise, as did his university textbook, *Gli elementi della teoria delle funzioni analitiche*, published by Zanichelli in 1922. To the Bolognese mathematician, we owe a masterful exposition of the principles of this theory, which Enriques judged to be "ingenious" [27, p. 125] because it bridged the gap between the mathematical notions taught in the first two years of university and the higher branches of analysis without omitting the

---

[58] [57, p. 806]: ... *nella quale s'inquadrano in modo unitario svariate teorie e concetti che prima erano stati svolti in modo autonomo ed indipendente gli uni dagli altri. Si ottiene in tal modo un'economia di pensiero ... sempre più necessaria di fronte al continuo allargarsi delle conoscenze e degli sviluppi matematici; e si riesce a cogliere la comune trama logica di svariate teorie, spesso sorte storicamente da dati intuitivi.*

[59] Thus, they conclude [57, p. 809]: *Già nella seconda metà del XIX secolo era stata pubblicata la grande Enciclopedia Matematica Tedesca. Ma si può ben dire che l'odierno tentativo di Bourbaki si presenta del tutto nuovo e significativo per il chiaro e rigido concetto informatore dell'intera opera; sicché sembra di poter sicuramente affermare che questa rappresenterà ognora un documento molto interessante delle matematiche dei nostri giorni, come ai suoi tempi il Formulaire de mathématique di Peano* (In the second half of the nineteenth century, the great *Enciclopedia Matematica* was published in Germany. Yet we may well argue that Bourbaki's enterprise today is wholly new and significant in the clear and rigid concept which informs the whole work. Hence, it seems that we can safely affirm that this will represent a lasting source of great interest on the mathematics of our time, as in its own day Peano's *Formulaire de mathématique* did).

elementary implications, which were of more direct utility to teachers. Of particular value are the development of the concepts of generalized dependence following Dirichlet and of analytic dependence, his treatment of functional equations for the exponential, of circular functions and of the classical procedure of recurrent series, for the solution of the differential equation of the logarithmic function.

The intertwining of scientific collaboration and personal relationships which underlies the publication of the *Enciclopedia delle Matematiche Elementari* enables us not only to analyse the dynamics of comparison between a single local mathematical community, such as that of Bologna, and the broader mathematical world at that time, but also to reconstruct how, in the terms of Enriques' telling metaphor:

> over the course of history, mathematicians, who were from time to time students and *Maestri*, have offered the spectacle of a universal reason which formulates eternal truth, one that rises above the differences and weaknesses of men. The *School* tends to extend itself beyond its own original environment, and its influence on the scholar is mingled with other, various forces which render it fertile. This is the fulfilment of the law of approach which Klein brought to light. That is to say, that the development of mathematical Schools, subject to alternating periods of progress and decadence within national limits, is revivified through the passage from one nation to another, almost as though the spirit of the world participated more broadly towards a common goal.[60]

Acknowledgements Translation of the entire paper was revised by Kim Williams. The transcriptions of the documents were done by the author. I would like to sincerely thank Prof. C.S. Roero, who first encouraged me to study this topic, read the first draft and supported me.

References

1. Amaldi, U. 1954. Della vita e delle opere di Salvatore Pincherle. In S. Pincherle, *Opere scelte*, ed. UMI, Roma: Cremonese: 3–16.
2. Archibald, R.C. 1950. Enciclopedia delle Matematiche Elementari III. *Bulletin of the American Mathematical Society* 56(6): 517–518.
3. Arrighi, G. (ed.). 1997. *Lettere a Mario Pieri*. Milano: Bocconi University, Quaderni Pristem n. 6.
4. Bennett, A. 1930. Enciclopedia delle Matematiche Elementari I[1]. *The American Mathematical Monthly* 37(7): 378–380.
5. Berzolari, L. 1924. Intorno alla pubblicazione di una Enciclopedia delle Matematiche Elementari, 20 December 1923. *Bollettino dell'UMI* s. 1, 3: 47.

---

[60][29, p. 181]: *Nella continuità della storia i matematici, a volta a volta successivamente scolari e Maestri, offrono lo spettacolo di una ragione universale che elabora la verità eterna, sopra alle differenze e alle debolezze degli uomini. La scuola tende ad allargarsi fuori del proprio ambiente d'origine, ed allora l'influenza sullo scolaro viene a comporsi con altri motivi diversi che la fecondano. Perciò si avvera la legge di avvicinamento che Klein ha messo in luce, cioè che lo sviluppo delle scuole matematiche, soggetto ad alternanze di progresso e di decadenza nei limiti di una nazione, si ravviva passando da una nazione ad un'altra, quasi a far partecipare più largamente all'opera comune lo spirito del mondo.*

6. Berzolari, L., and Bonola, R. 1909. Allegato E. Sopra una Enciclopedia di Matematiche Elementari da pubblicarsi sotto gli auspici della Mathesis, Relazione della Commissione Direttiva dell'opera. In *Atti del 2. Congresso della Mathesis, Società italiana di matematica, Padova, 20–23.9.1909*. Padova: Società Cooperativa Tipografica: 1–5, 43–45.
7. Berzolari, L., Vivanti, G., and Gigli, D. (eds.). 1930–1949. *Enciclopedia delle Matematiche Elementari*, Milano: Hoepli, I^1, 1930; I^2, 1932; II^1, 1937; II^2, 1938; III^1, 1949; III^2 1949; III^3, 1950.
8. Bompiani, E. 1937. Enciclopedia delle Matematiche Elementari II^1. *Bollettino dell'UMI* s. 1, 16: 155–156.
9. Bompiani, E. 1950. Commemorazione del Socio Luigi Berzolari. *Rendiconti Sc. Fis. Mat. e Nat. Lincei*, IX: 396–410.
10. Bonola, R. 1909. Sunto della relazione sul tema: Per la pubblicazione di una Enciclopedia delle Matematiche Elementari. *Bollettino della Mathesis*, I: 40–42.
11. Bonola, R. 1909, Per la pubblicazione di una Enciclopedia delle Matematiche Elementari. *Giornale di matematiche ad uso degli studenti delle Università Italiane* (Battaglini), 47 [2, 16], 343–344.
12. Borgato, M.T., and Pepe, L. (eds.). 1984. *Giuseppe Vitali. Opere sull'analisi reale e complessa. Carteggio*. Bologna: Cremonese.
13. Borgato, M.T., and Vaz Ferreira, A. 1987. Guerraggio, A. *Giuseppe Vitali: ricerca matematica e attività accademica dopo il 1918*. In Guerraggio A. (ed.). 1987: 43–58.
14. Brusotti, L. 1936. Luigi Berzolari. Cenni biografici. In *Scritti matematici offerti a Luigi Berzolari*, Pavia: Istituto matematico della R. Università: XXI–XXIX.
15. Cairns, S.S. 1939. Enciclopedia delle Matematiche Elementari II^2. *The American Mathematical Monthly* 46(1): 44.
16. Cantor, G. 1892. Sopra una questione elementare della teoria degli aggregati. *Rivista di matematica* 2: 165–167.
17. Cantor, G. 1895a. Sui numeri transfiniti. *Rivista di matematica* 5: 104–109.
18. Cantor, G. 1895b. Contribuzione al fondamento della teoria degli insiemi transfiniti. *Rivista di matematica* 5: 129–162.
19. Cassina, U. 1931. Enciclopedia delle Matematiche Elementari I^1. *Bollettino di Matematica* (A. Conti) 27: LIII–LVIII.
20. Cassina, U. 1932. Enciclopedia delle Matematiche Elementari I^2. *Bollettino di Matematica* (A. Conti) 28: LXVII–LXX.
21. Cassina, U. 1939. Enciclopedia delle Matematiche Elementari II^2. *Bollettino di Matematica* (A. Conti) 35: XXXI–XXXIV.
22. Chisini, O. 1937. Enciclopedia delle Matematiche Elementari II^1. *Bollettino di Matematica* (A. Conti) 33: LVII–LXIII, LXXVII–LXXXIII.
23. Cinquini, S. 1950. G. Vivanti. *Rendiconti dell'Istituto Lombardo. Accademia di Scienze e Lettere, Parte Generale e Atti Ufficiali* 78: 184–205.
24. Conforto, F. 1940. Enciclopedia delle Matematiche Elementari II^2. *Bollettino dell'UMI*. s. 2, 2: 506–507.
25. Dauben, J.W., and Scriba, C.J. (eds.). 2002. *Writing the history of mathematics: Its historical development*. Basel: Birkhauser.
26. Enriques, F. 1930. Enciclopedia delle Matematiche Elementari I^1. *Periodico di Matematiche* s. 4, 10: 39–41.
27. Enriques, F. 1932. Enciclopedia delle Matematiche Elementari I^2. *Periodico di Matematiche*, s. 4, 12, 124–125.
28. Enriques, F. 1937. Enciclopedia delle Matematiche Elementari II^1. *Periodico di Matematiche* s. 4, 17: 112–113.
29. Enriques, F., and Frajese, A. 1938. *Le matematiche nella storia e nella cultura*. Bologna: Zanichelli.
30. Fehr, H. 1929. Enciclopedia delle Matematiche Elementari I^1. *L'Enseignement mathématique* 28: 347–348.

31. Fehr, H. 1932. Enciclopedia delle Matematiche Elementari I^2. *L'Enseignement mathématique* 31: 319.
32. Fehr, H. 1937. Enciclopedia delle Matematiche Elementari II^1. *L'Enseignement mathématique* 36: 135.
33. Fehr, H. 1939. Enciclopedia delle Matematiche Elementari II^2. *L'Enseignement mathématique* 37: 115.
34. Feigl, G. 1930. Enciclopedia delle Matematiche Elementari I^1. *Jahrbuch über die Fortschritte der Mathematik* 56: 61.
35. Feigl, G. 1932. Enciclopedia delle Matematiche Elementari I^2. *Jahrbuch über die Fortschritte der Mathematik* 58: 1001.
36. Feigl, G. 1937. Enciclopedia delle Matematiche Elementari II^1. *Jahrbuch über die Fortschritte der Mathematik* 63: 1151.
37. Giacardi, L. (ed.). 2006. *Da Casati a Gentile: momenti di storia dell'insegnamento secondario della matematica in Italia*. Lugano: Lumières internationales.
38. Guerraggio, A. (ed.). 1987. *La matematica italiana tra le due guerre mondiali*. Bologna: Pitagora.
39. Hoepli, C. 1949. In Berzolari, L., Vivanti, G., Gigli, D., (eds.) III^2: nn.
40. Hornick, H. 1930. Enciclopedia delle Matematiche Elementari II^1 e II^2. *Monatshefte für Mathematik* 47(1): 21.
41. Lampe. 1909. Per la pubblicazione di una Enciclopedia delle Matematiche Elementari. *Jahrbuch über die Fortschritte der Mathematik* 40: 45.
42. Meschkowski, H., and Nilson, W. (eds.). 1991. *Georg Cantor Briefe*. Berlin: Springer.
43. Miller, G.A. 1932. Enciclopedia delle Matematiche Elementari I^2. *The American Mathematical Monthly* 39(3): 168–171.
44. Miller, G.A. 1932. Enciclopedia delle Matematiche Elementari I^1. *Bulletin of the American Mathematical Society* 38(3): 157–158.
45. Nastasi, P., and Scimone, A. (eds.). 1995. *Lettere a Giovanni Vacca*, Palermo: Quaderni Pristem n. 5 .
46. O. L. [Onofri, L.] 1929. Enciclopedia delle Matematiche Elementari I^1. *Bollettino dell'UMI* s. 1, 8, 276–277.
47. Palatini, A. 1931. Enciclopedia delle Matematiche Elementari I^1. *Scientia* 49: 224–225.
48. Palatini, A. 1933. Enciclopedia delle Matematiche Elementari I^2. *Scientia* 54: 50–51.
49. Palatini, A. 1938. Enciclopedia delle Matematiche Elementari II^1. *Scientia* 64: 37.
50. Palatini, A. 1939. Enciclopedia delle Matematiche Elementari II^2. *Scientia* 65: 176.
51. Palladino, F., and Palladino, N. (eds.). 2006. *Dalla "moderna geometria" alla "nuova geometria italiana". Viaggiando per Napoli, Torino e dintorni. Lettere di Sannia, Segre, Peano, Castelnuovo, D'Ovidio, Del Pezzo, Pascal e altri a Federico Amodeo*. Firenze: Olschki.
52. Peano, G. (ed.). 1895aa. *Formulaire de Mathématiques, tome 1 publié par la Rivista di matematica*. Torino: Bocca.
53. Peano, G. (ed.). 1895aa$^+$. *Formulaire de Mathématiques*. In Roero, C.S. (ed.) 2008.
54. Perna, A. 1950. L'Enciclopedia delle Matematiche Elementari di Luigi Berzolari. *Archimede* 2: 82–87.
55. Roero, C.S. (ed.). 2008. *L'opera omnia e i marginalia di Giuseppe Peano*. Torino: Dipartimento di Matematica, cd-rom N. 3b.
56. Sansone, G. 1932. Enciclopedia delle Matematiche Elementari I^2. *Bollettino dell'UMI* s. 1, 11: 47–49.
57. Severi, F., and Conforto, F. 1949. Caratteri e indirizzi della matematica moderna. In Berzolari, L., Vivanti, G., Gigli, D. (eds.) III^2: 751–814.
58. Sezione Lombarda Mathesis. 1909. Adunanza 23 Maggio 1909. *Bollettino della Mathesis* I: 35–37, 48.
59. Togliatti, E., Rollero, A., Castoldi, L., and Martinelli, E. 1950. Enciclopedia delle Matematiche Elementari III^2. *Bollettino dell'UMI* s. 3, 5: 368–376.
60. Vivanti, G. 1892. Sull'uso della rappresentazione geometrica nella teoria aritmetica dei numeri complessi. *Rivista di Matematica* 2: 167–176.

61. Vivanti, G. 1893. Lista bibliografica della teoria degli aggregati. *Rivista di Matematica* 3: 189–192.
62. Vivanti, G. 1894. Sulla parte VI del Formulario. *Rivista di Matematica* 4: 135–139.
63. Vivanti, G. 1895. Lista bibliografica fino a tutto l'anno 1893. In Peano, G. (ed.) 1895aa: 71–74.
64. Vivanti, G. 1900. Lista bibliografica della teoria degli aggregati 1893–1899. *Bibliotheca Mathematica* I: 160–165.
65. Vivanti, G. 1933. Duilio Gigli. *Periodico di Matematiche* s. 4, 13: 255–256.
66. Zacharias, M. 1938. Enciclopedia delle Matematiche Elementari II^2. *Jahrbuch über die Fortschritte der Mathematik* 64: 1255.

The Role of Salvatore Pincherle in the Development of Fractional Calculus

Francesco Mainardi and Gianni Pagnini

Abstract We revisit two contributions by Salvatore Pincherle (Professor of Mathematics at the University of Bologna from 1880 to 1928) published (in Italian) in 1888 and 1902 in order to point out his possible role in the development of Fractional Calculus. Fractional Calculus is that branch of mathematical analysis dealing with pseudo-differential operators interpreted as integrals and derivatives of non-integer order. Even if the former contribution (published in two notes on Accademia dei Lincei) on generalized hypergeomtric functions does not concern Fractional Calculus it contains the first example in the literature of the use of the so called Mellin–Barnes integrals. These integrals will be proved to be a fundamental task to deal with all higher transcendental functions including the Meijer and Fox functions introduced much later. In particular, the solutions of differential equations of fractional order are suited to be expressed in terms of these integrals. In the second paper (published on Accademia delle Scienze di Bologna), the author is interested to insert in the framework of his operational theory the notion of derivative of non integer order that appeared in those times not yet well established. Unfortunately, Pincherle's foundation of Fractional Calculus seems still ignored.

1 Some Biographical Notes of Salvatore Pincherle

Salvatore Pincherle was born in Trieste on 11 March 1853 and died in Bologna on 10 July 1936. He was Professor of Mathematics at the University of Bologna from 1880 to 1928. Furthermore, he was the founder and the first President of Unione Matematica Italiana (UMI) from 1922 to 1936.

F. Mainardi (✉)
Department of Physics, University of Bologna and INFN, Via Irnerio 46, 40126 Bologna, Italy
e-mail: francesco.mainardi@unibo.it

G. Pagnini
CRS4, Polaris Bldg. 1, 09010 Pula (CA), Italy
e-mail: pagnini@crs4.it

Pincherle retired from the University just after the International Congress of Mathematicians that he had organized in Bologna since 3–10 September 1928, following the invitation received at the previous Congress held in Toronto in 1924.[1]

Pincherle wrote several treatises and lecture notes on Algebra, Geometry, Real and Complex Analysis. His main book related to his scientific activity is entitled "Le Operazioni Distributive e loro Applicazioni all'Analisi"; it was written in collaboration with his assistant, Dr. Ugo Amaldi, and was published in 1901 by Zanichelli, Bologna, see [20]. Pincherle can be considered one of the most prominent founders of the Functional Analysis, as pointed out by J. Hadamard in his review lecture "Le développement et le rôle scientifique du Calcul fonctionnel", given at the Congress of Bologna (1928). A description of Pincherle's scientific works requested from him by Mittag-Leffler, who was the Editor of Acta Mathematica, appeared (in French) in 1925 on this prestigious journal [19]. A collection of selected papers (38 from 247 notes plus 24 treatises) was edited by Unione Matematica Italiana (UMI) on the occasion of the centenary of his birth, and published by Cremonese, Roma 1954.

2 Pincherle and the Mellin-Barnes Integrals

Here, we point out that the 1888 paper (in Italian) of S. Pincherle on the *Generalized Hypergeometric Functions* led him to introduce the afterwards named Mellin-Barnes integral to represent the solution of a generalized hypergeometric differential equation investigated by Goursat in 1883. Pincherle's priority was explicitly recognized by Mellin and Barnes themselves, as reported below.

In 1907 Barnes, see p. 63 in [1], wrote: "The idea of employing contour integrals involving gamma functions of the variable in the subject of integration appears to be due to Pincherle, whose suggestive paper was the starting point of the investigations of Mellin (1895) though the type of contour and its use can be traced back to Riemann." In 1910 Mellin, see p. 326ff in [15], devoted a section (Sect. 10: Proof of Theorems of Pincherle) to revisit the original work of Pincherle; in particular, he wrote "Before we are going to prove this theorem, which is a special case of a more general theorem of Mr. Pincherle, we want to describe more closely the lines L over which the integration preferably is to be carried out" [free translation from German].

---

[1]More precisely, as we know from the recent biography of the Swedish mathematician Mittag-Leffler by Arild Stubhaug [22]: *The final decision was to be made as to where the next international mathematics congress (in 1928) would be held; the options were Bologna and Stockholm. One strike against Stockholm was the strength of the Swedish currency; it was said that it would simply be too expensive in Stockholm. Mittag-Leffler was also in favor of Bologna, and in that context he had contacted both the Canadian J.C. Fields and the Italian Salvatore Pincherle. The latter even asked Mittag-Leffler whether he would preside at the opening meeting of what in reality would be the first international congress for mathematicians since 1912. This was because mathematicians from Germany and the other Central Powers would be invited to Bologna.*

The purpose of this section is, following our 2003 paper [8], to let know the community of scientists interested in special functions the pioneering 1888 work by Pincherle on Mellin–Barnes integrals, that, in the author's intention, was devoted to compare two different generalizations of the Gauss hypergeometric function due to Pochhammer and to Goursat. In fact, dropping the details on which the interested reader can be informed from our paper [8], Pincherle arrived at the following expression of the Goursat hypergeometric function

$$\psi(t) = \frac{1}{2\pi i} \int_{a-i\infty}^{a+i\infty} \frac{\Gamma(x-\rho_1)\Gamma(x-\rho_2)\ldots\Gamma(x-\rho_m)}{\Gamma(x-\sigma_1)\Gamma(x-\sigma_2)\ldots\Gamma(x-\sigma_{m-1})} e^{xt}\, dx, \quad (1)$$

where $a > \Re\{\rho_1, \rho_2, \ldots, \rho_m\}$, and ρ_k and σ_k are the roots of certain algebraic equations of order m and $m-1$, respectively. We recognize in (1) the first example in the literature of the (afterwards named) Mellin–Barnes integral. The convergence of the integral was proved by Pincherle by using his asymptotic formula[2] for $\Gamma(a+i\eta)$ as $\eta \to \pm\infty$. So, for a solution of a particular case of the Goursat equation, Pincherle provided an integral representation that later was adopted by Mellin and Barnes for their treatment of the generalized hypergeometric functions $_pF_q(z)$. Hince then, the merits of Mellin and Barnes were so well recognized that their names were attached to the integrals of this type; on the other hand, after the 1888 paper (written in Italian), Pincherle did not pursue on this topic, so his name was no longer related to these integrals and, in this respect, his 1888 paper was practically ignored.

In more recent times other families of higher transcendental functions have been introduced to generalize the hypergeometric function based on their representation by Mellin–Barnes type integrals. We especially refer to the so-called G and H functions introduced by Meijer [13] in 1946, and by Fox [4] in 1961, respectively, so Pincherle can be considered their precursor. For an exhaustive treatment of the Mellin–Barnes integrals we refer to the recent monograph by Paris and Kaminski [16].

In the original part of our 2003 paper, we have shown that, by extending the original arguments by Pincherle, we have been able to provide the Mellin–Barnes integral representation of the transcendental functions introduced by Meijer (the so-called G functions). In fact, we have shown how to formally derive the ordinary differential equation and the Mellin–Barnes integral representation of the G functions introduced by Meijer in 1936–1946. So, in principle, Pincherle could

---

[2] We also note the priority of Pincherle in obtaining this asymptotic formula, as outlined by Mellin, see e.g. [14], pp. 330–331, and [15], p. 309. In his 1925 "Notices sur les travaux" [19] (p. 56, Sect. 16) Pincherle wrote "Une expression asymptotique de $\Gamma(x)$ pour $x \to \infty$ dans le sens imaginaire qui se trouve dans [17] a été attribuée à d'autres auteurs, mais M. Mellin m'en a récemment révendiqué la priorité." This formula is fundamental to investigate the convergence of the Mellin–Barnes integrals, as one can recognize from the detailed analysis by Dixon and Ferrar [3], see also [16].

have introduced the G functions much before Meijer if he had intended to pursue his original arguments in this direction.

Finally, we like to point out that the so-called Mellin–Barnes integrals are an efficient tool to deal with the higher transcendental functions. In fact, for a pure mathematics view point they facilitate the representation of these functions (as formerly indicated by Pincherle), and for an applied mathematics view point they can be successfully adopted to compute the same functions. In particular we like to refer to our papers [9, 10] where we have derived the solutions of diffusion-wave equations of fractional order and their subordination properties by using the Mellin-Barnes integrals.

3 Pincherle's Foundation of Fractional Derivatives

The interest of S. Pincherle about Fractional Calculus was mainly motivated by the fact that literature definitions of derivation of not integer order, now called fractional derivation, were arbitrary introduced as generalization of some aspects of the ordinary integer order derivation. This lack of a rigorous foundation attracted him.

Remembering that one of the research field of S. Pincherle was the operational calculus, it seems straightforward to think that for him it was natural to apply his knowledge in this field to derive the most rigorous definition of derivation of not integer order. In fact, in the book [20] entitled *Le operazioni distributive* published in 1901 in collaboration with U. Amaldi, S. Pincherle analyzed the general properties of operators, and in particular of differential operators. This background inducted him to search for a rigorous foundation of Fractional Calculus, which overcomes the arbitrariness of literature definitions, deriving a generalized derivative operator which meets all the properties of differential operators in the most general sense. This problem was addressed by S. Pincherle in the memoir *Sulle derivate ad indice qualunque* [18].

In 1902, Fractional Calculus had put its basis with the works by Liouville [7], Riemann [21] Tardy [23], Holmgrem [6]. S. Pincherle was acquainted about these works that however he considered to have a paramount flaw because these fractional derivation was arbitrary defined.

In particular, with respect to Liouville definition of not integer derivation of order s

$$D^s e^{zx} = z^s e^{zx}, \qquad (2)$$

he observed that, from this arbitrary definition as an ingenuos extension for not integer s of the derivation of the exponential function, serious objections arise about the application of the distributive property of D^s for a sum of infinite terms. Arbitrariness has been highlighted also for Riemann definition of D^s, which was related to the coefficient of the term h^s when a function $f(x+h)$ is developed by a power series of terms $h^{\xi+n}$ with $n \in N$. Finally, also Holmgrem arbitrary assumed the derivation of not integer order s as the integral

$$D^s f(x) = \frac{1}{\Gamma(m-s)} D^m \int^x (x-z)^{m-s-1} f(z)\, dz, \tag{3}$$

which was presented by Riemann himself. The same integral was used by Hadamard [5] in his research on Taylor series.

The general properties of ordinary differential operator of integer order are

1. It is uniquely defined for any analytical fuction,
2. It is distributive,
3. It satisfies an index law,
4. It meets a composition law for $D^m(\phi\psi)$ by the application of D^n (with $n \le m$) to ϕ and ψ, as for example $D^m(x\phi) = xD^m\phi + mD^{m-1}\phi$.

Then, the generalization of the derivative operator for not integer order s is obtained by the construction of an operator $\mathcal{A}_s$ with the same properties of derivation of integer order. Hence, for the whole space of analytical functions or a part of it named $\mathcal{Q}$, the derivation of not integer order s meets the following constraints:

1. It is defined for any value of s, both real and complex, and for any function of $\mathcal{Q}$ generating at least one function belonging to the same space $\mathcal{Q}$,
2. It is distributive
$$\mathcal{A}_s(\phi + \psi) = \mathcal{A}_s(\phi) + \mathcal{A}_s(\psi),$$
3. It solves the equation
$$D\mathcal{A}_{s-1} = \mathcal{A}_s,$$
4. It solves the equation
$$\mathcal{A}_s(x\phi) = x\mathcal{A}_s(\phi) + s\mathcal{A}_{s-1}(\phi),$$
5. It reduces for integer value of $s = m$ to the operator D^m.

In this case, in the space $\mathcal{Q}$ it is defined the derivation of not integer order s and then $\mathcal{A}_s$ corresponds to D^s.

S. Pincherle derives such generalized operator D^s which emerges to be

$$x^s D^s \phi = \sum_{n=0}^{\infty} \binom{s}{n} \frac{x^n}{\Gamma(1-s+n)} D^n \phi, \tag{4}$$

where

$$\binom{s}{n} = \frac{s(s-1)\ldots(s-n+1)}{n!} = \frac{\Gamma(1+s)}{\Gamma(1+n)\Gamma(1+s-n)}, \tag{5}$$

is the binomial coefficient and $\Gamma(z)$, with $z \in C$, is the Euler gamma function that for integer argoument is related to the factorial by $\Gamma(1+n) = n!$. Furthermore, by adopting the Gamma function property $\Gamma(1+z) = z\Gamma(z)$ it results that

$$\Gamma(1-s+n) = \Gamma(1-s)(1-s)\ldots(n-s)$$
$$= (-1)^n \Gamma(1-s)(s-1)\ldots(s-n)$$
$$= (-1)^n \frac{\Gamma(1-s)\Gamma(1+s)}{\Gamma(1+s-n)} \frac{(s-n)}{s},$$

and by the definition of binomial coefficient (5) formula (4) can be rewritten as

$$x^s D^s \phi = \frac{s}{\Gamma(1-s)} \sum_{n=0}^{\infty} \frac{(-x)^n}{(s-n)n!} D^n \phi. \qquad (6)$$

S. Pincherle recognized that formula (6) was originally obtained by Bourlet [2]. However, S. Pincherle highlighted also that Bourlet obtained (6) assuming as definition of derivation of not integer order the same one given by Riemann, while S. Pincherle derived (6) as the necessery consequence of those properties that an operator must satisfy to be intended as derivation of not integer order.

Before concluding the memoir, S. Pincherle also shows that when s is a negative integer, i.e. $s = -m$, formula (4) becomes

$$D^{-m}\phi = \frac{x^m}{\Gamma(m)} \sum_{n=0}^{\infty} \frac{(-x)^n}{n!(m+n)} D^n \phi, \qquad (7)$$

which he recognized to be the generalized Bernoulli formula given in *Le operazioni distributive* [20].

Then the fractional derivation in the Pincherle sense is defined by (4) and it is emerged to be a series of weighted integer derivations up to the infinite order. This fractional derivation is not related to usual definitions in literature. But we remark that, differently from generally accepted definitions of fractional derivation, S. Pincherle derived an operator which meets all the general constraints that a derivation operator must satisfies. Unfortunately, in spite of the rigorous foundation, the fractional derivation in the Pincherle sense is not considered by the community of fractional analysts.

Moreover, we would like conclude this section stressing that, since Pincherle's fractional derivation has been obtained by strong analogy with ordinary differentiation satisfying all differential operator constraints, it could have a straightforward physical and geometrical interpretation, at variance with the actual literature of Fractional Calculus. This topic would be the argument of future analysis.

In fact, considering the interpretation of Pincherle fractional derivative as a weighted sum of infinity integer derivative of order n, $n = 0, 1, \ldots, \infty$, then

$$D^s \phi = \sum_{n=0}^{\infty} w(n; x, s) D^n \phi, \qquad (8)$$

where $w(n; x, s)$ is the weight of derivative of order n given the point x and the fractional order s and from normalization

$$\sum_{n=0}^{\infty} w(n; x, s) = 1.\qquad(9)$$

From (4) and (6) it follows that

$$w(n; x, s) = \binom{s}{n} \frac{x^{n-s}}{\Gamma(1-s+n)} = \frac{s\, x^{-s}}{\Gamma(1-s)} \frac{(-x)^n}{(s-n)n!},\qquad(10)$$

and the series (9) becomes

$$\sum_{n=0}^{\infty} w(n; x, s) = \frac{\Gamma(1+s)}{x^s} \sum_{n=0}^{\infty} \frac{x^n}{\Gamma(1+n)\Gamma(1+s-n)\Gamma(1-s+n)}$$
$$= \frac{s\, x^{-s}}{\Gamma(1-s)} \sum_{n=0}^{\infty} \frac{(-x)^n}{(s-n)n!}.\qquad(11)$$

Finally, computation of series (11) will give also a solution to the problem of the interpretation of fractional derivation.

4 Conclusions

We have revisited two contributions (in Italian) by Pincherle on generalized hypergeometric functions, dated 1888, and on derivatives of any order, dated 1902, in order to point out a possible role that he could have played in the development of the Fractional Calculus in Italy and abroad. As a matter of fact, unfortunately, these contributions remained practically unknown to the specialists of Fractional Calculus. However, we have recognized, since our 2003 paper [8], that the 1888 contribution if suitably continued could have led to the introduction of G functions before Meijer and used to deal with differential equations of fractional order. Up to nowadays the 1902 contribution has been ignored but in our opinion the approach by Pincherle is worth to be pursued in the framework of the modern theory of Fractional Calculus. It is interesting to note that even a former pupil of Pincherle at the University of Bologna, Antonio Mambriani, ignored the approach of his mentor in his papers on differential equations of fractional order, preferring the approach by Holmgren, see e.g. [11, 12].

Acknowledgements Research performed under the auspices of the National Group of Mathematical Physics (G.N.F.M. – I.N.D.A.M.) and partially supported by the Italian Ministry of University

(M.I.U.R) through the Research Commission of the University of Bologna. The authors are grateful to Prof. S. Coen for the discussions and the helpful comments.

References

1. Barnes, E.W. 1907. The asymptotic expansion of integral functions defined by generalized hypergeometric series. *Proceedings of the London Mathematical Society* (Ser. 2) 5: 59–116.
2. Bourlet, C. 1897. Sur les opérations en général et les équations différentielles linéaires d'ordre infini. *Annales de l'Éc. Normale* (Ser. III) 14: 133–190.
3. Dixon, A.L., and W.L. Ferrar. 1936. A class of discontinuous integrals. *The Quarterly Journal of Mathematics* (Oxford Series) 7: 81–96.
4. Fox, C. 1961. The G and H functions as symmetrical Fourier kernels. *Transactions of the American Mathematical Society* 98: 395–429.
5. Hadamard, J. 1892. Essai sur l'étude des fonctions données par leur développement de Taylor. *Journal de Mathématiques Pures et Appliquées* (Série IV) 8: 101–186.
6. Holmgren, Hj. 1865. Om differentialikalkylen med indices af havd natur som helst (Calcolo differenziale a indici qualunque). *Kongliga (Svenska) Vetenskaps-Akademiens Handligar* 5, No 11, 83, Stockholm, 7 Mars 1865.
7. Liouville, J. 1832. Mémoire sur le calcul des différentielles à indices quelconques. *J. Ecole Polytech.* 13 Cahier 21: 71–162.
8. Mainardi, F., and G. Pagnini. 2003. Salvatore Pincherle: the pioneer of the Mellin-Barnes integrals. *Journal of Computational and Applied Mathematics* 153: 331–342; [E-print:http://arxiv.org/abs/math/0702520].
9. Mainardi, F., Luchko, Yu., and G. Pagnini. 2001. The fundamental solution of the space-time fractional diffusion equation. *Fractional Calculus and Applied Analysis* 4(2): 153–192; [E-Print http://arxiv.org/abs/cond-mat/0702419].
10. Mainardi, F., Pagnini, G., and R. Gorenflo. 2003. Mellin transform and subordination laws in fractional diffusion processes. *Fractional Calculus and Applied Analysis* 6(4): 441–459; [E-print:http://arxiv.org/abs/math/0702133].
11. Mambriani, A. 1941. Derivazione d'ordine qualunque e la rísoluzione dell'equazione ipergeometrica. *Boll. Unione Mat. Ital.* (Ser. 2) 3: 9–18.
12. Mambriani, A. 1942. La derivazione parziale d'ordine qualunque e la rísoluzione dell'equazione di Euler e Poisson. *Annali R. Scuola Normale Superiore, Pisa,* Scienze Fis. Mat. (Ser. 2) 11: 79–97.
13. Meijer, G.S. 1946. On the G function, I–VIII, *Nederl. Akad. Wettensch. Proc.* 49: 227–237, 344–356, 457–469, 632–641, 765–772, 936–943, 1063–1072, 1165–1175; see also the translation to English in *Indagationes Math.* 8: 124–134, 213–225, 312–324, 391–400, 468–475, 595–602, 661–670, 713–723.
14. Mellin, H. 1891. Zur Theorie der linearen Differenzengleichungen erster Ordnung. *Acta Math.* 15: 317–384.
15. Mellin, H. 1910. Abriss einer einheitlichen Theorie der Gamma und der Hypergeometrischen Funktionen. *Mathematische Annalen* 68: 305–337.
16. Paris, R.B., and D. Kaminski. 2001. *Asymptotic and Mellin-Barnes integrals.* Cambridge: Cambridge University Press.
17. Pincherle, S. 1888. Sulle funzioni ipergeometriche generalizzate. *Atti R. Accademia Lincei, Rend. Cl. Sci. Fis. Mat. Nat.* (Ser. 4) 4: 694–700, 792–799 [Reprinted in *Salvatore Pincherle: Opere Scelte*, ed. UMI (Unione Matematica Italiana) 1, 223–230, 231–239. Roma: Cremonese, 1954.
18. Pincherle, S. 1902. Sulle derivate ad indice qualunque. *Rendiconti R. Accademia Scienze, Istituto di Bologna* (Ser. V) 9: 745–758.

19. Pincherle, S. 1925. Notices sur les travaux. *Acta Mathematica* 46: 341–362.
20. Pincherle, S., and U. Amaldi. 1901. *Le operazioni distributive*. Bologna: Zanichelli.
21. Riemann, B. 1892. Versuch einer allgemeinen Auffassung der Integration und Differentiation, 14 Janvier 1847. In *Bernhard Riemann's Gesammelte Mathematische Werke*, ed. Heinrich Weber with the assistance of Richard Dedekind, 353–362. Leipzig: Teubner Verlag [Reprinted in *The collected works of Bernhard Riemann*. New York: Dover, 1953].
22. Stubhaug, A. 2010. *Gösta Mittag-Leffler, A man of conviction*, 666. Berlin: Springer.
23. Tardy, P. 1868. Sui differenziali a indice qualunque. *Annali Mat. Pura Appl.* 1: 135–148.

Tullio Viola and his *Maestri* in Bologna: Giuseppe Vitali, Leonida Tonelli and Beppo Levi

Clara Silvia Roero and Michel Guillemot

Abstract The aim of this contribution is to illustrate the relationships between Tullio Viola (1904–1985) and his professors Beppo Levi (1875–1961), Giuseppe Vitali (1875–1932) and Leonida Tonelli (1885–1946) during his years as a student and just after graduation at the University of Bologna, in order to shed light on how they influenced his early research in analysis and his commitment to teaching. A transcription of Viola's manuscript concerning his studies of eight articles by Vitali is provided in the appendix.[1]

> *The mind does not need, like a vase, to be filled, but rather, like wood, it needs a spark that will kindle it and instill in it the impulse to research, and an ardent love for the truth.*
> *Plutarch, Moralia, On listening to lectures.*

Illustrious and dear Professor,
I send you cordial greetings from immense, wonderful Paris where I have been for three weeks now. I have given your greetings to Prof. Lebesgue and to the Eiffel Tower, who send theirs to you. (To be honest, the latter did not actually say that, because I only waved to it from afar, but I am almost sure that I have correctly guessed its intentions.)

---

[1] Translation of this paper from Italian is by Kim Williams. The transcripts of the documents and the appendix are by C. Silvia Roero. This research was performed within the Project *PRIN 2009 Scuole matematiche e identità nazionale nell'età moderna e contemporanea*, Unit of Turin University.

C.S. Roero (✉)
Department of Mathematics G. Peano, University of Turin, via Carlo Alberto 10,
I 10123 Turin, Italy
e-mail: clarasilvia.roero@unito.it

M. Guillemot
Department of Mathematics, Université Paul Sabatier, Toulouse, France
e-mail: guillemot@math.ups-tlse.fr

I have gradually begun to get settled, and to this end I have found no better means than getting intensely to work. At the Collège de France I attend lectures by Lebesgue (20 lessons *Sur quelques questions de construction géométrique*) and by Hadamard (25 lessons on *Analyses de Mémoires scientifiques*). I understand only a little of those of Lebesgue, to tell the truth, because I came in mid-course, but they are interesting because of the brilliant exposition.

The other professors have been very nice as well: some have been even quite cordial in welcoming me. Denjoy even did me the honour of taking an interest in my right-continuous functions and suggested some new ideas. That is how we came to find a new property (he stated it, and I proved it): the necessary and sufficient condition for a given aggregate H to be the aggregate of the points of discontinuity of a function whose right derivative is null everywhere. I hope to be able to carry on with this interesting research. However, my intention is to direct my activity to other areas, and so I have taken a look at the theory of functions of a complex variable: let's see what I am able to do!

When I said good-bye to you in Bologna I forgot to remind you that I remain at your disposal for the correction of the galleys of your book on the theory of functions of a real variable [96]. You may, if you believe my humble collaboration is useful, send me the galleys at the address that I send you. Today I received word of the award of the "Pincherle Prize". I immediately answered the Rector, thanking him. Please allow me, however, since I know that you were part of the jury commission, to tell you personally that I am very grateful to you for this, as I am for all the good that you have done many times in the past for me. Forgive me I have been too familiar here, and believe that I am your most devoted and affectionate Tullio Viola.[2]

Tullio Viola wrote these words on 14 January 1932 to Giuseppe Vitali, one of the *Maestri* at the University of Bologna most responsible for setting in motion Viola's

---

[2] T. Viola to G. Vitali, Paris 14 January 1932, in *Correspondence of G. Vitali*, U.M.I (Italian Mathematical Society) Archives, Department of Mathematics, University of Bologna:"*Illustre e caro Professore, Le mando un saluto cordiale dalla immensa e meravigliosa Parigi dove sono ormai da quasi tre settimane. Ho portati i Suoi saluti al prof. Lebesgue e alla torre d'Eiffel, i quali glieli ricambiano. (Quest'ultima veramente non me l'ha proprio detto, forse perché l'ho salutata soltanto da lontano, ma credo, così all'aspetto, di averne indovinata l'intenzione). Ho cominciato, a poco a poco, ad ambientarmi e, a questo scopo, non ho trovato nulla di meglio che di mettermi intensamente al lavoro. Frequento le conferenze al Collegio di Francia del Lebesgue (20 lezioni* Sur quelques questions de construction géométrique*) e dell'Hadamard (25 lezioni di* Analyses de Mémoires scientifiques*). Quelle del Lebesgue, veramente, le capisco un po' poco, perchè sono venuto a metà corso, ma sono attraenti per l'esposizione molto brillante. Anche gli altri professori sono stati molto gentili: alcuni mi hanno fatto un'accoglienza addirittura cordiale. Il Denjoy mi ha fatto l'onore persino d'interessarsi delle mie funzioni continue verso destra e mi ha suggerito nuove idee. Così abbiamo intanto trovata una nuova proprietà (lui l'ha enunciata ed io l'ho dimostrata): la condizione necessaria e sufficiente affinchè un dato aggregato H possa essere l'aggregato dei punti di discontinuità di una funzione avente derivata destra ovunque nulla. Spero di poter continuare queste interessanti ricerche. Ho però l'intenzione di rivolgere la mia attività ad altri campi ed ho quindi adocchiato la teoria delle funzioni di variabile complessa: vedremo che cosa riuscirò a fare! Nel salutarla a Bologna dimenticai di ricordarle che io mi tengo a Sua disposizione per la correzione delle bozze del Suo libro sulla teoria delle funzione di variabile reale. Ella, se crede di servirsi della mia umile collaborazione, può mandarmi le bozze senz'altro all'indirizzo che le trasmetto. Ho ricevuto oggi la comunicazione del conferimento del "Premio Pincherle". Ho immediatamente risposto al Rettore ringraziando. Mi permetta però, poichè so che Ella ha fatto parte della commissione giudicatrice, di ripeterle personalmente che Le sono grato di questo, come di tutto il bene che Ella ha più volte già fatto per me. Perdoni la confidenza che mi sono preso nella presente e mi creda, il suo dev.mo ed aff.mo Tullio Viola.*"

research in the field of analysis, and who contributed to the development of his sensitivity towards problems regarding education and teacher training.

Viola was born in Rome on 5 October 1904, the son of Carlo Maria Viola, full professor of mineralogy, and Clara Schneider, daughter of a jurisprudent who was an instructor at the University of Zurich. After having completed the classical high school in Parma, in accordance with his father's wishes he enrolled in the engineering school in Bologna, earning his degree in civil engineering on 17 November 1928.[3]

His passion for mathematics – as he loved to define it, "the purest and most exact of the sciences" – drove him to continue his studies and earn a second degree in mathematics from the University of Bologna, where he performed brilliantly in the courses of Higher Analysis with Leonida Tonelli and Beppo Levi, Elements of Theory of Functions with Beppo Levi, Higher Geometry with Enea Bortolotti, and Mathematical Physics with Pietro Burgatti.[4]

On 15 November 1930 Viola was awarded his degree with the highest score possible,[5] defending a thesis entitled *Studio intorno alle funzioni continue da una parte ed alla derivazione unilaterale*,[6] prepared under the advisement of Beppo Levi, which regarded a topic of analysis which the object of lively interest at the time, functions of a real variable. The research carried out for the thesis was published, with extensions and apposite in-depth analyses, in a series of notes and papers. This earned Viola the "Vittorio Emanuele II Prize" for 1930–1931 for best thesis for the Faculty of Sciences of the University of Bologna, the "Pincherle Prize" from the Salvatore Pincherle Foundation, which he mentions in the letter to Vitali, and a scholarship which took him to Paris.[7] Upon his return to Italy, various circumstances brought him first to Torino for a year at the Politecnico as assistant to Guido Fubini; then to Rome; next, as winner of the competition for the chair in Analysis, to the University of Bari in 1953; and finally in 1958 to Torino, where first he held the chair of complementary Mathematics and then, after Francesco G. Tricomi's retirement in 1968, the chair of Analysis [50, 51].

In 1982, on the 50th anniversary of the death of Vitali, Viola had this to say about his relationship with his teacher:

> I had just graduated a few months before in pure mathematics, and was a volunteer assistant in the course of theory of functions. Vitali was at the height of his scientific career, both in terms of the recognition he received – alas, so tardily! – by the highest Italian academies, and

---

[3] Historical Archive Bologna University (abbrev. HABU), Dossier n. 2076.

[4] HABU, Dossier No. 7224.

[5] HABU, Dossier No. 7224.

[6] The thesis is conserved in the form of a lithograph in Bologna (HABU, Dossier No. 7224) and in Torino in the library of the Department of Mathematics at the University of Torino, see [67].

[7] *Annuario della Regia Università di Bologna per l'anno accademico 1931–1932*, Bologna, Tip. Neri, 1932, pp. 69–70. The "Pincherle Prize" had been established in 1921 to celebrate Salvatore Pincherle's 40th year of teaching in Bologna. Cf. *Premio Salvatore Pincherle*, 9 January 1923, in HABU, BO II.F PINC S.

for having begun and, in part, concluded the preparation of his treatises, as well as for having begun, with renewed, surprising creative capacities, that research in stellar astronomy (...) Vitali was extraordinarily generous and good to me. Beginning with my degree thesis, at that time I was publishing my first works on functions of a real variable with a unilateral derivative. He came to see me, spoke with me at length, he presented me with extracts of his most beautiful, most famous publications on set theory, and on functions of a real variable, he gave me much good advice, he made it so I opened up to him about the whole of my studies and my projects for the future, and that I kept him informed about the results – alas, so modest! – of my research.[8]

Unfortunately, the dialogue between the two came to an abrupt end with Vitali's unexpected death on 29 February 1932, news of which reached Viola in Paris. Underlining the fact that he was Vitali's last student, he recalled that moment of his youth thus:

> The impression on my mind made by the announcement of his death can never be erased, an announcement that reached me in Paris, during a seminar I was taking part in. I heard people whispering "Vitali est mort!" A few days later I received the painful confirmation directly from Bologna: the Maestro had fallen, struck down suddenly while walking arm in arm with his colleague Ettore Bortolotti, under the porticos of that learned city in which I had spent the most beautiful years of the university studies, and in which were laid, and lie still, the remains of those who gave me life.[9]

Viola's research on the foundations of analysis, and his interest in the theory of integration as intended by Henri Lebesgue and developed by Arnaud Denjoy grew out of his conversations with Levi[10] and Vitali,[11] and from studying the lessons of Tonelli,[12] who from 1922 to 1930 taught at the Bologna University.

---

[8][83, p. 536]: *Io ero laureato da pochi mesi in matematica pura ed assistente volontario al corso di Teoria delle funzioni, Vitali era al vertice della Sua carriera scientifica, sia per i riconoscimenti giuntigli, ahimè tanto tardivamente!, dalle più alte accademie italiane, sia per aver Egli avviato e, in parte, concluso la redazione dei Suoi trattati, sia infine per aver iniziato, con rinnovate, sorprendenti capacità creative, quelle ricerche di astronomia stellare (...) Vitali fu straordinariamente generoso e buono verso di me. Partendo dalla mia tesi di laurea, io stavo allora pubblicando i miei primi lavori sulle funzioni di variabile reale unilateralmente derivabili. Mi venne a cercare, parlò a lungo con me, mi regalò gli estratti delle sue più belle, più famose pubblicazioni sulla teoria degli insiemi e su quella delle funzioni di variabile reale, mi diede molti buoni consigli, ottenne che io mi aprissi con Lui circa l'insieme dei miei studi e dei miei progetti a venire, che Lo tenessi al corrente dei risultati, ahimè tanto modesti!, delle mie ricerche.*

[9][83, p. 544]: *Incancellabile resterà nella mia mente l'impressione che mi fece l'annuncio della Sua morte, annunzio che mi raggiunse a Parigi, durante un seminario al quale assistevo. "Vitali est mort"!, sentii mormorare da più parti. Pochi giorni dopo ricevetti direttamente da Bologna la dolorosa conferma: il Maestro era caduto, fulminato improvvisamente mentre passeggiava al braccio del collega Ettore Bortolotti, sotto i portici di quella dotta città nella quale avevo passato i più belli anni dei miei studi universitari e nella quale riposavano, e tuttora riposano le spoglie di coloro che mi hanno dato la vita.*

[10]On Levi's scientific biography and works see [8, 39, 40, 57, 80] and on the relationships between Levi and Viola [19] and [8, p. XXXVI–XLIV].

[11]On Vitali's scientific biography, correspondence and works see [1, 2, 4, 62, 63, 66, 97].

[12]Viola mentions two notebooks (Cf. Appendix, note 42 and p. 404) containing his own notes on Tonelli's lectures on the Foundation of Calculus of Variations, but unfortunately these are now lost.

In the inaugural lecture that Vitali gave at the University of Bologna on 4 December 1930, he had this to say regarding students:

> To the young people who dedicate themselves to the study of mathematics, I say that the ability required to find the most suitable and most elegant devices is acquired through practise, through the examination of many examples, through the effort to imitate. To imitate, but not too much, if our discipline is not to become a marsh, a large one to be sure, but stagnant, with neither life nor movement. First imitate to learn, and then renew ourselves. (...) I love, at least when I am able, to regard science from a personal point of view, and always, again when possible, go beyond current opinions and look at the problem from a new perspective. I have the impression that some ways must be left behind, some mental habits must be abandoned, if we are not to clip the wings of progress. Even to science we must sometimes repeat Charon's cry: *By another way, by other ports, not here, you will find passage across the shore*. In my role as teacher I hope to be able to show you other ways, if not other ports.[13]

There are only a few elements at our disposal that allow us to reconstruct the relationship between Viola and Vitali in the brief span of time between winter 1930 and February 1932. We know that Vitali presented three Notes by Viola to the Accademia Nazionale dei Lincei [68–70], and that he supported Viola's nomination for the "Pincherle Prize". We also know that in his final years of life Viola dedicated some of his lectures in Modena, Cagliari and Milan to his distinguished Maestro,[14] recalling him with "esteem and affection"[83]. He had in mind a project for a lengthy article on Vitali's contributions, but was never able to carry through with it. The

---

On Tonelli's teaching in Bologna see [5, 59, 61] and L. CESARI, *L'opera di Leonida Tonelli e la sua influenza nel pensiero scientifico del secolo*, in [49, pp. 41–73].

[13][95], p. 2, 5–6 – *Opere*, p. 398, 401–402: *Ai giovani che si dedicano allo studio delle matematiche, io dirò che la abilità che si richiede per trovare gli artifici più convenienti e più eleganti si acquista con l'esercizio, coll'esaminare molti esempi, con lo sforzarsi di imitare. Imitare, ma non troppo, se non si vuole che la nostra scienza diventi una palude, immensa sì, ma stagnante, senza vita e senza movimento. Imitare dapprima per imparare, ma poi rinnovarsi.* (...) *Io amo, almeno quando vi riesco, trattare la scienza con una visione personale, e, sempre, quando mi è possibile, superare le opinioni correnti e prospettare i problemi sotto un aspetto nuovo. Ho l'impressione che alcune vie siano da lasciare, che alcuni abiti mentali siano da abbandonare, se non si vuole tarpare le ali al progresso. Anche alla scienza bisogna talvolta ripetere il grido di Caronte:* per altra via, per altri porti verrai a piaggia, non qui per passare. *Nelle mie funzioni di maestro spero di potervi indicare altre vie, se non altri porti.*

[14]On 26 September 1980, at the invitation of Emilio Baiada and Francesco Barbieri, Viola gave a lecture entitled *Giuseppe Vitali e l'integrale di Lebesgue* at the Mathematical Institute at the University of Modena which bears Vitali's name; the lecture was not however published. We have discovered among Viola's papers photocopies of four transparencies prepared for that occasion. On the 50th anniversary of Vitali's death, Viola commemorated his Maestro with the lecture *Ricordo di Giuseppe Vitali, a 50 anni dalla Sua scomparsa*, in Cagliari on 1 October 1982 (see [41, 83]). On 15 February 1983 at the Department of Mathematics at the University of Milan Viola gave the lecture *Come fu accolto e discusso in Italia l'integrale di Lebesgue nei primi decenni di questo secolo*, of which only the transparencies remain. According to his notebook no. 55 (1983), p. 46 he is supposed to have given a short lecture on the history of Zermelo's axiom and the opinions of Vitali, B. Levi and Tonelli in Cortona (Conference organized by the Istituto Nazionale di Alta Matematica, 26–29 April 1983), but this does not appear in the Symposia Mathematica XXVII, published in 1986, after Viola's death on 22 November 1985.

only texts that have come down to us are the *Ricordo di Giuseppe Vitali, a 50 anni dalla Sua scomparsa*, which appeared in the proceedings of the congress on the history of mathematics in Italy that took place in Cagliari in 1982, and the small notebook entitled *Giuseppe Vitali, L'integrale di Lebesgue. Lavori originali di Vitali. Riassunti ed Appunti* (Vitali's Lebesgue Integral, Original Works of Vitali, Abstracts and Notes), transcribed here in the Appendix.[15] Discovered in Viola's archives, it was probably assembled on the occasion of the invited lectures and the planned article just mentioned.

In the lecture entitled *How Lebesgue integral was received and discussed in Italy in the first decades of this century* given by Viola in Milan on 15 February 1983, he dwelt above all on the methodological aspect of the question [82, p. 48] and [52, pp. 76–77]. In this context he recalled the outcome of the competition for the chair in algebraic analysis of 20 October 1910, where the commission, comprised of Salvatore Pincherle, Luigi Bianchi, Gregorio Ricci Curbastro, Luigi Berzolari and Onorato Nicoletti, expressed this judgment regarding the contributions of Vitali, who was excluded from the trio of winners:

> The works of the candidate show the Author's through understanding of the various branches of analysis, his acumen in the treatment of questions that are delicate and not easy, arriving at interesting results, remarkable clarity and sobriety of composition. The fact itself of his having met with one of the creators of the new directions of integral calculus, Lebesgue, is proof that he stayed within the mainstream of this research, in which such results naturally presented themselves.[16]

He cited this report to show that at that time only Vitali, Tonelli and Levi grasped the significance of Lebesgue's new theory. In the lecture he also quoted the celebrated passage from the lecture entitled *Il teorema di riduzione per gli integrali doppi* given by Guido Fubini in 1942 at Princeton University and published posthumously in 1949 in the *Rendiconti* of the Mathematics Seminar of the University and the Politecnico of Torino, in which he reported Ulisse Dini's and Bianchi's opinions of Lebesgue's theory:

> When Lebesgue wrote his Thesis, Picard contacted Dini, then director of the *Annali di Matematica*, telling him that he had had a good thesis from one of his students, that this thesis studied the foundations of Calculus, that these foundations had been studied mainly

---

[15]The notebook is composed of 55 numbered pages. The notes and comments are mainly on odd-numbered pages, which some of the even-numbered pages are blank.

[16]T. VIOLA, Notebook no. 55 (1983), p. 14: *Giudizio complessivo su Vitali:"I lavori del concorrente dimostrano le buone cognizioni dell'Autore in vari rami dell'analisi, il suo acume nel trattare questioni delicate e non facili, giungendo a risultati interessanti, notevole chiarezza e sobrietà nella redazione. Il fatto stesso di essersi incontrato con uno dei creatori del nuovo indirizzo del calcolo integrale, il Lebesgue, prova come egli abbia saputo tenere in queste ricerche la via maestra, in cui tali risultati si presentavano naturalmente."* Terna: Levi Beppo, Cipolla Michele, Giambelli Giovanni Zenone. Commissari: Salvatore Pincherle, Luigi Bianchi, Gregorio Ricci Curbastro, Luigi Berzolari, Onorato Nicoletti. *Concorso alla cattedra di Analisi Algebrica nell'Università di Parma (20.10.1910)* (v. Bollettino Uff. del Ministero P. I. vol. XXXVIII, 1911, n. 3, pp. 2494–2508. Collocaz. Bibliot. Naz.le 16/1).

in Italy, and that it would therefore be best if the thesis were published in the *Annali*. I came to learn of this letter, and immediately thought that Picard did not appreciate this kind of research very much. Not even Dini was convinced of the importance of Lebesgue's thesis, but in order to fulfil the wishes of Picard, he published the work in the *Annali* [27]. This is how the first work of the new Calculus came to be published in an Italian periodical whose director didn't believe that the new methods were very important for the development of the science. For history's sake, I must add that Dini changed his opinion in his final years. This change was due to the definition of Lebesgue integration given by Perron, a definition that is very important because it does not require the notion, new, of the measure of a set, but was based solely on the more classic definitions. To comprehend the *forma mentis* of those times, by now long past, allow me to add that when I said to another great Italian mathematician, Bianchi, that the measure of the set of the rational numbers is null, he answered by mocking me and saying that I only studied the paradoxes of the infinite.[17]

Although it is often difficult to deduce the influence that teachers have on their students, in the case of young Viola it is possible to cite above all the teaching or the discussions he may have had with Tonelli, Levi and Vitali. He had attended Tonelli's lectures since 1928, and it is also possible that he listened to his lecture at the International Congress of Mathematicians held in Bologna from 3 to 10 September 1928, or at least heard echoes of it. In effect, the theme of Tonelli's lecture, *Il contributo italiano alla teoria delle funzioni di variabili reali* (The Italian contribution to the theory of the functions on real variables) can be linked to Viola's degree thesis. In both the lecture of Tonelli and the thesis of Viola we find the same references to U. Dini, P. Du Bois-Reymond, H. Lebesgue and A. Denjoy that appear in the introduction to Viola's thesis:

> The problem of research in the primitive functions, of fundamental importance for Integral Calculus, has induced modern analysts, such as for example Dini, Du Bois-Reymond, and in recent times Lebesgue and Denjoy, to study the operation of "limit of the incrementary ratio" for continuous functions.[18]

---

[17][14], p. 128 – *Opere scelte*, [15], vol. 3, p. 402: *Quando Lebesgue scrisse la sua Tesi, Picard si rivolse al Dini, allora direttore degli Annali di Matematica, dicendogli che aveva una buona tesi di uno dei suoi allievi, che questa tesi studiava i fondamenti del calcolo, che tali fondamenti erano studiati principalmente in Italia, e che perciò sarebbe stato meglio che la tesi fosse pubblicata negli Annali. Io venni a conoscenza di questa lettera, e subito pensai che Picard non apprezzasse molto questo genere di ricerche. Neppure il Dini era convinto dell'importanza della tesi di Lebesgue, ma per aderire al desiderio di Picard, pubblicò il lavoro negli Annali. In tal modo il primo lavoro sul nuovo Calcolo fu pubblicato in un periodico italiano, il cui direttore non credeva ai nuovi metodi molto importanti per lo sviluppo della scienza. Per la storia, devo aggiungere che il Dini mutò opinione nei suoi ultimi anni. Questo cambiamento si deve alla definizione degli integrali di Lebesgue data dal Perron, definizione molto importante perchè non esige la nozione, nuova, della misura di un insieme, ma si fonda soltanto sulle definizioni più classiche. Per comprendere la forma mentis di quei tempi ormai lontani permettetemi di aggiungere che quando dissi ad un altro grande matematico italiano, il Bianchi, che l'insieme dei numeri razionali ha misura nulla, egli mi rispose canzonandomi e dicendo che studiavo solo i paradossi dell'infinito.*

[18][67, p. 2]: *Il problema della ricerca delle funzioni primitive, di fondamentale importanza per il Calcolo Integrale, ha indotto analisti moderni, quali ad es. il Dini, il Du Bois-Reymond ed in epoca recente il Lebesgue e il Denjoy, allo studio della operazione di "limite del rapporto incrementale" per funzioni continue.* Cf. also [12, 20, 29, 30, 48, 53, 54, 56, 58].

Tonelli's statement regarding the axiom of choice is significant:

> Lebesgue's theory of integration is based on that of the measurement of groups of points and of measurable functions; this second theorem, in the general form Lebesgue has given it, requires for its carrying out the use of the so-called axiom of arbitrary choices, or Zermelo's postulate. With the auxiliary of this postulate, G. Vitali constructed, in 1905, the first example of a non-measurable set of points as Lebesgue intended it. Moreover, many analysts do not admit Zermelo's postulate; and even I maintain that the reasonings that use are insufficient. For this reason, I posed for myself the question of how to free the theory of the measurement of groups of points and of integration from this postulate; and in my *Foundations of Calculus of Variations*, I achieved this aim completely, introducing the sets that I called the *pseudointervals*, and the *quasi-continuous* functions.[19]

In fact, between Vitali, Tonelli and Levi, Viola found himself in a delicate situation regarding the axiom of choice,[20] since his teachers had adopted different stances: Vitali accepted it without difficulty,[21] while the other two rejected it, proposing some palliatives instead, Tonelli introducing new notions, and Levi stating a new postulate which he called the "approximation principle":

> Let there be given a deductive domain Ω, prime in which is the aggregate of the real numbers (at least the aggregate of the rational numbers); let $A, B, C, \ldots$ be aggregates defined in it and contained in prime aggregates within it (such that therefore it is permissible to consider an arbitrary element of them). Let E be instead a prime aggregate not belonging to deductive domain Ω, each element of which can be considered as constituted of infinite elements chosen arbitrarily in the aggregates $A, B, C, \ldots$ (possibly in only some of them): let there be given a function $f(x)$ with respect to which the domain D of x is contained in E, while the corresponding domain F of the function is contained in an aggregate G, prime or non prime, belonging or not belonging to Ω; finally, let there be defined a numeric function $d(y, z)$ of the couple of elements of G which is null always and only when $y = z$; let us suppose that a is an element of D such that, given an arbitrary number δ, among the elements of $A, B, C, \ldots$ that constitute a if there can be fixed a finite number n such that, indicating with a', a'' two elements whatsoever of D having in common with a said n elements, there will always be $d(f(a'), f(a'')) < \delta$. We then hold that the statement $f(a)$ exists as belonging to the natural extension of the deductive domain.[22]

---

[19][60, p. 248]: *La teoria dell'integrale di Lebesgue è fondata su quella della misura dei gruppi di punti e delle funzinì misurabili; e questa seconda teoria, nella forma generale datale dal Lebesgue, richiede per il suo svolgimento, l'uso del cosiddetto postulato delle scelte arbitrarie o postulato di Zermelo. Con l'ausilio di siffatto postulato, G. Vitali costruì, nel 1905, il primo esempio di insieme di punti non misurabile nel senso di Lebesgue. Peraltro, molti analisti non ammettono il postulato di Zermelo; ed anch'io ritengo insufficienti i ragionamenti che lo utilizzano. Per tale ragione, mi posi la questione di liberare la teoria della misura dei gruppi di punti e dell'integrazione dal postulato indicato; e nei miei* Fondamenti di Calcolo delle Variazioni *riuscii completamente allo scopo, introducendo gli insiemi, da me chiamati* pseudointervalli, *e le funzioni* quasi-continue.

[20]On the history of this axiom cf. [4, 19, 44, 46].

[21][89, p. 5]. Cf. also [19, p. 86].

[22][32, pp. 312–313]: *Sia assegnato un dominio deduttivo Ω, nel quale sia primo l'aggregato dei numeri reali (almeno l'aggregato dei numeri razionali); siano* A, B, C, ... *aggregati definiti in esso e contenuti in aggregati primi in esso (tali quindi che sia permesso considerarne un elemento arbitrario). Sia invece* E *un aggregato primo non appartenente al dominio deduttivo Ω, ciascun elemento del quale possa considerarsi costituito di infiniti elementi scelti in modo arbitrario negli aggregati* A, B, C, ... *(eventualmente in alcuni soltanto di essi): sia assegnata una funzione* $f(x)$

We have purposely cited in its entirety the principle formulated by Levi, so that the reader can immediately understand its complexity, and the various areas it leads to a consideration of. Here we are far from the axiomatic theory developed by Zermelo, which can assume the elementary form used by Vitali, that is, that "every set should be well-ordered". It should be noted that Levi's objectives are not always crystal clear, as Viola himself underlined in his obituary of Levi.[23]

In his degree thesis Viola distanced himself from the way paved by Tonelli, who positioned himself in the traditional field; nor did he follow Vitali. It is possible that his training as an engineer led him to adopt a position that was more constructive or that better reflected the reality. Perhaps at the time, the axiom of choice seemed to him to be an instrument reserved for more general studies:

> The consideration of individual examples of applications of Zermelo's postulate in the theory of aggregates suggests the classification of such examples into two categories: one comprises those in which we are dealing with aggregates in general, as for example, in the theory of transfinite numbers; the other in which we are dealing in particular with aggregates of real numbers (points in space of one or two dimensions). The first cannot be regularised with the approximation principle because they do not permit a consideration of distances, or, in general, or numeric functions $d(y, z)$ aimed at defining natural extensions. The second, *a priori*, are able to do this, but for the same reason require careful examination case by case of whether it is appropriate or not to adopt particular numeric regularising functions $d(y, z)$, whose geometric meaning is simple, expressive, and above all conform with the concept of *"approximation"* as used in everyday speech. It is not always possible to find such a function $d(y, z)$ and gives a closer look at the fact that the approximation principle has a selective nature and a more limited range than Zermelo's postulate, as was said earlier.[24]

---

rispetto alla quale il dominio D della x sia contenuto in E, mentre il corrispondente dominio F della funzione sia contenuto in un aggregato G, primo o non, appartenente o non a Ω; sia infine definita una funzione numerica d(y, z) *delle coppie di elementi di G la quale sia nulla sempre e solo quando* y=z*; supponiamo che* a *sia un elemento di D tale che, assegnato ad arbitrio un numero* δ*, fra gli elementi di A, B, C, ... che costituiscono* a*, se ne possa fissare un numero finito* n*, tale che, indicando con* a′, a″ *due elementi qualunque di D aventi comuni con* a *i detti* n *elementi, sia sempre* $d(f(a'), f(a'')) < \delta$. *Noi consideriamo allora l'affermazione* "f(a) esiste" *come appartenente all'ampliamento naturale del dominio deduttivo.*

[23][80, p. 515]: *Il Levi non fu altrettanto felice, a nostro parere, in altri lavori di filosofia matematica, e più precisamente di logica, nei quali si lasciò andare ad osservazioni eccessivamente sottili, tanto da apparire (in qualche dettaglio) addirittura oscure, quasi dei curiosi giuochi di parole!* (Levi was not as fortunate, in our opinion, in other works of mathematical philosophy, and more precisely, in logic, in which he allowed himself to make observations that were so excessively subtle that they even appeared (in some of their details) obscure, almost like curious plays on words!).

[24][71, p. 289]: Cf. [34, 40]. *La considerazione dei singoli esempi d'applicazione del postulato di Zermelo nella teoria degli aggregati ci suggerisce di classificare tali esempi in due categorie: l'una comprende quelli nei quali si tratta di aggregati in generale, come per es. nella teoria dei numeri transfiniti, l'altra quelli nei quali si tratta particolarmente di aggregati di numeri reali (punti di spazi a una o più dimensioni). I primi non possono venir regolarizzati col principio di approssimazione perchè non possono dar luogo a considerazione di distanze o, in generale, di funzioni numeriche* d(y, z) *atte a definire ampliamenti naturali. I secondi*, a priori, *lo possono, ma anche per questi occorre esaminare attentamente, caso per caso, la convenienza o meno di adottare particolari funzioni numeriche regolarizzatrici* d(y, z)*, aventi significato geometrico*

All of this leads to a re-examination of the approximation principle, situating it in the context of analysis, without attention being focussed on the purely logical aspect of the extension of the deductive domain, so dear to Beppo Levi. With a didactic purpose in mind, Viola considers the most natural example, that regarding the equivalence of the limit point and the convergence point:

> P is said to be the limit point of aggregate A of points if at every neighborhood of P there exists at least one point of the aggregate A that is different from P. In this regard a proposition frequently stated in which use is made of Zermelo's postulate says that there exists in A a succession of points $Q_1, Q_2, \ldots, Q_n, \ldots$ that tends to P.[25]

If we take, like Viola, the case of the real straight line, we can see how the axiom of choice works to link the abstract notion of limit point to the more concrete notion of convergence point. If we consider a series (I_n) of intervals centred in P whose lengths tend to zero, in each of these there exists at least one point that is distinct from P and belongs to A. In other words, using Q_n to indicate the element chosen in I_n, the series (Q_n) converges towards P and P is thus the convergence point. The approximation principle consists in justifying this choice in the following sense: after a finite number of choices, the others are of scant importance, because the difference between two elements within the same I_n does not have any effect on the nature of the series obtained. It is here that the approximation principle comes in. It must be possible to fix the degree of approximation by means of a certain distance and to determine it with the finite quantity of choices that have been made. Viola maintains that this distance must be formalised in a context that is as close as possible to that being studied, and in a certain sense represents the very foundation of the problem in question. In Paris, he wrote his paper *Sul principio di approssimazione di B. Levi nella teoria della misura degli aggregati e in quella dell'integrale di Lebesgue* (On the principle of approximation by B. Levi in the theory of the measure of aggregates and in Lebesgue integral theory), in which he proposed to regularise the fundamental theorem of the Lebesgue measure:

> The sum of a countable collection of measurable aggregates no two of which have any points in common, is a measurable aggregate, and its measure is the sum of the measures of the individual aggregates.[26]

Even if in the second edition of his *Leçons sur l'intégration et la recherche des fonctions primitives* Lebesgue thought it appropriate to add "From the definition of

---

semplice, espressivo e soprattutto conforme al concetto di "approssimazione" *nel parlare comune. Non sempre riesce trovare una tale funzione* d(y, z) *e ciò mostra più da vicino che il principio d'approssimazione ha carattere selettivo e una portata più limitata del postulato di Zermelo, come sopra si è detto.*

[25][71, p. 290]: *Si dice che* P *è punto limite di un aggregato* A *di punti se in ogni intorno di* P *esiste almeno un punto dell'aggregato* A *diverso da* P. *A questo riguardo una proposizione frequentemente affermata in cui si fa uso del postulato di Zermelo dice che esiste in* A *una successione di punti* $Q_1, Q_2, \ldots, Q_n, \ldots$ *che tende a* P.

[26][72, p. 75]: *La somma d'una infinità numerabile di aggregati misurabili, non aventi due a due nessun punto in comune, è un aggregato misurabile e la sua misura è la somma delle misure dei singoli aggregati addendi.*

measurable sets it results that ...",[27] the rest of the proof remained unchanged, and the axiom of choice was implicitly present in it, as Sierpinski revealed in 1928 [55, p. 223]. However, in his article *Sur les correspondances entre les points de deux espaces* – published in 1921 in the journal *Fundamenta Mathematicae* and cited by Viola [71, p. 294] – Lebesgue explained that "he did not think he had made a choice without a law!".[28] Even if Lebesgue's proof was quite clear at an intuitive level, it appeared that the difficulty was due to the external measure.

Thus Viola rightly focussed his efforts and proved, without using the axiom of choice, the part relative to the internal measure.[29] Then, however, the first part could be simplified by using only certain sets covers, and leaving the complementary ones aside. The notion proposed by Viola simplified comprehension with respect to that of Lebesgue.

After having covered the first set E_1 " at least by $\frac{\epsilon}{2}$" by means of the set S_1 the union of S_1 and E_2 is covered by S_2 " at least by $\frac{\epsilon}{2^2}$" and in general the union of S_n and E_{n+1} by means of S_{n+1} at least by "at least by $\frac{\epsilon}{2^n+1}$".

"At the limit" the series S_n furnishes the entity S, which needs to be justified, or at least regularised. As often happens, Viola sets the distance between S' and S'' as the measure of the set that is the union of the differences of these two sets. By means of some "slight touches" to the formation of S, it is possible to apply the approximation principle. In other words, in this proof Viola shows us the usefulness of this principle. Further, the application of the principle is not the only advantage of this proof. By comparing this and the one proposed by Lebesgue, Viola was able, on the one hand, to render one part independent of the axiom of choice and, on the other hand, to make another part easier to justify.

It often happens that first proofs are successively refined, but we want to underline that here, where the heuristic and didactic aspects come into play, the introduction of the approximation principle, which is apparently delicate, has led to a better grasp of the devices offered as proof by Lebesgue.

In 1933, Levi and Viola co-authored the article entitled *Intorno ad un ragionamento fondamentale nella teoria delle famiglie normali di funzioni*. In this case it is difficult to establish which part was written by which author. However, it seems to us that the in-depth study of Montel's work is attributable to Viola. In support of this claim we can cite on one hand Viola's doctoral thesis defended that same year, 1933, under the advisement of Montel [75, 76], and on the other hand, Viola's long review of one of Montel's books (Cf. [43], [77] and [35]). To this can be added the study of Montel's 1927 *Leçons sur les familles normales de fonctions analytiques*, a copy of which was owned by Viola. From Viola's autograph notes in the margins

---

[27][28, p. 115]: *De la définition des ensembles mesurables il résulte que...*

[28][28, p. 115]: *il ne croyait pas avoir fait de choix sans loi!*

[29][72, p. 78]: *La semplificazione introdotta ha poi il vantaggio di eliminare completamente il postulato della scelta dalla seconda parte della dimostrazione.* (The simplification introduced has then the advantage of completely eliminating the postulate of choice from the second part of the proof.)

of p. 3, we read "Infinite arbitrary choices?" and on pp. 35 and 36, regarding normal families, the observation, this time in French, "pour éviter les choix arbitraires on pourrait dire que les valeurs prises par une suite de fonctions appartenant à une famille normale dans un domaine sont bornées en un point fixe." (to avoid arbitrary choices we could say that the values taken from a series of functions belonging to a normal family within a domain are limited in a fixed point).

This is not too far from what Levi and Viola wrote on the definition of normal functions:

> It then appears appropriate to avoid this second-degree choice by slightly modifying the definition of normal families by asking that a family E is said to be normal when, arbitrarily assigned a succession of aggregates $E_n(n = 1, 2, ...)$ of elements of E, there exist successions of functions that are uniformly convergent to which belong elements contained in E_n of whatever index you wish.[30]

This suggests to us that the role played by the young student by the side of his teacher was one of inducing him to assume a more constructive attitude or to re-examine certain definitions, like Tonelli did. In 1934 the two allies, Maestro and student, went their separate ways, and each presented an article in the journal *Fundamenta Mathematicae* [38, 78]. As usual, Levi insisted on the deductive domains, while Viola placed himself on the terrain of applications. We can retrace in his article the lines of force of his thinking, which goes far beyond that of accepting or rejecting the axiom of choice. It is interesting to note that from the very beginning he places the accent on the aspect of reconciling the diverse positions:

> The point of view which we take could be called "conciliatory", inasmuch as we attempt to provide a justification of many mathematical procedures that make use of the well-known postulate of E. Zermelo.[31]

In other words, the approximation principle allows us to justify various choices. But Viola is aware that its application is not always possible, as in the case of reasoning that is too abstract:

> Instead, in other cases Zermelo's postulate is in effect applied in the abstract, that is, without specifying the deductive domain in which the arbitrary choices are being made, and it is then that its legitimacy appears doubtful to us. This is the case, for example, of almost all the propositions that are found in the theory of transfinite numbers.[32]

---

[30][36, p. 201]: *Pare allora conveniente di evitare questa scelta di secondo grado modificando leggermente la definizione delle famiglie normali col chiedere che si dica normale una famiglia* E *quando, assegnata arbitrariamente una successione di aggregati* E_n *(n = 1, 2, ...) di elementi di* E, *esistano successioni di funzioni uniformemente convergenti cui appartengono elementi contenuti in* E_n *di indice elevato quanto si vuole.*

[31][78, p. 75]: *Il punto di vista da cui ci mettiamo potrebbe chiarmarsi "conciliativo", in quanto cerca di dare una giustificazione di molti procedimenti matematici che fanno uso del noto postulato de E. Zermelo.*

[32][78, pp. 75–76]: *In altri casi, invece, il postulato di Zermelo viene effettivamente applicato in astratto, cioè senza precisazione del dominio deduttivo in cui si operano le scelte arbitrarie, ed è*

This is where Viola's attention to the didactic aspect emerges. The elimination or justification of the infinity of the choices can serve to increment simplicity and clarity:

> Extending a well-ordered deductive domain is, if you will, always permitted, and in many cases can even be considered advisable, for didactic purposes if nothing else, if the proofs gain in simplicity and clarity.[33]

Finally, Viola does not neglect the heuristic aspect of the question:

> On the other hand, from the heuristic point of view, that is, as a means of research, it is perhaps neither appropriate nor possible to eliminate Zermelo's postulate. Propositions such as that which we will examine in numbers 5, 7 as the lemma of Vitali's theorem (no. 10) and others have been found with proofs that are quite simple using Zermelo's postulate; and we will succeed in confirming them while liberating them completely from the use of Zermelo's postulate. The step forward that we want then to take in a proof in which the use of Zermelo's postulate is indispensable consists, if possible, in regularising the proof by means of the approximation principle. And it seems to me that from this present work it should become evident that Levi's approximation principle expresses in exact and satisfactory terms the intuitive yet somewhat vague concept which we usually attribute to the word 'approximation' (cf. Levi, p. 324) and is not an artful contrivance that disguises Zermelo's postulate itself.[34]

When the approximation principle intervenes, it is applied by means of a regularising function of the symmetric difference already used for the countable additivity of the Lebesgue measure. Basing himself on this simple and natural function, Viola is led to state that certain proofs cannot be regularised. Here is where the difficult problems of independence come in. The independence of the axiom of choice with regard to the other axioms of Zermelo-Fraenkel set theory [13, 98, 99] would only be solved many years later, in 1963 by Paul Cohen [9–11, 17, 44]. Thus Viola's statements about the fact of raising doubts about the impossibility or possibility of regularisation come as no surprise.

---

allora che la sua leggitimità ci pare discutibile. Tali sono ad esempio quasi tutte le proposizioni che s'incontrano nella teoria dei numeri transfiniti.

[33][78, p. 76]: *Ampliare un ben determinato dominio deduttivo è, se si vuole sempre lecito e in molti casi può essere anche consigliabile, se non altro a scopo didattico, se le dimostrazioni ne guadagnano in semplicità e chiarezza.*

[34][78, p. 76]: *D'altronde dal punto di vista euristico, cioè come mezzo di ricerca, non sarebbe opportuno né forse possibile eliminare il postulato di Zermelo. Proposizioni come quelle che esamineremo ai numeri 5, 7 come il lemma del teorema di Vitali (n. 10) ed altre, sono state trovate con dimostrazioni assai semplici facendo uso del postulato di Zermelo; e noi riusciremo a confermarle liberandone totalmente le dimostrazioni dal detto postulato. Il progresso che noi desideriamo poi realizzare in una dimostrazione in cui l'uso del postulato di Zermelo sia indispensabile, consiste, se è possibile, nel regolarizzare la dimostrazione mediante il principio di approssimazione. E mi pare che dal presente lavoro debba risultare evidente che il principio di approssimazione di Levi esprime in termini esatti e soddisfacenti il concetto intuitivo e d'altronde spesso un po' vago che noi siamo soliti attribuire alla parola "approssimazione" (Cfr. Levi, p. 324) e non è una convenzione artificiosa che dissimuli il postulato di Zermelo medesimo.* The reference is to the article [32], p. 324.

In conclusion, we can maintain that Viola was always aware of having used the axiom of choice in his works, and that it was not always possible for him to eliminate it, or to regularise the procedure in order to apply the approximation principle [19]. His various studies on this topic made it possible to show the profound nature, the limits of its applications and the heuristic consequences. Further, the young Viola had to adapt to the diverse opinions of his teachers. More than anyone else, he attempted to reconcile their various points of view. The benevolence with which Beppo Levi and Giuseppe Vitali received his early research show clearly that his efforts were not in vain.

Many years after that period, with the same characteristic modesty that was a distinguishing trait of Vitali, Viola confessed that he had not known how to profit from the valuable advice he received:

> At a distance of more than half a century, I understand that I did not know how to take advantage, as I should have done, from the occasion offered to me. I made the error that many young people make, a foolish error that I later bitterly regretted: that of letting haste to publish get the better of me instead of dedicating the most part of my time to broadening and deepening my knowledge. So it happened that I read, even with much interest and admiration, the publications of Vitali mentioned earlier, but I lacked the capacity to construct the possible connections with what I was looking for.[35]

In reality, he profited greatly from Vitali's knowledge and advise, which he described so vividly in the inaugural lecture quoted earlier. To give only one example, Mauro Picone had this to say about Viola's article *Sulla formula di integrazione per parti nell'integrazione doppia secondo Stieltjes* [79], which he presented for publication in Giuseppe Battaglini's *Giornale di Matematiche*:

> This is, in my opinion, a very valuable work which enriches mathematical analysis with theorems that are effectively and substantially new. I very much like, for example, the result according to which a function with several variables with finite variance according to Vitali belongs to the first Baire class.[36]

What is certain is that Viola was able to absorb from his illustrious Maestri in Bologna, along with their teachings, the examples they set, their accessibility and generosity towards their students, and the capacity to instil in them the inclination for research and an ardent love for the truth. From his teachers, and particularly from Vitali, he inherited a passion for civic duty and love for the school and for teachers.

---

[35][83, p. 536]: *A distanza di più di mezzo secolo, capisco che io non seppi approfittare, come avrei dovuto, dell'occasione che mi era stata offerta. Feci l'errore che fanno molti giovani, un errore stolto e del quale ebbi poi a pentirmi amaramente: quello di lasciarmi prendere dalla fretta di pubblicare, invece di dedicare la massima parte del mio tempo ad ampliare ed approfondire la mia cultura. Così accadde che lessi, ed anche con molto interesse ed ammirazione, le suddette pubblicazioni di Vitali, ma mi mancò la capacità di costruire i possibili collegamenti con quanto andavo cercando.*

[36]M. Picone to C. Miranda, Rome 2.6.1951, published in [18, p. 120]: *Trattasi secondo il mio parere di un lavoro assai pregevole che arricchisce l'Analisi matematica di teoremi effettivamente e sostanzialmente nuovi. Mi piace molto, per esempio, il risultato secondo il quale una funzione di più variabili a variazione finita secondo Vitali è della prima classe di Baire.*

Following the example of Vitali, who was for many years the regional president for the Ligurian section of the National Federation of Middle School Teachers [47], Viola was the national president of the Mathesis Association for teachers from 1959 to 1969 and of that same association's Torino section from 1959 to 1979 [16, 50, 51]. He did everything he could to see that the history of mathematics was introduced into secondary and university teaching, in the attempt to bridge the gap between the two cultures, humanist and scientific, caused by the Gentile reform. Among the quotation in his notebooks we find the words of the Hungarian geometer and pedagogist Ferenc Kàrteszi: "The most important problem is how to reawaken interest and activity in students, and lead them recognise the joy of work". There we also find the concepts of Charles Laisant, taken up by Giuseppe Peano and Giovanni Vailati as well, which were often inspirations for his lectures:

> He [Laisant] deplores the fact that also for mathematics (which is all talk), the school continues in spite of everything to be a kind of gymnasium for mnemonics rather than an institution for intellectual culture, that the student is too busy there with learning (learning, *accipere*) and too little with understanding (comprehending, *concipere*), that in short the student is considered more as a recipient to be filled than a field to be sown, a plant to cultivate, a fire to stoke.[37]

Acknowledgements To conclude, we wish to thank Giovanna and Carlo Viola for having put at our disposal the letters, notebooks and some books of their father's, Sandra Linguerri for having sent us material from the Archives of the University of Bologna regarding Viola's academic career and Sergio Garbiero, Sergio Console and Ernesto Buzano for their help with formulas and figures in the Appendix.

Appendix

TULLIO VIOLA, *Giuseppe Vitali, L'integrale di Lebesgue. Lavori originali di Vitali. Riassunti ed Appunti*, mss., f. 1–55.[38]

1. *"Sui gruppi di punti"* [39] (Rendic. Circ. Mat. di Palermo, 18, 1904, p. 116).

Richiamo (sugli insiemi E lineari misurabili secondo Lebesgue).

I) Insiemi limitati (cioè $\subseteq [a, b]$)

Lebesgue (perfezionata da C. J. de La Vallée Poussin, W. H. Young e altri):

---

[37][64], in [65, p. 259]: *Egli [Laisant] deplora che anche per la Matematica (il che è tutto dire), la scuola continui malgrado tutto a essere piuttosto una palestra mnemonica che non un istituto di cultura intellettuale, che l'allievo sia ivi occupato troppo a imparare (apprendere,* accipere*) e troppo poco a capire (comprendere,* concipere*), che lo scolaro insomma venga considerato più come un recipiente da riempire che non come un campo da seminare, una pianta da coltivare, un fuoco da eccitare.*

[38]The transcripts are done by Clara Silvia Roero, as well as the notes and references in square brakets, denoted by the abbreviation *(csr)*. The end of the page of the manuscript is indicated with ||.

[39] v. Pascal I_3, p. 1044. [*Ref.*: [84], cf. [25, 45], p. 1044. *(csr)*].

Sia A un insieme aperto (= unione di un numero finito o numerabilmente infinito d'intervalli aperti δ_i, due a due disgiunti). Si pone:

$$mis\, A = \sum_i \delta_i.$$

Sia C un insieme chiuso (= complementare, rispetto ad $[a,b]$ di un insieme aperto A). Si pone:

$$mis\, C = (b-a) - mis\, A.$$

Ciò posto, $\forall E$, si pone:

$$mis_e E = estr.inf.\{mis\, A\} \quad (\forall A \supseteq E),$$

$$mis_i E = estr.inf.\{mis\, C\} \quad (\forall C \subseteq E),$$

(v. Pascal I_3 p. 1044, Hobson I pp. 154–171, particolarmente a p. 166). H. Lebesgue, C. R. Paris 1901 vol. 132 p. 1025.[40] ||

È sempre: $mis_i E \le mis_e E$. Se $mis_i E = mis_e E$, si pone $mis\, E = mis_i E = mis_e E$.

Vitali definisce solo $mis_e E$. Passa poi a considerare il complementare E^* di E (cioè $E^* = [a,b] - E$), e osserva che è sempre

$$mis_e E + mis_e E^* \ge b - a.$$

Chiama la differenza

$$\mathcal{Z} = (mis_e E + mis_e E^*) - (b-a),$$

l'*allacciamento* dei due insiemi E, E^*. (Definisce "allacciamento" anche per due insiemi E_1, E_2 generici.) Seguono alcuni teoremi riguardanti gli allacciamenti.

Chiama poi *misurabile* ogni insieme E che ha allacciamento nullo col suo complementare E^*. Seguono numerosi teoremi sugli insiemi misurabili. ||

2. "*Sull'integrabilità delle funzioni*" [41] (Rendic. Ist. Lombardo, II, vol. 37, 1904, p. 69).

Riprende la definizione di misura di un insieme data nel lavoro precedente (n. 1), portandovi alcuni perfezionamenti e complementi.[42] Dimostra il teorema:

"*Condizione necessaria e sufficiente per l'integrabilità secondo Riemann*[43] *in un intervallo finito $[a,b]$, di una funzione limitata $f(x)$, è che l'insieme dei punti di discontinuità della $f(x)$ in $[a,b]$, abbia misura nulla.*"

Lebesgue dimostrò contemporaneamente questo teorema in "*Leçons sur l'intégration*" (I ediz. 1904, p. 29). Lo stesso Lebesgue nella II ediz. 1928, p. 29 rienuncia il teorema e, in nota, cita Vitali. Ivi, p. 26 dà un enunciato dello stesso teorema, che si riconduce alla misura secondo Peano-Jordan. In Pascal I_3 p. 1074, i due enunciati sono messi efficacemente a confronto.[44]||

---

[40][*Refs.*: [25,45], p. 1044; [22], vol. 1, pp. 159–171; [26] (*csr*)].

[41] v. Pascal I_3 [*Ref.*: [85], cf. [25,45], p. 1044. [*Ref.*: [42] (*csr*)].

[42] Complementi, varianti, osservazioni critiche in Carathéodory, pp. 460–475. [*Ref.*: [3]. (*csr*)].

[43] v. le 4 diverse forme in cui può essere espressa la condizione d'integrabilità secondo Riemann, nei miei appunti alle lezioni di Tonelli (Quaderno n. 1 pp. 1–14).

[44][*Refs.*: [28], p. 29, 2nd ed. 1928, p. 29; [25,45], p. 1074. (*csr*)].

3. *"Sulle funzioni integrali"*[45] (Atti Accad. Scienze Torino, vol. 40, 1904–1905, p. 1021).

Definizione di funzione $y = F(x)$, con $x \in [a, b]$, *assolutamente continua* in $[a, b]$:

"$\forall \epsilon$, $\exists \delta /$ Qualunque sia il gruppo d'intervalli parziali $[\alpha_i, \beta_i]$ di $[a, b]$, in numero finito e fra loro non sovrapponentisi, dalla limitazione $\sum_i (\beta_i - \alpha_i) < \delta$, segua:

$$\left| \sum_i \{F(\beta_i) - F(\alpha_i)\} \right| < \epsilon."$$

Vitali non dimostra qui il molto semplice

Teorema. *"Condizione necessaria e sufficiente affinché $F(x)$ sia assolutamente continua in $[a, b]$, è che $\forall \epsilon$, $\exists \delta /$ Qualunque sia il gruppo d'intervalli parziali $[\alpha_i, \beta_i]$ di $[a, b]$, in un numero finito e fra loro non sovrapponentisi, dalla limitazione*

$\sum_i (\beta_i - \alpha_i) < \delta$, *segua*: $\sum_i |\{F(\beta_i) - F(\alpha_i)\}| < \epsilon$."[46]

Dimostra che ogni funzione assolutamente continua è anche a variazione limitata,[47] ma dà esempio di una funzione || $F(x)$ continua in $[a, b]$, ivi non decrescente e non negativa, quindi a variazione limitata, ma non assolutamente continua.[48]

Teorema. *"Le variazioni totali di una funzione assolutamente continua sono assolutamente continue."*

Teorema. *"Una funzione assolutamente continua è la differenza di due funzioni assolutamente continue, non negative e non decrescenti."*

(Ma Vitali non scrive esplicitamente le due celebri relazioni:

$$F(x) = F(a) + P(x) - N(x),$$
$$V(x) = P(x) + N(x),$$

e non ne fa uso. V. i miei appunti alle lezioni di Tonelli, 2 p. 49).

I Teorema. Lemma. *"Le funzioni integrali sono assolutamente continue."*[49]

(Il concetto di funzione integrale è lo stesso del mio corso di analisi).[50]||

Dim. $f(x)$ è supposta sommabile, limitata o no, in $[a, b]$.

---

[45]Complementi, Varianti, Osservazioni critiche, Riferimenti al loc. cit. di Pascal (v. Pascal, I_3 p. 1088) Cfr. Tonelli, Calc. d. Variaz. I p. 165. [*Refs.*: [87]; [25, 45], p. 1088; [59], vol. 1, p. 165. (*csr*)].

[46]Una dimostrazione si trova in Tonelli, vol. 1 pp. 62–64. *Ref.*: [59, vol. 1, pp. 63–64]. (*csr*)].

[47]v. C. Jordan, C. R. 92 (1881) p. 228, e *"Cours d'analyse"* 2 ediz., 1 (1893) p. 54. Anche H. Hahn, pp. 483–513. [*Refs.*: [23]; [24, p. 54], [21, pp. 483–513]. (*csr*)].

[48]v. mio Taccuino 51.126, tema 218. [Viola's notebook no. 51, p. 126:"*218. Trovare una funzione crescente e continua in* [0, 1], *ma non assolutamente continua in nessun intervallo parziale di* [0, 1] *Generalizzare l'es. di Vitali,* Sulle funzioni integrali, *p. 1023 (Atti Torino, 40, 1904–05)."*(*csr*)].

[49]Dimostrazione *apparentemente* semplice in Tonelli (C. d. V. I, 165), perchè si basa su un lemma p. 153 relativo agl'integrali estesi a pseudointervalli. Altra dimostrazione *apparentemente* semplice in Lebesgue (II ediz. p. 158) perchè si basa su una teoria preliminare relativa alle funzioni additive d'insieme. *Refs.*: [59, vol. 1, p. 153–154, 165], [28, 2nd ed. 1928, p. 158]. (*csr*)].

[50]Cf. [81, vol. 1, pp. 390–394]. (*csr*)].

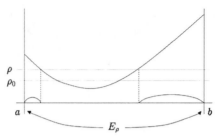

Limitazioni fondamentali:

$$\left|\int_E f(x)\,dx\right| \le \int_E |f(x)|\,dx \le \int_a^b |f(x)|\,dx.$$

Pongo:

$$E_\rho : \{x \in [a,b]/|f(x)| > \rho\}, \quad \psi(\rho) = \int_{E_\rho} |f(x)|\,dx \implies \lim_{\rho \to +\infty} \psi(\rho) = 0.$$

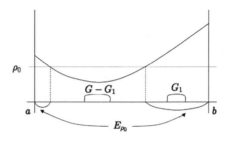

$$\forall \epsilon > 0, \quad \exists \rho_o / \forall \rho \ge \rho_o, \quad \psi(\rho) < \frac{\epsilon}{2}.$$

Prendo $G \,/\, mis\,G \le \frac{\epsilon}{2\rho_o}$. Sia $G_1 = G \cap E_{\rho_o}$.

$$\implies \int_{G-G_1} |f(x)|\,dx \le \rho_o \cdot mis\,(G-G_1) \le \rho_o \cdot mis\,G \le \rho_o \frac{\epsilon}{2\rho_o} = \frac{\epsilon}{2},$$

$$\int_{G_1} |f(x)|\,dx \le \int_{E_{\rho_o}} |f(x)|\,dx = \psi(\rho_o) < \frac{\epsilon}{2},$$

$$\int_G |f(x)|\,dx = \int_{G-G_1} |f(x)|\,dx + \int_{G_1} |f(x)|\,dx < \frac{\epsilon}{2} + \frac{\epsilon}{2} = \epsilon. \qquad \|$$

Sia ora

$$G = \bigcup_i [\alpha_i, \beta_i], \; F(x) = \int_a^x f(t)\,dt. \implies \sum_i \{F(\beta_i) - F(\alpha_i)\} = \int_G f(x)\,dx,$$

$$\left|\sum_i \{F(\beta_i) - F(\alpha_i)\}\right| = \left|\int_G f(x)\,dx\right| \le \int_G |f(x)|\,dx < \epsilon.$$

II Teorema. Lemma. *"Ogni funzione $F(x)$ assolutamente continua è un integrale."*

Dimostra precisamente che
$$F(x) = \int_a^x D_+ F(x)\,dx.$$

Ma un teorema di Lebesgue (*Leçons*, I ed. p. 124) dice che:

"*L'intègrale indefinito di una funzione sommabile limitata ammette questa funzione per derivata, salvo in punti di un insieme di misura nulla.*"[51] Vitali dunque dimostra il

Teorema Fondamentale. "*Condizione necessaria e sufficiente affinché una funzione $F(x)$ in $[a,b]$ sia ivi una funzione integrale, è che essa sia assolutamente con-* || *tinua in $[a,b]$. In tal caso $F'(x)$ esiste in quasi tutto $[a,b]$ e, posto*

$$\dot F(x) = \begin{cases} F'(x), & \forall\, x \text{ in cui esiste } F'(x), \\ 0, & \forall\, x \text{ in cui non esiste } F'(x), \end{cases}$$

risulta
$$F(x) = F(a) + \int_a^x \dot F(x)\,dx.$$

(cfr. Pascal I_3 p. 1088).[52] ||

4. "*Sul problema della misura dei gruppi di punti di una retta.*"[53] (Bologna, Edit. Gamberini e Parmeggiani 1905).

Secondo Lebesgue (I ediz. 1904, p. 103),[54] la *misura* $\mu(E)$ deve, per definizione assiomatica, esser tale che:

1. $\mu(E) \geq 0 \quad (\forall E)$
2. $\mu([0,1]) = 1$
3. $\mu(E') = \mu(E'')$, se E' congruente E'' (per spostamento)
4. $\mu(\bigcup_i E_i) = \sum_i \mu(E_i) \quad$ (se $E_i \cap E_j = \emptyset$)

Teorema. "*Se si ammette l'assioma di Zermelo, esistono insiemi non misurabili.*"

Dim. (Costruzione di un esempio).

$\forall x \in \mathbb{R}$, chiamo A_x l'insieme di *tutti* i punti $x + r$, con r razionale ($r \in \mathbb{Q}$). Prendo due punti $x_1 \in \mathbb{R}$, $x_2 \in \mathbb{R}$. Allora:

$$\text{se } x_1 - x_2 = \begin{cases} \text{raz.} \Longrightarrow A_{x_1} \equiv A_{x_2} \\ \text{irraz.} \Longrightarrow A_{x_1} \cap A_{x_2} = \emptyset. \end{cases} \quad ||$$

Sia $H \equiv (\alpha'\,\alpha''\,\alpha'''\ldots)$ l'aggregato di *tutti* gli A_x fra loro *effettivamente diversi*, e quindi anche *disgiunti*. Scelgo a piacere:

$$P_{\alpha'} \in \alpha', \quad P_{\alpha''} \in \alpha'', \quad P_{\alpha'''} \in \alpha''', \ldots$$

---

[51][28, pp. 124–125]:"*L'intégrale indéfinie d'une fonction sommable admet cette fonction pour dérivée sauf aux points d'un ensemble de mesure nulle.*" (*csr*)].

[52][*Ref.*: [25], p. 1088. (*csr*)].

[53]v. Enciclop. tedesca II_3, Parte II, p. 977 e Pascal I_3 p. 1042. Complementi e varianti in: H. Hahn pp. 575–580 e altri ivi p. 580 indicati; Carathéodory pp. 268, 349. v. i miei appunti allegati all'estratto. [*Refs.*: [88], [100, p. 977], [25,45, p. 1042] [21, pp. 575–580], [3, p. 268, 349]. (*csr*)].

[54][*Ref.*: [28, p. 103; 2nd ed. 1928, p. 110]. (*csr*)].

tutti in $[0, \frac{1}{2}]$. Essi sono tutti fra loro distinti. Pongo

$$G_0 : \{P_\alpha, \text{ al variare di } \alpha \text{ in } H\}.$$

Poi costruisco l'infinità numerabile $\{G_\rho\}$ d'insiemi

$$G_\rho : \{P_{\alpha'} + \rho, P_{\alpha''} + \rho, P_{\alpha'''} + \rho, \ldots\},$$

al variare di ρ in $\mathbb{Q}$.

Proprietà quasi immediate:

(I) Al variare di ρ in $\mathbb{Q}$, $P_\alpha + \rho$ descrive α.
(II) Se $\rho_1 \neq \rho_2$, è $G_{\rho_1} \cap G_{\rho_2} = \emptyset$.
Infatti non possono esistere due numeri razionali ρ_1, ρ_2 (distinti) tali che

$$P_{\alpha'} + \rho_1 = P_{\alpha''} + \rho_2,$$

perchè, se così fosse, si avrebbe $P_{\alpha'} = P_{\alpha''} + (\rho_2 - \rho_1)$, (assurdo).
(III) È $\mathbb{R} = \bigcup_{\rho \in \mathbb{Q}} G_\rho$. || Infatti, nell'ipotesi contraria, dovrebbe esistere un $x_0 \in \mathbb{R}$ estraneo a tutti i G_ρ (assurdo per la proprietà I).
(IV) Ogni G_ρ è congruo a G_0 per traslazione, e quindi (per l'assioma 3 di Lebesgue) dev'essere:

$$\mu(G_\rho) = \mu(G_0) \quad (\forall \rho \in \mathbb{Q}).$$

(V) È

$$\mu(\bigcup_{\rho \in \mathbb{Q}} G_\rho) = \infty \quad \text{(per la proprietà III)}.$$

(VI) È

$$\mu\left(\bigcup_{\rho \in \mathbb{Q}} G_\rho\right) = \sum_{\rho \in \mathbb{Q}} \mu(G_\rho) \quad \text{(per l'assioma 4 di Lebesgue}^{55}\text{)}$$

Per ottenere tale successione in $\mathbb{Q}$, considero G_0, $G_{\frac{1}{2}}$, $G_{\frac{1}{3}}$, $G_{\frac{1}{4}}$, ... che sono tutti contenuti in $[0, 1]$. ||

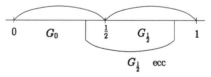

La VI si potrà dunque *specificare così*:

$$\mu\left(\bigcup_{i=2}^{\infty} G_{\frac{1}{i}}\right) = \sum_{i=2}^{\infty} \mu(G_{\frac{1}{i}}) = \lim_{n \to \infty} n \cdot \mu(G_0).$$

Posto

$$m = \mu\left(\bigcup_{i=2}^{\infty} G_{\frac{1}{i}}\right),$$

---

[55][Viola deleted here: *"almeno nel caso restrittivo che ρ assuma una successione di valori in $\mathbb{Q}$."* He added in margin: *"Va bene, perchè $\mathbb{Q}$ è per se stesso numerabile!"*] (csr).

dovrà essere $m \leq 1$, dunque $\mu(G_0) = 0$ ed anche $\mu(G_\rho) = 0$ ($\forall \rho \in \mathbb{Q}$), e ciò è assurdo per la proprietà VI. c.d.d.

P. S. Tabellazione dei numeri $P_\alpha + \rho$, alla p. 24 seguente.

Il termine generale $P_\alpha + \rho$ descrive il termine in testa a sinistra, se resta fermo α e varia ρ; descrive il termine in testa in alto, se resta fermo ρ e varia α.

Come si vede, si potrebbe sostituire alla successione $G_{\frac{1}{2}}, G_{\frac{1}{3}}, G_{\frac{1}{4}}, \ldots$, una qualunque altra $G_{\rho_1}, G_{\rho_2}, G_{\rho_3}, \ldots$ tale che tutti i numeri razionali $\rho_1, \rho_2, \rho_3, \ldots$ cadano in $[0, \frac{1}{2}]$. ||

Tabellazione, a doppia entrata, di tutti i numeri reali $P_\alpha + \rho$:

Descrive→ ↓	α'	α''	α'''	...	$\{\alpha\}$ (non numerab.)
G_{ρ_1}	$P_{\alpha'} + \rho_1$	$P_{\alpha''} + \rho_1$	$P_{\alpha'''} + \rho_1$	...	$\subset [\rho, \frac{1}{2} + \rho_1]$
G_{ρ_2}	$P_{\alpha'} + \rho_2$	$P_{\alpha''} + \rho_2$	$P_{\alpha'''} + \rho_2$	...	$\subset [\rho, \frac{1}{2} + \rho_2]$
G_{ρ_3}	$P_{\alpha'} + \rho_3$	$P_{\alpha''} + \rho_3$	$P_{\alpha'''} + \rho_3$	...	$\subset [\rho, \frac{1}{2} + \rho_3]$
...	...	...	...	...	
$\{G_\rho\}$ (numerab.)					

Ciascun $\alpha \in H$ è congruo a $\mathbb{Q}$ per traslazione. Esso è ovunque denso in $\mathbb{R}$. ||

5. *"Una proprietà delle funzioni misurabili."*[56] (Rend. Ist. Lombardo, s. II, 38 (1905), p. 599).

Dimostrazione di un teorema fondamentale che rivela un profondo legame fra le funzioni misurabili e le funzioni aventi classe di Baire:
"Una funzione di variabile reale, definita e misurabile in un intervallo $[a, b]$, si può spezzare nella somma di una funzione di classe ≤ 2, e di una funzione nulla dovunque all'infuori di un insieme di punti di misura nulla."

Nel corso della dimostrazione si ha occasione di dimostrare il seguente

Lemma I.[57] *"Se $f(x)$ è una funzione finita e misurabile in $[a, b]$ di lunghezza l, esiste per ogni numero $\epsilon > 0$, piccolo a piacere, un insieme $C[a, b]$, chiuso e di misura $> l - \epsilon$, tale che i valori di $f(x)$ nei punti di esso, formano una funzione continua."*

In nota, a fondo p. 601, Vitali cita Borel e Lebesgue, ma sembra che la prima dimostrazione del teorema sia sua. Tonelli (v. il Calc. d. V. I p. 131) capovolge il teorema assu||mendo la proprietà espressa dal teorema come definizione delle funzioni quasi continue.[58]

---

[56]v. Pascal I_3 p. 1065. In Pascal loc. cit. riferimenti vari. Cfr. anche H. Hahn p. 565; C. Carathéodory pp. 403–413, in particolare p. 410, e p. 279. [*Refs.*: [89], [25,45, p. 1065], [21, p. 565], [3, p. 279, 403–413]. (*csr*)].

[57]Per le funzioni di Baire e, in particolare, per la condizione necessaria e sufficiente affinchè una funzione sia di 1a classe di Baire, v. Pascal I_3 p. 1061. [*Ref.*: [25,45], p. 1061]. (*csr*)].

[58][*Ref.*: [59, vol.1, p. 131]. (*csr*)].

Nelle lezioni di Analisi Superiore (Anno 1927–1928, v. Quaderno n. 1, p. 15) di Tonelli, la definizione dell'integrale di Lebesgue è data per le funzioni *quasi-continue*, ivi intese come quelle che sono continue a prescindere da un *plurintervallo aperto* Δ avente *lunghezza* (cioè *mis* Δ) piccola a piacere.[59] Vitali passa poi al

Lemma II. *"Se $f(x)$ è una funzione misurabile in $[a, b]$ di lunghezza $l(= b - a)$, esiste una successione d'insiemi chiusi*

$$\Gamma_1, \Gamma_2, \Gamma_3, \ldots,$$

ciascuno contenuto nel successivo, le misure dei quali formano una successione avente per limite l, e tali che in ognuno di essi la $f(x)$ è continua."

Basandosi su questo II Lemma, Vitali arriva rapidamente alla dimostrazione del teorema enunciato in principio. ||

6. *"Sull'integrazione per serie."*[60] (Rendic. Circ. Mat. Palermo, 23, 1907, p. 137.)

Lebesgue (ediz. 1928), nota a p. 131, dice: *"C'est surtout pour le calcul effectif des intégrales de fonctions données par des développements en série qu'il importe de connaitre des cas d'intégration terme à terme. M. Vitali a écrit sur ce sujet un très important Mémoire que je ne puis ici que signaler."*

Cfr. Lezioni di Tonelli: Quaderno n. 1, p. 156; Quaderno n. 2, p. 131.

Nota preventiva di Vitali: *"Sopra l'integrazione di serie di funzioni di una variabile reale"*. (Bollettino Accad. Gioenia di Scienze Naturali a Catania, 86, maggio 1905.)[61]

(I) Generalizzazione del concetto di *assoluta continuità*, dimostrando che:

"Data una funzione $f(x)$ sommabile in un insieme E di misura finita,

$$\forall \epsilon, \exists \delta / \forall G \subseteq E \ con \ mis \ G < \delta,$$

risulta $| \int_G f(x)\, dx | < \epsilon$."

Def. *"Se $\mathcal{M} \equiv \{f(x)\}$ è un aggregato di funzioni sommabili in E, gli integrali $\int_G f(x)\,dx$ (al variare di $G \subseteq E$) si dicono equi-assolutamente continui in E, se*

$$\forall \epsilon, \exists \delta / \forall G \subseteq E \ con \ mis \ G < \delta, \ \forall \delta / \forall f(x) \in \mathcal{M} \ e \ \forall G \subseteq E \ con \ mis \ G < \delta,$$

risulta $| \int_G f(x)\, dx | < \epsilon$."

(II) *"Condizione necessaria e sufficiente affinché una serie*

$$\sum_{n=1}^{\infty} u_n(x)$$

di funzioni sommabili in E (insieme misurabile e di mis finita), sia integrabile || completamente per serie[62] *in E, è che la serie stessa*

---

[59]*Plurintervallo*: successione (finita o no) d'intervalli parziali di $[a, b]$, due a due non sovrapponentisi. Il plurintervallo è aperto, se ogni suo intervallo è considerato aperto. Dunque: $f(x)$ è quasi continua in $[a, b]$ (secondo le *"Lezioni"*) se, $\forall \epsilon > 0$, è possibile costruire un plurintervallo aperto $\Delta \subset [a, b]$, e di *mis* $\Delta < \epsilon$, a prescindere dal quale $f(x)$ è continua.

[60]Per vari riferimenti, v. Hobson [Refs.: [92], [22, vol. 1, p. 536; vol. 2, pp. 282, 289, 297, 309]. (*csr*)].

[61]Ivi afferma essersi ispirato al proprio lavoro: *"Sopra la serie di funzioni analitiche"*, Cap. 3, Annali di Matematica, s. 3, 10, 1904. [Refs.: [86,90]. (*csr*)].

[62]Cioè la serie sia tale che, $\forall G \subseteq E$, esistano e siano uguali

$$\sum_{n=1}^{\infty} u_n(x)$$

converga in E e che gli integrali delle sue somme parziali, cioè gli integrali

$$\int_G \sum_{n=1}^{p} u_n(x)\,dx \qquad (p = 1, 2, 3, \ldots)$$

siano equi-assolutamente continui in E."

(III) *"Se le funzioni $u_n(x)$ $(n = 1, 2, \ldots)$ sono sommabili in $[a, b]$, le condizioni del precedente teorema sono sufficienti per l' integrabilità per serie in senso ordinario[63] in $[a, b]$."*

N.B. Le condizioni ora dette *non sono* però *necessarie* all'integrabilità per serie in senso ordinario. ||

Esempio. Sia $\{x_n\}$ una successione di punti di (a, b), tali che[64]

$$x_1 < x_2 < x_3 < \cdots < x_n < x_{n+1} < \ldots \longrightarrow b.$$

Si pone:
$$y_n = \frac{x_n + x_{n+1}}{2}.$$

Inoltre:

$$\sigma_n(x) = \begin{cases} \dfrac{2k}{x_{n+1} - x_n} & \text{per } x_n < x < y_n, \\ -\dfrac{2k}{x_{n+1} - x_n} & \text{per } y_n < x < x_{n+1}, \\ 0 & \text{in tutti gli altri } x \in [a, b]. \end{cases}$$

$$\sum_{n=1}^{\infty} \int_G u_n(x)\,dx \quad \text{e} \quad \int_G \sum_{n=1}^{\infty} u_n(x)\,dx.$$

[63]Cioè la serie sia tale che, $\forall x \in (a, b]$, esistano e siano uguali

$$\sum_{n=1}^{\infty} \int_a^x u_n(x)\,dx \quad \text{e} \quad \int_a^x \sum_{n=1}^{\infty} u_n(x)\,dx.$$

[64]V. ivi n. 8. [*Refs.*: [92, p. 147–148; *Opere*, 247–248]. (*csr*)].

La serie
$$\sigma_1(x) + \sum_{n=1}^{\infty} \left[\sigma_{n+1}(x) - \sigma_n(x)\right] \tag{1}$$
ha per somma la costante 0 in tutto $[a,b]$. Per la generica somma parziale della serie, cioè
$$\sigma_1(x) + \sum_{n=1}^{p-1} \left[\sigma_{n+1}(x) - \sigma_n(x)\right] = \sigma_p(x) \quad (p > 1), \quad \|$$

si ha poi:
$$\int_a^x \sigma_p(t)\,dt = \begin{cases} 0 & \text{per } a \leq x \leq x_p \\ 2k\,\dfrac{x - x_p}{x_{p+1} - x_p} & \text{per } x_p \leq x \leq y_p \\ k - 2k\,\dfrac{x - y_p}{x_{p+1} - x_p} & \text{per } y_p \leq x \leq x_{p+1} \\ 0 & \text{per } x_{p+1} \leq x \leq b. \end{cases}$$

Tale è dunque il valore di
$$\int_a^x \sigma_1(t)\,dt + \sum_{n=1}^{p-1} \int_a^x \left[\sigma_{n+1}(t) - \sigma_n(t)\right] dt.$$

(v. il diagramma del tratto marcato nella figura appresso)

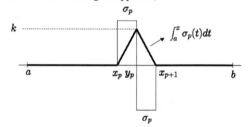

Segue:
$$\int_a^x \sigma_1(t)\,dt + \sum_{n=1}^{\infty} \int_a^x \left[\sigma_{n+1}(t) - \sigma_n(t)\right] dt = 0 \quad (\text{per ogni } x\,/\,a \leq x \leq b). \quad \|$$

Invece se si prende per G l'insieme unione di tutti gli intervalli $[x_n, y_n]$ ($n = 1, 2, \ldots$), risulta
$$\int_G \sigma_1(t)\,dt + \sum_{n=1}^{p-1} \int_G \left[\sigma_{n+1}(t) - \sigma_n(t)\right] dt = k,$$

e quindi anche
$$\int_G \sigma_1(t)\,dt + \sum_{n=1}^{\infty} \int_G \left[\sigma_{n+1}(t) - \sigma_n(t)\right] dt = k,$$

mentre invece:
$$\int_G \left\{\sigma_1(t) + \sum_{n=1}^{\infty} \left[\sigma_{n+1}(t) - \sigma_n(t)\right]\right\} dt = 0.$$

La serie (1) (v. p. 37) è dunque integrabile, in $[a,b]$, in senso ordinario, ma non è ivi completamente integrabile (e non lo è neppure nel sottoinsieme di $[a,b]$ indicato con G).

Per (II), naturalmente, gli integrali delle singole somme parziali della (1) non possono essere equi-assolutamente continui in $[a,b]$. Ed infatti, se ora si prefissa ad arbitrio un $\delta > 0$ e si prende poi p tanto grande che sia $y_p - x_p < \delta$, si avrà (assumendo per G_p l'intervallino $[x_p, y_p]$:

$$\int_G \sigma_p(t)\,dt = k \quad \text{(costante positiva)}. \quad \|$$

(IV) *"Se una serie*

$$\sum_{n=1}^{\infty} u_n(x)$$

è a termini tutti non negativi e sommabili in uno stesso insieme E, e se la serie numerica

$$\sum_{n=1}^{\infty} \int_E u_n(x)\,dx$$

converge, essa stessa converge in quasi tutto E ed è ivi integrabile completamente per serie."[65]

N.B. Se questo teorema si confronta con uno ben noto di B. Levi (Rendic. Ist. Lombardo, 39, 1906 p. 779, n. 4 e, più precisamente, p. 779[8-6], si trova una leggera generalizzazione nella precisazione:"integrabile *completamente per serie*."[66]

(V) *"Se la serie degli integrali dei valori assoluti dei termini di una serie converge, converge anche la serie stessa fuori che in un insieme di misura nulla ed è integrabile completamente per serie."*

Confronto critico fra i due concetti di equi-assoluta continuità (caso particolare di Giulio Ascoli) e di equi-continuità (di Giulio Ascoli).[67] Rileva che, da un lavoro di Arzelà (1899–1900),|| si deduce che, se le somme parziali di una serie convergente di funzioni sono equi-continue, la serie stessa è uniformemente convergente. Deduce il teorema

(VI) *"Affinché una serie $\sum_{n=1}^{\infty} u_n(x)$ sia integrabile completamente per serie, è necessaria l'equi-convergenza della serie degli integrali."* ||

7. *"Sui gruppi di punti e sulle funzioni di variabili reali."*[68] (Atti Accad. Scienze Torino, 43, 1907–1908, p. 229).

Nota preventiva: *"Sulle funzioni ad integrale nullo"* (Rendic. Palermo 1905, vol. 20).[69]

---

[65] V. Vitali p. 151[8-6], Hobson vol. II p. 536. [*Refs.*: [92, p. 151; *Opere*, p. 251], [22, vol. 2, p. 526]. (*csr*)].

[66] [*Ref.*: [31, p. 779]. (*csr*)].

[67] Mem. Acc. Bologna, s. V, vol. 8, pp. 3–58. Cfr. Tonelli, Calc. d. Variaz. I pp. 76–77. [*Ref.*: [59, vol. 1, pp. 76–77]. (*csr*)].

[68] v. Pascal I_3 p. 1043. Ivi Erich Kamke espone per $\mathfrak{R}$, con grande semplicità, quanto Carathéodory (*Reelle Funktionen* 1918 pp. 299–307) ha fatto per uno spazio euclideo $\mathfrak{R}_n$ a un numero qualunque n di dimensioni. (La questione di cui Carathéodory parla ivi, in nota a p. 304, è stata risolta?). Per i teoremi di Pincherle-Borel, di Borel-Lebesgue, e di Lindelöf, v. in Carathéodory, rispettivamente alle pp. 42, 45, 46. Pascal I_3. [*Refs.*: [93], [25, 45, pp. 1043–1044], [3, pp. 42, 45, 46, 299–307]. (*csr*)].

[69] [*Ref.*: [92]. (*csr*)].

Def. *Di un aggregato $\mathcal{M}$ di intervalli chiusi $I', I'', I''', \ldots$ dell'asse reale, è chiamato "corpo" $\gamma(\mathcal{M})$ l'insieme $\bigcup_{I \in \mathcal{M}} I$.*

(I) Lemma. *"Se $mis\, \gamma(\mathcal{M}) = \mu < +\infty$, $\forall \epsilon < \frac{\mu}{3}$, esistono in $\mathcal{M}$ un numero finito di intervalli $I_1, I_2, \ldots, I_N$, due a due disgiunti, tali che*

$$\sum_{n=1}^{N} mis\, I_n > \frac{\mu}{3} - \epsilon."$$

Def. *$\forall \epsilon$, sia $\mathcal{M}_\epsilon$ l'aggregato di tutti gli intervalli di $\mathcal{M}$, le cui ampiezze sono $< \epsilon$.[70] È chiamato "nucleo" $\nu(\mathcal{M})$ l'insieme $\bigcap_{\epsilon > 0} \gamma(\mathcal{M}_\epsilon)$.* ||

(II) *"Se $mis\, \nu(\mathcal{M}) = \mu < +\infty$, esistono in $\mathcal{M}$ un numero finito o numerabilmente infinito di intervalli $I_1, I_2, \ldots, I_n, \ldots$, due a due disgiunti, tali che $\sum_{n=1}^{\infty} mis\, I_n \geq m$."*

Alcune applicazioni di questo teorema:

(III) *"I punti in cui un numero derivato di una funzione continua e a variazione limitata non è finito, formano un insieme di misura nulla."*

(IV) *"Una funzione $f(x)$ sommabile in un insieme E e tale che*

$$\int_G f(x)\, dx = 0 \quad (\forall G \subseteq E)$$

è nulla in quasi tutto E."

Il teorema (II) è stato ripreso da C. Carathéodory (1918), trasformato effettivamente in un *teorema di copertura*, semplificato e generalizzato a spazi $\mathbb{R}_n$, con n qualunque.

Sull'asse $\mathbb{R}$, l'enunciato più semplice è dato da E. Kamke (*"Das Lebesguesche Integral"* 1925, pp. 68–70, poi in Pascal I_3 p. 1043), nella forma seguente: || Kamke (v. Pascal): ... *"von grosser Bedeutung"* Carathéodory (loc. cit. p. 299) *"Der Vitalische Überdeckungsatz verschärft in eigentümlicher Weise die Theorie des Inhalts einer Punktmenge."*[71]

Teorema di copertura di Vitali

"Sia E un insieme limitato, tale che $mis_e\, E > 0$. $\forall P \in E$ esista una successione infinita di intervalli $I_1(P), I_2(P), \ldots$, tutti contenenti P e tali che[72]

$$\lim_{n \to \infty} mis\, I_n(P) = 0.$$

Allora, $\forall \epsilon > 1$, si possono estrarre dall'aggregato di infiniti intervalli $\{I(P)\}$ (al variare di $P \in E$), un numero *finito* di intervalli $I^{(1)}, I^{(2)}, \ldots, I^{(N)}$, in modo che:

---

[70] Dunque $\mathcal{M}_\epsilon \subseteq \mathcal{M}$, $\forall \epsilon$. Inoltre, se $\epsilon' < \epsilon$, è $\mathcal{M}_{\epsilon'} \subseteq \mathcal{M}_\epsilon$.

[71] [Refs.: [25, 45, p. 1043], [3, p. 299]. (csr)].

[72] Carathéodory sostituisce questa condizione con altra (e il risultato è da lui chiamato *"merkwürdig"*). [Ref.: [3, p. 300]. (csr)].

$$mis_e\left\{E - E \cap \bigcup_{n=1}^{N} I^n\right\} < \epsilon \, mis_e \, E,$$

$$mis_e\left\{E \cap \bigcup_{n=1}^{N} I^n\right\} < (1-\epsilon) mis_e \, E." \quad \|$$

8. *"Sulla definizione di integrale delle funzioni di una variabile."* (Annali di Matematica, s. 4, 2, 1924–1925, p. 111)[73]

Richiama la definizione dell'integrale data da B. Levi nel 1923 (modificandola leggermente nella forma):[74]

"$f(x)$: *funzione limitata e > 0 in $[a, b]$. Si costruiscano due successioni (eventualmente finite) $(n = 1, 2, \ldots)$: $\{\delta_n\}$ di intervalli chiusi contenuti in $[a, b]$, e $\{h_n\}$ di numeri positivi, tali che $\forall x \in [a, b]$, esista un n tale che $x \in \delta_n$ e che $f(x) \leq h_n$. Corrispondentemente si formi la somma*

$$(1) \quad \sum_{n=1}^{\infty} (mis \, \delta_n) h_n.$$

Al variare, in tutti i modi possibili, delle due successioni $\{\delta_n\}$, $\{h_n\}$, la (1) descrive un insieme il cui estremo inferiore, per definizione, è l'integrale superiore

$$\overline{\int_a^b} f(x) \, dx." \quad \|$$

Se $f(x)$ è limitata (ma non più necessariamente > 0), si prenda una costante m tale che sia $f(x) + m > 0$ (in tutto $[a, b]$). Levi dimostra che l'espressione:

$$\overline{\int_a^b} [f(x) + m] \, dx - m(b - a),$$

non dipende da m. Tale espressione, per definizione, è l'integrale superiore $\overline{\int_a^b} f(x) \, dx$.

Infine, l'integrale inferiore è, per definizione,

$$\underline{\int_a^b} f(x) \, dx = -\overline{\int_a^b} [-f(x) \, dx].$$

Levi aveva dimostrato che, se $f(x)$ è misurabile in $[a, b]$, questa definizione coincide con quella di Lebesgue.

---

[73] [*Ref.*: [94]. (*csr*)].

[74] B. Levi, *Sulla definizione dell'integrale* (Annali di Matematica, s. 4, 1, 1923–1924, p. 58). [Cf. [33, pp. 61–66]. (*csr*)].

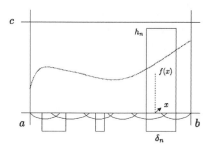

Vitali dimostra che effettivamente la presente definizione coincide con quella di Lebesgue.

References

1. Borgato, M.T., and L. Pepe (eds). 1984. *Lettere a Giuseppe Vitali*. In ed. G. Vitali *Opere*, 403–524.
2. Bortolotti, E. 1932–1933. *Giuseppe Vitali. Annuario dell' Università di Bologna*, 523–528.
3. Carathéodory, C. 1918. *Vorlesungen über reelle Funktionen*. Leipzig: Teubner.
4. Cassinet, J., and M. Guillemot. 1983. *L'axiome du choix dans les mathématiques de Cauchy (1821) à Gödel (1940)*. Thèse de mathématiques, Toulouse, Université Paul-Sabatier.
5. Cinquini, S. 1950. Leonida Tonelli. *Annali della Scuola Normale Superiore di Pisa (2)* 15: 1–37.
6. Coen, S. (ed.). 1991. *Geometry and complex variables, Proceedings of an international meeting on the occasion of the IX centennial of the University of Bologna*. New York: M. Dekker.
7. Coen, S. 1991. *Geometry and complex variables in the work of Beppo Levi*. In ed. S. Coen, 111–139.
8. Coen, S. 1999. *Beppo Levi: una biografia*. In Levi, *Opere*, vol. 1, pp. XIII–LIV.
9. Cohen, P. 1963. *The independence of the axiom of choice*. Palo Alto: Stanford University Press.
10. Cohen, P. 1963. The independence of the continuum hypothesis I. *Proceedings of the National Academy of Sciences* 50: 1143–1148.
11. Cohen, P. 1964. The independence of the continuum hypothesis II. *Proceedings of the National Academy of Sciences* 51: 105–110.
12. Denjoy, A. 1915. Mémoire sur les nombres dérivés des fonctions continues. *Journal de Mathématiques Pures et Appliquées* 1(7): 105–240.
13. Fraenkel, A. 1935. Sur l'Axiome du Choix. *L'Enseignement mathématique* 34: 32–51.
14. Fubini, G. 1949. Il teorema di riduzione per gli integrali doppi. *Rendiconti del Seminario Matematico dell'Università e del Politecnico di Torino* 9: 125–133; *Opere scelte* 3: 399–408.
15. Fubini, G. 1957–1962. *Opere scelte*. Roma: Cremonese, vol. 1, 1957, vol. 2, 1958, vol. 3, 1962.
16. Giacardi, L., and C.S. Roero (eds.). 2006. *Matematica, Arte e Tecnica nella Storia, In memoria di Tullio Viola*. Torino: Kim Williams Books.
17. Gödel, K. 1938. The consistency of the axiom of choice and of the generalized continuum hypothesis. *Proceedings of the National Academy of Sciences* 24: 556–557.
18. Guerraggio, A., M. Mattaliano, and P. Nastasi. 2009. *Carlo Miranda, Il dialogo epistolare con il suo maestro Mauro Picone*. Pristem/Storia, 25–26.
19. Guillemot, M. 2006. *Tullio Viola: assioma della scelta e principio di approssimazione*. In eds. L. Giacardi, and C.S. Roero, 81–117.
20. Hadamard, J. 1905. Cinq lettres sur la théorie des ensembles. *Bullettin de la S.M.F.* 33: 261–273.

21. Hahn, H. 1921. *Theorie der reellen Funktionen*, vol. 1. Berlin: Springer.
22. Hobson, E.W. 1907. *The theory of functions of a real variable and the theory of Fourier's series.* Cambridge: University Press; 2nd ed. New York, vol. 1, 1927, vol. 2, 1926.
23. Jordan, C. 1881. Sur la série de Fourier. *Comptes Rendus de l'Académie des Sciences Paris* 92: 228–230.
24. Jordan, C. 1893. *Cours d'Analyse de l'Ecole Polytechnique*, 2nd ed., vol. 1. Paris: Gauthier-Villars.
25. Kamke, E. 1929. *Neuere Theorie der reeelle Funktionen*. In eds. E. Pascal, and E. Salkowski, 1025–1095.
26. Lebesgue, H. 1901. Sur une généralisation de l'intégrale définie. *Comptes Rendus de l'Académie des Sciences Paris* 132: 1025–1027.
27. Lebesgue, H. 1902. Intégrale, Longueur, Aire. Thèse, Paris; *Annali di Matematica, (3)* 7: 231–359.
28. Lebesgue, H. 1904. *Leçons sur l'intégration et la recherche des fonctions primitives*. Paris: Gauthier-Villars, 2nd ed. 1928.
29. Lebesgue, H. 1921. Sur les correspondances entre les points de deux espaces. *Fundamenta Mathematicae* 2: 256–285.
30. Levi, B. 1902. Intorno alla teoria degli aggregati. *Rendiconti dell'Istituto Lombardo di Scienze e Lettere* 35: 863–868; *Opere* 1999, 1: 177–182.
31. Levi, B. 1906. Sopra l'integrazione delle serie. *Rendiconti dell'Istituto Lombardo di Scienze e Lettere* 39: 775–780; *Opere* 1999, 1: 177–182.
32. Levi, B. 1918. *Riflessioni sopra alcuni principii della teoria degli aggregati e delle funzioni*. In *Scritti matematici offerti ad Enrico d'Ovidio*, 305–324. Torino: Bocca; *Opere* 1999, vol. 2, 791–810.
33. Levi, B. 1923. Sulla definizione di integrale. *Annali di Matematica, (4)* 1: 57–82; *Opere* 1999, vol. 2, 879–904.
34. Levi, B. 1933. Nota di logica matematica. *Rendiconti dell'Istituto Lombardo di Scienze e Lettere, (2)* 66: 239–252.
35. Levi, B. 1933. Review P. Montel: Leçons sur les fonctions univalentes ou multivalentes. *Periodico di Matematiche, (4)* 13(5): 315–316.
36. Levi, B., T. Viola. 1933. Intorno ad un ragionamento fondamentale nella teoria delle famiglie normali di funzioni. *Bollettino U.M.I.* 12(4): 197–203.
37. Levi, B. 1934. Considerazioni sulle esigenze logiche della nozione del reale e sul principio delle definite scelte. In *Atti VIII Congr. Naz. di Filosofia*, Roma, 55–56.
38. Levi, B. 1934. La nozione di "dominio deduttivo" e la sua importanza in taluni argomenti relativi ai fondamenti dell'analisi. *Fundamenta Mathematicae* 23: 63–74.
39. Levi, B. 1999. *Opere 1897–1926*, 2 vols. Bologna: Cremonese.
40. Lolli, G. 1999. *L'opera logica di Beppo Levi*, In Levi, *Opere*, vol. 1, LXVII–LXXVI.
41. Montaldo, O., and L. Grugnetti. 1984. *La storia delle Matematiche in Italia Atti del Convegno, Cagliari 29-30.9-1.10.1982*. Bologna: Monograf.
42. Montel, P., and A. Rosenthal (eds.). 1923–1927. II C 9b *Integration und Differentiation, Encyklopädie der Mathematische Wissenschaften*, II.3, 1031–1063. Leipzig: Teubner.
43. Montel, P. 1927. *Leçons sur les familles normales de fonctions analytiques et leurs applications*. Paris: Gauthier-Villars.
44. Moore, G. 1982. *Zermelo's axiom of choice. It's origin, development and influence*. Berlin: Springer.
45. Pascal, E., and E. Salkowski (eds.). 1929. *Repertorium der höheren Mathematik*. 2nd ed., vol. 3. Leipzig: Teubner.
46. Peano, G. 1890. Démonstration de l'intégrabilité des équations différentielles ordinaires *Mathematische Annalen* 37: 182–228.
47. Pepe, L. 1983. Giuseppe Vitali e la didattica della matematica. *Archimede* 35: 163–176.
48. Picone, M., T. Viola. 1952. *Lezioni sulla teoria moderna dell'integrazione*. Torino: Einaudi.
49. Proceedings Lincei. 1986. *Convegno Celebrativo del centenario della nascita di Mauro Picone e di Leonida Tonelli (Roma 6-9 maggio 1985)*. Roma: Accad. Naz. Lincei.

50. Roero, C.S. 1999. *Tullio Viola*. In ed. C.S. Roero, *La Facoltà di Scienze Matematiche Fisiche Naturali di Torino 1848–1998*, vol. 2, *I docenti*. Torino: Deput. Sub. Storia Patria, 607–612.
51. Roero, C. S. 2006. *Tullio Viola ricercatore appassionato e maestro generoso*. In ed. L. Giacardi, and C.S. Roero, 31–42.
52. Skof, F. 2006. *Tullio Viola e l'Analisi matematica*. In ed. L. Giacardi, and C.S. Roero, 71–79.
53. Sierpinski, W. 1916. Sur le rôle de l'axiome de M. Zermelo dans l'analyse moderne. *Comptes Rendus de l'Académie des Sciences Paris* 163: 688–691.
54. Sierpinski, W. 1918. L'axiome de M. Zermelo et son rôle dans la théorie des ensembles et l'analyse. *Bulletin de l'Académie des Sciences de Cracovie Cl. Sci. Math. A*, 99–152; *Oeuvres choisies*. Warszawa, Éd. Scient. de Pologne, vol. 2, 1975, pp. 208–285.
55. Sierpinski, W. 1928. *Leçons sur les nombres transfinis*. Paris: Gauthier-Villars.
56. Solovay, R. 1970. A model of set theory in which every set of reals is Lebesgue mesurable. *Annals of Mathematics* 92: 1–56.
57. Terracini, A. 1963. Commemorazione del corrispondente Beppo Levi. *Rendiconti dell'Accademia Nazionale dei Lincei* 34: 590–606.
58. Tonelli, L. 1913. Sul valore di un certo ragionamento. *Atti della Reale Accademia delle scienze di Torino* 49: 4–14.
59. Tonelli, L. 1921–1923. *Fondamenti di Calcolo delle Variazioni*. Bologna: Zanichelli, vol. 1, 1921, vol. 2, 1923.
60. Tonelli, L. 1928. Il contributo italiano alla teoria delle funzioni di variabili reali. In *Atti Congr. Int. Matematici*, Bologna, 3–10 settembre 1928, vol. 1, 247–254, Bologna: Zanichelli.
61. Tonelli, L. 1960–1963. *Opere scelte*, vol. 4. Bologna: Cremonese.
62. Tonolo, A. 1932. Commemorazione di Giuseppe Vitali. *Rendiconti del Seminario Matematico della Università di Padova* 3: 67–81.
63. Tonolo, A. 1959. Giuseppe Vitali. *Archimede* 11: 105–110.
64. Vailati, G. 1898. Review *C. Laisant. La Mathématique: philosophie, enseignement*, Paris. Il Nuovo Risorgimento, 9, 8, 1899.
65. Vailati, G. 1911. *Scritti di G. Vailati (1863–1909)*. Firenze: Successori di Seeber.
66. Vaz Ferreira, A. 1991. *Giuseppe Vitali and the Mathematical Research at Bologna*. In ed. S. Coen, 375–395.
67. Viola, T. 1930. Studio intorno alle funzioni continue da una parte ed alla derivazione unilaterale. Tesi di laurea in Matematica pura, Università di Bologna, Litografia.
68. Viola, T. 1931. Proprietà notevoli di funzioni continue da una parte. *Rendiconti dell'Accademia Nazionale dei Lincei (6)* 13: 731–733.
69. Viola, T. 1931. Sulla derivata destra di una funzione continua verso destra e derivabile verso destra. *Rendiconti dell'Accademia Nazionale dei Lincei (6)* 13: 828–831.
70. Viola, T. 1931. Ancora sulla derivata destra di una funzione continua verso destra e derivabile verso destra. *Rendiconti dell'Accademia Nazionale dei Lincei (6)* 13: 918–920.
71. Viola, T. 1931. Riflessioni intorno ad alcune applicazioni del postulato della scelta di E. Zermelo e del principio di approssimazione di B. Levi nella teoria degli aggregati. *Bollettino U.M.I.* 10(5): 287–294.
72. Viola, T. 1932. Sul principio di approssimazione di B. Levi nella teoria della misura degli aggregati e in quella dell'integrale di Lebesgue. *Bollettino U.M.I.* 11(2): 74–78.
73. Viola, T. 1932. Sui punti irregolari di una famiglia non normale di funzioni olomorfe. Verhandl. Int. Math. Kongr. Zürich 1932, vol. 2, Zürich, Füssli, 1932, 36–37.
74. Viola, T. 1932. Una proprietà degli aggregati perfetti di punti, utile nello studio delle famiglie non normali di funzioni olomorfe. *Bollettino U.M.I.* 11(5): 287–290.
75. Viola, T. 1933. *Étude sur la détermination d'une fonction discontinue par sa dérivée unilatérale*, 1ère Thèse Faculté des Sciences de Paris. Paris: Gauthier Villars, 1–55; *Annales scientifique de l'École Normale Supérieure, (3)* 50, 1933, 71–125.
76. Viola, T. 1933. *Sur l'accumulation des valeurs des fonctions analytiques qui forment une suite uniformément convergente*. 2ème Thèse, Faculté des Sciences de Paris. Paris: Gauthier Villars, 1–52; *Journal de Mathématiques pures et appliquées, (9)* 12(2), 1933, 173–204.
77. Viola, T. 1933. Review P. Montel: Leçons sur les fonctions univalentes ou multivalentes. *Bollettino U.M.I.* 12(4): 261–265.

78. Viola, T. 1934. Ricerche assiomatiche sulle teorie delle funzioni d'insieme e dell'integrale di Lebesgue. *Fundamenta Mathematicae* 23: 75–101.
79. Viola, T. 1951. Sulla formula di integrazione per parti nell'integrazione doppia secondo Stieltjes. *Giornale di Matematiche (Battaglini), (4)* 80: 142–158.
80. Viola, T. 1961. Beppo Levi. *Bollettino U.M.I., (3)* 16: 513–516.
81. Viola, T. 1971–1976. *Lezioni di Analisi Matematica.* Torino: Levrotto-Bella, vol. 1, 1971, vol. 2, 1976.
82. Viola, T. 1982. Il contributo di Guido Fubini nell'approfondimento del concetto di integrale. *Atti del Convegno matematico in celebrazione di G. Fubini e F. Severi*, Torino: Accademia delle Scienze, 45–60.
83. Viola, T. 1984. *Ricordo di Giuseppe Vitali, a 50 anni dalla Sua scomparsa.* In eds. O. Montaldo, and L. Grugnetti, 534–544.
84. Vitali, G. 1904. Sui gruppi di punti. *Rendiconti del Circolo Matematico di Palermo* 18: 116–126; *Opere*, 139–149.
85. Vitali, G. 1904. Sull'integrabilità delle funzioni. *Rendiconti dell'Istituto Lombardo di Scienze e Lettere*, (2) 37: 69–73; *Opere*, 133–137.
86. Vitali, G. 1904. Sopra le serie di funzioni analitiche. *Annali di Matematica, (3)* 10: 65–82, *Opere*, 129–131.
87. Vitali, G. 1905. Sulle funzioni integrali. *Atti della Reale Accademia delle scienze di Torino* 40, 1021–1034; *Opere*, 205–220.
88. Vitali, G. 1905. *Sul problema della misura dei gruppi di punti di una retta.* Bologna: Tip. Gamberini, Parmeggiani; *Opere*, 231–235.
89. Vitali, G. 1905. Una proprietà delle funzioni misurabili. *Rendiconti del Istituto Lombardo, (2)* 38: 599–603; *Opere*, 183–187.
90. Vitali, G. 1905. Sopra l'integrazione di serie di funzioni di una variabile reale. *Bollettino dell'Accademia Gioenia di Scienze Naturali in Catania* 86: 3–9; *Opere*, 199–204.
91. Vitali, G. 1905. Sulle funzioni ad integrale nullo. *Rendiconti del Circolo Matematico di Palermo* 20: 136–141; *Opere*, 193–198.
92. Vitali, G. 1907. Sull'integrazione per serie. *Rendiconti del Circolo Matematico di Palermo* 23: 137–155; *Opere*, 237–255.
93. Vitali, G. 1907–1908. Sui gruppi di punti e sulle funzioni di variabili reali. *Atti della Reale Accademia delle scienze di Torino* 43: 229–246; *Opere*, 257–276.
94. Vitali, G. 1924–1925. Sulla definizione di integrale delle funzioni di una variabile. *Annali di Matematica (4)* 2: 111–121; *Opere*, 335–345.
95. Vitali, G. 1933. Del ragionare. *Bollettino U.M.I.* 12(2): 1–6; *Opere*, 397–402.
96. Vitali, G., G. Sansone. 1935. *Moderna teoria delle funzioni di variabile reale.* Bologna: Zanichelli, 2nd ed. 1943.
97. Vitali, G. 1984. *Giuseppe Vitali Opere sull'analisi reale e complessa. Carteggio.* Bologna: Cremonese.
98. Zermelo, E. 1904. Beweis, dass jede Menge wohlgeordnet werden kann. *Mathematische Annalen* 59: 514–516.
99. Zermelo, E. 1908. Untersuchungen über die Grundlagen der Mengenlehre I. *Mathematische Annalen* 65: 261–281; *Collected Works*, vol. 1, 160–229. Berlin: Springer.
100. Zoretti, L., and A. Rosenthal. 1923-1927. II C 9a *Die Punktmengen*, Encyklopädie der Mathematische Wissenschaften, II.3, 856–1030. Leipzig: Teubner.

Developement of the Theory of Lie Groups in Bologna (1884–1900)

Enrico Rogora

Abstract At the end of the nineteenth century some italian mathematicians got deeply interested in Lie's theory of transformations groups. The aim of this paper is to discuss an episode of this story which happened in Bologna. It is about the fruitful interactions between Tullio Levi-Civita, Salvatore Pincherle and Federigo Enriques, which finally led to the classification of infinite dimensional Lie groups acting on three dimensional space by Ugo Amaldi.

1 Introduction

One of the neglected aspects of the history of italian mathematics is the analysis of reactions and contributions to the theory of transformation groups, given by italian mathematicians, in particular to what is called today the theory of infinite dimensional pseudo groups of transformations or simply, in Lie's terminology, the theory of infinite transformation groups. The two main italian contributors to this theory, Paolo Medolaghi and Ugo Amaldi, are completely forgotten today. One of the reasons for this oblivion is that Paolo Medolaghi, after having published nine papers on infinite groups between 1887 and 1889, quit mathematics and become executive and then chief director of the italian social security department. As for Amaldi, he was a very shy person and he never fought to promote his works as they would have deserved.

The aim of this paper is to discuss an episode of this story which happened in Bologna. It is about the fruitful interactions between Tullio Levi-Civita, Salvatore Pincherle and Federigo Enriques during the few months that Levi-Civita spent in Bologna in 1894. We shall discuss the consequences of these interactions for Amaldi's research program on the classifications of transformation groups acting

---

E. Rogora (✉)
Sapienza, Università di Roma, Dipartimento di Matematica, p.le A. Moro 5, 00185 Roma, Italy
e-mail: rogora@mat.uniroma1.it

on three dimensional space and for Federigo Enriques' thoughts about the role of group theory in the foundations of geometry.

2 Levi-Civita, Pincherle and Amaldi

Between 1888 and 1893, Lie published the three volumes treatise *Theorie der Transformations gruppen* with Friedrich Engel [33–35]. This work spread Lie's ideas on transformation groups all over Europe and were quite popular in Italy, thanks to the affinity of italian heritage with Lie's conception of mathematics. An important effort for spreading Lie's ideas in Italy was made by Corrado Segre who had the idea to ask one of his students, Gino Fano to translate Klein's Erlangen Program [24] in order to reprint it in an italian Journal. Fano's translation [19] appeared in print in 1890. We read in Klein's preface that

> I comply with great pleasure with the proposal of Mr. Segre to publish in the "Annali" a translation of my 1872 Program, since the first volume of Lie's "Theorie der Transformationsgruppen" (Leipzig 1888) could arrange that the interest of geometers turned more steadily towards such discussions.[1] (Quoted by [19].)

Among the readers of Lie and Engels' treatise, young Levi-Civita was one of the most enthusiastic. We read in Ugo Amaldi's introduction to Levi Civita collected works [6, p. XI] that

> In his dissertation *on absolute invariants*,[2] Levi-Civita is the first to connect Ricci's views with Lie's theory and methods, of which he already shows complete knowledge and mastery. He studies form and properties of differential systems which define [...] differential invariants [...] with respect to any continuous group of transformations, not only, as in absolute differential calculus, with respect to that of all transformations.[3] (Quoted by [6], p. XI.)

In 1894, after his graduation, Levi-Civita moved to Bologna for few months to study with Pincherle

> In that environment of cozy speculative fervor, he had daily opportunities to meet new ideas and boost his innate versatility. He was attracted by Pincherle's researches on linear

---

[1] Alla proposta del sig. Segre di pubblicare negli Annali una traduzione del mio Programma del 1872 ho accondisceso tanto più volentieri, in quanto che il primo volume testè comparso della "Theorie der Transformationsgruppen" di Lie (Leipzig 1888) potrebbe far sì che l'interesse dei geometri si rivolgesse maggiormente a siffatte discusisioni.

[2] Published in [25].

[3] [Nella] Dissertazione di Laurea *Sugli invarianti assoluti*, [...] il Levi-Civita [ricollega] per primo le vedute del Ricci alle teorie e ai metodi di Sophus Lie, di cui fin d'allora palesa un pieno e sicuro possesso, [e] studia la forma e le proprietà dei sistemi differenziali, atti a definire [...] gl'invarianti differenziali ed integrali, non già, come nel Calcolo differenziale assoluto, rispetto al gruppo di tutte le possibili trasformazioni, bensì di fronte a un qualsiasi altro gruppo continuo (finito o infinito).

functionals which are representable by curvilinear integrals on the complex plane. In this class of operators Levi-Civita found a new field of applications for group theoretical views and he was able to classify, under suitable restrictions, all infinite [dimensional] continuous groups of such operators [26, 28]. From these results he characterized all ordinary differential equations reducible to linear forms by a suitable change of the unknown function and a general uniform inversion principle for definite integrals.[4] (Quoted by [27].)

During the period he spent in Bologna, Levi-Civita had many mathematical discussions with Pincherle. He was fascinated by Picherle's ideas to deal with functions as points in a "function space" and devised the idea of considering groups of transformations of functions and extend Lie's theory to these infinite dimensional geometries. In the introduction to [27], p. 529 he wrote

> From the concept of transformation one is naturally led to that of functional operation by imagining that the manifold on which one acts is not a manifold of points but a manifold of functions. [...] This genial idea is due to Prof. Pincherle who developed many aspects of it in his works in increasing degree of generality, so that it seems justifiable to hope that in a near future these developments will lead us to a complete theory.
> The appearance of such a wide and important class of transformations naturally rises group theoretic problems. Sophus Lie, in a now classical work, tracked a straight and secure approach. But for this new class of transformations, i.e. for functional operations, Lie's theories do not adapt smoothly. Some analogies survive, others do not, many useful facts are no longer true, some new intriguing relations appear for the first time. At any rate, this new field appears promising and despite the narrow goals of my venture I express the desire that someone else will tackle the challenge more boldly.[5]

From this we see that Levi-Civita had in mind a clear program for extending Lie's theory to functionals, at least to those functionals which were studied by

---

[4]In quell'ambiente di raccolto fervore speculativo trovò quotidiane occasioni a ricambi di idee e nuovi incentivi all'innata sua verstilità. Attratto dalle ricerche del Pincherle sulle operazioni funzionali lineari, rappresentabili per mezzo di integrali curvilinei sul piano complesso, e ravvisato in quella classe d'operazioni un nuovo campo per l'applicazione di vedute gruppali, ne determinò, sotto restrizioni imposte dalla natura del problema, tutti i gruppi continui infiniti [26, 28]; e dai risultati così conseguiti dedusse da un lato la caratterizzazione delle equazioni differenziali ordinarie riducibili con un cambiamento di funzione incognita alla forma lineare e dall'altro un principio generale ed uniforme per l'inversione degli integrali definiti.

[5]Dal concetto di trasformazione si [è] naturalmente condotti a quello di operazione funzionale, immaginando che il corpo degli enti, su cui si opera, invece che un varietà di punti, sia una varietà di funzioni. Dissi anche come questa concezione geniale, dovuta al Prof. Pincherle, venne svolgendosi per opera sua sotto aspetti molteplici e via via più generali, talchè forse è lecito sperare che ne sorgerà tra breve una completa dottrina. Di fronte al delinearsi di così estesa e importante classe di trasformazioni sorgeva spontaneo il problema gruppale. Sophus Lie in un'opera [33-35] [...] ha tracciata una via dritta e sicura; ma per questa nuova classe di trasformazioni, cioè per le operazioni funzionali le teorie dell'illustre autore non si adattano senz'altro; alcune analogie si mantengono, altre vengono a mancare, molti fatti di vantaggiosa applicazione scompaiono, qualche relazione non priva di interesse si presenta invece per la prima volta. In ogni modo il campo d'indagine non sembra infecondo e, malgrado l'esigua misura della mia iniziativa, mi permetto di esprimere il desiderio che altri in breve vi porti ben più valido impulso.

Pincherle. After the publication of [26–28] however he never came back to this project, attracted by other fruitful researches.

Pincherle's reaction to Levi-Civita suggestions was not immediate. He never wrote a paper about this program, but he was surely seduced by Levi-Civita's ideas. In fact, after Levi-Civita left Bologna, he gave some courses on the theory of transformation groups, as reported, for example, in Ugo Amaldi's necrology [3] of Roberto Bonola. One of the students who attended these courses was Amaldi himself, father of the famous italian physicist Edoardo. Amaldi was intrigued by Lie's theory of transformation groups and by Pincherle's theory of functionals and decided to study with Pincherle in order to pursue the connections. For a biography of Amaldi, see [38].

Amaldi was a student in Bologna from 1895 to 1898, when he graduated with a dissertation on Laplace Transform [1], inspired by Pincherle's work on functionals. During his student years in Bologna, he collaborated with Pincherle on the edition of [40] and began his lifelong collaboration with Enriques on a famous book on elementary geometry [17], reprinted in several editions until 1960. He also became a very good friend of Levi-Civita, with whom he wrote another famous treatise "Lezioni di Meccanica razionale" in three volumes [29, 30]. His ability as a writer of mathematical manuals is well established but his mathematical work is now completely forgotten.

Pincherle and Amaldi decided not to apply Lie's theory to classify groups of functionals, as Levi-Civita did, but to apply Pincherle's theory of functionals to put on firmer foundations Lie's theory, in particular his theory of transformation groups which depend on infinite parameters which, in Lie's conception, were at least as important as the theory of groups depending on a finite number of parameters. In fact the classification of finite dimensional groups of transformations in two variables is at the basis of Lie classification of systems of ordinary differential equations with respect to symmetries, while a classification of finite and infinite dimensional transformation groups acting on any number of variables play an analogous role for the classification of systems of partial differential equations. More on this can be found in Sect. 3.

Unfortunately infinite dimensional Lie groups are much harder to deal with than finite ones. The research program that Pincherle suggested to Amaldi was to use his theory of functionals in order to deal properly with them. After having several successes in the classification of infinite dimensional transformation groups acting on three dimensional space however, Amaldi became aware of the inadequacy of Pincherle's methods.[6] His disillusion is quite evident in the following excerpt taken from his article "group" for the "Enciclopedia Italiana"

> The infinitesimal transformations of an infinite group depend on arbitrary functions or, if we like, they make up a linear set with *infinite dimensions*, whence the aid of

---

[6]Amaldi was also one of the few mathematicians who studied thoroughly Elie Cartan' work on infinite dimensional transformation groups and realized their importance and profound novelty. The interesting letters which Cartan wrote to Amaldi are collected and commented in [38].

hyperspatial procedures and interpretations, which makes the consideration of infinitesimal transformations in the case of finite groups so fruitful, fails. In this field, moreover, the few results and methods that we have at our disposal about functions with infinite variables and spaces with infinite dimensions have not yet had any real application.[7] (Quoted by [5].)

Nonetheless, the results obtained by Amaldi are quite relevant. He completed the classification of punctual and contact transformation groups acting in three dimensional space, both finite and infinite dimensional. However his achievements were completely ignored. Nowadays the classification problem considered by Amaldi has begun to be considered important again for the algebraic theory of systems of partial differential equations. Peter Olver, an expert of the field, says.

> The general classification problem [of transformation groups acting on n dimensional manifolds] was investigated in great detail by Lie, who, at one time, viewed it as the principal goal of his mathematical career. He succeeded in completely classifying all nonsingular (i.e., without fixed points) finite-dimensional Lie group actions in one and two dimensions, determining a complete list of (local) canonical forms for Lie algebra of vector fields on one and two-dimensional manifolds [36]. Further, in the third volume of his treatise on transformation groups, [35], Lie claimed to have completed the three-dimensional classification, and present a large fraction thereof. Unfortunately he states that, while he has completed the rest of the classification, the results are too long to be included in the book. As far as I can determine, despite the evident importance of this problem (and despite unsubstantiated rumors appearing sporadically in the literature), the complete classification in three dimensions was never published by Lie, and, to this day, remains unknown! (Quoted by [39]).

It would be probably worthwhile to look again at Amaldi's work even if we may expect that considering his work from a modern point of view will reveal many omissions and unproved claims. Part of the responsibility for the oblivion of Amaldi's work is due to Amaldi himself. At a certain point of his career he decided to accept a position at the "School of Architecture" in Rome, which, contrary to his expectations, never became a university institution. Hence he had been out of the mainstream of italian mathematical research for years. Amaldi also realized the loss of centrality of the classification problem for the mathematics of his period, due to the necessity to put Lie's theory of transformation groups on firmer basis and he always considered his work with excessive modesty, as is shown, for example, in the excerpt of the following letter to Levi-Civita, where he thanks his friend for his considerations on his most important work [4].

> I heartful thank you for your kind words about my huge memoir. By experience I perfectly know that I need to soften your judgments about my work because of your great benevolence toward myself. Even so, I have greatly appreciated your good words, since now that I am in

---

[7]Le trasformazioni infinitesime di un gruppo infinito dipendono da funzioni arbitrarie o, se si vuole, costituiscono un insieme lineare *a infinite dimensioni*, onde vien meno il sussidio di quelle interpretazioni e di quei procedimenti iperspaziali, che rendono così feconda la considerazione delle trasformazioni infinitesime nel caso dei gruppi finiti; nè, per ora, hanno trovato in questo campo effettiva applicazione i risultati e i metodi, ancora alquanto scarsi, che sino ad oggi si possiedono sulle funzioni di infinite variabili e sugli spazi a infinite dimensioni.

front of this huge volume, I feel concerned, especially with respect to the Society of LX,[8] by the responsibility to publish such a work for which I have honestly to admit the umbalance between size and interest. After the half committement I took in a previous work, it was a kind of point of honor for me to complete this classification: in any case, this work is the end of my researches on the classification of continuous groups, on which I have already insisted too much.[9] (Quoted by [38], letter to Levi-Civita, 2.VI.1918.)

A more objective evaluation of Amaldi's work is, in our opinion, expressed in the following letter by Paolo Medolaghi, a now almost forgotten mathematician who gave important contributions to Lie's theory of transformation groups [41].

> I have received your memoir on the determination of all kind of transformation groups, both finite and infinite, and my thoughts went back to many years ago when I was completely seduced by the researches on the theory of groups. In those times such a substantial piece of work would have appeared to me like a distant dream. Although not actively involved any more, I am able to understand your sharp arguments, your genial and zealous researches at the basis of your work, which remain as a document and a monument of your considerable scientific activity.
> I have just had time to read some part of it, having received the parcel yesterday morning, and I have already found some chapters of the greatest interest.[10] (Quoted by [38] lettera to Amaldi, 31.V.1918.)

3 Some Notes on Lie's Theory of Differential Equations

The goal of this section is to give a rough idea of the reason why *Lie fundamental classification problem* i.e. the problem to classify all transformation groups acting on n dimensional space, is so important for the algebraic theory of differential

---

[8]The "Academia Nazionale delle Scienze, detta dei XL", was founded in 1782 in Verona. It was named after the number of its original members. Today there are 65 Italian members and 22 non italian members.

[9]Di gran cuore mi affretto a ringraziarti delle gentili parole a proposito di quella mia mastodontica memoria. Per esperienza so benissimo di dover fare nei tuoi giudizi sulle cose mie, la debita parte alla grande tua benevolenza a mio riguardo: ma, pur così (attenuate) da mia parte, le tue buone parole mi sono riuscite molto gradite, perché, ora che mi vedo innanzi quel volume piramidale, mi sento turbato dalla responsabilità assunta, sopra tutto di fronte alla Società dei XL, pubblicando un lavoro, nel quale debbo riconoscere una sproporzione tra la mole e l'interesse. Dopo il mezzo impegno preso in una memoria precedente, era per me una specie di punto d'onore il venir a capo di quella classificazione: e ad ogni modo, con questa ho definitivamente finito le mie ricerche di determinazione di gruppi continui, sulle quali già troppo ho insistito.

[10]Ho ricevuto la sua memoria sulla determinazione di tutti i tipi di gruppi finiti e infiniti di trasformazioni di contatto dello spazio, ed il mio pensiero è tornato ai tempi ormai lontani in cui il fascino delle ricerche sulla teoria dei gruppi mi aveva tutto conquistato. Allora una opera così considerevole come quella che Ella ha compiuto mi sarebbe parso un sogno lontano! Sebbene passato alla riserva, sono al caso di intendere la serie di sottili accorgimenti, di trovate geniali e di coscienziose ricerche che è rappresentata dal suo lavoro, il quale resta come monumento e documento considerevole della sua attività scientifica.

Ho appena avuto il tempo di leggere alcune parti, avendo ricevuto il plico ieri mattina, e già ho trovato dei capitoli che mi hanno altamente interessato.

equations. We start from ordinary differential equations and recall two basic facts [37].

Fact 1. All first order equations $y' = f(x, y) = 0$ are equivalent to the simplest one $y' = 0$. For example, the Riccati equation $y' + y^2 = 0$ is transformable to $Y' = 0$ by means of the coordinate transformation $y = \frac{1}{Y+X}, x = X$.

Fact 2. All *linear* second order equations are equivalent to $y'' = 0$ by means of invertible point transformations.

We will not discuss several delicate questions, like what we mean exactly by equivalent differential equations, what kind of transformations we allow to realize the equivalence and how to deal properly with singularities. We just note that in order to transform a differential equation we can consider, in general, transformations acting on variables and derivatives. The simplest transformations are those which act only on variables, like the one considered in the example of Riccati equation. These transformations are called *point transformations*. To transform a differential equation with a point transformation one needs to *extend* its action to derivatives, and this is done by suitably deriving the transformation itself. The procedure is simple in principle but the actual computations are quite cumbersome. To deal properly with this extension procedure and with more complicated transformations, Jet spaces and their calculus has been introduced by Ehresmann [8]. This is the modern algebro-geometric framework for Lie's theories of differential equations, see [39] and the references therein.

Many questions about the classification of differential equations arise naturally at this point, and Lie considered many of them in his works, as well as their manifold applications. For example, in [32], he considered several questions about the classification of second order ordinary differential equations under the action of point transformations.

In order to deal with these problems, Lie defined the *symmetry algebra* of an analytic differential equation as the Lie algebra of complex vector fields in the plane which transform the differential equation in itself[11]. Let us denote by L_r the r dimensional complex symmetry algebra of a second order ordinary differential equation. He showed that r can be 0, 1, 2, 3 or 8 and proved that a second-order equation cannot admit a (maximal) four-, five-, six- or seven-dimensional symmetry Lie algebra. He also showed that if a second-order equation admits an eight-dimensional algebra, it is linearizable by a point transformation and equivalent to the simplest equation $y'' = 0$. In order to prove his results he classified complex all Lie algebra of vector fields in the plane. He selected from his classification all algebras which are symmetry algebra for a second order ODE and for each of these he gave a canonical representative. For this classification we refer to [39], p. 476 or

---

[11]Lie did not use the words "Lie algebra of vector fields" but "group of infinitesimal transformations".

to [37], p. 2005, where one can find the adaptions to the real analytical case, for this classification. We just copy one of the entries in the classification table of [39] to make some general comments.

Symmetry group	Dim	Type	Invariant equation
$\partial_x, \partial_u, x\partial_x + (x+u)\partial_u$	3	1.8	$u_{xx} = ce^{-u_x}$

The symmetry algebra (or symmetry group in Lie's notation) is given by its infinitesimal generators. There are three infinitesimal generators in the example and each defines a vector field in the plane, hence the dimension of the symmetry algebra is 3. The type refers to Olver classification table in [39], p. 472 and is not the same as in the classification of simple Lie algebras. In fact, non-isomorphic Lie algebras of vector fields may be isomorphic as abstract Lie algebras, like, for example,

$$\partial_x, x\partial_x - u\partial_u, x^2\partial_x - 2xy\partial_u$$

and

$$\partial_x, x\partial_x - u\partial_u, x^2\partial_x - (2xy+1)\partial_u$$

which are both isomorphic to $\mathfrak{sl}(2)$ but cannot be transformed one into the other by an invertible change of coordinates.

As Lie himself already remarked, in order to solve similar classification problems for partial differential equations one need to classify infinite dimensional algebras of vector fields in n dimensional space and also to consider *contact transformations* [23, 39]. For example, for the classification of second order partial differential equations in two variables, one should first classify all finite and infinite dimensional algebras of vector fields in $\mathbb{R}^3$ both of point and contact transformations. This is precisely the problem that Amaldi claimed to have solved in his works.

4 Levi-Civita and Enriques

In 1891, Federigo Enriques, a student at the Scuola Normale Superiore in Pisa, graduated in Mathematics and got a one year fellowship in Pisa . Next year, he got a fellowship in Rome, to study with Luigi Cremona. He moved to Rome in autumn 1892, where he met Guido Castelnuovo, with whom he begun a well known and fruitful scientific collaboration. In November 1893, he went to Torino to study with Corrado Segre and in January 1894 he was appointed at the university of Bologna for teaching projective and descriptive geometry.

His first papers [9–11] concerned the theory of homographies, and were inspired by papers by Bertini, Segre and Pradella. During his year in Rome with Cremona he published some papers connected with Lie's theory of transformation groups [12–15]. In [12] he solved one of the problems raised by Klein in his Erlangen's program, namely to classify all algebraic surfaces admitting infinite projective transformations. Lie had already solved the problem in [31], but Enriques did not know that. The paper was prepared during his fellowship in Rome and it is possible that it was suggested by Cremona himself, who was an enthusiastic supporter of Lie's ideas and suggested other works of this kind to his students and colleagues, for example Medolaghi and Pittarelli.

In 1896, a student of Corrado Segre, Gino Fano generalized the results of [12] and published two important works on the classification of hypersurfaces of $\mathbb{P}^4$ with a continuous group of projectivities [20, 21]. Fano, unlike Enriques, followed an approach which was closer in spirit to the theory of representations of Lie algebras. In Fano's works, one can find the first correct proof of Study's theorem on the complete reducibility of the representations of $\mathfrak{sl}(2, \mathbb{C})$ [22, 23].

Later Fano and Enriques gave original and important contributions to the classifications of Cremona groups, where Lie's ideas were combined with an algebraic geometric approach [14, 18]. The relevance of these researches have been well expressed in a conference of Ugo Amaldi for the 1907 congress of the "Società Italiana per il Progresso delle Scienze" [2]

> The work of Enriques and Fano on continuous groups on one side and on surfaces which admit such a group on the other, is one of the most interesting chapter in continuous group theory, not only for the importance and the singular difficulties of the problems which they solve, but also for the deep views and the richness and geniality of expedients that they employ. So that one can say that they contain, at least implicitly, the fundamentals of the algebraic geometry of finite continuous groups, that the work of Lie left completely untouched.[12]

When Enriques moved to Bologna in 1894 his knowledge of Lie's theory was good enough to allow profitable conversations with Levi-Civita when he met him in november 1894 (see [7], letter n. 150, p. 151). We know from Enriques' letters to Castelnuovo that among the topics which they discussed together there was the group theoretic foundations of geometry proposed by Klein in his Erlangen Program. These discussions turned out to be quite fruitful for Enriques who sketched a program for enlarging the original scope of Erlangen program to include situations commonly met in algebraic geometry, but not directly under the scope of Klein's program. The following quotations from the letters to Castelnuovo clarify the genesis and the content of this sketch

> Some students ask me to give a course of Higher Geometry: I am not against the idea to please them, at least in part, with a series of weekly conferences, that however I shall begin later (after january). In case, I will give you a sketch of these: they will be based on a general principle which will complete Klein's one (Program) in order to embody under that point of view, many other researches (e.g. the geometry over an algebraic manifold) which are not directly contained in it. [13] (Quoted by [7], lettera 149, p. 151.)

---

[12]I lavori di Enriques e Fano, su gruppi continui da una parte e sulle superficie che ammettono un gruppo siffatto dall'altra, costituiscono indubbiamente uno dei più bei capitoli della teoria dei gruppi continui, non soltanto per l'importanza e la singolare difficoltà dei problemi risoluti, ma anche per la profondità di vedute e la ricchezza e genialità di spedienti che vi furono profuse. Talchè si può dire ormai che siano poste, almeno implicitamente, le prime basi di quella geometria algebrica dei gruppi continui finiti, che l'opera del Lie aveva lasciato completamente in ombra.

[13]Alcuni giovani mi chiedono che faccia un corso di Geometria Superiore: non sono alieno dall'idea di contentarli, almeno in parte con un seguito di conferenze settimanali che però

These ideas, that were already in the back of my mind, have become concrete by talking to Levi-Civita. [14] (Quoted by [7], lettera 151, p. 154.)

As Enriques explained, a bit obscurely, in [7], letter 151 and much more clearly in [16], when one considers a geometrical object X immersed in a space whose geometry is given by a certain group of transformations G (for an algebraic variety in projective space S_n whose geometry is given by birational transformations) then the properties of X which are preserved by G may very well be preserved by a bigger group G_X. In the example of birational transformations, each *rational* transformation which become birational between X and its image, preserves the same properties, hence X and its images may be not birationally equivalent under a birational transformation of S_n but they may be birational equivalent if we abstract from the relations between them and the ambient space. Enriques claims that this generalization enlarges the scope of Klein's program to include the *geometry over a manifold* which did not fit the original Erlangen program.

Enriques points out the usefulness of considering *abstract* geometric objects, i.e. independent on their immersions and *abstract* groups, i.e. independent on the geometric space on which they manifest themselves as transformation groups. He observes that any group of transformations gives rise to other groups, that can be thought of as different manifestations of the original one. For example, the group of projective transformations of the projective line induces a group of transformations in the space whose elements are all pairs of points of the line, which, in turn, may be shown to be the same as the projective transformations of the plane which fix a conic curve (the points of the conic corresponds to pairs of coincident points on the line). Hence the projective geometry of the line is the same as that of the plane with respect to a fixed conic curve, i.e. the geometry of lines in $\mathbb{R}^3$ through a point with respect to rotations. This implies the classification of the finite subgroups of the projectivities of the line as those corresponding to finite groups of rotations.

Enriques remarks therefore that it is sometimes useful that a group lose its concreteness as a group of transformations of a given space and be considered with respect to all its possible realizations, among which, he remarks, there is one which is *canonical*. This is simply obtained by considering the action of the group on itself given by multiplication.

These views, which anticipates some of the view of modern algebra, have not been pursued any further by Enriques, whose research concentrated on the classification of algebraic surfaces, where he got the great successes for which he is famous today.

---

comincerei solo più tardi (dopo Gennaio). Ti dirò in caso il piano di queste: s'informerebbero ad un principio generale che completa quello di Klein (Program) per far rientrare in quell'ordine di idee vari altri tipi di ricerche (per esempio la geometria sull'ente) che ad esso sfuggono, almeno direttamente.

[14] Queste idee che avevo in germe, mi son venute concrete parlandone col Levi-Civita.

5 Conclusions

The interest of italian mathematicians for Lie's theory of transformations group, especially the infinite dimensional ones, has been largely neglected. In this paper, we have shown an example of the deep and lively interaction between this theory and the mathematical ideas that circulated in Bologna at the end of XIX century, both analytical and geometrical. Other interactions with Lie's ideas are very interesting to track in the works of Medolaghi, Bianchi, Fubini, Fano, Elia L. Levi, etc. We shall come back to this theme in future works.

References

1. Amaldi, U. 1898. Sulla trasformazione di Laplace, *Rendiconti dell'Accademia Nazionale dei Lincei*, (5), VII$_2$, 117–124.
2. Amaldi, U. 1908. Sui principali risultati ottenuti nella teoria dei gruppi continui dopo la morte di S. Lie, *Ann. di matem*, (3), XV, 293–328.
3. Amaldi, U. 1911. *Roberto Bonola. 14 novembre 1874–16 maggio 1911*. Boll della "Mathesis" 3(127): 145–152.
4. Amaldi, U. 1918. Sulla classificazione dei gruppi continui di trasformazioni di contatto nello spazio, *Mem. della Soc. Ital. dei Quaranta*, (3), XX, 167–350.
5. Amaldi, U. 1933. *Gruppo*. Bologna: Zanichelli.
6. Amaldi, U. 1954. *Tullio Levi-Civita*. Bologna: Zanichelli.
7. Bottazzini, U., A. Conte, and P Gario, editors. 1996. *Riposte armonie: lettere di Federigo Enriques a Guido Castelnuovo*. Torino: Bollati Boringhieri.
8. Ehresmann, C. 1953. Introduction à la thèorie des structures infinitèsimales et des pseudo-groupes de Lie. *Geometrie Differentielle, Colloq. Inter. du Centre Nat. de la Recherche Scientifique, Strasbourg, 97–127.*.
9. Enriques, F. 1890. Alcune proprietà dei fasci di omografie negli spazi lineari ad n dimensioni, *Rom. Acc. L. Rend.* (4), VI$_2$, 63–70.
10. Enriques, F. 1892. Le omografie armoniche negli spazi lineari ad n dimensioni. XXX: 319–325.
11. Enriques, F. 1892. Le omografie cicliche negli spazii ad n dimensioni. XXX: 311–318.
12. Enriques, F. 1893. Le superficie con infinite trasformazioni proiettive in sè stesse. *Ven Ist Atti, Ser VII, Tomo* 4: 1590–1635.
13. Enriques, F. 1893. Sopra un gruppo continuo di trasformazioni di Jonquières nel piano. *Rom. Acc. L. Rend.* (5), II$_1$, 532–538.
14. Enriques, F. 1893. Sui gruppi continui di trasformazioni cremoniane nel piano. *Rom. Acc. L. Rend.*, (5), II$_1$, 468–473.
15. Enriques, F. 1894. Intorno alla Memoria: "Le superficie con infinite trasformazioni proiettive in sè stesse". *Ven Ist. Atti, Ser VII, Tomo* 5, 636–642.
16. Enriques, F. 1895. Conferenze di geometria. Autographirt.
17. Enriques, F., and U. Amaldi. 1902. *Elementi di geometria ad uso delle scuole secondarie superiori*. Bologna: Zanichelli. XXII u. 655 S. 16°.
18. Enriques, F., and Fano, G. 1897. Sui gruppi continui di trasformazioni cremoniane dello spazio. *Annali di Mat.* (2) 26, 59–98.
19. Fano, G. 1890. Considerazioni comparative intorno a ricerche geometriche recenti. *Annali mat. Pura e appl.* (2) *17, 307–342.*
20. Fano, G. 1896. Sulle varietà algebriche con un gruppo continuo non integrabile di trasformazioni proiettive in sè. *Mem Accad Sci Torino, Cl Sc fis, Mat, Nat, Ser II*, **46**: 187–218.

21. Fano, G. 1896. Sulle varietà algebriche dello spazio a quattro dimensioni con un gruppo continuo integrabile di trasformazioni projettive in sè. *Ven. Ist. Atti.*, (7) 7, 1069–1103.
22. Hawkins, T. 1994. Lie groups and geometry: The italian connection. *Rendiconti del Circolo Matematico di Palermo, Supplemento*, s. II(36): 185–206.
23. Hawkins, T. 2000. *Emergency of the theory of Lie Groups*. New York: Springer..
24. Klein, F. 1872. Vergleichende Betrachtungen über neuere geometrische Forschungen. Programm zum Eintritt in die philosophische Facultät und den Senat der Universität zu Erlangen. Erlangen. A. Deichert..
25. Levi-Civita, T. 1894. Sugli invarianti assoluti. *Ven. Ist. Atti* (7) 77.
26. Levi-Civita, T. 1895. Alcune osservazioni alla nota "Sui gruppi di operazioni funzionali". *Lomb. Ist. Rend.* (2) XXVIII, 864–873.
27. Levi-Civita, T. 1895. I gruppi di operazioni funzionali e l'inversione degli integrali definiti. I and II. *Lomb. Ist. Rend.* (2) XXVIII, 529–544, 565–577.
28. Levi-Civita, T. 1895. Sui gruppi di operazioni funzionali. *Lomb. Ist. Rend.* (2) XXVIII, 458–468..
29. Levi-Civita, T., and U. Amaldi. 1923. *Lezioni di meccanica razionale. Vol. 1. Cinematica. Principi e statica.* Bologna: Zanichelli, XIII u. 743 S. 8° .
30. Levi-Civita, T., and U. Amaldi. 1926, 1927. *Lezioni di meccanica razionale. Vol. II. Dinamica dei sistemi con un numero finito di gradi di liberta.* Part. 1 (1926), 2 (1927). IX + 527 p., IX + 685 p. Bologna: Zanichelli.
31. Lie, S. 1882. Ueber Flöchen, die infinitesimale und lineare Transformationen gestatten. *Lie Arch* VII: 179–193.
32. Lie, S. 1888. Klassification und Integration von gewöhnlichen Differentialgleichungen zwischen x, y, die eine Gruppe von Transformationen gestatten. *Math Ann* XXXII: 213–281.
33. Lie, S. 1888. *Theorie der Transformationsgruppen. Erster Abschnitt. Unter Mitwirkung von F. Engel bearbeitet.* Leipzig: Teubner. X + 632 S. gr. 8° .
34. Lie, S. 1890. *Theorie der Transformationsgruppen. Zweiter Abschnitt. Unter Mitwirkung von F. Engel.* Leipzig: B.G. Teubner. IV + 554 S. 8°.
35. Lie, S. 1893. *Theorie der Transformationsgruppen. Dritter (und letzten) Abschnitt. Unter Mitwirkung von Fr. Engel bearbeitet.* Leipzig: B.G. Teubner. XXVII + 830 S. 8°.
36. Lie, S. 1934. *Gesammelte Abhandlungen. Herausgegeben von dem Norwegischen Mathematischen Verein durch Friedrich Engel und Poul Heegaard. 5. Bd.: Abhandlungen über die Theorie der Transformationsgruppen. 1. Abteilung, herausgegeben von Friedrich Engel.* Leipzig: B. Teubner u. Kristiania: H. Aschehong & Co., XII u. 776 S. gr. 8°.
37. Mahomed, F.M. 2007. Symmetry group classification of ordinary differential equations: Survey of some results. *Math. Meth. Appl. Sci.* 30, 1995–2012.
38. Nastasi, P., and E. Rogora. 2007. *Mon cher ami, Illustre professore.* Roma: Edizioni Nuova Cultura.
39. Olver, P. 1995. *Equivalence, invariants and symmetry.* Cambridge: Cambridge University Press (London Mathematical Society Lecture Notes).
40. Pincherle, S., and U. Amaldi. 1901. *Le operazioni distributive e le loro applicazioni all' analisi.* Bologna: Zanichelli. XII u. 490 S. 8°.
41. Pommaret, J.F. 1978. *Systems of partial differential equations and Lie pseudogroups.* New York: Gordon and Breach Science Publishers.

Difference Equations in Spaces of Regular Functions: a tribute to Salvatore Pincherle

Irene Sabadini and Daniele C. Struppa

Abstract In [14], Pincherle studies the surjectivity of a difference operator with constant coefficients in the space of holomorphic functions. In this paper, we discuss how this work can be rephrased in the context of modern functional analysis and we conclude by extending his results and we show that difference equations act surjectively on the space of quaternionic regular functions.

1 Introduction

Salvatore Pincherle was one of the pioneers in the study of what we now call difference equations, as well as infinite order differential equations. Among his works in this direction we should mention [14], which has been the inspiration for this article. For a more complete view of Pincherle's work in this and related areas, we refer the reader to his collected works published by Edizioni Cremonese and curated by the Unione Matematica Italiana [16]. We should also mention our earlier work [19] on the evolution of the theory of infinite order differential operators in Pincherle.

In [14], Pincherle studies the surjectivity of a difference operator with constant coefficients, as well as the surjectivity of what we would now call infinite order differential operators. This paper is probably one of the most important early works in the theory of such operators, and, as the editor of [16] explains in his notes to [14], its importance was recognized by Mittag-Leffler who, in 1926, reprinted in French this article in his journal Acta Mathematica [15].

I. Sabadini (✉)
Politecnico di Milano, Dipartimento di Matematica, Via E. Bonardi, 9, 20133 Milano, Italy
e-mail: irene.sabadini@polimi.it

D.C. Struppa
Schmid College of Science, Chapman University, Orange, CA 92866, USA
e-mail: struppa@chapman.edu

In what was the typical style of late nineteen century, Pincherle studies the surjectivity of a variety of operators in the space of analytic functions without the sophisticated machinery from the theory of topological vector spaces that is now available, but rather uses a very direct approach. Specifically, in [14], he first considers the problem of formally solving an infinite order differential equation of the form

$$\sum_{0}^{+\infty} \frac{a_n}{n!} \varphi^{(n)}(x) = f(x). \tag{1}$$

In this process he finds at least some of its solutions; for example he states that (1) admits the solution

$$\Sigma C_\nu e^{\beta_\nu x}, \tag{2}$$

where the β_ν are the roots of the characteristic function $a(t) = \sum_0^{+\infty} \frac{a_n t^n}{n!}$, and the C_ν are arbitrary constant. While this is certainly true (at least as far as the series is convergent), we know that if some of the roots have multiplicity larger than one, than the constants C_ν are to be replaced by polynomials of suitable degree. Moving further, Pincherle analyzes the special case in which the function $f(x)$ in (1) is an entire function of exponential type, and he shows that the formal solutions that one can derive by formal integration of (1) are indeed entire solutions.

Pincherle also considers more general equations, namely difference equations of the form

$$\sum_{r=1}^{m} h_r \varphi(x + \alpha_r) = f(x)$$

and shows that, in the space of entire functions, such equations are always solvable.

In Sect. 2 of this paper, we introduce some modern terminology, we will discuss how the results of Pincherle can be placed in the context of modern functional analysis, and how the issue of surjectivity actually lends itself to some rather interesting considerations on the nature of entire functions and their growth at infinity. Section 2 will also provide the necessary background for the more novel part of this paper, namely the study of difference equations in the quaternionic setting, namely on the space of regular functions defined (see below) as infinitely differentiable solutions of the Cauchy–Fueter system.

In Sect. 2 we discuss the notion of AU-spaces: these are spaces of generalized functions introduced by Ehrenpreis in [10], for which many general results hold, including the surjectivity of linear constant coefficients differential operators. As it is known if X is an AU-space, and $\vec{P}(D)$ is a system of linear partial differential operators with constant coefficients, then the kernel of $\vec{P}(D)$ in X is an AU-space itself, and thus, in principle at least, partial differential operators should be surjective on the space of regular functions.

This would definitely be true if the space of regular function were the space of solutions to a single differential equation, or even solutions to a system with a single unknown. In our case, however, we actually have a square system of partial differential equations and the situation is a little more complicated. In our previous work [7] we have, for example, shown, see Corollary 4.8, that indeed any linear constant coefficient partial differential operator is surjective on the space of regular functions on open convex sets, but the argument required a degree of algebraic analysis of the system rather than a simple invocation of the theory of analytically uniform spaces.

Since difference equations, as well as infinite order differential equations, are special cases of convolution equations, one can wonder whether any of the results that hold for differential equations on AU-spaces hold true in the case of convolution equations. That this is the case (and to which extent it is the case) is the object of the work of Berenstein–Taylor [6], and successively Struppa [18], and Berenstein–Kawai–Struppa [2]. A rather comprehensive survey of the results in this area is given in [5].

In this paper, we prove that difference equations act surjectively on the space of regular functions on an open convex set in $\mathbb{H}$ by extending the arguments from [7].

2 Convolution Equations in Analytically Uniform Spaces

In this section, we will recall briefly some of the general facts regarding the theory of convolution equations in Analytically Uniform spaces.

To begin with, we will give a short introduction to Analytically Uniform spaces (AU-spaces in short). These spaces were introduced by Ehrenpreis in [10], in order to give the appropriate setting for the so-called Ehrenpreis–Palamodov Fundamental Principle (see [10, 13]), which gives an integral exponential representation for solutions of homogeneous systems of linear constant coefficients differential equations. The actual definition of AU-spaces is rather convoluted (and we refer the interested reader to the first chapter of [10] for the details) but it can be intuitively described as follows. A topological space X of generalized functions is Analytically Uniform if its dual X' is topologically isomorphic, through some transformation which usually looks like the Fourier transform, to a space of entire functions whose growth at infinity is bounded by some suitable weight functions. These weight functions (specific examples will follow) are usually referred to as the AU-structure of the space.

We realize that the way we have given the definition is rather ambiguous, but it will be enough for most of our purposes, and it will be clarified by the next two examples.

Example 1. The space $\mathcal{E}(\mathbb{R}^n)$ of infinitely differentiable functions on $\mathbb{R}^n$ is an AU-space. Its dual is the space $\mathcal{E}'$ of compactly supported distributions, and the Fourier transform is an isomorphism between $\mathcal{E}'$ and the space

$$\hat{\mathcal{E}}' = \{F \in \mathcal{H}(\mathbb{C}^n) : \exists A, B > 0 \text{ s.t. } |F(z)| \leq A\,(1+|z|)^B \exp(B|\text{Im}(z)|)\}$$

of entire functions of exponential type whose growth on the real axis is polynomial. More generally, the space of infinitely differentiable functions on any open convex subset $\Omega \subseteq \mathbb{R}^n$ is also an AU-space; its dual is the space of distributions with compact support contained in Ω, and this last space is isomorphic (via the Fourier transform) to the space of entire functions F satisfying, for some positive constants A, B

$$|F(z)| \leq A (1 + |z|)^B \exp H_K(\mathrm{Im}(z)),$$

where H_K is the indicator function of the convex hull of the compact set K.

Example 2. The space $\mathcal{H}(\mathbb{C}^n)$ of entire functions on $\mathbb{C}^n$ is an AU-space. Its dual is the space $\mathcal{H}'(\mathbb{C}^n)$ of analytic functionals, and the Fourier–Borel transform is an isomorphism between $\mathcal{H}'(\mathbb{C}^n)$ and the space

$$\hat{\mathcal{H}}'(\mathbb{C}^n) = \{F \in \mathcal{H}(\mathbb{C}^n) : \exists A > 0 \text{ s.t. } |F(z)| \leq A \exp(A|z|)\}$$

of entire functions of exponential type. Once again, the space of holomorphic functions on any open convex subset of $\mathbb{C}^n$ is also an AU-space.

Whenever X is an AU-space, an element $\mu \in X'$ can be used as a convolutor on X by setting

$$\mu * f := \langle \mu, \tau_t f \rangle,$$

where τ_t denotes the translation of t, and μ acts on $\tau_t f$ as a generalized function of the variable t. Thus convolution equations in X can be written as

$$\mu * f = 0,$$

and the symbol of the convolution operator is the Fourier transform of μ defined as the element in $\hat{X}'$ given by

$$\hat{\mu}(z) := \langle \mu_\zeta, \exp(\zeta \cdot \bar{z}) \rangle.$$

The theory of differential equations in AU-spaces is described in great detail in [10, 13], and in particular we know of the surjectivity of linear constant coefficients differential operators on such spaces. The theory of convolution equations in such spaces is more delicate, and a good survey can be found in [5]. In particular, however, we know the conditions under which convolution operators are surjective. While the general conditions are sometimes complex, the general idea for the proof of a surjectivity result is quite simple.

Let T be a continuous operator acting from a suitable topological vector space X to itself. Then, [9], T is surjective if and only if its adjoint $T' : X' \to X'$ is injective and has closed image. Usually it is easy to prove injectivity, while the key of the proof lies in establishing the closure of the image. Because of the nature of AU-spaces, such closure can be interpreted in terms of a division theorem in the space of entire functions which are Fourier transforms of the objects in X'. This specifically relies on finding appropriate conditions for the following to be true: if $\mu, \nu \in X'$ and $\hat{\nu}/\hat{\mu}$ is an entire function, then $\hat{\nu}/\hat{\mu}$ is the Fourier transform of an object in X'. If this is the case, one usually says that μ is invertible.

Let us now see how, from this point of view, one can demonstrate that for any finite sets $\{\alpha_1, \ldots, \alpha_m\}$ and $\{h_1, \ldots, h_m\}$ with $\alpha_i, h_i \in \mathbb{C}$, the difference equation

$$\sum_{r=1}^{m} h_r \varphi(z + \alpha_r) = f(z)$$

always admits an entire solution φ in the space $\mathcal{H}(\mathbb{C})$ of holomorphic functions in the complex plane, for any f in the same space $\mathcal{H}(\mathbb{C})$. The proof of this result is quite standard (see e.g. [3, 5, 12]) and goes as follows. Denote by D the operator acting on $\mathcal{H}(\mathbb{C})$ which maps a function φ to the function

$$D(\varphi) = \sum_{r=1}^{m} h_r \varphi(z + \alpha_r).$$

The space $\mathcal{H}(\mathbb{C})$ is a bornological, barreled, topological vector space and therefore in view of the well-known theory of Dieudonné and Schwartz [9] we have that the operator D is surjective if and only if its transpose $D^t \colon \mathcal{H}(\mathbb{C})' \to \mathcal{H}(\mathbb{C})'$ is both injective and has closed range. Since the Fourier–Borel transform is a topological isomorphism from $\mathcal{H}(\mathbb{C})'$ to the space

$$\mathrm{Exp}(\mathbb{C}) = \{f \in \mathcal{H}(\mathbb{C}) \mid \text{for some } A > 0 \, |f(z)| \leq A \exp(A|z|)\},$$

the surjectivity of this is equivalent to the request that the multiplication map $\hat{D}^t \colon \mathrm{Exp}(\mathbb{C}) \to \mathrm{Exp}(\mathbb{C})$ is injective and has closed range. Since

$$\hat{D}^t = \widehat{\sum_{r=1}^{m} h_r \delta_{-a_r}} = \sum_{r=1}^{m} h_r e^{-iza_r},$$

it is obvious that the map is injective. In order to demonstrate that the range is closed it is enough to show that the ideal generated by the exponential sum $\sum_{r=1}^{m} h_r e^{-iza_r}$ in $\mathrm{Exp}(\mathbb{C})$ is the intersection of $\mathrm{Exp}(\mathbb{C})$ with the ideal generated by the exponential sum $\sum_{r=1}^{m} h_r e^{-iza_r}$ in $\mathcal{H}(\mathbb{C})$. In other words, one has to prove that if F is a function in $\mathrm{Exp}(\mathbb{C})$ and if

$$\frac{F}{\sum_{r=1}^{m} h_r e^{-iza_r}}$$

is entire, then

$$\frac{F}{\sum_{r=1}^{m} h_r e^{-iza_r}}$$

belongs to $\mathrm{Exp}(\mathbb{C})$. That this is the case is essentially a variation of the Ritt theorem, see [5].

This same approach can be used if instead of difference equations, one considers infinite order differential operators. In Pincherle's work, the distinction between

these operators is not always clear. For example, Pincherle writes that the translation operator
$$D(f)(x) := f(x+1),$$
which is clearly a difference operator, can be written as
$$D(f)(x) := \sum_{n=0}^{+\infty} \frac{f^{(n)}(x)}{n!}.$$

Though this way of representing the translation is technically incorrect (in fact the translation operator is not even a local operator), this particular representation is still commonly used by physicists, see e.g. [1], who also apply approximations based on this series expansion. From our point of view, however, infinite order differential operators are local operators which are defined by infinite series of derivatives whose coefficients are such that the operator they define acts locally on germs of holomorphic functions. Specifically, a differential operator of infinite order is an operator of the form
$$D := \sum_{n=0}^{+\infty} a_n \frac{d^n}{dz^n},$$
where the coefficients a_n satisfy the condition
$$\lim(|a_n|n!)^{1/n} = 0.$$

Under this condition the operator D is a local operator, which maps the space of entire functions to itself. Its symbol is the holomorphic function
$$P(z) = \sum_{n=0}^{+\infty} a_n z^n,$$
which can be proved to be of exponential type zero (or of infraexponential type, as the Japanese school would call it), namely such that for every $\epsilon > 0$, there is a positive constant A_ϵ such that
$$|P(z)| \leq A_\epsilon \exp(\epsilon |z|).$$

One can show, see e.g. [6], that infinite order differential operators are invertible, and thus surjectivity follows once again in a rather immediate way.

The situation is considerably more complicated when one considers difference equations acting on spaces of holomorphic functions defined on convex subsets of $\mathbb{C}^n$. In this case, the domain and the range of the operators do not coincide, and the conditions for invertibility become significantly more complex, and involve the growth at infinity of the Fourier–Borel transform of the operator, as well as the geometry of the support K of the operator itself. We will not discuss this case, but we will refer the reader to the comprehensive analysis in [3, 5, 11, 12].

We conclude this section with a brief reference to the situation that occurs when we study difference equations on spaces of infinitely differentiable functions. This is necessary if we want to study difference equations on regular functions, since such functions are vectors of infinitely differentiable functions satisfying a suitable system of differential equations (the Cauchy–Fueter system), but the components of the vectors are definitely not holomorphic functions. Thus, while the theory described above offers a modern interpretation of Pincherle's work, it does not help us when we want to extend his results to regular functions.

The situation is indeed somewhat different in this case, because unlike what happens in the case of entire functions, it is not true that every convolutor is invertible, even when we are looking at infinitely differentiable functions defined on the entire Euclidean space $\mathbb{R}^n$. The reason for this lies in the AU-structure for the space $\mathcal{E}$, which uses a weight (namely $|\text{Im} z| + \log(1 + |z|^2)$) which is not radial, unlike the AU-structure for $\mathcal{H}$, whose weight is $|z|$. As a consequence, not every convolutor is invertible, and specific conditions are necessary in order to ensure the invertibility. These conditions (known as *slow decrease* conditions) are expressed in terms of the growth of the Fourier transform of the convolutors involved, and in particular of the size of the areas of $\mathbb{C}^n$ where such Fourier transforms are small. Finally, we note that infinitely differential operators cannot be defined on infinitely differentiable equations, since, unlike analytic functionals, distributions are locally defined as finite order distributions. It is nevertheless possible to study some classes of infinite order differential operators acting on regular functions (as showed in [17]), though we will not focus on this aspect in this paper.

3 Regular Functions of a Quaternionic Variable

In this section, we will follow some of the ideas in [7] to show that difference equations are surjective on the space of regular functions in one quaternionic variable. Functions on quaternions have been of interest for many years, and in the first decades of the last century a notion of holomorphicity, called regularity, was developed for them. In recent years, there has been a renewed interest in the theory of regularity in the space of quaternions. The theory officially began with the work of Fueter and, independently, Moisil and Theodorescu in the 1930s (see e.g. [20], and the more recent [8] for a historical perspective), and has gained interest and success over the last few decades, see again [8] and its references.

Let $\mathbb{H}$ denote the skew field of quaternions. From a vector-space point of view, one can regard $\mathbb{H}$ as $\mathbb{R}^4$, but of course the elements of $\mathbb{H}$ can be represented as

$$q = x_0 + ix_1 + jx_2 + kx_3$$

where each x_t is real. The elements i, j, k are three imaginary units (i.e. $i^2 = j^2 = k^2 = -1$) such that $ij = k, jk = i, ki = j$. As a consequence, the space $\mathbb{H}$ of quaternions admits a non-commutative multiplicative structure, that differentiates it from $\mathbb{R}^4$ and makes it an algebra over the real numbers.

Definition 1. Let Ω be an open subset of $\mathbb{H}$, and let $f : \Omega \to \mathbb{H}$ be a real differentiable function. We say that f is *Cauchy–Fueter regular* (or simply *regular* in this paper, since no confusion with other notions of regularity can arise) if it satisfies the so-called *Cauchy–Fueter equation*

$$\frac{\partial f}{\partial \bar{q}} := \frac{\partial f}{\partial x_0} + i\frac{\partial f}{\partial x_1} + j\frac{\partial f}{\partial x_2} + k\frac{\partial f}{\partial x_3} = 0. \tag{3}$$

One should note that even though (3) appears as a single equation, we can rewrite it by taking advantage of the explicit expression of the $\frac{\partial}{\partial \bar{q}}$ operator and of the quaternionic representation of f as $f = f_0 + if_1 + jf_2 + kf_3$. From this point of view, (3) becomes the following 4×4 system of differential equations in the unknown variables f_0, f_1, f_2, f_3:

$$\begin{cases} \dfrac{\partial f_0}{\partial x_0} - \dfrac{\partial f_1}{\partial x_1} - \dfrac{\partial f_2}{\partial x_2} - \dfrac{\partial f_3}{\partial x_3} = 0 \\ \dfrac{\partial f_0}{\partial x_1} + \dfrac{\partial f_1}{\partial x_0} - \dfrac{\partial f_2}{\partial x_3} + \dfrac{\partial f_3}{\partial x_2} = 0 \\ \dfrac{\partial f_0}{\partial x_2} + \dfrac{\partial f_1}{\partial x_3} + \dfrac{\partial f_2}{\partial x_0} - \dfrac{\partial f_3}{\partial x_1} = 0 \\ \dfrac{\partial f_0}{\partial x_3} - \dfrac{\partial f_1}{\partial x_2} + \dfrac{\partial f_2}{\partial x_1} + \dfrac{\partial f_3}{\partial x_0} = 0. \end{cases}$$

This is clearly a system of linear constant coefficients partial differential equations (in fact, a rather easy system, since every equation is a first order equation) to which the Fundamental Principle therefore applies. The theory of regular functions is very well developed, and standard reference is probably the paper of Sudbery [20]. More recently, the authors and some of their collaborators have also developed in full the theory for the case of several quaternionic variables (defined in the natural way) [8].

In order to study difference equations in the space of regular functions on $\mathbb{H}$, we begin by recalling that a general theory of systems of convolution equations in AU-spaces, and in particular for infinitely differentiable functions, was developed in [18]. As it turns out, and this was already demonstrated in [6], it is impossible to develop a theory for systems of convolution equations that completely parallels the theory of Ehrenpreis and Palamodov for systems of constant coefficients differential equations. The reason for this is the fact that it is impossible to bound, in general, the multiplicity of the varieties associated to the Fourier transforms of convolutors which are not polynomials. The complexities associated with this issue are too elaborated to be discussed in this context, but it will be sufficient to recall that in order for the theory of Ehrenpreis and Palamodov to be extended to convolution equations, it became necessary to limit the attention to what are known as *slowly decreasing* systems of convolution equations. The definition is too complicated to be replicated here, it obviously includes all linear constant coefficients differential operators, and in its full form it was given (for systems) first in [18] and later refined in [2]. What is important, however, is the following result (Example 4.2 in [18]):

Theorem 1. *Let $M = [\mu_{ij}]$ be a matrix of $q \times s$ convolutors on the space $\mathcal{E}(\mathbb{R}^n)$, and assume that each one of them is associated to a difference equation with integral delays. Assume furthermore that the characteristic variety of the common zeroes of the maximal minors of the Fourier transform of the matrix M is discrete. Then M is a slowly decreasing system of convolution equations.*

The importance of this theorem lies in the fact that slowly decreasing systems are exactly those systems for which the Ehrenpreis–Palamodov theory can be extended. In particular, a Fundamental Principle for exponential representations for solutions of homogeneous slowly decreasing systems holds [18, Theorem 5.4], and a theorem on the solvability of non-homogeneous equations can be stated (Theorem 5.6 in [18]):

Theorem 2. *Let M be a slowly decreasing system of convolutors. Then the non-homogeneous system*
$$M * \vec{f} = \vec{g},$$
with given $\vec{g} \in \mathcal{E}^q$ admits a solution $\vec{f} \in \mathcal{E}^s$ if and only if
$$N * \vec{g} = 0,$$
for all matrices N of convolutors such that $M^t * N^t = 0$.

Let us now consider a quaternionic valued function f defined on the space $\mathbb{H}$ of quaternions. We then consider a difference equation of the form
$$\sum_{r=1}^{m} h_r f(q + \alpha_r) = g(q)$$
with complex coefficients h_r, and integral delays α_r, and we are interested in studying its surjectivity in the space of regular functions.

Before we can actually demonstrate that these conditions are sufficient to prove surjectivity, we will recall the analysis we carried out in [7].

To this purpose, let **R** denote a ring of symbols of operators. There are many important examples. For example, **R** could be the ring of polynomials, when the operators are constant coefficients linear partial differential operators, or could be the space of infraexponential type functions, when the operators are constant coefficients linear partial differential operators of infinite order, or could be the space of entire functions of exponential type, if the operators are convolution operators with analytic functionals, or, finally, could be the space $\hat{\mathcal{E}}'$, when we are interested in convolution operators with compactly supported distributions. This is actually the case we are interested in. We then denote by $\mathcal{R}_n$ the ring of $n \times n$ square matrices with coefficients in **R** (when n is irrelevant, or clear from the context, we will omit the index). We then proved (Corollary 4.1 in [7]) the following result:

Theorem 3. *Let* **R** *be a ring of symbols of operators for an AU-space X. Let P_1, P_2 be two commuting matrices forming a regular sequence in $\mathcal{R} = \mathrm{Mat}_n(\mathbf{R})$, and denote by D_1, D_2 the matrices of operators for which P_1, P_2 are symbols. Then the system*

$$\begin{cases} D_1(f) = g_1 \\ D_2(f) = g_2, \end{cases}$$

with $g_1, g_2 \in X^n$ has a solution $f \in X^n$ if and only if $D_1(g_2) = D_2(g_1)$.

In other words, we can compute the syzygies of the system associated to D_1 and D_2 as if the matrices were elements of **R**, at least under some reasonably mild assumptions. Note that this is not the case in more general situations. We are now ready to prove the key result of this paper.

Theorem 4. *Let g be a regular function on $\mathbb{H}$. Then the system*

$$\sum_{r=1}^m h_r f(q + \alpha_r) = g(q),$$

with $h_r \in \mathbb{R}$ and $\alpha_r \in \mathbb{Z}$, always has a solution f, regular on $\mathbb{H}$.

Proof. The conclusion of the theorem is equivalent to studying whether or not there are infinitely differentiable solutions to the system:

$$\begin{cases} \dfrac{\partial f}{\partial \bar{q}} = 0 \\ \sum_{r=1}^m h_r f(q + \alpha_r) = g(q). \end{cases}$$

This system can be written in a symbolic form as follows:

$$\begin{cases} P(D)\vec{f} = 0 \\ M * \vec{f} = \vec{g}, \end{cases}$$

where $P(D)$ is the 4×4 matrix associated to the Cauchy–Fueter system, M is the 4×4 convolution matrix associated to the difference equation, $\vec{f}$ is the 4-vector $\vec{f} = (f_0, f_1, f_2, f_3)^t$, and $\vec{g}$ is the 4-vector $\vec{g} = (g_0, g_1, g_2, g_3)^t$. Note, in particular, that the way in which we have represented the vectors, and since the coefficients h_r are real numbers, the matrix M is actually a diagonal matrix, whose diagonal entry is nothing but $\sum_{r=1}^m h_r \delta_{\alpha_r}$. This is a consequence of the immediate fact that convolving a function f with the translation of the Dirac delta is exactly the same as translating the function f itself.

Thus, in view of the results on difference equations on differentiable functions we see, [2, 18], that the system given by the matrix $Q = [P(D), M]$ is a slowly decreasing system and therefore if we set $\vec{g} = (0, 0, 0, 0, g_0, g_1, g_2, g_3)^t$, the system of convolution equations

$$Q \times f = \vec{g},$$

has a solution $\vec{f}$ if and only if $\vec{g}$ satisfies the systems $R \times \vec{g} = 0$, for any matrix R in the syzygies of Q. But now, we can use the same argument as described in Sect. 4 of [7] to show that the matrices $P(D)$ and M form a regular sequence in the ring $\mathcal{R}$, which concludes the proof. □

Remark 1. Two comments come to mind. The first is that the method we are using is forcing us to consider difference equations with integral delays. This is not necessary when dealing with difference equations on holomorphic functions, and thus one may wonder whether the limitation is due to the technique used in the proof (specifically its reliance on [18]) rather than on an intrinsic need. We do not know how to answer this question. The second comment deals with the fact that the theorem can be extended to the case of regular functions on any convex set $U \subseteq \mathbb{H}$. That this is true is not simple to demonstrate, as one needs to show that difference operators are slowly decreasing convolutors when considered in the space of infinitely differentiable functions on convex sets. Though the difference seems minor, in reality the study of convolution equations on convex sets is rather complex. However, in [4] the authors study exactly that case, and demonstrate again that difference operators satisfy the conditions necessary to obtain the surjectivity result.

Remark 2. In the statement of Theorem 4 we require that the coefficients h_r are real numbers in order to have that M ia a diagonal matrix. It is also possible to extend the result to the case in which the coefficients are complex. It is sufficient to consider the functions f and g with values in the space of complex quaternions, also called biquaternions, i.e. $\mathbb{H} \otimes \mathbb{C}$. The proofs remains the same, since complex numbers $a + \iota b$ commute with quaternions $q = x_0 + i x_1 + j x_2 + k x_3$.

Acknowledgement The authors are grateful to Professor S. Coen for inviting them to contribute to the volume.

References

1. Aharonov, Y., and D. Rohrlich. 2005. *Quantum paradoxes: quantum theory for the perplexed.* Weinheim: Wiley.
2. Berenstein, C.A., T. Kawai, and D.C. Struppa. 1996. Interpolating varieties and the Fabry–Ehrenpreis–Kawai gap theorem. *Adv Math* 122(2): 280–310.
3. Berenstein, C.A., and D.C. Struppa. 1987. A remark on: "Convolutors in spaces of holomorphic functions" by A. Meril and Struppa. Complex Analysis, II (College Park, Md., 198586), 276–280. In *Lecture notes in mathematics*, 1276. Berlin: Springer.

4. Berenstein, C.A., and D.C. Struppa. 1987. Solutions of convolution equations in convex sets. *Am J Math* 109: 521–43.
5. Berenstein, C.A., and D.C. Struppa. 1989. Complex analysis and convolution equations. In *Several complex variables V: Encyclopaedia of mathematical sciences*, ed. G.M. Khenkin, 54, 5–111, Moscow: Akad Nauk SSSR, Vsesoyuz Inst Nauchn i Tekhn Inform (Russian). English transl.: 1–108, Berlin-Heidelberg: Springer, 1993.
6. Berenstein, C.A., and B.A. Taylor. 1980. Interpolation problems in $\mathbb{C}^n$ with applications to harmonic analysis. *J Analyse Math* 38: 188–254.
7. Colombo, F., A. Damiano, I. Sabadini, and D.C. Struppa. 2005. A surjectivity theorem for differential operators on spaces of regular functions. *Compl Var Theor Appl* 50: 389–400.
8. Colombo, F., I. Sabadini, F. Sommen, and D.C. Struppa. 2004. *Analysis of Dirac systems and computational algebra. Progress in mathematical physics*, 39. Boston: Birkhäuser.
9. Dieudonné, J., and L. Schwartz. 1950. La dualité dans les espaces $\mathcal{F}F$ et $(\mathcal{L}\mathcal{F})$. (French) *Ann Inst Fourier Grenoble* 1(1949): 61–101.
10. Ehrenpreis, L. 1970. *Fourier analysis in several complex variables*. New York: Wiley.
11. Krivosheev, A.S., and V.V. Napalkov. 1992. Complex analysis and convolution operators. (Russian) *Uspekhi Mat Nauk* 47(288): 3–58; translation in Russian Mathematical Surveys 47(6): 1–56.
12. Meril, A., and D.C. Struppa. 1987. Convolutors in spaces of holomorphic functions, Complex analysis II (College Park, MD, 1985–86), 253–75. In *Lecture Notes in Mathematics*, 1276. Berlin: Springer.
13. Palamodov, V.P. 1970. *Linear differential operators with constant coefficients*. Translated from the Russian by A. A. Brown. Die Grundlehren der mathematischen Wissenschaften, Band 168. New York, Berlin: Springer.
14. Pincherle, S. 1888. Sulla risoluzione dell'equazione funzionale $\Sigma h_\nu \varphi(x + \alpha_\nu) = f(x)$ a coefficienti costanti. *Mem R Accad Sci Ist Bologna* 9: 45–71.
15. Pincherle, S. 1926. Sur la résolution de l'équation fonctionnelle $\sum h^\nu \varphi(\alpha^\nu) = f(x)$ à cofficients constants. *Acta Math* 48: 279–304.
16. Pincherle, S.1954. *Opere Scelte. Unione Matematica Italiana*, Edizioni Cremonese, Roma vol. I, II.
17. Sabadini, I., and D.C. Struppa. 1996. Topologies on quaternionic hyperfunctions and duality theorems. *Compl Var Theor Appl* 30: 19–34.
18. Struppa, D.C. 1981. The fundamental principle for systems of convolution equations. *Memoir Am Math Soc* 273: 1–64.
19. Struppa, D.C., and C. Turrini. 1991. Pincherle's contribution to the Italian school of analytic functionals. Conference on the history of mathematics, Italian, Cetraro, 1988, 551–560, Sem Conf 7, EditEl, Rende.
20. Sudbery, A. 1979. Quaternionic analysis. *Math Proc Camb Phil Soc* 85: 199–225.

The Work of Beniamino Segre on Curves and Their Moduli

Edoardo Sernesi

Abstract In this paper, we discuss part of the mathematical production of B. Segre, focusing on his contributions on families of plane curves and their moduli, especially in the years 1928/1930. Our purpose is to illustrate the relevance of his work to today's research activity on the subject. We will take the opportunity to survey recent development on these areas of research.

1 The Context

The purpose of this article is to overview some of B. Segre's early work on algebraic curves and to outline its relevance to today's research on the subject, hoping that this can be of some interest both to historians and to researchers. The papers we will consider represent only a very small part of the impressive scientific production of B. Segre (see [37] for the list of its publications). In particular I will concentrate on few papers published at the beginning of his career, in the years 1928/30, when he served as an assistant professor of Severi in Roma. His scientific production of those few years opened the way to his appointment on the chair of *Geometria Superiore* in Bologna in 1931.

Recall that Beniamino Segre studied with Corrado Segre in Torino, where he was born in 1903. After spending one year in Paris with E. Cartan, he arrived in Roma in 1927. There he found the three greatest italian algebraic geometers, Castelnuovo, Enriques and Severi, and his research interests immediately focused on some of the hottest problems of contemporary algebraic geometry. He was only 24 years old.

E. Sernesi (✉)
Dipartimento di Matematica, Università Roma Tre, Largo S.L. Murialdo 1, 00146 Roma, Italy
e-mail: sernesi@mat.uniroma3.it

2 Severi's Vorlesungen

In order to put Segre's work in the appropriate perspective it is necessary to review the advancement of curve theory at that time. Severi's *Vorlesungen* [47] had appeared in 1921. Many authors have already criticized and commented this well known work, so we don't need to discuss it in detail. We only want to highlight some parts relevant to Segre's work. This book contained some incomplete proofs and even some false statement, but nevertheless it indicated the state of the art on algebraic curves and their moduli. His main asserted achievement, supported by confused and unconvincing arguments, was the positive solution of *Zeuthen's Problem* asking whether, for each $r \geq 3$, every irreducible nonsingular and nondegenerate curve in $\mathbb{P}^r$ can be flatly degenerated to a *polygon* (also called a *stick figure*), i.e. to a curve which is a reduced nodal union of lines. This statement had been proved previously in some special cases [5] but we now know it to be false in general: a counterexample was found by Hartshorne [27]. The problem was motivated by the need of having some discrete invariants, finer than just degree and genus, that could distinguish among distinct irreducible components of the Hilbert scheme of curves. As of today such invariants have not been discovered yet. It is worth observing that the flat degeneration to a stick figure is correctly described by Severi in the case of linearly normal curves of degree $n = g + r$ in $\mathbb{P}^r$, and that there are no counterexamples known in the case of curves of degree n and genus g in $\mathbb{P}^r$ such that the inequality $\rho(g, r, n) \geq 0$ holds, where

$$\rho(g, r, n) := g - (r + 1)(g - n + r)$$

is the *Brill–Noether number*.

In a previous series of two notes [46] Severi had sketched his results and indicated new directions of research. Here he outlined several ideas and made remarkable conjectures. Among the topics he considered we find:

(a) *Riemann's existence theorem (RET) and algebraic geometry.*

He outlined [46, n. 4] an idea, which was expanded in [47], Anhang F n. 8, for a purely algebro-geometrical statement and proof of RET. He later gave a modified argument in [48]. The lack of such a proof has been remarked by Mumford ([34], p. 15) and apparently the idea of Severi, based on a degeneration argument, still awaits for a modern critical reconsideration.

(b) *Rationality and unirationality properties of moduli spaces of curves.*

He formulated [46, n. 2] the famous conjecture about the unirationality of the moduli space M_g of curves of any genus g, and sketched a proof for genus $g \leq 10$ using families of plane curves. We will come back on his proof below, when discussing the content of [40]. The problem of unirationality of M_g is mentioned as a very important and difficult one in [34, p. 37]. The conjecture has been confirmed up to genus 14 [14, 45, 51] and disproved in general [18, 20, 26]. It is still unsettled for $15 \leq g \leq 21$ (even though we know that the moduli space of *stable* curves of genus 15 is rationally connected [9] and that M_{16} has negative Kodaira dimension [15]).

(c) *Families of plane curves,* especially of nodal curves.

In [47], Anhang F, he considered the family $\mathcal{V}_{n,g}$ of plane nodal irreducible curves of given degree n and geometric genus g and sketched a proof of its irreducibility. His argument is incomplete (see [22] for a discussion), and the irreducibility of $\mathcal{V}_{n,g}$ has remained unsettled until 1986, when Harris proved it [25].

(d) *Brill–Noether theory.*

He proposed several approaches to the proof of the main statements of the so-called *Brill–Noether theory* concerning the existence and dimension of the varieties W_n^r parametrizing special linear series of dimension r and degree n on a general curve. Such statements had been made without proof, because considered as evident by the authors, in the seminal paper [6]. Severi proposed at least two proofs of them by degeneration. The first one, less known, is sketched in [46, n. 3]: it is by flatly degenerating a nonsingular curve of genus g to an irreducible nodal curve of genus $g - 1$, and then proceeding by induction. It has never been fixed or disproved. The second proof is outlined in [47], Anhang G, and it uses a degeneration of a nonsingular curve of genus g to an irreducible g-nodal rational curve. These problems, and the second argument of Severi, have attracted much attention in post-Grothendieck times. The existence statement of Brill–Noether theory, which is far from trivial, has been proved independently and almost simultaneously by Kempf [29] and by Kleiman–Laksov [30]. The statement about the dimension of the varieties W_n^r has remained unsettled for many years and finally proved by Griffiths–Harris [24] using a degeneration argument which follows quite closely Severi's idea.

(e) *Classification of maximal families of projective curves,* i.e. what today goes under the name of *Hilbert and Chow schemes.*

Severi claimed something equivalent to the statement that, whenever $n \geq g + r$ the Hilbert scheme of $\mathbb{P}^r$ has a unique irreducible component $\mathbf{I}_{g,r,n}$ whose general point parametrizes an irreducible nonsingular and non-degenerate curve of genus g and degree n. The existence of $\mathbf{I}_{g,r,n}$ is a slightly stronger assertion that the existence statement in Brill–Noether theory (see (d) above), but it is trivial in this case. What we know today is that there is a unique component $\mathbf{I}_{g,r,n}$ of Hilb^r which parametrizes *general curves* (in the sense of moduli): this has been proved by Fulton and Lazarsfeld [23] using the irreducibility of M_g for all g, r, n such that $\rho(g, r, n) \geq 0$. The irreducibility statement of Severi has been proved in some special cases [17] but it is false in general. Harris found a series of counterexamples, i.e. of irreducible components different from $\mathbf{I}_{g,r,n}$, reported in [17]; more examples have been found by Mezzetti and Sacchiero [33]. The components found by Harris generically parametrize nonsingular trigonal curves of degree n and genus g which are *not linearly normal*. The same happens for the components described in [33]. No components different from $\mathbf{I}_{g,r,n}$ and generically consisting of linearly normal curves have been found yet for any value of $\rho(g, r, n) \geq 0$, so that Severi's irreducibility conjecture is still open for the extended set of values of g, r, n such that $\rho(g, r, n) \geq 0$, if we interpret it as a statement about components of Hilb^r generically parametrizing *linearly normal* curves.

It is interesting to observe that Severi was constantly considering curves from the point of view of their moduli spaces. Brill and Noether had the same point of view when they spoke of *the general curve,* but with Severi all questions were clearly related to moduli, even though there was no definition of M_g as an algebraic variety at that time. With the language of today we could say that Severi had a *stacky* point of view, because he was looking at those properties of moduli spaces which could be detected by means of properties of families of curves, so that for him M_g was just the target of the functorial morphisms induced by such families. Another notable fact is the systematic use of degeneration methods. Such methods derived from Schubert, and had already been applied by Castelnuovo [10] to enumerative problems on algebraic curves, but now Severi tried to use them to prove irreducibility statements, where very delicate monodromy-type of phenomena take place. We can say that, more than proving theorems, Severi raised questions and made conjectures, and they had a long lasting influence on curve theory.

3 On Riemann's Existence Theorem

This scenery must have been very stimulating for the young and talented Segre, and certainly Severi exerted a strong influence on him. As we will see next, Beniamino worked on moduli of curves, but he never ventured into the use of degeneration techniques: he rather relied on a solid knowledge of the geometry of plane curves and Cremona transformations.

The first paper I will discuss is [38]. Here, Segre considered d-gonal curves of genus g with $d < \frac{1}{2}g + 1$ and showed that the general such curve has only finitely many linear series of degree d and dimension 1 (i.e. g_d^1's). This is equivalent to showing that the locus $M_{g,n}^1 \subset M_g$ of such curves has the expected dimension $2g + 2d - 5$. The result was obtained by means of the construction of plane models of d-gonal curves. The question of counting the dimension of $M_{g,n}^1$ had been already considered in [48], but the estimate given there, even though correct, overlooked the finiteness question considered in [38]. In [3], this paper has been analyzed, commented and generalized. Arbarello and Cornalba improved Segre's result by proving that a general d-gonal curve as above has a *unique* g_d^1. Moreover, using the plane models constructed in [38], they obtained a new proof of the unirationality of M_g for $g \leq 10$. It is interesting to note that Severi expressed the codimension of $M_{g,n}^1$ in M_g at a point C as $\dim[H^1(L^2)]$, where L is the line bundle defining the (complete) g_n^1. Today, using cohomological methods, we understand the meaning of this estimate because we know that $H^1(L^2)$ is the conormal space of $M_{g,n}^1$ in M_g at the point C. On the other hand the point of view of B. Segre is related with an interesting and still unsolved problem concerning the so-called *Petri loci* in M_g. For given g, n such that $n \geq \frac{1}{2}g + 1$ one defines the Petri locus $P_{g,n} \subset M_g$ as the set of curves having a line bundle L of degree n such that the so-called *Petri map*

$$\mu_0(L) : \ker[H^0(L) \otimes H^0(K - L) \longrightarrow H^0(K)]$$

is not injective. This locus is not well understood yet, and the first question one can ask is whether it has pure codimension one. This would be true if a general curve C in any component of $P_{g,n}$ had only finitely many L's such that $\mu_0(L)$ is not injective. This is exactly what Segre's theorem says in the case $n < \frac{1}{2}g + 1$; unfortunately the problem is of a completely different nature in the case $n \geq \frac{1}{2}g+1$, and Segre's proof does not extend. For partial results in this direction we refer to [7, 8, 13, 19, 21, 50]. These examples show that the problems considered by Severi and Segre are still meaningful and interesting today.

4 On Moduli of Plane Curves

Another interesting, but almost forgotten, paper is [40]. Here Segre consider a problem that has been suggested to him by Severi, as he says in the introduction, namely to try to extend Severi's proof of the unirationality of M_g for $g \leq 10$, to higher values of the genus. Before describing Segre's paper we briefly recall the proof of Severi.

The degree n of a plane model of a *general* curve of genus g must satisfy the inequality

$$n \geq \frac{2}{3}g + 2.$$

Taking n minimal with this property and assuming that such a model C has only nodes as singularities, one easily computes that when $g \leq 10$ the number δ of nodes of C satisfies:

$$3\delta \leq \frac{1}{2}n(n + 3). \tag{1}$$

The right hand side of this inequality is the dimension of the linear system of plane curves of degree n, while the left hand side is an upper bound for the number of linear conditions imposed to such curves if we want them to have nodes at given distinct points $P_1, \ldots, P_\delta$. Therefore (1) implies that there exist nodal curves of degree n and genus g with nodes at any *general* set of δ distinct points $P_1, \ldots, P_\delta$. Therefore, Severi concludes, the family W of plane nodal curves of degree n and genus g is a union of linear systems parametrized by an open set of $(\mathbb{P}^2)^{(\delta)}$, the δ-th symmetric product of $\mathbb{P}^2$, and therefore it is rational because it is a union of projective spaces parametrized by a rational variety: it follows that M_g is unirational because it is dominated by W. This argument, as it stands, is not complete because some of its steps need to be justified: for example it might be a priori possible that the family of irreducible curves of degree n and genus g has two components, and that the one having its singular points in general position does not consist of general curves. This and other minor objections to this argument can be fixed easily, so that the proof can be made to satisfy modern standards (see [4] for a discussion).

In [40], Segre began by showing that there exists a *linear system* of plane irreducible curves of genus g having general moduli, i.e. containing the general curve of genus g, if and only if $g \leq 6$. The existence part of his argument is elementary; nevertheless it is worth reading his elegant argument in the case $g = 5$. The proof of non-existence for $g > 10$ is based on the observation that a linear system containing the general curve must have dimension at least $3g - 3$, and therefore strictly larger than $2g + 7$. A theorem of Castelnuovo [11] (see [12] for a modern discussion) implies that a linear system of plane curves of genus g and dimension larger than $2g + 7$ consists of hyperelliptic curves and therefore it cannot have general moduli if $g > 10$. The remaining cases $7 \leq g \leq 10$ are treated with a special, and non-obvious, argument.

In the second part of [40] Segre investigated whether there exist irreducible families of plane curves of genus $g > 10$, not necessarily linear systems, having general moduli and whose singularities are general points in $\mathbb{P}^2$; in other words, whether the proof of Severi could be extended by allowing the singularities of the curves of the system to be arbitrary, and not just nodes. The conclusion of Segre is that such a family W cannot exist, under the following assumption. The general element C of the family W will be of degree n and will have singularities of multiplicities $v_1 \geq v_2 \geq \cdots \geq v_r \geq 2$ say, at general points $P_1, \ldots, P_r$; then the assumption is that the linear system $\mathcal{L}(n; v_1 P_1, \ldots, v_r P_r)$ of curves of degree n and having multiplicity $\geq v_i$ at P_i is regular. Assuming this fact Segre gives an elaborated argument to show that $v_1 + v_2 + v_3 > n$ and from this fact the conclusion follows immediately because then the degree of the curve C can be lowered by means of the quadratic transformation centered at P_1, P_2, P_3: it follows that the family W can be replaced by an analogous one consisting of curves of lower degree and this leads to a contradiction. This interesting proof has not been rewritten in modern language yet. Moreover, it must be remarked that today the regularity of the linear system $\mathcal{L}(n; v_1 P_1, \ldots, v_r P_r)$ with general base points $P_1, \ldots, P_r$ is known to be true only in special cases, but not for any choice of $n, v_1, \ldots, v_r$; the question of regularity of such linear systems, which originates from Segre [44], has generated a great deal of research in the last few years, and many partial results are known. We refer to [16] for an overview about these problems. It follows that the truth of Segre's result is conditioned by the validity of the regularity assumption made. On the other hand, from his argument it follows that if a plane curve of genus $g \geq 11$ has general moduli, has degree n and has singularities of multiplicities $v_1 \geq v_2 \geq \cdots \geq v_r \geq 2$ say, at general points $P_1, \ldots, P_r$, then the linear system $\mathcal{L}(n; v_1 P_1, \ldots, v_r P_r)$ is superabundant.

In the last part of his paper Segre drops the regularity assumption on the linear systems $\mathcal{L}(n; v_1 P_1, \ldots, v_r P_r)$ and considers a similar problem. He arrives at the conclusion that *every family of curves of genus $g > 36$ having singular points* *in general position cannot consist of curves with general moduli.* The proof of this result is long and elaborate and it has not been reconsidered in recent times. It would deserve a critical screening.

In [2], assuming a technical lemma used in [40], Arbarello gives a generalization of the last result of Segre to rational surfaces. To my knowledge this is the only relatively recent paper taking up the earlier work [40].

Also, in this case, the work of Segre turns out to be strongly related to problems of contemporary research. It has to do with the problem of deciding about the unirationality or uniruledness of the moduli space M_g for low values of g, which is still unsettled when $16 \leq g \leq 21$.

5 On Plane Curves with Nodes and Cusps

As we have seen, B. Segre had a deep knowledge of plane algebraic curves, a subject he certainly had learned from his master C. Segre. No surprise then if he also ventured into difficult problems on families of plane curves with nodes, cusps and higher singularities. This subject was already classical, after the work of Lefschetz [32], Albanese [1] and, of course, Enriques and Severi. Let's briefly recall the set-up in modern language.

If we fix n, δ, κ, then there is a universal, possibly empty, family (called a "maximal" family in Segre's language):

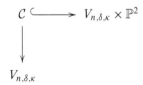

parametrizing all plane curves of degree n with δ nodes, κ cusps and no other singularities. Here $V_{n,\delta,\kappa} \subset \mathbb{P}^{n(n+3)/2}$ is a locally closed subscheme. The existence of $V_{n,\delta,\kappa}$ has been proved in modern standards by Wahl [52]. One of the main tools available at that time, and still now, in the study of $V_{n,\delta,\kappa}$ is the so-called "characteristic linear series" which can be introduced as follows. Given an irreducible $[C] \in V_{n,\delta,\kappa}$ one considers the linear system of curves of degree n which is the projective tangent space to $V_{n,\delta,\kappa}$ at $[C]$: then the induced linear series on the normalization $\widetilde{C}$ of C is the *characteristic linear series* of $V_{n,\delta,\kappa}$ at $[C]$. With a suggestive terminology the classical geometers said that the characteristic linear series is cut on $\widetilde{C}$ by the curves of the family $V_{n,\delta,\kappa}$ which are "infinitely near" C. This series turns out to be defined by sections of the sheaf $\mathcal{O}_{\widetilde{C}}(K + 3H - p_1 - \cdots - p_\kappa)$, where $p_1, \ldots, p_\kappa \in \widetilde{C}$ are the inverse images of the cusps and H is the pullback of a line section under the normalization morphism $\widetilde{C} \longrightarrow C$; therefore the Zariski tangent space $T_{[C]} V_{n,\delta,\kappa}$ is naturally a subspace of $H^0(\widetilde{C}, \mathcal{O}_{\widetilde{C}}(K + 3H - p_1 - \cdots - p_\kappa))$. On the other hand a standard count of constants shows that $V_{n,\delta,\kappa}$ has dimension at least

$$3n + g - 1 - \kappa = h^0(\widetilde{C}, \mathcal{O}_{\widetilde{C}}(K + 3H - p_1 - \cdots - p_\kappa))$$
$$- h^1(\widetilde{C}, \mathcal{O}_{\widetilde{C}}(K + 3H - p_1 - \cdots - p_\kappa))$$

at $[C]$, where $h^i(-)$ means $\dim[H^i(-)]$ and

$$g = \frac{(n-1)(n-2)}{2} - \delta - \kappa$$

is the geometric genus of C, i.e. the genus of $\widetilde{C}$. The lower bound $3n + g - 1 - \kappa$ is called the *virtual dimension* of $V_{n,\delta,\kappa}$. The above facts imply:

1. If
$$H^1(\widetilde{C}, \mathcal{O}_{\widetilde{C}}(K + 3H - p_1 - \cdots - p_\kappa)) = 0, \tag{2}$$
then $V_{n,\delta,\kappa}$ is nonsingular at $[C]$ of dimension equal to the virtual dimension. We then say that $V_{n,\delta,\kappa}$ is *regular at* $[C]$. If this happens at all $[C]$ in an irreducible component $W \subset V_{n,\delta,\kappa}$ (resp. in $V_{n,\delta,\kappa}$) we say that W (resp. $V_{n,\delta,\kappa}$) is *regular*.
2. If $\kappa < 3n$ then (2) holds and therefore $V_{n,\delta,\kappa}$ is regular.

The main problems concerning the varieties $V_{n,\delta,\kappa}$ are:

(a) Establish for which values of n, δ, κ we have $V_{n,\delta,\kappa} \neq \emptyset$
(b) Prove or disprove irreducibility. In the negative case prove or disprove equidimensionality
(c) Prove or disprove the existence of non-regular components
(d) Investigate the existence of singular points of $V_{n,\delta,\kappa}$, and give examples

All these problems (except irreducibility, as we saw before) had been completely solved by Severi [47] in the case $\kappa = 0$ (no cusps), but they were widely open in the presence of cusps.

In [41], after introducing some terminology, Segre gave examples of non-regular, but everywhere nonsingular, components of $V_{n,\delta,\kappa}$ for infinitely many values of n, δ, κ. He then went on by giving, in [42], examples of reducible $V_{n,\delta,\kappa}$ both in distinct regular components and in components of different dimensions. The simplest of these examples are sextics with six cusps: $V(6, 0, 6)$ consists of two components, both regular hence of dimension 15, distinguished by the condition that the cusps do/do not belong to a conic; in fact the sextics with six cusps on a conic appear as branch curves of general projections of cubic surfaces. With these examples Segre took care of problems (b) and (c). His examples are beautifully simple, and are essentially the only one known. They have been quoted by Zariski in [53, p. 220 and 223]. For a modern treatment we refer to [49].

In the second part of [41] Segre suggested a procedure to construct new components starting from given ones, in the attempt to solve problem (a). He claimed, at the end, the non-emptiness of $V(2n, 0, n^2)$, for all $n \geq 3$ and of $V(2n + 1, 0, n^2 + n - 1)$ for all $n \geq 1$. These claims are not clearly explained, the inductive procedure he proposes involves smoothing arguments of non-reduced curves, and the all construction needs a careful inspection. Certainly his conclusions, as they stand, are incorrect. In fact Zariski [53, p. 222] showed the non-existence of curves of degree 7 with 11 cusps, even though they appear in the above list of Segre. Nevertheless Segre's procedure seems to be correct when he constructs new *regular* components by deforming cusps to nodes and smoothing nodes in regular components [41, n. 7]. This idea still awaits for a modern treatment. Note that the existence problem (a) is not completely solved yet. For recent partial results see [28, 31].

Recall that branch curves of multiple planes are curves with nodes and cusps, and not only nodal. This fact introduces a whole variety of cases which are partly responsible for the many possibilities that can occur in problems (a)–(c), as opposed to the case of families of curves with only nodes, that are always irreducible and regular. Segre took up this point of view in [43], where he characterized those plane curves which are branch curves of general projections of nonsingular surfaces of $\mathbb{P}^3$. His treatment of this problem is entirely algebro-geometrical and extremely elegant, and the characterization is given in terms of degrees of curves containing the singular points.

Problem (d) has remained out of reach for the classical geometers, essentially due to lack of technique. Only with the help of scheme theory the study of families has reached a level of sophistication sufficient to understand such questions. The first example of a singularity of a family $V(n, \delta, \kappa)$ or, in modern terminology, of an *obstructed* curve, has been given by Wahl [52]: it is a curve of degree 104 with 3,636 nodes and 900 cusps.

Finally, I would like to mention the interesting Note [39], which has been completely forgotten so far, where Segre showed that the characteristic linear series of a family of curves with an ordinary tacnode is always incomplete. He raised the question of giving a geometrical interpretation of the defect of completeness of this series. It is a question which deserves to be studied.

The all subject of families of plane curves with nodes and cusps is discussed at length in Chap. VIII of [53], to which we refer the reader for other classical references.

6 Final Considerations

From the above discussion it should hopefully emerge how much Segre's work was inspired, if not guided, by the scientific figure of Severi. At the same time, despite his young age and the presence near him of such an influential personality, his independent creativity emerged quite strongly. We have seen this in the way he went beyond Severi's [48] in computing the dimension of $M_{g,n}^1 \subset M_g$ in [38], or in the way he explored moduli of plane curves using his technique that, we can guess, came to him directly from C. Segre. The papers on plane curves with nodes and cusps reflect perhaps another influence, coming from Castelnuovo and especially Enriques, even though I was not able to find direct confirmation about any scientific relation of B. Segre with them. We should also note that everything he did on algebraic curves at the time was restricted to plane curves, a very classical and safe point of view. He did not venture into the geometry of curves in higher projective spaces, and in this respect he did not take up the spirit of [47]. Nevertheless, his work is far from being forgotten today, since the problems he considered were deep and difficult and, in fact, still unsolved or related with important open problems. In closing observe that in the 1920s there has been another important development in curve theory, represented by the nowadays well known papers by Petri [35, 36], who was the last student of M. Noether. It is interesting to see how different the points of view of Severi/Segre and of Petri were. The work of Petri was bringing the equations to the center of attention, mixing up projective geometry with elegant methods of an algebraic-homological nature, thus reflecting traditions and developments of the German and British schools of algebra and invariant theory. The italians insisted in carrying on a purely synthetic point of view, no algebra being allowed to contaminate their world. This attitude, while coherent with their tradition and style, would eventually be a cause of isolation and decline for the headquarters of italian algebraic geometry.

Acknowledgements The author is member of GNSAGA-INDAM.

References

1. Albanese, G. 1923. Sui sistemi continui di curve piane algebriche. doctoral thesis, Nistri, Pisa.
2. Arbarello, E. 1975. Alcune osservazioni sui moduli delle curve appartenenti ad una data superficie algebrica, *Rend Accad Naz Lincei* (8) 59: 725–732.
3. Arbarello, E., and M. Cornalba. 1981. Footnotes to a paper of Beniamino Segre. *Math Ann* 256: 341–362.
4. Arbarello, E., and E. Sernesi. 1979. The equation of a plane curve. *Duke Math J* 46: 469–485.
5. Brill, A. 1907. Uber algebraische raumkurven. *Math Ann* 64: 289–324.
6. Brill, A., and M. Noether. 1873. Über die algebraischen funktionen und ihre anwendung in geometrie. *Math Ann* 7: 269–310.
7. Bruno, A., and E. Sernesi. 2011. A note on the Petri loci. *Manuscripta Math* 136: 439–443.
8. Bruno, A., and E. Sernesi. 2011. The symmetric square of a curve and the Petri map., arXiv 1106.3190.
9. Bruno, A., and A. Verra. 2005. M_{15} is rationally connected. In *Projective varieties with unexpected properties*, 51–65. edited by C. Ciliberto, A.V. Geramita, B. Harbourne, R.M. Miro'-Roig, K. Ranestad. New York: Walter de Gruyter .
10. Castelnuovo, G. 1889. Una applicazione della geometria enumerativa alle curve algebriche. *Rend Circolo Mat Palermo* 3: 27–37.
11. Castelnuovo, G. 1892. Ricerche generali sopra i sistemi lineari di curve piane. *Mem R Accad Sci Torino Cl Sci Fis Mat Nat* (2), 42: 137–188.
12. Castorena, A., and C. Ciliberto. 2010. On a theorem of Castelnuovo and applications to moduli. arXiv: 1008.3248.

13. Castorena, A., and M. Teixidor i Bigas. 2008. Divisorial components of the Petri locus for pencils. *J Pure Appl Algebra* 212: 1500–1508.
14. Chang, M.C., and Z. Ran. 1984. Unirationality of the moduli spaces of curves of genus 11, 13 (and 12). *Invent Math* 76: 41–54.
15. Chang, M.C., and Z. Ran. 1991. On the slope and Kodaira dimension of $\overline{M}_g$ for small g. *J Diff Geom* 34: 267–274.
16. Ciliberto, C., and R. Miranda. 2001. The Segre and Harbourne–Hirschowitz Conjectures. In: *Applications of algebraic geometry to coding theory, physics and computation (Eilat 2001)*, 37–51, edited by C. Ciliberto, F. Hirzebruch, R. Miranda, M. Teicher. *NATO Science Series II: Mathematics, Physics and Chemistry*, 36. Dordrecht: Kluwer.
17. Ein, L. 1987. The irreducibility of the Hilbert scheme of smooth space curves. In *Algebraic geometry, Bowdoin, 1985 (Brunswick, Maine, 1985)*, 83–87, edited by S. Bloch, *Proceedings of symposia in pure mathematics, Part 1*, 46. Providence, RI: American Mathematical Society.
18. Eisenbud, D., and J. Harris. 1987. The Kodaira dimension of the moduli space of curves of genus $\geq$ 23. *Invent Math* 90: 359–387.
19. Farkas, G. 2005. Gaussian maps, Gieseker–Petri loci and large theta-characteristics, *J Reine Angew Math* 581: 151–173.
20. Farkas, G. 2005. $\overline{M}_{22}$ is of general type. Manuscript.
21. Farkas, G. 2010. Rational maps beween moduli spaces of curves and Gieseker–Petri divisors. *J Algebr Geom* 19: 243–284.
22. Fulton, W. 1983. On nodal curves. In *Algebraic geometry—open problems (Ravello, 1982)*, edited by C. Ciliberto, F. Ghione, F. Orecchia. *Lecture Notes in Mathematics*, 997, 146–155. Berlin: Springer. .
23. Fulton, W., and R. Lazarsfeld. 1981. On the connectedness of degeneracy loci and special divisors. *Acta Math* 146: 271–283.
24. Griffiths, P., and J. Harris. 1980. The dimension of the space of special linear series on a general curve. *Duke Math J* 47: 233–272.
25. Harris, J. 1986. On the Severi problem. *Invent Math* 84: 445–461.
26. Harris, J., and D. Mumford. 1982. On the Kodaira dimension of the moduli space of curves. *Invent Math* 67: 23–86.
27. Hartshorne, R. 1997. Families of curves in $\mathbb{P}^3$ and Zeuthen's problem. *Memoir Am Math Soc* 130(617). p. viii, 96
28. Hirzebruch, F. 1986. Singularities of algebraic surfaces and characteristic numbers. *Contemp Math* 58: 141–155.
29. Kempf, G. 1971. *Schubert methods with an application to special divisors*. Amsterdam: Publications of Mathematical Centrum.
30. Kleiman, S., and D. Laksov. 1972. On the existence of special divisors. *Am J Math* 94: 431–436.
31. Langer, A. 2003. Logarithmic orbifold Euler numbers of surfaces with applications. *Proc London Math Soc* (3) 86: 358–396.
32. Lefschetz, S. 1913. On the existence of loci with given singularities. *Trans AMS* 14: 23–41.
33. Mezzetti, E., and G. Sacchiero. 1989. Gonality and Hilbert schemes of smooth curves. In *Algebraic curves and projective geometry (Trento, 1988)*, edited by E. Ballico, C. Ciliberto. *Lecture Notes in Mathematics*, 1389. Berlin: Springer.
34. Mumford, D. 1975. *Curves and their Jacobians*. Ann Arbor, Mich: The University of Michigan Press.
35. Petri, K. 1922. Über die invariante darstellung algebraischer funktionen einer veränderlichen. *Math Ann* 88: 242–289.
36. Petri, K. 1925. Über Spezialkurven I. *Math Ann* 93: 182–209.
37. Segre, B. 1987. *Memorie Scelte*, vol. I. Roma: Cremonese.
38. Segre, B. 1928. Sui moduli delle curve poligonali e sopra un complemento al teorema di esistenza di Riemann. *Math Ann* 100: 537–551.
39. Segre, B. 1929. Sui sistemi continui di curve con tacnodo.*Rend Accad Naz Lincei* (6) 9_1: 970–974.
40. Segre, B. 1929–30. Sui moduli delle curve algebriche. *Ann di Mat* (4) 7: 71–102.

41. Segre, B. 1929. Esistenza e dimensioni di sistemi continui di curve piane algebriche con dati caratteri. *Rend Accad Naz Lincei* (6) 10_2: 31–38.
42. Segre, B. 1929. Esistenza di sistemi continui distinti di curve piane algebriche con dati numeri plückeriani. *Rend Accad Naz Lincei* (6) 10_2, 557–560.
43. Segre, B. 1930. Sulla caratterizzazione delle curve di diramazione dei piani multipli generali. *Memor Reale Acc d'Italia* 1, 5–31.
44. Segre, B. 1961. Alcune questioni su insiemi finiti di punti in geometria algebrica. In *Atti Convegno Intern di Geom Alg di Torino*, 15–33.
45. Sernesi, E. 1981. L'unirazionalità della varietà dei moduli delle curve di genere dodici. *Ann Sc Norm Super Pisa* (4), 8: 405–439.
46. Severi, F. 1915. Sulla classificazione delle curve algebriche e sul teorema di esistenze di Riemann. *Rend Accad Naz Lincei* (5) 241: 877–888, 1011–1020.
47. Severi, F. 1921. *Vorlesungen über algebraische Geometrie: Geometrie auf einer Kurve, Riemannsche Flächen, Abelsche Integrale*. Leipzig: Teubner.
48. Severi, F. 1922. Sul teorema di esistenza di Riemann. *Rend Circolo Mat Palermo* 46: 105–116.
49. Tannenbaum, A. 1984. On the classical characteristic linear series of plane curves with nodes and cuspidal points: two examples of Beniamino Segre. *Compos Math* 51: 169–183.
50. Teixidor, M. 1988. The divisor of curves with a vanishing theta null. *Compos Math* 66: 15–22.
51. Verra, A. 2005. The unirationality of the moduli spaces of curves of genus 14 or lower. *Compos Math* 141: 1425–1444.
52. Wahl, J. 1974. Deformations of plane curves with nodes and cusps. *Am J Math* 96: 529–577.
53. Zariski, O. 1971. *Algebraic surfaces*, Ergebnisse der Mathematik und ihrer Grenzgebiete, Bd. 61, 2nd supplemented ed. Berlin, New York: Springer.

Lamberto Cattabriga and the Theory of Linear Constant Coefficients Partial Differential Equations

Daniele C. Struppa

Abstract This article focuses on the contributions of Cattabriga and De Giorgi to the study of surjectivity of linear constant coefficients partial differential equations on spaces of real analytic functions. Their contributions are placed in the context of the concurrent development of the general theory of Analytically Uniform spaces due to Ehrenpreis.

1 Introduction

Lamberto Cattabriga was one of the most important contributors to the development of the theory of linear constant coefficients partial differential equations which, in the aftermath of the creation of the Schwartz theory of distributions, underwent a dramatic evolution in the years between 1950 and 1980. It would be impossible to discuss in detail these developments in a short article, but what I would like to do here is to outline the impact of the work of Cattabriga on the understanding of the issue of surjectivity of linear constant coefficients differential operators.

More specifically, I will concentrate on the early developments of the theory of surjectivity for partial differential operators, which was fully developed through the works of Ehrenpreis [8–12], Hormander [13–16], Malgrange [23], and Palamodov [25] among others. All of these developments were related to those spaces which Ehrenpreis called Analytically Uniform (or AU-spaces in short), and the theory of such spaces provided a wonderfully unifying setup for the study of constant coefficients operators. However, it turned out that there was at least one extremely important space that could not be studied within this framework, namely the space $\mathcal{A}$ of real analytic functions. It was one of the major accomplishments of Cattabriga, in a series of joint papers with Ennio De Giorgi, to show that despite the fact that $\mathcal{A}$

D.C. Struppa (✉)
Schmid College of Science and Technology, Chapman University, Orange, CA 92866, USA
e-mail: struppa@chapman.edu

was not an Analytically Uniform space, it was still possible to study surjectivity in it and actually prove it at least in some cases.

I will attempt to recreate the intellectual landscape of the early 1960s into the 1970s, and I will describe the evolution of the study of the division problem (an equivalent formulation of surjectivity, as we will soon see) in this context. This will give us an opportunity to then address some of the fundamental contributions of Cattabriga. I am fully aware that it is not possible to do justice to the importance of Cattabriga's work in a short paper, and I hope the reader will find my comments sufficiently interesting to stimulate his/her desire to delve more deeply in the work of the Italian school.

2 Analytically Uniform Spaces and Surjectivity of Linear Constant Coefficients Operators

In the 1960s and 1970s, Ehrenpreis [12] and Palamodov [25] developed a rather comprehensive theory for the study of global properties of solutions of systems of linear constant coefficients partial differential equations.

Maybe the most important result of this theory is the celebrated Ehrenpreis–Palamodov Fundamental Principle, which states that every solution (in suitable spaces, see below) of a homogeneous system of such differential equations can be expressed as a sum of integrals of exponential solutions (and their derivatives). In this sense, the Fundamental Principle is a very far reaching extension of the classical result of Euler for ordinary linear differential equations with constant coefficients. In order to describe this result, we will begin with some notations.

If we are working with (generalized) functions on an n-dimensional space (e.g. $\mathbb{R}^n$ or $\mathbb{C}^n$), we use the symbol D to indicate the formal vector $(\frac{\partial}{\partial x_1}, \ldots, \frac{\partial}{\partial x_n})$, where $x_1, \ldots, x_n$ are the variables in the n-dimensional space. Thus, if $P = \Sigma_{|I|=0}^{m} a_I z^I$ is a polynomial in n variables, with $z = (z_1, \ldots, z_n)$, $I = (i_1, \ldots, i_n)$, and $z^I = z_1^{i_1} \cdot \ldots \cdot z_n^{i_n}$, we have that $P(D)$ defines the linear constant coefficients differential operator

$$P(D) = \Sigma_{|I|=0}^{m} a_I D^I.$$

More generally, instead of a single polynomial (and operator), we will consider a rectangular matrix of polynomials so that $P(D) = [P_{ij}(D)]$ is an $q \times s$ matrix of linear constant coefficients differential operators, and if $\vec{f} = (f_1, \ldots, f_s)$ is a vector of generalized functions, the equation $P(D)\vec{f} = 0$ represents the homogeneous system

$$\Sigma_{j=1}^{s} P_{ij}(D) f_j = 0, \quad i = 1, \ldots, q. \tag{1}$$

The Ehrenpreis–Palamodov Fundamental Principle states that, under suitable conditions on the space of generalized functions, it is possible to represent the solutions of (1) in an integral form. This is indeed the case for a very large class of spaces, which Ehrenpreis called Analytically Uniform spaces, or AU-spaces in short. The actual definition of AU-Spaces is rather convoluted (see Sect. 3 for details) but it can be intuitively described as follows. A topological space X of generalized functions is Analytically Uniform if its dual X' is topologically isomorphic, through some transformation which usually looks like the Fourier transform, to a space of entire functions whose growth at infinity is bounded by some suitable weight functions. These weight functions are usually referred to as the AU-structure of the space. Let us clarify this notion with two examples.

Example 2.1. The space $\mathcal{E}(\mathbb{R}^n)$ of infinitely differentiable functions on $\mathbb{R}^n$ is an Analytically Uniform space. Its dual is the space $\mathcal{E}'$ of compactly supported distributions, and the Fourier transform is an isomorphism between $\mathcal{E}'$ and the space

$$\hat{\mathcal{E}}' = \{F \in \mathcal{H}(\mathbb{C}^n) : \exists A, B > 0 \text{ s.t. } |F(z)| \leq A(1+|z|)^B \exp(B|\operatorname{Im}(z)|)\}$$

of entire functions of exponential type whose growth on the real axis is polynomial. More generally, the space of infinitely differentiable functions on any open convex subset $\Omega \subseteq \mathbb{R}^n$ is also an Analytically Uniform space; its dual is the space of distributions with compact support contained in Ω, and this last space is isomorphic (via the Fourier transform) to the space of entire functions F satisfying, for some positive constants A, B

$$|F(z)| \leq A(1+|z|)^B \exp H_K(\operatorname{Im}(z)),$$

where H_K is the indicator function of the convex hull of some compact set K contained in Ω.

Example 2.2. The space $\mathcal{D}'(\mathbb{R}^n)$ of Schwartz distributions on $\mathbb{R}^n$ is an Analytically Uniform space. Its dual is the space $\mathcal{D}(\mathbb{R}^n)$ of infinitely differentiable functions with compact support, and the Fourier transform is an isomorphism between $\mathcal{D}$ and the space

$$\hat{\mathcal{D}} = \{F \in \mathcal{H}(\mathbb{C}^n) : \exists A > 0 \text{ s.t. } |F(z)| \leq A(1+|z|)^{-p} \exp(A|\operatorname{Im}(z)|) \text{ for all } p > 0\}$$

of entire functions of exponential type, which decrease, on the real axis, faster than the inverse of any polynomial. Once again, the space of Schwartz distributions on any open convex subset of $\mathbb{R}^n$ is also an Analytically Uniform space.

We are now ready for the statement of the Fundamental Principle, which, for simplicity, we only present in the case $s = 1$.

Theorem 2.3. *Let X be an n-dimensional AU-space, and let $P_1, \ldots, P_q$ be q polynomials in n variables. Then there is a finite number, say p, of algebraic varieties V_j, differential operators ∂_j, supported by the varieties V_j, as well as Radon measures μ_j on V_j, such that every solution $f \in X$ of the system*

$$P_1(D)f = \ldots = P_q(D)f = 0$$

can be written as

$$f(x) = \Sigma_{j=1}^{p} \int_{v_j} [\partial_j (\exp(iz \cdot x))] \, d\mu_j(z),$$

where the integrals converge in the topology of X.

Remark 2.4. The integral representation given by the Fundamental Theorem is only symbolic when the space X is a space of generalized functions, for example distributions. If, on the other hand, the space is a space of functions for which a pointwise value is defined, for example differentiable or holomorphic functions, the integral representation yields the actual value of the solution at any given point.

Remark 2.5. The varieties and the differential operators which appear in Theorem 2.3 are collectively referred to as the characteristic variety of the system, while the operators are often referred to as the Noetherian operators of the system.

The theory of differential equations in AU-spaces is described in great detail in [1, 12, 25, 27], and in particular we know of the surjectivity of linear constant coefficients differential operators on such spaces.

The general argument for surjectivity goes as follows: let T be a continuous operator acting from a suitable topological vector space X to itself. Then T is surjective if and only if its adjoint $T' : X' \to X'$ is injective and with closed image [7]. Usually it is easy to prove injectivity, while the key of the proof lies in the establishment of the closure of the image. Because of the nature of AU-spaces, such closure can be interpreted in terms of a division theorem in the space of entire functions which are Fourier transforms of the objects in X'. This specifically relies on finding appropriate conditions for the following to be true: if P is a polynomial, F is an entire function in $\hat{X}'$, and F/P is an entire function, then F/P is the Fourier transform of an object in X'. If this is the case, one usually says that the division problem in X can be solved. In particular, if T is a linear partial differential operator with constant coefficients $P(D)$, where P is the polynomial in n variables which denotes its symbol, then the solvability of the division problem in X is equivalent to the surjectivity of $P(D)$ on X, namely. the solvability of the division problem in X is equivalent to

$$P(D)X = X.$$

3 The Case of Real Analytic Functions

In Chap. 5 of [12], Ehrenpreis offers a series of concrete examples of his newly developed AU-spaces. In addition to the examples described above, he study the case of what he calls *quasianalytic classes*. In the context of these spaces, Ehrenpreis poses the following problem (Problem 5.6 at p. 174 of [12]):

Problem: Study solutions of systems of constant coefficient linear partial differential equations in these spaces. In particular, does the analog of the solvability theorem hold for the space of real analytic functions? (We don't know the result, says Ehrenpreis, even for single equation systems.)

The first thing to ask is, naturally, whether the space of real analytic functions is an AU-space, since in this case the solution would be affirmative. This seems quite plausible for a variety of reasons. On one hand, complex analytic functions (holomorphic functions) form an AU-space. In addition, one can think of $\mathcal{A}$ as the intersection of the so-called quasi-analytic classes, which are AU as well, so even though a proof was lacking for a while, it seemed a reasonable assumption. However, we will now follow [2] to show that the space $\mathcal{A}$ of real analytic functions in *not* an AU-space.

We will begin by giving a more precise definition for the notion of AU-spaces, and of convolutors on such spaces.

Definition 3.1. A Hausdorff locally convex topological vector space X is said to be an AU-space of dimension n if:

1. X is the strong dual of a locally convex topological space Y with topology $\mathcal{T}_Y$.
2. There exists a continuous analytic embedding $\imath : \mathbb{C}^n \to X$ such that the linear combinations of the elements in $\imath(\mathbb{C}^n)$ are dense in X. Moreover, for every $T \in Y$ the function $\hat{T} := \langle T, \imath(z) \rangle$ is entire on $\mathbb{C}^n$. The map $\mathcal{F}$ that maps T to $\hat{T}$ is an isomorphism of topological spaces.
3. There exists a family $\mathcal{K}$ of positive continuous functions on $\mathbb{C}^n$ such that

$$\mathcal{F}(Y) := \hat{Y} = \{f \in \mathcal{H}(\mathbb{C}^n) : ||f||_k := sup_{\zeta \in \mathbb{C}^n} \frac{|f(\zeta)|}{k(\zeta)} < \infty \; \forall k \in \mathcal{K}\}$$

and its topology is given through the norms $||\cdot||_k$.
4. There is such a family $\mathcal{K}$ such that for every positive integer N the family

$$\mathcal{K}_N := \{k_N(z) := \max_{|z-z'| \leq N}[k(z')(1+|z'|)^N]\}$$

satisfies those same properties.
5. There exists a family $\mathcal{M}$ of positive continuous functions on $\mathbb{C}^n$ such that for each $m \in \mathcal{M}$ and each $k \in \mathcal{K}$ the family $\{A(m,\alpha)\}_{\alpha > 0, m \in \mathcal{M}}$ of the sets

$$A(m,\alpha) := \left\{ S \in Y : sup_{z \in \mathbb{C}^n} \frac{|\hat{S}(z)|}{m(z)} < \alpha \right\}$$

defines a fundamental system of bounded sets in the space Y.
6. There is such a family $\mathcal{M}$ such that for every positive integer N the family

$$\mathcal{M}_N := \{m_N(z) := \max_{|z-z'| \leq N}[m(z')(1+|z'|)^N]\}$$

satisfies those same properties.

Remark 3.2. Any family $\mathcal{K}$ as in 3 and 4 above is called an Analytically Uniform structure (or AU-structure in brief) for X. Any family $\mathcal{M}$ as in 5 and 6 above is called a Bounded Analytically Uniform structure (BAU-structure in brief) for X.

It is immediate to note that the definition we just offered is such that multiplication by a polynomial defines a continuous map of $\hat{Y}$ into itself. Since, in all cases when X is a space of generalized functions, multiplication by a polynomial $P(z)$ in $\hat{Y}$ corresponds to the action of the partial differential operator $P(D)$ on X, we can naturally define the notion of convolutor on an AU-space as follows.

Definition 3.3. Let F be an entire function such that multiplication by F is a continuous map of $\hat{Y}$ into itself. We then say that F is a multiplier for Y. The operator defined as the adjoint of the operator that maps S to $\mathcal{F}^{-1}(F(\mathcal{F}(S)))$ is said to be a convolutor for X. A convolutor such that the corresponding multiplication by F is an open endomorphism of $\hat{Y}$, is said to be invertible in X.

It is easy to verify the following sufficient condition to insure invertibility of a convolutor.

Theorem 3.4. *Let F be a multiplier for a space Y. Suppose there are positive constants A and B such that for all $z \in \mathbb{C}^n$ there is ρ_z with $0 < \rho_z \leq B$ such that*

$$\min_{|z'-z|=\rho_z} |F(z')| \geq \frac{A}{(1+|z|)^B}. \qquad (2)$$

Then the convolutor $\mathcal{F}^{-1}(F(\mathcal{F}))$ is invertible in X.

Proof. Note that for any $k \in \mathcal{K}$, condition 4 in the definition of an AU-structure shows that, for every $B > 0$ there exists $\tilde{k} \in \mathcal{K}$ and $C > 0$ such that

$$\{f \in \hat{Y} : ||f||_{\tilde{k}_B} \leq 1\} \subset \{f \in \hat{Y} : ||f||_k \leq C\}.$$

Thus, if $F\hat{S} \in \{f \in \hat{Y} : ||f||_{\tilde{k}} \leq 1\}$, by the inequality above we obtain that

$$|\hat{S}(z)| \leq A^{-1} \max_{|z'| \leq B}(|F(z+z')\hat{S}(z+z')|(1+|z|)^B) \leq A^{-1}\tilde{k}_B(z),$$

which shows that $\hat{S}$ belongs to $\{f \in \hat{Y} : ||f||_k \leq C/A\}$, and therefore implies that multiplication by F is an open mapping of $\hat{Y}$ to itself. $\square$

As a consequence of these definitions, and in particular of this last result, one immediately obtains that linear constant coefficients partial differential operators act surjectively on any AU-space.

Theorem 3.5. *Every linear constant coefficients partial differential operator is always surjective on any AU-space.*

Proof. This follows immediately from the so-called Ehrenpreis–Malgrange lemma (see e.g. [8, 23]) which shows that every polynomial satisfies (2).

Let us now conclude this section by giving the proof, from [2], of the fact that the space of real analytic functions on $\mathbb{R}$ is *not* an AU-space.

Definition 3.6. Given $\epsilon > 0$, and an interval $I_n = [-n, +n]$, we denote by $\mathcal{A}_{\epsilon,n}$ the Banach space of functions holomorphic on $\{z \in \mathbb{C} : dist(z, I_n) \leq \epsilon\}$. Then the space $\mathcal{A}$ of real analytic functions on $\mathbb{R}$ is defined by

$$\text{proj}\lim\nolimits_{n \to +\infty} \text{ind}\lim\nolimits_{\epsilon \to 0} \mathcal{A}_{\epsilon,n}.$$

Theorem 3.7. *The space $\mathcal{A}$ is not an AU-space.*

Proof. Assume there is a family $\mathcal{K}$ which describes $\hat{\mathcal{A}}'$. We begin by proving that for each $k \in \mathcal{K}$, and every integer j, there exists $\epsilon_j > 0$ such that

$$\exp(\epsilon_j |x| + j|y|) = O(k(z)). \tag{3}$$

In fact, if this were not the case, we could find sequences $\{\epsilon_n\}$ and $\{z_n\}$, with $\{\epsilon_n\} \to 0$, and $\{|z_n|\} \to +\infty$, such that

$$\exp(\epsilon_n |x_n| + n|y_n|) \geq nk(z_n). \tag{4}$$

Now set

$$f(z) := \prod_{k=1}^{+\infty} \left(1 - \frac{z}{z_n}\right),$$

$$\varphi(z) := \sum_n \frac{\exp(\epsilon_n |z_n|)}{f'(z_n)} \frac{f(z)}{z - z_n} \left(\frac{z}{z_n}\right)^{[\epsilon_n |z_n|]+1},$$

where $[x]$ denotes as usual the largest integer less than or equal to x, and finally

$$F(z) := \exp(-i(j + \epsilon_1)z)\varphi(z).$$

Then one can verify (see e.g. [12] or [24]) that $F \in \hat{\mathcal{A}}'$. This implies that, for some $C > 0$, one has $|F(z)| \leq Ck(z)$ which contradicts (4). This shows that (3) holds. But in fact, see [2] for details, one can in fact establish that (3) holds for some ϵ independent of j. Thus, for every j and every z one has, for example,

$$\exp(1/3(\epsilon_1 |x| + j|y|)) = O(k(z)).$$

Let now h be an arbitrary real analytic function. Since every function in an AU-space can always be written as a Fourier integral, we can write

$$h(w) = \int_\mathbb{C} \exp(iwz)\, d\mu(z)/k(z).$$

This integral is convergent in particular if $\text{Im}(w) \leq \epsilon_1/3$ and $\text{Re}(w)$ is bounded. Thus, we have shown that every real analytic function on $\mathbb{R}$ can be extended analytically to a strip of fixed size along the real axis, which is clearly false. This concludes the proof. □

4 Cattabriga and De Giorgi: Surjectivity Results

As we have seen in the previous sections, the issue of surjectivity is fully settled, in a positive sense, for Analytically Uniform spaces; furthermore the notion of P-convexity introduced by Malgrange in [23] and its subsequent variations allow the description of necessary and sufficient conditions for surjectivity in spaces of infinitely differentiable functions or Schwartz distributions on open sets in $\mathbb{R}^n$.

We have also seen, in the last section, that the space of real analytic functions is not an Analytically Uniform space, despite the fact that it is the intersection of an infinite family of Analytically Uniform spaces (the so-called quasianalytic classes discussed in detail by Ehrenpreis in [12]).

Therefore, it was surprising when De Giorgi and Cattabriga proved, in [5], the following result:

Theorem 4.1. *Let $P(D)$ be a linear partial differential operator with constant coefficients. Then $P(D)$ is surjective on the space $\mathcal{A}(\mathbb{R}^2)$ of real analytic functions in the real plane.*

The result is important because it shows that surjectivity is not necessarily due to the existence of an Analytically Uniform structure. It is also very interesting because it uses in a specific fashion the bidimensionality of the problem. Finally, it is a brilliant result, because it does not use any of the sophisticated machinery of the theory of distributions. Rather, it is an ingenious (but for this reason even more admirable) application of simple elementary properties of the Fourier transform and of the theory of ordinary constant coefficients differential equations.

The key ingredient of the proof can be found in another (immediately precedent) paper of De Giorgi and Cattabriga who proved in [4] the following result:

Lemma 4.2. *Let f be a real analytic function in the real plane, and let φ be a positive non-increasing function defined on $\mathbb{R}$ such that*

$$\int_{\mathbb{R}^2} \varphi(\alpha^2 + \beta^2)\, d\alpha\, d\beta < +\infty.$$

Then there exist an infinitely differentiable function g on $\mathbb{R}^3$ and an infinitely differentiable, positive, non-increasing function ψ on $\mathbb{R}$, such that

$$\text{supp}(g) \subset \{(\alpha, \beta, \gamma) : \psi(\alpha^2 + \beta^2) \leq \gamma \leq 2\psi(\alpha^2 + \beta^2)\},$$

$$\int_0^{+\infty} |g(\alpha, \beta, \gamma)|\, d\gamma \leq \varphi(\alpha^2 + \beta^2),$$

and

$$f(x, y) = \int_{\mathbb{R}^3} \frac{g(\alpha, \beta, \gamma)}{[(x-\alpha)^2 + (y-\beta)^2 + \gamma^2]^2} \, d\alpha \, d\beta \, d\gamma.$$

By using this lemma, De Giorgi and Cattabriga show that if $v(x, y)$ is a suitable solution of an auxiliary differential equation

$$\frac{\partial^n v}{\partial y^n} + \sum_{i=0}^{n-1} c_i \frac{\partial^n v}{\partial y^i \partial x^{n-i}} + \sum_{0 \leq i+j \leq n-1} c_{ij} \frac{\partial^{i+j} v}{\partial y^i \partial x^j} = \frac{1}{[(x-\alpha)^2 + (y-\beta)^2 + \gamma^2]^2},$$

then the function

$$u(x, y) = \int_{\mathbb{R}^2} d\alpha \, d\beta \int_0^{+\infty} g(\alpha, \beta, \gamma) v(x, y, \alpha, \beta, \gamma) \, d\gamma$$

is a solution of the differential equation

$$P(D)u = f$$

in all the real plane. This proves the surjectivity of constant coefficients differential operators in the space of real analytic functions in the plane.

Remark 4.3. Independently of De Giorgi and Cattabriga, and with the use of substantially different tools, Kawai, from the Kyoto's school of algebraic analysis, had concurrently developed a theory for the surjectivity of differential operators in spaces of real analytic functions. These results do not diminish the importance and novelty of those due to De Giorgi and Cattabriga, whose techniques are quite different from those used by Kawai in [18–22], and whose approach relies on the newly developed theory of hyperfunctions, and microfunctions see e.g. [17]. I hope I will be forgiven if I insert here a rather lengthy quote from [22], since in it Kawai describes the difference between the two methods with his usual clarity and depth of thought. Specifically Kawai makes explicit reference to the fact that "the topological structure of the space of real analytic functions on an open set is rather complicated, hence even Professor Ehrenpreis, who initiated and completed the general theory of linear differential equations *with constant coefficients* in the framework of distributions with Professors Malgrange, Hormander and Palamodov, seems at present to have abandoned the attempt to attack the problem of global existence of real analytic solutions...But we can treat this problem without much difficulty by the aid of the theory of hyperfunctions and that of the sheaf $\mathcal{C}$, at least when we restrict ourselves to the consideration of the operators satisfying suitable regularity conditions which allow us to consider the problems geometrically. In a sense our method can be regarded as "method of algebraic analysis" contrary to "method of functional analysis", which is developed, for example, in Hormander [14], Palamodov [25], Ehrenpreis [12], etc. (The word "algebraic analysis" seems to go back to Euler but it has recently been endowed with positive meanings by Professor Sato, who aims at the Renaissance of classical analysis.)"

Note that the result of [5] only works for the case in which the differential equation is considered on the entire real plane. No comments are made there concerning the case in which real analytic functions are considered on open sets in the real plane. In [20], however, it is shown the following important condition:

Theorem 4.4. *If Ω is a bounded open subset in $\mathbb{R}^2$ such that its intersection with every characteristic line of the operator $P(D)$ is an interval, then $P(D)$ is surjective, i.e.*

$$P(D)\mathcal{A}(\Omega) = \mathcal{A}(\Omega).$$

Remark 4.5. As Cattabriga points out, the condition in this theorem is necessary and sufficient for surjectivity in the space of infinitely differentiable functions, but is only sufficient in the case of real analytic functions.

The article [5] is short and to the point, and its last section opens the discussion for the case of real analytic functions in more than two variables. Indeed, De Giorgi and Cattabriga point out that they *have serious doubts* that even a very simple differential operator (in two variables) such as

$$\frac{\partial^2}{\partial x^2} + \frac{\partial^2}{\partial y^2}$$

be surjective in the space of real analytic functions in three variables. Similarly, they express doubts as to the surjectivity of the operator

$$\frac{\partial}{\partial t} + \frac{\partial^2}{\partial x^2} + \frac{\partial^2}{\partial y^2}$$

in the space of real analytic functions in three variables.

That the suspicions of De Giorgi and Cattabriga were well founded was seen quite rapidly after their first landmark paper, as shown for example in [3, 26]. Thus, the next natural step consisted in identifying necessary and sufficient conditions for the surjectivity, or equivalently relationships between the open sets in which the equations are considered and the operator which is involved. This was done in their next paper, [6], where (inspired by the examples above) they offer a sufficient condition for the compatibility between subsets of $\mathbb{R}^{n+1}$ and polynomials in n variables. To clarify the result, we need to rephrase Lemma 4.2 from [4], in the case of several variables (as De Giorgi and Cattabriga underline, the formulation we will give below is slightly weaker than the original, but simpler and sufficient for the applications).

Lemma 4.6. *Let $f : \mathbb{R}^n \to \mathbb{C}$ be a real analytic function. Then there exist a closed set $E \subset \mathbb{R}^{n+1}$, a non-negative completely additive function μ defined on the set of*

all the Borel sets of $\mathbb{R}^{n+1}$, and a Baire function $\varphi : \mathbb{R}^{n+1} \to \mathbb{C}$ such that
$$E \cap \{(x,t) \in \mathbb{R}^{n+1} : t = 0\} = \emptyset,$$
$$\mu(\mathbb{R}^{n+1}) = \mu(E) < +\infty, |\varphi(x,t)| = 1,$$

and finally
$$f(x) = \int_{\mathbb{R}} (|x-\alpha|^2 + \beta^2)^{-(n+1)/2} \varphi(\alpha, \beta) d\mu(\alpha, \beta).$$

This lemma leads naturally to define the notion of compatible polynomial. Given a polynomial P in n complex variables, and a closed set E in $\mathbb{R}^{n+1}$, we say that E is P-compatible if for every μ and φ as in the previous lemma, there exists an analytic solution u in $\mathbb{R}^n$ to the equation
$$P(D)u = f,$$
where f is given by the representation in the lemma.

The main result of [6] is then the following:

Theorem 4.7. *Let P be a complex polynomial in $\mathbb{C}^n$ such that*
$$\frac{\partial^{m+1} P(p,\lambda)}{\partial \lambda^{m+1}} \equiv 0,$$
and
$$\left| \frac{\partial^{m+1} P(p,\lambda)}{\partial \lambda^{m+1}} \right| \geq 1 \quad \forall p \in \mathbb{R}^{n-1}.$$
Denote by $\lambda_j(p)$ the roots of the equation $P(p,\lambda) = 0$ and assume that there exists a positive constant C such that
$$|\lambda_j(p)| \leq C(1 + |p|)$$
and that there are four positive constants $C_1 < 1, C_2, C_3, C_4$ such that for every p with $|p| \geq C_2$ one of the two following inequalities hold:
$$\mathrm{Im}(\lambda_j(p)) \leq -2C_3|p|$$
or
$$\mathrm{Im}(\lambda_j(p)) \geq -C_4|p|^{C_1}.$$
Finally, assume there exists a continuous positive function defined on $\mathbb{R}$ such that
$$E = \{(\alpha, \beta) \in \mathbb{R}^{n+1} : \alpha_n \geq 0, \beta \geq g(\alpha_n)\}.$$
Then, E is P-compatible.

We conclude by pointing out how the importance of the work of De Giorgi and Cattabriga was immediately very well understood and appreciated by the community of experts. So, for example, Hormander was able, in [16], to analyze and expand the ideas contained in [3] to obtain the following necessary and sufficient conditions for the solvability of linear constant coefficients partial differential equations in $\mathcal{A}$.

Theorem 4.8. *Let* $P(z) = \sum_{|I| \leq m} a_I z^I$ *be a polynomial in n variables, and let* $P_m(z) = \sum_{|I|=m} a_I z^I$ *be its principal part. Then* $P(D)$ *is surjective on the space* $\mathcal{A}$ *of real analytic functions on* $\mathbb{R}^n$ *if and only if there exists a constant C such that every plurisubharmonic function* φ *in* $\mathbb{C}^n$ *which satisfies*

$$\varphi(z) \leq |z| \quad \text{on } \mathbb{C}^n; \quad \varphi(x) \leq 0 \quad \text{if } x \in \mathbb{R}^n, \quad P_m(x) = 0,$$

also satisfies

$$\varphi(z) \leq C|Im(z)| \quad \text{if } P_m(z) = 0.$$

A more sophisticated version of this theorem is also available, again from [16], to give necessary and sufficient conditions for solvability in the space of real analytic functions on convex subsets of $\mathbb{R}^n$.

Acknowledgements The author is grateful to Professor S. Coen for inviting him to participate in this celebration of the birthday of the "Dipartimento di Matematica" of the University of Bologna.

References

1. Berenstein, C.A., and D.C. Struppa. 1993. Complex analysis and convolution equations. In Several complex variables V, Encyclopaedia of Mathematical Sciences, ed. G.M. Khenkin, 54, 5–111, Akad. Nauk SSSR, Vsesoyuz Inst Nauchn i Tekhn Inform, Moscow, 1989 (Russian). English transl.: 1–108. Berlin-Heidelberg: Springer.
2. Berenstein, C.A., and M.A. Dostal. 1972. *Analytically uniform spaces and their applications to convolution equations. Springer lecture notes in mathematics* 256. Berlin-Heidelberg-New York: Springer.
3. De Giorgi, E. 1971–1972. Solutions analytiques des équations aux dérivées partielles à coefficients constants, École Polytechnique. Séminaire Goulaouie-Schwartz, exposé 24.
4. De Giorgi, E., and L. Cattabriga. 1971. Una formula di rappresentazione per funzioni analitiche in $\mathbb{R}^n$. *Boll Un Mat Ital* 4: 1010–1014.
5. De Giorgi, E., and L. Cattabriga. 1971. Una dimostrazione diretta dell'esistenza di soluzioni analitiche nel piano reale di equazioni a derivate parziali a coefficienti constanti. *Boll Un Mat Ital* 4: 1015–1027.
6. De Giorgi, E., and L.Cattabriga. 1972. Sull'esistenza di soluzioni analitiche di equazioni a derivate parziali a coefficienti costanti in un qualunque numero di variabili. *Boll Un Mat Ital* (4) 6: 301–311.
7. Dieudonné, J., and L. Schwartz. 1950. La dualité dans les espaces $\mathcal{F}F$ et $(\mathcal{LF})$. (French) *Ann Inst Fourier Grenoble* 1(1949): 61–101.
8. Ehrenpreis, L. 1954. Solutions of some problems of division. I. Division by a polynomial of derivation. *Am J Math* 76: 883–903.

9. Ehrenpreis, L. 1955. Solutions of some problems of division. II. Division by a punctual distribution. *Am J Math* 77: 286–292.
10. Ehrenpreis, L. 1956. Solutions of some problems of division. III. Division in the spaces $\mathcal{D}'$, $\mathcal{H}$, $\mathcal{Q}_A$, $\mathcal{O}$. *Am J Math* 78: 685–715.
11. Ehrenpreis, L. 1960. Solutions of some problems of division. IV. *Am J Math* 82: 522–588.
12. Ehrenpreis, L. 1970. *Fourier analysis in several complex variables*. New York: Wiley.
13. Hormander, L. 1955. On the theory of general partial differential operators. *Acta Math* 94: 161–248.
14. Hormander, L. 1963. *Linear partial differential operators*. Berlin: Springer.
15. Hormander, L. 1971. On the existence and the regularity of solutions of linear pseudo-differential equations. *Enseignement Math* 17: 99–163.
16. Hormander, L. 1973. On the existence of real analytic solutions of partial differential equations with constant coefficients. *Invent Math* 21: 151–182.
17. Kato, G., and D.C. Struppa. 1999. *Fundamentals of microlocal algebraic analysis*. New York: Marcel Dekker.
18. Kawai, T. 1971. Construction of a local elementary solution for linear partial differential operators I. *Proc Jpn Acad* 47: 19–23.
19. Kawai, T. 1971. Construction of a local elementary solution for linear partial differential operators II. *Proc Jpn Acad* 47: 147–152.
20. Kawai, T. 1972. On the global existence of real analytic solutions of linear differential equations I. *J Math Soc Jpn* 24: 481–517.
21. Kawai, T. 1973. On the global existence of real analytic solutions of linear differential equations II. *J Math Soc Jpn* 25: 644–647.
22. Kawai, T. 1973. On the global existence of real analytic solutions of linear differential equations. *Springer Lecture Notes in Mathematics* 287; 99–121.
23. Malgrange, B. 1955. Existence et approximation des solutions des équations aux dérivées partielles et des équations de convolution. *Ann Inst Fourier Grenoble* 6: 271–355.
24. Martineau, A. 1963. Sur les fonctionelles analytiques et la transformation de Fourier–Borel. *J d'Analyse Math* 9: 1–144.
25. Palamodov, V.P. 1970. *Linear differential operators with constant coefficients*. Translated from the Russian by A. A. Brown. Die Grundlehren der mathematischen Wissenschaften, Band 168, New York-Berlin: Springer.
26. Piccinini, L.C. 1973. Non surjectivity of the Cauchy–Riemann operator on the space of the analytic functions on $\mathbb{R}^n$. Generalization to the parabolic operators. *Boll Un Mat Ital* 7: 12–28.
27. Struppa, D.C. 1981. The fundamental principle for systems of convolution equations. *Mem Am Math Soc* 273: 1–167.

New Perspectives on Beltrami's Life and Work – Considerations Based on his Correspondence

Rossana Tazzioli

Abstract Eugenio Beltrami was a prominent figure of nineteenth century Italian mathematics. He was also involved in the social, cultural and political events of his country. This paper aims at throwing fresh light on some aspects of Beltrami's life and work by using his personal correspondence. Unfortunately, Beltrami's Archive has never been found, and only letters by Beltrami – or in a few cases some drafts addressed to him – are available. In this paper, some letters addressed by Beltrami to his foreign colleagues are published and annotated for the first time in order to give a more exhaustive picture of Beltrami's personality and to point out the impact of his work in Italy and abroad.

1 Introduction

Eugenio Beltrami (1835–1900) was one of the most influential mathematicians of the nineteenth century in Italy. One of the protagonists of the Risorgimento (the political movement preceding the Italian unification of 1861), he participated, together with other Italian mathematicians such as Enrico Betti (1823–1892), Francesco Brioschi (1824–1897), Felice Casorati (1835–1890), and Luigi Cremona (1830–1903), in the political and cultural life of post-unification Italy. Section 2 gives a picture of the involvement of Italian mathematicians in the Risorgimento and of their institutional commitments in the period immediately after unification.

Beltrami is well-known mainly for his works on differential geometry and non-Euclidean geometry, but he also made important contributions to mathematical physics over roughly 30 years of his scientific career. On his scientific work and the connection between geometry and mathematical physics several studies have

---

R. Tazzioli (✉)
U.F.R. de Mathématiques, Laboratoire Paul Painlevé U.M.R. CNRS 8524
Université de Sciences et Technologie de Lille, 59655 Villeneuve d'Ascq Cedex, France
e-mail: rossana.tazzioliatuniv--lille1.fr

already been published.[1] This paper does not provide an exhaustive scientific biography of Beltrami; it aims rather at giving a fresh perspective on his life and work by considering his unpublished correspondence with some of his foreign colleagues.

As far as we know, no Beltrami Archive exists; however, many of his letters to Italian and foreign colleagues have already been published.[2] In these letters new aspects of Beltrami's life and work are brought forth, such as his engagement with mathematics teaching and with institutional issues, his ideas on non-Euclidean geometry and its relations with Gaussian theory of surfaces, and his strong interest in mathematical physics.

In this paper, we give an account of Beltrami's letters to Richard Dedekind (1831–1916), Pierre Duhem (1861–1916), Wilhelm Killing (1847–1923), Felix Klein (1849–1925), Rudolph Lipschitz (1832–1903), and Gösta Mittag-Leffler (1846–1927),[3] where it becomes very clear that Beltrami attributed great importance to mathematical physics both as a researcher and as a teacher charged with giving lectures on various areas of mathematical physics – on mechanics, elasticity, potential theory, the theory of electricity and magnetism (including Maxwell's equations), and the theory of heat. His interest in the *new physics* – as the Maxwell theory was called at the time[4] – is often confirmed in this correspondence, as well as his troubles due to the complexity and obscurity of this theory, which indeed had not yet been rigorously founded. These letters also show the strong influence of Beltrami's methods and results not only on geometry but also on mathematical physics. Furthermore, thanks to his high reputation in Italy and abroad, Beltrami's advice was generally followed by his colleagues. He often recommended skilled young mathematicians for positions or grants and he was often gratified to see his recommendation accepted.

Section 3 introduces Beltrami as a mathematician and as a political figure, and in particular points out what is new in the letters to his foreign colleagues published in the Appendix. In more detail, Sect. 3.1 sketches Beltrami's life and scientific activity; Sect. 3.2 aims at describing his opinion on Maxwell's electromagnetic theory, and how he managed to solve problems coming from this theory as a researcher and as a teacher; Sect. 3.3 concerns his high reputation – some remarkable examples from his correspondence are shown, such as his engagement as a teacher, the influence of his research abroad, and his requests to participate in

---

[1] For a scientific biography of Beltrami and especially for his work on non-Euclidean geometry see the Introduction to [16]; for his contribution to mathematical physics see [65, 67].

[2] See, for example, Beltrami's letters to Hoüel in [16], to Betti, Tardy and Gherardi in [39], to Cesaro in [62], and to Chelini in [36].

[3] These letters are unpublished except for the letter to Klein dated 17 April 1888 and the letter to Lipschitz dated 13 March 1877; both of them are published but not annotated in [16] on pp. 211–213.

[4] Maxwell published his electromagnetic theory in the *Treatise on Electricity and Magnetism* in 1873 [59].

international undertakings. We attempt to express what Beltrami's letters highlight and how they contribute to a clarification of some aspects of his life and work.

Appendices 4.1–4.6 contain the unpublished correspondence between Beltrami and Dedekind (Appendix 4.1), and Beltrami's letters to Duhem, Killing, Klein, Lipschitz, and Mittag-Leffler (Appendices 4.2–4.5). Appendix 4.7 reproduces some letters exchanged between Cremona and Betti, and two letters from Beltrami to Betti concerning Sonya Kovalevskaya's affair of 1886, when several mathematicians tried to induce the Stockholm Academy of Sciences to accept Sonya Kovalevskaya (1850–1891) – though a woman – as a member of the Academy. Appendix 4.8 contains a Report written by Beltrami in 1889 and addressed to the University of Stockholm in support of Kovalevskaya. She was appointed full professor at the University of Stockholm the same year.

All the documents in the Appendices are annotated; the footnotes contain short biographies, bibliographical references, and some explanations when they are necessary.

2 The Risorgimento Generation

The history of modern Italy starts in 1861. In that year the various smaller states into which Italy had been politically and administratively divided were unified in a process called the Risorgimento. The government of Piedmont (the northwestern region of Italy, whose capital is Turin) obtained significant popular support which allowed it to become the leader of the idealistic and democratic mobilization against Austria. After popular uprisings, two wars of independence against Austria, and intense diplomatic activity, the Unification of Italy was finally achieved. Only in 1870, after the defeat of the papal army, did Rome become the capital of unified Italy.

Many Italian mathematicians – such as Beltrami, Betti, Brioschi, Casorati, Cremona, and others – participated as soldiers in the wars of independence, and as politicians in a general sense after unification. They raised the level of mathematical studies and, more generally, saw themselves as the builders of the new Italy. Betti was a volunteer and fought in Curtatone and Montanara in the battalion of the students of the University of Pisa; in 1848 Cremona participated in the defence of Venice against the Austrian regime, where he attained the ranks of corporal and then sergeant; Brioschi participated in the insurrection of Milan in 1848 against the Austrians and in 1870 he was involved in the taking of Rome. Beltrami and his family shared the ideals of the Risorgimento from the popular uprisings of 1848–49, as Beltrami wrote to Charles Hermite (1822–1901) in November 1887:

> [...] as a consequence of the political events of 48–49, the members of my family could no more expect to become civil servants of the Austrian government. [...] At the end of my studies – I had to renounce my doctorate for lack of means – [...] I had the good luck to become and remain for six years the private secretary of M. Didier, chief engineer of mines and the director of railways of the North Italy, an excellent man taking an almost fatherly

care of me. In this period my vocation – a bit uncertain – turned to mathematical research. Since I was badly prepared by the feeble courses of that time, I began to learn the arithmetic of M. Serret. (My translation)[5]

In 1853, Beltrami had indeed started to study mathematics at the University of Pavia, where he had attended Brioschi's lectures. After the working period described in this letter to Hermite, Beltrami continued his mathematical education in Milan where he met Cremona, who drove his interests to the geometry of curves and surfaces. Very soon Beltrami began to publish, and produced several interesting works on this subject. In 1862, thanks to Brioschi's intervention (Brioschi was the secretary of the Ministry of Public Education at that time), Beltrami became professor of algebra and analytic geometry at the University of Bologna, though he had never obtained his doctorate in mathematics.

In 1858, Betti, Brioschi and Casorati visited the universities of Göttingen, Berlin, and Paris, and came into contact with the most important mathematicians and institutes in Europe. They met Dedekind, Peter Gustav Lejeune Dirichlet (1805–1859), Bernhard Riemann (1826–1866), Leopold Kronecker (1823–1891), Karl Weierstrass (1815–1897), Joseph Bertrand (1822–1900), Hermite, and others. By tradition, Italian mathematics is supposed to have had its origins in this trip. The argument that the Risorgimento created a new starting point in mathematics – as well as in politics – obviously fits a patriotic ideology which emphasizes that the Unification of Italy had given wings to the aspirations and enthusiasm of the best minds in the country, in science no less than in other domains.

In 1908, in his inaugural address to the *International Congress of mathematicians* in Rome, Vito Volterra (1860–1940) claimed:

> [...] it is then not surprising if, in the progress of science, a sudden change in Italian thought occurred, provoked by the quick development and spreading of this thought, and by the new qualities which it acquired and which made it richer in the years following the period of the political Risorgimento. (My translation) [69, p. 59]

In reality, it is inconceivable that a mathematical school should have sprung up from nothing, or simply from a documentary trip. Even so, this view – extreme as it may appear – can still be taken as a suitable starting point, from which a deeper analysis may begin.[6]

---

[5]"[...] les événements politiques de 48–49 ayant mis ma famille au nombre de celles dont les membres ne pouvaient pas espérer un emploi public par le gouvernement autrichien, [...] à la fin de mes études (en renonçant au doctorat, faute de moyens) [...] j'ai eu le bonheur de devenir et de rester, pendant six ans, secrétaire particulier de M. Diday, Ingénieur au chef des mines, Directeur de l'exploitation des chemins de fer de la Haute-Italie, excellent homme qui a eu pour moi des soins presque paternels. C'est dans cet intervalle que ma vocation un peu flottante, s'est décidée pour les études mathématiques et que, me trouvant décidément trop mal préparé par les faibles cours de ce temps là, j'ai entrepris de refaire toute mon instruction, à commencer par l'arithmétique de M. Serret." Letter from Beltrami to Hermite, Venise 1st November 1887; in Betti's Archive, Archivio della Biblioteca della Scuola Normale Superiore, Pisa.

[6]On the commitment of Italian mathematicians during the Risorgimento, see [18, 63].

Italian mathematicians were therefore in the front line of this process of nation-building, through the public and political offices they held. In particular, Beltrami and Casorati were senators; Brioschi and Betti were deputies, senators, and also under-secretaries in the Ministry of Education (in the period 1861–62 and 1874–76, respectively); Cremona was appointed Minister of Education in 1898, albeit only for one month, and was the director of the *Scuola degli Ingegneri* in Rome.

Brioschi was a key figure in the formation and guidance of the new entrepreneurial middle class of Northern Italy. This emerging class, which took shape in reaction against a sluggish and passive landowning class, expected much from the progress of science and technology. Therefore the mentality and outlook of many scientists were in perfect harmony with the ideas of this middle class and its desire to further and accelerate the growth of Italian industry. As an expression of this shared project – matching the interests of the middle class and the orientation of mathematicians regarding their research and teaching – Brioschi founded the *Polytechnic* in Milan in 1863, with the intention of training engineers able to contribute to the development and modernization of Italy.[7]

The collaboration between Betti and Brioschi was the true driving force behind the initial development of Italian mathematics; it was extremely fruitful both because of its organizational impact and in consequence of the high-level research to which it led. The influence of Riemann – whom Betti met at the University of Göttingen for the first time – was crucial for Betti's scientific activity.[8] Betti later invited Riemann to spend two years (1863–1865) at the University of Pisa, when Beltrami was professor of geodesy at the same university. Betti deeply involved himself in promoting Riemann's ideas not only in Italy but also in Europe at large.

Generally speaking, Italian mathematicians tried to build Italian national identity by following several directions:

1. Developing new research fields
2. Creating new mathematical chairs and new research institutes – such as the Polytechnic of Milan
3. Starting the publication of new mathematical high-level journals able to compete with German or French mathematical journals – such as the *Annali di Matematica pura ed applicata* in 1858 and the *Giornale di Matematiche* in 1863[9]
4. Increasing the number of grants for young mathematicians who aimed at specializing in foreign universities, and especially in Germany and France
5. Reforming university and secondary school
6. Writing or translating good mathematical treatises for mathematical teaching

Just after the celebrated trip throughout Europe, Betti translated Riemann's *Inauguraldissertation* on the theory of complex functions [64] and spread interest in

---

[7]On Brioschi and the Polytechnic of Milan, see [58].

[8]For Betti and his mathematical school at the University of Pisa, see [17].

[9]On these mathematical journals see [19, 23].

this field of research in Italy, to which especially Casorati made important contributions. Beltrami published his important works on non-Euclidean geometry [5, 6] employing the methods of differential geometry he had learned when he was a student in Pavia. Cremona's geometrical research led to the emergence of the so-called Italian school of algebraic geometry.

They also engaged themselves in writing new treatises on different subjects – the theory of determinants (Brioschi, [21]), real and complex analysis (Ulisse Dini (1845–1918) [28] and Casorati [22]), rational mechanics (Domenico Chelini (1802–1878) [24]), graphical statics (Cremona [26]), potential theory (Betti [15]), and many others. Beltrami did not publish any treatise but, as it appears from reading his letters (see Sect. 3.2), he often drafted his courses on mathematical physics. One of these course contains his lectures on the theory of elasticity held at the University of Rome; it was handwritten by one of his student, Alfonso Sella (1865–1907), and is held by the Library of the Department of Mathematics in Genoa. Other lectures on mathematical physics held by Beltrami at the University of Pavia in the period 1881–1883 are contained in the *Dipartimento di Matematica* in Pavia. These lectures concern heat theory, magnetism and electromagnetism, potential theory, electrodynamics, and elasticity of solid bodies.

Betti was the director for about 30 years of the celebrated Scuola Normale Superiore of Pisa, where many mathematicians studied and were trained in mathematical research. Among them we may mention Dini, Salvatore Pincherle (1853–1936), Luigi Bianchi (1859–1928), Gregorio Ricci Curbastro (1853–1925), and Volterra. Pincherle spent a year (1877–1878) at the University of Berlin with Weierstrass; Bianchi and Ricci Curbastro attended Klein's lectures at the University of Munich in 1879–1880 and in 1877–1878, respectively.

The Risorgimento generation had the great merit of creating the conditions for the development of the *second generation* of mathematicians, who made Italy a great power in mathematics, ranking only after France and Germany. This period can be called the golden age of Italian mathematics; the mathematicians of the next generation would have to measure up to this high standard.

3 New Perspectives

3.1 A Biographical Sketch

Beltrami studied with Brioschi at the University of Pavia from 1853 to 1856, when he had to leave Pavia before graduating because of his Risorgimento ideals.[10] Just after the unification of Italy he was offered the chair of complementary algebra and

---

[10]The director of the Collegio Ghislieri in Pavia expelled Beltrami for political reasons; see, for example, [57].

analytic geometry at the University of Bologna, which he held during the academic year 1862–63. In 1863, Ottaviano Mossotti (1791–1863), professor of geodesy at the University of Pisa, died; therefore Betti asked Beltrami to come to Pisa to take over Mossotti's chair and Beltrami accepted. From 1866 to 1873 he was back in Bologna, where he taught rational mechanics. In 1873, he accepted an offer to join his old colleague and friend Cremona in Rome, where Cremona was the director of the *Scuola degli Ingegneri* and Beltrami held the chair of rational mechanics at the university. In 1876, Beltrami moved to Pavia where he taught mathematical physics. In 1891, he returned to Rome, where he taught until his death. In 1898, he became the president of the Accademia dei Lincei and, the following year, a senator of the Italian Kingdom.

Beltrami's correspondence allows us to highlight academic motivations and personal reasons which led him to change his university and his teaching direction several times.[11]

Beltrami is mainly known for his contribution to differential geometry, which continued the works of Gauss and Riemann, and to non-Euclidean geometry. These two research fields were indeed deeply connected, as Beltrami himself often claimed. In his *Saggio di interpretazione della geometria non euclidea* [5], Beltrami shows that hyperbolic geometry could be represented on a particular surface of the Euclidean space having constant negative curvature (called the *pseudospherical surface*). By using a suitable lexicon, where "geodesic line on the pseudospherical surface" stands for straight line of the hyperbolic plane" and "angle between two geodesic lines" for "angle between two straight lines", Beltrami translates all the theorems that are valid in the Lobachevskij geometry to theorems valid on the pseudospherical surface. However, his model of the hyperbolic plane in Euclidean space is valid only locally.

His second fundamental paper on non-Euclidean geometry, *Teoria fondamentale degli spazi di curvatura costante* [6], originates from a deep analysis of Riemann's work on differential geometry and rigorously proves some important results concerning Riemannian geometry. [5] and [6] were soon translated into French and gave an impulse to the development of non-Euclidean geometry in all Europe.

In the 1870s, Beltrami changed his research field and directed his interests to mathematical physics – first to potential theory and then, from 1882 onward, to the theory of elasticity. However, he also published many works concerning rational mechanics, electricity and magnetism. The mathematical tools employed in his research on mathematical physics are those of differential geometry. Beltrami, like other mathematicians and physicists of the nineteenth century, tried to explain propagation of physical phenomena in space by imagining an ether filling the whole universe – to a particular strain system of the ether corresponds a phenomenon

---

[11] In particular, see Beltrami's letters to Hoüel in [16] and his (unpublished) letters to Betti contained in Betti's Archive, Archivio storico della Scuola Normale Superiore di Pisa; Beltrami's letters to Betti will be published in [39]. On Beltrami's life and work, see also [67].

which is transmitted through space by contact (and not at-a-distance) from one ether particle to an immediately contiguous one.

It was also a (natural-)philosophical motivation which led Beltrami to prove new and interesting results, such as the Laplace–Beltrami equation in potential theory (which generalizes the classical Laplace equation to an n-dimensional Riemannian manifold); the equations of elastic equilibrium in a space with constant curvature; and the necessary and sufficient conditions for six given functions to constitute the components of an actual elastic deformation – these formulae led Beltrami to deduce the equations of elastic equilibrium expressed by elastic stress, which are nowadays called "Beltrami–Mitchell's equations".[12]

Beltrami and Betti are the two key figures of mathematical physics in Italy in the second half of the nineteenth century. Both were members of university and ministerial commissions, engaged in politics, education, and mathematical teaching as well as in mathematical research. As already mentioned Betti was the director of the Scuola Normale Superiore of Pisa and trained many mathematicians in research. On the contrary, Beltrami did not found an actual school of geometry and/or of mathematical physics – his frequent changes of town and university probably prevented him from creating a stable group of students and researchers working in the same university and on similar subjects. Nevertheless, his research on geometry and on mathematical physics had a great influence in Italy and abroad, as these letters show. In Italy, Valentino Cerruti (1850–1909), Ernesto Cesaro (1859–1906), Gian Antonio Maggi (1856–1937), Ernesto Padova (1845–1896), and Carlo Somigliana (1860–1955) followed Beltrami's ideas and researches, even though sometimes they were not his students in the usual sense. His letters to his foreign colleagues reproduced in the Appendices show how his works – not only on non-Euclidean geometry, but also on mathematical physics – were well-known and appreciated also abroad.

3.2 The New Physics

As we have already pointed out, since the 1870s Beltrami turned his interests from geometry to mathematical physics, and published about 50 papers on potential theory, the theory of elasticity, hydrodynamics, electrostatics, electrodynamics, magnetism, and electromagnetism. In his papers Beltrami often combined mathematical physics and non-Euclidean geometry, and obtained remarkable results. In particular, he successfully applied his geometrical research to the generalization of some aspects of mechanics, potential theory and theory of elasticity to non-Euclidean spaces. Beltrami's ideas are well described in the following passage by Maurice Lévy:

---

[12]On this subject see [65].

M. Beltrami is indeed a follower of Faraday and Maxwell; that is to say, he aims at suppressing any assumption of action at-a-distance and at explaining physical phenomena by means of a medium connecting all natural bodies, so that one can not touch a body without producing a more or less strong effect on the others. (My translation)[13]

In Beltrami's opinion, in fact, an ethereal medium existed, which was able to propagate forces in space by means of its own deformations. Strains and tensions of the ether were submitted to the laws of the theory of elasticity. As already mentioned, Beltrami published various papers on Maxwell's equations and tried to give a mechanical interpretation of electromagnetic phenomena. For example, in a paper published in 1882 [9], he described a system of elastic deformations of ether "in accordance with" the ideas of Faraday and Maxwell. Explaining electric and magnetic phenomena by means of an ethereal and elastic fluid allowed the reduction of electromagnetism to mechanics – and that was considered as an important achievement by mathematicians and physicists of the nineteenth century.

Between 1884 and 1886, Beltrami published three papers [10–12] on Maxwell's electromagnetic theory and its mechanical interpretation. On January 4th, 1885 Beltrami explained to Ludwig Schläfli (1814–1895) that he was very interested in Maxwell's electromagnetic theory and also his university lectures dealt with this subject. In particular, he tried to connect together electric and elastic phenomena. He wrote to Schläfli:

> This year I am especially interested in the theory of elasticity, both in my lectures and in my research. This theory is very interesting to me, also from the point of view of the method. In particular, I am investigating a question connected to Maxwell's ideas, which tends to establish a deep connection between electric and elastic phenomena. Until now my deductions seem to be rather contrary to the possibility of this connection, but I do not dare to formulate conclusive statements – Maxwell himself did not want to attribute a precise form to his interpretation and hoped for a long time, unfortunately in vain, to get more concrete results. (My translation)[14]

Beltrami referred to his attempt to justify Maxwell's electromagnetic theory by deducing the ether deformations able to produce Maxwell's equations. In a

---

[13]"Mr Beltrami, en effet, est un disciple de Faraday et de Maxwell, c'est–dire que sa tendance est d'arriver à la suppression de l'hypothèse des actions à distance et à l'explication des phénomènes physiques par la présence d'un milieu servant de lien entre tous les corps de la nature, de telle sorte qu'on ne puisse pas toucher à l'un d'eux, sans que tous les autres s'en ressentent plus ou moins." This unpublished report of Maurice Lévy concerns Beltrami's election as a corresponding member of the Paris Academy of Sciences, and is contained at the Archives de l'Académie des Sciences in Paris.

[14]"In questo anno mi occupo specialmente, tanto nelle lezioni, quanto nel mio studio privato, della teoria dell'elasticità, la quale mi interessa molto, anche dal lato del metodo. In particolare sto studiando una questione che si collega con quelle vedute di Maxwell le quali tenderebbero a stabilire un nesso intimo fra i fenomeni elettrici e gli elastici. Finora le mie deduzioni sarebbero piuttosto contrarie alla possibilità di questo nesso, ma non ardirei ancora di formulare conclusioni assolute, tanto più che lo stesso Maxwell non ha voluto dare una forma precisa alla sua interpretazione ed ha sperato, disgraziatamente invano, di ottenere da tempo risultati più concreti." Letter by Beltrami to Schläfli, in [40], pp. 120–121.

paper published in 1886 [12] Beltrami found a negative answer: he proved that no elastic medium could produce strains and tensions in accordance with Maxwell's electromagnetic theory. Nevertheless, he believed in the validity of Maxwell's theory and was one the first Italian scientists who actively studied the new physics.

In a letter to Duhem dated July 25th, 1891, Beltrami confessed he feared that he misunderstood the ideas of the illustrious "English physicist". In another letter to his French colleague (15 April 1892) Beltrami wrote how his aim was to justify Maxwell's results, though "maybe there does not exist in the history of science another example of a doctrine so badly digested and explained as the theory contained in the famous Treatise". (My translation)[15]

In various papers Beltrami extended potential theory, the theory of elasticity, and classical mechanics to non-Euclidean spaces. He indeed considered the possibility that physical phenomena could propagate in a non-Euclidean space. This idea was shared by the German mathematicians Killing, Lipschitz, and Ernst Schering (1824–1889), who tried to extend certain classical theories of mathematical physics – such as Hamiltonian mechanics and potential theory – to a Riemannian manifold with constant curvature.[16] The long letter by Beltrami to Killing (March 22nd, 1888) shows that his German colleague essentially shared Beltrami's research program. Beltrami suggested him to study physical laws concerning mutual interactions between electric particles in non-Euclidean spaces that could replace Ampère's law in the usual space. Beltrami thought that this research, and more in general research concerning electro-dynamic theory in non-Euclidean spaces, could lead to "very interesting results".

However, in Italy mathematical physics was not very popular at that time, as Beltrami complained in some of his letters. On July 25th, 1891 Beltrami wrote to Duhem that he felt himself "very isolated" in his research on mathematical physics. And in a letter to Klein (April 17th, 1888) he deplored the attitude of Italian physicists towards mathematical physics. According to Beltrami, "only two or three physicists" knew the fundamental theories of mathematical physics in Italy. And he added:

> As to the fear you seem to feel concerning an increasing divorce between mathematics and physics, I would dare to say that Germany is, in my opinion, the country where this divorce seems to me the less advanced. (My translation)[17]

In the same letter, Beltrami remarked that it could be profitable if a journal of mathematical physics were to exist. In fact, such a journal could help those who are usually considered as mathematicians among physicists and as physicists among mathematicians. It is evident that Beltrami referred to his own experience.

---

[15]The original French is: "il n'y a peut-être pas d'exemple, dans l'histoire de la science, d'une doctrine aussi mal digérée et exposée, que celle du célèbre Treatise".

[16]On these works see [66].

[17]"Quant à la crainte que Vous semblez concevoir d'un divorce toujours croissant entre les mathématiques et la physique, j'oserais vous dire que l'Allemagne est, à mon avis, le pays où ce divorce me paraîtrait le moins avancé". Letter by Beltrami to Klein, in [16], p. 213.

3.3 Beltrami's Reputation

Beltrami had strong international relationships and often met foreign colleagues, as it is shown by his letters. For example, he met Schläfli several times between 1871 and 1873; in 1873 he met Hermann von Helmholtz (1821–1894) for the first time and again in 1877; during his summer holidays he met Kronecker, Georg Ferdinand Frobenius (1849–1917), Thomas Archer Hirst (1830–1892), Carl Friedrich Geiser (1843–1934), Jacob Rosanes (1842–1922) and others. In a letter to Betti (30 August 1879) Beltrami was happy to write that he has joined "Schläfli, Geiser, Frobenius, Meyer, Hirst, Cremona, Casorati" and that at that moment "they were waiting for Zeuthen". In another letter to Betti (17 September 1880) he wrote that he had to keep Hermann Amandus Schwarz (1843–1921) company.[18]

In spite of the isolation he felt, Beltrami was a reference figure in Italy as well as abroad. I shall show some examples. Dedekind asked Beltrami for information about an Italian mathematician – Aureliano Faifofer (1843–1909) – who intended to translate Dirichlet's treatise on number theory into Italian. This work had been edited by Dedekind himself and published in 1863. Beltrami's answer convinced Dedekind to accept Faifofer's proposal and the Italian translation of Dirichlet's classical book appeared in 1881 (see the correspondence between Dedekind and Beltrami in Appendix 4.1).

Beltrami was asked to participate in prestigious celebrations: he was invited by Schwarz to take part in the centenary celebration of Gauss's birth held in Göttingen in 1877. In the end Beltrami had to forego participation in the celebrations.[19] Beltrami was also asked to participate in the scientific meeting celebrating the 70th birthday of Charles Hermite (see Beltrami's letter to Mittag-Leffler dated June 27th, 1892). Mittag-Leffler often invited Beltrami to publish his papers in *Acta Mathematica*, the well-known Swedish journal which Mittag-Leffler had founded and directed for many years.

As we have already mentioned, his frequent changes of university and teaching area prevented Beltrami from creating his own mathematical school. Nevertheless, his letters show his commitment to raise the level of Italian mathematics. He always

---

[18] See Beltrami's letters to Schläfli dated 6 August 1871, 17 August 1871, and 20 September 1873, in [40], pp. 113–116; Beltrami's letter to Mittag-Leffler dated 19 July 1883; Beltrami's letter to Schwarz dated 5 April 1877. Beltrami's letters to Schwarz are in the Archiv der Berlin-Brandeburgischen Akademie der Wissenschaften of Berlin, Nachlass Schwarz.

[19] That can indeed be read in Beltrami's letter to Schwarz dated April 25th, 1877: "Le circostanze alle quali alludevo nella mia lettera precedente mi hanno definitivamente obbligato a privarmi del piacere di assistere alla solennità scientifica del 30 aprile. Nel mentre mi reco a dovere di ciò significarLe, per ringraziarLa dei cortesi eccitamenti venutami da Lei, faccio appello alla di Lei benevolenza per pregarLa di attestare ai geometri tedeschi la viva partecipazione che io prendo, benché lontano, ai sentimenti di ammirazione e di gratitudine che li fanno convenire a Gottinga per onorare la memoria del grand'uomo che, sebbene avesse applicato a sé stesso il motto Nil actum reputans, si quid superesset agendum, ha pur tuttavia lasciato tanto di actum da rendere ardua ai posteri la via dell'agendum." This letter is unpublished (see footnote 18).

supported the best students, and considered mathematical teaching an important issue. In a letter to Betti (20 February 1882) he praised Enrico d'Ovidio (1843–1933) "because he is able to train good students for research; something which did not occur in Turin for a long time". (My translation)[20] In another letter to Betti (13 March 1881) he recommended Gian Antonio Maggi, one of his students at the University of Pavia, who possessed all the requisites necessary for "becoming a good physicist".

His high reputation allowed Beltrami to write to his colleague Felix Klein in order to introduce to him some young Italian mathematicians who aimed at doing further specialized work in Germany. In a letter dated 7 December 1883 Beltrami introduced to Klein Giacinto Morera (1856–1909) and in another letter (5 January 1885) he wrote to him: "I take this opportunity to thank you for your kindness towards M. Morera and the other young Italian people who have preceded him in Leipzig".[21]

Beltrami addressed Klein again in order to introduce another young Italian researcher, the physicist Alfonso Sella, who was the son of Quintino Sella. The latter was a mineral engineer and an "eminent man", a well-known politician who also became Ministry of Finance.

Furthermore, Mittag-Leffler asked for Beltrami's intervention on a delicate question concerning the Russian mathematician Sonya Kovalevskaya. She had difficulty finding an academic position, though she was en excellent mathematician – in 1874 she obtained her PhD with a dissertation supervised by Weierstrass, and in 1888 she won the Prix Bordin proposed by the Academy of Sciences of Paris. In 1884, she obtained a position of extraordinary professor at the University of Stockholm for five years. Mittag-Leffler asked Beltrami to support the cause of Kovalevskaya in order to obtain for her a permanent position. Answering Mittag-Leffler's request, Beltrami wrote the Report some days later.[22] Finally, Kovalevskaya obtained a professorship at the University of Stockholm in 1889. Unfortunately she died two years later, in 1891. The letter addressed by Beltrami to the University of Stockholm, where he referred on Kovalevskaya's scientific works, is reproduced in Appendix 4.8.

We point out that another international undertaking had already taken place in 1886, when Weierstrass proposed Kovalevskaya for a vacancy at the Academy of Sciences of Stockholm. However only men were accepted as members of the Stockholm Academy. Though many mathematicians supported Sonya Kovalevskaya, the project was finally abandoned [25, p. 109], [1, pp. 162–163].

Appendix 4.7 contains two letters from Beltrami to Betti, written in 1886: as Beltrami explained, Angelo Genocchi (1817–1889) and Betti asked him and

---

[20] The original Italian is: "In quanto al D'Ovidio debbo dire [...] che egli merita lode anche perché gli riesce di fare buoni allievi, addestrandoli alla ricerca; cosa che a Torino da molto tempo non si vedeva."

[21] "Je saisis cette occasion pour vous remercier aussi de la bienveillance que vous avez bien voulu témoigner à M.r Morera, ainsi qu'aux autres jeunes gens italiens qui l'ont précédé à Leipzig."

[22] See Beltrami's letter to Mittag-Leffler dated 16 May 1889.

Casorati to sign a document supporting the cause of Kovalevskaya. But he hesitated and, apparently, Casorati shared his opinion. Indeed, as Beltrami himself confessed, he did not know what this document contained exactly and he had never read Kovalevskaya's works carefully. Furthermore, Beltrami thought that German mathematicians, who better knew Kovalevskaya and her mathematical skill, had to support her more strongly. Finally, Beltrami and Casorati preferred to avoid any personal involvement in this matter.

We may point out that other Italian mathematicians, such as Betti and Cremona, supported Kovalevskaya's cause since 1886. A letter from Betti to Cremona (26 March 1886) and Cremona's answer are reproduced in the same Appendix 4.7. Betti informed Cremona about an international undertaking in favour of Kovalevskaya, and said that many French mathematicians – such as Hermite, Camille Jordan (1838–1922), Gaston Darboux (1842–1917), Paul Appel (1855–1930), Henri Poincaré (1854–1912), Emile Picard (1856–1941), and Félix Tisserand (1845–1896) – joined it. In his answer (2 April 1886) Cremona wrote that he intended to support Kovalevskaya, together with Betti, Genocchi and other Italian mathematicians.

4 Appendix

4.1 The correspondence Beltrami–Dedekind

The Beltrami–Dedekind[23] correspondance is contained in the Archive of the Niedersächsische Staats-und Universitätsbibliothek, Göttingen, Cod. Ms. Dedekind XIII: 2, XIV: 6.

Letter n. 1
Draft by Richard Dedekind to Eugenio Beltrami

An Herrn Ritter Eugenio Beltrami Professor an der K. Universität in Rom

Hochgeehrter Herr Professor! Hiermit erlaube ich mir, Ihnen eine Bitte um gefällige Auskunft vorzutragen, deren Veranlassung die folgende ist. Gestern habe ich ein Schreiben des Herrn Faifofer[24] Professor am Lyceum zu Venedig, welcher mir seine Absicht zu erhennen giebt, eine italienische Übersetzung der "Vorlesun-

---

[23]Richard Dedekind (1831–1916) studied at the universities of Göttingen and Berlin. At the University of Göttingen he followed Dirichlet's lectures on number theory, which he collected and published in 1863. In 1858 he began teaching at the Polytechnic of Zürich and in 1862 came back to his native Braunschweig, where he taught at the Technische Hochschule. He made important contributions to abstract algebra, algebraic number theory and the foundations of the real numbers.

[24]Aureliano Faifofer (1843–1909) studied at the University of Padua, where he spent some years as assistant professor. In 1868, he was appointed professor at the Liceo Foscarini in Venice. He published several mathematical textbooks, translated into many languages.

gen über Zahlentheorie von Lejeune Dirichlet" herauszugeben, und mir zugleich mittheilt,[25] dass Sie, hochgeehrter Herr Professor, und ebenso Herr Professor Bellavitis[26] zu Padua seinen Plan gebilligt haben. Die Ausführung derselben wird mir, dem deutschen Herausgeber, sehr angenehm sein, aber nur in der Voraussetzung, dass die Übersetzung sich mit vollem Verständniss und möglichst treu dem Originale anschliesst. Da mir nun der Name des Herrn Professor Faifofer bisher noch nicht bekannt gewesen ist, so werden Sie es natürlich finden, wenn ich noch vor der erforderlichen Verhandlung mit dem hiesigen Verleger Vieveg bitte mir vertraulich mitzutheilen, ob Sie Herrn Faifofer näher kennen und für vollkommen fähig halten, ein solches, keineswegs leichtes Unternehmen gut auszuführen; es würde mich auch besonders interessieren zu erfahren, ob Herr Faifofer, der mir über seine eigene Person gar nichts schreibt, sich bisher schon mit der *Theorie der Zahlen* ernstlich beschäftigt hat, oder ob er erst jetzt ein eigentliches Studium derselben beginnen will. Dies Letztere würde zwar in meinen Augen kein absolutes Hinderniss sein, wenn Sie ihn im Übrigen als einen guten *Mathematiker* kennen, aber es würde mich veranlassen die Bedingung zu stellen, dass ich die Übersetzung vor dem Druck genau durchsehen und eventuell auch Abänderungen derselben verlangen dürfte.

Die Ausführung des Plans, welchen ich mit Freude begrüsse, würde für mich zugleich eine willkommene Veranlassung sein, die italienische Sprache gründlicher als bisher zu erlernen. Freilich bin ich wohl im Stande, die ausgezeichneten Arbeiten der jetzigen italienischen Mathematiker zu verstehen, und ich erlaube mir bei dieses Gelegenheit, Ihnen, hochgeehrter Herr Professor, meinen herzlichen Dank für die höchst interessanten und werthvollen Geschenke auszusprechen, durch welche Sie mich hoch erfreut haben; allein weiter gehen meine Kenntnisse bisher nicht, und aus diesem Grunde habe ich mich in meiner Muttersprache an Sie gewendet, deren Sie, wie aus Ihren Arbeiten hervorgeht, vollständig mächtig sind. Sie werden mich zu grossem Danke verpflichten, wenn Sie meine Bitte erfüllen und mich mit einer italienischen Antwort beehren wollen.

Mit ausgezeichneter Hochachtung verbleibe ich Ihr ganz ergebener

R. Dedekind, Professor
Braunschweig
7 Juni 1876
Petrithorpromenade 24

---

[25] Johann Peter Gustav Lejeune Dirichlet (1805–1859) was professor at the University of Berlin from 1831 to 1855, when after Gauss's death he became professor at the University of Göttingen. His lectures on number theory were edited and published by Dedekind, who was one of his students at the University of Göttingen: Dirichlet, P.G., *Vorlesungen ber Zahlentheorie*, Braunschweig, Vieweg, 1863. The Italian translation of this book was actually published by Aureliano Faifofer, *Lezioni sulla teoria dei numeri di P. G. Lejeune Dirichlet, pubblicate e corredate di appendice da R. Dedekind*, translated and edited by A. Faifofer, Venezia, Tipografia Emiliana, 1881.

[26] Giusto Bellavitis (1803–1880) was professor at the University of Padua from 1845 onwards, though he did not attend any establishment of higher education. Faifofer was one of his students. Bellavitis is well-known for his works on equipollencies, a kind of vector calculus.

Letter n. 2
Eugenio Beltrami to Richard Dedekind

Roma 13 Giugno 1876

Chiarissimo Signor Professore
   Mi affretto a soddisfare al desiderio manifestato nella graditissima di Lei lettera del 7 corr., jeri soltanto pervenuta nelle mie mani.
   Io conosco il Sig.r D.r Faifofer già da qualche anno, senza però avere (anche per la lontananza delle nostre rispettive sedi) una grande intimità con lui, vedendolo io di quando in quando soltanto nelle ferie autunnali.
   Rispetto però allo scopo delle informazioni da Lei desiderate, credo di poterLe fornire alcuni elementi abbastanza sicuri, che dividerò in due parti, cioè in quella che si riferisce alla cognizione della lingua tedesca, ed in quella che si riferisce alla competenza matematica.
   In quanto alla prima parte, credo di potere assicurare che il prof. Faifofer conosce bene la lingua anzidetta, e la parla correntemente, avendo anche più volte viaggiato in Germania. D'altronde la precisa e ristretta fraseologia matematica non presenta alcuna delle difficoltà che lo stile letterario può presentare talvolta anche a chi possiede profondamente una lingua straniera.
   In quanto alla conoscenza dell'argomento delle ormai celebri Vorlesungen del Dirichlet, di cui Ella ha così largamente accresciuto il valore pedagogico e scientifico colla propria illuminata collaborazione, mi riesce più difficile pronunziare un giudizio esplicito. Il Sig.r Faifofer è conosciuto in Italia come autore di due recenti trattati elementari, l'uno di Aritmetica, l'altro di Algebra, i quali, senz'avere alcuna pretesa di originalità, anzi essendo forse troppo rigidamente calcati sui programmi e sulle istruzioni Ministeriali (relative all'insegnamento matematico nei Ginnasi e Licei), sono però da collocarsi fra i migliori di quelli fin qui pubblicati per uso delle scuole, ed hanno già trovato buona accoglienza.[27] Inoltre, il medesimo Sig.r Faifofer è considerato, e giustamente, come uno dei migliori insegnanti di matematica elementare, tanto dal punto di vista della chiarezza quanto da quello della esattezza e del rigore delle sue lezioni.
   Ma altre prove del suo sapere in matematica, e specialmente nelle parti più elevate della scienza, non esistono (almeno a mia conoscenza); né, da quanto egli stesso mi comunicò intorno agli studii universitarii da lui fatti, mi risulta ch'egli abbia in essi approfondita particolarmente qualche disciplina matematica superiore. Mi ricordo perfettamente d'avere udito, dalla sua bocca, che stava da ultimo studiando le Vorlesungen, e che, per suo esercizio, le traduceva. Più tardi poi mi parlò delle difficoltà che aveva dovuto superare nell'andare piïnnanzi, e dell'intenzione che cominciava a formare di dare pubblicità alla sua traduzione, intenzione alla quale io non ho certamente mancato di far plauso. Sono ora circa

---

[27] In 1876, when Beltrami wrote the letter, Faifofer had already published the following books for secondary schools: *Elementi di geometria*, Venezia, 1878; *Elementi d'algebra e trigonometria : ad uso dei licei*, Venezia, 1878.

due anni che ebbe luogo la prima di queste comunicazioni; e, sebbene io non creda che il Faifofer abbia allargato i suoi studii sulla teoria dei numeri (ricorrendo ad altri autori più o meno speciali), inclino tuttavia a ritenere, specialmente dopo il passo fatto presso di Lei, che egli abbia esattamente compreso il classico libro da lui tradotto, e che, riservandosi Ella la revisione dei fogli (come non posso a meno di consigliarLe), la versione italiana sia per riescire soddisfacente.

Io Le sono grato a ogni modo delle di Lei gentili righe, che mi hanno fornito l'occasione di dirigerLe questa mia, e mi professo con tutta stima e devozione

obbl.mo E. Beltrami

Letter n. 3
Eugenio Beltrami to Richard Dedekind

Pavia 9 Marzo 1879

Illustrissimo Signore

La ringrazio distintamente del gentilissimo dono ch'Ella si è compiaciuto di farmi della III ed. delle Vorlesungen ueber Zahlentheorie, divenute ormai classiche per il felice innesto delle di Lei proprie elucubrazioni sul tronco originale delle Lezioni di Dirichlet. Io ammiro queste profonde e delicate ricerche, quantunque non abbia finora avuto il coraggio di affrontare i problemi.

Con questa di Lei pubblicazione, e con quelle dei Sig.ri Meyer e Grube,[28] credo sia presso a poco esaurito il campo così largamente fecondato dalle lezioni orali di Dirichlet. Credo però anche che, per coronare l'opera, bisognerebbe fare per lui quello che è stato fatto per Riemann[29] e per altri eminenti matematici, cioè ripubblicarne in un sol corpo le Memorie e gli scritti sparsi. Ho ferma opinione che una tal pubblicazione sarebbe estremamente gradita a tutto il mondo matematico. La perfezione di forma che distingue i lavori di Dirichlet li rende eminentemente adatti ad essere, dirò così, popolarizzati fra gli studiosi, e resi accessibili anche a chi non ha opportunità di fruire delle pubbliche Biblioteche.

RinnovandoLe i miei più sinceri ringraziamenti, La prego di aggradire l'espressione della mia profonda stima e di credermi sempre di Lei div.mo obbl.mo

Eugenio Beltrami

---

[28] He refers to Dirichlet's lectures edited by Franz Grube (1835–1893) with the title *Vorlesungen über die im umgekehrten Verhältnisse des Quadrats der Entfernung wirkenden Kräfte*, Leipzig, Teubner, 1876. Another edition of Dirichlet's lectures has been published by G. Arendt, *Vorlesungen über die Lehre von den einfachen und mehrfachen Bestimmten Integralen*, Braunschweig, Vieweg, 1904. Oscar Emil Meyer (1834–1909) edited Franz Neumann's lectures on the theory of elasticity: *Vorlesungen über die Theorie der Elasticität der festen Körper und des Lichtäthers gehalten in der Universität Königsberg*, Leipzig, B. G. Teubner, 1885.

[29] He refers to Riemann's collected works edited by H. Weber and published in 1876 for the first time (Leipzig, Teubner). Bernhard Riemann (1826–1866) – a personal friend of Dedekind – was professor at the University of Göttingen from 1859 until his death.

4.2 Letters by *Eugenio Beltrami* to *Pierre Duhem*

The letters by Eugenio Beltrami to Pierre Duhem,[30] Dossier Duhem, B-10, Archives de l'Académie des Sciences, Paris

Letter n. 1
Eugenio Beltrami to Pierre Duhem

[Undated letter]

Eugenio Beltrami Professore nella R. Università di Roma offre ses remerciements les plus sincères à M. le prof. Duhem pour son précieux cadeau.

Letter n. 2
Eugenio Beltrami to Pierre Duhem

[Undated letter]

Eugenio Beltrami exprime à M. Duhem ses meilleurs souhaits pour la nouvelle anné, le félicitant de son infatigable activité scientifique et faisant ses meilleurs voeux pour l'heureux succès de son oeuvre de savant et de professeur.
E. B.

Letter n. 3
Eugenio Beltrami to Pierre Duhem

Pavia 25 Juillet 1891

Monsieur et très honor Collègue

Je viens de recevoir, aujourd'hui même, par l'entremise de M. Gauthier-Villars, le beau volume, formant la 1.ère partie d'un Traité d'électricité et de magnétisme, que Vous avez bien voulu m'adresser.[31] Je tiens à Vous remercier de la manière la plus distinguée de cet offre si bienveillant et si considérable, et qui m'est d'autant plus

---

[30]Pierre Duhem (1861–1916) studied at the Ecole Normale Supérieure in Paris. He was professor at the universities of Lille (1887–1893), Rennes (1893–1894) and Bordeaux (1894–1916). He strongly contributed to mathematics, and to history and philosophy of science. His scientific works concern thermodynamics, hydrodynamics, elasticity and chemical physics. His works on history of science mainly concern history of mechanics and, more generally, history of mathematical physics.

[31]See [29].

précieux que son sujet est celui sur lequel se concentre mon attention en ce moment. D'après ce que Vous dîtes dans Votre préface, touchant les méthodes et les points de vue qui Vous sont propres, j'ai la certitude que j'apprendrai beaucoup de choses par Votre livre, surtout du côté thermodynamique, que, je Vous l'avoue, je n'ai jamais eu l'occasion d'approfondir, autant que l'importance su sujet le mérite. Je me trouve malheureusement très-isolé depuis longtemps, par rapport aux études physico-mathématiques, et l'occasion me manque d'éclaircir mes idées sur plusieurs points délicats, que je suis obligé, par là, à laisser de côté.

Etant sur le point de partir pour la campagne, j'ai fait à la hâte un paquet de quelques unes de mes publications antérieures et je vais Vous l'adresser, en Vous priant de vouloir bien les agréer comme une faible marque de mon estime et de ma reconnaissance. Il y en aurait bien d'autres que je voudrais Vous faire parvenir, mais je ne les ai pas sous ma main en ce moment: j'en tiendrai note pour mon retour, si tant est que ce retour ait lieu: car c'est presque décidé que j'irai à Rome, pour la nouvelle année scolaire, et dans ce cas j'aurai sur le bras la grosse affaire d'un déménagement complet!

Vous ne parlez pas des théories de Maxwell, et en effet Vous n'aviez pas besoin d'en parler, au point de vue des sujets que Vous traitez dans ce premier volume. C'est dans le second que Vous la rencontrerez, et je désire beaucoup voir la place que Vous lui ferez. Je dis cela, car mon éternelle préoccupation est de bien comprendre, si c'est possible, la pensée de l'ingénieux physicien anglais.

Veuillez agréer, Monsieur et honoré collègue, l'expression de mes sentiments de considération la plus distinguée.

Votre bien obligé E. Beltrami

P. S. Voulez-Vous bien me permettre une rectification méticuleuse? Dans la 2.e Note à la page 3 de Votre Traité c'est la 1.ère série du Journal de Physique qu'il faut indiquer.

Letter n. 4
Eugenio Beltrami to Pierre Duhem

Rome 27 Décembre 1891

Monsieur et très honoré Collègue

Les vicissitudes de mon changements de résidence m'ayant amené à force retenir à Pavie tous les envois d'imprimés, qui m'ont été adressés dans ces derniers mois, ce n'est qu'aujourd'hui même que j'ai reçu les nombreuses publications, en partie nouvelles, que Vous avez eu la bonté de m'envoyer à différentes reprises. Je regrette ce retard à double titre: d'abord parce que je n'ai pu Vous présenter plus tôt mes remerciements les plus sincères, ce dont je Vous demande bien pardon; ensuite parce que j'aurais pu tirer grand avantage de quelques unes de ces publications, pour une étude sur l'électromagnétisme, que je viens de rédiger pour l'Académie de

Bologne.³² Mais je tâcherai de suppléer à cet inconvénient dans l'intervalle assez long qui s'écoulera avant l'impression.

Je n'ai presque vu que les titres et les sujets principaux de Vos ouvrages, et j'aurai l'occasion de Vous en parler plus au long dans la suite, car plusieurs de ces sujets m'intéressent au plus haut degré. Permettez-moi cependant de Vous offrir, dès à présent, mes compliments les plus sincères pour cette imposante masse de travail accumulée en aussi peu d'années, dans plusieurs directions différentes, dont quelques unes ont été frayées la première fois par Vous.

Parmi Vos gracieux envois il y a plusieurs doubles. Ce sont les Mémoires suivants:
Sur l'équivalence des courants et des aimants;
Sur les dissolutions d'un sel magnétique;
Sur les équations générales de la thermodynamique;
Des corps diamagnétiques;
Sur la continuité entre l'état liquide et l'état gazeux et sur la théorie générale des vapeurs.³³

Si ce n'est que par mégarde que ces Mémoires m'ont été adressés en double exemplaire, je Vous les renverrai tout-de-suite: Vous n'avez qu'à me le dire. Si c'est avec intention, j'en ferai l'usage que Vous voudrez bien m'indiquer. J'ai même pensé que Vous m'aviez peut-être écrit à ce sujet: mais je n'ai reçu aucune lettre de Vous. Malheureusement les nombreuses tournées que j'ai dû faire récemment, légitiment le doute que quelques lettres se soit perdu.

Je n'ai pas encore de logement fixe proprement dit: je suis encore à la recherche. En attendant mon adresse est: Institut Royal de Physique, Panisperna 89.

Veuillez bien agréer, excellent collègue, mes sentiments les plus distingués et les plus affectueux.

E. Beltrami

Letter n. 5
Eugenio Beltrami to Pierre Duhem

Rome 15 Avril 1892

Monsieur et Collègue très-honoré!

Je ne suis que trop en retard avec les remerciements que je Vous dois, et que je Vous offre de la manière la plus distinguée, pour l'envoi du troisième volume de votre Traité,³⁴ ainsi que pour celui de Votre Discours. Veuillez pardonner ce retard, qui n'est dû qu'à une suite malheureuse de préoccupations douloureuses que j'ai eu, depuis le commencement de cette année, pour la santé toujours chancelante de ma

---

[32] Beltrami refers to [2].
[33] See references [31–35].
[34] See [29].

compagne et qui m'ont d'autant plus affecté que c'était surtout à cause de cela que je m'étais déterminé à accepter la nouvelle résidence de Rome, jugée plus favorable à la santé de M.me Beltrami.

La manière tout-à-fait nouvelle dont Vous posez, dans Votre nouveau volume, la question fondamentale des actions électrodynamiques mérite un examen sérieux, que je n'ai pu jusqu'ici qu'initier. J'ai pu, au contraire, puiser des connaissances précieuses dans divers chapitres détachés de ce volume, ayant rapport à certains sujets particuliers dont j'avais besoin de m'occuper. J'espère pouvoir me familiariser un peu plus tard avec les considérations d'ordre plus général qui caractérisent Votre ouvrage: mais je voudrais auparavant en finir avec la révision des épreuves (qui [...] on ne peut plus en longueur) d'un Mémoire sur l'électromagnétisme[35] où, comme Vous verrez, j'ai laissé dominer des tendances en quelques sorte opposées, soit en ne considérant que des systèmes continus de courants, soit en renonçant le plus possible à l'invocation des lois dites élémentaires. Je dois ajouter cependant que mon cadre était bien plus restreint que le Vôtre, et que mon but principal était de justifier plusieurs résultats de Maxwell, dont moi-même j'avais longtemps été sans ne rendre bien compte. Les confirmations que j'ai trouvées me disposent, je l'avoue, à croire que l'on pourra également justifier d'autres parties d'une théorie que Vous me semblez juger moins favorablement. Mais je puis Vous assurer que je suis tout prêt à me raviser: et je Vous accorde, en tous cas, qu'il n'y a peut-être pas d'exemple, dans l'histoire de la science, d'une doctrine aussi mal digérée et exposée, que celle du célèbre Treatise.[36]

En attendant je vais Vous adresser quelques petites brochures, dont l'une se rapporte de nouveau au principe de Huygens.[37] J'ai notablement simplifié la démonstration que Vous savez, de manière, surtout, à rendre inutile toute discussion touchant l'existence des intégrales: il me semble que le nouveau procédé se prête encore mieux aux exigences didactiques. J'y ai rattaché quelques développements ultérieurs, dont Vous saisirez facilement l'esprit.

Veuillez agréer, Monsieur et excellent collègue, l'expression renouvelée de ma reconnaissance, ainsi que celle de mon estime la plus sincère et la plus affectueuse.

E. Beltrami

Letter n. 6
Eugenio Beltrami to Pierre Duhem

Rome 6 Xbre 1898

Monsieur et Collègue très-honoré!

J'éprouvais depuis longtemps le plus vif désir de Vous offrir mes remerciements les plus sincères et les plus empressés pour le gracieux envoi de Votre 3.me

---

[35] See [2].

[36] He refers to Maxwell's treatise [59].

[37] He refers to his paper [3].

volume de la Mécanique chimique,[38] et ce sentiment s'est encore augmenté à la suite de l'arrivée d'un nouveau paquet de Vos travaux les plus récents, dont la seule inspection des sujets suffit pour faire apprécier de plus en plus la bonté de Vos choix, d'un côté, et l'extraordinaire puissance de travail, de l'autre, puissance qui se révèle aussi dans l'oeuvre dont Vos assistants puisent la conception et les méthodes dans l'étude et l'imitation de la Vôtre.

Malheureusement depuis quelque temps ma santé, qui s'était bien rétablie, ou paraissait l'être, s'est troublée de nouveau, par la répétition, quoique atténuée, des accidents qui l'avaient fortement compromises il y a deux ans, et j'ai dû me mettre tout-à-fait au repos, en renonçant non seulement à la reprise de mes leçons, mais à toute espèce de travail intellectuel, et ce n'est que depuis quelques jours que les médecins me permettent ma correspondance, que je dois borner, cependant, d'après leur terminologie, à sa partie non-scientifique. Je commence cependant à espérer, si tout continue de marcher comme à présent, de pouvoir reprendre en Janvier au moins une partie de mes occupations.

Quoiqu'il en soit [...] pourquoi il ne m'a pas été donné de Vous faire parvenir plus tôt l'expression de mes sentiments de reconnaissance, ainsi que de Vous prier de vouloir bien présenter mes remerciements bien distingués à Votre excellent disciple M. Marchis.[39] Je serai bien heureux si je pourrai encore recouvrer la faculté de suivre autrement que par le désir et les souhaits les développements et les progrès que Vous apportez à nos connaissances de la mécanique des phénomènes.

Veuillez agréer les salutations cordiales de Votre bien obligé

E. Beltrami

Letter n. 7
Eugenio Beltrami to Pierre Duhem

Rome 6 Juillet 1899

Monsieur et Collègue très-honoré

Je regrette infiniment d'avoir dû tant retarder à Vous exprimer mes remerciements, et plus encore mon admiration sincère et profonde, pour le gracieux envoi du dernier volume de Votre Ouvrage sur la mécanique chimique.[40] Je suis malheureusement trop peu versé dans la chimie moderne pour être à même d'apprécier le détail de Vos recherches sur les différents sujets que Vous avez traités, mais je puis assez suivre la ligne générale de Vos raisonnements et admirer la souplesse que Vous avez su donner aux procédés d'analyse et de déduction rationnelle, pour acquérir la conviction que Vous avez rendu un grand service à la science et fait faire de beaux progrès à sa coordination et à son enseignement!

---

[38] See [30].

[39] Lucien Marchis (1863–1941), professor of Aviation at the Faculty of Sciences of Paris.

[40] See [30].

Comme je Vous je le disais tout à l'heure, je regrette bien vivement de ne pas avoir eu moyen de prendre plus tôt connaissance de Votre important volume, mais d'abord ma santé n'est plus la même qu'au temps passé et tout travail un peu suivi me fatigue beaucoup, et ensuite mes fonctions académiques, dans les deux derniers mois, m'ont donné bien des tracas, et m'en donnent encore. En outre (et cela Vous surprendra beaucoup) depuis quelques semaines le Roi a voulu me nommer Sénateur, et quoique mes connaissances ne me donnent pas beaucoup d'avantages pour l'exercice de cette nouvelle fonction, il n'en est pas moins vrai qu'elle me prend assez de temps et qu'elle détourne matériellement (pour ainsi dire) mon attention de ce qui a trait aux études et à la science pure. Quoique il en soit je suis bien heureux d'arriver encore à temps pour Vous féliciter de Votre oeuvre savante et géniale et Vous souhaiter les meilleurs succès dans le monde scientifique.

Votre bien obligé et bien dévoué
E. Beltrami

4.3 A Letter by Beltrami to Wilhelm Killing

This letter by Eugenio Beltrami to Wilhelm Killing[41] is contained in Nachlass Killing, N. 8, Universitäts-und Landesbibliothek, Münster

Pavia 22 Marzo 1888

Chiarissmo collega

Una disgrazia che ho avuto nella famiglia,[42] mi ha impedito di rispondere prima d'ora alla gradita di Lei lettera del 2 di questo mese, e di darLe, in ispecie, l'assicurazione che non solo io, ma tutti gli studiosi delle scienze matematiche, hanno per Lei tutta quella considerazione e quella simpatia che vengono giustamente tributate a chi lavora con costanza e con successo per il progresso della scienza.

Rilevo dalla di Lei lettera che Ella avrebbe intenzione, dopo il termine delle difficili ricerche in cui è attualmente impegnato e di cui ho ricevuto con gratitudine i saggi già pubblicati, di studiare la teoria dell'attrazione degli ellissoidi nello spazio non euclideo. Questo sarà certamente un interessante soggetto di studio: tuttavia, conoscendosi già la forma che assume nel detto spazio la legge newtoniana,[43]

---

[41]Wilhelm Killing (1847–1923) studied at the universities of Münster and Berlin. In 1892, he was appointed professor at the University of Münster. His first research concerns the theory of surfaces and non-Euclidean geometry; later on he contributed to the theory of transformation groups and to Lie algebras.

[42]Most probably Beltrami refers to the death of his aunt living in Padoa.

[43]Some of Beltrami's papers concern potential theory in non-Euclidean spaces; in particular, he explicitly deduced the potential function in a space with constant negative curvature in [10]; on the subject see [67].

è probabile che l'interesse del lavoro si riporti soltanto sulla natura analitica delle nuove formole che Ella incontrerà nel risolvere i problemi già trattati nello spazio euclideo.

Senza quindi volerLa distogliere da questa intrapresa, io mi permetterei di indicargliene un'altra, la quale potrebbe forse condurre a conseguenze più inaspettate, e sarebbe la ricerca della legge che, nello spazio non euclideo, fa le veci di quella d'Ampère, relativamente all'azione mutua delle correnti elettriche, ed in generale lo studio della teoria elettrodinamica nel detto spazio.[44] Non sarei alieno dal credere che in questa via si potessero incontrare dei risultati molto interessanti. L'unico tentativo che è stato già fatto (a mia conoscenza) in questo genere di argomenti è quello del Sig.r Prof.re Ernesto Schering,[45] il quale si è occupato di ricerche molto generali intorno alla legge di Weber.[46] Ma, come Ella sa, la validità di questa ultima legge è molto controversa, mentre la legge d'Ampère e quelle trovate dal venerando F. Neumann sono universalmente riconosciute come inattaccabili, od almeno come basi necessarie di qualunque teoria: cosicché sarebbero queste che si dovrebbero, innanzi tutto, riportare allo spazio non euclideo.[47] Potrebbe darsi, del resto, che Ella conoscesse qualche lavoro, a me ignoto, già pubblicato sull'argomento, nel qual caso ritirerei la mia proposta e la convertirei nella preghiera di darmi, con Suo comodo, qualche notizia su tali lavori già compiuti da altri.

Gradisca i sensi della mia affettuosa stima e mi creda sempre

di Lei dev.mo
Eugenio Beltrami

---

[44] Ampère's law, discovered by André-Marie Ampère in 1826, relates the magnetic field around a closed loop to the electric current passing through the loop.

[45] Possibly Beltrami referred to the paper by Ernst Christian Julius Schering (1824–1889), professor at the University of Göttingen, "Zur mathematischer Theorie elektrisch. Ströme" published in the fourth volume (1857) of *Annalen der Physik und Chemie*.

[46] Starting from Ampère's law, Wilhelm Eduard Weber (1804–1891) obtained a general fundamental law of electrical action, expressing the force between moving charges. His fundamental electrical law depends upon the relative velocities and the relative accelerations of the particles. The Coulomb electrostatic law thus becomes a special case of Weber general law, when the particles are at relative rest. Weber's contributions to electrodynamics appeared in the *Elektrodynamische Maasbestimmungen* (Berlin, Springer, 1893) which collect seven long works published from 1846 to 1878. See [27].

[47] Franz Ernst Neumann (1798–1895), the father of Carl Neumann, studied at the universities of Berlin and Jena. In 1825, he became Privatdozent and in 1829 professor at the University of Königsberg. Neumann established mathematically the laws of induction of electric currents in some papers published in the 1840s. As a starting point he took Lenz's proposition (1834) according to which the current induced in a conductor moving in the vicinity of a current or a magnet will flow in the direction that tends to oppose the motion. Considering Ampère's law for a closed circuit, Neumann was led to what it is known as the mutual potential of two circuits – the amount of mechanical work that must be performed against the electromagnetic forces in order to carry the two circuits to an infinite distance, when the currents are maintained unchanged. See [27].

4.4 Letters by Eugenio Beltrami to Felix Klein

The letters by Eugenio Beltrami to Felix Klein,[48] Cod. Ms. Klein 8, 79, Niedersächsische Staats-und Universitätsbibliothek, Göttingen). Letter 3 is published, but not annotated, in [16, p. 213].

Letter n. 1
Eugenio Beltrami to Felix Klein

Pavie 7 Xbre 83

Monsieur et vénéré collègue

Je prends la liberté de Vous présenter M.r le Docteur G. Morera,[49] qui vient à Leipzig pour y faire des études mathématiques. Je vous serai infiniment obligé si vous voudrez bien lui faire un accueil bienveillant et lui donner les secours et les conseils dont il pourra avoir besoin. Je suis sûr qu'il fera tout son possible pour mériter vos soins. Il s'est occupé surtout, avec succès, des recherches sur les équations de la dynamique: aussi j'espère que vous pouvez le recommander à M.r le prof.r Mayer,[50] que je n'ai pas l'honneur de connaître personnellement.

M.r Gerbaldi[51] m'a dit que Vous avez été incommodé dans la santé, il y a quelques mois. Je veux espérer que vous vous portiez tout-à-fait bien maintenant et je vous souhaite de tout mon coeur un rétablissement complet, si votre état de santé vous laisse encore quelque chose à désirer.

Je Vous prie de vouloir bien présenter mes hommages à M.me votre femme, à laquelle ma femme aussi fait ses compliments, et j'ai l'honneur de me dire

votre bien obligé E. Beltrami

---

[48] Felix Klein (1849–1925) studied at Göttingen, Berlin, and Paris. He was appointed professor at the University of Erlangen in 1872; later on he moved to the Technische Hochschule of Munich (1875–1880), and then to the universities of Leipzig (1880–1886) and Göttingen (1886–1913). His main research concerns geometry, theory of algebraic equations and function theory. He also published important treatises on elementary mathematics and history of mathematics.

[49] Giacinto Morera (1856–1909) studied at Turin, Pisa, Pavia, Leipzig and Berlin. Beltrami, Klein and Weierstrass were among his teachers. In 1886 he became professor of rational mechanics at the University of Genoa, and in 1901 succeeded Vito Volterra at the University of Turin. His main research concerns complex analysis, but he also gave contributions to analytical mechanics, thermodynamics, and elasticity.

[50] Christian Gustav Adolph Mayer (1839–1908) studied at the Universities of Leipzig, Göttingen, Heidelberg and Königsberg. He taught mathematics at the University of Heidelberg all his life. He studied the theory of differential equations, the calculus of variations and mechanics. From 1871 to 1907 he exchanged many letters with Klein.

[51] Francesco Gerbaldi (1858–1934) studied at the University of Turin, and specialized at the University of Pavia and in Germany. In 1890 he was appointed professor of geometry at the University of Palermo, where he collaborated with Giuseppe Bagnera, Michele De Franchis, and Michele Cipolla on questions of geometry and algebraic geometry.

Letter n. 2
Eugenio Beltrami to Felix Klein

Pavie, 5 Janvier 85

Monsieur et cher collègue

Je vous prie de vouloir bien m'excuser si je ne Vous ai pas encore offert mes remerciements pour l'obligeant envoi des vos Vorlesungen ueber das Ikosaeder.[52] Mais j'ai été pendant plusieurs mois préoccupé à cause de la santé de ma femme, et ce n'est que depuis peu que je suis heureusement délivré de toute inquiétude.

Votre livre a sans doute fait beaucoup de plaisir à ceux qui sont au courant des théories élévées auxquelles il se rapporte: mais il en fait un plus grand encore à tous ceux qui, comme moi, ont eu la malheur de rester en dehors de ce grand courant, car il va leur permettre de s'y orienter, à l'aide d'un guide dont on ne pouvait pas trouver le meilleur. Aussi peu de livres auront ils un plus sympathique accueil.

Je saisis cette occasion pour vous remercier aussi de la bienveillance que vous avez bien voulu témoigner à M.r Morera, ainsi qu'aux autres jeunes gens italiens qui l'ont précédé à Leipzig.

Je vous prie de vouloir bien présenter mes hommages à M.me Klein, de la part de ma femme aussi, et d'agréer l'expression de mes sentiments de reconnaissance et d'admiration.

votre bien dévoué E. Beltrami

Letter n. 3
Eugenio Beltrami to Felix Klein

Pavie 17 Avril 1888

Très cher et honoré collègue[53]

J'ai reçu avec grand plaisir votre chère lettre du 15, et je me suis aussitôt occupé de prendre des mesures afin que le programme que Vous m'avez envoyé soit inséré dans le *Cimento* et dans les Comptes Rendus des *Lincei*: j'espère que cela se fera au plus tôt.[54] Cependant si Vous étiez à même de m'envoyer quelques autres exemplaires du programme, je m'en servirais peut-être utilement pour le faire connaître à des professeurs qui pourraient à leur tour exciter quelques uns de leurs meilleurs élèves à s'occuper du sujet.

J'ai été bien content d'apprendre que Vous vous intéressez à présent aux questions de physique mathématique: elles ne pourront que gagner en passant par vos mains et je m'attends à quelques uns de vos heureux tours de force.

---

[52]See [42].

[53]Je vous écris en français dans la pensée de vous causer moins de fatigue. (Footnote by Beltrami).

[54]Most probably Beltrami refers to Klein's *Erlangen Program* which was translated into Italian by Gino Fano [37].

Quant à la crainte que Vous semblez concevoir d'un divorce toujours croissant entre les mathématiques et la physique, j'oserais vous dire que l'Allemagne est, à mon avis, le pays où ce divorce me paraîtrait le moins avancé: peut-être certaines parties de la science n'y sont pas encore dévenues assez populaires, pour ainsi dire, parmi les physiciens, mais en revanche il y en a plusieurs autres qui sont entrées assez bien dans le domaine commun. Que diriez Vous donc si vous voyiez de près ce qui se passe en Italie, où deux ou trois physiciens seulement ont une connaissance approfondie des théories fondamentales!

Une chose qui m'a toujours surpris c'est qu'au milieu de tant de publications périodiques, dont plusieurs ont un caractère très-spécial, personne n'ait jamais songé à fonder un Recueil pour la physique mathématique: ce qui pourrait être très-avantageux surtout pour les physiciens, qui ont souvent quelque répugnance à consulter les journaux mathématiques, et dont quelques uns n'en ont pas même le moyen. Seulement il faudrait qu'un pareil Recueil eusse une allure un peu moins rigide que de coutume: qu'on y fit de la place à des bonnes Monographies sur les sujets les moins généralement connus, à des Recensions dûment développées, à des discussions scientifiques, destinées surtout à détruire les préventions, quelques fois même assez fondées, contre l'emploi de l'analyse en général, ou d'une certaine analyse en particulier. Un tel Recueil contribuerait peut-être à rendre moins désagréable la position de beaucoup de ceux qui s'adonnent à de semblables études, et dont le sort est assez souvent celui de passer pour physicien parmi les mathématiciens, et pour mathématicien parmi les physiciens, et de n'être écouté, en définitive, ni par les uns ni par les autres.

La nouvelle que Vous me donnez de la présence chez Vous d'un des fils du regretté Sella m'a beaucoup intéressé[55]: j'aurai probablement connu ce jeune homme lorsqu'il était encore enfant, car j'allais souvent, à Rome, chez son père, un homme éminent dont je garde un souvenir des plus chers et dont la perte a été un véritable malheur pour notre pays. Je fais des voeux sincères pour que ce fils fasse honneur à sa naissance.

Veuillez agréer, mon excellent ami, mes salutations les plus affectueuses, et présenter mes hommages à M.me Klein: veuillez aussi offrir mes compliments à vos honorables collègues Riecke et Voigt.[56]

votre bien dévoué E. Beltrami

---

[55]He refers to Alfonso Sella (1865–1907) who studied elasticity, radioactivity, acoustics and crystallography. His father was the Italian politician Quintino Sella.

[56]Eduard Riecke (1845–1915) studied mathematics and physics at the universities of Stuttgart, Tübingen and Göttingen. In 1873, he was appointed professor of physics in Göttingen. He was interested in different fields of physics and mathematical physics, such as magnetism, hydrodynamics, thermodynamics, electricity, and crystallography. He also published an important textbook of physics (*Lehrbuch der Physik*, 1896). Woldemar Voigt (1850–1919) studied at the universities of Leipzig and Königsberg, where he became extraordinary professor of physics in 1875. In 1883, he was appointed full professor at the University of Göttingen, with the promise that he and Riecke were to have a new physical institute (which was not ready until 1905). Voigt studied crystallography and its connection to the theory of elasticity. In 1908, he introduced the word *tensor* in the vocabulary of mathematical physics in relation to his research on crystals.

4.5 Letters by Beltrami to Rudolph Lipschitz

The letters by Eugenio Beltrami to Rudolph Lipschitz[57] are contained in Nachlass Lipschitz, Mathematisches Institut, Bonn. Letter 3 is published, but not annotated, in [16], pp. 211–212.

Letter n. 1
Eugenio Beltrami to Rudolph Lipschitz

Bologna 18 Novembre 1872

Chiar.mo Sig. Professore

Mi prendo la libertà di inviarLe una mia fotografia, nella speranza ch'Ella voglia accoglierLa con quella gentilezza con cui si compiace di accettare i miei piccoli lavori, e di favorirmi le tanto importanti sue produzioni. Se poi Le fosse possibile, ora o più tardi, di farmi un dono della sua propria fotografia, io me ne terrei onorato come di favore veramente singolare. Voglia continuarmi la di Lei benevolenza ed aggradire i sentimenti della mia più profonda considerazione.

Dev.mo aff.mo Eugenio Beltrami

Ormai tutti piangono la morte immatura del grande Clebsch.[58] Possa, questo universale compianto, confortare la sua povera famiglia!

Letter n. 2
Eugenio Beltrami to Rudolph Lipschitz

Bologna 12 Luglio 1873

Egregio signor professore

A suo tempo ebbi la graditissima sua lettera del 25 Maggio, e, posteriormente, l'Estratto francese delle di Lei memorie sulle funzioni differenziali, di cui sommamente La ringrazio.[59]

---

[57]Rudolph Lipschitz (1832–1903) studied at the universities of Königsberg and Berlin. He taught at secondary schools in Königsberg and from 1857 at the University of Berlin. In 1862, he was appointed professor at the University of Breslau and in 1864 moved to the University of Bonn. His publications concern different field of mathematics and mathematical physics: analysis, mechanics, number theory, theory of differential equations, potential theory and elasticity. His well-known treatise on the foundations of analysis, *Grundlagen der Analysis* (Bonn, 1877–1880), is mentioned in the letters.

[58]Rudolf Friedrich Alfred Clebsch (1833–1872) studied in Königsberg and Berlin. His scientific career started at the University of Berlin and ended at the University of Göttingen. Together with Carl Neumann he founded the journal *Mathematische Annalen*. His main works concern the theory of differential equations, invariant theory, and algebraic geometry.

[59]See [51].

Io posso assicurarLa che l'onore derivante all'Accademia dalla nomina di Lei a socio corrispondente non è stato sentito ed apprezzato soltanto da me[60]; ma da tutti quelli che si interessano alle scienze esatte nell'Accademia stessa, e l'unanimità colla quale fu accolta la relativa proposta, mostra che questo sentimento era penetrato anche negli altri soci. Io deploro che gli Statuti attualmente in vigore, e non ancora informati ai bisogni del nostro tempo, limitino i diritti dei Corrispondenti al ricevimento del solo volumetto dei rendiconti annuali. Ma ho fondata speranza che alla prima revisione dei detti Statuti (che risalgono ancora all'epoca del Governo pontificio!) si introdurrà, fra molte altre, anche la regola di mandare ai Corrispondenti esteri la collezione delle Memorie in 4to.

Ho già fatto introdurre, nella nota dei Soci, la correzione relativa al di Lei prenome. Credo che un consimile errore sia stato commesso, e parimenti corretto, riguardo al di Lei collega sig. prof. Clausius.[61]

La prego di aggradire i miei cordiali e rispettosi saluti, e di volermi ricordare al sig. Maestro Wasiliewsky.[62]

Di Lei Dev.mo
aff.mo E. Beltrami

Letter n. 3
Eugenio Beltrami to Rudolph Lipschitz

Pavia 13 Marzo 1877

Illustre collega

Ho ricevuto con vera gratitudine il cospicuo dono delle sei Memorie che Ella si compiacque inviarmi, e alcune delle quali mi riescono in questo momento utilissime. Ella riceverà fra non molto una mia Nota su argomento di elettrostatica,[63] provocata in parte da una certa discussione che regna da molti anni in Italia sui principii fondamentali di questa scienza, per opera principalmente di Volpicelli, professore di fisica matematica a Roma. Per verità le obbiezioni di costui non meriterebbero molta attenzione; ma siccome non manca chi se ne lascia preoccupare, così non è forse inutile totalmente il combatterle.[64] Ma io lo farò senza nominare nessuno, e senza neppure accennare alla controversia.

---

[60]Lipschitz was indeed appointed corresponding member of the Academy of Lincei.

[61]Rudolf Clausius (1822–1888) studied at the universities of Berlin and Halle. In 1855 he became professor of mathematical physics at the new Polytechnicum of Zürich. In 1867, he accepted a professorship at the University of Würzburg and in 1869 moved to the University of Bonn. His main contributions concern heat theory and, in particular, the foundations of thermodynamics.

[62]Wilhelm Joseph von Wasielewski (1822–1896) was a musician. He studied at the Conservatory of Leipzig and then moved to Bonn. He published an important biography of R. Schumann (1858), the volume *Die Violine und ihre Meister* (1869) and other books.

[63]Beltrami refers to [8].

[64]Paolo Volpicelli (1804–1879) studied philosophy at the University of Rome, where he taught at secondary schools from 1832. In 1872 he was appointed professor of mathematical physics at the

Applaudisco di gran cuore al felice pensiero, che Ella ha cominciato ad attuare, di una propedeutica matematica agli studii di fisica razionale. Ella sarà benemerito di questa scienza in doppia guisa, colle ricerche originali, e colle agevolezze procurate a chi deve studiarla, ed io faccio voti perché mi sia dato fruire presto di tale ajuto, e farne fruire l'insegnamento.

Non Le sembra che gioverebbe molto un lavoro analogo, ma in fondo inverso, da farsi da un fisico, cioè l'esposizione dei fenomeni accertati, in forma schematica, senza l'attiraglio *störend* dei minuti procedimenti sperimentali? Io desidero sempre un tal lavoro. Mi creda sempre

Obbl.mo affez.mo suo
E. Beltrami

Letter n. 4
Eugenio Beltrami to Rudolph Lipschitz
Pavia 28 Febbrajo 79

Illustre Sig.r Professore

Nel ringraziarLa del recente invio dell'interessante Memoria sul principio di Gauss,[65] mi reco a dovere di avvertirLa che io mi sono trasferito dall'Università di Roma a quella di Pavia, e che quindi è bene ch'Ella mi indirizzi in quest'ultima città le future pubblicazioni che vorrà favorirmi.

Al Sig.r Kaiser, che mi chiese, dietro una di Lei indicazione, una copia della breve nota sulla cinematica pseudosferica,[66] ho tosto fatto la chiesta spedizione.

Sto qui attendendo ad un corso di lezioni sulla fisica matematica, e precisamente, per quest'anno, sull'elettrostatica. Mi sarebbe molto gradito d'avere le di Lei pregiate Memorie sull'argomento;[67] e, se mai Ella ne conservasse qualche esemplare, io oserei (fidando nella ben nota cortesia di Lei) di farLene richiesta. Posso bensì consultarle nel *Journal der Mathematik*, la cui collezione esiste nella Biblioteca dell'Università: ma non avendole in separato esemplare, mi riesce meno facile il trovarle sempre a disposizione.

Mi perdoni, e mi creda sempre

dev.mo aff.mo
E. Beltrami

University of Rome. He was the author of about 200 publications on number theory, experimental physics, and history of science. In his treatise *Analisi e rettificazione di alcuni concetti e di alcune sperienze che appartengono alla elettrostatica* (Roma, Tipografia Belle Arti, 1866) Volpicelli explained his ideas on electricity. He was against the "recent" electric theory according to which a unique electric force exists.

[65] See [52].
[66] See [7].
[67] See [53, 54].

Letter n. 5
Eugenio Beltrami to Rudolph Lipschitz

Pavia 20 Giugno 1880

Chiarissimo Sig.r Professore

Ricevo or ora il 2 volume del *Lehrbuch der Analysis*, che Ella si è compiaciuto di farmi pervenire.[68]

Sono veramente confuso di tanta Sua gentilezza, alla quale io non posso corrispondere se non col riconoscere tutto il pregio della nuova produzione dovuta alla Sua penna. Mi chiamo ben fortunato di esser messo da Lei, così cortesemente, a immediata cognizione di un libro che sarà premurosamente accolto da tutti gli studiosi, i quali da lungo tempo desiderano un trattato di Calcolo che sia in armonia collo stato attuale della scienza e collo spirito delle più importanti sue applicazioni.

Unicamente per attestarLe in qualsiasi modo la mia gratitudine, Le invio alcune recenti pubblicazioni, che avevo messe da parte per unirLe ad altre di cui aspetto gli esemplari. Voglia gradirLe colla Sua solita bontà e credermi sempre

di Lei Ch.mo Sig.r Professore
obbl.mo dev.mo E. Beltrami

Letter n. 6
Eugenio Beltrami to Rudolph Lipschitz

Roma 18 Marzo 1898

Chiarissimo Collega

La ringrazio veramente di cuore delle benevole espressioni contenute nella cara Sua lettera del 13 corrente. Esse sono una nuova prova della di Lei buona e cordiale amicizia. Quanto all'onore conferitomi dall'Accademia dei Lincei, io ho dovuto accettarlo per deferenza alla volontà dei Colleghi, ma pur troppo non so se le mie forze saranno proporzionate al bisogno.[69]

Nello scorso anno ebbi un grave attacco di malattia dal quale ho potuto fortunatamente salvarmi, ma non posso dire di trovarmi interamente ristabilito in perfetta salute. Comunque sia io Le presento l'espressione della mia sincera riconoscenza, insieme con quella dei voti che io faccio per la di Lei prosperità.

Dev.mo aff.mo E. Beltrami

Letter n. 7
Eugenio Beltrami to Rudolph Lipschitz

13/8/[...] Macugnaga (Monte Rosa)

Chiarissimo collega

---

[68] See [55].
[69] Beltrami was appointed President of the Academy of Lincei.

Ricevo qui la di Lei lettera e il primo volume dell'Opera di cui Ella mi scrisse già in addietro.[70] Le sono oltremodo obbligato del prezioso regalo, che mi propongo di ben presto esaminare e di studiare, e intanto mi affretto a dargliene ricevuta.
Sono qui da una quindicina a fare la vita di montagna. Passerò poi i mesi di Settembre e di Ottobre a Venezia, e colà spero potermi occupare piò utilmente dello studio del di Lei libro.

Mi creda, con devota attenzione
obbl.mo Beltrami

4.6 Letters by Beltrami to Gösta Mittag-Leffler

The letters by Eugenio Beltrami to Gösta Mittag-Leffler[71] are contained in Mittag-Leffler's Archive, Djursholm, Stockholm.
Letter n. 1
Eugenio Beltrami to Gösta Mittag-Leffler

Pavie 8 Janvier 83

Cher Monsieur

J'ai reçu les deux exemplaires du 1er cahier des *Acta mathematica* et je vous remercie bien vivement de l'honneur que vous avez bien voulu me faire en m'en offrant un. J'ai déjà envoyé l'autre exemplaire à l'Institut de Venise,[72] avec un article de présentation, qui sera lu par un Membre qui se charge de faire cela en mon nom. Cette communication sera faite dans la premiïre séance que l'on tiendra, et je vous ferai connaître, à son temps, les détails de cette communication, ainsi que le compte rendu qui en sera donné.

Je vous remercie ainsi de la demande, que vous avez bien voulu me faire, de contribuer par quelque Article à votre Journal, qui sort sous d'excellents auspices. Je me tiens honoré par cette demande et je me propose d'en tenir compte. Dans ce moment je suis excessivement absorbé par le travail de préparation de deux cours que je fais: mais sitôt que je serai un peu plus libre, je n'hésiterai pas à vous adresser quelque Note qui me paraîtrait moins indigne de figurer dans votre Recueil.

---

[70]See [56].

[71]Gösta Mittag Leffler (1846–1927) studied at Uppsala, Paris, Göttingen, and Berlin. In 1876, he was appointed professor at the University of Helsinki and, in 1881, moved to the new University of Stockholm. He mostly contributed to complex analysis by following Weierstrass's ideas, but also to analytic geometry and theory of probability. In 1882 he founded and then directed the journal *Acta Mathematica*. Mittag-Leffler engaged himself in finding a position for the Russian mathematician Sonya Kovalevskaya, who in 1884 obtained a position at the University of Stockolm for five years. Some of these letters are devoted to the case of Sonya Kovalevskaya. See also the Appendices 4.7 and 4.8.

[72]The *Istituto Veneto di Scienze, Lettere ed Arti*.

J'ai présenté à ma femme vos salutations et tous les deux nous vous félicitons pour votre récent mariage, pour le bonheur duquel nous faisons tous vos voeux. Veuillez présenter nos hommages à votre épouse, en attendant que l'occasion se présente d'en faire la connaissance personnelle.[73]

J'ai toujours votre nom noté parmi ceux des personnes auxquelles je dois ma photographie. Malheureusement ma paresse et le manque d'occasion (car il n'y a pas de bons ateliers photographiques ici) m'ont fait trop négliger ce devoir. Soyez sûr que je m'empresserai de satisfaire à votre obligent désir, aussitôt que je réussirai à me procurer ce qui me manque encore depuis notre entrevue.

Veuillez agréer, cher collègue, l'expression de mon sincère dévouement.

E. B.

Letter n. 2
Eugenio Beltrami to Gösta Mittag-Leffler

Pavie 10 Mai 1883

Monsieur et honoré collègue

J'ai l'honneur d'expédier à votre adresse, dans un paquet sous bande, quelques exemplaires de la communication par laquelle j'ai annoncé à l'Institut Royal de Venise l'offre, de votre part, du 1er Cahier des *Acta*. À la suite de cette communication, l'Institut a voté un remerciement accompagné de l'offre de ses Comptes rendus,[74] en échange de votre Journal. Peut être le Secrétaire de l'Institut vous a déjà donné participation de cette délibération. Dans le cas où la chose souffrait quelque retard, veuillez avoir l'obligeance de me le marquer, afin que je puisse faire des démarches, pour la solliciter.

Je saisis cette occasion pour vous confirmer mes sentiments de considération très-distinguée, pour vous prier de vouloir bien présenter mes hommages à M.me votre épouse et pour me dire

Votre bien dévoué
E. Beltrami

Letter n. 3
Eugenio Beltrami to Gösta Mittag-Leffler

Pavie 23 Juin 1883

Monsieur et honoré collègue

Profitant de vos invitations flatteuses, je prends la liberté de vous envoyer un Article pour votre Journal de Mathématiques. Le manuscrit de cette communication

---

[73]In 1882 Mittag-Leffler had married Signe af Lindfors who came from a wealthy family.
[74]Beltrami refers the *Atti dell'Istituto Veneto di Scienze, Lettere ed Arti*.

vous parviendra sous-bande recommandée, que je remettrai à la Poste en même temps que la présente.[75]

Vous devez me dire sans scrupule si vous croyez que l'Article en question, à cause de son sujet, puisse entrer convénablement dans le cadre de vos publications.

Dans le cas où il vous paraissait admissible, je désirerais, lorsque le moment sera venu, de pouvoir faire au moins la *dernière* révision des épreuves. Je n'aurai probablement rien à changer, mais il pourrait m'être échappé quelque impropriété de rédaction en français, ou l'on pourrait n'avoir pas exactement reproduit quelque mot ou quelque formule.

Je resterai ici encore un mois, mais si vous voudrez bien me dire à quelle époque, à peu près, les épreuves pourraient arriver, je pourrai vous indiquer, de mon côté, l'adresse qu'il serait bon d'adopter.

Je Vous prie de vouloir bien présenter mes hommages et ceux de ma femme à Mme Mittag-Leffler et agréer l'expression de ma considération et de mon dévouement.

Votre bien obligé
E. Beltrami

Letter n. 4
Eugenio Beltrami to Gösta Mittag-Leffler

Pavie 19 Juillet 1883

Très honoré collègue

J'adresse la présente à Stockholm, puisque je pense que, si vous n'y êtes pas revenu, on vous la renverra à votre adresse actuelle, que je n'ai pas bien comprise.

Je vous remercie de l'empressement que vous avez bien voulu mettre à agréer l'article que je vous ai envoyé. Je m'arrangerai de manière à me faire renvoyer les épreuves que vous comptez adresser à Pavie. La chose serait mieux assurée si ces épreuves étaient expédiées en enveloppe fermée (le manuscrit ne m'est point nécessaire), car j'ai l'habitude de faire retenir ici tous les imprimés qui arrivent sous bande ouverte, à cause des chances d'égarement qu'il y a lorsqu'on se déplace souvent, comme je fais pendant les vacances. Mais en tous cas j'ai ici un ami qui veut bien se charger de passer en revue ce qui arrivera à mon adresse (car ma maison reste vide).

Si cela n'est pas en dehors des règles que vous avez établi pour le tirage à part, je serais bien aise d'avoir au moins 50 exemplaires de ma Note, car je voudrais bien en faire connaître le contenu à ceux qui s'occupent du sujet. Il est inutile d'ajouter que je tiens à être informé de toutes les dépenses qui seront nécessaires soit pour le tirage, soit pour l'envoi du paquet des exemplaires séparés, soit pour celui des épreuves, notamment si vous consentez à le faire par lettre fermée. En vous renvoyant les épreuves, je vous écrirai pour vous indiquer le lieu où il me serait le plus agréable de recevoir les exemplaires.

---

[75]Beltrami refers to his paper later published in *Acta*: [13].

Je vais partir dans trois ou quatre jours pour la Suisse, mais je ne peux vous indiquer exactement, dans ce moment, le lieu où je me fixerai. Si, dans une semaine, je pourrai prévoir de rester longtemps dans le même endroit, je vous ferai connaître mon adresse, et alors il n'y aura plus la difficulté dont je vous parlais tout à l'heure.

Il est très-probable que M. Casorati[76] sera avec moi, en Suisse. Dans ce moment il ne se porte pas tout-à-fait bien, mais j'espère qu'il ne s'agisse que d'une indisposition passagère. Je vous fais ses salutations les plus cordiales.

J'espère que vous aussi vous pensez, quelque autre fois, à venir passer quelque temps en Suisse avec M.me votre femme. Vous aurez là l'occasion de rencontrer plusieurs professeurs, notamment de l'Allemagne. Il y a quelques annïes, j'ai pu jouir pendant plusieurs jours de la compagnie de MM. Zeuthen, Frobenius, Hirst, Schläfli, Geiser, Rosanes,[77] qui se trouvaient tous en même temps dans le village où je m'étais établi avec MM. Cremona[78] et Casorati. L'année passée, aussi, étant avec M. Cremona à Pontresina, nous y avons trouvé MM. Kronecker et Helmholtz, avec leurs familles.[79]

Veuillez agréer mes salutations cordiales ainsi que celles de ma femme, et n'oubliez pas de présenter nos hommages à M.me votre épouse, si elle est avec vous en ce moment.

C'est toujours avec la plus grande estime que je vous prie de me croire Votre bien obligé et dévoué

E. Beltrami

---

[76]Felice Casorati (1835–1890) was professor at the University of Pavia. He mainly studied real and complex analysis.

[77]Hieronymus Georg Zeuthen (1839–1920) studied mathematics at the University of Copenhagen, where he was appointed professor in 1871. His main research topics were algebraic geometry and history of mathematics. Georg Ferdinand Frobenius (1849–1917) was one of Weierstrass' students at the University of Berlin. He taught at the University of Berlin and at the Polytechnic of Zürich. His research concerns group theory and the theory of differential equations. Thomas Archer Hirst (1830–1892) was born in England, studied mathematics and physics in Germany and then lived in Paris and Rome. In 1860, he came back to England were he started his academic career in London. He published many papers and book on geometry, algebraic geometry, and physics (mechanics and thermodynamics). Ludwig Schläfli (1814–1895) studied theology at the University of Bern, and then accepted a post as a teacher of mathematics and science at secondary school in Thun. In 1848, he became Privatdozent at Bern, in 1853 extraordinary professor and in 1868 full professor. His works concern n-dimensional geometry and invariant theory. Carl Friedrich Geiser (1843–1934) studied and taught at the Polytechnic of Zürich. His research concerns algebraic geometry, differential geometry and invariant theory. Jakob Rosanes (1842–1922) studied mathematics and physics at the University of Breslau, where he taught from 1873 onwards. His main research subjects were algebraic geometry and invariant theory.

[78]Luigi Cremona (1830–1903) was professor at the University of Bologna, taught at the Polytechnic of Milan, and directed the *Scuola degli Ingegneri* in Rome. He especially contributed to the theory of surfaces and curves.

[79]Leopold Kronecker (1823–1891) taught at the University of Berlin from 1862 onwards, but achieved an official position only in 1883 when Kummer retired. He mainly contributed to analysis and number theory. Hermann von Helmholtz (1821–1894) was professor of physiology at the universities of Königsberg (1849), Bonn (1855) and Heidelberg (1858). In 1871, he became professor of physics at the University of Berlin. He contributed to medicine, physiology, physics, and mathematics.

Letter n. 5
Eugenio Beltrami to Gösta Mittag-Leffler

Silvaplana (Engadina) 25 Juillet 83

Monsieur

Arrivé ici ce matin, et ayant rencontré des conditions de logement assez convenable, je puis vous dire que je resterai ici au moins trois semaines. Par suite, vous pouvez donner l'ordre qu'on m'adresse ici les épreuves que je vous savez, par envoi ordinaire sous-bande (Drucksache): ainsi aucun retard n'aura lieu pour le retour de ces épreuves après ma correction. J'attends M. Casorati demain ou après demain: je vous avance ses compliments en même temps que je Vous renouvelle mes salutations les plus cordiales.

Votre bien obligé E. Beltrami

Letter n. 6
Eugenio Beltrami to Gösta Mittag-Leffler

Silvaplana (Suisse) 21 Août 83

Monsieur et excellent collègue

J'ai reçu et renvoyé à M.r Eneström, il y a quelque jours, les épreuves de ma Note pour les *Acta*, et, après avoir remercié M.r Eneström pour son obligeance et pour ses soins, je ne veux pas tarder plus longtemps à vous remercier vous même pour la promptitude avec laquelle vous avez bien voulu pouvoir à l'impression de la susdite Note.

Comme je quitterai ces lieux dans une semaine pour me rendre à Padoue, après une autre semaine que j'emploierai à faire un petit tour dans la Suisse, j'ai donné mon adresse dans la ville susdite pour l'envoi des exemplaires à part: je vous transcrirai cette adresse ci-dessous, quoique je l'aie déjà indiquée à M.r Eneström.

Je vous fais les salutations amicales de M.r Casorati, qui est ici avec nous, et qui est notablement amélioré dans son état de santé depuis qu'il a quitté Pavie.

Je vous présente aussi, au nom de ma femme et au mien, les hommages les plus sincères pour M.me votre épouse et je vous prie de me croire toujours

Votre bien obligé
E. Beltrami

Adresse pour les exemplaires à part:
Sig. Professore Eugenio Beltrami
Padova (Italia) Via S. Fermo 1251

Letter n. 7
Eugenio Beltrami to Gösta Mittag-Leffler

Venise 2 8bre 83

Monsieur et très honoré collègue

Arrivé à Padoue après une absence assez longue, j'ai retrouvé là, en bon ordre, le paquet des exemplaires à part de la Note que vous avez eu la bonté d'accueillir dans les *Acta*.

Je saisis cette occasion pour vous remercier encore une fois de toutes vos bontés et de tous vos soins et pour vous prier de vouloir bien exprimer ma reconnaissance à M. Eneström pour toutes les peines qu'il a eues à cause de moi.

J'ai à vous prier de vouloir bien me faire connaître le montant de ma dette envers l'Administration des *Acta*, pour le tirage à part et pour les frais de poste et autres. Je ne reviendrai définitivement *à Pavie* que dans un mois, mais, comme il est probable que je devrai me déplacer dans l'intervalle, il est bon de m'adresser toujours les lettres à Pavie, d'où on me les renvoie là où je me trouve.

Veuillez présenter mes compliments les plus respectueux à Madame votre épouse et agréer l'expression de mon affectueux dévouement.

Votre bien obligé E. Beltrami

Letter n. 8
Eugenio Beltrami to Gösta Mittag-Leffler

Venise 18 8bre 83

Cher Monsieur

Je vous remercie infiniment du portrait d'Abel dont vous avez voulu me faire cadeau. A cause d'une erreur de la Poste il avait d'abord été renvoyé à une autre personne de mon nom, mais fort heureusement il est revenu hier dans mes mains.

Je n'insiste pas, d'après ce que vous avez l'obligeance de m'écrire, dans ma demande. Quant à votre désir bienveillant, vous pouvez être assuré que je tiendrai à honneur de vous envoyer quelque article, lorsque je penserai qu'il puisse paraître dans les Acta sans faire de tort à la réputation de ce nouveau Recueil.

Ma femme et moi nous remercions M.r et M.me Mittag Leffler de leur bon souvenir et présentons leur nos salutations les plus distinguées.

Votre bien dévoué
E. Beltrami

Letter n. 9
Eugenio Beltrami to Gösta Mittag-Leffler

Pavie 17 Mars 1889

Cher collègue et ami

Au retour d'une courte absence, qui a été motivée par des affaires de famille, je viens de recevoir votre aimable lettre du 11 de ce mois ainsi que le Mémoire de M.me Kowalewsky, que vous avez l'obligeance de m'envoyer et que je reçois avec le plus grand plaisir.[80] J'étais très-désireux de savoir comment cette excellente géomètre avait pu remporter de nouvelles victoires sur un domaine qui avait été le champ d'aussi grandes batailles analytiques, et je vient de reconnaître, d'une

---

[80]Sonya Kovalevskaya (1850–1891), pupil of Weierstrass, became extraordinary professor of mathematics at the University of Stockholm in 1884 and ordinary professor at the same university in 1889. In 1888, she won the Prix Bordin (proposed by the Academy of Sciences of Paris) with the paper Sur le problème de la rotation d'un corps solide autour d'un point fixe [49]. Most probably Beltrami refers to this paper.

manière encore très-sommaire, qu'il s'agit d'un exploit des plus originels et de plus brillants. Aussi vous fais-je mes compliments d'avoir, en bon Rédacteur en chef, assuré aux *Acta* la prémisse de cet admirable travail.

Il y a une discontinuit très-extraordinaire à vous parler, après ça, de l'invitation bienveillante que vous me faites de vous envoyer, pour plus tard, quelque Article pour les Acta; mais il me [...] mal de ne pas vous en remercier, de tout mon coeur, et de vous assurer que je ferai mon possible pour vous satisfaire, si je réussirai à vous offrir quelque chose qui me semble peu digne de vous et de votre Journal.

Je dois vous remercier de même de l'intention excessivement obligeante que vous avez de m'envoyer le grand mémoire couronné de M. Poincaré,[81] dont les travaux acquièrent une encore plus grande importance, pour moi en particulier, depuis que leur Auteur se rapproche de plus en plus au domaine de la physique mathématique.

Vos renseignements au sujet de mon allée à Rome ne sont pas tout-à-fait exacts. Il ne s'agit pas de la chaire d'analyse supérieure, qui d'ailleurs a toujours existé, peut-on dire, car elle a été occupée pendant plusieurs années par M. Battaglini, et l'est maintenant par un suppléant: il s'agissait de celle de physique mathématique, qui est vacante depuis plus de dix ans (savoir après la mort de Volpicelli, l'ancien secrétaire perpétuel de l'Académie des Lincei qui lui doit de ne pas avoir été tout à fait supprimée par le Pape). Je n'ai pas absolument renoncé à occuper cette position et si je ne m'y suis pas encore décidé c'est, indépendamment de quelques circonstances de famille, parce que je ne crois pas que je ferais grande chose, à moi-seul, c'est-à-dire sans des cours bien et fortement organisés sur les autres parties des hautes mathématiques. Tout cela serait bien facile si M. Cremona revint aux études avec l'ardeur d'autrefois: mais malheureusement je ne puis pas y compter.

Ma femme vous est bien obligée de vos compliments et nous vous prions tous les deux de vouloir bien présenter les nôtres à M.me Mittag-Leffler. M.r Casorati aussi Vous fait ses salutations affectueuses: je voudrais pouvoir vous dire que sa santé est bonne, mais peut-être laisse-t-elle quelque chose à désirer, quoique il ne s'en plaigne pas: probablement, après le mariage de sa seconde fille, il n'a pas encore trouvé un système de vie qui lui convienne. Veuillez agréer, très-cher ami, mes salutations les plus dévouées.

E. Beltrami

Letter n. 10
Eugenio Beltrami to Gösta Mittag-Leffler

Pavie 16 Mai 1889

Mon cher ami

Je viens de recevoir votre aimable lettre et j'aime à vous répondre tout de suite, soit pour vous assurer de la réception de vos envois, soit pour vous dire que je vais

---

[81] Beltrami refers to the work by Henri Poincaré (1854–1912) – awarded by the King of Sweden Oscar II on January 21st, 1889 – Sur le problème des trois corps et les équations de la dynamique published in *Acta Mathematica in* 1890 (vol. 13, pp. 1–278).

me mettre tout de suite à préparer ma réponse à la lettre officielle de l'Université. Je vous suis bien obligé du témoignage d'estime que vous m'avez procuré et que je serais bien heureux de mériter: mais je ne puis vous cacher ma conviction que M. Betti ou M. Brioschi[82] auraient pu formuler un jugement bien plus autorisé que le mien. Quoiqu'il en soit ce sera pour moi un plaisir que de contribuer à une cause aussi juste que celle de mettre M.me Kowalewsky à l'abri des attaques des personnes malveillantes dont vous me parlez.

Je saisis aussi cette occasion pour vous offrir la photographie que vous m'aviez demandé depuis si longtemps: elle n'est pas d'un aussi bon travail que celle que vous avez bien voulu m'envoyer, et dont je vous remercie bien vivement, mais il me semble que la ressemblance soit assez bien gardée.

Je ferai au plus tôt votre commission à M.r Casorati, que je ne vois depuis quelques jours et que je crois toujours plongé dans certaines recherches sur la courbure des surfaces dont il m'a parlé la dernière fois que je l'ai vu, et dont il était profondément intéressé. Peut-être même la proposition qu'il vous a fait a-t-elle rapport avec ce sujet: mais il ne m'en a rien dit. Je crois que je le verrai demain.

Croyez vous bien de faire aussi vite le voyage de Paris, pour visiter l'Exposition?[83] D'après les correspondances qu'envoi à Notre journal le mieux informé, la *Perseveranza*,[84] un très-habile et très-intelligent correspondant, qui demeure à Paris depuis 30 ans, il résulterait qu'il faut encore un bon mois avant que l'Exposition puisse être considérée comme suffisamment complète: ce correspondant conseille décidemment ses lecteurs à attendre le Juillet. Il est vrai qu'on aurait alors l'inconvénient du chaud.

Je pense que je pourrai vous adresser la lettre pour l'Université après demain, au plus tard: ainsi vous la recevrez sans doute avant votre départ.

Je joins à mes salutations très-affectueuses celles que ma femme me charge de vous faire, et celles que moi même je vous prie de vouloir bien présenter à M.me Mittag-Leffler. Je vous souhaite un bon voyage et je désire aussi qu'il soit profitable pour Votre santé.

Votre bien dévoué et affectionné
Eugène Beltrami

P. S. Il me vient à l'esprit que M.me Kowalewsky pouvait bien essayer de voir si le nouveau cas d'intégration qu'Elle a découvert pour le problème du corps tournant, ne trouverait pas aussi une interprétation élégante dans la théorie de la tige élastique, d'après le rapprochement signalé par Kirchhoff (Mechanik, XXVIII Vorles. §6)

---

[82]Enrico Betti (1823–1892) was professor at the University of Pisa from 1857 onwards. He gave important contributions to algebra, differential geometry and mathematical physics. Francesco Brioschi (1824–1897) was professor at the University of Pavia. He founded (1863) and directed the Polytechnic of Milan. He contributed to the theory of determinants and to analysis.

[83]Beltrami refers to the *Exposition Universelle* held in 1889 in Paris.

[84]*Perseveranza* was an Italian journal founded at the eve of the Unification of Italy (1859).

entre les deux questions.[85] Mais ce n'est qu'une idée, laquelle d'ailleurs s'est peut-être déjà présentée à cette Analyste distinguée.

E. B.

Letter n. 11
Eugenio Beltrami to Gösta Mittag-Leffler

Pavie 18 Mai 89

Mon cher collègue

Je vous écris deux mots pour vous dire plusieurs choses. D'abord que j'ai déjà remis à la Poste, en lettre recommandée, ce matin de bonne heure, le Rapport pour l'Université. Naturellement je l'ai adressé à votre nom. Je crains maintenant qu'il ne soit trop long: mais je ne pouvais me dépouiller des mes habitudes *rapporteuses*.

Ensuite je vous demande bien excuse dans le cas où j'aurais oublié de signer la photographie que je vous ai adressée. Si j'ai été coupable de cet oubli, j'y remédierai à la première occasion où je pourrai vous en envoyer une autre, meilleure.

Enfin je vous prie de me dire si je dois renvoyer à Stockholm la Note des Astr. Nachr. de Mad.me Kowalewsky.[86] Il me serait sans doute très-agréable de pouvoir bien la garder, mais il peut se faire que l'Autrice n'aie pas à sa disposition d'autres exemplaires. Ainsi j'attends vos indications.

M. Casorati m'a dit que vous lui avez écrit directement et qu'il va vous répondre, ou envoyer la traduction dont il s'agissait.

Avec mille salutations cordiales
E. Beltrami

Enclosed letter by Eugenio Beltrami to Sophia Kovalevskaya

Pavie 26 Avril 1890

Madame

À mon retour d'une absence de Pavie, je viens de retrouver ici l'exemplaire, que vous avez eu la complaisance de me faire parvenir, d'un Mémoire où j'entrevois un complément analytique de votre grand travail couronné.[87] Je m'empresse de vous offrir l'expression de ma reconnaissance bien sincère pour ce précieux cadeau, et je vous demande pardon si je ne pourrai pas de sitôt vous prier d'agréer quelque publication nouvelle. J'ai bien des matériaux assez étendus: mais les préoccupations cruelles auxquelles je suis en proie depuis plusieurs mois, à cause de la santé de

---

[85]Gustav Robert Kirchhoff (1824–1887) studied at the University of Königsberg, where he had Jacobi and Franz Neumann as teachers. In 1850, he was appointed professor at the University of Breslau. In 1854, he moved to the University of Heidelberg and in 1875 accepted the chair of theoretical physics in Berlin. He contributed to the fundamental understanding of electrical circuits and spectroscopy. In the letter, Beltrami refers to Kirchhoff's book on mechanics [41].

[86]See [43].

[87]Beltrami refers to the paper: [44]. This paper is a complement of the memoir already cited [49].

ma femme, m'ôtent le calme et la tranquillité d'esprit dont j'aurais besoin pour les mettre en ordre et les rédiger d'une manière convenable.

Veuillez agréer, Madame et très-honorée Collègue, l'expression de mes sentiments de considération la plus distinguée.

E. Beltrami

Letter n. 12
Eugenio Beltrami to Gösta Mittag-Leffler

Pavie 21 Juin 1890

Monsieur et collègue très-honoré

J'ai reçu votre nouveau Mémoire sur la représentation analytique des intégrales et des invariants des éq.ns diff.les et je vous présente mes meilleurs remerciements pour ce joli cadeau. J'ai pu déjà apprécier, par une première lecture (ce qui est encore bien loin d'une première étude), toute l'importance et tout l'intérêt des vos recherches, dont le point de départ est une représentation conforme qui rentre dans un type ayant déjà rendu des services considérables en physique mathématique.

Je regrette de ne rien avoir à vous offrir de mon côté. Cette année a été excessivement peu favorable à mes études. Une longue maladie que ma femme a eu cet hiver et dont elle est encore bien loin d'être guérie aussi complètement qu'il y avait bien de l'espérer, m'a profondément troublé et c'est à peine si j'ai eu la force de m'occuper de mes leçons à l'université.

Ce n'est pas, aussi, un petit chagrin pour moi que celui d'assister, sans ne pouvoir rien faire, à la lente destruction de la santé de mon excellente ami M. Casorati. Il vous aura peut-être écrit lui-même de se trouver incommodé depuis quelque temps: mais il s'est toujours fait, et il se fait encore, une illusion complète sur son état. Il est, depuis quelques jours, à la campagne, où il désirait vivement d'aller, et pour où on l'a laissé partir, vu qu'il n'y avait absolument rien à faire pour lui. Il est victime d'une affection cancéreuse aux intestins, maladie terrible dont sont morts son père et sa mère. Qui sait si je pourrai le revoir encore avant que je part pour les vacances; puisque, dans la sécurité où il vit, ma visite de ma part lui semblerait extraordinaire et pourrait lui réveiller des soupçons! Vous me croirez facilement si je vous dis que tout cela me cause une grande douleur.

Je vous prie d'agréer, pour vous et pour votre honorable famille, l'expression de mes voeux sincères pour votre bonheur aussi que celle de ma considération bien distinguée et de mon affectueux dévouement.

E. Beltrami

Letter n. 13
Eugenio Beltrami to Gösta Mittag-Leffler

Pavie 25 Novembre 1890

Monsieur et collègue très-honoré

Je viens vous présenter mes remerciements les plus empressées, pour la bonté que vous avez eue de me faire parvenir une exemplaire du grand Mémoire couronné de M. Poincaré. Si je suis un peu en retard avec vous pour cet objet, c'est que j'ai dû m'absenter dans ces derniers jours, soit pour prendre part aux élections politiques, soit pour satisfaire au désir de mon vieil et excellent ami M. Cremona, qui m'avait prié d'assister, en qualité de témoin pour la fiancée, au mariage de sa dernière fille avec un médecin distingué de Naples.

Je ne puis juger jusqu'à présent de l'importance et du contenu du Mémoire de M. Poincaré que d'après la préface et la table des matières, ce qui suffit cependant à prévoir qu'il s'agit d'un travail fondamental; comme vous me l'aviez déjà assuré, au reste, à la suite de la lecture que vous aviez faite de la rédaction primitive. Mais je vais en prendre une connaissance plus détaillée, au moins en tant que cela me sera permis par mes occupations obligatoires, qui sont plus absorbantes au commencement des cours que dans la suite.

Je suis encore sous le coup du malheur qui nous a frappés, dans la personne de notre excellent ami M. Casorati. M. Brioschi est occupé à écrire une commémoration scientifique de ce savant, qui a été d'abord son élève et puis son ami, peut-être le plus intime. Cet écrit paraîtra, je crois, dans les *Annali di Matematica* et comprendra la liste complète des articles et des Mémoires publiés par M. Casorati.[88] Il pourrait vous être très-utile dans le cas où vous voudriez consacrer quelques pages des Acta au souvenir d'un de ses collaborateurs étrangers.[89] La place que M. Casorati a laissé vacante ici va être occupée par un jeune analyste très-distingué, M. Ernest Pascal,[90] qui a fait tout récemment une série de travaux sur les fonctions hyperelliptiques.

J'ai passé une bonne partie des vacances en Tyrol, à cause d'un traitement minéral que M.me Beltrami devait suivre pour améliorer son état de santé. Cet état est maintenant assez satisfaisant: mais je ne suis pas sans craintes à cause de l'hiver, qui a été très-doux jusqu'ici, mais qui devient d'ordinaire très-rigide dans les mois de Décembre et de Janvier. C'est la saison la plus défavorable à la santé de mon épouse et celle qui exige le plus de précautions.

Je vous prie de vouloir bien, cher collègue et ami, présenter mes compliments les plus distingués a votre honorable famille, ainsi qu'à votre beau-frère M.r Del Pezzo,[91] s'il habite encore Stockholm en ce moment: veuillez bien aussi recevoir

---

[88] In the volume 18 (1890) of *Annali di matematica pura ed applicata* Brioschi shortly communicated the death of Felice Casorati and promised to publish Casorati's obituary in the next volumes of *Annali*, but it never appeared.

[89] In *Acta Mathematica* no obituary of Felice Casorati has been published.

[90] Ernesto Pascal (1865–1940) studied at the universities of Naples, Pisa and Göttingen with Klein. In 1890, he was appointed professor of calculus at the University of Pavia. In 1907, he move to the University of Naples. His works deal with various research fields; in differential geometry he developed a kind of generalization of Ricci's tensor calculus which, however, did not lead to any interesting consequence.

[91] Pasquale del Pezzo (1859–1936) came from a noble family of Naples. He studied mathematics at the University of Naples, where he was appointed professor of projective geometry in 1884. He married the writer Anne Charlotte Leffler, sister of the mathematician Gösta. Del Pezzo was many times chancellor of the University of Naples, mayor of the town and senator of Italy. He introduced the *Del Pezzo surfaces*, well-known in modern algebraic geometry, as the surfaces having elliptic curves as plane sections.

mes salutations les plus affectueuses, ainsi que l'expression renouvelée de ma reconnaissance.

Votre bien obligé E. Beltrami

Letter n. 14
Eugenio Beltrami to Gösta Mittag-Leffler

Pavie 28 Décembre 1890

Cher et honoré collègue

J'ai beaucoup retardé à répondre à votre bonne lettre du 26 du mois passé (qui s'est croisée avec celle que je vous écrivais pour vous remercier de l'envoi du grand Mémoire couronné de M. Poincaré), parce que j'étais très-contrarié de ne pas pouvoir satisfaire a votre demande d'un exemplaire à part de mes écrits sur l'hydrodynamique.[92] Je m'en trouvais en effet tout-à-fait dépourvu, le tirage en ayant été assez faible. Cependant je me suis adressé à plusieurs de mes amis pour voir si quelqu'un d'entre eux, pour qui le sujet de ces Mémoires n'avait pas un grand intérêt, consentirait à m'en faire la restitution; et j'ai enfin obtenu de l'un d'eux la promesse de me les renvoyer. J'en ferai relier les feuilles en brochure, avec le frontispice dont je conserve encore quelques tirages, et je vous en ferai l'envoi dans quelques jours, en y joignant, conformément à votre désir, le peu d'autres ancien articles que je pourrai retrouver. Je dois vous dire, encore à propos des mémoires sur l'hydrodynamique, qu'à présent je suis très-peu content de ce travail, qui a paru nécessairement par fragments, et dont le plan s'est pour ainsi dire changé dans le cours de la rédaction, ainsi que l'indique assez la différence du titre que j'ai fini par lui assigner, d'avec celui des volumes de l'Académie de Bologne. J'ai eu aussi l'occasion de constater plusieurs erreurs d'impression, plusieurs inadvertances, et même quelques inexactitudes: je me repens, en outre, d'avoir traîné en longueur plusieurs recherches de peu d'importance. Mais, enfin, péché confessé, dit-on, est à moitié pardonné.

Je Vous remercie bien de l'intérêt que vous prenez à la santé de M.me Beltrami. Le Dieu merci, je suis assez content de ce côté: malgré la rigueur du froid de cet hiver, jusqu'à présent il ne s'est produit aucun phénomène grave comme ceux de l'hiver passé et je commence à espérer que la mauvaise saison puisse s'écouler sans accident.

Je vous prie de vouloir bien présenter mes hommages les plus respectueux à M.me Kowalewsky. Je vous prie aussi de vouloir bien agréer, pour vous et pour M.me Mittag-Leffler, les souhaits les plus sincères que M.me Beltrami et moi nous faisons pour votre bonheur, dans la nouvelle année qui va commencer.

Veuillez accueillir l'expression de mes sentiments de considération la plus distinguée. Votre bien dévoué

E. Beltrami

---

[92]See [14].

Letter n. 15
Eugenio Beltrami to Gösta Mittag-Leffler

Pavie 20 Mars 1891

Collègue très honoré
Après un bien long retard, je viens satisfaire au désir par vous exprimé d'avoir un exemplaire de mon Hydrodynamique. Je n'en avais plus chez moi, comme je vous l'ai dit, et je comptais m'en procurer un chez un ami: mais celui que j'ai reçu était si délabré que je n'ai pas osé vous l'offrir. Aussi ai-je dû faire d'autres recherches et j'en ai enfin trouvé un en bon état de conservation: c'est celui que je vous adresse, après y avoir fait quelques corrections à la plume. Il y en aurait bien d'autres à faire; mais, ainsi que j'ai déjà eu l'occasion de vous l'écrire, c'est un ancien travail qu'il faudrait revoir et refaire de fond en comble.

Vous désiriez aussi des exemplaires d'autres anciens Mémoires de moi et j'ai toujours toute la meilleure volonté de vous contenter. Mais les circonstances sont souvent plus fortes que nos volontés. J'ai passé une très-longue période d'appréhensions et de souffrances à cause de la santé toujours imparfaite de M.me Beltrami, et je n'ai eu la tête à rien. A présent les choses vont en peu mieux, mais il surgit des préoccupations d'un autre genre. Il s'agirait pour moi de me transférer à Rome (je crois que vous m'avez parlé une fois de cela, n'approuvant pas ma rénitence à me décider à ce passage): c'est un grand tracas qui va me frapper, mais auquel je pourrai me d'aider en vue de transporter M.me Beltrami en un climat plus doux. La chose n'est pas tout-à-fait décidée, mais elle se décidera assez vite. A cette occasion je devrai faire une grande révision de tous mes papiers et de tous mes livres, et je crois que je vais retrouver beaucoup de choses qui depuis bien longtemps ont disparu ou se sont soustraites à mes recherches. J'espère alors pouvoir former un paquet pour vous, beaucoup moins incomplet que je ne pourrais le faire à présent.

J'ai été énormément affligé de la nouvelle, tout-à-fait imprévue pour moi, de la mort de M.me Kowalewsky. C'est un bien grande perte pour les sciences mathématiques, et pour vous en particulier qui étiez à même d'apprécier de tout fait le mérite exceptionnel de cette femme illustre. J'espère qu'on en publiera une biographie détaillée dans les *Acta*, afin que les savants de tous pays apprennent les circonstances parmi lesquelles s'est éclore et s'est développée cette intelligence si élevée et si fine, que l'on aurait cru destinée à une bien plus longue carrière![93]

Veuillez, Monsieur et excellent Collègue, agréer pour vous-même et pour Mad.me votre épouse l'expression de mes sentiments les plus distingués, ainsi que l'assurance de mon parfait dévouement.

E. Beltrami

---

[93] See [61].

Letter n. 16
Eugenio Beltrami to Gösta Mittag-Leffler

Rome 27 Juin 1892

Monsieur et excellent collègue

Je m'empresse de vous adresser ci-jointe la Circulaire, que vous avez bien voulu me transmettre après y avoir apposé une signature et mon adresse. Je suis tout-à-fait heureux de m'associer à la fête que l'on veut faire à l'éminent analyste et je regrette seulement de ne pas pouvoir, à cause de mon prochain départ pour les vacances, contribuer autrement que par une offre personnel. Aussi suis-je bien content que M.r Guccia,[94] dont l'activité m'est bien connue, soit disposé à prendre sur lui la plus grande partie de la peine. Je vais me mettre en rapport avec lui à cet objet.

Comme vous voyez, je me suis enfin décidé à venir à Rome, où j'ai déjà fait l'année scolaire 91–92. Malheureusement avant que j'eusse eu le temps de m'installer convénablement dans cette ville, M.me Beltrami a été attaquée par l'influence, à la suite de la quelle elle a eu une longue série de souffrances qui ne sont pas même aujourd'hui tout-à-fait disparues, et qui nous ont obligé de vivre ici en touristes. Ce n'est qu'en Septembre que nous pourrons faire le déménagement complet. J'espère pouvoir alors reprendre mon travail plus régulièrement, et réparer au temps perdu. Nous espérons aussi, M.me Beltrami et moi, que le jour viendra où nous vous reverrons à Rome, ce qui sera un grand plaisir pour nous. Veuillez agréer, en attendant, nos salutations les plus cordiales ainsi que l'expression des sentiments de profonde et affectueuse estime de

Votre bien dévoué E. Beltrami

This letter is enclosed
Circolare per l'omaggio a Hermite

Palermo, 26/6/1892

Monsieur

Dans quelques mois, l'un des plus éminents géomètres qu'il y ait jamais eu, M. Hermite, va avoir 70 ans. La vie tout entière a été consacrée à la Science. Depuis ces précoces travaux qui attiraient sur un jeune écolier, l'attention de Jacobi, jusqu'à son récent mémoire sur le propriétés du nombre e, il a sans cesse marché

---

[94]Giovan Batti Guccia (1855–1914) studied at the University of Rome, where Luigi Cremona directed his first research in geometry. He was appointed professor at the University of Palermo and in 1884 founded the *Circolo Matematico di Palermo*. This institution became more and more important in Italy and abroad. The journal of the Mathematical Circle, the *Rendiconti del Circolo Matematico di Palermo*, was an international mathematical journal which published fundamental papers in the history of mathematics. See [20].

de découverte en découverte. De tous ces efforts il se croyait assez récompensé par les progrès de ses deux sciences de prédilection l'arithmétique et l'Analyse et il ne recherchait ni les honneurs, ni la gloire.

Mais s'il fut une notoriété bruyante, il ne reposera pas sans doute un témoignage sincère de reconnaissance et de respect. C'est, pourquoi un groupe d'élèves et d'admirateurs de M. Hermite croit devoir faire appel à ceux qui ont suivi ses leçons, comme à ceux qui l'ont approché ou qu'ont d'une manière quelconque subi son influence. Et nous en effet, nous lui devons beaucoup; non seulement sa parole, ses ouvrages et ses conseils ont guidé nos premiers pas, mais sa vie nous a donné un grande exemple; elle nous a appris à aimer la science d'un amour désintéressé.

Puisse notre concours lui prouver que cette leçon n'a pas été perdue et combler un de ses voeux le plus chers en lui donnant l'espoir que d'autres recolleront un jour, au profit de l'Analyse, la moisson qu'il a si libéralement semée.

Nous espérons, Monsieur, que Vous êtes dans les mêmes sentiments et que Vous penserez comme nous que le meilleur moyen de prouver à M. Hermite notre respectueuse admiration, c'est de lui offrir à l'occasion de son 70 anniversaire, une médaille commémorative avec une adresse portant les signatures de nombreux amis de la science.

Eugène Beltrami
Professeur de physique mathématique
Rome, Institut Royal de Physique, rue de Panisperna 89

4.7 Some letters concerning Sonya Kowalewskaia's affair of 1886

Letter n. 1
Enrico Betti to Luigi Cremona
This unpublished letter is contained in the Istituto Mazziniano di Genova

Pisa 26 Marzo 1886

Caro Cremona

Mi scrive Genocchi[95] che ha ricevuto una lettera di Hermite colla quale gli chiede di fare adesione insieme con i suoi amici matematici d'Italia a una testimonianza in favore della Sig.ra Sofia Kowalewski, alla quale hanno fatto già adesione Jordan, Darboux, Appel, Poincaré, Picard et Tisserand.[96] Genocchi ha già aderito

---

[95] Angelo Genocchi (1817–1889) was professor at the University of Turin from from 1859. Peano served as his assistant in 1881–82. He contributed to number theory, series and integral calculus.

[96] On March 19th, 1886, Hermite wrote to Genocchi: "Permettez-moi mon cher Président de vous faire part d'une circonstance qui intéresse Madame de Kowalevski, l'éminente analyste, et dont m'a informé M. Mittag-Leffler. Plusieurs membres de l'Académie des Sciences de Stockholm, auraient

unitamente a D'Ovidio[97] e prega me a mandare la mia adesione, e a scrive a te, Beltrami, Casorati, etc. per sentire se acconsentite a unirvi a noi.

La ragione di questa testimonianza a favore della Sg. Kowalewski è il rigetto della proposta di nominarla Socia dell'Accademia di Stokholm il quale ha destato mali umori e malevolenza verso di Lei per la quale è stato anche posto in dubbio il suo valore matematico.

Genocchi avrebbe scritto da sè se la cataratta progrediente e altri incomodi di salute non gli rendessero troppo grave scrivere lettere.

Ricevi i più cordiali saluti dal

Tuo aff Amico e Collega

Letter n. 2
Eugenio Beltrami to Enrico Betti
Beltrami's letters to Betti are contained in Betti's Archive, Archivio della Scuola Normale Superiore di Pisa.[98]

Pavia 28 Marzo 86

Carissimo Betti

La proposta che tu hai fatto, al Casorati ed a me, a nome di Genocchi, ci ha alquanto imbarazzati.

Sotto qual forma si dovrebbe fare la richiesta *adesione*? Trattasi di formulare, ciascuno personalmente per proprio conto, una dichiarazione di stima, o trattasi di aderire, in genere, ad una dichiarazione redatta da altri, e di cui non si possa conoscere il tenore? – Questa è una prima questione. Una seconda è quest'altra. Cosa fanno i matematici tedeschi? Sarebbe importante saperlo, perché avendo la

---

désiré qu'elle fût appelée à remplir une place vacante dans cette Académie, mais une opposition forte vive s'est produite contre la proposition d'admettre une femme à siéger, quelque fût d'ailleurs son talent scientifique, et Madame Kowalevski n'a pas été elue. Non seulement elle n'a pas été elue, mais une sorte de malveillance s'est attachée à sa personne, et on a été jusqu'à contester son mérite mathématique. M. Mittag-Leffler me demande de prendre sa défense, et dans cette intention, j'ai obtenu de MM. Camille Jordan, Darboux, Appell, Poincaré, Picard, Tisserand, l'autorisation de joindre au mien leur témoignage en sa faveur. Je viens vous demander, sous les plus expresses réserves de votre convenance, de vous réunir aux géomètres français, et dans le cas où ce ne serait point contraire à votre sentiment, de m'obtenir les adhésions de vos amis mathématiques d'Italie", in [60], p. 179. Other interesting notices on this subject are in [1], pp. 161–163. As it is emphasized in Sect. 3.3, though many mathematicians signed the letter, Sonya Kovalevskaya did not become a member of the Academy. Charles Hermite (1822–1901), Camille Jordan (1838–1922), Gaston Darboux (1842–1917), Paul Appel (1855–1930), Henri Poincaré (1854–1912), and Emile Picard (1856–1941) were important French mathematicians; Félix Tisserand (1845–1896) was an astronomer, who in 1886 held the chair of celestial mechanics at the Sorbonne.

[97]Enrico D'Ovidio (1843–1933) was appointed professor at the University of Turin in 1872. He contributed to (Euclidean and non-Euclidean) geometry. In 1905 he became senator of the Italian Kingdom.

[98]They will be published in [39].

signora Kowalewsky studiato in Germania, dev'essere colà conosciuta da molti professori valenti, i quali dovrebbero pei primi sentire la necessità di manifestare la loro opinione, dato che una manifestazione cosiffatta sia opportuna e motivata.

A dire la verità è una faccenda che al Casorati ed a me sembra alquanto delicata. Posto anche che l'opposizione dell'Accademia di Stoccolma sia eccessiva, è pur nondimeno desiderabile che gli scrupoli di quelli Accademici vengano rispettati e che non si eserciti dal di fuori una pressione che, in altro momento ed in altre condizioni, potrebbe essere prodotta con fini meno buoni. D'altronde poniamo che quei signori fossero ostili in genere all'ammissione di donne, e non già a quella della signora Sofia in particolare: come si fa ad imporre loro un'altra opinione? Giacché, sebbene si tratti di affermare genericamente il merito della Signora, è certo che in pratica questo verdetto va a colpire gli Accademici. Il Casorati poi osserva giustamente che, essendo stata creata per la Kowalewsky una cattedra apposita, non può essere mancata a lei l'occasione di farsi conoscere al pubblico studioso come insegnante di elevate dottrine matematiche, cosicché, se il suo insegnamento è realmente buono, non è ammissibile che un fatto tanto comune, quale è quello di una cattedra accademica non riuscita, possa scuotere la riputazione che essa ha potuto crearsi per questa via.

Mia moglie ricambia i tuoi cordiali saluti.

Aff.mo tuo E. Beltrami

Letter n. 3
Luigi Cremona to Enrico Betti
The letters by Cremona to Betti are contained in Betti's Archive, Archivio della Scuola Normale Superiore di Pisa, and published in [38], pp. 7–90; the cited letter is on p. 89.

[Roma] 2 aprile 1886

Carissimo Betti

Ho ricevuto la tua del 26, e ti chieggio scusa se non sono stato più sollecito nel rispondere. Ben volentieri aderisco alla manifestazione o testimonianza in onore della Sig. Kowalewski. Perciò, nel rispondere per tuo conto, ti prego di aggiungere al tuo nome il mio, sebbene la mia adesione abbia ben poco valore in confronto alla tua.

Abbimi sempre per tuo affezionatissimo amico

L. Cremona

Letter n. 4
Eugenio Beltrami to Enrico Betti

Pavia 11 Aprile 86

Carissimo Betti

Stando le cose come tu scrivi, il Casorati ed io pensiamo di astenerci dall'interloquire nella nota faccenda Kowalewsky, tanto più che realmente né l'uno, né l'altro abbiamo conoscenza sufficiente dei lavori di lei. Del resto i matematici italiani sono già più che distintamente rappresentati fra gli aderenti alla dichiarazione di stima.

Rilevo oggi dal Fanfulla[99] che il 18 si raduna la Giunta del Consiglio. Ciò mi spiega come la Commissione del Pinto sia stata convocata per il 19: ma io ci ho risposto che non vado.

Mia moglie ricambia i tuoi saluti ed io ti stringo affettuosamente la mano.

Aff.mo tuo E. Beltrami

4.8 Beltrami's Report on Sonya Kowalewskaia (1889)

Eugenio Beltrami to the University of Stockholm
This unpublished Report is contained in Mittag-Leffler's Archive, Djursholm, Stockholm[100]

Pavie 18 Mai 1889

À l'Université de Stockholm

Invité par l'Université de Stockholm à exprimer mon avis sur les travaux scientifiques de Madame Kowalewsky et sur les titres que ces travaux peuvent lui donner pour être nommée professeur ordinaire d'Analyse supérieure en cette Université, je m'empresse de répondre par ce qui suit à cette invitation qui m'honore, en même temps qu'elle me donne l'occasion bien agréable de rendre justice à un véritable mérite.

Les travaux dont il s'agit sont tous des pièces d'analyse la plus élevée et la plus exquise: mais tandis que les deux premiers (la Dissertation inaugurale et le Mémoire inséré au J.M. des *Acta Mathematica*) se apportent par leurs sujets aux mathématiques pures, les trois autres ont trait à des applications physico-mécanique aussi importantes que difficiles.[101] M'étant moi-même, depuis quinze ans, presque entièrement dévoué à ce genre de recherches, on me permettra de ne m'arrêter un peu que sur ces derniers travaux, d'autant plus que le temps me manquerait de revoir avec assez d'attention les deux premiers dont cependant je me rappelle parfaitement

---

[99] An Italian daily newspaper published from 1870 to 1886.

[100] As is pointed out in Sect. 3.3, Sonya Kovalevskaya was extraordinary professor of mathematics at the University of Stockholm since 1884. Also thanks to Beltrami's report, in 1889 she was appointed ordinary professor at the same university. Also Hermite and the Norwegian physicist and meteorologist Vilhelm Friman Koren Bjerknes (1862–1951) addressed their letters of recommendation to the Academy of Science of Stockholm; see [1], p. 131.

[101] The first two papers which Beltrami refers to are: [45, 46]. The other three papers concerning mathematical physical subjects – and which Beltrami refers to – are: [47–49].

bien l'appréciation ne peut plus honorable qu'en faisait un de mes confrères les plus éminents et les plus compétents, M.r le Sénateur Betti, de Pise. E d'ailleurs un fait bien avéré que la solution exacte des questions mécaniques exige le plus souvent les ressources les plus cachées de l'analyse, ce qui en fait des sujets très propres à donner la mesure des forces et des aptitudes de ceux qui les attaquent.

Je commencerai par le Mémoire très-considérable qui a paru aux J.M. des *Acta Mathematica* et qui est intitulé *Ueber die Brechung des Lichtes cristallinischen Mitteln.*[102] Quiconque a étudié la théorie de la double réfraction exposée par Lamé a pu voir combien est fatigante et enveloppée sa longue recherche finale sur la possibilité d'un seul centre d'ébranlement, et plus d'un lecteur a du penser que cette recherche n'était peut-être pas placée sur la véritable voie. Mme Kowalevsky a conçu l'idée de trancher la difficulté en entreprenant l'intégration directe des équations différentielles du mouvement de l'éther et, ce que l'on doit encore plus admirer, elle a eu le mérite de réussir de la manière la plus complète dans cette entreprise aussi hardie que compliquée. Elle a pu sans doute s'avantager des procédés que Poisson et Cauchy avaient depuis longtemps fait connaître (pour des questions analogues mais beaucoup plus simples) et que M. Weierstrass avait grandement perfectionnés et préparés, en quelque sorte, pour ce dernier effort: mais il y avait encore bien du chemin à faire, bien des obstacles à vaincre. Mme Kowalewsky a su parcourir la carrière jusqu'au bout, sans jamais renoncer ni à l'élégance des formules, ni à la rigueur des déductions. Ce chapitre important de l'optique mathématique a désormais acquis, par elle, sa forme la plus parfaite, et les conséquences des hypothèses qui lui servent de base pourront être maintenant facilement poussées jusqu'aux derniers détails.

Le Mémoire intitulé *Zusätze und Bemerkungen zu Laplace's Untersuchungen ueber die Gestalt der Saturnringe,*[103] inséré, d'après la proposition de l'illustre Mr. Gyldon, au J.M. des *Astronomyschen Nachrichten* est beaucoup moins étendu que le précédent, mais se rapporte à un point de la mécanique céleste qui n'est pas moins important, dans cet autre domaine de la science. Il s'agit des conditions d'équilibre du fluide répandu à la surface d'un anneau tournant autour d'un noyau central. Sous certaines hypothèses très-restrictives, cette question avait été résolue par Laplace, mais par des considérations qui, entre autres choses, ne laissaient pas reconnaître bien nettement le degré d'approximation que l'on devait attribuer, en réalité, au résultat formulé par lui. Mme Kowalewsky a repris toute cette question, pour la traiter d'une manière bien plus large et plus complète ; et par une analyse très-habile elle est arrivée non seulement à bien fixer ce degré d'approximation, mais elle a donné une méthode générale à l'aide de laquelle on pourrait satisfaire aux conditions du problème avec telle approximation que l'on voudrait. On doit surtout admirer, dans ce morceau, la sûreté par laquelle la savante analyste, ne se laissant

---

[102] See [47]. Beltrami did not realize that Kovalevskay committed a fatal mistake. This mistake was pointed out and corrected by Volterra in the paper: [68]. On this subject see [1].

[103] See [48].

pas rebuter par de très-réelles complications de calcul, retrouve toujours son chemin et arrive au but qu'elle s'était proposée. Pour ce qui est du dernier Mémoire *Sur la rotation d'un corps pesant autour d'un point fixe*, T. 12 des *Acta*[104] le jugement que vient d'en prononcer l'Académie des Sciences de Paris me dispenserait tout-à-fait d'en parler. Je ne puis cependant m'empêcher de remarquer qu'il s'agit d'un problème qui a été l'objet d'à peu près un siècle d'efforts, et qui, après cela, n'avait été complètement résolu que dans deux cas, auxquels se rattachent les noms d'Euler et de Lagrange. Les brillants aperçus géométriques de Poinsot, que l'on pouvait croire destinés d'abord à élargir la voie, n'ont malheureusement abouti à rien, en ce qui concerne la véritable difficulté, qui est celle de l'intégration, et ont plutôt laissé craindre que le chemin resterait désormais fermé. Mme Kowalewsky a eu le bonheur de découvrir un nouveau cas, intégrable par des fonctions bien connues. J'ai dit le bonheur, mais ce n'est pas le mot, car tout bonheur passe, tandis que le résultat de Mme Kowalewsky restera, et aura sa place marquée dans les traités de mécanique, après les solutions classiques dont j'ai nommé les auteurs.

Je dois maintenant répondre à la seconde partie de la question qui m'a été adressée et dire si les travaux dont je viens de parler peuvent être considérés comme des titres valables pour une nomination de professeur ordinaire à l'Université. Cette question étant en quelque sorte administrative, ou juridique si l'on veut, plutôt que scientifique, ne peut pas comporter, évidemment, une solution absolue. Ayant toutefois été membre, dans mon pays, d'un grand nombre de Commissions de concours et de promotion, avec mandat de juger les titres des candidats aux différentes chaires de mathématiques supérieures, et ayant aussi été membre, pendant quelques années, du Conseil supérieur de l'Instruction publique, qui exerce sa révision sur les Rapports des Commissions et fait ensuite ses propositions définitives au Ministre, je suis porté pour ma parte à assimiler la question dont il s'agit à une de celles que j'ai eu bien souvent à traiter : je veux dire à l'appréciation des titres pour la promotion d'un professeur extraordinaire (nommé d'année en année par le Ministre) au rang de professeur ordinaire (nommé à vie par le Roi). Or je puis déclarer en toute conscience que dans une occasion semblable, les titres à l'appui étaient de la force de ceux de Mme Kowalewsky, la Commission la plus exigeante et la plus sévère n'hésiterait par un instant à donner une approbation unanime et le Conseil n'hésiterait pas non plus, à ratifier pleinement ce jugement: je puis même ajouter que ce Conseil autoriserait le Ministre à excéder d'une unité le nombre des chaires disponibles, s'il se trouvait que ce nombre fût déjà au complet. En me plaçant ainsi au point de vue de ce qui se passerait en Italie, le seul dont je puis parler avec compétence, il n'y a pas de doute, pour moi, qu'en nommant Mme Kowalewsky à une chaire permanente d'analyse, on ferait non seulement acte de justice mais on satisferait d'une manière splendide aux exigences de l'enseignement supérieure.

Ayant ainsi répondu de mon mieux, mais en toute sincérité, aux questions que l'honorable Université de Stockholm a bien voulu m'adresser, je n'ai plus qu'à

---

[104]See [49].

exprimer à ses dignes représentants les sentiments de profond respect et de très-vive reconnaissance avec lesquels j'ai l'honneur de me signer

Eugène Beltrami
Professeur à l'Université Royale de Pavie (Italia)

Acknowledgments I warmly thank Michèle Audin, Livia Giacardi, Jens Høyrup, and Pietro Nastasi for their precious suggestions and Thomas Archibald for checking my English.

References

1. Audin, M. 2008. *Souvenirs sur Sofia Kovalevskaya*. Montrouge: Calvage et Mounet.
2. Beltrami, E. 1892. Considerazioni sulla teoria matematica dell'elettromagnetismo. *Memorie della Reale Accademia delle Scienze dell'Istituto di Bologna* 5(2): 313–378; in [4] 4: 436–498.
3. Beltrami, E. 1892. Sull'espressione analitica del principio di Huygens. *Rendiconti della Reale Accademia dei Lincei* 1: 99–108; in [4] 4: 499–509.
4. Beltrami, E. 1902–1920. *Opere matematiche* (ed. by Facoltà di Scienze della R. Università di Roma), 4 vols., Milano, Hoepli.
5. Beltrami, E. 1869. Saggio di interpretazione della geometria non euclidea. *Giornale di matematiche* 6: 285–315; in [4] 1: 374–405; French Transl. Essai d'interprétation de la géométrie non euclidienne. *Annales scientifiques de l'Ecole Normale Supérieure* 6: 251–288.
6. Beltrami, E. 1869. Teoria fondamentale degli spazii di curvatura costante. *Annali di matematica pura ed applicata* (2) 2: 232-255, in [4] 1: 406–429; Frenche Transl. Théorie fondamentale des espaces de courbure constante. *Annales scientifiques de l'Ecole Normale Supérieure* 6: 347–377.
7. Beltrami, E. 1876. Formules fondamentales de Cinématique dans les espaces Annali di matematica pura ed applicata (2) 2: 232–255, de courbure constante. *Bulletin des sciences mathématiques et astronomiques* 11: 233–240; in [4] 3: 23–29.
8. Beltrami, E. 1877. Intorno ad alcune questioni di elettrostatica, *Rendiconti del Reale Istituto Lombardo* 2(10): 171–185; in [4] 3: 73–88.
9. Beltrami, E. 1882. Sulle equazioni generali dell'elasticità. *Annali di matematica pura ed applicata* (2)10: 188–211 in [4] 3: 383–407.
10. Beltrami, E. 1884. Sull'uso delle coordinate curvilinee nelle teorie del potenziale e dell'elasticità. *Memorie della R. Accademia delle Scienze dell'Istituto di Bologna* (4) 6: 401–448; in [4] 4: 136–179.
11. Beltrami, E. 1884. Sulla rappresentazione delle forze newtoniane per mezzo di forze elastiche. *Rendiconti del R. Istituto Lombardo* (2) 17: 581–590; in [4] 4: 95–103.
12. Beltrami, E. 1884. Sull'interpretazione meccanica delle formole di Maxwell. *Memorie della R. Accademia dell'Istituto di Bologna* (4) 7: 1–38; in [4] 4: 190–223.
13. Beltrami, E. 1884. Sur les couches de niveau électromagnïtique. *Acta Mathematica* 3: 141–152; in [4] 4: 77–86.
14. Beltrami, E. 1871. Sui principi fondamentali dell'idrodinamica razionale. *Memorie della Reale Accademia delle Scienze dell'Istituto di Bologna* 3(1): 431–476 (2 (1872), 381–437; 3 (1873), 349–407; 5 (1874), 443–484; these works are reprinted in [4] 2: 202–379 with the title *Ricerche sulla cinematica dei fluidi*).
15. Betti, E. 1879. *Teorica delle forze newtoniane e sue applicazioni all'elettrostatica e al magnetismo*. Pisa: Nistri.

16. Boi, L., L. Giacardi, and R. Tazzioli (eds.) 1998. *La découverte de la géométrie non euclidienne sur la pseudosphère*. Paris: Blanchard.
17. Bottazzini, U. 1982. Enrico Betti e la formazione della Scuola Matematica Pisana. Atti del Convegno *La Storia delle Matematiche in Italia*, Cagliari, 29–30 settembre –1 ottobre 1982, pp. 229–275.
18. Bottazzini, U. 1994. *Va' pensiero. Immagini della matematica nell'Italia dell'Ottocento*. Bologna: Il Mulino.
19. Bottazzini, U. 2000. Brioschi e gli Annali di Matematica, in [50], pp. 71–84.
20. Brigaglia, A., G. Masotto. 1982. *Il Circolo Matematico di Palermo*. Bari: Dedalo.
21. Brioschi, F. 1854. *La teorica dei determinanti e le loro principali applicazioni*. Pavia: Fratelli Fusi.
22. Casorati, F. 1868. *Teorica delle funzioni di varibili complesse*. Pavia: Fratelli Fusi.
23. Castellana, M., and F. Palladino (ed.) 1996. *Giuseppe Battaglini. Raccolta di lettere (1854–1891) di un matematico al tempo del Risorgimento d'Italia*. Bari: Levante.
24. Chelini, D. 1860. *Elementi di meccanica razionale con Appendice sui principii fondamentali delle matematiche*. Bologna: G. Legnani.
25. Cooke, R. 1984. *The mathematics of Sonya Kovalevskaya*. New York: Springer.
26. Cremona, L. 1872. *Le figure reciproche della statica grafica*. Milano: Hoepli.
27. Darrigol, O. 2000. *Electrodynamics from Ampère to Einstein*. Oxford: Oxford University Press.
28. Dini, U. 1878. *Fondamenti per la teorica delle funzioni di variabili reali*. Pisa: Nistri.
29. Duhem, P. 1891–1892. *Leçons sur l'électricité et le magnétisme*, 3 vols. Paris: Gauthier-Villars.
30. Duhem, P. 1897–1899. *Traité élémentaire de mécanique chimique, fonde sur la thermodynamique*, 4 vols. Paris: Hermann.
31. Duhem, P. 1889. Sur l'équivalence des courants et aimants. *Annales scientifiques de l'Ecole Normale Supérieure* 3(6): 297–326.
32. Duhem, P. 1890. Sur les dissolutions d'un sel magnétique. *Annales scientifiques de l'Ecole Normale Supérieure* 3(7): 289–322.
33. Duhem, P. 1891. Sur les équations générales de la thermodynamique. *Annales scientifiques de l'Ecole Normale Supérieure* 3(8): 231–266.
34. Duhem, P. 1889. *Des corps diamagnétiques*. Lille: Au Siège des Facultés.
35. Duhem, P. 1891. *Sur la continuité entre l'état liquide et l'état gazeux et sur la théorie générale des vapeurs*. Lille: Au Siège des Facultés.
36. Enea, M.R. (ed.) 2009. *Il Carteggio Beltrami-Chelini (1863–1873)*. Milano: Mimesis.
37. Fano, G. 1889. Considerazioni comparative intorno a ricerche geometriche recenti. *Annali di Matematica* 2(17): 307–343.
38. Gatto, R. (ed.) 1996. Lettere di Luigi Cremona a Enrico Betti (1860–1890). In *La corrispondenza di Luigi Cremona (1830–1903)*, vol. III, ed. by M. Menghini, 7–90. Palermo: Quaderno Pristem N. 9.
39. Giacardi, L., and R. Tazzioli (ed.) 2011. *Pel lustro della Scienza italiana e per progresso dell'alto insegnamento. Le lettere di Beltrami a Betti, Tardy e Gherardi*. Milano: Mimesis, to appear.
40. Graf, J.H. (ed.) 1915. *La correpondance entre Ludwig Schläfli et des Mathématiciens Italiens de son époque*. Bollettino di Bibliografia e Storia delle Scienze Matematiche 17: 81–86, 113–122.
41. Kirchhoff, G. 1876. *Vorlesungen über mathematischen Physik. Mechanik*. Lepzig: Teubner.
42. Klein, F. 1884. *Vorlesungen über das Ikosaeder und die Auflösung der Gleichungen vom 5. Grade*. Leipzig: Teuber.
43. Kovalevskaya, S. 1885. Zusätze und Bemerkungen zu Laplace's Untersuchung über die Gestalt der Saturnringe. *Astronomische Nachrichten* 111: 37–48.
44. Kovalevskaya, S. 1890–91. Sur une propriété du système d'équations différentielles qui définit la rotation d'un corps solide autour d'un point fixe. *Acta Mathematica* 14: 81–93.
45. Kovalevskaya, S. 1874. *Ueber die Teorie di Partielle Differentialgleichungen*. Dissertation, (also published in *Journal für die reine und angewandte Mathematik* 80, 1875, pp. 1–32).

46. Kovalevskaya, S. 1884. Ueber die Reduction einer bestimmten Klasse Abel'scher Integrale dritten Ranges auf elliptische Integrale. *Acta Mathematica* 4: 393–414.
47. Kovalevskaya, S. 1885. Ueber die Brechung des Lichtes cristallinischen Mitteln, *Acta Mathematica* 6: 249–304.
48. Kovalevskaya, S. 1885. Zusaetze und Bemerkungen zu Laplace's Untersuchungen ueber die Gestalt der Saturnringe. *Astronomische Nachrichten* 111: 37–48.
49. Kovalevskaya, S. 1889. Sur le problème de la rotation d'un corps solide autour d'un point fixe. *Acta Mathematica* 12: 177–232.
50. Lacaita, C.G., and A. Silvestri (ed.) 2000. *Francesco Brioschi e il suo tempo (1824–1897)*. Milano: Franco Angeli.
51. Lipschitz, R. 1874. Extrait de six mémoires dans le journal de mathématiques de Borchardt. *Bulletin des sciences mathématiques et astronomiques* 4: 97–110, 142–157, 212–224, 297–307, 308–320, 414–420.
52. Lipschitz, R. 1877. Bemerkungen zu dem Princip des kleinsten Zwanges. *Journal für die reine und angewandte Mathematik* 82: 316–342.
53. Lipschitz, R. 1861. Beiträge zur Theorie der Vertheilung der statischen und dynamischen Electricität in leitenden Körpern. *Journal für die reine und angewandte Mathematik* 58: 1–53.
54. Lipschitz, R. 1863. Untersuchungen über die Anwendung eines Abbildungsprincips auf die Theorie der Vertheilung der Electricität. *Journal für die reine und angewandte Mathematik* 61: 1–21.
55. Lipschitz, R. 1880. *Lehrbuch der Analysis. Zweiter Band: Differential-und Integralrechnung*. Bonn: Cohen.
56. Lipschitz, R. 1877. *Lehrbuch der Analysis. Erster Band: Grundlagen der Analysis*. Bonn: Cohen.
57. Loria, G. 1901. Eugenio Beltrami e le sue opere matematiche. *Bibliothca Mathmatica* 2(3): 392–440.
58. Maiocchi, R. 2000. Il Politecnico di Francesco Brioschi, in [50], pp. 51–69.
59. Maxwell, J.C. 1873. *A treatise on electricity and magnetism*. 2 vols. Oxford: Clarendon.
60. Michelacci, G. (ed.) 2003. Le lettere di Charles Hermite ad Angelo Genocchi (1868–1887). *Quaderni matematici, Università degli Studi di Trieste* 546.
61. Mittag-Leffler, G. 1892–93. Sophie Kowalewsky. Notice biografique. *Acta Mathematica* 16: 385–392.
62. Palladino, F., and R. Tazzioli. 1996. Le lettere di Eugenio Beltrami nella corrispondenza di Ernesto Cesìro. *Archive for History of Exact Sciences* 49: 321–353.
63. Pepe, L. 2002. *Universitari italiani nel Risorgimento*. Bologna: Clueb.
64. Riemann, B. 1859. Fondamenti di una teorica generale delle funzioni di una variabile complessa (transl. by E. Betti). *Annali di Matematica Pura ed Applicata* 1(2): 288–304, 337–356.
65. Tazzioli, R. 1993. Ether and theory of elasticity in Betrami's work. *Archive for History of Exact Sciences* 46: 1–37.
66. Tazzioli, R. 1994. Rudolph Lipschitz's work on differential geometry and mechanics. In *The history of modern mathematics*, ed. E. Knobloch, and D.E. Rowe, 113–138. Boston: Academic Press.
67. Tazzioli, R. 2000. *Beltrami e I matematici relativisti. La meccanica in spazi curvi nella seconda metà dell'Ottocento*. Quaderni dell'Unione Matematica Italiana, 47, Bologna, Pitagora.
68. Volterra, V. 1892. Sur les vibrations lumineuses dans les milieux biréfringents. *Acta Mathematica* 16: 153–215.
69. Volterra, V. 1908. Le matematiche in Italia nella seconda metà del secolo XIX, in [70], 55–79.
70. Volterra, V. 1920. *Saggi scientifici*. Bologna: Zanichelli.

On Cimmino Integrals as Residues of Zeta Functions

Sergio Venturini

Abstract The following paper is a variation on a theme of Gianfranco Cimmino on some integral representation formulas for the solution of a linear equations system.

Cimmino was probably motivated for giving a representation formula suitable not only for theoretical investigations but also for applied computation.

In this paper, we will prove that the Cimmino integrals are strictly related to the residues of some zeta-like functions associated to the linear system.

1 Introduction

Gianfranco Cimmino was born in Naples on March 12, 1908.

He received his Laurea degree in Mathematics at the University of Naples under the direction of Mauro Picone (1885–1977) in 1927.

At the end of 1939 Cimmino moved permanently to the University of Bologna to occupy the chair of Mathematical Analysis.

Cimmino died in Bologna on May 30, 1989.

Gianfranco Cimmino (1908–1989) was a student of Renato Caccioppoli (1904–1959) and Mauro Picone toghether with Giuseppe Scorza Dragoni (1907–1996) and Carlo Miranda (1912–1982).

Mauro Picone founded the "Istituto per le Applicazioni del Calcolo" (IAC) in 1927. Indeed, long before the introduction of digital computers, Picone had the intuition of the potential impact on real-life problems of the combination of computational methods with mathematical abstraction.

S. Venturini (✉)
Dipartimento Di Matematica, Università di Bologna, Piazza di Porta S. Donato 5, I-40127, Bologna, Italy
e-mail: venturin@dm.unibo.it

Probably the influence of Picone ideas and Cesari papers [2, 3] on numerical solution of linear systems leads Cimmino to be interested to the numerical treatment of the solutions of linear systems of algebraic equations.

Among his researches he also obtained an interesting representation of the solution of a linear system of real linear equations that now we describe.

Let

$$Ax = b \tag{1}$$

be a linear system of n equation and n unknown, where the unknown values $x_1, \ldots, x_n$ are the components of the column vector $x \in \mathbb{R}^n$ and A is non singular matrix with real coefficient of order n.

The well known Cramer's rule say that if $D = \det(A) \neq 0$ then

$$x_i = \frac{D_i}{D}, \quad i = 1, \ldots n, \tag{2}$$

where D_i is the determinant of the matrix obtained replacing the i-th column of the matrix A with the column vector b.

The alternative representation of (2) given by Cimmino is

$$x_i = \frac{C_i}{C}, \quad i = 1, \ldots n, \tag{3}$$

where

$$C = \int_{S^{n-1}} \|A^t u\|^{-n} \, du, \tag{4}$$

and

$$C_i = n \int_{S^{n-1}} \|A^t u\|^{-n-2} \langle b, u \rangle \langle A^t x, e_i \rangle \, du. \tag{5}$$

The integration here is made with respect to the standard $(n-1)$-dimensional measure on S^{n-1}, the boundary of the Euclidean unit ball in $\mathbb{R}^n$, $\|\cdot\|$ and $\langle \cdot, \cdot \rangle$ stand respectively for the Euclidean norm and the Euclidean inner product on $\mathbb{R}^n$ and $e_1, \ldots, e_n$ is the canonical basis of $\mathbb{R}^n$, and A^t denote the transpose of the matrix A.

In [4, 5, 7], Cimmino gives a probabilistic argument which justify the existence of such kind of formulas and in [6] he also give an elementary but not trivial proof of (3). See also [8] for some applications. We refer to [1] for a detailed discussion and a background information on Cimmino's papers in the field of the numerical analysis.

The purpose of this paper is to show that the Cimmino ideas fit nicely into the theory of the residues of zeta-like functions, a powerful tool coming from the analytic number theory.

Namely, we will show that (4) and (5) are the integral representation of the residues of suitable zeta-like function associate to the matrix A and the vector b of the linear system (1); see Theorem 8 for the complete statement.

It should be noted that formulas 4 and 5 actually are due to Jacobi: cf. [10].

2 Theta functions and their Mellin transform

Let us begin with the following (almost trivial) observation:

Proposition 1. *Let f be a continuous complex function defined on the real interval $[0, 1]$. Assume that for some constants $R, \alpha, \beta \in \mathbb{R}$ with $\alpha > \beta$ we have*

$$f(t) = Rt^{-\alpha} + O(t^{-\beta}), \quad t \to 0^+.$$

Then, given $s \in \mathbb{C}$, the integral

$$g(s) = \int_0^1 f(t) t^s \frac{dt}{t}$$

converges absolutely on the half space $\mathrm{Re}(s) > \alpha$ and extends to a meromorphic function on the half space $\mathrm{Re}(s) > \beta$ having a simple pole at $s = \alpha$ with residue R.

Proof. Since $f(t) = O(t^{-\alpha})$ then $g(s)$ converges and is holomorphic when $\mathrm{Re}(s) > \alpha$. If we denote

$$h(t) = f(t) - Rt^{-\alpha},$$

then

$$g(s) = \int_0^1 f(t) t^s \frac{dt}{t} = \frac{R}{s - \alpha} + \int_0^1 h(t) t^s \frac{dt}{t}.$$

Since $h(t) = O(t^{-\beta})$ the last integral define a holomorphic function on the half space $\mathrm{Re}(s) > \beta$ and we are done. □

Thus, the singularities of the analytic function $g(s)$ describe the behaviour of the function $f(t)$ as $t \to 0^+$.

Let us recall that $C(\mathbb{R}^n)$ and $L^1(\mathbb{R}^n)$ denotes respectively the space of the continuous complex function on $\mathbb{R}^n$ and the space of the absolutely integrable complex functions with respect to the Lebesgue measure on $\mathbb{R}^n$.

Given $f \in L^1(\mathbb{R}^n)$ the *Fourier transform* of f is the function $\hat{f}$ defined by the formula

$$\hat{f}(y) = \int_{\mathbb{R}^n} f(x) e^{-2\pi i \langle x, y \rangle} \, dx.$$

Observe that

$$\hat{f}(0) = \int_{\mathbb{R}^n} f(x)\,dx.$$

We denote by $\mathcal{S}(\mathbb{R}^n)$ is the Schwartz space of smooth functions rapidly decreasing at the infinity together with their derivatives of all orders. Given $\alpha, \beta > 0$ two positive real number we denote by

$$\mathcal{S}_{n+\alpha}^{n+\beta}(\mathbb{R}^n)$$

the space of all measurable function such that

$$\sup_{x \in \mathbb{R}^n} |f(x)|\,(1 + \|x\|^{n+\alpha}) < +\infty$$

and

$$\sup_{y \in \mathbb{R}^n} \left|\hat{f}(y)\right|(1 + \|y\|^{n+\beta}) < +\infty.$$

Of course $\mathcal{S}_{n+\alpha}^{n+\beta}(\mathbb{R}^n) \subset L^1(\mathbb{R}^n)$ and $\mathcal{S}(\mathbb{R}^n)$ is the intersection of all the spaces $\mathcal{S}_{n+\alpha}^{n+\beta}(\mathbb{R}^n)$ when α and β varies on all the positive real numbers.

Since

$$\mathcal{S}(\mathbb{R}^n) = \bigcap_{\alpha,\beta>0} \mathcal{S}_{n+\alpha}^{n+\beta}(\mathbb{R}^n)$$

is useful to set

$$\mathcal{S}_{n+\infty}^{n+\infty}(\mathbb{R}^n) = \mathcal{S}(\mathbb{R}^n).$$

Let $\alpha, \beta > 0$, finite or infinite, be fixed.
Let $f \in \mathcal{S}_{n+\alpha}^{n+\beta}(\mathbb{R}^n)$ be an arbitrary function.
By the very elementary approach to the (Riemann) integration theory we have

$$\hat{f}(0) = \int_{\mathbb{R}^n} f(x)\,dx = \lim_{t \to 0^+} t^n \sum_{\omega \in \mathbb{Z}^n} f(t\omega).$$

It is then natural to define for $t > 0$ and $d > 0$

$$\theta_d(f,t) = \sum_{\omega \in \mathbb{Z}^n} f(t^{1/d}\omega),$$

$$\theta_d^*(f,t) = {\sum_{\omega \in \mathbb{Z}^n}}' f(t^{1/d}\omega) = \theta_d(f,t) - f(0),$$

(where $\sum'_{\omega \in \mathbb{Z}^n}$, as usual, stands for $\sum_{\omega \in \mathbb{Z}^n \setminus \{0\}}$) and considering the *Mellin transform*.

$$\xi_d(f,s) = \int_0^{+\infty} \theta_d^*(f,t) t^s \frac{dt}{t}$$

(The introduction of the constant d will simplify some computations).

Our hope is to study the Riemann approximation (6) looking at the residues of $\xi_d(f,s)$.

Indeed we have

Theorem 1. *Let $\alpha, \beta > 0$ be given and let $f \in \mathcal{S}_{n+\alpha}^{n+\beta}(\mathbb{R}^n)$ be an arbitrary function. Then the integral defining $\xi_d(f,s)$ converges in the strip $n/d < \mathrm{Re}(s) < (n+\alpha)/d$ and the function $\xi_d(f,s)$ extends to a meromorphic function on the strip $-\beta/d < \mathrm{Re}(s) < (n+\alpha)/d$ having exactly two simple poles respectively at $s = n/d$ with residue $\hat{f}(0)$ and at $s = 0$ with residue $-f(0)$. Moreover, we have the functional equation*

$$\xi_d(f,s) = \xi_d\left(\hat{f}, n/d - s\right), \quad -\frac{\beta}{d} < \mathrm{Re}(s) < \frac{n+\alpha}{d}.$$

The *proof* of the theorem above follows closely the lines of (some of) the standard proof of the functional equation for the classical zeta functions used in number theory, but the *statement* of the theorem in the form above is not so common. We refer to [11] for a detailed proof.

Observe that $\theta_d^*(f,t) = \theta_1^*(f, t^{1/d})$ and hence, by a simple change of integration variable, $\xi_d(f,s) = d\xi_1(f,ds)$. It follows that the function $\xi_d(f,s)$ has a pole at $s = n/d$ if, and only if, the function $\xi_1(f,s)$ has a pole at $s = n$ with the same residue.

3 Zeta functions machinery

The following proposition describe the hearth of our approach to the treatment of Cimmino integrals.

Proposition 2. *Let $\alpha, \beta, d > 0$ be given and let $f \in \mathcal{S}(\mathbb{R}^n)$ be an arbitrary function. Assume that in the strip $-\beta/d < \mathrm{Re}(s) < (n+\alpha)/d$ the function $\xi_d(f,s)$ decomposes as*

$$\xi_d(f,s) = G_f(s) Z_f(s)$$

with $Z_f(s)$ being meromorphic with exactly a single simple pole at $s = n/d$.

Then

$$Z_f(0) \operatorname*{Res}_{s=0} G_f(s) = -f(0),$$

$$G_f(n) \operatorname*{Res}_{s=n/d} Z_f(s) = \hat{f}(0).$$

Proof. By the previous theorem $\xi_d(f,s)$ has a simple pole at $s=0$ and

$$-f(0) = \operatorname*{Res}_{s=0} \xi_d(f,s) = \operatorname*{Res}_{s=0} G_f(s) Z_f(s).$$

Since $Z_f(s)$ is holomorphic at $s=0$ then

$$-f(0) = Z_f(0) \operatorname*{Res}_{s=0} G_f(s).$$

Replacing f with $\hat{f}$, we have

$$-\hat{f}(0) = \operatorname*{Res}_{s=0} \xi_d\left(\hat{f},s\right).$$

Using the functional equation $\xi_d(f,s) = \xi_d\left(\hat{f}, n/d - s\right)$ we obtain

$$\hat{f}(0) = -\operatorname*{Res}_{s=0} \xi_d\left(\hat{f},s\right) = -\operatorname*{Res}_{s=0} \xi_d(f, n/d - s) = \operatorname*{Res}_{s=n} \xi_d(f,s).$$

Since $\xi_d(f,s)$ and $Z_f(s)$ have a simple pole at $s=n$ then necessarily $G_f(s)$ is holomorphic at $s=n$ and hence

$$\hat{f}(0) = G_f(n) \operatorname*{Res}_{s=n/d} Z_f(s),$$

as desired. □

In the sequel, we will apply the above proposition to functions f which gives a decompositions $\xi_d(f,s) = G_f(s) Z_f(s)$ where the function $G_f(s)$ is an algebraic combination of elementary functions and functions of the form $\Gamma(as+b)$, being $\Gamma(z)$ the Euler gamma function, and $Z_f(s)$ is representable when $\operatorname{Re}(s) \gg 0$ as (generalized) Dirichlet series

$$Z_f(s) = \sum_{k=1}^{\infty} c_k \lambda_k^{-s},$$

where $c_k, \lambda_k \in \mathbb{R}$, $\lambda_1 < \lambda_2 < \cdots$, and $\lambda_k \to +\infty$ as $k \to +\infty$.

4 Quadratic forms

We begin considering the well known classical example of the Gaussian integrals associated to quadratic form.

Let us recall that $\Gamma(s)$ denotes the Euler gamma function; it is a meromorphic function on $\mathbb{C}$ which for $\operatorname{Re}(s) > 0$ satisfies

$$\Gamma(s) = \int_0^{+\infty} e^{-t} t^s \frac{dt}{t}.$$

We assume the reader knows all the properties of such a function.

Let Q be a real symmetric matrix of order n. We set

$$q_Q(x) = \langle Qx, x \rangle.$$

Assume that Q is positive definite and consider the function

$$g_Q(x) = e^{-\pi q_Q(x)},$$

then $g_Q \in \mathcal{S}(\mathbb{R}^n)$ and a standard argument yields

$$\hat{g}_Q(y) = \frac{1}{\sqrt{\det Q}} g_{Q^{-1}}(y).$$

In particular,

$$\hat{g}_Q(0) = \frac{1}{\sqrt{\det Q}}.$$

Since $q_Q(tx) = t^2 q_Q(x)$ then, choosing $d = 2$,

$$\theta_2^*(g_Q, t) = {\sum_{\omega \in \mathbb{Z}^n}}' e^{-t\pi q_Q(\omega)}$$

and hence, when $\operatorname{Re}(s) > n/2$,

$$\xi_2(g_Q, s) = {\sum_{\omega \in \mathbb{Z}^n}}' \int_0^{+\infty} e^{-t\pi q_Q(\omega)} t^s \frac{dt}{t}.$$

The change of variable $u = t\pi q_Q(\omega)$ yields

$$\xi_2(g_Q, s) = \pi^{-s} \left(\int_0^{+\infty} e^{-u} u^s \frac{du}{u} \right) {\sum_{\omega \in \mathbb{Z}^n}}' q_Q(\omega)^{-s}$$

$$= \pi^{-s} \Gamma(s) {\sum_{\omega \in \mathbb{Z}^n}}' q_Q(\omega)^{-s}.$$

The function defined for $\operatorname{Re}(s) > n/2$ by

$$\zeta(q_Q, s) = {\sum_{\omega \in \mathbb{Z}^n}}' q_Q(\omega)^{-s}$$

is the *Epstein zeta function* associated to the quadratic form q_Q.

Since $g_Q(x) \in \mathcal{S}(\mathbb{R}^n)$ then, by Theorem 1, $\xi_2(g_Q, s)$ is a meromorphic function on $\mathbb{C}$ having exactly two simple poles at $s = n/2$ and 0 with residues respectively $R = 1$ and -1.

Since the function $\pi^{-s} \Gamma(s)$ never vanishes on $\mathbb{C}$ and the function $\Gamma(s)$ has a simple pole at $s = 0$ with residue $R = 1$ it follows that $\zeta(q_Q, s)$ extends to a meromorphic function on $\mathbb{C}$ having exactly a simple pole at $s = n/2$.

By Proposition 2 and the relation

$$\Gamma(s+1) = s\Gamma(s),$$

we obtain that the residue of $\zeta(q_Q, s)$ at $s = n/2$ is

$$\operatorname*{Res}_{s=n/2} \zeta(q_Q, s) = \frac{\pi^{n/2}}{\Gamma\left(\frac{n}{2}\right)} \frac{1}{\sqrt{\det Q}} = \frac{n}{2} \left(\frac{\pi^{n/2}}{\Gamma\left(\frac{n}{2}+1\right)} \right) \frac{1}{\sqrt{\det Q}}$$

and that

$$\zeta(q_Q, s) = -1.$$

All that is well known to number theorists.

But now consider two quadratic forms, q_Q and q_S with Q and S symmetric matrices with Q positive definite and set

$$g_{Q,S}(x) = q_S(x) e^{-\pi q_Q(x)}.$$

Then $g_{Q,S} \in \mathcal{S}(\mathbb{R}^n)$.

Proposition 3. *The Fourier transform of* $g_{Q,S}(x)$ *is*

$$\hat{g}_{Q,S}(y) = -\frac{1}{\sqrt{\det Q}} g_{Q^{-1}, Q^{-1}SQ^{-1}}(y) + \frac{\operatorname{Tr}(Q^{-1}S)}{2\pi\sqrt{\det Q}} g_{Q^{-1}}(y)$$

$$= -\frac{1}{\sqrt{\det Q}} q_{Q^{-1}SQ^{-1}}(y) e^{-\pi q_{Q^{-1}}(y)} + \frac{\operatorname{Tr}(Q^{-1}S)}{2\pi\sqrt{\det Q}} e^{-\pi q_{Q^{-1}}(y)},$$

where $\operatorname{Tr}(Q^{-1}S)$ *denotes the trace of the matrix* $Q^{-1}S$.
In particular, we have

$$\int_{\mathbb{R}^n} g_{Q,S}(x)\,dx = \hat{g}_{Q,S}(0) = \frac{\operatorname{Tr}(Q^{-1}S)}{2\pi\sqrt{\det Q}}.$$

Proof. The function

$$e^{-\pi\|x\|^2},$$

where

$$\|x\|^2 = \langle x, x\rangle = \sum_{i=1}^n x_i^2,$$

coincide with its Fourier transform. Set

$$h_{ij}(x) = x_i x_j e^{-\pi\|x\|^2}.$$

Then, the standard properties of the Fourier transform yields

$$\hat{h}_{ij}(y) = -\frac{1}{4\pi^2} D_{y_i} D_{y_j} e^{-\pi\|y\|^2},$$

where D_{y_i} denote the operator of derivation with respect to the variable y_i.
We compute

$$h_{ij}(y) = y_i y_j e^{-\pi\|y\|^2},$$

$$D_{y_j} h_{ij}(y) = -2\pi y_j e^{-\pi\|y\|^2},$$

$$D_{y_i} D_{y_j} h_{ij}(y) = -2\pi D_{y_i}\left(y_j e^{-\pi\|y\|^2}\right) = 4\pi^2 e^{-\pi\|y\|^2} - 2\pi\delta_{ij} e^{-\pi\|y\|^2},$$

$$\hat{h}_{ij}(y) = -\frac{1}{4\pi^2} D_{y_i} D_{y_j} e^{-\pi\|y\|^2} = -e^{-\pi\|y\|^2} + \frac{1}{2\pi}\delta_{ij} e^{-\pi\|y\|^2}.$$

Let denote by I_n the identity matrix of order n. Let $C = (c_{ij})$ be a real symmetric matrix of order n and set

$$h_C(x) = g_{I_n, C}(x) = q_C(x) e^{-\pi \|x\|^2} = \sum_{i,j=1}^n c_{ij} h_{ij}(x).$$

Then,

$$\hat{h}_C(y) = \sum_{i,j=1}^n c_{ij} \hat{h}_{ij}(y)$$

$$= \sum_{i,j=1}^n -c_{ij} e^{-\pi \|y\|^2} + \sum_{i,j=1}^n \frac{1}{2\pi} c_{ij} \delta_{ij} e^{-\pi \|y\|^2}$$

$$= -q_C(y) e^{-\pi \|y\|^2} + \frac{1}{2\pi} \operatorname{Tr}(C) e^{-\pi \|y\|^2}.$$

Let $Q^{\frac{1}{2}}$ be the unique positive definite symmetric matrix such that

$$\left(Q^{\frac{1}{2}}\right)^2 = Q,$$

and let $Q^{-\frac{1}{2}}$ be its inverse.
Set

$$f(x) = h_C(x),$$

where $C = Q^{-\frac{1}{2}} S Q^{-\frac{1}{2}}$. We have

$$\operatorname{Tr}(C) = \operatorname{Tr}(Q^{-\frac{1}{2}} S Q^{-\frac{1}{2}}) = \operatorname{Tr}(Q^{-\frac{1}{2}} Q^{-\frac{1}{2}} S) = \operatorname{Tr}(Q^{-1} S)$$

and hence

$$\hat{f}(y) = -q_C(y) e^{-\pi \|y\|^2} + \frac{1}{2\pi} \operatorname{Tr}(Q^{-1} S) e^{-\pi \|y\|^2}.$$

We also have

$$q_C(Q^{\frac{1}{2}} x) = q_S(x),$$
$$q_C(Q^{-\frac{1}{2}} y) = q_{Q^{-1} S Q^{-1}}(y),$$
$$\left\| Q^{\frac{1}{2}} y \right\| = q_Q(y).$$

It follows that
$$g_{Q,S}(x) = f(Q^{\frac{1}{2}}x)$$

and hence
$$\hat{g}_{Q,S}(y) = \frac{1}{\sqrt{\det Q}} \hat{f}(Q^{-\frac{1}{2}}y)$$
$$= -\frac{1}{\sqrt{\det Q}} q_{Q^{-1}S^{-1}}(y) e^{-\pi q_{Q^{-1}}(y)} + \frac{\text{Tr}(Q^{-1}S)}{2\pi \sqrt{\det Q}} e^{-\pi q_{Q^{-1}}(y)},$$

as desired. □

Let now compute $\theta_2^*(g_{Q,S}, t)$ and $\xi_2(g_{Q,S}, s)$.
Since $q_Q(tx) = t^2 q_Q(x)$ and $q_S(tx) = t^2 q_S(x)$ we have
$$\theta_2^*(g_{Q,S}, t) = {\sum_{\omega \in \mathbb{Z}^n}}' t q_S(\omega) e^{-t\pi q_S(\omega)},$$

and
$$\xi_2(g_{Q,S}, s) = \int_0^{+\infty} {\sum_{\omega \in \mathbb{Z}^n}}' t q_S(\omega) e^{-t\pi q_S(\omega)} t^s \frac{dt}{t}$$
$$= {\sum_{\omega \in \mathbb{Z}^n}}' \int_0^{+\infty} q_S(\omega) e^{-t\pi q_S(\omega)} t^{s+1} \frac{dt}{t}.$$

By the change of variable $u = t\pi q_S(\omega)$ we obtain
$$\xi_2(g_{Q,S}, s) = \pi^{-(s+1)} \left(\int_0^{+\infty} e^{-u} u^{s+1} \frac{dt}{t} \right) {\sum_{\omega \in \mathbb{Z}^n}}' q_S(\omega) q_Q(\omega)^{-(s+1)}$$
$$= \pi^{-(s+1)} \Gamma(s+1) {\sum_{\omega \in \mathbb{Z}^n}}' q_S(\omega) q_Q(\omega)^{-(s+1)}.$$

Set
$$\zeta(q_Q, q_S, s) = {\sum_{\omega \in \mathbb{Z}^n}}' q_S(\omega) q_Q(\omega)^{-s}$$

The meromorphic function $\pi^{-(s+1)} \Gamma(s+1)$ never vanishes on $\mathbb{C}$. Since $g_{Q,S}(0) = 0$ by Theorem 1 the function $\xi_2(g_{Q,S}, s)$ is a meromorphic function on $\mathbb{C}$ having exactly a simple pole at $s = n/2$ and hence the sum defining $\zeta(q_Q, q_S, s)$ converges for $\text{Re}(s) > n/2$ and extends to a meromorphic function on $\mathbb{C}$ having exactly a simple pole at $s = n/2$. Then proposition 2 and proposition 3 give:

Theorem 2. *Let Q and S be two symmetric matrices of order n with Q positive definite. Then the sum defining the zeta function $\zeta\left(q_Q, q_S, s\right)$ converges absolutely for $\mathrm{Re}(s) > n/2 + 1$.*

The zeta function $\zeta\left(q_Q, q_S, s\right)$ extends to a meromorphic function on $\mathbb{C}$ having exactly a simple pole at $s = n/2 + 1$ with residue

$$\operatorname*{Res}_{s=n/2+1} \zeta\left(q_Q, q_S, s\right) = \frac{1}{2}\left(\frac{\pi^{n/2}}{\Gamma\left(\frac{n}{2}+1\right)}\right) \frac{\mathrm{Tr}(Q^{-1}S)}{\sqrt{\det Q}}.$$

From the equality

$$\hat{g}_{Q,S}(y) = -\frac{1}{\sqrt{\det Q}} g_{Q^{-1}, Q^{-1}SQ^{-1}}(y) + \frac{\mathrm{Tr}(Q^{-1}S)}{2\pi\sqrt{\det Q}} g_{Q^{-1}}(y)$$

proved in Proposition 3 we obtain

$$\xi_2\left(\hat{g}_{Q,S}, s\right) = -\frac{1}{\sqrt{\det Q}} \pi^{-(s+1)} \Gamma(s+1) \zeta\left(q_{Q^{-1}}, q_{Q^{-1}SQ^{-1}}, s+1\right)$$

$$+ \frac{\mathrm{Tr}(Q^{-1}S)}{2\sqrt{\det Q}} \pi^{-(s+1)} \Gamma(s) \zeta\left(q_{Q^{-1}}, s\right).$$

Inserting such expressions in the functional equation

$$\xi_2\left(g_{Q,S}, \frac{n}{2} - s\right) = \xi_2\left(\hat{g}_{Q,S}, s\right)$$

and dividing by π we obtain:

Theorem 3. *Let Q and S be as in Theorem 2. The zeta function $\zeta\left(q_Q, q_S, s\right)$ satisfies the functional equation:*

$$\pi^{-\left(\frac{n}{2}-s\right)} \Gamma\left(\frac{n}{2}+1-s\right) \zeta\left(q_Q, q_S, \frac{n}{2}+1-s\right)$$

$$+ \frac{1}{\sqrt{\det Q}} \pi^{-s} \Gamma(s+1) \zeta\left(q_{Q^{-1}}, q_{Q^{-1}SQ^{-1}}, s+1\right)$$

$$= \frac{\mathrm{Tr}(Q^{-1}S)}{2\sqrt{\det Q}} \pi^{-s} \Gamma(s) \zeta\left(q_{Q^{-1}}, s\right).$$

5 Lattices

A *lattice* in $\mathbb{R}^n$ is a set of the form

$$L = \{A\omega \mid \omega \in \mathbb{Z}^n\},$$

where $A \in GL(n, \mathbb{R})$ is a real invertible matrix of order n.

If A_1 is an other matrix such that $A_1(\mathbf{Z}^n) = L$ then $U = AA_1^{-1}(\mathbf{Z}^n) \subset \mathbf{Z}^n$. It follows that the invertible matrix U has integral coefficients and hence $|\det U| = 1$, that is $|\det A| = |\det A_1|$. We define the volume of the lattice L as

$$|L| = |\det A|.$$

By the argument given above the definition does not depend on the choice of the matrix A.

If $L = A(\mathbf{Z}^n)$ is a lattice the *dual lattice* is the lattice $L^\wedge$ associate to the inverse of the transpose of the matrix A. For convenience we also set

$$\hat{A} = (A^t)^{-1}.$$

Given a lattice $L \subset \mathbb{R}^n$ and a positive definite symmetric matrix Q of order n we define

$$\zeta_L(q_Q, s) = {\sum_{\omega \in L}}' q_Q(\omega)^{-s},$$

and if S is any symmetric matrix we also define

$$\zeta_L(q_Q, q_S, s) = {\sum_{\omega \in L}}' q_S(\omega) q_Q(\omega)^{-s}.$$

Of course we have

$$\zeta_{\mathbf{Z}^n}(q_Q, s) = \zeta(q_Q, s)$$

and

$$\zeta_{\mathbf{Z}^n}(q_Q, q_S, s) = \zeta(q_Q, q_S, s).$$

If $L = A(\mathbf{Z}^n)$ with $A \in GL(n, \mathbb{R})$, then we have

$$\zeta_L(q_Q, s) = \zeta(q_{A^tQA}, s),$$

where A^t denotes the transpose of the matrix A, and

$$\zeta_L(q_Q, q_S, s) = \zeta(q_{A^tQA}, q_{A^tSA}, s).$$

We also have

$$\sqrt{\det(A^tQA)} = |\det A| \sqrt{\det Q} = |L| \sqrt{\det A}$$

and

$$\mathrm{Tr}\big((A^tQA)^{-1}(A^tSA)(A^tQA)^{-1}\big) = \mathrm{Tr}\big(A^{-1}(Q^{-1}SQ^{-1})A\big)$$
$$= \mathrm{Tr}(Q^{-1}SQ^{-1}).$$

Applying the theta–zeta machinery to the function

$$g_{A,Q,S}(x) = g_{Q,S}(Ax) = q_S(Ax)e^{-\pi q_Q(Ax)},$$

observing that the Fourier transform of $g_{A,Q,S}(x)$ is

$$\hat{g}_{A,Q,S}(y) = \frac{1}{|\det A|}\hat{g}_{Q,S}(\hat{A}y),$$

it follows that the results of the previous section generalize:

Theorem 4. *Let $L \subset \mathbb{R}^n$ be a lattice and Let Q and S be two symmetric matrices of order n with Q positive definite. Then the sum defining the zeta functions $\zeta_L(q_Q, s)$ and $\zeta_L(q_Q, q_S, s)$ converges absolutely respectively for $\mathrm{Re}(s) > n/2$ and $\mathrm{Re}(s) > n/2 + 1$.*

The zeta functions $\zeta_L(q_Q, s)$ and $\zeta_L(q_Q, q_S, s)$ extends to a meromorphic function on $\mathbb{C}$ having exactly a simple pole respectively at $s = n/2$ and $s = n/2+1$ with residues

$$\operatorname*{Res}_{s=n/2} \zeta_L(q_Q, s) = \frac{n}{2}\left(\frac{\pi^{n/2}}{\Gamma(\frac{n}{2}+1)}\right)\frac{1}{|L|\sqrt{\det Q}}$$

and

$$\operatorname*{Res}_{s=n/2+1} \zeta_L(q_Q, q_S, s) = \frac{1}{2}\left(\frac{\pi^{n/2}}{\Gamma(\frac{n}{2}+1)}\right)\frac{\mathrm{Tr}(Q^{-1}S)}{|L|\sqrt{\det Q}}.$$

Theorem 5. *Let $L \subset \mathbb{R}^n$, Q and S be as in Theorem 4. The zeta functions $\zeta_L(q_Q, s)$ and $\zeta_L(q_Q, q_S, s)$ satisfy the functional equations, respectively*

$$\pi^{-(\frac{n}{2}-s)}\Gamma\left(\frac{n}{2}-s\right)\zeta_L\left(q_Q, \frac{n}{2}-s\right)$$
$$= \frac{1}{|L|\sqrt{\det Q}}\pi^{-s}\Gamma(s)\zeta_{L^\wedge}(q_{Q^{-1}}, s)$$

and

$$\pi^{-(\frac{n}{2}-s)}\Gamma\left(\frac{n}{2}+1-s\right)\zeta_L\left(q_Q, q_S, \frac{n}{2}+1-s\right)$$
$$+ \frac{1}{|L|\sqrt{\det Q}}\pi^{-s}\Gamma(s+1)\zeta_{L^\wedge}(q_{Q^{-1}}, q_{Q^{-1}SQ^{-1}}, s+1)$$
$$= \frac{\mathrm{Tr}(Q^{-1}S)}{2|L|\sqrt{\det Q}}\pi^{-s}\Gamma(s)\zeta_{L^\wedge}(q_{Q^{-1}}, s).$$

6 Integral representation

Let q_Q and q_S be two quadratic form with Q and S symmetric matrices and Q positive definite.

We now will give such residues as integrals over the boundary of the unit ball in $\mathbb{R}^n$.

We already observed that

$$\int_{\mathbb{R}^n} e^{-\pi q_Q(x)} \, dx = \frac{1}{\sqrt{\det Q}}$$

and by Proposition 3 we also have

$$\int_{\mathbb{R}^n} q_S(x) e^{-\pi q_Q(x)} \, dx = \frac{\operatorname{Tr}(Q^{-1}S)}{2\pi \sqrt{\det Q}}.$$

If $A \in GL(n, \mathbb{R})$ then a simple change of variable gives

$$\int_{\mathbb{R}^n} e^{-\pi q_Q(Ax)} \, dx = \frac{1}{|\det A| \sqrt{\det Q}}$$

and

$$\int_{\mathbb{R}^n} q_S(Ax) e^{-\pi q_Q(Ax)} \, dx = \frac{\operatorname{Tr}(Q^{-1}S)}{2\pi |\det A| \sqrt{\det Q}}.$$

Let us recall that given $f \in L^1(\mathbb{R}^n)$ the integration by polar coordinates gives

$$\int_{\mathbb{R}^n} f(x) \, dx = \int_{S^{n-1}} \left(\int_0^{+\infty} f(ru) r^{n-1} \, dr \right) du,$$

where du is the Euclidean (hyper-)surface measure on the unit sphere S^{n-1}.

Using such formula we obtain

$$\frac{\operatorname{Tr}(Q^{-1}S)}{2\pi |\det A| \sqrt{\det Q}} = \int_{\mathbb{R}^n} q_S(Ax) e^{-\pi q_Q(Ax)} \, dx$$

$$= \int_{S^{n-1}} \left(\int_0^{+\infty} q_S(rAu) e^{-\pi q_Q(rAu)} r^{n-1} \, dr \right) du$$

$$= \int_{S^{n-1}} q_S(Au) \left(\int_0^{+\infty} e^{-\pi q_Q(Au) r^{n+2}} \frac{dr}{r} \right) du.$$

By the substitution $\pi r^2 q_Q(Au) = t$ in the inner integral we obtain

$$\int_{S^{n-1}} q_S(Au) \left(\int_0^{+\infty} e^{-\pi q_Q(Au)} r^{n+2} \frac{dr}{r} \right) du$$

$$= \frac{1}{2} \int_{S^{n-1}} q_S(Au) \left(\int_0^{+\infty} e^{-t} \left(\frac{t}{\pi q_Q(Au)} \right)^{n/2+1} \frac{dt}{t} \right) du$$

$$= \frac{1}{2} \pi^{-(n/2+1)} \Gamma\left(\frac{n}{2}+1\right) \int_{S^{n-1}} q_S(Au) q_Q(Au)^{-(n/2+1)} du$$

and hence

$$\int_{S^{n-1}} q_S(Au) q_Q(Au)^{-(n/2+1)} du = \left(\frac{\pi^{n/2}}{\Gamma\left(\frac{n}{2}+1\right)} \right) \frac{\mathrm{Tr}(Q^{-1}S)}{|\det A| \sqrt{\det Q}}.$$

When $S = Q$ we obtain

$$\int_{S^{n-1}} q_Q(Au)^{-n/2} du = n \left(\frac{\pi^{n/2}}{\Gamma\left(\frac{n}{2}+1\right)} \right) \frac{1}{|\det A| \sqrt{\det Q}}.$$

Replacing s with $s/2$ in the formulas of the residues in Theorem 4 we obtain

$$\mathop{\mathrm{Res}}_{s=n} \zeta_L\left(q_Q, s/2\right) = n \left(\frac{\pi^{n/2}}{\Gamma\left(\frac{n}{2}+1\right)} \right) \frac{1}{|L| \sqrt{\det Q}}$$

and

$$\mathop{\mathrm{Res}}_{s=n+2} \zeta_L\left(q_Q, q_S, s/2\right) = \left(\frac{\pi^{n/2}}{\Gamma\left(\frac{n}{2}+1\right)} \right) \frac{\mathrm{Tr}(Q^{-1}S)}{|L| \sqrt{\det Q}}.$$

Thus we obtained:

Theorem 6. *Let q_Q and q_S be two quadratic form with Q and S symmetric matrices and Q positive definite. Let $A \in GL(n, \mathbb{R})$ and set $L = A(\mathbb{Z}^n)$. Then we have*

$$\mathop{\mathrm{Res}}_{s=n} \zeta_L\left(q_Q, s/2\right) = \int_{S^{n-1}} q_Q(Au)^{-n/2} du$$

and

$$\mathop{\mathrm{Res}}_{s=n+2} \zeta_L\left(q_Q, q_S, s/2\right) = \int_{S^{n-1}} q_S(Au) q_Q(Au)^{-(n/2+1)} du.$$

7 Linear systems

We recall that $\hat{A}$ denotes the inverse of the transpose of the matrix A.
Given $A \in GL(\mathbb{R}^n)$ we define

$$\zeta(A, s) = \sum_{\omega \in n}{}' \|A\omega\|^{-2s},$$

and given also $b \in \mathbb{R}^n$ we define the vector value zeta function

$$\zeta(A, b, s) = \sum_{\omega \in n}{}' \|A\omega\|^{-2s} \langle b, \omega \rangle A\omega.$$

Of course we have

$$\zeta(A, s) = \zeta_L(q_{I_n}, s)$$

with $L = A(\mathbb{Z}^n)$, I_n the identity matrix of order n and hence, by Theorem 4, we have

$$\operatorname*{Res}_{s=n/2} \zeta(A, s) = \frac{n}{2} \left(\frac{\pi^{n/2}}{\Gamma\left(\frac{n}{2}+1\right)} \right) \frac{1}{|\det A|}.$$

But we also have:

Theorem 7. *The series defining $\zeta(A, b, s)$ converges for $\operatorname{Re}(s) > n/2 + 1$ and the function $\zeta(A, b, s)$ extends to a (vector value) meromorphic function on $\mathbb{C}$ having only a simple pole at $s = n/2 + 1$ with residue*

$$\operatorname*{Res}_{s=n/2+1} \zeta(A, b, s) = \frac{1}{2} \left(\frac{\pi^{n/2}}{\Gamma\left(\frac{n}{2}+1\right)} \right) \frac{1}{|\det A|} \hat{A}b.$$

Moreover, if $c \in \mathbb{R}^n$ then we have the functional equation

$$\pi^{-\left(\frac{n}{2}-s\right)} \Gamma\left(\frac{n}{2}+1-s\right) \left\langle \zeta\left(A, b, \frac{n}{2}+1-s\right), Ac \right\rangle$$
$$+ \frac{\pi^{-s}}{|\det A|} \Gamma(s+1) \left\langle \hat{A}b, \zeta\left(\hat{A}, c, s+1\right) \right\rangle$$
$$= \frac{\langle b, c \rangle}{2|\det A|} \pi^{-s} \Gamma(s) \zeta\left(\hat{A}, s\right).$$

Proof. Given $u, v \in \mathbb{R}^n$, with $u = (u_1, \ldots, u_n)$ and $v = (v_1, \ldots, v_n)$ we denote by $u \otimes v$ the symmetric matrix (w_{ij}) of order n with entries given by $w_{ij} = u_i v_j$. Then we have

$$\mathrm{Tr}(u \otimes v) = \langle u, v \rangle.$$

Now observe that

$$\langle \zeta(A, b, s), Ac \rangle = \zeta_L(q_{I_n}, q_S, s),$$

where $L = A(\mathbb{Z}^n)$, I_n is the identity matrix of order n and

$$S = \hat{A}b \otimes Ac.$$

We have

$$\mathrm{Tr}(\hat{A}b \otimes Ac) = \langle \hat{A}b, Ac \rangle = \langle b, c \rangle$$

and hence, the formulas for the residue and the functional equation of the function $\langle \zeta(A, b, \frac{n}{2} + 1 - s), Ac \rangle$ follows from Theorem 4 and Theorem 5. $\square$

We are now ready to state and prove the reformulation of Cimmino's results.

Theorem 8. *Let*

$$Ax = b$$

be a linear system of n equation and n unknown, where the unknown values $x_1, \ldots, x_n$ are the components if the column vector $x \in \mathbb{R}^n$ and A is non singular matrix with real coefficient of order n.

Then we have

$$x_i = \frac{R_i}{R}, \quad i = 1, \ldots n,$$

where

$$R = \operatorname*{Res}_{s=n} \zeta(A^t, s/2)$$

and

$$R_i = n \operatorname*{Res}_{s=n+2} \langle \zeta(A^t, b, s/2), e_i \rangle, \quad i = 1, \ldots n,$$

where $e_1, \ldots, e_n$ is the canonical basis of $\mathbb{R}^n$.

Moreover, we have the identities

$$R = \int_{S^{n-1}} \|A^t u\|^{-n} \, du$$

and

$$R_i = n \int_{S^{n-1}} \|A^t u\|^{-n-2} \langle b, u \rangle \langle A^t x, e_i \rangle \, du.$$

Proof. By assumption

$$x = A^{-1} b$$

and we have

$$x_i = \langle x, e_i \rangle, \quad i = 1, \ldots, n,$$

where $e_1, \ldots, e_n$ is the canonical basis of $\mathbb{R}^n$. As previously observed we have

$$R = n \left(\frac{\pi^{n/2}}{\Gamma\left(\frac{n}{2}+1\right)} \right) \frac{1}{|\det A^t|} = n \left(\frac{\pi^{n/2}}{\Gamma\left(\frac{n}{2}+1\right)} \right) \frac{1}{|\det A|},$$

and for $i = 1, \ldots, n$ we have

$$R_i = n \frac{1}{2} \left(\frac{\pi^{n/2}}{\Gamma\left(\frac{n}{2}+1\right)} \right) \frac{1}{|\det A^t|} \langle \hat{A}^t b, e_i \rangle$$

$$= \frac{n}{2} \left(\frac{\pi^{n/2}}{\Gamma\left(\frac{n}{2}+1\right)} \right) \frac{1}{|\det A|} \langle A^{-1} b, e_i \rangle,$$

and hence

$$\frac{R_i}{R} = \langle A^{-1} b, e_i \rangle = x_i,$$

as required.

The last assertions of the Theorem follows from Theorem 6. □

Our tour around Cimmino's ideas is completed.

References

1. Benzi, M. 2005. Gianfranco Cimmino's contribution to numerical mathematics. Atti del Seminario di Analisi Matematica di Bologna, pp. 187–109.

2. Cesari, L. 1937. Sulla risoluzione dei sistemi di equazioni lineari per approssimazioni successive. *Rendiconti della Classe di Scienze Fisiche, Matematiche e Naturali dell'Accademia Nazionale dei Lincei, VI* 25: 422–428.
3. Cesari, L. 1937. Sulla risoluzione dei sistemi di equazioni lineari per approssimazioni successive. *La Ricerca Scientifica II* 8: 512–522.
4. Cimmino, G. 1967. Un metodo Monte Carlo per la risoluzione numerica dei sistemi di equazioni lineari. *Rendiconti dell'Accademia delle Scienze dell'Istituto di Bologna, XII* 2: 39–44.
5. Cimmino, G. 1972. Su uno speciale tipo di metodi probabilistici in analisi numerica. *Symposia Mathematica, Istituto Nazionale di Alta Matematica, X,* 247–254. London and New York: Academic Press.
6. Cimmino, G. 1987. La regola di Cramer svincolata dalla nozione di determinante. *Atti della Accademia Scienze Istitut Bologna Classe di Scienze Fisiche Rendiconti (14)* 3: 115–138, 1985/86.
7. Cimmino, G. 1986. An unusual way of solving linear systems. *Rendiconti della Classe di Scienze Fisiche, Matematiche e Naturali dell'Accademia Nazionale dei Lincei, VIII* 80: 6–7.
8. Cimmino, G. 1989. On some identities involving spherical means. *Atti della Accademia Nazionale dei Lincei Rendiconti Classe di Scienze Fisiche, Matematiche e Naturali (8)* 83: 69–72 (1990).
9. Cimmino, G. 2002. *Opere Scelte*. In ed. C. Sbordone, and G. Trombetti, Accademia di Scienze Fisiche e Matematiche della Società Nazionale di Scienze Lettere e Arti in Napoli, Giannini, Napoli.
10. Jacobi, C.G. 1834. Dato systemate n equationum linearium inter n incognitas, valores incognitarum per integralis definita (n-1)-tuplicia exhibentur. *Crelle Journal* 14: 51–55.
11. Venturini, S. 2008. Volumes, traces and zeta functions. *arXiv: 0812.2754v1*, pp. 1–24.

Index of Names and Locations

Note: Page numbers in italics indicate reference pages.

A

Abel, Niels H.   79, 80, 103, 500
Adams, Clarence Raymond   *51*
Adwa   114
Agostini, Amedeo   247, 253, *314*, 351, 359
Aharonov, Yakir   *437*
Ahlfors, Lars V.   25, *29*
Akivis, Maks A.   *173*
Albanese, Giacomo   445, *448*
Alexander, Waddel   129, *136*
Aliprandi, Giuseppe   47
Almagià, Francesco   251, 254
Amaldi, Edoardo   418
Amaldi, Ugo   v, 127, 128, 134, *139*, *141*, 145, 152, *176*, 192, 198, 203, *208*, 226, 239–242, 249, 257, 266, 267, *272–275*, 358, 363, *369*, 374, 376, *381*, 415–416, 418–420, 422, 423, *425*, *426*
Amodeo, Federico   107, 113, 114, 118, *136*, *139*, 217, 346, *371*
Amsterdam   *173*, *449*
Ann Arbor   *449*
Appell, Paul Émile   33, 35, *50*, 198, 477, 509, 510
Arbarello, Enrico   442, 445, *448*
Archibald, Raymond C.   *369*
Archibald, Thomas   515
Archimedes   242, 253
Arcozzi, Nicola   1–30
Arendt, G.   480
Aristote   255
Armellini, Giuseppe   254
Aronhold, Siegfried H.   121, 122
Arrighi, Gino   *369*
Artom, Emilio   247, 351, 353, 359
Arzelà, Cesare   *vii*, 32, 33, 36–38, 41, *50*, 108, 207, 291, 356, 362, 367, 407
Aschieri, Ferdinando   114
Asciano   309
Ascoli, Giulio   37, 407
Ascoli, Guido   367
Audin, Michèle   515
Avellone, Maurizio   *270*
Aymerich, Giuseppe   191

B

Babbit, Donald   *136*
Baccelli, Guido   109, 114
Badoglio, Pietro   309
Bagley Stouffer, Ellis   156
Baglioni, Silvestro   251, 252, 254
Bagnera, Giuseppe   292–294, 488
Baiada, Emilio   301, 387
Bain, William A.   335
Baire, René-Louis   37, 41, *51*, 396, 403, 461
Baldassarri, Francesco   328
Baldoni, Rina   47
Baltzer, Arnim R.   89, 226
Banach, Stefan   42, 45, 46, *52*, *53*, 457
Banfi, Carlo   191, 192
Barcelona   254
Bari   385
Barnes, Ernest William   374, 375, *380*
Baroni, Ettore   127, 226
Bartels, Martin   162
Basel   71, *176*, *315*, *370*

Battaglini, Giuseppe 11, 346, 396, 501, *516*
Battelli, Angelo 267
Bauer, Heinz 328, *331*
Baxter, Andrew 254
Bayle, Lionel *136*
Beauville, Arnaud *136*
Bellavitis, Giusto 478
Beltrami, Eugenio v, *vii*, 1–28, *29*, *30*, 46, 79, 82, 85–87, *103*, *104*, 119, 121, 126, *136*, 179, 189, 191, 193, 197, 203, 207, 212, 218, 238, 261, *273*, 281, 339, 465–514, *515*, 515, *516*, *517*
Benedetti, Claudia viii
Benedetti, Piero 349, 352, 353
Benenti, Sergio 206, *208*
Bennett, Albert A. *369*
Bentley, Richard 254
Benzi, Michele *537*
Berenstein, Carlos Alberto *437*, *438*, *462*
Bergamo *315*
Bergman, Stefan 20, 21
Bergson, Henri 336
Berlin 58, 91, 101, 117, 214, 254, 334, 368, 468, 470, 475, 477, 478, 486–488, 491, 492, 495, 498, 503
Bern 498
Bernardini, Gilberto *314*
Bertini, Eugenio 33, 101, *103*, *104*, 107, 108, 114, 129–131, *136*, 156, 343, 422
Bertrand, Joseph 468
Berwald, Ludwig 168
Berzolari, Luigi 48, 107, 147, 151, 170, 171, *173*, 215, 217, *270*, 310, 343, 345–350, 352–355, 357–366, *369–371*, 388
Bettazzi, Rodolfo 214, 346, 367
Betti, Enrico 46, 57, 80, 87, 101, 179, 183, 189, 207, 281, 313, 465–472, 475–477, 502, 509–512, *515–517*
Betti, Mario 296
Bezout, Étienne 95
Bianchi, Luigi 32–35, 46, 108, 119, 126, 128, 154, 158–160, 163, 167, 170, *173*, *175*, 305, 306, 313, 388, 389, 425, 470
Biggini, Carlo Alberto 307–309
Biggiogero, Giuseppina 270, 349, 351, 355
Bischoff, Johan N. 87
Biscottini, Umberto 307, 308
Blaschke, Wilhelm 155, 164, 325
Boggio, Tommaso 198, 202, *208*, 351, 352, 355, 360
Bohm, David 288
Boi, Luciano *516*

Bologna v–viii, 1, 32, 33, 35, 38, 46, 49, 58, 60, 70, 73–103, 105–135, 143–173, 185, 187–191, 193, 194, 197, 203, 207, 214–222, 224, 227, 239, 240, 243, 245, 253, 255, 256, 263–266, 269, 289–298, 300, 301, 304, 311, 317, 318, 334, 343, 355–366, 368, 373, 374, 379, 383–410, 415–425, 439, 468, 471, 491, 498, 519
Bolondi, Giorgio 270, 274
Bolyai, János 1, 2, 6, 7, 9, 16, 21, 126
Bolyai, Wolfgang (Farkas) 6
Bolzano, Bernhard 7
Bombelli, Rafael v, 253
Bompiani, Arturo 145
Bompiani, Enrico v, *vii*, 46, 47, 143–173, *173–177*, 192, 254, 256, 296, 310, 350, 354, *370*
Bonfiglioli, A. *331*
Bonghi, Ruggero 211
Bonino, Giovanni Battista 302, *315*
Bonn 491, 492, 498, *517*
Bonnesen, T. 246
Bonola, Roberto 3, 5, 6, *29*, 126, 127, 135, *136*, 215, 226, *270*, 343–347, *370*, 418, *425*
Bony, Jean Michel 328, *331*
Boole, George 337, 338
Bordeaux 481
Bordoni, Antonio Maria 80
Borel, Émile 37, 38, 40–42, *51*, *52*, 62, 314, 461
Borgato, Maria Teresa 31–50, *50*, *51*, 297, *370*, *410*
Borio, Agostino 352
Born, Max 287
Bortolotti, Enea 46, 47, 49, *53–55*, 156, 168, 169, *175*, 385, *410*
Bortolotti, Ettore v, vi, *viii*, 35, 247, 252, 253, 300, 343, 348, 358–361, 386
Boschi, Pietro 87, 89
Boselli, Paolo 212
Bosello, Carlo Alberto 189
Boston 282
Bottazzini, Umberto 57–70, *70*, *71*, *136*, *270*, *342*, *425*, *516*
Bourlet, Carlo 378, *380*
Boyer, Carl B. *29*
Braunschweig 477, 478, 480
Brera 98
Breslau 491, 498, 503
Bridson, Martin R. 3, *29*
Brigaglia, Aldo 73–103, *103*, *136*, *173*, 252, 259, *270*, 299, *516*

Index of Names and Locations 541

Brill, Alexander von   129, *136*, 442, *448*
Brill Leiden   *138*
Brillouin, Léon   280
Brioschi, Francesco   78, 80, 83, 85, 89, 91, 357, 465, 467–470, 502, 505, *516*, *517*
Brizzi, Gian Paolo   259
Broad, Charles D.   334, *342*
Brouwer, Luitzen E.   122
Brown, A.A.   *438*, *463*
Bruno, Andrea   *448*
Brusotti, Luigi   349, 351, 352, 355, *370*
Brussels   91, 222, 224
Budapest   254
Burali Forti, Cesare   198–202, 207, *208*, 340, 341, *342*, 351, 366
Burgatti, Pietro   v, vi, *vii*, 189–193, 197–207, *208*, 385
Burkhardt, Heinrich F.   125
Bussotti, Paolo   *137*, 259, *274*
Buzano, Ernesto   397

C

Caccioppoli, Renato   305, 519
Cagliari   87, 127, 190, 240, 293, 318, 387, 388
Calabri, Alberto   89, *137*
Calcutta   290
Calò, Benedetto   127, 226, 347
Cambridge   146, 154
Cambridge (Mass.)   290, 314
Campedelli, Luigi   *142*, *270*, 301
Cannon, James W.   3, 28, *29*
Cantat, Serge   *137*
Cantelli, Francesco Paolo   299, 360
Cantimori, Delio   *314*
Cantor, Georg   37, 38, 122, 182, 367, *370*, *371*
Capelli, Alfredo   121
Capellini, Giovanni   84
Caprioli, Luigi   184, 191
Caratheodory, Costantin   42, *52*, 398, 401, 403, 407, 408, *410*
Carbone, Luciano   *103*, *270*, *315*
Cardano, Girolamo   v
Carducci, Giosuè   78, 86
Carnap, Rudolf   334, *342*
Carruccio, Ettore   252, 253
Cartan, Élie   144, 159, 162, 168, *175*, 418, 439

Casanova   310
Casati, Gabrio   77, 210
Casini, Paolo   255, *270*
Casorati, Felice   57–59, *71*, 80, 100, 354, 465, 467–470, 475, 477, 498, 505, 510, 511, *516*
Cassina, Ugo   247, 347, 351, 352, 362, *370*
Cassinet, Jean   *410*
Cassini, Giandomenico   v, 65
Castelfidardo   78
Castellana, Mario   *516*
Castellani, F.   302
Castelnuovo, Emma   258, *270*
Castelnuovo, Gino   48, 252, 253, *273*
Castelnuovo, Guido   101, 106–109, 112–114, 116, 118, 125–127, 129–132, 134, 135, *136–138*, *141*, *142*, 143–147, 150–154, 158, 163, 167, 169, 171, 172, *175*, 214, 216–222, 224–226, 242, 243, 246, 249, 250, 253–255, 258, *270*, 293–295, 299, 312, 334, *342*, 346, 360, 362, *371*, 422, 423, *425*, 439, 442, 444, 448, *448*
Castoldi, Luigi   *371*
Castorena, Abel   *448*, *449*
Catania   127, 404
Catania, Sebastiano   362
Cattabriga, Lamberto   v, 451–462, *462*
Cattaneo, Carlo   187
Cauchy, Augustin-Louis   7, 57, 59, 63, 65–67, 69, 204, 513
Cavalieri, Bonaventura   v
Cayley, Arthur   91, 119, 121
Cebyšëv, Pafnutij L'vovič   291
Cech, Eduard   159, *175*, *177*
Cecioni, Francesco   312
Cento   197
Cermeli, L.   268
Cerroni, Cinzia   *103*
Cerruti, Valentino   197, 472
Cesari, Lamberto   185, 190, 290, 299, 301, 387, 520, *538*
Cesàro, Ernesto   197, 362, 466, 472
Cetraro   *438*
Ceyne, George   254
Chang, Mei Chu   *449*
Charrier, G.   358
Chasles, Michel   92, 93, 97, 115, 116
Chelini, Domenico   82, 84–87, *103*, 466, 470, *516*
Cheng, Shiu-Yuen   *175*
Chern, Shiing-Shen   146, 159, *175*

Chersoni, Cristina viii, 259
Chiarugi, Alberto 308
Chiarugi, Giulio 302
Chicago 146, 155
Chisini, Oscar 116, *142*, 222, 229, 247, 268, 270, *272*, 352, *370*
Chow, Wei-Liang 441
Ciani, Edgardo 107, 114
Cicenia, Salvatore *29*
Ciliberto, Ciro *103*, 105–135, *136*,*137*, 143–173, *173*, *175*, 216, 252, *270*, *448*, *449*
Cimmino, Gianfranco v, *vii*, 193, 317, 519–537, *537*, *538*
Cinquini, Silvio 290, 296, 301, *314*, *370*, *410*
Cipolla, Michele 348, 349, 351, 388, 488
Cisbani, Renzo 298, 299, 302
Clairaut, Alexis 253, 258
Clarke, Samuel 254
Clarkson, James A. *51*
Clausius, Rudolf 492
Clebsch, Rudolf F.A. 59, 92, 100, 113, 115, 116, 122, 128, *137*, 491
Clifford, William Kingdon 5, 9
Cocchi, Chiara viii
Codazzi, Delfino 7, 47, 158
Coen, Salvatore v–viii, *viii*, 4, *50*, *71*, *137*, 380, *410*, *412*, 437, 462
Cohen, Paul 395, *410*
Colombo, Bonaparte 351
Colombo, Fabrizio *438*
Comessatti, Annibale 48, 52, 152, 343
Conforto, Fabio 132, 133, *142*, 252, 301, 360, 368, *370*, *371*
Console, Sergio 397
Conte, Alberto *136*, *270*, *342*, *425*
Conti, Alberto 127, 226, *272*, 346, *370*
Cooke, Roger *516*
Coolidge, Julian Lowell 148, *175*
Copenhagen 498
Corbino, Orso Mario 216, 302
Cornalba, Maurizio 442, *448*
Cosserat, Eugène 146
Courant, Richard 205, *208*
Cournot, Antoine A. 360
Credaro, Luigi 242
Crelle, August L. 87, 91
Cremona Cozzolino, Itala 74
Cremona, Elena 79
Cremona, Gaudenzio 74
Cremona, Giovanni 74
Cremona, Giovannina 74
Cremona, Giuseppe 74, 75
Cremona, Giuseppe (Gaudenzio's father) 74

Cremona, Luigi v, *vii*, 2, 33, 55, 73–103, *103*, *104*, 115, 129, 130, 133, 134, *137*, 143, 163, 212, 213, 224, 226, *270*, 361, 422, 423, 465, 467–471, 475, 477, 498, 501, 505, 508–511, *516*
Cremona, Luisa 86
Cremona, Tranquillo 74
Cremona, Vittorio 79
Crocco, Gaetano Arturo 303
Croce, Benedetto 211, 215, 230, 234, 243, *271*, *294*, 334, *342*
Cugiani, Marco *314*
Cuneo 347, 348

D

Dainelli, Dino 299
D'Alambert, Jean-Baptiste Le Rond 123
Dall'Acqua, Aurelio 205, 206
Dall'Aglio, L. *208*
d'Ambra, Lucio 303
Daniele, Ermenegildo 127, 226, 312, 355
Dantoni, Giovanni 299
Darboux, Jean Gaston 149, 160–162, 165, 167, *175*, 477, 509, 510
Darrigol, Olivier *516*
Dauben, Joseph Warren *370*
De Broglie, Louis 277–279, 287, 288
De Castro, Ercole 191
De Dominicis, Saverio 213, *270*
De Finetti, Bruno 360
De Franchis, Michele 488
De Francisci, Pietro 252
De Giorgi, Ennio 290, 451, 458–462, *462*
De Jonquières, Jean-Philippe-Ernest Fauque 97, 100
de Lisio, Carlo *176*
De Paolis, Riccardo 107, 110, *137*, 367
Decroly, Ovide 257
Dedekind, Richard 37, 111, 253, 336, 466–468
Dedò, Modesto *270*
Dehn, Max 110
Del Chiaro, Adolfo 299
del Ferro, Scipione v
del Pezzo, Pasquale 130, 149, 505
Del Re, Alfonso 107, 109, 114
Delboeuf, Joseph 335
Deligne, Pierre *55*
Della Seta, Fabio *270*
Demazure, Michel 123, *137*
Demidov, Sergei S. *71*

Index of Names and Locations 543

Demoulin, Alphonse 165, 166
Denjoy, Arnaud 42, *52*, 384, 386, 389, *410*
Derham, William 254
Deserti, Julie *104*
Di Sieno, Simonetta 73–103, *103*, *136*, *173*, *176*, *270*, *314*
Dieudonné, Jean 42, 431, *438*
Dini, Ulisse 32, 33, 37, 38, 40, *51*, 108, 305, 306, 367, 388, 389, 470, *516*
Dirac, Paul A.M. 277, 288, *288*, 437
Dirichlet, Johann Peter G. Lejeune 57, 60, 64, 79, 80, 292, 318, 319, 369, 468, 475, 477–480, 524
Dixon, Arthur Lewis 375, *380*
Doetsch, G. 321, *331*
Dolgachev, Igor 101, *104*, *137*
Donati, Luigi vii
Doob, Joseph Leo 328
Dordrecht 449
Dornig, Mario 152
Dostal, Milos A. *462*
Dover 29, *71*, *288*, *381*, 420
D'Ovidio, Enrico 112, *137*, 476, 510
Du Bois Reymond, Paul 37, 336, 389
Dugac, Pierre *71*
Duhem, Pierre 466, 467, 474, 481–486, *516*
Dunham, William *50*

E

Ehrenpreis, Leon 428, 429, 434, 435, *438*, 451, 452, 454, 455, 458, 459, *462*, *463*
Ehresmann, Charles 421, *425*
Ein, Lawrence 449
Einstein, Albert 47, *53*, 157, 198, 246, 278, 334
Eisenbud, David 449
Ellero, Pietro 79
Emanuelli, Pio 252
Enea, Maria Rosaria *104*, *516*
Eneström Gustav 366, 367, 499, 500
Engel, Friedrich 416
Enriques, Federigo v, *vii*, 3, *29*, 32, 33, 35, 101, 103, 105–135, *136–142*, 151, 153, 154, 156, 171, 189, 192, 207, 209–269, *270*, *271*, *272*, *273*, *274*, *275*, 295, 299, 301, 302, 333–342, *342*, 347, 349, 352, 353, 357–359, 362, 367–369, *370*, 415, 416, 418, 422–424, *425*, 439, 445, 448

Ering, Ewald 219
Erlangen 124, 351, 416, 422–424, 488, 489
Euclid 4, 95, 119, 126, 253, 258, 260, 262, 271, *275*
Euler, Leonhard 377, 452, 460, 514, 524, 525
Evangelisti, Giuseppe 189
Evans, L.C. 327, *331*

F

Faà di Bruno, Francesco 211
Fabrizio, Mauro 179–195
Faedo, Alessandro (Sandro) 290, 293, 297–299, 301, 302, 305, 311
Faifofer, Aureliano 475, 477–480
Fano, Gino 113, 114, 118, 123, 124, *137*, *141*, 210, 214–217, 224, *272*, 351, 367, 416, 423, 425, *425*, *426*, 489, *516*
Fantappié, Luigi vi, *vii*, 48, 207, 310
Faraday, Michael 473
Farini, Luigi Carlo 77, 81
Federzoni, Luigi 303
Fedoryuk, Mikhail V. *288*
Fehr, Henri 226, *272*, *370*, *371*
Fenaroli, Giuseppina *103*
Fermi, Enrico 46, 247, 257, 298
Ferrar, William L. 375, *380*
Ferrara 83, 185, 197
Ferrari, Elisa 79, 81, 82, 86, 91
Ferrari, Italo 191
Ferrari, Ludovico v
Ferrari, Nicolao 81
Ferrarotto, M. *314*
Ferrazzano *176*
Ferri, Luigi 78, 89
Ferrière, Adolphe 257
Feynman, Richard 288
Fichera, Gaetano 180, 193, *315*
Fields, John Charles 374
Finzi, Aldo 348, 352
Finzi, Bruno 195, 203, *208*
Finzi, Cesare 33
Fiorentino, Francesco 79
Fiorini, Matteo v, 82
Florence 219, 224, 226, 243, 248, 282, 309
Floyd, William J. 3, *29*
Foà, Emanuele vi, 190, 192
Föppl, August 198
Forlì 317
Forti, Umberto 247, 252, 253, *272*

Fourier, Jean Baptiste J.   26, 36, 63, 186, 295,
        429, 430, 433–435, 453, 454, 457,
        458, 521, 527, 532
Fox, Charles   375, *380*
Fraenkel, Abraham A.   *410*
Frajese, Attilio   252, 253, *370*
Francesconi, Stefano   *viii*, *71*
Franchi, Franca   191
Franchi, Oliviero   267
Frattini, Giovanni   357
Fréchet, Maurice René   37, 45, *53*,
        182, 291
Frege, F.L. Gottlob   337
Freguglia, Paolo   197–207,
        208, 236
Frenet, Jean   162, 168
Freud, Sigmund   334, 478
Freudenthal, Hans   *137*, *138*
Fribourg   258
Frisone, Rosetta   352
Fröbel, Friedrich   257
Frobenius, Ferdinand Georg   60, 475, 498
Fubini, Guido   33, 34, 44, 46–48, 144, 145,
        147–149, 156–164, 166–168, 171,
        172, *175*, 190, 292, 293, 297, 362,
        385, 388, *410*, 425
Fueter, Karl R.   433
Fulks, W.   324, *331*
Fulton, William   441, *449*
Funaioli, Giovan Battista   308
Furinghetti, Fulvia   *272*
Furlani, Giacomo   248, 266–269
Fürth, Reinhard   288

G

Gabba, Luigi   355
Gaeta   81
Galilei, Galileo   253
Galletti, Alfredo   246
Gallipoli   290
Galuzzi, Massimo   *104*, 236, *270*
Gandino, Giovanni Battista   79
Garbasso, Antonio   302
Garbiero, Sergio   397
Gardner, J. Helen   *104*
Garibaldi, Giuseppe   77, 81, 82, 92
Gariepy, L.F.   327, *331*
Gario, Paola   105–135, *136*, *138*, 216, *270*,
        272, *425*
Gatto, Romano   *103*, *104*, *270*, *516*

Gauss, Carl F.   1, 6, 7, 9, 11–13, 16, 20, 22,
        47, 63, 64, 79, 80, 161, 167, 323,
        335, 360, 375, 471, 475, 478, 493
Geiser, Carl F.   475, 498
Gemona   290
Genoa   33, 34, 37, 43, 44, 74, 83, 91, 156,
        311, 355, 362, 470, 488
Genocchi, Angelo   2, 74, 76, 78–80, 82, 83,
        85, 100, 226, 476, 477, 509, 510
Gentile, Giovanni   vi, 43, 44, 79, 192, 234,
        239, 240, 243, 248–252, 256, 257,
        290, 293–298, 300, 302, 304–307,
        309–311, 397
Gerbaldi, Francesco   107, 150, 151, 345, 367,
        488
Gergonne, Joseph Diaz   91, 339
Gherardi, Silvestro   74, 78, 82, 83, 89, 466
Ghersi, Italo   *104*
Ghione, Franco   247
Giacardi, Livia   *52*, *104*, *140*, 209–269, *272*,
        273, *274*, 335, *371*, *410*, *412*, 515,
        516
Giaccardi, Fernando   351, 360
Giacomini, Amedeo   127, 226
Gianturco, Emanuele   114
Gigli, Duilio   48, 229, 345, 348, 349, 353,
        355, 364, *370*–*372*
Gini, Corrado   294, 349
Giordani, Francesco   303
Giorgi, Giovanni   303, 304, 310, 352
Giudice, Francesco   211, 214, 367
Giuliano, Landolino   301, 307, 308, 312
Gliozzi, Mario   360
Gödel, Kurt   *410*
Goldberg, Vladislav V.   *173*
Gonella, Guido   313
Gonseth, Ferdinand   254, 255
Goodstein, Judith   *136*
Gordan Paul   59, 122, 128, *137*
Gorenflo, Rudolf   *380*
Gotti, Vincenzo   *104*
Göttingen   91, 113, 126, 150, 154, 211, 214,
        223, 224, 241, 259, 263–266, 468,
        469, 475, 477, 478, 480, 487, 488,
        490, 491, 495, 505
Goursat, Edouard   374, 375
Graf, J.H.   *516*
Graffi, Dario   v, *vii*, 179–195
Graffi, Michele   189
Graffi, Sandro   197–207, 277–288
Grassmann, Hermann G.   97, 199–201
Gray, Jeremy   3, 6, 17, *29*, *71*
Green, Gabriel M.   148, 160, 162, *175*, 321,
        323

Index of Names and Locations

Grenoble *331*, *438*, *462*, *463*
Gribodo, Giovanni 107
Griffiths, Phillip 159, 172, *175*, *449*
Groningen 171
Grossmann, M. 226, *273*
Grothendieck, Alexander 42, 441
Grube, Franz 480
Grugnetti, Lucia *411*, *413*
Grygor'yan, A. 323
Guarducci, Alfredo 127, 128, 226
Guccia, Giovan Battista 508
Guerraggio, Angelo *50*, *173*, *176*, *270*, *273*, 289–314, *314*, *315*, *370*, *371*, *410*
Guglielmini, Domenico v
Guidi, Camillo 303
Guillemot, Michel 383–413
Gunning, Robert C. *55*
Gurtin Morton E. 183

H

Hadamard, Jacques Solomon 58, 70, *71*, 182, 283, 320–322, 328, 331, *331*, *331*, 374, 377, *380*, 384, *410*
Häfliger, André 3, *29*
Hahn, Hans 42, 399, 401, 403, *411*
Hale, Jack 185
Halle 110, 492
Hallet, Michael *138*
Halphen, Georges H. 147, 158, *176*
Hankel, Hermann 37, 356
Harnack, Carl G.A. von 26–28, 317–331, *331*, *332*
Harris, Joe 172, *175*, 441, *449*
Hartshorne, Robin 440, *449*
Hausdorff, Felix 24, 39, 455
Hawkins, Thomas *138*, *426*
Heegaard, Paul *426*
Heiberg, Johan L. 252, 253, *273*
Heidelberg 148, 254, 488, 498, 503
Heine, Heinrich Eduard 37, 42, 60, 62, 69
Heisenberg, Werner K. 280
Helmholtz, Hermann von 117, 119, 124, 126, *138*, 219, 220, 335, 475, 498
Helsinki 495
Herbart, Johann F. 335
Hermann, Robert 159, *176*, 254
Hermite, Charles 79, 80, 467, 468, 475, 477, 508–510, 512

Hilbert, David 10, 28, 110, 119, 120, 126, 127, *138*, *141*, 182, *208*, 241, 277, 279, 280, 291, 340, 342, 440, 441
Hildesheim 72
Hirst, Thomas 91, 97, 98, 475, 498
Hirst, Thomas A. 97, 98, *104*, 475, 498
Hirzebruch, Friedrich *449*
Hobson, Ernest William 40, *51*, 398, 404, 407, *411*
Hoepli, Carlo 349, *371*
Hoepli, Ulrico 348, 357, 363, 366, *370*
Holmgren, Hjalmar 379, *380*
Hormander, Lars 451, 459, 462, *463*
Hornick, H. *371*
Hoüel, Jules 2, 23, *29*, 466, 471
Hoyrup, Jens 515
Hudson, Hilda Phoebe *104*
Hugoniot, Pierre Henri 283

I

Iitaka, Shigeru *138*
Iskovskikh, Vasily A. *137*
Isnenghi, Mario *315*
Israel, Giorgio *138*, *176*, 236, 239, *273*
Ivey, Thomas A. *176*

J

Jackiw, Roman *288*
Jacobi, Carl Gustav Jacob 79, 80, 287, 503, 508, 521, *538*
James, William *29*
Jena 487
Jerusalem 282
Jordan, Camille 38, 280, 399, *411*, 509, 510

K

Kàarteszi, Ferenc 397
Kaminski, David 375, *380*
Kamke, Erich 407, 408, *411*
Kant, Immanuel 6, 7, *29*, 117, 254
Kantor, Seligmanri 129, 130, *138*
Karachalios, Andreas *315*
Kassmann, M. 323, *331*
Kato, Goro *463*

Kato, Tosio 288
Kawai, Takahiro 437, 459, *463*
Kempf, George 441, *449*
Kenyon, Richard 3, *29*
Kiev 184, 185
Killing, Wilhelm 466, 467, 474, 486–487
Kingston *136*, *173*, *270*
Kirchhoff, Gustav 502, 503, *516*
Kirner, Giuseppe 43
Kleiman, Steven 101, *104*, *449*
Klein, Felix 3, 11, 13, 20, 43, 66, *71*, 113, 119, 121, 124–127, 135, *138*, 210, 211, 214, 215, 217, 218, 223–229, 234, 241, 243, 263–266, *271*, *273*, *274*, 349, 361, 362, 369, 416, 422–424, *426*, 466, 467, 470, 474, 476, 488–490, 505, *516*
Königsberg 480, 487, 488, 490, 491, 498, 503
Koszul, Jean-Louis 169
Kowalevski, Sonya (Sophie) 206, 467, 476, 477, 495, 500, 503–504, 509–512, *515–517*
Krall, Giulio 310
Kramers, Hendrik A. 280
Krank, Philipp 334, *342*
Krazer, Adolf 265
Kristiania *426*
Krivosheev, A.S. *438*
Kronecker, Leopold 58, 59, 468, 475, 498
Kummer, Ernst E. 79, 80, 498
Kyoto 459

L

La Spezia *137*, *139*, *272*, *274*
La Vallée Poussin, Charles-Jean de 34, 45, *52*
Labriola, Antonio 294
Lacaita, Carlo *517*
Lagrange, Joseph-Louis 8, 123, 203, 514
Laisant, Charles 397
Laksov, Dan 441
Lambert, Johann 6, 62, 337
Lamé, Gabriel 80, 513
Lampe, Emil *371*
Lamy, Stephane 123, *137*
Lanaro, Giorgio *273*, *342*
Lanconelli, Ermanno 317–*331*
Landau, Edmund 36
Landis, E.M. 327
Landsberg, Joseph M. 172, *176*

Lane, Ernest P. 146, 147, 156, 160, 161, 164
Langer, Adrian 447, *449*
Laplace, Pierre Simon 165, 360
Lazarsfeld, Robert 441
Lazzarino, Orazio 312
Lazzeri, Giulio 246, 269, 346
Lebesgue, Henri 122, 386
Lecce 290
Lecornu, Léon François Alfred 198
Leffler, Anne C. 505
Lefschetz, Solomon 185, 193, 445
Legendre, Adrien-Marie 126, 261
Lehto, Olli 58, *71*
Leibniz, Gottfried W. 126, 337, 339
Leipzig 91, 226, 416, 476, 480, 488–490, 492
Lenzi, Enrico 346
Leone, Enrico 294
Leray, Jean 159, *175*
Letta, Giorgio 36, *50*
Levi, Beppo v, vi, *vii*, 39, 41, 44, *52*, 207, 241, 277–288, 294, 296, 297, 341, *342*, 358, 383–410, *410–413*
Levi, Eugenio Elia 44, 149, *176*, 425
Levi-Civita Tullio 34, 46, 145, 189, 192, 198, 202, 203, 205, 207, *208*, 281, 290, 292, 294, 295, 415–420, 422–424, *426*
Lévy, Maurice 472, 473
Li, P. (Peter) 146, 159, *175*
Libois, Paul 254
Liceni, Rita 47
Lie, Sophus 119, 121, 123, 146, 159, 168, 172, 415–425
Lille, 481
Lindelöf, Ernst Leonard 40
Lindemann, Ferdinand C.L. von 242
Linguerri, Sandra 252, 259, 397
Liouville, Joseph 16, 91, 205, 206, 376
Liporesi, Alfeo 311
Lipschitz, Rudolph 466, 467, 474, 491–495
Livorno 120, 269, 301
Lo Surdo, Antonino 304
Lobachevskij, Nikolaj 1, 2, 6, 7, 12, 21, 126, 335, 471
Lodi, E. 358
Lolli, Gabriele 227, 259, 333–342, *411*
Lombardini, Maria 253
Lombardo Radice, Lucio 243, 249, *273*
London 91, 154, 193, 254, 498
Loria, Gino 102, 127, 128, 156, 211, 215, 247, 344, 346, 351, 355, 367
Lotze, Rudolph H. 335
Luchko, Yuri 376, *380*

Index of Names and Locations 547

Luciano, Erika 215, 343–372
Lugano 248
Lugli, Aurelio 214, 264, 266, 491
Lüroth, Jacob 37, 129, 261
Luzin, Nikolai 41, 49

M

Maccaferri, Eugenio 347
Mach, Ernst 219, 246, 334
Macugnaga (Monte Rosa) 494
Madelung, Erwin 288
Magenes, Enrico 309
Maggi, Gian Antonio 33, 198, 354, 472, 476
Magni, Francesco 79, 86
Magnus, Ludwig I. 97–99
Mahomed, Fazal *426*
Mainardi, Francesco 373–381, *380*
Mainardi, Gaspare 158
Maiocchi, Roberto *517*
Maiorana, Quirino 188, 189, 304, 311
Majer, Ulrich *138*
Malgrange, Bernard 451, 458, 459, *463*
Mambriani, Antonio 296, 301, 379, *380*
Mamiani, Terenzio 73, 76, 78
Manara, Carlo Felice 118
Manarini, M. 191, 358
Mancini, Augusto 308, 309
Manfredi, Gabriele v
Maniá, Basilio 310
Mantova 317
Marchis, Lucien 485
Marchisotto Elena A. *138*
Marcolongo, Roberto 198–203, 207, *208*, 249, 310, 351
Marconi, Guglielmo 170
Marino, G.C. *315*
Marsala 77
Martineau, Andre *463*
Martinelli, Enzo *176*, *371*
Marziani, Marziano 191
Maslov, Victor P. *288*
Masotto, Guido *516*
Mattaliano, Maurizio 171, *410*
Matteotti, Giacomo 295
Matteucci, Carlo 86
Maxwell, James C. 186, 201, 466, 473, 474, 482, 484, *515*, *517*
Mayer, Christian G.A. *72*, 488
Mayoux, Jean-Jacques *273*
Mazurkiewicz, Stefan 45, *53*

Mazzini, Giuseppe 75, 81, 82, 92
Mazzola, Annibale 309
Medolaghi, Paolo 416, 420, 423, 425
Mehmke, Rudolf 148, 166, 542
Meijer, Cornelis Simon 373, 375, 376, 379, *380*
Meldola 81
Mellin, Hjalmar 373–376, *380*, 521, 523
Menghini, Marta *104*, *273*, *516*
Mengoli, Pietro 5, 305
Mercuri, L. *315*
Mercurio, Anna Maria *104*
Meril, Alex *437*, *438*
Merri Mamarini, Marisa 191
Merrik, Teri *29*
Meschkowski, Herbert *371*
Messina 190, 197, *208*, 293, 356, 362, 363
Metzger, Héléne 254, 255, *274*
Meusnier, Jean Baptiste *173*
Meyer, Friedrich Wilhelm Franz 125
Meyer, Oscar Emil 475, 480
Mezzetti, Emilia 441, *449*
Michel, Ch. 115, 198
Michelacci, Giacomo *517*
Michelson, Albert A. 246
Mieli, Aldo 251, 254
Milan 75, 81, 83, 86, 90, 92, 100, 150, 154, 248, 294, 355, 362, 387
Milesi, Luigi 367
Millán Gasca, Ana Maria *104*
Miller G.A. *371*
Milnor, John 3, *29*
Minghetti, Marco 81
Miranda, Carlo 396, *410*, 519
Miranda, Rick *449*
Mises, Richard von 360
Mitropolsky, Yuri 185
Mittag-Leffler, Gösta 58, 374, 427, 466, 467, 475, 476, 495–509, 512–515, *517*
Möbius, August F. 97
Modena 32, 34, 114, 127, 128, 152, 292, 318, 355, 387
Moisil, Grogore C. 433
Monge, Gaspard 16, *29*, 88, 147, 162, 167, 169, *176*
Montaldo, Oscar 322, 324, *332*, *411*
Montalenti, Giuseppe 252, 254
Montanari, Antonio 81–83
Montel, Paul 34, 36, 393, *411*
Montesano, Domenico 108, 114
Moore, Gregory H. *411*
Morera, Giacinto 476, 488

Moretti, M. 247, *274*
Morghen, Raffaello 171
Moscow *438*
Moser, Jürgen 323, *332*
Mossa, Lorenzo 308
Mossotti, Ottaviano Fabrizio 87, 471
Moufang, Ruth 110
Moulton, Forest Ray 110
Moutard, Théodore Florentin 165
Mumford, David 135, *138*, *449*
Munich 87, 488
Münster 486–487
Mussolini, Benito 297, 300, 303, 310, 313

N

Nabonnand, Philippe 259
Nalli, Pia 46
Napalkov, V.V. *438*
Naples *vii*, 78, 108, 114, 216, 226, 248, *274*, 305, 309, *314*, *315*, 355, 505, 519, *538*
Nardini, Renato 181, 191, 192
Nastasi, Pietro *104*, *139*, 145, 171, *173*, *176*, *208*, 259, *259*, *273*, *274*, 289–315, *315*, *371*, *410*, *426*, 515
Nastasi, Tina *139*, *274*
Natoli, Giuseppe 85
Natucci, Alpinolo 247
Nelson, Edward 288
Nendeln *175*
Neumann, Carl 60, 64, 487, 491
Neumann, Franz E. 480, 487, 503
Neumann, John von 277, 288, *288*
Neurath, Otto 334
New York 29, *50*, *70–72*, *104*, *140*, 177, *208*, *273*, *288*, *331*, *342*, *381*, *410*, *411*, *426*, *438*, *448*, *450*, *463*, *516*, *538*
Newton, Isaac *50*, 93, 97, 252–254, *272*
Nicoletti, Onorato 249, 388
Nikodým, Otton M. 46
Nilson, Winfried *371*
Nirenberg, Louis 322
Noether, Max 100, *104*, 115, 129, 132, *136*, *139*, *142*, 442, 448
Nonni, Giuseppe 32
Notari Cuzzer, Vittoria 256
Novara, Domenico Maria v
Novi, Giovanni 87
Nuesnschwander, Erwin *71*
Nuremberg *140*
Nurzia, Laura *104*, *273*, *274*

O

Olver, Peter 419, 422, *426*
Onesti, Carla, 259
Orange 427, 451
Orestano, Francesco 303, 304
Orlando, Luciano 292, 293
Orsini, Felice 81
Osgood, William Fogg 36
Ossermann Robert 24, *29*
Ostwald, Wilhelm 246
Ottaviani, G. 299
Oxford *288*, *380*, *516*, *517*

P

Pacioli, Luca v
Padoa, Alessandro 120, *139*, 215, 229, *273*, 294, 346, 348, 352, 353, 486
Padova, Ernesto 472
Padua 32, 34, 49, 127, 128, *140*, 215, 224, *270*, *273*, 295, 312, 345, 346, 355, *370*, 478, 499
Pagnini, Gianni 373–379, *380*
Painlevé, Paul 198
Palamodov, Viktor P. 434, 435, *438*, 451, 452, 459, *463*
Palatini, Attilio 355, *371*
Palermo 35, 78, *104*, 301, *315*, *371*, 407, 508, *516*
Palladino, Dino *139*
Palladino, Franco *103*, *104*, *270*, *371*, *516*, *517*
Palladino, Nicla *104*, *139*, *371*
Palo Alto *410*
Panetti, M. 302
Paoloni, Giovanni *176*
Papa, Emilio R. 294
Papini, Pier Luigi *50*
Paris 29, 34, 49, *50*, *51*, *52*, *53*, *71*, *72*, 91, 120, *139*, *175*, *176*, 181, 193, 243, 254, *273*, 334, *342*, 383–386, 392, *411*, *412*, 439, 468, 473, 476, 481, 485, 488, 495, 498, 500, 502, 514, *516*
Paris, Richard B. 375, *380*
Parma 33, 292, 293, 295, 296, 385
Parry, Walter R. 3
Pascal, Ernesto 48, *139*, 349, 355, *371*, 397–399, 401, 403, 407, 408, *411*, 505

Index of Names and Locations

Pascal, Mario   292–294
Pasch, Moritz   110, 111, 119, *139*, 340, *342*
Pavia, 57, *71*, 74, 75, 78, 80, 85, 89, 90, *104*, 150–153, 163, 203, *208*, 290, 343, 345–347, 354–359, 362–366, 368, *370*, 470, 471, 480–482, 486, 488, 489, 492–497, 499–501, 503, 504, 506, 507, 510–512, *516*
Pazzini, Adalberto   252
Peano, Giuseppe   10, 28, 38, *51*, 111, 112, 119, 120, *139*, *140*, 198–201, 215, 238, 333, 337, 339–341, 346–349, 353, 354, 356, 357, 362, 366–368, *371*, *372*, 397, *411*, 509
Pedrini, Antonio   299
Pedrini, Claudio   *103*
Pende, Nicola   303
Pentimalli, Francesco   309
Pepe, Luigi   32, *50*, 292, 297, *315*, *370*, *410*, *411*, *517*
Perna, Alfredo   *274*, 346, *371*
Perrono, Oskar   389
Persico, Enrico   190, 247, 282
Pesaro   290, 294
Pession, Giuseppe   304
Pestalozzi, Johann H.   257
Petralia, Stefano   299
Petri, K.   448, *449*
Petrocchi, Giuseppe   313
Pettini, Giovanna   191
Piacenza   241
Piani, Domenico   83
Picard, Émile   135, *139*, 246, 388, 389, 477, 509, 510
Piccinini, L.C.   *463*
Pick, George   24, 25, *29*
Picone, Mauro   170, 193, 292, 293, 295, 299, 301–306, 310–313, *314*, *315*, 396, *410*, *411*, 519, 520
Pier, Jean-Paul   *50*
Pieri, Mario   33, 107–109, 111, 113, 114, 120, *138*, *139*, *140*, 217, *270*, 341, 346, *369*
Pincherle, Salvatore   v, vi, *vii*, 34, 38, 48, 57–70, *71*, *72*, 108, 127, 153, 207, 211, 214, 222, *274*, 291–296, 300, 317, 343, 354, 356, 358, 359, 362, 364, 365, 368, *369*, 373–379, *380*, *381*, 384, 385, 387, 388, 415–418, *426*, 427, 428, 429, 430, 431–433, 434–437, *438*, 470
Pini, Bruno   v, 317–331, *332*
Pini de Socio, Marialuisa   191
Pirio, Luc   *176*

Pisa   32–35, 37, 46, *50*, *51*, 57, 78, 87, 101, 107, 108, 124, 127, 128, *139*, 152, 156, 158, 226, 247, *274*, 293, 295–302, 306–314, *315*, 355, *380*, 422, *448*, 468, 470–472, 488, 505, 509, *515*, *516*
Pistoia   *271*
Pistolesi, Enrico   308
Pittarelli, Giulio   144, 145, 151, *176*, 294, 423
Pittsbourg   155
Pizzetti, E.   299
Pizzocchero, Livio   *176*
Planck, Max   198
Platon   255
Plotinus   233
Plutarch   383
Pochhammer, Leo A.   375
Poggiorusco   317
Poincaré, Henri   3, 16, 25, *29*, 62–64, 66, 67, *72*, 334, 360, 477, 501, 509, 510
Poinsot, Louis   514
Poisson, Siméon-Denis   360
Pommaret, Jean-François   *426*
Pompeo Faracovi, Ornella   *137*, *139*, 259, *270*, *272*, *274*, *275*
Pompilj Giuseppe   349
Poncelet, Jean-Victor   116
Pontresina   498
Pordenone   *272*, *273*
Posidonius   5
Pradella, Piro   422
Prato   128, 226
Priestley, Joseph   254
Princeton   *50*, *55*, 193, *288*, 388
Prouhet, Pierre-Marie E.   92
Providence   *104*, *176*, *449*
Prym, Friedrich E.   58
Pucci, Carlo   *315*
Puccianti, Gigi   307
Pula   373

Q

Quilici, Leana   *274*

R

Racah, Giulio   282
Racinaro, Roberto   *315*

Raffaele, Federico   251
Ragghianti, Renzo   *274*
Ramenghi, Sante   88
Ran, Ziv   *449*
Rasetti, Franco   302
Ravenna   32
Razzaboni, Amilcare   108, 109
Reed, Michael   *288*
Rennes   481
Respighi, Lorenzo   84
Retali, Virginio   349, 351
Reykjavik   *272*
Ricasoli, Bettino   77
Riccati, Jacopo   *173*
Ricci Curbastro, Gregorio   34, 46, 157, 198, 388, 470
Ricci, Giovanni   *314*, *315*, 416
Ricci, P.E.   *315*
Richiardi, Sebastiano   212
Riecke, Eduard   150, 490
Riemann, Bernhard G.F.   1, 2, 5, 7, 10, 13, 14, 16, 17, 20, 22, 23, 28, *30*, 37, 46, 57, 67, 87, 122, 126, *139*, *140*, 374, 376–378, *381*, 468, 469, 471, 480, *517*
Rietti, Guido   253
Rignano, Eugenio   243, 246, *273*
Rionero, Salvatore   184
Roero, Clara Silvia   369, *371*, 383–413, *410*, *412*
Roghi, Gino   301
Rogora, Enrico   *139*, *176*, *274*, 415–426, *426*
Rohlfing, Helmut   259, 265
Rohn, Karl   355
Rohrlich, D.   *437*
Rollero, Aldo   *371*
Rome   *vii*, *viii*, 33, *55*, *72*, 82, 86, 90, 92, 100, *104*, 107, 109, 110, 112, 120, 126, 127, 132, *139*, *140*, 144–147, 150–154, 156, 163, 169–171, 173, *176*, *177*, 187, 189, 197, 203, 212, 217, 224, 226, 239, 248, 250–254, 256, 259, 266, 268, *270*–*274*, 282, 290, 293, 295, 296, 299, 301, 302, 304, 306, 307, 311, *314*, *315*, 334, *342*, 344, 355, *369*, 374, *380*, 385, 396, *410*, *411*, 419, 422, *425*, *426*, *438*, 439, *449*, 467–471, 479, 482–485, 490, 492–494, 498, 501, 507–509, 511, *515*
Rosanes, Jacob   475, 498
Rosenthal, Arthur   *411*, *413*
Rossi, Bruno   189, 282

Rossi, Pellegrino   81
Rossi, Vincenzo   308
Rovigo   189
Royce, Josiah   336
Rufini, Enrico   253
Russell, Bertrand   246, 334, 338, *342*, 346
Russo, Luigi   307, 308, 309, 311–313, *315*

S

Sabadini, Irene   427–437, *438*
Sabbatini, Alfredo   256
Saccheri, Girolamo   5, 6, 9, 261
Sacchiero, Gianni   441, *449*
Sacilotto, Ines   47
S.Agata Bolognese   359
Saks, Stanislaw   42, 317, 325
Sala Bolognese   *273*
Sallent Del Colombo, Emma   143–177, 259
Salò   307–309, 312, 313
Saloff-Coste, L.   323, *332*
Salkowski, Erich   *411*
Salvemini, Gaetano   43, 246
Sannia, Gustavo   34, *139*, 292, *371*
Sansone, Giovanni   34, 48, *50*, *54*, 310, *315*, *371*, *413*
Santacroce, C.   299, 310
Santo Stefano Magra   33
Saporetti, Antonio   88
Sassari   33, 37
Sato, Mikio   460
Sbordone, Carlo   *315*, *538*
Scanlan, Michael J.   21, 23, *30*
Scarantino, Luca M.   *274*
Scarpis, Umberto   229
Schepp, Adolf   37, *51*
Schering, Ernst   487
Schiaparelli, Giovanni Virgilio   98, 99, *104*
Schläfli, Ludwig   29, 473, 475, 498, *516*
Schlick, Moritz   334
Schneider, Clara   385
Schouten, Jan Arnoldus   53
Schröder, Ernst   337
Schröter, Heinrich Eduard   92, 97
Schubert, Hermann   442, *449*
Schubring Gert   259
Schumann, Robert   492
Schwartz, Laurent   431, *438*, *462*
Schwarz, Karl Hermann Amandus   29, 37, *72*, 475
Scimone, Aldo   *315*, *371*
Scorza Dragoni, Giuseppe   519

Index of Names and Locations 551

Scorza, Gaetano   170, 171, 246, 299, 519, *519*
Scriba, Christoph J.   *370*
Segre, Beniamino   v, vi, 46, 48, *55*, 207, 343, 358, 359, 361, *377*, 439, 440–448, *448–450*
Segre, Corrado   101, 107, 109, 111–116, 122, 124, 129, 131, 132, *137*, *138*, *139*, *140*, 144, 147, 148, 149, 150, 158, 160, 162, 164, 165, 171, 172, *175*, *176*, 210, 216, 217, 218, 221, 238, *274*, 334, 354, *371*, 416, 342, 423, 439–448
Sella, Alfonso   470, 476, 490
Sella, Quintino   476, 490
Senff, Carl Eduard   162
Serini, Rocco   345, 355
Sernesi, Edoardo   *175*, *448*, 439–448, *450*
Serret, Joseph Alfred   468
Serrin, James   184
Severi, Francesco   46, 48, 101, 135, *137*, *141*, 144, 145, 249, 250, 252, *270*, *272*, *274*, 294, 295, 297, 299, 301–304, 310, 311, 345, 346, 355, 360, 367, 368, *371*, *413*, 439–448, *450*
Seydewitz, Franz   97
Shouten, Jan   168
Sibirani, Filippo   v, 343, 355, 358–360
Sierpinski, Waclaw   39, 298, 299, 393, *412*
Signorini, Antonio   202, 310
Silvaplana (Engadina)   499
Silvestri, Andrea   *517*
Simart, George   135, *139*
Simili, Raffaella   *274*
Simon, Barry   *288*
Simoncelli, P.   311, *315*
Simson, Robert   102
Singapore   *175*
Skof, Fulvia   *412*
Smith, J.T.   *138*, *140*
Solovay, Robert Martin   *51*, *412*
Somigliana, Carlo   293, 303, 304, 310, 472
Sommen, Franciscus   *438*
Spencer, Herbert   115, 119
Speranza, Francesco   *139*, *270*, *274*
Stäckel, Paul G.   205, *208*
Staudt, Karl G.C. von   111, 113, 115, 116, *140*, 217, 339
Steiner, Jacob   93, 97, 98, 102, 115, 116, 134
Stendhal (Marie-Henri Beyle)   92, 93, *104*
Stillwell, John   *140*
Stockholm   374, *380*, 467, 476, 495, 497, 500, 503, 505, 509, 512–515

Struick, Ruth   253
Struik, Dirk   *53*, 160, 253
Struppa, Daniele C.   427–437, *437*, *438*, 451–462, *462*, *463*
Stubhaug, Arild   374, *381*
Sturm, Jacques C.F.   198
Stuttgart   *342*, 490
Sudbery, A.   434, *438*
Supino, Giulio   vi, 190
Swansea   154
Sylvester, James J.   115, 122

T

Tacchella, Giuseppe   348
Taeklind, Sven   322, *332*
Taine, Hippolyte   335
Tannenbaum, Allen   *450*
Tannery, Jules   246
Tardy, Placido   78, 81, 84, 85, *103*, *273*, 376, *381*, 466, *516*
Taylor, B. Allan   429, *438*
Tazzioli, Rossana   86, *208*, *273*, 465–515 *516*, *517*
Tealdi, Alice   *274*
Tedeschi, Amalia   189, 360
Teixidor i Bigas, Montserrat   *449*, *450*
Terquem, Olry   92, *104*
Terracini, Alessandro   147–150, 158, 160, 163, 164, *177*, *412*
Teza, Emilio   79
Thomsen, Gerhard   165, *177*
Thun   498
Tian, Gang   *175*
Tisserand, Félix   477, 509, 510
Tedeschi, Bruno   360
Togliatti, Eugenio   147, 149, 352, *371*
Toland, John   254
Tomasi, Tina   118, *275*
Tonelli, Alberto   250
Tonelli, Leonida   *vii*, 33–35, 39, 41, 44, 45, *52*, *53*, 144, 145, 154, 189, 192, 207, 250, 289–314, *314*, *315*, 362, 383, 385–391, *398*, *399*, 403, 404, 407, *410–412*
Tonolo, Angelo   47, 49, *50*, *412*
Torelli, Gabriele   34, 292
Toronto   289, 290, 374
Toulouse   *51*, 383, *410*
Treccani, Giovanni   256
Trento   248, 267, *449*
Tricomi, Francesco   190, 193, 362, 385

Trieste   248, 266–269, 355, *372*, *517*
Trombetti, Guido   *538*
Truesdell, Clifford A.   184
Tübingen   490
Turi, Gabriele   *315*
Turin   107–109, 111–114, 127, 131, 132,
   147, 157, 160, 167, 190, 197, 209,
   282, 309, *315*, 334, 343, 347–349,
   353–356, 358, 361, 383, 467, 476,
   488, 509, 510
Tzitzeica, Gheorge   *177*

U

Udine   *270*
Uguzzoni, F.   *331*
Umemura, Hiroshi   *140*
Uppsala   495
Urbino   *208*

V

Vacca, Giovanni   251, 254, 290, 295, *315*,
   337, 346, 353, 360, *371*
Vaccaro, Giuseppe   *177*
Vailati, Giovanni   118, 222, 226, 238, 241,
   243, *273*, *275*, 336, 337, *342*, 360,
   362, 397, *412*
Vallauri, Giancarlo   303, 304, 309
Van der Waerden, Bartel Leendert   *288*
Vaz Ferreira, Arturo   36, *50*, *370*, *412*
Veneroni, Emilio   345, 347
Venice   75, 282, 467, 477
Venturi, Luigi   85, 86
Venturini, Sergio   519–538
Vergerio, Attilio   293
Verona   420
Veronese, Giuseppe   90, *104*, 112, 114, 126,
   139, *140*, 168, 200, 224, 367
Verra, Alessandro   *448*, *450*
Vetro, Pasquale   328
Viareggio   154, 269
Vienna   254, 334
Vigevano   354
Viglezio, Elisa   352
Villa, Mario   360
Villari, Pasquale   212
Viola, Carlo   397

Viola, Carlo Maria   385
Viola, Giovanna   397
Viola, Tullio   v, 49, *50*, 258, *275*, 383–410,
   410–413
Vitali, Giuseppe   v, vii, *viii*, 31, *50–55*, 127,
   128, 226, 291, 292, 293, 297, *315*,
   343, 357, 358, 359, 362–366, *370*,
   410–413
Vitali, Luisa   32, 86
Vivanti, Giulio   v, 48, 247, 343, 345–349,
   355–368, *370*, *371*, *372*, 392
Voghera   32, 33, 37, 362
Voigt, Woldemar   490
Volpicelli, Paolo   492, 493, 501
Volterra, Vito   33, 34, 37, 44, 58, 108, 127,
   144, 145, 150, 170, 171, 180–183,
   189, 207, 238, 246, 249, 257, 281,
   290, 291, 294–296, 300, 302, 345,
   354, 468, 470, 488, 513, *517*

W

Wahl, Jonathan   445, 447, *450*
Wallis, John   6
Warszawa   *53*, *412*
Washburne, Carleton   258, 259, *275*
Wasielewski, Wilhelm Joseph von   492
Watson, N.A.   324, 325, *332*
Weber, Heinrich   349, *381*, 480, 487
Weierstrass, Karl Theodor Wilhelm   7, 37,
   58–60, 62, 63, 66, 67, *72*, 115, 368,
   468, 470, 476, 488, 495, 498, 500,
   513
Weil, André   35, *50*, 96, *104*
Weingarten, Julius   159
Weinheim   *437*
Weitzenböck, Roland   47, *53*
Weizel, Wilhelm   288
Wentzel, Gregor   280
Weyl, Hermann   168
Whiston, William   254
Whittaker, Edmund T.   205, *208*
Wiechert, Emil J.   150
Wiener, Hermann   100, *140*
Wilczynski, Ernest J.   67, 146–149, 159–161,
   165, *176*, *177*
Williams, Kim   259, 269, 369, 383, *410*
Wilson, Robin J.   *104*, 356
Winger, R.M.   *30*
Wölffing, Ernst   148
Wundt, Wilhelm   219, 220, 335, *342*

Y

Young, William H. 397

Z

Zac, Fyodor *177*
Zacharias, Johann Martin *372*
Zapelloni, Maria Teresa 247, 253

Zappa, Guido 299
Zappulla, Carmela *270*
Zariski, Oscar 98, 253, 445–447, *450*
Zecca, G.B. 358
Zermelo, Ernst 387, 390–392, 394, 395, 401, *413*
Zeuthen, Hieronymus G. 475, 498
Zoretti, Ludovic *413*
Zuccheri, Luciana 266, *275*
Zudini, Verena 266, *275*
Zurich 254, 290, 300, 385